The Remote Sensing of Tropospheric Composition from Space

Physics of Earth and Space Environments

The series *Physics of Earth and Space Environments* is devoted to monograph texts dealing with all aspects of atmospheric, hydrospheric and space science research and advanced teaching. The presentations will be both qualitative as well as quantitative, with strong emphasis on the underlying (geo)physical sciences. Of particular interest are

- contributions which relate fundamental research in the aforementioned fields to present and developing environmental issues viewed broadly
- concise accounts of newly emerging important topics that are embedded in a broader framework in order to provide quick but readable access of new material to a larger audience

The books forming this collection will be of importance for graduate students and active researchers alike.

For other titles published in the series, go to
www.springer.com/series/5117

John P. Burrows • Ulrich Platt • Peter Borrell
Editors

The Remote Sensing of Tropospheric Composition from Space

With 158 Figures and 23 Tables

Prof. Dr. John P. Burrows
Universität Bremen
Institut für Umweltphysik (IUP)
Otto-Hahn-Allee 1
28359 Bremen
Germany
burrows@iup.physik.uni-bremen.de

Prof. Dr. Ulrich Platt
Universität Heidelberg
Institut für Umweltphysik
Im Neuenheimer Feld 229
69120 Heidelberg
Germany
ulrich.platt@iup.uni-heidelberg.de

Dr. Peter Borrell
P & PMB Consultants
6 Berne Avenue
Newcastle-under-Lyme
ST5 2QJ, United Kingdom
peter@ppmborrell.co.uk

ISSN 1610-1677 e-ISSN 1865-0678
ISBN 978-3-642-14790-6 e-ISBN 978-3-642-14791-3
DOI 10.1007/978-3-642-14791-3
Springer Heidelberg Dordrecht London New York

Cover design: eStudio Calamar S.L.

Printed on acid-free paper

Springer is part of Springer Science+Business Media (www.springer.com)

Preface

The impact of anthropogenic activities on our atmospheric environment is of much public concern, and the economic and technical solutions needed to provide a sustainable environment require reliable observations, coupled with a proper scientific understanding. Satellite-based techniques now provide an essential component of observational strategies on regional and global scales.

It is now some 15 years since the launch of GOME, the first satellite instrument designed specifically to retrieve the composition of trace gases and pollutants in the troposphere. Since then the number of satellite instruments has increased steadily, and the availability of satellite data is providing the capability of monitoring the state of the global atmosphere. It is also radically changing the field of atmospheric chemistry.

The purpose of this book is to summarise the state of the art in the field; to describe the technology and techniques used; and to demonstrate the key findings and results. The book has its origins in TROPOSAT, a project initiated within the EUROTRAC framework, to encourage the use and usability of satellite data for tropospheric research; the project was continued within the EU air quality project, ACCENT. Two of the book's editors were proposers of SCIAMACHY and the smaller scale GOME, which initiated European-based remote sensing of tropospheric trace gases from space. The third has coordinated the various TROPOSAT activities, having previously been the Executive Scientific Secretary of the EUROTRAC project. All the contributing authors to this volume are senior scientists actively involved in the field – in satellite data retrievals, in the validation of tropospheric data, in the interpretation of the global and regional results and in the modelling, which relies on the data; most are part of the TROPOSAT community.

The book opens with an historical perspective of the field together with the basic principles of remote sensing from space. Three chapters follow on the techniques and on the solutions to the problems associated with the various spectral regions in which observations are made.

The particular challenges posed by aerosols and clouds are covered in the next two chapters. Of special importance is the accuracy and reliability of remote sensing data and these issues are covered in a chapter on validation.

The final section of the book is concerned with exploitation of the data for scientific and operational applications. These include investigations using individual data products and synergistic studies using a variety of data products. Comparison of global and regional observations with chemical transport and climate models are discussed and the potential added value from the synergetic interaction of model and measurements identified.

The book concludes with scientific needs and likely future developments in the field, and the necessary actions to be taken if we are to have the global observation system that the Earth needs in its present, deteriorating state.

The appendices provide a comprehensive list of satellite instruments, global representations of some ancillary data such as fire counts and light pollution, a list of abbreviations and acronyms, and a set of colourful timelines indicating the satellite coverage of tropospheric composition in the foreseeable future.

The recent impact of volcanic ash on European air transport (Chapter 10) has provided a forceful reminder of the utility of satellite observations in monitoring and understanding the tropospheric constituents in the atmosphere. Thus the book provides a timely account of the developments in a new area of much utility to sustaining a healthy atmosphere.

Bremen, Germany and NERC CEH, Wallingford, UK — John P. Burrows
Heidelberg, Germany — Ulrich Platt
Newcastle-under-Lyme, UK — Peter Borrell

Acknowledgements

We would like to thank our co-contributing authors, for their excellent contributions and for their patience with the editing process; our contributors, Cathy Clerbaux, Klaus Kunzi and Gerrit de Leeuw for their thoughtful reading of our own two chapters; Christian Caron and his colleagues at Springer for their patient encouragement; our many colleagues and friends in TROPOSAT, in ACCENT and elsewhere, for their continued encouragement and support; and Dr Patricia Borrell for her thorough reading of the manuscript and many appreciable contributions to the content and form of this book.

University of Bremen
Bremen, Germany
and
NERC Centre for Ecology and Hydrology
Wallingford, United Kingdom

John P. Burrows

University of Heidelberg
Heidelberg, Germany

Ulrich Platt

P&PMB Consultants
Newcastle-under-Lyme, UK

Peter Borrell

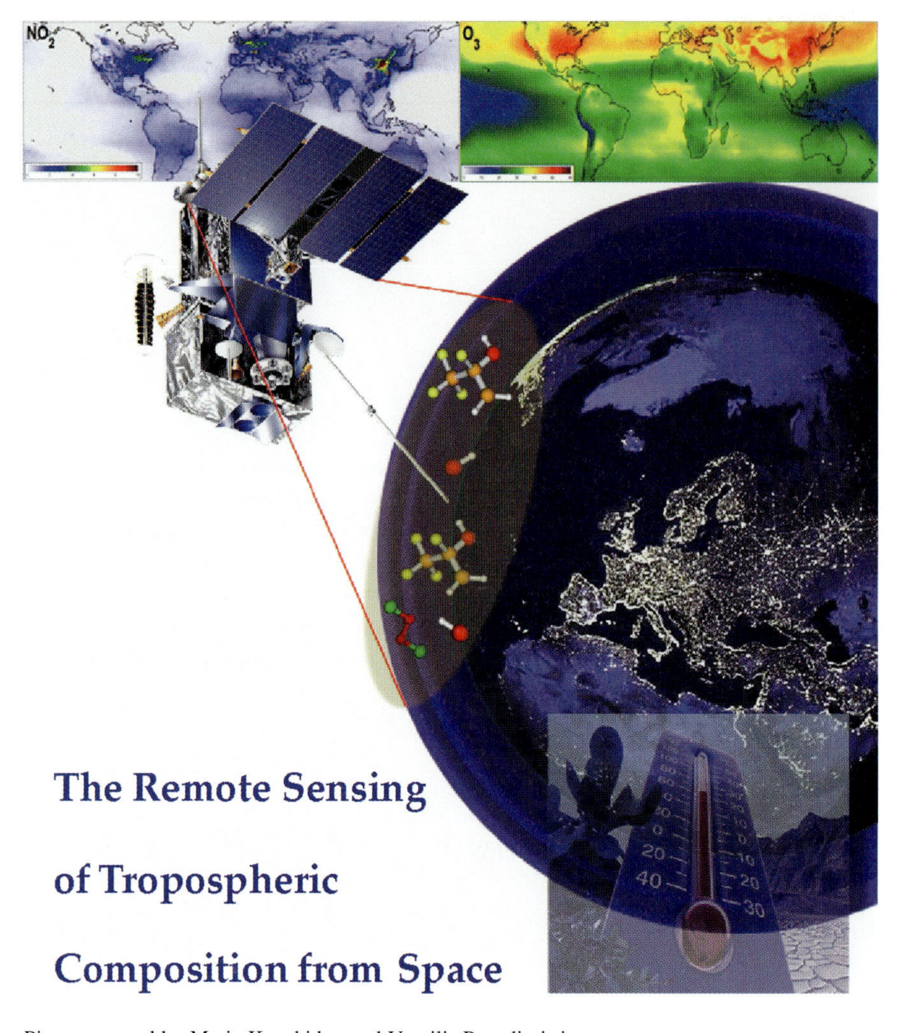

Picture created by Maria Kanakidou and Vassilis Papadimitriou

Contents

Contributors

Dr. Peter Bauer European Centre for Medium-Range Weather Forecasts (ECMWF), Reading, UK

Dr. Matthias Beekmann Laboratoire Interuniversitaire des Systèmes Atmosphériques (LISA) CNRS, Université Paris, Est et Paris 7, Créteil, France

Dr. Steffen Beirle Max-Planck-Institut für Chemie, Mainz, Germany

Dr. Peter Borrell P&PMB Consultants, Newcastle-under-Lyme, Uinted Kingdom

Dr. Dominik Brunner Laboratory for Air Pollution Technology, Empa, Swiss Federal Laboratories for Materials Testing and Research, Dübendorf, Switzerland

Dr. Brigitte Buchmann Laboratory for Air Pollution Technology, Empa, Swiss Federal Laboratories for Materials Testing and Research, Dübendorf, Switzerland

Prof. John P. Burrows Institute of Environmental Physics (IUP), University of Bremen, Germany; NERC Centre for Ecology and Hydrology, Wallingford, United Kingdom

Dr. Cathy Clerbaux UPMC Univ. Paris 06; CNRS/INSU, LATMOS-IPSL, Paris, France

Prof. Ronald C. Cohen Department of Chemistry, University of California, Berkeley, CA, USA

Prof. James R. Drummond Department of Physics and Atmospheric Science, Dalhousie University, Halifax, Canada

Dr. Hendrik Elbern Rhenish Institute for Environmental Research at the University of Cologne, Köln, Germany

Dr. Reima Eresmaa European Centre for Medium-Range Weather Forecasts (ECMWF), Reading, UK

Dr. Patrick Eriksson Department of Earth and Space Science, Chalmers University of Technology, Gothenburg, Sweden

Prof. Jean-Marie Flaud Université Créteil Paris 12, CNRS UMR 7583, LISA-IPSL, Paris, France

Dr. Sean Healy European Centre for Medium-Range Weather Forecasts (ECMWF), Reading, UK

Dr. Wolfgang von Hoyningen-Huene Institute of Environmental Physics, University of Bremen, Bremen, Germany

Dr. Michael D. King Laboratory for Atmospheric and Space Physics, University of Colorado, Boulder, CO 80309, USA

Dr. S. Kinne MPI-Meteorology, Hamburg, Germany

Dr. Alexander A. Kokhanovsky Institute of Environmental Physics, University of Bremen, Germany

Dr. Igor B. Konovalov Institute of Applied Physics, Russian Acadamy of Sciences, Nizhnig, Novgorod, Russia

Prof. Klaus Kunzi University of Bremen, Institute of Environmental Physics, Bremen, Germany

Dr. Jean-Christopher Lambert Belgian Institute for Space Aeronomy (BIRA-IASB), Brussels, Belgium

Prof. Gerrit de Leeuw Climate Change Unit, Finnish Meteorological Institute, Helsinki, Finland; Department of Physics, University of Helsinki, Helsinki, Finland; TNO Environment and Geosciences, Utrecht, The Netherlands

Dr. Jean-Francois Léon LOA, Lille, France

Dr. Nathaniel Livesey Microwave Atmospheric Science Team, Jet Propulsion Laboratory, Pasadena, CA, USA

Dr. Maarten Krol Meteorology and Air Quality, Environmental Sciences Group, Wageningen University, Wageningen, The Netherlands

Prof. Maria Kanakidou Environmental Chemical Processes Laboratory, Department of Chemistry, University of Crete, Heraklion, Greece

Prof. Martin Dameris Deutsches Zentrum für Luft- und Raumfahrt, Institut für Physik der Atmosphäre, Oberpfaffenhofen, Germany

Prof. Paul S. Monks Department of Chemistry, University of Leicester, Leicester, United Kingdom

Dr. Alberto Mugnai Istituto di Scienze dell'Atmosfera e del Clima (ISAC), CNR, Roma, Italy

Dr. Lars Nieradzik Rhenish Institute for Environmental Research at the University of Cologne, Köln, Germany

Prof. Johannes Orphal Institute for Meteorology and Climate Research, Karlsruhe Institute of Technology (KIT), Germany

Dr. Jacques Pelon Université Pierre et Marie Curie, Paris, France

Dr. Ankie J.M. Piters Royal Netherlands Meteorological Institute (KNMI), De Bilt, The Netherlands

Dr. Steven Platnick NASA Goddard Space Flight Center, Greenbelt, MD, USA

Prof. Ulrich Platt Institute of Environmental Physics (IUP), University of Heidelberg, Heidelberg, Germany

Dr. Catherine Prigent CNRS, Observatoire de Paris, Paris, France

Dr. Andreas Richter Institute of Environmental Physics, University of Bremen, Bremen, Germany

Prof. Daniel Rosenfeld The Hebrew University of Jerusalem, Jerusalem, Israel

Dr. Martijn Schaap TNO Environment and Geosciences, Utrecht, The Netherlands

Dr. Eric A. Smith NASA/Goddard Space Flight Center, Greenbelt, MD, USA

Dr. Piet Stammes Royal Netherlands Meteorological Institute (KNMI), De Bilt, The Netherlands

Dr. Graeme Stephens Colorado State University, Fort Collins, CO, USA

Dr. Achim Strunk Rhenish Institute for Environmental Research at the University of Cologne, Köln, Germany

Dr. Pepijn J. Veefkind Royal Netherlands Meteorological Institute (KNMI), De Bilt, The Netherlands

Dr. Ben Veihelmann ESA/ESTEC, European Space Agency, Noordwijk, The Netherlands

Prof. Thomas Wagner Max-Planck-Institute for Chemistry, Mainz, Germany

Dr. Michiel van Weele Royal Netherlands Meteorological Institute (KNMI), De Bilt, The Netherlands

Dr. David M. Winker NASA Langley Research Center, Hampton, USA

Dr. Folkard Wittrock Institute of Environmental Physics, University of Bremen, Germany

List of Tables

List of Figures

Chemical Names and Molecular Formulae

Oxygen and hydrogen containing molecules and radicals	
Oxygen	O_2
Oxygen atom	O
Oxygen atom (ground state)	$O(^3P)$
Oxygen atom (first excited state)	$O(^1D)$
Ozone	O_3
Water (Ice, liquid,vapour)	H_2O
Water (Partially deuterated)	HDO
Hydrogen peroxide	H_2O_2
Hydroxyl radical	OH
Hydroperoxy radical	HO_2
Nitrogen compounds	
Nitrogen	N_2
Nitric oxide	NO
Nitrogen dioxide	NO_2
Nitrous oxide	N_2O
Nitrate radical	NO_3
Nitric acid	HNO_3
Dinitrogen pentoxide (nitric acid anhydride)	N_2O_5
Peroxynitric acid	HNO_4
Ammonia	NH_3
Hydrogen cyanide	HCN
Oxidised carbon	
Carbon monoxide	CO
Carbon dioxide	CO_2

(continued)

Organic compounds	
Methane	CH_4
Ethyne (acetylene)	C_2H_2
Ethane	C_2H_6
Ethene (ethylene)	C_2H_4
Methanol	CH_3OH
Formaldehyde	HCHO
Formic acid	HCOOH
Glyoxal	CHOCHO
Acetone	CH_3COCH_3
Peroxyacetyl nitrate (PAN)	$CH_3COO_2NO_2$
Halogen compounds	
Chlorine nitrate	$ClONO_2$
Hypobromous acid	HOBr
Hypochlorous acid	HOCl
Bromine nitrate	$BrONO_2$
Hydrogen fluoride	HF
Hydrogen chloride	HCl
Methyl chloride	CH_3Cl
Halogen radicals	
Chlorine monoxide	ClO
Bromine monoxide	BrO
Iodine monoxide	IO
CFCs	
CFC-11	$CFCl_3$
CFC-12	CF_2Cl_2
CFC-113	$Cl_2FCCClF_2$

(continued)

HCFCs	
HCFC-142b	ClF_2CCH_3
HCFC-22	$CHClF_2$
Sulfur compounds	
Sulfur dioxide	SO_2
Hydrogen Sulfide	H_2S
Dimethyl Sulfide DMS	CH_3SCH_3
Carbon disulfide	CS_2
Sulfuric acid	H_2SO_4
Carbonyl sulfide	OCS
Sulfur hexafluoride	SF_6

(continued)

A Full list of Abbreviations and Acronyms is given in Appendix C.

Chapter 1
Tropospheric Remote Sensing from Space

John P. Burrows, Ulrich Platt and Peter Borrell

1.1 Remote Sensing and the Scope of the Book

The study of the distributions and amounts of trace constituents in the troposphere, using satellite instruments orbiting some 800 km above in space, is a science that has developed within the past twenty or so years and which is having a large influence on both monitoring the global and regional atmospheric environment, and within the research field of atmospheric chemistry. The field is a relatively new scientific discipline within the field of Earth Observation.

Remote sensing is the act of obtaining information about an object from a distance. In the context of our atmosphere, it comprises the detection of changes in electromagnetic radiation reaching the observer and the subsequent determination of the abundance of trace constituents in the atmosphere, such as trace gases, including water vapour, and particles such as aerosols, clouds, smoke and dust.

Our own eyes and ears evolved as remote sensors, and we all use remote sensing when looking at clouds, smoke, dust and hazes. Chemical constituents are not so evident, but used to be seen in the orange plumes of nitrogen dioxide from reactors in chemical factories and, unhappily, from the greenish tinge of chlorine over the battlefields of the First World War. We are able to see, i.e. use remote sensing, through the atmosphere because the principal gases, nitrogen, N_2 and oxygen, O_2, are largely transparent in the visible region of the spectrum.

Remote sensing principally utilizes the interaction of electromagnetic radiation with the trace constituents; thus the absorption, emission and scattering of radiation by molecules or particles results in changes in the spectral or temporal

J.P. Burrows (✉)
Institute of Environmental Physics (IUP), University of Bremen, Germany
and
NERC Centre for Ecology and Hydrology, Wallingford, United Kingdom

U. Platt
Institute of Environmental Physics (IUP), University of Heidelberg, Heidelberg, Germany

P. Borrell
P&PMB Consultants, Newcastle-under-Lyme, United Kingdom

J.P. Burrows et al. (eds.), *The Remote Sensing of Tropospheric Composition from Space*,
Physics of Earth and Space Environments, DOI 10.1007/978-3-642-14791-3_1,

characteristics of the radiation. These changes are then interpreted to reveal information about the object.

In the context of this book, remote sensing refers to obtaining information on trace constituents using spectroscopic instruments mounted on satellite platforms. These generally give a global coverage and provide data on global and regional scales.

As indicated below, remote sensing has been extensively used for studying the atmosphere, particularly air pollutants such ozone, O_3, whose existence was discovered by Christian Schönbein (Schönbein 1840), nitrogen dioxide, NO_2, and sulfur dioxide, SO_2 both at ground level and vertically using purpose-built lidar instruments.

In addition, measurements of the global distribution of trace constituents are needed for both environmental monitoring and to test our scientific understanding of the transport and transportation of pollution, air quality and global climate change.

The creation by the governments of the world of the Global Observing System of Systems, GEOSS (GEOSS 2009), which builds on initiatives such as the Integrated Global Observing System, IGOS, is recognition that accurate data is needed by the international policymakers attempting to manage the Earth's resources and control atmospheric pollution from the local to the global scale. Measurements of tropospheric constituents from space will play an important part in this system.

The progress in the remote sensing of the troposphere has been dramatic in the past two decades. The advances have been facilitated in part by technological improvements, the creation of the programmes by space agencies and also by the scientific community through individual work and through national and international projects (Borrell et al. 2003).

So we have moved from an age of scepticism about the value of remote sounding of trace tropospheric constituents from space, to a golden age of first observations and exploitation for science. We are now at the beginning of an age of operational applications and services. The measurements will, we hope, be used to achieve sustainable development delivering an environment having its ecosystem services intact and maximising human well being

The purpose of this book is to describe the state of the art in the field, to explain the technology and techniques used and to demonstrate some key recent findings and results. The book begins in this chapter with an historical perspective of the field together with some the basic physical principles and knowledge required to understand remote sensing from space. The following chapters present the techniques and solutions to the problems associated with the various spectral regions (ultraviolet, visible, infrared, microwave) in which observations are made. Aerosols and clouds, which are covered in the next two chapters, are of particular interest – they not only pose problems for making observations of the lower troposphere from space but they offer intrinsic interest themselves in the fields of climate change, air quality and atmospheric chemistry. Of special importance is the accuracy and reliability of remote sensing data; these issues are covered in a chapter on validation. The final section of the book is concerned with the exploitation of data. There is a chapter on observational results, giving not only results of current interest from individual studies, but also examples of the use of data from two or more satellites to obtain insights into particular atmospheric environmental problems. A chapter follows on the comparison of global and regional observations with chemistry-transport models and climate models, and the added value that the interaction brings to both. The book concludes by summarising our progress and indicating likely future developments in the field. It also includes suggestions for the actions needed if we are to have the global observation system that the Earth needs to assess its present, deteriorating state.

Also included in the book is a list of satellite instruments used for tropospheric composition measurements (Appendix A), an Atlas with a global view of several complementary parameters useful in interpreting satellite data (Appendix B) and, in Chapter 8, an Atlas of global distributions of most common tropospheric trace substances. This is also accessible as a pull-out section in the rear cover. Appendix C is a list of the abbreviations and acronyms which prevail in this field. Appendix D gives a set of time lines for future missions which highlight the likely gaps in our coverage in future years, unless action is taken soon.

Overall the book is addressed to anyone interested in Earth observation and the details of remote sensing of trace gases, clouds and aerosols, in particular students of the environmental sciences, and scientists working in the field. It gives a timely account of the recent developments in this emerging area, which are proving to be of much utility in the atmospheric environment.

1.2 Earth Observation and Remote Sensing

The science of Earth observation, the study of the Earth and its atmosphere using space-based instrumentation is a new field, of much importance to the modern world. Since the industrial revolution, both the population of the Earth, and its standard of living, have been increasing nearly exponentially, as a result of the availability of low cost energy and materials from fossil fuels. However such activity results in pollution and the growth has resulted in air pollution intensifying on local scales and expanding regionally and globally. Such anthropogenic activity is now changing the conditions in much of the lower atmosphere and at the Earth's surface globally and, it is generally agreed, is changing the state of the atmosphere leading to climate change and global warming.

In order to assess these changes detailed measurements on all scales, both to monitor our changing atmosphere and to test and verify our current knowledge and understanding of the Earth–atmosphere system, are needed, and those from space potentially meet this need on regional and global scales.

Over the last 50 years, remote sensing has developed at a great pace through technical advances in scientific instrumentation, optics and rocketry. The period has seen a pioneering voyage of discovery, and Earth observation has come of age during this period. These advances have been made possible by the ingenuity of mankind, powered by the availability of cheap energy from fossil fuel combustion.

Much public attention has been focused on man's exploration of space, and the greatest use of satellites is probably for military surveillance, communications and television. However the most important outcome of the space age has been perhaps the development of remote sensing techniques for studying our atmosphere. Techniques often developed initially for ground based and aircraft platforms, have been adapted to exploit the unique advantages of the observational geometry available from space, with both passive and active remote sensing techniques being exploited for Earth observation.

Passive remote sensing exploits the measurement and analysis of the electromagnetic radiation coming from the sun and the Earth itself, after it has passed through the atmosphere. In active remote sensing, artificial sources of radiation are used to detect the back scattered radiation and the temporal evolution of changes in intensity.

Examples of parameters or objects, which are targets for Earth Observation include:

- surface colour (vegetation; "land use", ocean colour),
- albedo and/or surface spectral reflectance,
- cloud cover,
- aerosol properties,
- thunderstorms (flash frequency),
- fire parameters,
- surface and air temperature,
- wind velocity and direction,
- ocean waves,
- sea ice,
- column densities of trace gases,
- vertical profiles of trace gases.

The production of meaningful data requires instrument development, data acquisition, retrieval theory, data analysis and interpretation. The results in the geophysical sciences provide "fields", and the goal is to determine these fields and observe their changes in time and space:

- pressure and temperature fields (e.g. meteorology),
- gravity fields (e.g. geophysics, planetology),
- electrostatic fields (e.g. geophysics, planetology),
- electromagnetic fields (e.g. geophysics, geology, geography).

The spatial and temporal variations in the radiation field are measured by sensors yielding the following quantities:

- radiometric parameters for electromagnetic radiation: geophysical calibrated intensities, polarisation etc.,
- spectral dependence of radiometric parameters.

The aim of tropospheric work is to produce column densities and concentration profiles of gases and aerosols. Since the troposphere can only be observed from space through the stratosphere, then knowledge of the stratospheric components is also needed. In addition, the tropospheric concentrations are directly related to surface sources, both natural and anthropomorphic, and are strongly influenced by air movements, clouds and precipitation. Thus a variety of satellite-based information is needed for a full interpretation of the results.

1.3 Atmospheric Remote Sensing from Space

This section provides some aspects of the programmatic development of remote sensing, together with a short introductory review of some of the most relevant developments in remote sensing in the solar, thermal infrared and microwave spectral regions. These are expanded upon and developed in detail in the subsequent chapters.

A full list of satellite instruments for observing tropospheric chemical composition with solar back-scattered radiation is given in Appendix A.

1.3.1 Pre-Satellite Days

Remote sensing from elevated platforms is not a new concept; such measurements were foreseen by Benjamin Franklin, who, in 1783 as the American ambassador to France and Belgium, observed Montgolfier Balloons flying over Paris and predicted their use for military reconnaissance purposes. The remote sensing of the Earth's surface from above began in the nineteenth century with the first aerial photographs. A little over a decade after the invention of photography, the French photographer Gerard Felix Tournachon (1820–1910) obtained the first aerial photographs over Paris in October 1859. He later used the same technique to map the countryside. The military applications of remote sensing from balloons were recognized immediately with aerial photography being used by the French in 1859 and by the Northern Army under the command of General McClelland in the American civil war.

Following the invention of manned flight by the Wright brothers at the beginning of the twentieth century, aircraft were exploited extensively for civil and military reconnaissance. In addition low spectral resolution spectroscopic methods, including colour photography, enabled different types of surfaces (vegetation, deserts, etc.) to be characterized. Remote sensing of the surface from aircraft and balloon platforms still remains an important element in Earth observation.

Observations of atmospheric trace constituents using remote sensing also have a long history. The aurora borealis has long been a source of curiosity for the Nordic peoples initially being attributed to divine powers (Roach and Gordon 1973; Bone 1991; Brekke and Egeland 1994). The development of our understanding of atomic and molecular spectroscopy in the nineteenth and twentieth centuries enabled the observation of air glows to be used to retrieve the amounts and abundances of atoms, molecules and ions in the mesosphere and thermosphere. Another early success of ground based atmospheric remote sensing was the discovery of the stratospheric ozone layer: in 1879 Cornu discovered that absorption of solar ultraviolet radiation was occurring in the upper atmosphere and Hartley in 1880 attributed this absorption to O_3.

1.3.2 Some Historical Milestones in Satellite Remote Sensing

In the following some historical organisational milestones are presented, some of which are expanded in the subsequent chapters of this book. In 1947, captured German V2 rockets were modified by the U.S. military to take pictures of clouds from altitudes of 110–160 km. In July 1955, President Eisenhower announced the intention of the United States to launch satellites for earth observation within the International Geophysical Year, IGY. With the launch of Sputnik by the Soviet Union in October 1957 mankind's exploration of space began.

The first American satellite, Explorer 1, was launched into orbit in January 1958 and carried experiments to study the magnetosphere which led to the discovery of the Van Allen radiation belts. July 1958 saw the creation of the US National Aeronautics and Space Administration, NASA, which aimed to study all aspects of the civilian exploration of space. Within the IGY activity, the Vanguard 2 was successfully launched in January 1959 and measured the Earth's albedo. Canada became the third country to launch a man-made satellite into space with the launch of Alouette-1 in 1962, whose payload was used to study the ionosphere.

In these early days, primitive remote sensing payloads were often carried by astronauts in the first satellite missions. The resultant collection of photographs of the Earth and spectrometric measurements, collected by the NASA Mercury and Gemini astronauts, was of pioneering significance. It also led to the development of missions to explore systematically the Earth and its atmosphere from space and in the 1960, resulted in the successful Nimbus series of satellites, and subsequently the NASA Mission to Planet Earth.

In the USA, not only NASA but also the Department of Defence (DOD), and the National Oceanic and Atmospheric Administration (NOAA) developed satellite systems for meteorological applications. NOAA operates two types of satellite systems: geostationary satellites and polar-orbiting satellites. Geostationary satellites constantly monitor the western hemisphere from an altitude of ~36,000 km, and polar-orbiting satellites circle the Earth and provide global information from ~800 km above the Earth. The instruments on board these address the evolving needs to supply data for numerical weather prediction, NWP. In 1994 the National Polar-orbiting Operational Environmental Satellite System (NPOESS), which will be the new generation of low earth orbiting environmental satellites, was created with contributions from DOD, NOAA and NASA. Recently the system has been separated again into independent but complementary parts.

European space activity began in the early 1960s with the formation of ESRO (European Space Research Organisation) and ELDO (European Launcher Development Organisation). In 1975 ESRO and ELDO merged to form ESA the European Space Agency. ESA developed its programmes in Earth Observation for research and operational meteorological purposes. In addition, ESA developed the first European Geostationary satellite, Meteosat, whose operation was transferred to the agency EUMETSAT (European Organisation for the Exploitation of Meteorological Satellites), after its creation in 1986. ESA and EUMETSAT collaborate to provide Europe with explorer and operational meteorological Earth observation missions. The European MetOp series of polar orbiting platforms, complements the Meteosat Second Generation for operational meteorology, and began its operation in late 2006.

The EU has created its Global Monitoring of Environment and Security programme, GMES. GMES has many aims including the provision of an adequate space segment for environmental monitoring. The EU, ESA and EUMETSAT are contributing to GMES. GMES has become Europe's contribution to GEOSS.

In addition to the activities in Europe and USA, the Canadian Space Agency has launched atmospheric constituent monitoring missions as have the Japanese Aerospace Space Agency, JAXA. The Chinese, Indian, and Korean space agencies are also developing instrumentation for atmospheric remote sensing from space.

1.3.3 Tropospheric Remote Sensing Using Back-Scattered Solar Radiation

A full list of satellite instruments for observing tropospheric chemical composition with solar back-scattered radiation is given in Appendix A.

Much of this book describes results and applications from the use of passive techniques in the solar spectral range. The technique of measuring the spectral absorption structures of trace gases was pioneered by Noxon (1975) using a ground based spectrometer, and the first balloon measurements of this type were made by Pommereau, Goutail and colleagues at CNRS (Pommereau 1982; Pommereau and Piquard 1994a; 1994b). A ground based network followed using the same approach. Independently, the Differential Optical Absorption Spectroscopy technique, DOAS, was developed initially for long path active remote sensing measurements in the troposphere (Perner and Platt 1979; Platt and Stutz 2008) and later applied to measurements with ground based, aircraft and satellite borne instrumentation. Long path infrared spectroscopy has also been used under atmospheric conditions to detect species in chambers and outside (Tuazon et al. 1980).

In the first satellite missions, observations of mesospheric emissions and stratospheric trace gases were targeted, an important scientific emphasis being to understand the global behaviour of the upper atmospheric ozone, O_3, and the dynamics of the middle atmosphere. The first measurements of O_3 were made within the Soviet programme, and NASA began a substantial research programme in the early 1970s. One focus for measurements within the Nimbus series of satellites was the use of back scattered ultraviolet radiation, BUV instruments, to determine the vertical profile and total column amount of O_3. The Solar Backscattered Ultraviolet, SBUV, instrument and Total Ozone Mapping Spectrometer, TOMS were launched together on Nimbus 7 in 1979 (Heath et al. 1973; 1975). Subsequently an improved SBUV called SBUV-2 became part of the NOAA operational meteorological system and a series of TOMS instruments were built and flown by NASA (see Appendix A).

The recognition by Fishman that combining the retrieved vertical profile of O_3, derived from SBUV measurements, with the total O_3 column amount, retrieved from TOMS measurements, could be used to separate tropospheric from stratospheric O_3 (Fishman et al. 1990) for the remote sensing of tropospheric gases. O_3 was the first trace gas to be retrieved in the troposphere from space. The approach was later extended to combine TOMS with information from the Stratospheric Aerosol and Gas Experiment, SAGE, which provided retrievals of aerosol, O_3 and water vapour, H_2O, in the stratosphere and upper troposphere under cloud free conditions.

In Europe, the SCIAMACHY (Scanning Imaging Absorption spectroMeter for Atmospheric CHartographY) concept was developed between 1984 and 1988 by Burrows and colleagues at the Max Planck Institute in Mainz and elsewhere (Burrows et al. 1990; 1995; Bovensmann et al. 1999). In contrast to the NASA instruments, which used selected wavelengths, SCIAMACHY aimed to measure the entire solar spectrum from the ultraviolet to the short wave infrared at a spectral resolution appropriate to the retrieval of the spectroscopic absorptions and emissions of trace gases in the atmosphere (Gottwald and Bovensmann 2010).

SCIAMACHY has a long history, which illustrates the vicissitudes that a satellite instrument must undergo before it is launched. As proposed, SCIAMACHY was to make simultaneous observations of limb and nadir back-scattered and reflected radiation from the atmosphere, as well as solar and lunar occultation, as part of a mission proposed for launch in 1988. As well as exploiting the unique capability of the nadir limb and occultation measurements, the nadir and limb measurement strategy was selected to enable stratospheric and mesospheric amounts of trace constituents, determined in limb, to be synergistically subtracted from the total columns determined from nadir observations. Thus SCIAMACHY was the first instrument intended to study tropospheric trace gas distributions. It was successfully launched on ENVISAT in February 2002 and is continuing to deliver data. Thus nearly 20 years elapsed between the first ideas and the launch.

In the meantime however, in December 1988, in response to an ESA call for instrumentation to measure atmospheric constituents for the second European Research Satellite, ERS-2, the SCIA-mini concept was proposed by Burrows, Crutzen and colleagues. SCIA-mini was a smaller scale version of SCIAMACHY and comprised instruments measuring in limb and nadir the UV visible and near-IR at the top of the atmosphere. SCIA-mini, although selected for investigation, was then de-scoped to make only measurements in nadir viewing geometry, being renamed GOME, Global Ozone Monitoring Experiment (Burrows et al. 1999). GOME was launched on ERS-2 in April 1995. Using the ideas developed for SCIAMACHY, GOME became the first instrument to fly in space, dedicated to the study of the amounts and distributions of tropospheric trace gas constituents.

The OMI instrument, proposed and developed in the Netherlands, was launched as part of the NASA Aura payload in 2004. OMI like GOME is a nadir viewing instrument. An operational follow-on of GOME was selected by EUMETSAT and ESA as part of the MetOp series of platforms in the late 1990s. The first GOME-2 instrument was launched aboard MetOp-A in October 2006 and is continuing to make successful measurements.

A more thorough review of remote sensing of troposphere constituents and parameters using solar radiation is given in Chapters 2, 5 and 6 of this book. Validation of these data products are discussed in Chapter 7 and their use for science explained in Chapters 8 and 9.

1.3.4 Remote Sensing Using Thermal Infrared in the Troposphere

The potential of using thermal infrared (TIR) and microwave radiation emerging from the atmosphere for the measurements of atmospheric parameters was recognised by the pioneers of atmospheric remote sensing. Instruments such as correlation radiometers (e.g. pressure modulators), spectrometers, interferometers and instruments utilising the heterodyne technique at longer wavelengths have all flown successfully aboard space based platforms. During the Nimbus programme the first correlation radiometers were tested in space. The MAPS experiment, launched by NASA and flown aboard the space shuttle in 1984 and again in 1994, successfully measured CO from space in limited duration missions. In 1996 JAXA launched as part of its ADEOS-I satellite, the nadir sounding FTIR instrument IMG. Although the satellite ADEOS-1 failed after 9 months, IMG data were used successfully to retrieve the tropospheric abundance of CO and several other species.

MOPITT proposed in 1988, is a Canadian contribution to the NASA Terra mission. MOPITT was launched in December 1999 aboard NASA Terra and has now made over a decade of measurements of carbon monoxide, CO. The AIRS spectrometer, flying aboard the Aqua satellite, is a spectrometer having, as its primary mission, the retrieval of water, and temperature profiles. AIRS however also retrieves CO and SO_2 in the troposphere.

The Microwave Limb Sounder, MLS, was launched in September 1991 as part of the highly successful Upper Atmospheric Research Satellite, UARS, mission (Livesey et al. 2003). MLS measured temperature, O_3, H_2O, and chlorine monoxide, ClO.

The Michelson Interferometer for Passive Atmospheric Sounding instrument, MIPAS, was launched by ESA on ENVISAT in March 2002. MIPAS is a Fourier transform infrared, FTIR, spectrometer for the detection of limb emission spectra in the middle and upper atmosphere (Fischer and Oelhaf 1996). It observes a wide spectral interval with high spectral resolution. The primary geophysical parameters of interest are vertical profiles of atmospheric pressure, temperature, and the volume mixing ratios of over 25 trace constituents.

The Tropospheric Emission Spectrometer instrument, TES, a high resolution nadir sounding FTIR, was launched on the NASA Aura platform. It has been used to retrieve a variety of tropospheric data products including tropospheric O_3 and NH_3. Solar occultation and limb scanning instruments in the TIR are able to retrieve trace gases in the upper tropospheric region.

The CNES/EUMETSAT instrument IASI is an FTIR instrument which has, as its primary operational objective, the retrieval of vertical profiles of temperature and water vapour. However it has also successfully retrieved tropospheric amounts and distributions of other tropospheric trace gases. In addtion to the thermal emission experiments, there are also solar occultation experiments, which target the upper troposphere and above. The NASA ATMOS experiment aboard the shuttle, which flew several times as part of the ATLAS mission was the first using a FTIR interferometer. The Canadian stand alone SCISAT-1 mission with the ACE instrument, which is also an FTIR interferometer has now been making measurements successfully since is launch in 2003.

Further details of the thermal infrared sounding of the troposphere are provided in Chapter 3. The developments for the measurements of trace gases cloud and aerosol data products in the microwave and sub mm regions are described in Chapter 4. A full list of satellite instruments for observing tropospheric chemical composition with thermal IR radiation is given in Appendix A and with microwave radiation in Chapter 4.

1.3.5 TROPOSAT and AT2

One aspect, often overlooked in the development of large technological projects such as the launch of satellite instruments, is the contributions made by individual scientists to the formulation of the initial scientific needs and ideas, and to the development of the spectroscopic instruments themselves. In the field of the remote sensing of tropospheric composition a user group has been formed to encourage the exploitation of data derived from satellite measurements and to suggest future missions that could or should be mounted.

The TROPOSAT project (Use and Usability of Satellite Data for Tropospheric Research) (Borrell et al. 2003) was initiated within the EUROTRAC framework (a EUREKA Environmental Project) by the editors of this book, in order to foster the development of instrumentation and retrieval algorithms and their validation and the utilization for scientific and operational applications. Initially the project received some modest support towards coordination from ESA. The aim was the exploitation of the space-based measurements of back scattered solar radiation from the atmosphere, using particularly the measurements from GOME and SCIAMACHY. Later, the project was incorporated within the framework of the EU ACCENT network of excellence as ACENT-TROPOSAT-2 (AT2) and was extended to include the utilization of measurements of thermal infrared radiation and the retrieval of aerosols and clouds. Many of the scientists supporting AT2 have contributed to the development of this book.

1.4 The Atmosphere, Tropospheric Chemistry and Air Pollution

Satellite instruments observe the atmosphere from high above and so knowledge of the structure of the atmosphere is essential to the interpretation of the results. Furthermore the trace constituents observed are part of a complex chemistry involving primary sources, rapidly reacting chemical intermediates, and products which are eventually rained out. In this section an outline will be given of the atmosphere and the chemistry of the troposphere.

1.4.1 The Physical Structure of the Atmosphere

The Earth was formed approximately 4.54 billion years ago. The Earth's atmosphere comprises the thin envelope of gas surrounding the Earth, which is held in place by gravitation. The primordial atmosphere, produced by out gassing from volcanic eruptions and the Earth's surface, had a very different composition to that of the present day, being a reducing atmosphere, quite in contrast to our present day oxidising atmosphere. About 3.8 billion years ago life appears to have started on Earth and, as a result, the biosphere began to change the composition of the atmosphere. Since this time the sun, the Earth's atmosphere and surface comprise a complex system, essential for maintaining our environment and life as we know it.

The bulk constituents of the Earth's atmosphere, molecular nitrogen, N_2, and oxygen, O_2, contrast with those of its nearest neighbours Mars and Venus, whose atmospheres are primarily composed of carbon dioxide, CO_2. The conditions within these atmospheres are effectively closer to an inorganic photochemical equilibrium, compared to the Earth's atmosphere, which is better described as a complex biogeochemical reactor.

The chemical processing within the Earth's atmosphere depends on the conditions in the atmosphere itself and at its surface. The pressure within the Earth's atmosphere obeys the barometric equation and falls off approximately exponentially as a function of height as shown logarithmically in Fig. 1.1.

The atmosphere up to ~120 km can be divided conveniently into four regions of positive and negative temperature gradient, the troposphere (Greek: well mixed region), the stratosphere (Greek: stratified region), the mesosphere (Greek: middle) and the thermosphere (Greek: heated region), as shown in Fig. 1.2. The regions are separated by the tropopause, the stratopause and the mesopause respectively.

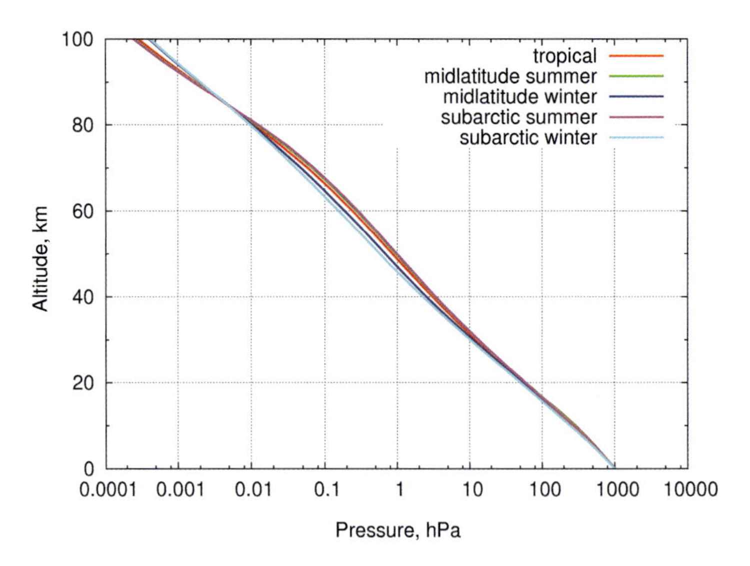

Fig. 1.1 The change in pressure as a function of altitude in the Earth's atmosphere for tropical regions, mid latitudes in summer and winter, and high latitude sub arctic summer and winter. The diagram uses data from the US Standard Atmosphere (1976), and was constructed by J.P. Burrows and S. Noel IUP, Bremen.

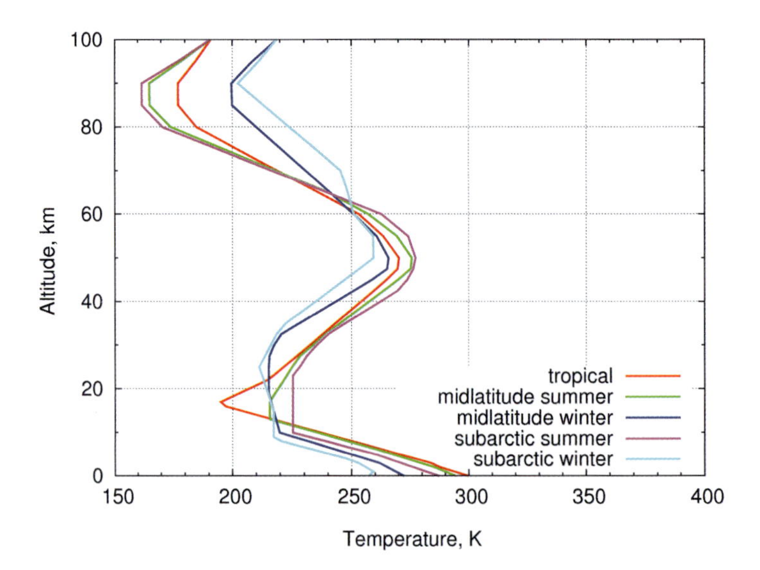

Fig. 1.2 The temperature variation in the atmosphere as a function of altitude for different latitudes and seasons. The regions of the atmosphere having different temperature gradients are: troposphere, below ~12–20 km; stratosphere, ~20 to ~50 km; mesosphere; ~50 to ~90 km; thermosphere, above ~90 km. The diagram uses data from the US Standard Atmosphere (1976), and was constructed by J.P. Burrows and S. Noel IUP, Bremen.

Two regions, the thermosphere and the stratosphere, have positive temperature gradients (or temperature lapse rates) indicating that they are being heated. The thermosphere is the region where short wavelength solar ultraviolet radiation is absorbed by atoms and molecules. Additionally interstellar dust and coronal mass ejections from the sun result in energy and matter being deposited in this region. Photo-ionisation takes place initiating complex physics and chemistry. Overall these processes result in the heating of the thermosphere.

In the mesosphere the photochemistry is considerably less, so the solar warming is diminished and the temperature gradient is negative, as expected from an adiabatic expansion of rising air.

In the stratosphere, which was designated by Teisserenc de Bort and Aßman (1929) the O_2 concentration increases rapidly towards lower altitudes, and absorption by O_2, below about 214 nm in the Schumann Runge bands and continuum, results in O atoms, $O(^3P)$, being formed. These in turn react in a third body reaction to produce ozone, O_3. The O_3 production depends approximately on the square of the O_2 concentration, as suggested by Chapman in 1929 (Chapman 1930). He coupled this with loss of O_3 by photolysis between 220 and 320 nm, and reaction with O atoms to explain the variation of O_3 in the atmosphere as a function of height.

Later it was shown that the reaction of ground state O atoms, $O(^3P)$, with O_3 is too slow to provide the O_3-loss required to bring the model into agreement with observation, but a series of catalytic cycles resulting in an odd oxygen equivalent removal of O_3 in the stratosphere (Wayne 2000) was later identified. The catalytic destruction mechanisms, by which the amount of O_3 is controlled, is complex and have been affected by human activity, namely the tropospheric release of chlorofluorocarbons and bromine-containing species, which are sufficiently long lived to be transported to the stratosphere. As a result of the strong ultraviolet absorption of O_3, the temperature increases in the stratosphere above the tropopause.

The sign of the temperature gradient changes at the tropopause and the region from the surface of the Earth to the tropopause is called the troposphere, which means well mixed. As a result of its negative temperature gradient the troposphere is a region of convective and turbulent mixing; it is this gradient that provides our weather. The lower part of the troposphere is referred to as the planetary boundary layer and is frequently separated from the free troposphere by a further temperature inversion, which has a strong diurnal cycle over land.

Biogenic emissions and most anthropogenic emissions are released directly into the planetary boundary layer; most aircraft emissions are released in the troposphere and the lower stratosphere. These emissions are processed in the atmosphere.

1.4.2 Tropospheric Chemistry

Tropospheric chemistry describes the complex myriad of reactions determining the composition of the troposphere. The chemistry and dynamics of the atmosphere as a whole are complex and coupled. While changes in the extra terrestrial solar radiation, coronal mass ejections and dust modulate the chemistry and radiation balance in the upper reaches of the atmosphere, the Earth's surface and the planetary boundary layer are the dominant source of bulk and trace constituents for the stratosphere and troposphere (Holloway and Wayne 2010).

Within the troposphere the natural sources of atmospheric constituents (gases and aerosols) include direct release from the biosphere, exchange at the surface, (land, ocean and cryosphere) lightning, natural fires, and stratospheric-tropospheric exchange. Anthropogenic activity, such as biomass burning, the combustion of fossil fuels and changes in land usage, has and is modifying tropospheric chemistry. Figure 1.3a indicates some of the processes diagramatically.

Atmospheric pollution refers to the trace constituents (chemical and aerosol) added to the atmosphere by anthropogenic activity (primary pollutants), and to constituents resulting from the reactions of the primary pollutants in the atmosphere (secondary pollutants). In general during pollution episodes, air masses contain elevated amounts of O_3, aerosol, acids, and other noxious chemical species in comparison with unperturbed air masses. The term *air quality* usually describes the chemical composition of trace constituents close to the surface of the earth, which impact on humans.

Remote sensing of tropospheric composition from space potentially provides global and regional information about the amounts and distributions of key constituents. To use this information to constrain our knowledge of the sources and sinks of atmospheric constituents requires a good knowledge of the chemical processing within the troposphere.

a Free Radical Reactions in the Troposphere

A large part of tropospheric chemistry is associated with the oxidation and subsequent transformation of chemical species released naturally from the biosphere or by anthropogenic activity to the troposphere. Oxidation is often initiated by the attack on a trace substance by the hydroxyl radical, OH, and other free radicals. Hydroxyl is a *free radical*, i.e. a molecule having an unpaired electron, and it is a particularly reactive species. OH can be regarded as the scavenger of the atmosphere, the purpose of which is to remove trace gases and pollutants and so keep the atmosphere clean. OH scavenges hydrocarbons and organic molecules by removing a hydrogen atom to form water. A single simple example is provided by methane, CH_4:

$$OH + CH_4 \rightarrow CH_3 + H_2O \tag{1a}$$

The new species, the methyl radical CH_3, is itself a free radical and reacts immediately with O_2 to form the reactive methylperoxy radical CH_3O_2:

$$CH_3 + O_2 + M \rightarrow CH_3O_2 + M \tag{1b}$$

a

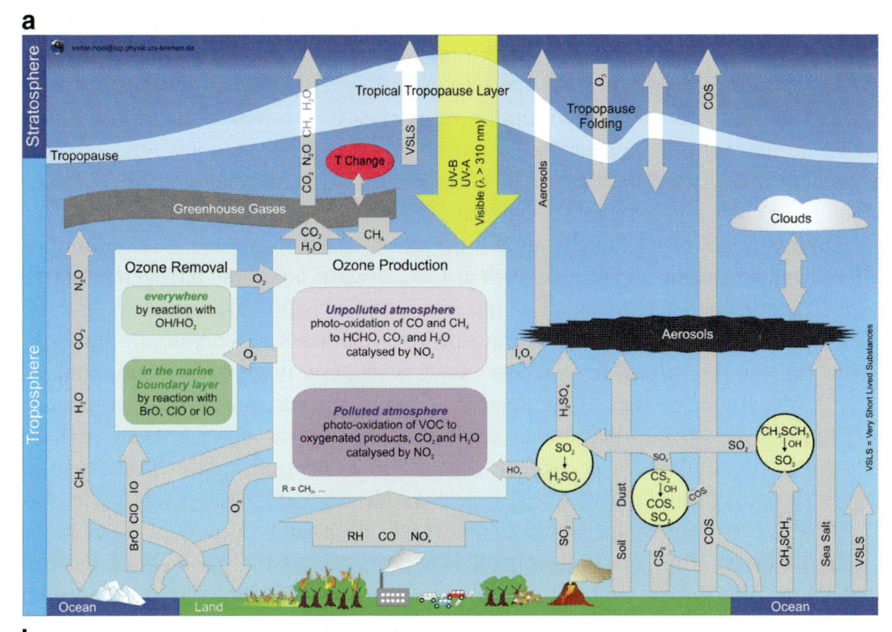

b

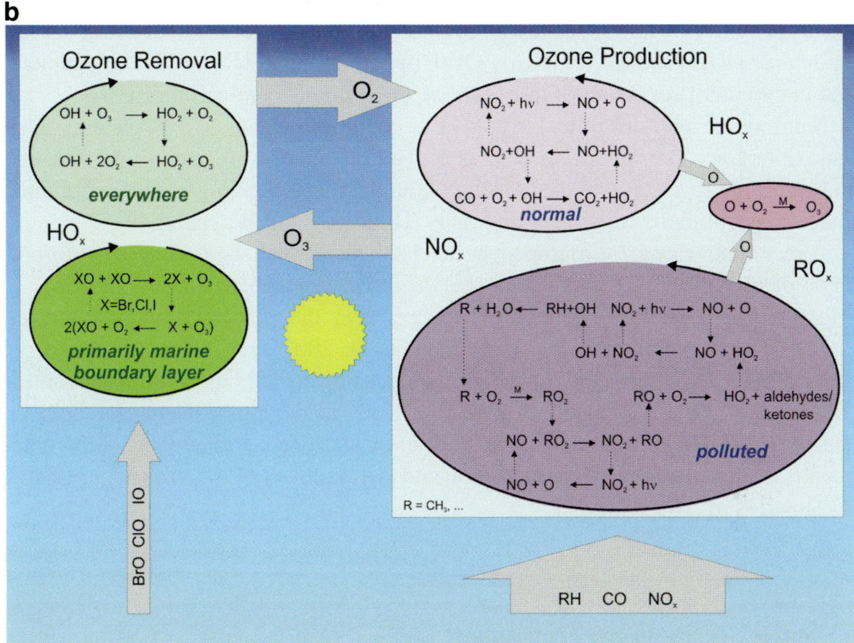

Fig. 1.3 (**a**) A schematic diagram, depicting the complex chemical and physical interactions in the troposphere. The Earth's surface is a natural and anthropogenic source of volatile organic compounds, VOC (also shown as RH), CO, NO_x and other species. Aerosols are produced by physical and chemical processes. Species are removed from the troposphere by deposition to the surface or transport up to the stratosphere. (**b**) An expansion of the boxes in (a) showing some of the chemical processes involved in the production and removal of troposphic ozone. Figures: J.P. Burrows and S. Noel, IUP, Bremen.

where M is any other molecule. Further reactions then occur generating still more species most of which can react with one another and any other reactive species present. During these processes the OH radical is re-formed so, although it only present at tiny concentrations (mixing ratio of ~10^{-11}), it is always present and active in the daytime atmosphere.

Since virtually all species react with OH, tropospheric chemistry is of much complexity, some of which is illustrated Figs 1.3a and b.

The net result of all the chemical processes in the troposphere is the formation of more stable, oxidised products such as CO_2, organic acids and nitrogen products, which result from reactions involving nitric oxide, NO, NO_2, and many others. These are then removed from the troposphere by both wet and dry deposition or by transport into the stratosphere, where they can be photolysed by UV radiation.

For many trace species of intermediate lifetime, the transport and transformation of longer-lived species represents an important removal process from polluted regions but is simultaneously a source of pollution for pristine regions. These species, of which peroxyacetyl nitrate, PAN, is an example, are termed temporary reservoirs.

Prior to the 1970s the global source of tropospheric O_3 was thought to be transport down from the stratosphere. However, one of the products of the free radical reactions, following further reactions involving NO and NO_2, is the secondary pollutant, O_3. Thus the anthropogenic and natural emissions from the surface to the troposphere result in catalytic cycles (Fig. 1.3b) which, depending on the amount of NO_x, produce or destroy O_3 (Crutzen 1973; 1974). So it is now thought that the greater proportion of tropospheric O_3 is photochemical in origin.

Summer or photochemical smog is a condition where O_3, PAN and other obnoxious trace substances, as well as aerosol precursors, are produced efficiently and locally or regionally in the planetary boundary layer leading to severe pollution under stagnant meteorological conditions (Holloway and Wayne 2010).

A recent development in tropospheric chemistry is the recognition of the importance of chemistry at the boundary between the atmosphere and the cryosphere. Snow and ice interact chemically with gases in the planetary boundary layer. Photochemical reactions in snow (Grannas et al. 2007), result in emissions from the surface. The release of inorganic halogens from ice, in particular bromine (Barrie et al. 1988; Simpson et al. 2007), leads to depletion events for both O_3 and mercury, Hg, in polar regions (Steffan et al. 2008). Release of iodine compounds results in new particle formation and is the subject of much current research (O'Dowd et al. 1998; 2002).

b Stable Species in the Troposphere

The OH radical cannot attack stable oxidised species, so CO_2 produced both by combustion processes and by photochemical oxidation largely remains in the atmosphere, where it has an extremely long chemical lifetime and, since the beginning of the Anthropocene (see below), an increasing concentration. Over land it is removed by plant photosynthesis and the small signal from the seasonal variation can be seen imposed on the Mauna Loa records, and can be observed from space (see Chapter 8). At the ocean surface, CO_2 participates in an inorganic equilibrium with bicarbonate, HCO_3^-. The net effect is acidification of the oceans and the gradual removal of CO_2.

Another group of compounds unreactive to OH are the chlorofluorocarbons, CFC. These are thus inert in the troposphere and were seen as ideal products for refrigerants, and as blowing agents for plastics. As they don't dissolve appreciably in water, their only fate is very slow diffusion up into the stratosphere; there, they are photolysed into atoms and free radicals which are very active in reducing stratospheric O_3 concentrations. The stratospheric ozone hole, which appears in spring over the Antarctic, leads to increased intensities of surface UV irradiation with deleterious effects on surface ecosystems.

As with CFCs, transport to the stratosphere is the fate of all longer lived species released from the Earth's surface, such as the greenhouse gases, CO_2, CH_4, and nitrous oxide, N_2O. They are mixed by transport, turbulence and molecular diffusion within the troposphere, prior to their exchange to the stratosphere. Once in the stratosphere, they are removed either by photolysis due to the increasing amounts of shorter wavelength solar UV radiation, or reaction with atoms and free radicals, which are more abundant in the stratosphere. The oxidized products are eventually removed from the stratosphere by slow diffusion down to the troposphere and subsequent rain-out.

1.4.3 Air Pollution and Environmental Policy

Perhaps the earliest recorded environmental policy was that of Edward 1 of England in 1306 when he banned the burning of coal in London. The quality of city air remained a matter of public concern over the centuries and was, for example, addressed by Evelyn in "the inconvenience of the aer and smoake of London" in the seventeenth century.

In 1905 the word smog, combining fog and smoke, was coined by Dr. des Voeux to describe the air conditions often observed in large cities. This smog was a characteristic of the boundary layer conditions of London in winter at this time, where the emissions from the burning of coal for domestic heating and industrial processes, coupled with boundary layer humidity, resulted in conditions of low visibility which had adverse health effects. Winter smog, which includes carbon particulates and sulfur dioxide, SO_2, is chemically reducing in character.

A new kind of smog was observed for the first time in the 1950s in southern California, and was termed summer, or photochemical, smog (Haagen-Smit 1952; Haagen-Smit et al. 1952). It is oxidizing in character and also detrimental to health. In the early part of the twentieth century, the population in the region of Los Angeles grew rapidly, as did the large fleet of automobiles. The resultant emissions from transportation and also power generation, coupled with the intense sunshine (i.e. actinic radiation), the local topography and related local meteorology, yielded particularly favourable conditions for the production of summer smog.

During the past 50 years this type of smog has become an increasingly important environmental issue for urban areas around the globe, particularly in the tropical and semi-tropical regions. While smog control measures have led to reduced pollution in the USA and western Europe, in the summer of 2003 pollution in Europe reached some of the highest levels ever recorded. Severe photochemical smog is now endemic in the vast population centres of developing countries throughout the world.

As a result of global climate change, occasional episodes of pollution, currently considered as extreme, are likely to become more frequent later in the century (Schär and Jendritzky 2004; Schär et al. 2004). The recent attempts to minimise air pollution during the Olympic Games in Beijing (Wang et al. 2009) have further highlighted the potential adverse health effects of photochemical smog, and the drastic measures required to control pollution levels in megacities.

Since the industrial revolution, the rapid growth of the human population and its standard of living has been facilitated by the availability of relatively inexpensive energy, generated in large part from the combustion of fossil fuels. Large changes in land usage have also occurred, where the temperate and tropical forests have been cut or burned down to provide agricultural land. To provide power and to feed the world's population, agriculture has become increasingly intensive. The agro-chemical industrial sector now supplies artificial fertilisers, pesticides and other products to the world's agricultural industry on an unprecedented scale. This has resulted in persistent organic pollutants being distributed globally, because they are resistant to environmental degradation through chemical, biological, and photolytic processes. Concern about the global impact of anthropogenic activities on the Earth system has led to the definition of a new geological age, the Anthropocene (Crutzen and Stoermer 2000), which is considered to have begun with the industrial revolution at the end of the eighteenth century.

The research community, initially driven by scientific curiosity, has become involved in quantifying the impact of anthropogenic activity on our environment by both measurement and modelling activities. Comprehensive assessments of our knowledge about ozone, O_3, have evolved since the first concerns about anthropogenically induced destruction of stratospheric O_3 were raised in the 1970s. In agreement with the parties to the Vienna Convention for the Protection of the Ozone Layer and the Montreal Protocol and its amendments on substances that deplete the ozone layer, the WMO, UNEP and their partners now sponsor Ozone Assessments, which appear at four yearly intervals, the latest being the Scientific Assessment of Ozone Depletion: 2006 (WMO 2007); the next one, WMO 2010, is due for publication in 2011.

Twenty years ago, the Intergovernmental Panel on Climate Change, the IPCC, was created by UNEP and the WMO to provide a scientific consensus for policy-makers about our knowledge and understanding of climate change. The IPCC publishes specialist reports on topics relevant to the implementation of the United Nations Framework Convention on Climate Change (UNFCCC), an international treaty that acknowledges the possibility of harmful climate change. The most recent report by the IPCC (2007) confirmed the findings of the previous reports and concluded that there is convincing evidence that climate change is being driven by human activity. The success of the IPCC was recognised in the award of the Nobel Peace Prize in 2007.

The Executive Body of the 1979 UNECE Convention on Long-range Transboundary Air Pollution (LRTAP 2004) established the Task Force on Hemispheric Transport of Air Pollution (HTAP 2007) to develop a fuller understanding of intercontinental transport of air pollution in the northern hemisphere. LRTAP has been important in establishing national and European directives about air pollution. The current debates about the use of agricultural land for the production of bio-fuels and the potential impact of ship emissions are other examples of the challenge facing policymakers in modern times. Overall the assessment process has become embedded in international environmental policy related to air pollution.

1.4.4 Environmental Issues of Relevance to the Troposphere

In the post war period scientific research was linked to the public recognition of a series of environmental issues, which are driven by anthropogenic activity.

(a) Global increase of tropospheric O_3 and degradation of air quality
(b) The transport and transformation of pollution
(c) Biomass burning, and fire
(d) Persistent organic pollutants
(e) Acid deposition
(f) Fine particulate matter with dry diameters smaller than 1, 2.5 or 10 μm, (PM1, PM2.5 or PM10)
To these the following must be added.
(g) Global climate change
(h) Stratospheric ozone depletion, which have tropospheric dimensions

These issues have also led to the recognition of the complex coupling and feedback within the Earth atmosphere system and the need to quantify the mechanisms and processes driving it.

a Global Increase of Tropospheric Ozone and the Effect on Air Quality

Following its discovery in Los Angeles, summer smog is now observed in many urban and rural locations in the northern and southern hemispheres, particularly tropical and sub-tropical conurbations. Emissions from fossil fuel combustion, biomass and waste burning are considered to be the causes of poor air quality. Under typically anticyclonic conditions O_3, aerosols and other toxic compounds can be generated in amounts which are biologically hazardous. The development of the

Asian economies in the past 30 years has been coupled with increasing emissions of a variety of gases and particulate matter. Parts of Asia now have some of the most dramatic episodes of atmospheric pollution and poorest air quality in the world. As noted previously, the heat wave during the European summer of 2003 produced a severe health impact and it is thought that global climate change will result in a higher frequency of such summers in the coming years (Schär et al. 2004).

The recognition of the detrimental effects of smog conditions has led to air pollution and air quality becoming important national and international issues. Air quality addresses the boundary layer condition of air, in particular that experienced close the surface by the biosphere. O_3 is toxic and considered to have negative health impacts at boundary layer mixing ratios above 60 ppb, which is only about 50% higher than the typical mixing ratio observed in summer in Europe. Fine particulate matter is considered to have both short and long term health effects leading to premature death and reduced life expectancy. While at the outset of the 1960s, the troposphere was still perceived to be relatively chemically inactive and globally tropospheric O_3 was considered to originate from the stratosphere and to be removed by dry deposition at the Earth's surface, Crutzen (1973) pointed out that O_3 can be catalytically produced and destroyed in the troposphere. Later analyses of older and more recent measurements indicate a global increase in the amount of tropospheric O_3 (Volz and Kley 1988; Horowitz 2006) by a factor of two to three, with which Models agree (Pavelin et al 1999).

b The Transport and Transformation of Pollutants

As already indicated the importance of transboundary pollution was recognised in the LRTAP Convention (HTAP 2007), and substantial efforts are being made to develop a fuller understanding of intercontinental transport of air pollution in the northern hemisphere. The transport and transformation of pollution around the globe has been recognized in the past decade to be an important part of the anthropogenic modification of the troposphere. For example, in the northern hemisphere, Europe is affected by American emissions, European pollution is transported to the Arctic and Asian pollution to North America. Similarly, in the southern hemisphere, plumes from biomass burning can be tracked from South America to Africa and Africa to Australia.

c Biomass Burning and Fire

Biomass burning and fire has been identified as a potential source of atmospheric pollutants (Watson et al. 1990). Biomass burning and fire are naturally occurring phenomena but there is an increasing anthropomorphic contribution, particularly in the tropics. It is now a major global environmental issue in its own right (Crutzen and Andreae 1990; Levine 1991). Biomass burning is one of the dominant sources for anthropogenic aerosol (Seinfeld and Pandis 1998) and an important source of absorbing aerosol which contributes to warming the atmosphere, thus reinforcing the effect of greenhouse gases and opposing the cooling effect of most other aerosol species. In addition, fires used for cooking in developing countries contribute to the high concentrations of black carbon with an associated low single scattering albedo.

d Persistent Organic Pollutants

Another important environmental issue relates to persistent organic pollutants, POPs. The United Nations Environment Programme Governing Council started to investigate POPs, beginning with a short list of the following twelve POPs, known as the "dirty dozen": aldrin, chlordane, DDT, dieldrin, endrin, heptachlor, hexachlorobenzene, mirex, polychlorinated biphenyls, polychlorinated dibenzo-*p*-dioxins, polychlorinated dibenzofurans, and toxaphene. Since then, the list has been extended generally to include such substances as carcinogenic polycyclic aromatic hydrocarbons, PAHs, and certain brominated flame-retardants, as well as some organometallic compounds such as tributyltin.

It is unlikely that such substances can be observed directly from space or the ground by remote sensing techniques, as they have low volatility. However knowledge of meteorological parameters, and aerosols as well as biological vectors is required to understand their behaviour. One related example of this general type, is the deposition of mercury, Hg, at high latitudes by a natural process involving halogen oxides.

e Acid Deposition

Acid deposition is a phenomenon of the natural world. Dimethyl sulfide, DMS, and other sulfur containing compounds are emitted by a variety of sources including oceans and volcanoes. These sulfur compounds are oxidised to form sulfur dioxide, SO_2, which reacts in the gas and liquid phases with OH and hydrogen peroxide, H_2O_2, to produce sulfuric acid, H_2SO_4 (Wayne 2000). This has a low vapour pressure and forms cloud condensation nuclei, CCN.

NO_2 reacts with OH to form nitric acid, HNO_3, in the gas phase. The acid anhydride, dinitrogen pentoxide, N_2O_5, formed by the reaction of NO_2 with the nitrate radical, NO_3, readily produces HNO_3 on reaction with H_2O on surfaces or in the liquid phase. HNO_3 has a high solubility in water and, therefore, accumulates in aerosols and cloud droplets, and is rained out.

In an unpolluted atmosphere the natural sources of SO_2 and NO_2 provide a mechanism by which the pH of aerosol and rain are expected to be slightly acidic. In tropical rain forests organic acids produced as a result of biogenic emissions provide a further natural source of acidity.

However, large amounts of NO and NO_2 (together denoted as NO_x) and SO_2 are produced during fossil fuel combustion, which makes the precipitation still more acidic and results in so-called acid rain. In Europe public recognition of the consequences of acid deposition peaked in the middle of the 1980s with the concern over forest die-back and the acidification of the Scandinavian lakes. This led to the introduction of sulfur scrubbers in the power plants in western Europe.

However acid deposition remains an important environmental issue and research area (Heij and Erisman 1995). Similarly measures, such as catalytic converters in cars, improved design of furnaces in power stations, chemical removal of acid gases, are now also taken to reduce the emissions of NO_x.

f Global Climate Change

Global climate change resulting from man's activity is currently receiving much public attention, having been recognised as a threat to both the biosphere and mankind. Le Treut et al. (2007) have recently provided an interesting review of the development of our understanding of global climate change and greenhouse gases using previous studies by Fleming (1998) and Weart (2003). The first studies of relevance for global climate change date back to the seventeenth century. In the eighteenth century, the mathematician Fourier identified the basic processes of the atmosphere which lead to the greenhouse effect, namely solar radiation is transmitted whereas thermal infrared radiation is absorbed. Tyndall identified the gases CO_2 and H_2O as having this characteristic property, whereas O_2 and N_2 do not. Arrhenius (1896) and others discussed the issue of global warming, caused by the injection of the greenhouse gases such as CO_2 into the atmosphere on a quantitative basis and pondered its impact on glaciers. In the twentieth century progress has been rapid in developing our knowledge of the carbon cycle.

Other gases are often more effective than CO_2, and this has led to the definition of the "global warming potential" of trace gases. The list of greenhouse gases now comprises many species, including water vapour, CO_2, CH_4, N_2O, CFCs and tropospheric O_3. Aerosols have an appreciable effect on the atmospheric radiative balance, as do clouds which have an overall cooling impact that partly compensates for the warming effect of greenhouse gases; however, locally, they are often warming.

The publication of the fourth Assessment Report of the IPCC (2007) and the award of the Nobel Peace prize for 2007 to the IPCC and former USA Vice-President Al Gore for their championing of the issue of climate change has brought an unprecedented but much-needed focus on global warming. The recent assessment of the IPCC has pointed out again the importance of the combustion of fossil fuels as the main driver of global climate change. The United Nations Convention for Climate Change, UNFCCC, has tried to establish targets for reducing emissions but there is still no general international acceptance, even for the necessity of such reductions.

g Stratospheric Ozone Depletion and Its Impact on the Troposphere

A dramatic example of the role of science in the understanding and prediction of an environmental issue was the discovery that stratospheric O_3 is depleted by the tropospheric release of chlorofluorocarbon compounds, CFCs. Lovelock et al. (1973) made the first measurements of CFCs in the troposphere. Molina and Rowland (1974) then proposed that their presence might lead to a global depletion of the stratospheric O_3 layer in the mid and upper stratosphere. About a decade later large depletions of O_3, measured above the British Antarctic Survey measurement station at Halley in Antarctica, were reported during the Austral spring (Farman et al. 1985). Subsequently the satellite measurements of O_3 from the NASA TOMS and SBUV instruments showed O_3 to be depleted throughout the polar vortex in spring and this behaviour was termed the "Ozone Hole". No atmospheric models had predicted such an effect, because the key chemical processes were not then recognised. During autumn the polar vortex in the lower stratosphere splits off and thereby generates a special set of conditions. The mechanism of the ozone hole formation within the vortex is complex comprising heterogeneous and homogeneous chemical reactions in the lower stratosphere (WMO 2007).

The consequences of a reduction of the stratospheric O_3 layer affect both the stratosphere and the troposphere. In particular the amount of UV-B radiation at the surface and in the troposphere is increased. This biologically damaging radiation also initiates tropospheric photo-oxidation and the generation of tropospheric O_3 is increased. It has been recognized that global warming influences both the intensity and duration of the stratospheric O_3 loss (Shindell et al. 1998; Dameris et al. 1998; WMO 2007). Recently it has been shown that the increasing tropospheric burden of nitrous oxide, N_2O, in the troposphere results from an anthropogenic modification of the Earth's surface environment (Ravishankara et al. 2009). Although the role of N_2O as the source of stratospheric NO_x is well understood, this new study highlights the long term impact of anthropogenic activity changing the surface emission of N_2O. So here is a new aspect of chemistry – climate change – coupling the importance for both the stratosphere and troposphere. In summary, to understand the chemistry of the troposphere accurate knowledge of the following stratospheric parameters is required.

- Stratospheric O_3, which determines the amount of photochemically and biologically active UV radiation in the troposphere.
- Exchange of constituents between the upper troposphere and the lower stratosphere.

These various environmental issues, impacting as they do on local, regional and global scales, are of both national and international importance. To a large extent the issues are the consequence of the higher standards of living and the increase in the world population over the last two centuries and particularly the last 50 years, which is resulting in dramatic changes in the nature of the Earth's surface and increasing emissions of many trace gases to the atmosphere. Management of the anthropogenic influence on global change is a challenging task for the future. Global measurements of key tropospheric constituents are required to be able to attribute appropriately the cause and assess the longer term potential impact of anthropogenic activity on the environment.

1.5 Measuring Atmospheric Composition

Measurements of atmospheric parameters are of central importance for improving our understanding of our atmosphere and environment. However this goal can be achieved in a number of quite different ways. Measurements are characterised by their time scale; i.e. there may be specialised short term measurement campaigns aimed at investigating certain phenomena; on the other hand, long term studies follow the evolution of chemical changes. In addition the spatial scale of the investigation may be local, regional or global. In addition to the measurements made for primarily scientific research purposes, a large number of measurements of atmospheric trace gases are made every day to document the state of the atmosphere, and also to alert the public to extreme pollution events, in order to fulfil legal requirements.

1.5.1 Long Term Observations

Observing and monitoring the changes in atmospheric composition is of great scientific interest. A famous example is the detection of the continuously rising (annual mean) CO_2 mixing ratio directly observed at Manau Loa since 1960 (Keeling et al. 1976; 2003). As the measurement technology has developed, the temporal behaviour of many other important trace gases are now monitored at specific locations around the globe. Long term observations are aimed at monitoring gradual changes in the trace composition of the atmosphere, and include the following trend observations.

- The evolution of tropospheric amounts of CFCs, HCFCs, halons, methyl bromide, and other species which supply halogens to the stratosphere.
- The stratospheric ozone trend.
- The change of stratospheric chemistry (as realised in NDACC).
- The trend in the tropospheric O_3 mixing ratio (GAW).
- Trends in greenhouse gases such as CO_2, CH_4, and N_2O.
- Trends of gases indicating the atmospheric oxidation capacity (i.e. the ability of the atmosphere to remove trace gases). For instance O_3, CH_3CCl_3 or ^{14}CO are monitored for this purpose.
- Trends in aerosol properties.

In this context the "operator dilemma" should be noted: the success of the measurement series hinges on the careful calibration and execution of the measurement procedure over a long period of time. However making the required measurements with the same technique over an extended period of time is often not considered much of a scientific challenge. Thus the psychological side of the project may be as important as the technological aspects in obtaining reliable long term measurements.

1.5.2 Regional and Episodic Studies

Regional and episodic atmospheric studies seek to investigate causes, extent, and consequences of air pollution. While routine monitoring is an issue many fundamental questions can only be investigated by observations made on a regional scale. Typical measurements tasks in this context are:

- monitoring of air pollutants such as O_3, SO_2, NO, NO_2, hydrocarbons, and aerosols,
- investigation of urban plume evolution (e.g. with respect to O_3 formation downwind of source regions),
- mapping of continental plumes,
- observation of the Antarctic stratospheric O_3 hole and
- polar boundary-layer ozone loss events (the "tropospheric ozone hole" (Platt and Lehrer 1996)).

1.5.3 Investigation of Fast In Situ Photochemistry

Studies in smog chambers (also called reaction chambers or photo-reactors) allow of transport processes to be surpressed so that the effect of chemistry alone can be investigated. In fact the phenomenon of tropospheric O_3 formation was observed in smog-chambers long before the chemical mechanism was discovered. The main disadvantage of smog chambers is the presence of surfaces, and so care has to be taken to avoid artefacts, which may arise from chemical processes at the chamber walls. In order to minimise these problems very large smog chambers with volumes exceeding 100 m^3 have been built in recent years, offering surface/volume ratios below 1 m^{-1}.

However, investigation of fast (time scale of seconds) *in situ* chemical and photochemical processes (see Section 2.5) in the open atmosphere allows one to neglect the effect of transport, since transport takes place only at longer time scales. Thus it is possible to study chemical processes directly in the atmosphere. In particular this is true for free radical (OH/HO_2) photochemistry, where the lifetime of the reactive species is of the order of seconds.

1.5.4 In Situ Observational Techniques

A straightforward way to analyse the trace gas composition of the atmosphere is to take a sample of air either in a suitable container for later analysis or inside an instrument. In either case the instrument or sampling apparatus has to be brought to the place in the atmosphere where the measurement is desired. These "*in situ*" measurements come close to the ideal of determining trace gas concentrations at a "point" in space i.e. usually very close to the instrument. A large variety of measurement techniques for atmospheric trace gases (and other atmospheric parameters) is available; a few examples of *in situ* techniques and instruments are given below.

(a) Gas Chromatography (GC, e.g. for the quantification of organic species).
(b) Chemical Ionisation – Mass Spectrometry (CIMS).
(c) Gas-phase Chemiluminescence (e.g. for the detection of NO_2 or O_3).
(d) Chemical amplifiers for the detection of peroxy radicals.
(e) Electrochemical techniques (e.g. used in ozone sondes).
(f) Matrix Isolation – Electron Spin Resonance (MI-ESR).
(g) Derivatisation – HPLC or Hantzsch reaction (e.g. for aldehyde speciation and detection).
(h) "Bubblers" combined with wet chemical, colourimetric, or ion-chromatographic (IC) analysis.
(i) Photoacoustic spectroscopic detection (e.g. ethene).
(j) Short-path non-dispersive absorption of radiation (e.g. for the measurement of O_3 and CO_2).
(k) Folded-path (either by multi-reflection cell or optical resonator) absorption spectroscopy. This technique comes in many varieties, e.g. using tunable diode lasers or broad-band light sources combined with spectrometers.

In situ techniques represent the traditional approach to study and monitor the atmospheric trace gas composition; their strength is the conceptual simplicity; their main weakness the effort and cost required, in particular when spatial distributions of relatively short lived species (like air pollutants) are to be observed over large areas and over long periods of time. Moreover, *in situ* observations require supporting infrastructure on the ground, which is difficult to obtain in remote areas and over the oceans.

1.5.5 Remote Sensing Versus In Situ Techniques

Remote sensing techniques allow one to detect properties of an object from a distance. Applied to the atmosphere, the trace gas composition can be measure at a point which is remote from the probing instrument. Examples are lidar instruments (see Section 1.12), Multi-Axis DOAS measurements, or observation of trace gas distributions from space. In contrast *in situ* instrumentation (see Section 1.5.4) measures trace gas concentrations (or other parameters) at a particular location. There are a series of applications where localised measurements are desirable, for instance for the determination of strong spatial gradients. Since many trace gases have a strong vertical gradient close to the ground, the observation of these gradients requires measurements that are localised in the vertical dimension. It should be noted, however that *in situ* measurements frequently require relatively long integration times, i.e. they average the concentration over a period t_m of time. As a consequence t_m will average over the distance d_m given by:

$$d_m = t_m \times v_w$$

where v_w denotes the wind speed.

Thus an integration time of 5 min at $v_w = 2$ m/s already translates into a spatial averaging over a distance of $d_m = 600$ m (and more at higher wind speeds). This property of *in situ* measurements is important when comparing them with remote measurements from space, which exhibit spatial resolutions approaching 10 km, and probably less for future instruments. Moreover, space-based remote sensing techniques in principle allow measurements at any point on Earth, including remote continental and marine areas.

1.5.6 The Need for Global Tropospheric Measurements from Space

Most chemical compounds regarded as pollutants, with a few exceptions such as CFCs and HFCs, are present in trace amounts in the natural atmosphere. Anthropogenic activity increases their quantities and changes their distributions. The assessment of the impact and consequences of increasing anthropogenic emissions of trace constituents into the atmosphere is not trivial – because of the inherent non-linear and complex nature of the processes in the atmosphere, a detailed knowledge of the elementary atmospheric processes is required. Thus the measurement of the composition and changing trends in the amounts and distribution of atmospheric constituents (gases, aerosols and clouds) and meteorological parameters provides the data needed to test our understanding of the biogeochemical cycles within the atmosphere.

For long-lived atmospheric species, such as the greenhouse gases, CO_2, CH_4 and N_2O, measurement stations around the globe provide a monitoring network. The network makes highly accurate measurements, suitable for the assessment of the increasing background amounts. However such networks are sparse. It is now recognised that an accurate knowledge of the regional sources and sinks of both short lived and long lived pollutants is needed, and that this can be provided by a mixture of ground based measurements of different types together with satellite based remote sensing (Barrie et al. 2004).

For short-lived species and species having sources that exhibit variability temporally and spatially, the global measurement of constituents from remote sounding instrumentation aboard orbiting space-based platforms provides a unique opportunity to augment our knowledge of atmospheric pollution and biogeochemical cycling. The challenge for the remote sensing of long lived gases from space is to retrieve data products that have sufficient precision and accuracy to test our current understanding of the sources and sinks of these gases.

The interpretation of satellite observations of tropospheric constituents requires the separation of stratospheric contributions, if they are significant, and must take into account the changes in radiative transfer created by changes in surface spectral reflectance, the changing burden of aerosols, and the vertical profile of the trace constituents. In this context, constraints from *a priori* climatologies and atmospheric models provide essential information.

A hierarchy of atmospheric models has been developed to simulate the current state of the atmosphere, to predict its future behaviour and to estimate its response to both natural and anthropogenically induced change. As indicated in Chapter 9, using data assimilation to combine model and measurements is playing an increasingly important role in determining the distribution of trace constituents. In the future the coupling of models and measurements from the local to the global scale will be of increasing importance.

1.6 Electromagnetic Radiation and Molecular Energy Levels

Remote sensing and earth observation depend on the interaction of electromagnetic radiation with matter, and the accuracy of the derived data products depends on our knowledge of the interaction of light and matter.

1.6.1 Electromagnetic Radiation

Electromagnetic radiation of different types is distinguished by its wavelength or frequency as shown in Fig. 1.4. The wavelength λ is connected to the frequency ν by the speed of light through a medium

$$c = \nu \times \lambda$$

where c is the speed of light and its value is 2.998×10^8 m/s for vacuum. For air under standard conditions c is about 0.03% smaller.

Figure 1.4 illustrates the range of wavelengths of interest and gives the names for different types of electromagnetic radiation. At very short wavelengths ($\lambda < 30$ nm) the scale begins with γ-radiation and X-radiation. At longer wavelengths the UV and IR ranges bracket the spectral range of visible radiation (ca. 380–750 nm). Near IR and short wave IR radiation (ca. 1 μm wavelength) is then followed by the mid IR and far IR regions. Radiation of even longer wavelength is known as sub-mm wave ($\lambda < 1$ mm), microwave, and radio-wave radiation.

While either wavelength or frequency can be used to characterise electromagnetic radiation, normally the wavelength, λ, in nm, is used in the UV-visible, the wavenumber, $1/\lambda = \nu/c$ expressed in cm^{-1}, is used in the IR and the frequency, MHz/GHz, in the microwave regions.

It is a result of quantum mechanics that some aspects of electromagnetic radiation can be described as behaviour of waves while others can only be understood

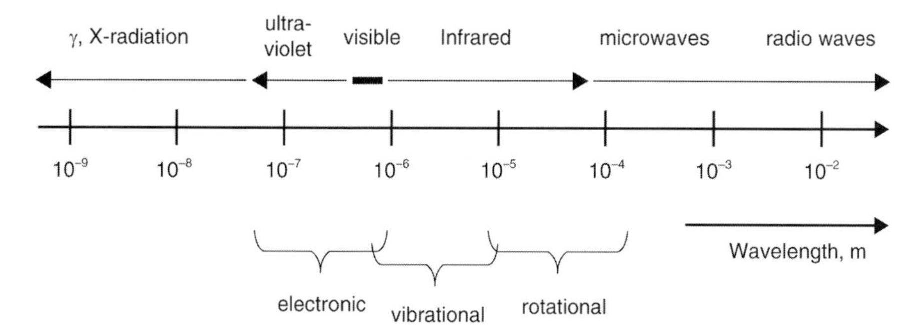

Fig. 1.4 Sketch of the electromagnetic spectrum – γ-rays to radio waves. Also indicated are the types of molecular spectroscopic transitions (electronic, vibrational and rotational) induced by radiation of the respective wavelength ranges. In this book we concentrate on the UV, visible, IR, and sub-mm wave spectral ranges.

when describing radiation as a stream of particles, or photons. The energy E of the photons depends on the frequency (or the wavelength) of the radiation:

$$E = \mathrm{h}\nu = \mathrm{h}c/\lambda$$

where Planck's constant $\mathrm{h} = 6.626 \times 10^{-34}$ Js.

a Scattering and Absorption of Radiation

Electromagnetic radiation passing through a gas may either be scattered or absorbed. When a photon is scattered, the scattering molecule is unaffected but the photon changes direction. Such scattering is known as Rayleigh scattering. The scattering depends on the wavelength of the light and is responsible for the blue skies in the atmosphere (blue light preferentially scattered from light passing above) and red sunsets (red light remaining in the sunlight after losses of blue scattered light).

Scattering also occurs from aerosols, particles and clouds and in this case is known as Mie scattering. Mie scattering depends on wavelength and the particle properties and so can be used to determine aerosol characteristics. Mie theory can be solved exactly for spherical particles. Rayleigh and Mie scattering are examples of elastic scattering (Section 1.9.1).

Rather than scattering the photon of light, the molecule may absorb it; the molecule undergoes an internal change and the energy of the photon promotes the molecule to a higher energy state. Analysis of the interactions, the science of spectroscopy, provides information on the energy states of the molecule normally allowing simple molecules to be easily identified, and their concentrations to be determined.

b Spontaneous Emission, Stimulated Absorption and Emission

Absorption is just one of three processes, identified by Einstein (1917), by which a molecule can interact with radiation while undergoing a change in energy state: stimulated absorption, emission and stimulated emission. These are illustrated in Fig. 1.5.

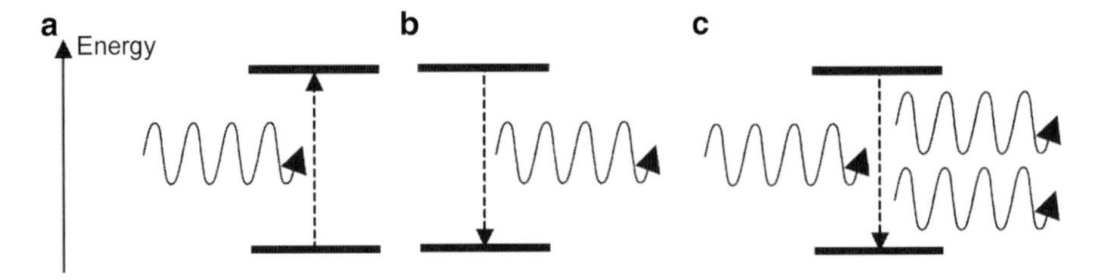

Fig. 1.5 Interaction of radiation with matter: (**a**) stimulated absorption: upon absorption of a photon with the appropriate energy the molecule changes to an excited state; (**b**) spontaneous emission: the molecule returns from the excited to the lower state emitting a photon; this process occurs in fluorescence or phosphorescence; molecules may also lose energy by collision with other molecules; (**c**) stimulated emission: a photon of the appropriate energy interacts with a molecule which is already in an excited state. The molecule is stimulated to return to the ground state emitting a further photon. This coherent emission is the basis of the amplification inherent in laser operation.

- In stimulated absorption, the process already mentioned, a molecule in a lower energy state is excited by the absorbed radiation to an upper excited state.
- In spontaneous emission, a molecule in an upper excited state undergoes a transition to a lower state, losing its energy as an emitted photon of light. This process is the basis of fluorescence and all normally seen light sources.
- Stimulated emission: a molecule, already in an upper excited state interacts with an incoming photon of the correct wavelength and undergoes a transition to a lower energy state so that two photons of the same wavelength are produced. The new photon has the same phase, frequency, polarisation, and direction of travel as the original, so that the radiation emitted is *coherent*. Such a system can act as an amplifier (laser: light *amplification* by stimulated emission of radiation).

Both absorption and emission are used to identify and determine the concentrations of trace species by remote sensing. Stimulated emission is the basis of laser technology used to provide light beams in active systems.

c Raman Scattering

The Raman Effect is a weak interaction of radiation with matter that lies between scattering and absorption. When passing through air, a small fraction of light is elastically scattered, as mentioned above, by Rayleigh scattering, with the scattering molecules remaining unchanged. However a much smaller fraction is also scattered by molecules, which however, undergo an internal transition between energy states. The scattered photon thus has a different energy from the exciting photon, so that an analysis of this very weak radiation can be used to identify molecules and determine their molecular concentrations. The Raman Effect is an example of inelastic scattering.

Raman scattering, though weak, can provide useful information in passive remote sensing as well as in active systems with intense laser light sources: see Sections 1.9.1 and 1.11.

1.6.2 Molecular Energy States

Molecules, the smallest particle of a chemical compound consisting of chemically bound groups of atoms, are quantum mechanical multi-particle systems. As such their internal energy is quantised and they have discrete energy states. It is the transitions between these states that provide the basis for the interaction with electromagnetic radiation, and the spectroscopic methods by which the molecules can be characterised. There are three molecular energy states which concern us here.

- Rotation of the entire molecule; it is the angular momentum of the entire molecule that gives rise to rotational energy states.
- Vibration of the atoms within the molecule relative to each other gives rise to vibrational energy states.
- The changed configuration of the electrons in the molecule gives rise to electronic energy states.

Typical transition energies ΔE for the different types of transition are:

- Rotational transitions: ΔE of the order of 0.1–1 kJ/mol (or 10^{-3}–10^{-2} eV); the corresponding wavelengths are in the sub-mm or microwave range;
- Vibrational transitions: ΔE of the order of 10 kJ/mol (or 0.1 eV); corresponding wavelengths are in the IR spectral range. Note that a change in vibrational energy will be accompanied by a change in rotational energy, so these transitions may have a rotational "fine structure" when observed;
- Electronic transitions: ΔE of the order of 96 kJ/mol (or 1 eV), corresponding wavelengths are in the visible or UV spectral ranges. Here also a change in electronic energy may well be accompanied by changes in vibrational and rotational energy, so these transitions may well have a vibrational structure with a rotational "fine structure" when observed.

a Rotational Energy Levels and Transitions

Quantum mechanically, the rotational energy levels in a molecule are given by:

$$\mathrm{E_j} = B \cdot J(J+1) \quad \text{with } B = \frac{\hbar^2}{2\Theta} \tag{1.1}$$

where J is the rotational quantum number, B denotes the rotational constant of the particular molecule and rotation mode (rotation axis) with the moment of inertia Θ

with respect to this axis. To a first approximation, Θ is assumed to be independent of J (rigid rotor model). However molecules are not rigid i.e. the atoms within a molecule can change their relative positions (see discussion of vibrational transitions below) due to the centrifugal force leading to a slight increase of Θ at higher values of J compared to its value at low rotational levels. Thus $B \propto 1/\Theta$ will be somewhat lower at higher J values.

Selection Rules for Rotational Transitions

For pure rotational transitions the molecule must have a dipole moment – i.e. an unevenly distributed charge. Then, the difference in the angular momentum quantum number of initial and final state, $\Delta J = \pm 1$, since the photon exchanged with the atom/molecule has a spin (intrinsic angular momentum) of unity. The $\Delta J = -1$ transitions are denoted as P-branch transitions, $\Delta J = +1$ as the R-branch, and $\Delta J = 0$ as the Q-branch where transitions can occur if an electronic transition takes place at the same time, or in Raman transitions. Consequently the photon energy $h\nu = \Delta E$ of allowed transitions is given by the energy difference of two consecutive states:

$$\Delta E_j = E_{j+1} - E_j = B \cdot [(J+1) \cdot (J+2) - J \cdot (J+1)] = 2B(J+1) \propto J \quad (1.2)$$

Thus a rotational band (a set of observed transitions) consists of a series of equally spaced lines, the difference between two lines being 2B. The value of 2B is of the order of 10^{-3}–10^{-2} eV (about 242–2,418 GHz or 8.1–81 cm^{-1}), so that wavelengths of photons exchanged in such "purely rotational" transitions occur in the sub-mm or microwave ranges.

The rotational lines which accompany vibrational and electronic transitions have a similar structure.

The energies of rotational states are of the order of the thermal kinetic energy of molecules at room temperature, so that most molecules are rotationally excited under ambient conditions (see below).

b Vibrational Energy Levels and Transitions

Vibrations of molecules can be treated approximately as harmonic oscillations, the energy levels of a quantum mechanical oscillator being given by:

$$E_v = \left(v + \frac{1}{2}\right) \cdot \hbar\omega_0 \quad (1.3)$$

with $v = 0, 1, 2, \ldots$ denoting the vibrational quantum number (vibration level) and $\frac{1}{2}\hbar\omega_0 = \frac{1}{2}h\nu_0$ is the zero-point energy of the molecular oscillator. Thus energies of different vibrational states are proportional to their associated vibrational quantum number, v. Energies of $\hbar\omega_0$ are of the order of 0.1 eV (~24,000 GHz or 800 cm^{-1}), corresponding to wavelengths in the IR spectral range.

Molecules only absorb in the IR if they have a permanent dipole moment, or if the vibration generates a dipole moment. Thus N_2 and O_2 are not IR active while CO is. Of the three vibrations of CO_2, two are IR active. The selection rule for pure vibrational transitions is $\Delta v = \pm 1$.

In addition to vibrational excitation, a molecule is likely to be rotationally excited at ambient temperatures. Thus each vibrational state splits into a series of ro-vibrational states, and the transitions between states are observed as a series of rotational lines (see Fig. 1.6).

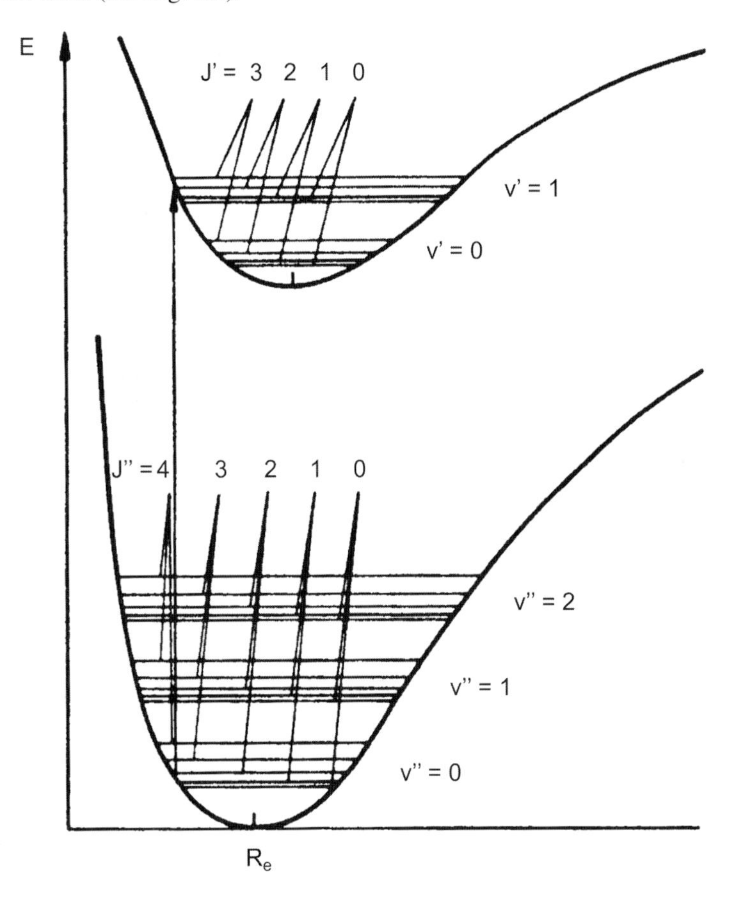

Fig. 1.6 A sketch of the ro-vibrational energy levels (with energy E) of two electronic states of a molecule. Electronic energies are given as function of the distance (R) between the nuclei of the atoms within the molecule, they are at a minimum at a certain distance R_e, which usually differs for the different electronic states. The equidistant, horizontal groups of lines denote vibrational states (ground state: $v'' = 0, 1, 2$, excited state: $v' = 0, 1$) with their attendant lowest rotational levels. Note that the energy separation of the rotational states is exaggerated in comparison to the vibrational energies. (Platt and Stutz 2008).

c Electronic Energy States and Transitions

Electronic energy states result from differing configuration of the electrons in the molecule. Each electronic state has its own set of vibrational and rotational states, these differing from those in the lowest of ground state because both the molecular bond length and strength changes with the electronic configuration.

A potential energy diagram depicting two electronic states in a molecule is shown in Fig. 1.6. The vibrational constant ω, and rotational constant B, can differ in the two states. As a consequence the rotational line spacing in the spectrum is no longer equidistant; rather the lines tend to group towards a certain wavelength either below ("blue shaded" band) or above ("red shaded" band). Electronic states are designated according to their angular momentum (electronic and rotational combined) and their electronic spin. The reader is referred to a spectroscopic text for further details (e.g. Herzberg 1950).

There are no simple selection rules for electronic transitions: the transitions between electronic states are determined by the quantum mechanical characteristics of the respective states. One finds that some transitions are allowed, that for IO in Fig. 1.7 for example. Some are forbidden and so have no spectrum, and some are very weak such as the absorption in the near IR at 762 nm by O_2 itself.

The Frank-Condon principle provides a guide to the intensity of the vibrational transitions for a given electronic transition: during the transition the bond length, or inter-nuclear distance, cannot change, so in terms of the diagram, transitions must be vertical. Since the inter-nuclear distance differs between the states then the observed transitions usually result in a change in the vibrational quantum number.

d The Populations of Molecular Energy States

The population, i.e. the number of molecules in a particular state with energy E, N(E) of different states of a molecule with the energies E_1, E_2 above the ground state (where $E_0 = 0$) as a function of temperature T is given by the Boltzmann–Distribution:

$$\frac{N(E_2)}{N(E_1)} = \frac{G_2}{G_1} \cdot e^{-\frac{E_2 - E_1}{kT}} \tag{1.4}$$

Here k denotes the Boltzmann constant, while G_1, G_2 are the statistical weights or degeneracy factors (i.e. the number of different states with the same energy E) of the respective states. Vibrational excitation in molecules is not degenerate i.e. $G_v = 1$, while rotational states have a degeneracy of $G_J = 2J + 1$, thus the population of rotational states is given by:

$$\frac{N(E_2)}{N(E_1)} = \frac{2J_2 + 1}{2J_1 + 1} \cdot e^{-\frac{E_2 - E_1}{kT}} = \frac{2J_2 + 1}{2J_1 + 1} \cdot e^{-\frac{B(J_2(J_2+1) - J_1(J_1+1))}{kT}} \tag{1.5}$$

With E(J) = B × J(J + 1), where B is the rotational constant of the particular molecule and rotational axis as given in Eq. 1.1. The kinetic energy of a molecule at room temperature of around 290 K is comparable to or larger than the lowest rotational energy level, but smaller than the lowest vibrational energy level of most molecules. Therefore, under ambient conditions several rotational states are usually populated, but only a small fraction of the molecules are vibrationally excited.

1.7 Molecular Spectra and Line Broadening

An example of an electronic spectrum is provided in Fig. 1.7, which shows the spectrum of the IO molecule. The bands seen are vibrational bands the intensities of which are, as explained above, governed by the Frank-Condon principle.

The rotational structure is not generally resolved, but can indeed be observed under high resolution in some of the IO electronic vibrational bands. As indicated below, the linewidth of the rotational lines is governed by the lifetime of the electronic state. For stable states, the lines are sharp, but for shorter lived electronic states the lines are broader. As will be explained, lines are broadened by pressure and by the Doppler effect.

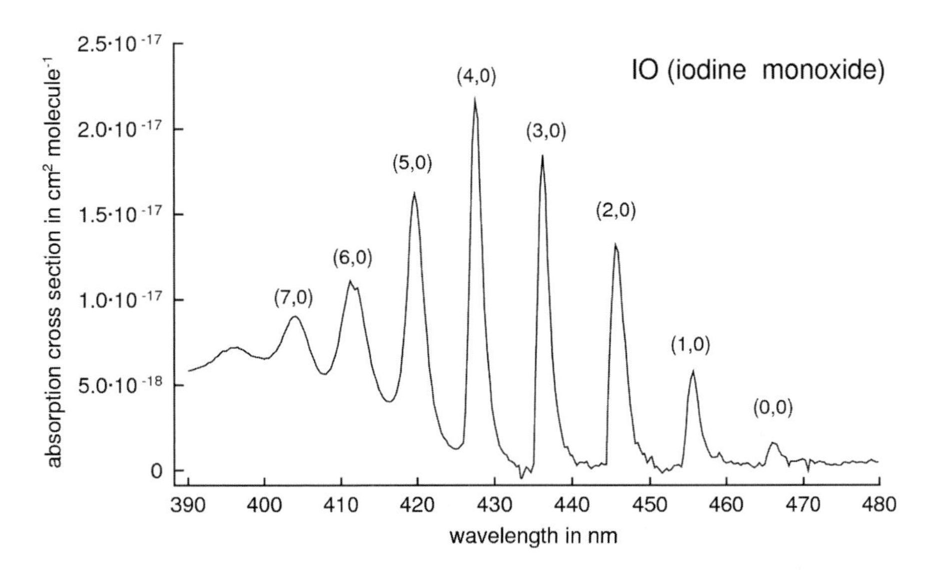

Fig. 1.7 Example of the vibrational structure of the spectrum of IO (electronic transition: $A^2\Pi \rightarrow X^2\Pi$). The bands correspond to transitions to different vibrational levels; the designation (v′, v″) gives the vibrational quantum numbers of the upper and lower levels, v′ and v″. The rotational fine structure is not resolved. Figure: J.P. Burrows, A. Schoenhardt, J.C. Gomez Martin and P. Spietz, IUP, Bremen.

In general, thanks to the line broadening effects and to the mass of the molecules, only diatomic molecules and a few small polyatomic molecules exhibit resolved rotational structure in their electronic spectrum. Larger molecules usually have continuous spectra. Thus only a limited range of molecules have useful spectra in the UV-visible region and can be detected by satellite instruments. However some of these are the most important environmental species.

In the IR, a greater range of molecules display a rotational fine structure and so thermal IR measurements are useful for a range of molecules that cannot be seen in the UV-visible.

1.7.1 Line Broadening Mechanisms and the Width of Absorption Lines

According to classical electrodynamics, for an excited molecule with energy E_0 above the ground state, the energy of the excited electrical dipole will decrease exponentially with time *t* as a result of emission of electromagnetic radiation:

$$E(t) = E_0 e^{-\delta t} \tag{1.6}$$

Here δ denotes a damping constant, where $\tau = 1/\delta$ as the time in which $E(t)$ decays to 1/e of E_0. Accordingly the corresponding amplitude of the electric field $\widehat{E}(t)$ of the emitted radiation will decrease exponentially with time *t*:

$$\widehat{E}(t) = \widehat{E}_0 e^{-\delta t/2} \cdot \cos(\omega_0 t + \varphi) \tag{1.7}$$

Here ω_0 denotes the frequency of the oscillation, which is related to the energy E_0 of the excited state $\omega_0 = 2\pi E_0/h$ (with h = Planck's constant), while φ is the phase of the oscillation. Since the amplitude $\widehat{E}(t)$ is damped, the radiation cannot be truly monochromatic, as can be seen from the Fourier transform of $\widehat{E}(t)$:

$$\widehat{E}(t) = \frac{1}{\sqrt{2\pi}} \int_{-\infty}^{\infty} a(\omega) e^{i\omega t} d\omega \tag{1.8}$$

with the spectral function $a(\omega)$:

$$a(\omega) = \frac{E_0}{2} \left[\frac{e^{i\varphi}}{i(\omega - \omega_0) - \delta/2} - \frac{e^{-i\varphi}}{i(\omega + \omega_0) + \delta/2} \right] \tag{1.9}$$

Since $\delta << \omega_0$ at frequencies ω near the resonance frequency ω_0, we can neglect the second term in the square bracket of Eq. 1.9 and obtain the radiation intensity $I(\omega)$ as the square of $a(\omega)$ (and normalising the integral over $I(\omega)$ to unity):

$$I(\omega) d\omega = I_0 \frac{\delta^2/4}{(\omega - \omega_0)^2 - \delta^2/4} d\omega \tag{1.10}$$

Equation 1.10 is known as the Lorentz distribution, its maximum is at $\omega = \omega_0$ while $I(\omega) = 0.5\ I(\omega_0)$ at $\omega = \omega_0 \pm \delta/2$, i.e. δ denotes the full width at half maximum (FWHM) of the distribution.

1.7.2 The Natural Linewidth

The classical image of an oscillating dipole cannot directly be applied to an excited atom or molecule, but the correspondence principle suggests that, at least at large quantum numbers, Eq. 1.10 should be an approximation, with which to describe the profile of an emission line of an isolated molecule, using an empirical δ. In addition there might be more than one level to which an excited molecule can decay: thus $\delta_L = \Sigma\delta_i$ has to be used instead of δ, where the individual δ_i describe the transition frequency to the individual states, to which the excited state under consideration can decay. The natural line shape of an isolated molecule is thus given by:

$$I_L(\omega)\, d\omega = I_0 \cdot \frac{\delta_L^2/4}{(\omega - \omega_0)^2 - \delta_L^2/4}\, d\omega \tag{1.11}$$

with the natural line width δ_L being the FWHM. It essentially depends on the lifetime of the excited state of the molecule, but not directly on the energy difference of the transition and thus the frequency of the emitted or absorbed radiation. For the usual observed transitions the natural lifetime is of the order of 10^{-8} s corresponding to $\delta_L = 10^8$ Hz. For radiation of $\lambda = 400$ nm wavelength ($\omega = 2\pi c/\lambda \approx 4.7 \times 10^{15}$ Hz) this would amount to $\Delta\omega/\omega \approx 2.1 \times 10^{-8}$ or a linewidth, $\Delta\lambda \approx 8.5 \times 10^{-6}$ nm (≈ 0.01 pm). This is usually negligible compared with pressure or Doppler broadening.

1.7.3 Pressure Broadening (Collisional Broadening)

Collisions of the molecules in a gas will reduce the lifetime of the excited state below the value given by radiative transitions alone Eq. 1.10 which determines the natural line width. The shape of a pressure broadened line is Lorentzian, as given by Eq. 1.10 with the width of the pressure broadened line, $\delta = \delta_P$. In principle the collisional damping constant, δ_C, is given by the product of the gas kinetic collision frequency z_{AB} and the deactivation probability p_{AB} per collision. Since p_{AB} can differ for each molecular species, the amount of pressure broadening depends not only on the pressure of the molecular species itself but also on the nature of the other species present. Of particular interest are the self broadening (the species itself is the pressure gas, $p_{AB} = p_{AA}$) and air broadening (species A is occurring as a trace species in air, which is species B). In air, deactivation occurs almost exclusively in collisions between molecule A and air molecules. The collision frequency is directly proportional to the product of gas density (i.e. pT^{-1} where p is the pressure) and also to the average molecular velocity (and thus to $T^{1/2}$). The expression for the pressure – broadened line width δ_P is then:

$$\delta_P(p,T) = \delta_P(p_0,T_0)\cdot\frac{p}{p_0}\cdot\sqrt{\frac{T_0}{T}} = \delta_0\cdot\frac{p}{p_0}\cdot\sqrt{\frac{T_0}{T}} \tag{1.12}$$

Here $\delta_p(p_0, T_0) = \delta_0$ denotes the pressure broadening at some reference pressure and temperature. Typical values at one atmosphere for pressure broadening of light molecules in the near UV are $\Delta\lambda \approx 1$ pm.

1.7.4 Doppler Broadening

The Brownian motion of the molecules in the atmosphere has two main consequences: pressure broadening and Doppler broadening.

1. The energy E is not only dissipated by radiation but also by collisions with other molecules present in the gas. This effect manifests itself as an increased $\delta_P > \delta_L$, the "pressure broadening" discussed above. Since the effect can be described in terms of a damped oscillation, the line shape is also given by Eq. 1.10.
2. The Doppler Effect will change the frequency ω_0 of the emitted radiation, as a first approximation, to:

$$\omega = \omega_0\cdot\left(1+\frac{v_x}{c}\right) \tag{1.13}$$

where v_x denotes the velocity component of the emitting system along the direction of propagation and c the speed of light. The distribution of an individual component of the velocity $N(v_x)dv_x$ denoting the number N of molecules, where the x-component of their velocity is in the range of v_x to $v_x + dv_x$, is Gaussian (not be confused with the distribution of the absolute value of the velocity, which is given by the Maxwell–Boltzmann distribution). The resulting Gaussian intensity distribution, centred at ω_0, is given by:

$$\begin{aligned} I_D(\omega)\,d\omega &= I_0\cdot\exp\left(\left(Mc^2/2RT\right)\cdot(\omega_0-\omega)^2/\omega^2\right)d\omega \\ &= I_0\cdot\exp\left((\delta_D/2)\cdot(\omega_0-\omega)^2\right)d\omega \end{aligned} \tag{1.14}$$

Here M denotes the molar weight of the molecule and R the universal gas constant. The FWHM of a purely Doppler broadened line is given by:

$$\delta_D = \omega \frac{2\sqrt{2R \ln 2}}{c} \cdot \sqrt{\frac{T}{M}} = \omega \cdot C_D \cdot \sqrt{\frac{T}{M}} \tag{1.15}$$

The constant $C_D = 2\sqrt{2R \ln 2}/c$ has the value of 2.26×10^{-8} $(kg/mol)^{1/2}$ $K^{-1/2}$. For radiation with a wavelength of $\lambda = 400$ nm, at $T = 300$ K and for air ($M \approx 0.029$ kg/mol) we obtain: $\delta_D/\omega_0 = \Delta\omega/\omega \approx 2.26 \times 10^{-8}$ or a Doppler width of $\Delta\lambda \approx 9 \times 10^{-4}$ nm (about 1 pm).

1.7.5 Atmospheric Spectral Line Shapes in Different Spectral Ranges

The three broadening mechanisms discussed above have very different magnitudes. The natural linewidths are often very small and can only be observed at very low pressures. Pressure broadening and Doppler broadening have line shape described by Lorentzian and Gaussian functions, and so have different dependencies on wavelength (or frequency).

1. For Lorentzian (i.e. natural or pressure) broadening, the half-width is independent of the frequency ω (Eq. 1.10), while Doppler broadening (Eq. 1.15) is proportional to ω.
2. The intensity (or absorption cross section) of a Lorentzian line (Eq. 1.10) decays only with the square of the deviation from the centre frequency $1/(\omega-\omega_0)^2$, so that a large fraction of the total intensity (or total absorption) is in the wings of the line. In contrast, the Gaussian profile (Eq. 1.14) decays exponentially, so that there is little emission (or absorption) in the wings of the line.

Thus the spectral line shape depends on the spectral frequency and atmospheric temperature and pressure. Doppler broadening is usually negligible at low frequencies, i.e. in the microwave and far IR regions; it becomes noticeable in the near IR, and dominates in the short-wavelength UV. In the visible and UV spectral range Doppler broadening is comparable to pressure (Lorentzian) broadening. The resulting line shape, usually called the Voigt shape, is obtained by convoluting the Gaussian and Lorentzian line shapes:

$$I_V(\omega, \delta_D, \delta_P) = \int_{-\infty}^{\infty} I_D(\omega', \delta_D) \cdot I_P(\omega - \omega', \delta_P) d\omega' \tag{1.16}$$

Fig. 1.8 shows a comparison of the three line shapes.

As absorption is the inverse process to spontaneous emission, the equations above also describe the line shapes of atomic or molecular absorption lines, and so the absorption cross section $\sigma(\omega)$ (or $\sigma(\lambda)$) will show the same wavelength dependence as $I(\omega)$ in Eq. 1.16 above.

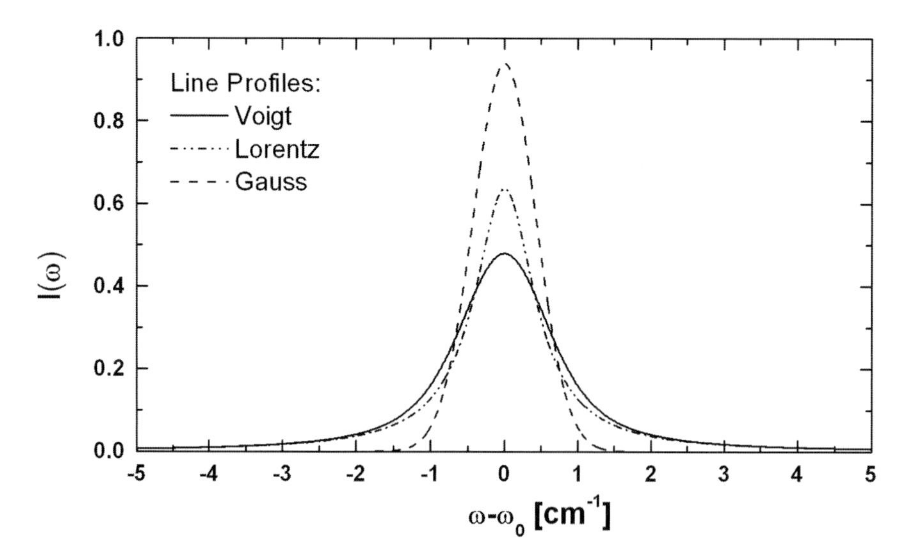

Fig. 1.8 Sketch of the different types of line profiles: Lorentz shape (collision broadening), Doppler (Gaussian) shape (thermal motion of molecules), and Voigt shape resulting from simultaneous Doppler and collision broadening. Platt and Stutz (2008).

1.8 Spectroscopic Techniques for Chemical Analysis

The interaction of radiation with matter provides a powerful tool for a wide variety of investigations, being used for both the investigation of molecular structure and the analysis of the chemical composition of complex mixtures such as are found in the atmosphere. We must first distinguish between two different experimental approaches to spectroscopy.

1.8.1 Absorption Spectroscopy

This spectroscopic technique makes use of the absorption of electromagnetic radiation by matter (Fig. 1.9). Quantitatively the absorption of radiation is expressed by Beer–Lambert law (or Bouguer–Lambert law):

$$I(\lambda) = I_0(\lambda) \cdot \exp(-\sigma(\lambda) \cdot c \cdot L) \tag{1.17}$$

where $I_0(\lambda)$ denotes the initial intensity emitted by a source of radiation and $I(\lambda)$ is the radiation intensity after passing through the medium of thickness L. The species to be measured is present at the concentration (number density) c. The quantity $\sigma(\lambda)$ denotes the absorption cross section at wavelength λ.

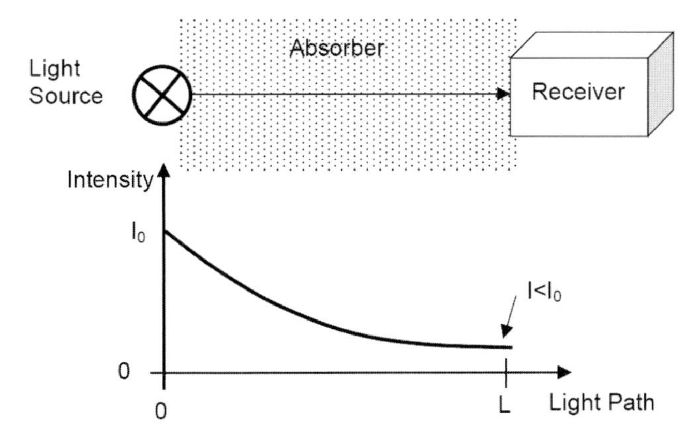

Fig. 1.9 The basic principle of absorption spectroscopic trace gas detection: *top*: an experimental arrangement; *below*: the fall in light intensity as the light passes through the absorber.

The absorption cross section is a characteristic property of any species. $\sigma(\lambda)$ can be measured in the laboratory, while the light-path length L is derived from radiation transport considerations. In the atmosphere, a more general formulation of the Lambert–Beer law involves the column density S:

$$I(\lambda) = I_0(\lambda) \cdot \exp\left(-\sigma(\lambda) \cdot \int c(x) \cdot dx\right) = I_0(\lambda) \cdot \exp(-\sigma(\lambda) \cdot S) \quad (1.18)$$

Once the intensities I, I_0, and the cross section, σ, are known the trace gas concentration, c can be calculated from the measured ratio, $I_0(\lambda)/I(\lambda)$:

$$c = \frac{\log(I_0(\lambda)/I(\lambda))}{\sigma(\lambda) \cdot L} = \frac{D(\lambda)}{\sigma(\lambda) \cdot L} \quad (1.19)$$

If L is unknown the column density, S, can be deduced instead:

$$S = \frac{\log(I_0(\lambda)/I(\lambda))}{\sigma(\lambda)} = \frac{D(\lambda)}{\sigma(\lambda)} \quad (1.20)$$

The expression:

$$D(\lambda) = \log\left(\frac{I_0(\lambda)}{I(\lambda)}\right) = S \cdot \sigma(\lambda) \quad (1.21)$$

is known as the *optical density* of a layer of a given species (Note that in the definition of the optical density the decadic can be used instead of the natural logarithm; here we shall always use the natural logarithm).

In thermal equilibrium, the emission of any object (solid, liquid, or gas) will be proportional to the Planck function:

$$I(\nu, T) = \varepsilon_x \cdot B(\nu, T) = \frac{2kc^2\nu^3}{e^{hc\nu/kT} - 1} \tag{1.22}$$

where ε_x denotes the emissivity of the object, which, by Kirchhoff's law, is equal to the absorptivity of the same object.

1.8.2 Emission Spectroscopy

The spontaneous emission of radiation by excited atoms or molecules is called fluorescence. As the intensity of the emitted radiation is generally proportional to the number of molecules present in a particular volume, emission spectroscopy can be used for the quantitative determination of atmospheric trace gas concentrations. The radiation is emitted at a characteristic set of wavelengths with the photon energy corresponding to the energy difference of the participating states (see Section 1.6). In practice, geometrical factors, the absorption cross section and quenching processes have to be taken into account.

The energy required to transfer the molecule from its ground state to its excited state can be supplied by various mechanisms including collisional excitation. In the TIR and microwave spectral regions, excitation occurs at ambient temperatures in the atmosphere and it is this radiation which is detected by the satellite instruments (Chapters 3 and 4).

Emission spectra in the laboratory can be produced by electron bombardment, chemical reactions, or absorption of radiation. Laser light is a common excitation source in the laboratory and in ground-based observation techniques. Laser induced fluorescence, LIF, is employed in several instruments for the measurement of transient atmospheric trace gases such as the OH radical.

1.9 Atmospheric Scattering and Radiation Transfer

The radiation detected by a satellite instrument has passed from the source, perhaps the sun, to the Earth's surface or a cloud top, through the atmosphere. During the passage some radiation is lost by scattering by molecules or aerosol particles, and some radiation from other sources, also scattered by the atmosphere, reaches the instrument. Thus knowledge of scattering phenomena and the equations for atmospheric radiative transfer are essential in the estimation of trace constituents from space. This section provides a brief introduction.

1.9.1 *Scattering*

a Rayleigh Scattering

In the atmosphere scattering by molecules and particles is an important process. This is simply demonstrated by looking at the blue colour of a cloud free sky or the white or grey colour of a cloudy sky. Elastic scattering (i.e. scattering without change of the photon energy) by air molecules is called Rayleigh scattering. Another important elastic scattering process, called Mie scattering, is due to aerosol particles (see below). The Rayleigh scattering cross section $\sigma_R(\lambda)$ is given by (Rayleigh 1899):

$$\sigma_R(\lambda) = \frac{24\pi^3}{\lambda^4 N_{air}^2} \cdot \frac{\left(n_0(\lambda)^2 - 1\right)^2}{\left(n_0(\lambda)^2 + 2\right)^2} \cdot F_K(\lambda)$$

$$\approx \frac{8\pi^3}{3\lambda^4 N_{air}^2} \cdot \left(n_0(\lambda)^2 - 1\right)^2 \cdot F_K(\lambda) \tag{1.23}$$

where:

$n_0(\lambda)$ is the wavelength dependent index of refraction of air (real part);

N_{air} is the number density of air (e.g. 2.4×10^{19}molecules/cm^3 at 20°C, 1 atm);

$F_K(\lambda) \approx 1.061$ is a correction for anisotropy (polarisability of air molecules).

However since $n_0(\lambda)^2 - 1 \approx 2(n_0(\lambda) - 1) \propto N_{air}$, $n_0(\lambda) \approx 1$ (e.g. n_0(550 nm) = 1.000293), and $n_0(\lambda) - 1 \propto N_{air}$, $\sigma_R(\lambda)$ is essentially independent of N_{air}. On that basis a simplified expression for the Rayleigh scattering cross section was given by Nicolet (1984):

$$\sigma_R(\lambda) \approx \frac{4.02 \times 10^{28}}{\lambda^{4+x}} \tag{1.24}$$

with $x = 0.04$ (for $\lambda > 550$ nm) and $x = 0.389\lambda + 0.09426/\lambda - 0.3228$ for 200 nm $< \lambda <$ 550 nm.

For simple estimates the Rayleigh scattering cross section can be written as:

$$\sigma_R(\lambda) \approx \sigma_{R0} \times \lambda^{-4} (\sigma_{R0} \approx 4.4 \times 10^{-16} \text{cm}^2\text{nm}^4 \text{ for air}) \tag{1.25}$$

The extinction coefficient due to Rayleigh scattering $\varepsilon_R(\lambda)$ is then given by:

$$\varepsilon_R(\lambda) = \sigma_R(\lambda) \times N_{air} \tag{1.26}$$

The Rayleigh scattering phase function is given by:

$$\Phi(\cos\vartheta) = \frac{3}{4}\left(1 + \cos^2\vartheta\right) \tag{1.27}$$

Taking the anisotropy of the polarisability into account (Penndorf 1957): the above equation becomes

$$\Phi(\cos\vartheta) = 0.7629 \cdot \left(0.9324 + \cos^2\vartheta\right) \tag{1.28}$$

b Raman Scattering

Inelastic scattering occurs if the scattering molecule changes its state of excitation during scattering, a process first identified by Raman. A part of the photon's energy passes from the photon to the molecule to excite either a vibration or rotation or both.

Rotational Raman scattering (RS) only changes the rotational excitation state (Stokes lines, $\Delta J = +2$, S-branch) or vice versa (Anti-Stokes, $\Delta J = -2$, O-branch) ($\Delta v = 0$). If the vibrational state also changes, the term rotational-vibrational RS is used ($\Delta v = \pm 1$). Only discrete amounts of energy given by the energy difference between the excitation states can be absorbed or emitted.

For O_2 and N_2 nuclear spin effects impact on the allowed rotational states and their statistical weighting. A rotational, RS frequency shift of up to ± 200 cm^{-1} occurs. There is also a vibrational shift of $\pm 2{,}331$ cm^{-1} for nitrogen and $\pm 1{,}555$ cm^{-1} for oxygen. The rotational-vibrational RS cross sections are an order of magnitude weaker than the rotational RS cross sections, and the rotational RS, in turn, is roughly one magnitude weaker than Rayleigh scattering.

However, while Raman backscatter from the atmosphere is weak, backscattered radiation from the oceans, rivers and lakes is appreciably enhanced by vibrational RS from liquid water.

c Mie Scattering

Mie scattering (Mie 1908) is defined as the interaction of light with particulate matter with a dimension comparable to the wavelength of the incident radiation. It can be regarded as the radiation resulting from a large number of coherently excited elementary emitters (molecules for example) in a particle. Since the linear dimension of the particle is comparable to the wavelength of the radiation, interference effects occur. The most noticeable difference to Rayleigh scattering is, generally, the much weaker wavelength dependence and a strong dominance of the forward direction in the scattered light.

The calculation of the Mie scattering cross section, which involves summing over slowly converging series, is complicated even for spherical particles; it is worse for particles of an arbitrary shape. However, the Mie theory for spherical particles is well developed and a number of numerical models exist to calculate scattering phase functions and extinction coefficients for given aerosol types and particle size distributions (Van de Hulst 1980; Wiscombe 1980). The computational effort is substantially reduced by the introduction of an analytical expression for the scattering phase function, which only depends on a few observable parameters. The Henyey-Greenstein parameterisation is most commonly used:

$$\Phi(\cos\vartheta) = \frac{(1 - g^2)}{(1 + g^2 - 2g\cos\vartheta)^{3/2}} \tag{1.29}$$

which is only dependent on the asymmetry factor g (average cosine of the scattering function):

$$g = \langle\cos\vartheta\rangle = \frac{1}{2}\int_{-1}^{1} P(\cos\vartheta)\cdot\cos\vartheta \, d\cos\vartheta \tag{1.30}$$

For isotropic scattering ($P(\cos\vartheta) =$ constant) the asymmetry factor $g = 0$; for tropospheric aerosol a typical value might be as large as 10.

Tropospheric aerosol is either emitted from the surface (examples are sea salt, mineral dust and soot and particles from biomass burning) or formed in the gas phase by condensation of chemically derived hygroscopic species, principally sulfate, nitrates or oxidized organic material. The physical and chemical properties of aerosols in the atmosphere depends on the aerosol origin and history (see Chapter 6). Parameters for typical aerosol scenarios (urban, rural, maritime, background) can be found in the data base for the radiative transfer model LOWTRAN, (Isaacs et al. 1987), which includes the extinction coefficients and the asymmetry factors as well as their spectral dependence. A radiative transfer model, including all cloud effects known to date, was developed by Funk (2000).

d Total Scattering

Mie scattering is only partly an absorption process but, by similar arguments to those for Rayleigh scattering, it can be treated for narrow beams, such as an absorption process with the extinction coefficient:

$$\varepsilon_M(\lambda) = \varepsilon_M(\lambda_0)\times(\lambda/\lambda_0)^{-\alpha} \tag{1.31}$$

where the Angström Exponent α is inversely related to the mean aerosol particle radius. Typically α is found in the range 0.5–2.5 with an "average" value of $\alpha = 1.3$ (Angström 1929).

In short a more comprehensive description of atmospheric extinction in the presence of a single trace gas species, having a concentration, c, and an absorption cross section, σ, can be expressed as:

$$I(\lambda) = I_0(\lambda)\exp[-L\,(\sigma(\lambda)\times c + \varepsilon_R(\lambda) + \varepsilon_M(\lambda)] \tag{1.32}$$

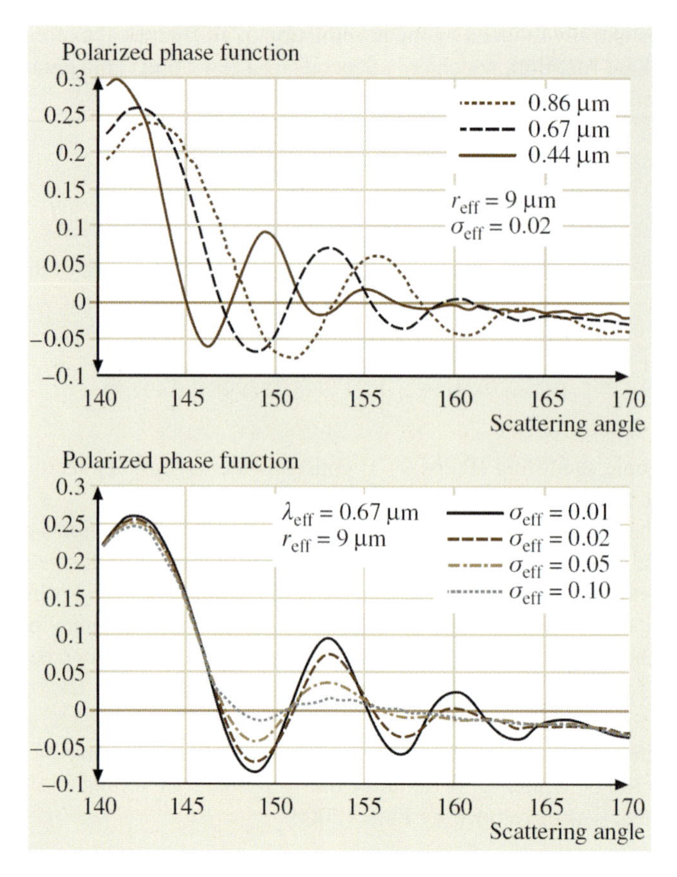

Fig. 1.10 Polarised Mie scattering phase function as a function of scattering angle for cloud droplets having a lognormal particle size distribution with an effective radius r_{eff} = 9μm. Upper panel: phase function as a function of wavelength with fixed σ_{eff} = 0.02 effective size variance; lower panel as a function of effective size variance (adapted from Platt et al. (2007)).

Typical extinction coefficients due to Rayleigh and Mie scattering in the UV at 300 nm are 1.3×10^{-6} cm^{-1} and 1–10 $\times$ 10^{-6} cm^{-1}, respectively. In the IR Rayleigh scattering can usually be neglected. Fig. 1.10 shows the Mie theory scattering function for polarised radiation as a function of scattering angle. With an appropriate selection of scattering angle the polarisation information can be used to retrieve information about particles and aerosols.

1.9.2 Atmospheric Radiative Transfer

Radiation transfer in the atmosphere is an important area of scientific study for several reasons.

- Atmospheric radiation transport is central to quantifying the greenhouse effect and thus climate change.
- Atmospheric radiation transport determines the efficiency of photochemical reactions in the Earth's atmosphere.
- Most importantly for this book: the retrieval of trace gas, aerosol or cloud parameters from remote sounding experiments requires an accurate knowledge of the sources of electromagnetic radiation and its transport through the atmosphere.

Accurate radiative transfer models are needed to invert tropospheric trace constituent data products. In the following we describe the physical processes described in radiative transfer. In practice, radiative transfer models are complex and usually require numerical integration, as the resulting system of equations cannot be solved analytically. Monte Carlo methods can be used but remain computationally expensive, preventing their routine use at this stage.

Consider a stream of photons travelling through the atmosphere. The radiant flux, Φ, per unit space angle, Ω, and wavelength, λ, i.e. the incoming intensity I_λ(Eq. 1.33) is attenuated within the distance, ds, by absorption and scattering. The subscripts for the absorption, a, and scattering coefficients, s, are used where the respective coefficients are $\varepsilon_a(\lambda) = n \times \sigma_a(\lambda)$ and $\varepsilon_s(\lambda) = n \times \sigma_s(\lambda)$; n is the number of absorber or scatterers per unit volume and $\sigma_i(\lambda)$ are the absorption or scattering cross sections; they consist of contributions from all the scattering species. The sum of absorption and scattering is commonly referred to as the extinction. The continuity equation for the intensity of the incoming radiation traversing the distance, ds, for a single species is then:

$$dI_\lambda = -[\varepsilon_a(\lambda) + \varepsilon_s(\lambda)] \cdot I_\lambda \cdot ds = -[\sigma_a(\lambda) + \sigma_s(\lambda)] \cdot n \cdot I_\lambda \cdot ds \qquad (1.33)$$

ε_a = absorption coefficient,
ε_s = scattering coefficient,
$\sigma_a(\lambda)$ = absorption cross section of the absorbing molecule,
$\sigma_s(\lambda)$ = scattering cross section of the absorbing molecule,
n = number of absorbers/scatterers per unit volume.

There are three primary sources of radiation in the atmosphere: external radiation from the sun (or other celestial bodies), thermal emission by molecules in the atmosphere or from the ground, and scattering, i.e. the radiation removed from the primary beam due to scattering (Eq. 1.33) reappearing as a radiation source. The outgoing light receives some intensity, $dI_{s,\lambda,}$ by scattering from all angles, θ and φ. We introduce a dimensionless scattering function $S(\theta, \varphi)$:

$$S_\lambda(\vartheta, \phi) = \frac{4\pi}{\sigma_s} \cdot \frac{d\sigma_s(\lambda)}{d\Omega} \qquad (1.34)$$

which we integrate over all angles weighted with the incoming intensity $I_{s,\lambda,}(\theta, \varphi)$

$$dI_S(\lambda) = \varepsilon_s(\lambda)ds \int_0^{\pi}\int_0^{2\pi} I(\lambda, \vartheta, \varphi) \cdot \frac{S(\vartheta, \varphi)}{4\pi} d\varphi \cdot \sin\vartheta d\vartheta \tag{1.35}$$

to obtain the intensity added to the outgoing intensity dI_s.

The intensity due to thermal emission, $dI_{th}(\lambda,T)$, from the volume element, $dV = A \times ds$, is added to the outgoing intenstity:

$$dI_{th}(\lambda, T) = \varepsilon_a(\lambda) \cdot I_p(\lambda, T)\ ds = \varepsilon_a(\lambda) \cdot F_p(\lambda, T) \cdot A\ ds \tag{1.36}$$

where $F_p(\lambda,T)$ is the flux of radiation at wavelength λ through the area A and, as before, ε_a denote the absorption coefficient and $I_p(\lambda,T)$ the Planck function respectively:

$$dI_p(\lambda, T) = \frac{2hc^2}{\lambda^5} \cdot \frac{d\lambda}{e^{hc/\lambda kT} - 1} \tag{1.37}$$

Combining all the processes above, we obtain the radiation transport equation:

$$\frac{dI(\lambda)}{ds} = -[\varepsilon_a(\lambda) + \varepsilon_s(\lambda)] \cdot I(\lambda) + \varepsilon_a(\lambda) \cdot I_p(\lambda, T) + \varepsilon_s(\lambda) \int_0^{\pi}\int_0^{2\pi} I(\lambda, \vartheta, \varphi) \cdot \frac{S(\vartheta, \varphi)}{4\pi} d\varphi \cdot \sin\vartheta d\vartheta \tag{1.38}$$

In general the radiation transport Eq. 1.38 cannot be solved by analytical methods, and numercial modelling is required. However simplifications are frequently possible. For example at short wavelengths (UV-vis) the Planck term can usually be neglected so that:

$$\frac{dI(\lambda)}{ds} = -(\varepsilon_a(\lambda) + \varepsilon_s(\lambda)) \cdot I(\lambda) + \varepsilon_s(\lambda) \int_0^{\pi}\int_0^{2\pi} F(\lambda, \vartheta, \varphi) \cdot \frac{S(\vartheta, \varphi)}{4\pi} d\varphi \cdot \sin\vartheta d\vartheta \tag{1.39}$$

Similarly Rayleigh scattering and Mie scattering by aerosol particles can be neglected for thermal infrared radiation, due to its long wavelength. If there are no clouds (i.e. in the absence of any scattering) we have:

$$\frac{dI(\lambda)}{ds} = \varepsilon_a(\lambda) \cdot \left(I_p(\lambda, T) - I(\lambda)\right) \tag{1.40}$$

With the definition of the optical density $d\tau = \varepsilon_a(\lambda) \cdot ds$ Eq. 1.40 simplifies to:

$$\frac{dI(\lambda)}{d\tau} = I_p(\lambda, T) - I(\lambda) \tag{1.41}$$

This equation is also known as Schwarzschild equation. Even further simplification, i.e. setting the thermal emission to zero ($I_p(\lambda,T) = 0$) leads to:

$$\frac{dI(\lambda)}{ds} = -\varepsilon_a(\lambda) \cdot I(\lambda) = -c \cdot \sigma_a(\lambda) \cdot I(\lambda) \tag{1.42}$$

or

$$\frac{dI(\lambda)}{I(\lambda)} = -c \cdot \sigma_a(\lambda) \cdot ds \tag{1.43}$$

Integration of Eq. 1.43 (taking $\ln(I_0)$ as integration constant) over distance L yields the Beer–Lambert law, Eq. 1.17. Thus the Beer–Lambert law can be identified as the most simple solution of the radiation transport equation.

In Chapters 2 to 6 the particular approaches and simplifications used to solve the atmospheric radiative transfer equation in the different spectral regions are discussed.

1.10 Remote Sensing: Images and Spectroscopy

Using the knowledge of the interactions of electromagnetic radiation with matter and molecular spectroscopy as described above, remote sensing systems can provide data products about atmospheric constituents and surface parameters. The essential components of any remote sensing system comprise the following:

- radiation source (passive: sun, moon, Earth, stars or active: laser, Lamp etc.);
- radiation path (e.g. the atmosphere);
- object (e.g. trace gases in the atmosphere or the Earth's surface);
- sensor (e.g. spectrometer, scanner, radiometer, camera).

1.10.1 Satellite Images

The development over the last 25 years of solid state detector devices and the improvements in fibre optics have revolutionised remote sensing. The modern image sensors determine the spatial distribution of the radiation intensity (radiance) in only one wavelength band, i.e. they produce monochromatic images. Colour image sensors determine the spatial distribution of the radiation intensity in a small number (typically three) of wavelength bands which are typically those corresponding to the human eye's colour impressions, red, green, and blue. An example is shown in Fig. 1.11, which shows a volcanic plume from Etna. However, such images often contain spectral information from UV and IR sensors outside the visible region, and are thus false colour images.

Fig. 1.11 Smoke plume of the Etna volcano – Outbreak on 23rd July 2001.
Source: DWD, NOAA.

Multi-spectral images, such as those used for vegetation detection, are produced by a Hyper-Spectral Scanner. An example of the different reflectivities of various types of ground cover in the spectral range of 400–2,600 nm is given in Fig. 1.12.

Using hundreds or thousands of wavelength intervals, trace gas column densities can be evaluated, for example using DOAS, and reproduced as images. Many examples of this type of image are found in the later chapters of this book.

1.10.2 Spectroscopic Techniques in Remote Sensing

No single spectroscopic technique fulfils every need so that, for a particular application, the technique selected will be based on the particular task to be performed. Questions to be answered in selecting a method are: which species are to be measured; is the simultaneous determination of several species necessary; what is the required accuracy, time resolution, horizontal and vertical resolution?

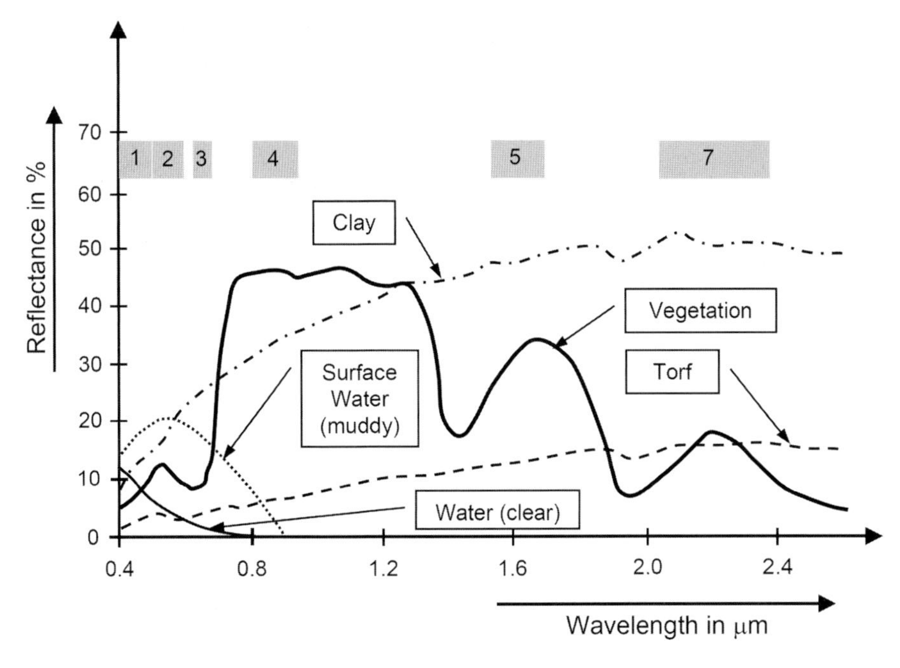

Fig. 1.12 Reflectivity of various ground covers and river water in the spectral range of 0.4–2.6 μm. *Gray, numbered boxes* denote the spectral position of the Landsat-TM channels.

Spectroscopic techniques can be broadly divided into methods relying on absorption of radiation directed to the sample from some source, and into spectroscopic analysis of radiation emitted by the sample itself (see Section 1.8).

a Microwave Spectroscopy

Spectroscopy in the mm and sub-mm wavelength range can, in principle, be used in active and passive configurations. To our knowledge detection of atmospheric gases with active microwave instruments (i.e. with instruments employing their own radiation sources) are not currently made, with the exception of water vapour which can be retrieved with the use of GPS signals, see Section 4.8. However measurements of atmospheric parameters (clouds, precipitation, and turbulence) with active radar techniques are common.

Sensors measuring the thermal radiation at mm and sub-mm wave are mostly applied to the quantification of stratospheric species, the technique registering the thermal emission due to rotational transitions of the atmospheric molecules [Janassen et al, 1993]. Many minor atmospheric constituents have emission lines at mm and sub-mm wavelengths. These lines are dominated by pressure broadening which is much larger than Doppler broadening. An upper limit for the profile occurs at an altitude where the pressure and Doppler broadening become comparable.

Since Doppler broadening is frequency dependent, low frequencies are suitable for measuring mixing ratio profiles at higher altitudes. Transitions at mm waves yield the profile well into the mesosphere, examples being O_3, water vapour and CO. At low altitudes the lines become very broad so the present sensor technology limits the profile retrieval to altitudes above 10 km. Observations using the limb sounding geometry yield profile information down to the upper troposphere. Microwave radiometers also provide information on total water vapour, liquid water content and for ice clouds (see Chapter 4).

Measurements by passive microwave and sub-mm wave emission are mostly applied to the quantification of stratospheric species, the technique registering the thermal emission due to rotational transitions of the atmospheric molecules (Janssen 1993). Taking the chlorine monoxide, ClO, molecule as an example, it radiates at $\nu_o = 649.448$ GHz (wavelength: 0.46 mm; $18_{1/2} \rightarrow 17_{1/2}$ transition of ^{35}ClO) (Klein et al. 2002). The line width is dominated by collisional (pressure) broadening, which strongly dominates Doppler broadening ($\Delta\nu_D/\nu_0 \approx v_{molec}/c \approx 10^{-6}$). While the strong variation in line-width with pressure, and thus with the altitude of the absorbing molecule, allows the retrieval of vertical profiles from a thorough analysis of the recorded line shape, it also limits the detection sensitivity to atmospheric pressures exceeding a few mbar. Thus for ClO the best sensitivity is reached in the upper stratosphere, while detection at lower stratospheric or upper troposphere is difficult.

The microwave and sub-mm wavelength range is presently not used for measurements of trace gases in the mid or lower troposphere, due to the large pressure broadening and strong absorption. It is however used in limb and for the retrieval of H_2O and other parameters (see Chapter 4).

b IR Spectroscopy

This spectral range is characterized by thermal emission from the Earth's atmosphere and the ground and encompasses the interval from approximately 3.5–30 μm. In addition, IR solar occultation is employed to probe the upper troposphere and above. Modern IR spectrometers are based on FT techniques and are able to measure a large number of species relevant to atmospheric chemistry, such as CO, CO_2 and O_3 (see Chapter 3).

In the TIR, the averaging kernel (i.e. the region to which the measurement is sensitive, see Chapter 3) depends on the temperature difference between the atmosphere and the Earth's surface. If this difference is low, as in the lower troposphere, there is then little sensitivity and the information content in the observation is primarily in the middle and upper troposphere.

An advantage offered by TIR, as well as microwave and sub mm regions, is that passive operation is not restricted to daylight and observing can continue in the Earth's shadow. Thus one can obtain double the number of measurements from the satellite instrument. Furthermore it is possible to study atmospheric chemical processes in the absence of daytime photochemistry.

c UV/Visible/Short-Wave IR Absorption Spectroscopy

This region encompasses the UV spectrum from about 300 to 400 nm, the visible range (400–700 nm) and the IR range from 700 to about 2,400 nm (2.4 μm). Stratospheric absorption by O_3 precludes observation of the troposphere at shorter UV wavelengths.

The basic optical arrangement is shown in Fig. 1.9. A light source, the reflected light from the Earth, and a receiving system are separated by the vertical extent of the atmosphere. The strength of the technique in this region lies in good specificity and sensitivity for a series of species relevant to atmospheric chemistry and climate. The absence of thermal emission in this range simplifies the radiation transport calculation appreciably (see Section 1.9). A particular advantage is the potential for high sensitivity throughout the atmosphere including the atmospheric boundary layer. In particular, at relatively long wavelengths such as the short wavelength IR, the sensitivity allows the precise determination of total column amounts of important greenhouse gases such as CO_2 and CH_4, with relative errors typically below 1%.

The normal mode of observation is passive, but active techniques have the potential to combine the advantages of the passive approach with the capability of providing altitude-resolved measurements and measurements in the absence of daylight. Two techniques have been proposed: (a) the use of artificial light sources on the ground, and (b) aerosol lidar and Differential Absorption Lidar (DIAL) on satellite platforms (see Section 1.12.2). While lidar techniques in space look promising for the determination of trace gas profiles, only aerosol lidar instruments have so far been deployed (see Chapter 6).

1.10.3 Passive and Active Remote Sensing

A schematic description of active and passive remote sensing scenarios is illustrated in Fig. 1.13 and active and passive remote sensing are discussed in more detail below.

1.10.4 Nadir, Limb and Occultation Views

For passive remote sensing, the viewing geometry is a defining parameter and enables particular aspects of atmospheric composition to be probed. There are basically three viewing geometries being used, which are shown schematically in Fig. 1.14.

a Nadir view

Looking down from space towards the nadir direction or close to it, the sunlight reflected from the Earth's surface and the atmosphere (called the "earthshine" in

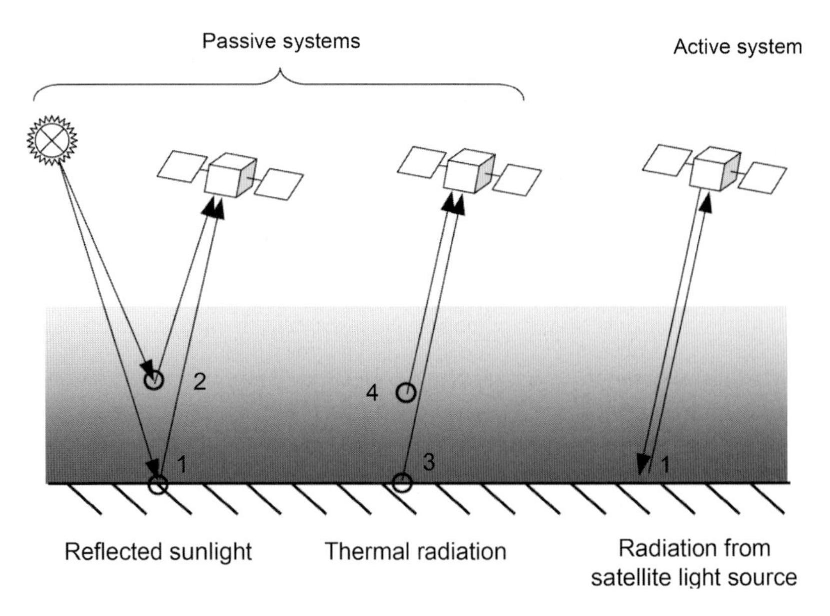

Fig. 1.13 Active and passive remote sensing systems. Possible pathways of the radiation from source to object and sensor are shown. Direct solar radiance can be reflected from the Earth's surface (1), or scattered in the atmosphere (2). Long-wave radiation is emitted from the Earth's surface (3) or from atmospheric gases (4). In summary all processes lead to a modification of the information content.

analogy to "moonshine") is utilised to obtain information on atmospheric trace gas column densities. In this geometry the radiation observed from the atmosphere originates from either the sun, back-scattered from the Earth's surface and the atmosphere or, from the thermal emission in the Earth's surface and atmosphere, or from emissions within the atmosphere. Examples of satellite instruments using this geometry include, for solar measurements: TOMS, SBUV, SBUV-2, GOME, SCIAMACHY, GOME-2 and OMI for UV/vis/NIR spectra, and for thermal IR measurements: MOPITT, TES, IASI. See Appendix A for a full list of satellite instruments.

b Multiple Views

Instruments with several viewing directions offer, in addition to the nadir view, the advantage of multiple paths through the atmospheric boundary layer, which is especially useful for aerosol retrieval. Not only is the AOD higher along a longer atmospheric path, thus providing an extra constraint to the retrieval results but, in particular, the additional view(s) provide a means to eliminate the effect of the reflectance of land surfaces which often dominates the total reflectance. Only a few instruments have more than one view over a particular pixel: (A)ATSR, POLDER/

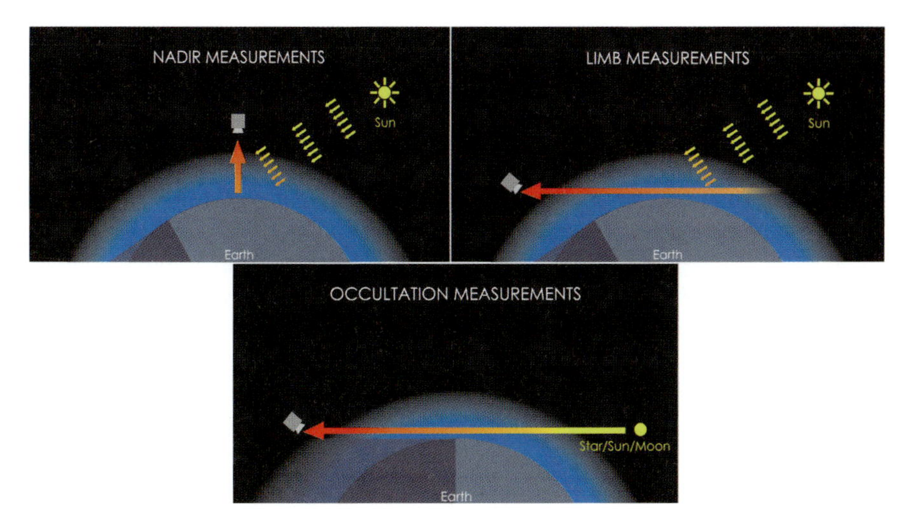

Fig. 1.14 A schematic representation of nadir, limb and occultation viewing geometries used by SCIAMACHY. Figure: J.P. Burrows and S. Noel, IUP, Bremen.

PARASOL and MISR. Hence these instruments provide more constraints on the path radiance and surface effects. In addition to this decoupling of the surface effects, the multi-view instruments can also provide information about the aerosol phase function (Martonchik et al. 1998), and the height of aerosol plumes (Kahn et al. 2007; 2008).

c Limb Mode

In this geometry light scattered from the Earth's rim is analysed. By observing the limb at different tangent altitudes vertical trace gas concentration profiles can be inferred (see Fig. 1.14). Examples of satellite instruments using this geometry include SCIAMACHY, CLAES, ILAS, MLS, ACE-FTS, and OSIRIS. See Appendix A.

d Occultation

By observing the light of the rising or setting sun, moon, or stars through the atmosphere at different tangent altitudes, vertical trace gas concentration profiles can be deduced. Examples of satellite instruments using this geometry include SCIAMACHY, GOMOS, ACE-FTS, HALOE, SAM-II, SAGE-I,II,III, and OSIRIS. See Appendix A.

1.10.5 Active Techniques

In contrast to passive remote sensing which uses the sun, moon, stars and the Earth and its atmosphere as the source of radiation, active remote sensing instruments illuminate the object of interest with a radiation source as part of the instrument. A particular example, which has been used on satellites, is lidar (Rothe et al. 1974; Hinkley 1976; Svanberg 1992; Sigrist 1994). In principle short pulses of a strong, collimated light source, typically from a pulsed laser, are emitted into the atmosphere. By analyzing the temporal evolution of the intensity back-scattered from the atmosphere, the spatial distribution of scattering and extinction along the direction of the emitted (and received) radiation can be determined; this is illustrated in Fig. 1.15.

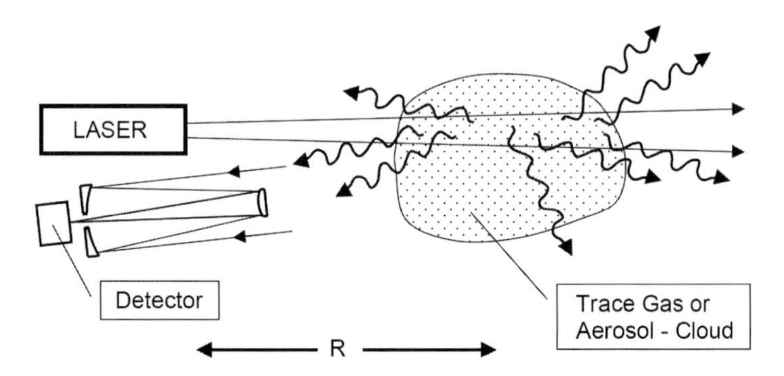

Fig. 1.15 The principle of the lidar technique.

The amout of energy returned is given by the general lidar equation:

$$E(\lambda, R, \Delta R) = K \cdot E_0 \cdot c_S(R)\, \sigma_{SR} \frac{\Delta R}{R^2} \cdot \exp\left(-2\int_0^R [\sigma_A(r) \cdot c_A(r) + \sigma_S(r) \cdot c_S(r)]\, dr\right) \tag{1.44}$$

where R denotes the distance from which the radiation intensity (under consideration) is scattered back; $R = (c \times t)/2$ where c = speed of light and t = time after emission of the laser pulse at which the signal reaches the detector.

$E(\lambda, R, \Delta R)$ is the radiation energy received from a volume of air between R and $R + \Delta R$ where:

ΔR = distance interval to be averaged over (corresponds to $\Delta t = \frac{2\Delta R}{c}$);

K = constant of the system describing the particular properties and size of the receiving system;

E_0 = radiation energy emitted by the LASER;

$c_s(R)$ = concentration of backscattering centres (molecules, aerosol particles)

σ_{SR} = backscatter cross section;

σ_A = absorption cross section (aerosol or gas);
$c_A(R)$ = concentration of absorbers (i.e. gas molecules); and
σ_S = total scattering cross section (note that $\sigma_S \neq \sigma_{RS}$).

Note that here the returned lidar signal is proportional to R^{-2} (Eq. 1.44); in contrast, a returned radar signal varies with R^{-4}. This is due to the fact that the scattering volume for lidar increases with R^2, while in the case of radar the object (e.g. an aircraft) is of constant size and so both the transmitted signal arriving at the target and the returned signal reaching the receiver decrease with R^{-2}.

Lidar techniques have the unique capability of making range-resolved measurements along the line of sight. In the form described so far lidar systems have been successfully used to map aerosol distributions. There are, however, two central problems with lidar systems.

- The signal (Eq. 1.44) depends on the back-scattering as well as on the total scattering (or extinction) properties of the atmosphere. In the case of Mie scattering, these are difficult to separate, since the ratio of σ_S/σ_{SR} is usually unknown and must be estimated from *a priori* information, However if Raman lidar technology is used, the scattering at the Raman wavelength can be corrected with the observed scattering at the excitation wavelength.
- The back scattered signal is usually very weak, so that, even when a high energy laser is used, only a relatively small number of photons are received.

Thus the advantages of the active system are obtained at the expense of sensitivity. However one lidar system, CALIOP, is now being successfully operated in space for aerosol and cloud measurements; for details see Chapter 6 and Appendix A.

a Differential Absorption Lidar (DIAL)

In order to measure the distribution of trace gases at least two different wavelengths are required (Svanberg 1992; Sigrist 1992). The two wavelengths are chosen so that, while the difference in their wavelengths is as small as possible, the difference in the absorption cross sections, $\Delta\sigma = \sigma_A(\lambda_2)–\sigma_A(\lambda_1)$, is as large as possible; normally, one wavelength is at the centre of an absorption line, the other close to the line.

The DIAL equation is obtained by dividing two general lidar equations at two different wavelengths λ_1 and λ_2,

$$\frac{E(\lambda_2,\ R)}{E(\lambda_1,\ R)} = \exp\left(-2(\sigma_A(\lambda_2) - \sigma_A(\lambda_1)) \cdot \int_0^R c_A(r)\,dr\right) \tag{1.45}$$

and it is assumed that σ_{SR} and σ_S are the same for both λ_1 and λ_2, which is justified as long as $\Delta\lambda = \lambda_2 - \lambda_1$ is sufficiently small (a few nm).

The proposed mission, the Water Vapour Lidar Experiment in Space (WALES), will use this technique (Wirth et al. 2009) as will the new Franco-German MERLIN mission.

1.11 Satellite Orbits

The Earth is now orbited by a series of satellites carrying instruments for remote sensing of atmospheric composition. Appendix A summarises the instruments, the spectral ranges used, main species analysed, the type of satellite orbit and other features.

The orbit of the satellite carrying the instrument largely determines its observation capabilities. For instance the "classical" orbit with a small inclination to the equator allows only low latitudes to be probed. A more useful orbit is the polar orbit, which allows the whole earth to be studied. A sun-synchronous orbit always observes the Earth at the same local time during each orbit. In contrast satellites on geo-stationary orbits see only part of the Earth, but they can observe diurnal variations.

1.11.1 Low Earth Orbits (LEO)

In LEO, the satellite circles earth at a relatively low altitude (around 800 km). Sun-synchronous polar orbits are particularly advantageous, where the satellite instrument can observe the whole earth every day or within a few days. The time for the orbit can be adjusted so that the instrument will see the equator at essentially the same local time on each traverse, Fig. 1.16.

In a sun-synchronous orbit the non-sphericity of Earth combined with a small tilt of the orbit (the inclination with respect to the equator) produces a precession so keeping the plane of the orbit at a fixed angle relative to the sun–Earth line. By selecting the inclination of the orbit to the planet's equator, the orbit will precess at the same rate as the planet goes around the sun. For the Earth, the sun-synchronous inclination in LEO is about 98°: it is a retrograde, near-polar orbit, which will always cross the equator at the same local time.

An example of an instrument in a sun-synchronous orbit is provided by the GOME instrument on the ERS-2 satellite; Fig. 1.17. (Burrows et al. 1991; Burrows 1999). The satellite is on a descending orbit crossing the equator at 10.30 a.m. local time. The re-visit time to the same location on the Earth's surface is 3 days.

Sunlight, back-scattered from the Earth, is collected by a scan mirror and then focussed on the entrance slit of a spectrometer, which observes the entire spectral range between 232 and 793 nm with a resolution of between 0.2 nm and 0.33 nm. The instrument is optimised for the collection of the radiation from earth (nadir view) directly by the scan mirror. Fig. 1.18 shows the "whisk broom" scanning scheme used by GOME and other satellite instruments.

The direct, extraterrestrial solar irradiance (the Fraunhofer reference spectrum) is also measured, using a diffuser plate to reduce its intensity prior to it being reflected by the scan mirror into the instrument. The spectra and other signals recorded by GOME are transmitted to the ESA ground station at Kiruna, and then

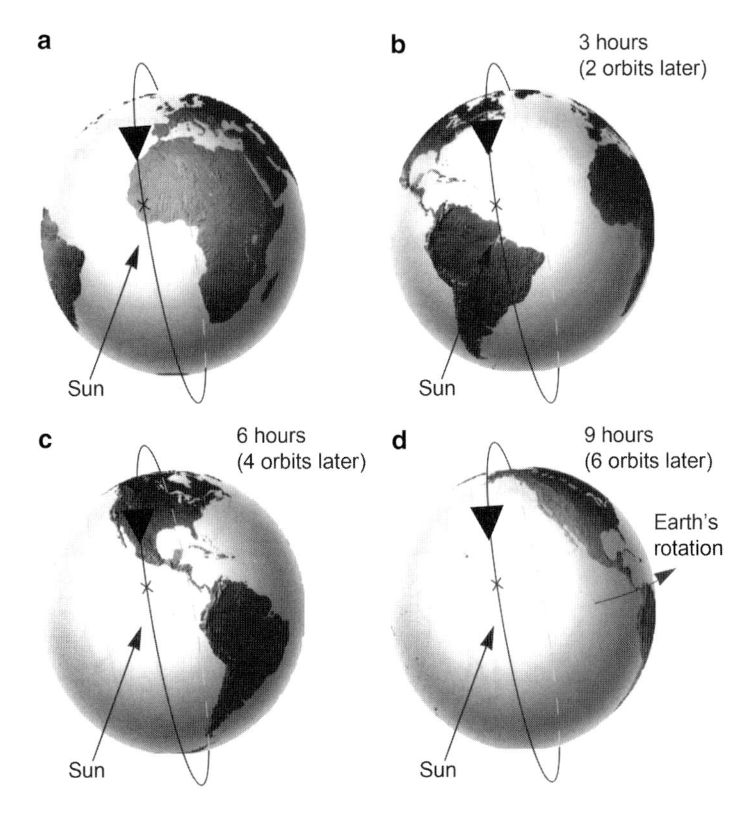

Fig. 1.16 Many satellites circle the Earth on polar orbits (800 km altitude, 100 min per orbit). In a sun synchronous polar orbit, the satellite always crosses the equator at the same local time. Instruments on satellites with these orbits include GOME, SCIAMACHY, AVHRR3, IASI and GOME-2, all of which are all in a descending node traversing from the north to south poles. On the other hand MODIS, OMI and TES are on satellites in an ascending mode. Others are given in Appendix A.

distributed to the waiting laboratories. Details for the other satellite instruments observing the troposphere are given in the table in Appendix A.

1.11.2 Geostationary Orbits (GEO)

Instrumentation placed in the GEO is approximately 36,000 km from the earth and rotates with the earth. As the meteorological observations have demonstrated, this enables optimal temporal sampling and yields diurnal variations. In order to meet the needs of tropospheric chemistry and air quality, a number of missions have been proposed by the community to measure trace gases from a geostationary orbit: for example, the GeoSCIA and GeoFIS instruments and the GeoTROPE concept (Bovensmann et al. 2002; 2004; Burrows et al. 2004).

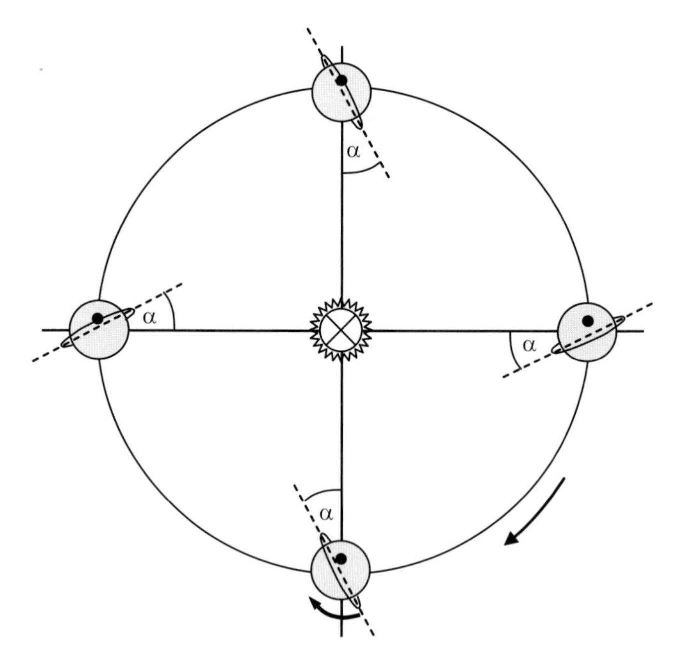

Fig. 1.17 In a sun synchronous polar orbit the satellite always crosses the equator at the same local time. Thus the angle α between the plane of the orbit and the direction towards the sun stays constant, i.e. the plane of the orbit makes a full rotation once a year. The required change in angular momentum is supplied by gravitational forces due to the "equator bulge" of Earth.

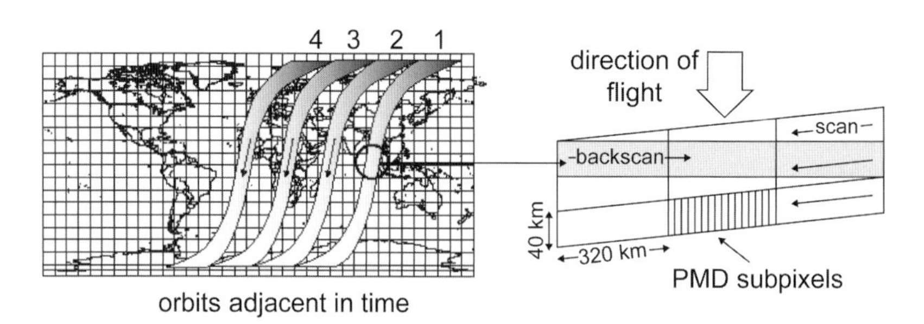

Fig. 1.18 Whisk broom scanning scheme of the GOME instrument. Left panel: Subsequent satellite orbits (numbered 1...4) on a world map. Right panel: Each scan consists of three ground pixels plus one "back-scan pixel". Later instruments (SCIAMACHY, GOME-2) divide a scan into more pixels. Fig. from Leue et al. (2001).

Based on this innovative work Europe has now decided, as part of its Meteosat Third Generation and GMES Sentinel 4 programmes, to fly a UVN and an IRS instrument, which meet some of the objectives of the GeoTROPE mission.

A full list of satellite instruments for observing tropospheric chemical composition with solar back-scattered radiation is given in Appendix A.

1.12 Summary

This chapter has provided the historical highlights in the development of tropospheric remote sounding from space, which is now an essential part of the modern science of Earth observation. The scientific background required to understand why the remote sounding of trace constituents is important and how it has evolved has been presented, together with an introduction to the approaches needed to retrieve trace gases, cloud and aerosol parameters. The following chapters expand on these topics and present the remarkable results that have been obtained so far. Some outlines of future needs and activities are given in the concluding chapter, Chapter 10.

References

Angström A., 1929, On the atmospheric transmission of sun radiation and on dust in the air, *Geogr. Ann.* Stockholm **11**, 156–166.

Arrhenius S., 1896, On the influence of carbonic acid in the air upon the temperature of the ground, *The London, Dublin, and Edinburgh Philosophical Magazine and Journal of Science* **41**, 237–276.

Barrie L.A., Bottenheim J.W., R.C. Shnell, P.J. Crutzen and Rasmussen R.A., 1988, Ozone destruction and photochemical reactions at polar sunrise in the lower Arctic atmosphere, *Nature* **334**, 138–141.

Barrie, L.A., P. Borrell and J. Langen, 2004, The Integrated Global Atmospheric Chemistry Observations Theme (IGACO) Report on Monitoring of our Environment from Space and from Earth, European Space Agency (ESA) SP1235; Global Atmospheric Watch (GAW) 159, *World Meteorological Organisation (WMO) TD* 1235, pp 54.

Bone N., 1991, The Aurora: Sun–Earth Interactions, Ellis Horwood, Chichester.

Borrell P, Borrell P.M., Burrows J.P., and Platt U. (Eds.), 2003, Sounding the Troposphere from Space: a new Era for Atmospheric Chemistry, The TROPOSAT Final Report, Springer Verlag, Heidelberg, 446 pp.

Brekke A. and Egeland A., 1994, The northern lights: their heritage and science, Grøndahl Dreyer, ISBN-13: 9788250421059 ISBN: 8250421051.

Bovensmann, H., J. P. Burrows, M. Buchwitz, J. Frerick, S. Noël, V. V. Rozanov, K. V. Chance, A. P. H. Goede, 1999, "SCIAMACHY- mission objectives and measurement modes", *Atmos Sci* 56 127–150.

Bovensmann H., S. Noel, P. Monks, A.P.H. Goede, J.P. Burrows, 2002, The Geostationary Scanning Imaging Absorption Spectrometer (GEOSCIA) Mission: Requirements and Capabilities, *Adv. Space Res.*, 29(11), pp 1849–1859.

Bovensmann H., K. U. Eichmann, S. Noel, J. M. Flaud, J. Orphal, P.S. Monks, G. K. Corlett, A. P. Goede, T, von Clarmann, T. Steck, V. Rozanov and J. P. Burrows, 2004, The Geostationary scanning imaging absorption spectrometer (GeoSCIA) as part of the Geostationary pollution exlorer (GeoTROPE) mission: requirements concepts and capabilities. *Adv. Space Research* **34**(2004) 694–699.

Burrows J. P., W. Schneider, J. C. Geary, K. V. Chance, A. P. H. Goede, H. J. M. Aarts, J. de Vries, C. Smorenburg and H. Visser, 1990, "Atmospheric remote sensing with SCIAMACHY.", Digest of Topical Meeting on Optical Remote Sensing of the Atmosphere 1990 (Optical Society of America, Washington D.C. 1990) 4 71 1990.

Burrows J. P., E. Hölzle, A. P. H. Goede, H. Visser and W. Fricke, 1995, "SCIAMACHY - Scanning Imaging Absorption Spectrometer for Atmospheric Chartography." *Acta Astronautica* **35(7)** 445–451

Burrows J.P., 1999, "Current and future passive remote sensing techniques used to determine atmospheric constituents", in *Developments in Atmospheric Sciences* **24**: Approaches to Scaling Trace Gas Fluxes in Ecosystems, Ed A. F. Bouwman Elservier Amsterdam pp 315–347. ISBN: 0-444-82934-2.

Burrows J.P., Chance K.V., Crutzen P.J., Fishman J., Fredericks J.E., Geary J.C., Johnson T.J., Harris, G.W., Isaksen I.S.A., Kelder H., Moortgat G.K., Muller C., Perner D., Platt, U., Pommereau J.-P., Rodhe H., Roeckner E., Schneider W., Simon P., Sundqvist H., Vercheval J., 1991, SCIAMACHY Phase A Study – *Scientific Requirements Specification.*

Burrows J.P., Weber M., Buchwitz M., Rozanov V., Ladstätter-Weißenmayer A., Richter A., DeBeek, R., Hoogen R., Bramstedt K., Eichmann K. -U., Eisinger M., and Perner D., 1999, The Global Ozone Monitoring Experiment (GOME): Mission Concept and First Scientific Results, *J. Atmos. Sci.* **56**, 151–171.

Burrows J.P., H. Bovensmann, G. Bergametti, J.M. Flaud, J. Orphal, S. Noel, P.S. Monks, G.K. Corlett, A.P. Goede, T. v. Clarmann, T. Steck, H. Fischer, F. Friedl-Vallon, 2004, The geostationary tropospheric pollution explorer (GeoTROPE) missions: objects, requirements and mission concept, *Adv. Space Research*, **34**, 682–687, doi. 10.1016/j.asr. 2003.08.067.

Chapman S., 1930, On ozone and atomic oxygen in the upper atmosphere, *Phil. Mag.* **S7**, 369–383.

Climate Change 2007 – Impacts, Adaptation and Vulnerability, Contribution of Working Group II to the Fourth Assessment Report of the IPCC, (978 0521 88010-7 Hardback; 978 0521 70597-4 Paperback).

Climate Change 2007 – Mitigation of Climate Change, Contribution of Working Group III to the Fourth Assessment Report of the IPCC, (978 0521 88011-4 Hardback; 978 0521 70598-1 Paperback). Climate Change 2007 – Synthesis Report.

Crutzen P.J., 1973, A discussion of the chemistry of some minor constituents in the stratosphere and troposphere, *Pure Appl. Geophys.* **106,** 1385–1399.

Crutzen P.J., 1974, Photochemical reactions initiated by and influencing ozone in unpolluted tropospheric air, *Tellus* **26**, 47–57.

Crutzen P.J. and Andreae M.O., 1990, Biomass burning in the tropics: Impact on atmospheric chemistry and biogeochemical cycles, *Science* **250**, 1669–1678.

Crutzen P.J., and Stoermer E.F., 2000, The "Anthropocene". *Global Change Newsletter*. **41**: 17–18.

Dameris M., Grewe V., Hein R., Schnadt C. Brühl C., and Steil B., 1998, Assessment of the future development of the ozone layer, *Geophys. Res. Lett.* **25 (19)**, 3579–3582.

Einstein, A., 1917 Zur Quantentheorie der Strahlung, *Phys Z.*, **18**, 121–128.

Farman J.C., Gardiner B.G., Shanklin J.D., 1985, Large losses of total ozone in Antarctica reveal seasonal ClO_X/NO_x interaction, *Nature* **315**, 207–210.

Finlayson-Pitts B.J. and Pitts J.N., 1986, Chemistry of the Upper and Lower Atmosphere: Theory, Experiments, and Applications, Academic Press, 1st edition (November 1999), 969 pp., ISBN-10: 012257060X, ISBN-13: 978-0122570605, and references therein.

Fischer H. and Oelhaf H., 1996, Remote sensing of vertical profiles of atmospheric trace constituents with MIPAS limb emission spectrometers, *Applied Optics* **35 (16)**, 2787–2796.

Fishman, J., C.E. Watson, J.C. Larsen and J.A. Logan, 1990, Distribution of tropospheric ozone determined from satellite data. *J. Geophys. Res.,* **95**:3599–3617.

Fleming J.R., 1998, Historical Perspectives on Climate Change. Oxford University Press, New York, 208pp.

Funk O., 2000, Photon path length distributions for cloudy skies; oxygen a-band measurements and radiative transfer calculations, Doctoral Thesis, University of Heidelberg, Heidelberg, Germany.

GEOSS 2009, The Global Earth Observation System of Systems, http://earthobservations.org/

Gottwald M., Bovensmann H., (eds.), 2010, SCIAMACHY - Exploring the Changing Earth's Atmosphere. Springer Heidelberg, ISBN: 978-90-481-9895-5, doi: 10.1007/978-90-481-9896-2.

Grannas A.M., Jones A.E., Dibb J., Ammann M., Anastasio C., Beine H.J., Bergin M., Bottenheim J., Boxe C.S., Carver G., Chen G., Crawford J.H., Dominé F., Frey M.M., Guzmán M.I., Heard D.E., Helmigh D., Hoffmann M.R., Honrath R.E., Huey L.G., Hutterli M., Jacobi H.W., Klán P., Lefer B., McConnell J., Plane J., Sanders R., Savarino J.,

Shepson P.B., Simposon W.R., Sodeau J.R., von Glasow R., Weller R., Wolff E.M.and Zhu T., 2007, An overview of snow photochemistry: evidence, mechanisms and impacts, *Atmos. Chem. Phys.* **7**, 4329–4373.

Haagen-Smit A.J., 1952, Chemistry and physiology of Los Angeles smog, *Ind. Engin. Chem.* **44**, 1342–1346.

Haagen-Smit, A.J., Darley, E.F., Zailin, M., Hull, H., and Noble, W., 1952, Investigation on injury to plants from air pollution in the Los Angeles area. *Plant Physiology* **27**:18.

Heath D.F., Mateer C. L. and Krueger A. J., 1973, The Nimbus -4 Backscatter Ultraviolet (BUV) Atmospheric Ozone Experiment – Two Years of Operation, *Pure and Applied Geophysics* **106–108** 1238–1253.

Heath D.F., Krueger A.J, Roeder H.A., and Henderson B.D., 1975, "The Solar backscatter Ultraviolet and the Total Ozone Mapping Spectrometere (SBUV/TOMS) for Nimbus G. *Opt. Eng.* **14** 323–332

Heij G.J. and Erisman J.W. (Eds.) 1995, Acid rain research: Do We Have Enough Answers?, Elsevier, Amsterdam, 502pp., ISBN 0-444-82038-9.

Herzberg, G., 1950, Spectra of Diatomic Molecules, van Nostrand, New York.

Hinkley E.D. (Ed.) 1976, Laser Monitoring of the Atmosphere, Topics in Appl. Physics, Vol. 14, Springer Berlin, Heidelberg.

Holloway, A.M. and Wayne, R.P., 2010, Atmospheric Chemistry, Royal Society of Chemistry, London, ISBN 9781847558077.

Horowitz, L.W., 2006, Past, present, and future concentrations of tropospheric ozone and aerosols: Methodology, ozone evaluation, and sensitivity to aerosol wet removal. *J. Geophys. Res.* **111** (D22), D22211.

HTAP 2007, Hemispheric Transport of Air Pollution 2007, Air Pollution Studies No. 16, Interim report prepared by the Task Force on Hemispheric Transport of Air Pollution acting within the framework of the Convention on Long-range Transboundary Air Pollution, Economic Commission for Europe, Geneva, United Nations, New York and Geneva, 2007

IPCC 4th Assessment Report, 2007 available at http://www.ipcc.ch/Climate Change 2007 and comprising the elements: The Physical Science Basis. Contribution of Working Group I to the Fourth Assessment Report of the IPCC, (ISBN 978 0521 88009-1 Hardback; 978 0521 70596-7 Paperback).

Isaacs, R. G., W.-C. Wang, R. D. Worsham, and S. Goldberg, 1987. Multiple Scattering Lowtran and Fascode models. *Appl.* Opt. **26**, 1272–1281.

Janssen M.A. (1993), An Introduction to the Passive Remote Atmospheric Remote Sensing by Microwave Radiometry, J. Wiley & Sons Inc., New York, pp.1–36.

Kahn, R.A. W.H. Li, C. Moroney, D. J. Diner, J. V. Martonchik, and E. Fishbein, 2007 Aerosol source plume physical characteristics from space-based multiangle imaging, *J. Geophys. Res.*, **112**, D11205, doi:10.1029/2006JD007647

Kahn, R., A. Petzold, M. Wendisch, E. Bierwirth, T. Dinter, M. Esselborn, M. Fiebig, B. Heese, P. Knippertz, D. Műller, A. Schladitz and W. von Hoyningen-Huene, 2008 Desert dust aerosol air mass mapping in the western Sahara, using particle properties derived from space-based multi-angle imaging. *Tellus* **61**, 239–251. Doi:10.1111/j.1600-0889.2008.00398.x

Keeling C.D., Barcastow R.B., Bainbridge A.E., Ekdahl C.A., Guenther P.R., Waterman L.S., 1976, Atmospheric carbon dioxide variations at Mauna Loa observatory, Hawaii, *Tellus* **28,** 538–551.

Keeling, R.F., S.C. Piper, A.F. Bollenbacher and J.S. Walker, 2003, CDIAC, DOI: 10.3334/CDIAC/atg.035

Klein U., Wohltmann I., Lindner K., and Künzi K.F., 2002, Ozone depletion and chlorine activation in the Arctic winter 1999/2000 observed in Ny-Ålesund, *J. Geophys. Res.* **107(D20)**, 8288, doi:10.1029/2001JD000543.

Le Treut, H., R. Somerville, U. Cubasch, Y. Ding, C. Mauritzen, A. Mokssit, T. Peterson and M. Prather (2007), Historical Overview of Climate Change. In: Climate Change 2007: The Physical Science Basis. Contribution of Working Group I to the Fourth Assessment Report of the Intergovernmental Panel on Climate Change.

Leue C., Wenig M., Wagner T., Platt U. and Jähne B., 2001, Quantitative analysis of NOX emission from Global Ozone Monitoring Experiment satellite image sequences, *J. Geophys. Res.* **106**, 5493–5505.

Levine J.S., 1991, Global Biomass Burning: Atmospheric, Climatic, and Biospheric Implications. MIT Press, Cambridge, MA.

Livesey N.J., Read W.G., Froidevaux L., Waters J.W., Santee M.L., Pumphrey H.C., Wu D.L., Shippony Z., and Jarnot R.F., 2003, The UARS Microwave Limb Sounder version 5 data set: Theory, characterization, and validation, *J. Geophys. Res.* **108(D13)**, 4378, doi:10.1029/2002JD002273.

Lovelock J.E., Maggs R.J., and Wade R.J., 1973, Halogenated hydrocarbons in and over the Atlantic, Nature **241**, 194–196.

LRTAP 2004, Handbook for the 1979 Convention on Long-range Transboundary Air Pollution and its protocols, United Nations, New York and Geneva. (see also: http://www.unece.org/env/lrtap/welcome.html)

Martonchik, J.V., D. J. Diner, B. Pinty, M. M. Verstraete, R. B. Myneni, Y. Knyazikhin, H. R. Gordon, 1998, Determination of land and ocean reflective, radiative, and biophysical properties using multiangle imaging, http://citeseerx.ist.psu.edu/viewdoc/summary? doi: 10.1.1.32.5688

Mie, Gustav, 1908, Beiträge zur Optik trüber Medien, speziell kolloidaler Metallösungen, *Annalen der Physik, Vierte Folge,* Band **25, No. 3,** 377–445.

Molina, M.J. and Rowland, F.S., 1974, Stratospheric sink for chlorofluoromethanes: chlorine atomc-catalysed destruction of ozone, *Nature* **249**, 810–812; doi:10.1038/249810a0

Nicolet M., 1984, On the Molecular Scattering in the Terrestrial Atmosphere: An Empirical Formula for its Calculation in the Homosphere. *Planet. Space Sci.* **32**(11) 1467–1468.

Noxon J.F., 1975, Nitrogen dioxide in the stratosphere and troposphere measured by ground-based absorption spectroscopy, *Science* **189**, 547–549

O'Dowd, C. D., Geever, M., and Hill, M. K. (1998), New particle formation: Nucleation rates and spatial scales in the clean marine coastal environment, *Geophys. Res. Lett.*, **25(10)**, 1661–1664.

O'Dowd, C. D., Jimenez, J. L., Bahreini, R., Flagan, R. C., Seinfeld, J. H., Hameri, K., Pirjola, L., Kulmala, M., Jennings, S. G., and Hoffmann, T., 2002, Marine aerosol formation from biogenic iodine emissions, *Nature,* **417,** 632–636.

Penndorf R., 1957, Tables of the refractive index for standard air and the Rayleigh scattering coefficient for the spectral region between 0.2 and 20.0 μ and their application to atmospheric optics, *J. Opt. Soc. Amer.* **47**, 176–182.

Perner D. and Platt U., 1979, Detection of nitrous acid in the atmosphere by differential optical absorption, *Geophys. Res. Lett.* **6**, 917–920.

Platt U. and Lehrer E., (Eds.) 1996, Arctic Tropospheric Halogen Chemistry, *Final Report to EU.*

Platt U., Pfeilsticker K. and Vollmer M., 2007, Radiation and Optics in the Atmosphere, Ch. 19 in: "Springer Handbook of Lasers and Optics", F. Träger Ed., Springer, Heidelberg, ISBN-10: 0-387-95579-8, pp. 1165–1203.

Platt, U. and J. Stutz, 2008, Differential optical absorption spectroscopy: principles and applications, Springer Verlag, Heidelberg, ISBN 978-3540211938, pp597.

Pommereau, J.P., 1982, First balloon flight of vis spectrometer for NO2. Observation of NO_2 diurnal variation in the stratosphere, *Geophys. Res. Lett.*, 850.

Pommereau, J.P. and J. Piquard, 1994a, First publications relative to SAOZ balloon flights in Kiruna, Ozone. Nitrogen dioxide and Aerosol vertical distributions by UV-visible solar occultation from balloons, *Geophys. Res. Lett,* **21**, 1227–1230.

Pommereau, J.P. and J. Piquard, 1994b, Observations of the vertical distribution of stratospheric OClO, *Geophys. Res. Lett.*, **21**, 1231–1234.

Rayleigh, Lord, 1899, On the transmission of light through an atmosphere containing many small particles in suspension, and on the origin of the blue of the sky, *Phil. Mag.* **41**, 447–454. Also: in 'the scientific papers of Lord Rayleigh', Vol. 4, Dover, New York, 1964.

Ravishankara, A.R., J.S. Daniel and R.W. Portmann. 2009, Nitrous Oxide (N_2O): The dominant ozone-depleting substance emitted in the 21st century Published online 31 August 2009. DOI: 10.1126/science.1176985

Roach, F.E.and Gordon J.L., 1973, The Light of the Night Sky, Reidel, Dordrecht.

Rothe, K.W., Brinkmann U., Walther H., 1974, Applications of tunable dye lasers to air pollution detection: Measurements of atmospheric NO_2 concentrations by differential absorption, *Appl. Phys.* **3**, 115–119.

Schär C. and Jendritzky G., 2004, Hot news from summer 2003, *Nature* **432**, 559–560.

Schär C., Vidale P.L., Lüthi D., Frei C., Häberli C., Liniger M.and Appenzeller C., 2004, The role of increasing temperature variability for European summer heat waves, *Nature* 427, 332–336.

Seinfeld, J.H., S.N. Pandis, 1998, Atmospheric Chemistry and Physics, from Air Pollution to Climate Change, *J. Atmos. Chem.*, **37**, 212–214.

Shindell D.T., Rind D. and Lonergan P., 1998, Increased polar stratospheric ozone losses and delayed eventual recovery owing to increasing greenhouse-gas concentrations. *Nature* **392**, 589–592, doi:10.1038/33385.

Sigrist M.W. (Ed.) 1994, Air monitoring by spectroscopic techniques, *Chemical Analysis Series,* Vol. **127,** John Wiley & Sons, Inc.

Simpson W.R., R. von Glasow, K. Riedel, P. Anderson, P. Ariya, J. Bottemheim, J. Burows, L. J. Carepnter, U. Frieß, M. Goodsite, D. Heard, M. Hutterli, H.-W. Jacobi, L. Kaleschke, B. Neff, J. Plane, U. Platt, A. Richter, H. Roscoe, R. Sander, P. Shepson, J. Sodeau, A. Stefann, T. Wagner and E. Wolff, 2007, Halogens and their role in polar boundary-layer ozoen depletion, *Atmos Chem. Phys.* **7**, 4375–4418.

Steffan A., Douglas T., Amyot M., Ariya P., Aspmo K., Berg T., Bottenheim J., Brooks S., Corbett F., Dastoo A., Dommergue A., Ebinghaus R., Ferrari C., Gardefeldt K., Goodsite M.E., Lean D., Poulain A.J., Scherz C., Skov H., Sommar J.and Temme C., 2008, A synthesis of atmospheric mercury depletion event chemistry in the atmosphere and snow, *Atmos. Chem. Phys.* **8**, 1445–1482.

Svanberg S., 1992, Atomic and Molecular Spectroscopy, 2nd Edition, Springer Series on Atoms and Plasmas, Springer Berlin, Heidelberg.

U.S. Standard Atmosphere, 1976, U.S. Government Printing Office, Washington, D.C.

Van de Hulst, H.C., 1980, Multiple Light Scattering, Tables, Formulas and Applications, Volume 1 and 2. London: Academic Press.

Volz A. and Kley D., 1988, Ozone Measurements in the 19th century: An Evaluation of the Montsouris series, *Nature* **332**, 240–242.

Wang J., Zhang X., Keenan T. and Duan Y., 2009, Air-quality management and weather prediction during the 2008 Beijing Olympics, *WMO Bulletin* **58 (1),** 31–40.

Watson, C.E.; Fishman, J.; Reichle, H.G.Jr., 1990, The significance of biomass burning as a source of carbon monoxide and ozone in the southern hemisphere tropics: A satellite analysis, *J. Geophy. Res.,* **95 (D10),** 16,443–16, 450.

Wayne, R.P. (2000), Chemistry of atmospheres (3rd Ed.). Oxford University Press. ISBN 0-19-850375-X

Weart, S., (2003), The Discovery of Global Warming. Harvard University Press, Cambridge, MA, 240 pp.

Wirth M., Fix A., Mahnke P., Schwarzer H., Schrandt F., and Ehret G., 2009, The airborne multi-wavelength water vapour differential absorption lidar WALES: system design and performance. *App. Phys. B-Lasers & Optics*, **96**, 201–213.

Wiscombe, W. J. (1980). Improved Mie scattering algorithms. *Appl. Op*t. **19**, 1505–1509.

WMO (World Meteorological Organization) 2007, Scientific Assessment of Ozone Depletion: 2006, *Global Ozone Research and Monitoring Project—Report* **50**, 572 pp., Geneva, Switzerland.

Chapter 2
The Use of UV, Visible and Near IR Solar Back Scattered Radiation to Determine Trace Gases

Andreas Richter and Thomas Wagner

2.1 Basics and Historical Background

Satellite remote sensing in the near-IR, visible and UV spectral range makes use of absorption and emission processes of electromagnetic radiation corresponding to electronic transitions, combined with simultaneous rotational-vibrational molecular transitions. One important difference compared to atmospheric observations in the microwave and thermal IR spectral range is that, usually thermal emission can be neglected at short wavelengths (there might, however, be emissions from, for example, excited gases in the high atmosphere). Thus the observed spectral signatures can be directly related to *absorption* spectra of atmospheric constituents. The neglect of emission terms makes the spectral analysis in the UV/vis spectral range usually reasonably straight forward. Another important and related advantage is that from satellite observations in the UV/vis spectral region, information from all atmospheric height layers (including the near surface layers) can be obtained. This makes UV/vis satellite observations a powerful tool for the monitoring of atmospheric pollution and for the characterisation and quantification of emission sources which are usually located close to the ground. It should, however, also be noted that, in contrast to observations in the microwave or thermal IR, usually little or no information on the vertical distribution of a trace gas is obtained.

The analysis procedures applied to satellite observations in the UV/vis/NIR spectral range have evolved from those of ground-based, ship, aircraft or balloon-borne observations, which have been successfully performed over several decades before instruments were placed on orbiting satellites in space. A brief overview of the major developments in this field during the last century is given below.

A. Richter (✉)
Institute of Environmental Physics, University of Bremen, Bremen, Germany

T. Wagner (✉)
Max-Planck-Institute for Chemistry, Mainz, Germany

J.P. Burrows et al. (eds.), *The Remote Sensing of Tropospheric Composition from Space*, Physics of Earth and Space Environments, DOI 10.1007/978-3-642-14791-3_2,

In 1879, Marie Alfred Cornu proposed that the short wavelength limit of the solar radiation on the Earth's surface must be caused by an absorber located in the Earth's atmosphere. Just a year later, Sir Walther Noel Hartley described the strong UV absorptions between 200 and 300 nm and it became obvious that the absorbing properties of ozone, O_3, fulfilled the requirements of the postulated atmospheric absorber. Additional electronic bands of the O_3 absorption spectrum were discovered in the following years by J. Chappuis and Sir William Huggins (see Fig. 2.1). In 1902 Leon Teisserenc de Bort and Richard Aßmann discovered that the atmospheric temperature starts to increase above about 10 km and in 1908 Teisserenc de Bort called this layer between about 10 km and 50 km the stratosphere. This temperature increase was consistent with the assumption that an absorbing (i.e. O_3) layer exists at these altitudes. In 1925 G. Dobson developed a new very stable spectrophotometer (a double monochromator using quartz prisms and a photomultiplier) for the quantification of the vertically integrated atmospheric concentration or vertical column density, VCD, of O_3. Today the "thickness" of the atmospheric O_3 layer is still expressed in Dobson units (DU), which are defined as the thickness of the atmospheric O_3 vertical column density under standard conditions for temperature and pressure (STP: 293.15K, 1013.25 hPa), measured in 10^{-5} m. Thus, 1 DU of O_3 is equivalent to a column density of 2.68×10^{16} molecules/cm^2. Dobson spectrometers use a simple (but stable) spectroscopic method: the direct or scattered solar intensity is measured in different narrow (~1 nm) spectral intervals of which some are located inside and some outside the O_3 (Huggins) absorption bands (see Fig. 2.1). From the ratio of the radiation at different wavelengths (and a geometric correction factor accounting for the effect of varying solar zenith angle, SZA) the thickness of the O_3 layer is determined. The impact of absorption and scattering by aerosols or SO_2 on the O_3 retrieval can be accounted

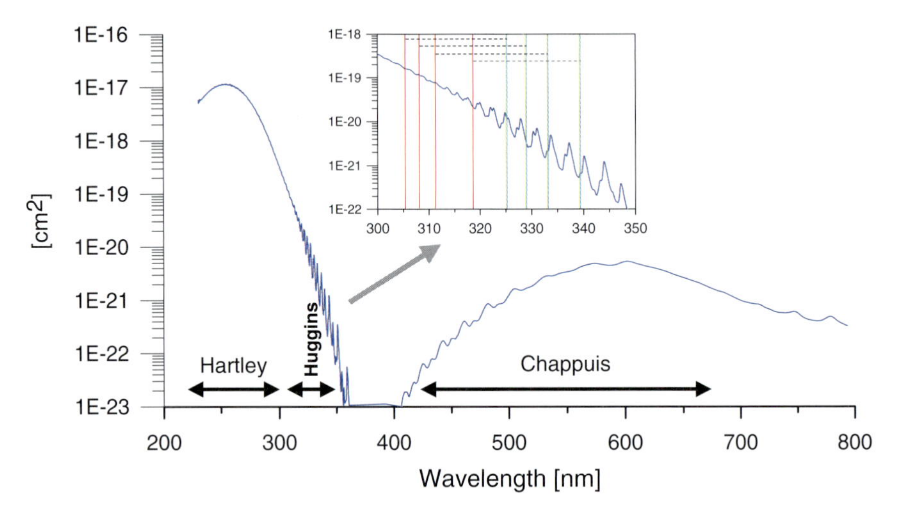

Fig. 2.1 The O_3 absorption cross-section (Bogumil et al. 2003). The highly structured O_3 Huggins Bands are displayed in the panel with an expanded wavelength scale. Also shown in the panel are the wavelength pairs used by Dobson spectrometers (see text).

for by combining different wavelength pairs, for which the influence of O_3 and such interfering effects are different. Dobson spectrometers became widely used and, as a result of an initiative in the international Geophysics Year in 1957 by the International Ozone Commission, a global network of more than 100 instruments was established (Komhyr et al. 1989): the majority still being in use. These instruments provide data over long time, an important basis for the investigation of the long term evolution of the atmospheric O_3 layer.

In 1973 Alan Brewer used a similar method to measure the atmospheric nitrogen dioxide, NO_2, column density from the ground (Brewer et al. 1973). His instrument observes light in different narrow wavelength intervals around 450 nm, for which the NO_2 absorption varies strongly with wavelength. Although the atmospheric NO_2 absorptions are by far weaker than those of O_3, it was possible to identify them with this new spectrometer. It was, in particular, possible to confirm the permanent presence of NO_2 in the stratosphere. One peculiarity of the Brewer spectrometers is that they can also measure O_3 using the light of a full moon when the sky is clear; this is useful in the polar winter. Similar to the Dobson instruments, a network of Brewer spectrometers was established which is still in operation.

Two significant advances with respect to the retrieval of "weak atmospheric absorbers", were introduced in 1975 by Noxon (1975). First, he measured the sky radiances over a continuous spectral interval covering the targeted atmospheric absorption, in this case around 440 nm for NO_2. Secondly he removed the strong structures of the solar Fraunhofer lines (Kirchhoff 1859) (see Fig. 2.3) by dividing a spectrum taken either during sunset or sunrise by another spectrum measured at noon for which the atmospheric light path is much shorter. The difference in the NO_2 absorption of both spectra is a good approximation for the stratospheric NO_2 amount.

In 1976 U. Platt and D. Perner introduced the DOAS (Differential Optical Absorption Spectroscopy) method (Perner et al. 1976; Platt et al. 1979; Perner and Platt 1979), which was first applied to tropospheric observations using artificial light sources. The key concept of DOAS is the simultaneous fit of several trace gas absorption spectra to the measured atmospheric spectrum using only the high frequency part (Platt 1994). This concept is an elegant way to deal with possible spectral interference between absorption structures of different trace gases (e.g. between sulfur dioxide, SO_2, and O_3 for Dobson measurements) and the effects of atmospheric scattering.

After the discovery of the stratospheric O_3 hole in 1985 (Farman et al. 1985), DOAS observations of scattered solar radiation in zenith viewing direction became very important for the monitoring of stratospheric NO_2 and the halogen oxides, BrO and OClO (Solomon et al. 1987). The weak absorption features of these species (typically $< 0.1\%$) could be identified during twilight, when the light paths through the stratosphere are especially long. Today, DOAS observations of scattered sun light are often also performed at a variety of additional viewing directions. The so called Multi AXis (MAX) DOAS observations collect the light at slant elevation angles making such observations especially sensitive to tropospheric trace gases and aerosols.

2.1.1 *Satellite Observations in the UV/vis/NIR Spectral Range*

Shortly after the launch of the first satellite (Sputnik in 1957), the new space borne platforms were used for observing our planet. In 1960, the first UV/vis image of the Earth's surface and atmosphere was taken by TIROS-1; a few years later, composite images of the whole Earth and the Earth seen as a blue planet from the moon enabled a completely new view of the Earth in space. Satellite images are now widely used, today's instruments typically providing measurements at various spectral intervals, thus allowing the retrieval of a variety of atmospheric and surface properties.

After early conceptual studies and prototype measurements, long-term spectroscopic UV/vis observations from space started in 1970 on board the US research satellite Nimbus 4 (Heath et al. 1973). These instruments [Backscatter Ultraviolet, BUV, later also called Solar BUV or SBUV, see (Frederick et al. 1986)] operated in nadir geometry i.e. pointing downward to the surface and measuring the solar light reflected from the ground or scattered from the atmosphere (Fig. 2.2). The atmospheric penetration depth of the observed light depends on the optical thickness of the atmosphere. Typically most of the observed photons have traversed the stratosphere twice and thus the sensitivity to stratospheric absorbers is high. In the UV, the sensitivity for trace gases close to the surface is usually much lower (except over ice and snow) and for trace gases located below thick clouds it can become almost zero.

In a similar way to the Dobson instruments, the BUV/SBUV also measured the intensity in narrow spectral intervals, for which the atmospheric O_3 absorption differs. From the BUV/SBUV data it is possible to retrieve information on the atmospheric O_3 concentration profile, because the penetration depth into the atmosphere strongly depends on the absolute value of the O_3 absorption cross-section and thus on wavelength (see Fig. 2.1). Typically, the received light at the shortest wavelengths has only "seen" the highest parts of the O_3 layer whereas the longest wavelengths have seen the total column.

Like the BUV/SBUV instruments, the TOMS instruments [Total Ozone Mapping Spectrometer, first launched in 1979 on Nimbus 7, (Heath et al. 1975)] observe the backscattered light in distinct wavelength intervals, but at longer wavelengths, for which the light can penetrate the whole atmosphere. TOMS observations thus yield the total O_3 VCD, constituting a major breakthrough in UV/vis satellite remote sensing. The TOMS instrument on board Nimbus 7 has yielded the longest continuous global data set on the O_3 layer (1979–1992) (McPeters et al. 1996), covering in particular the formation and evolution of the ozone hole. Since then, several further TOMS instruments were launched on other satellites.

A completely new quality of the observed spectral information became available in 1995 with the launch of the first DOAS-type instrument, the Global Ozone Monitoring Experiment (GOME) on the European research satellite ERS-2 (ESA 1995; Burrows et al. 1999 and references therein). Similar to SBUV and TOMS, GOME is a nadir viewing instrument, but it contiguously measures a large spectral

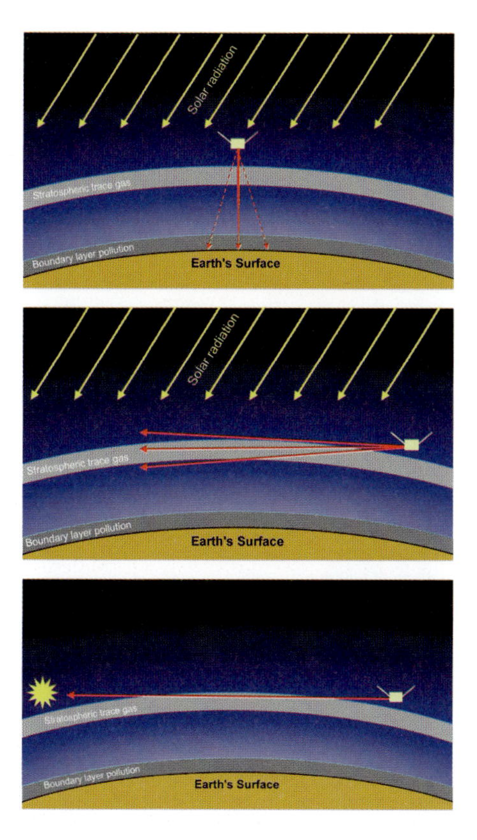

Fig. 2.2 Different viewing geometries for satellite observations in the UV/vis/NIR spectral range (thickness of the atmosphere and altitude of the satellite are not to scale). *Top*: in nadir mode (often the actual viewing angles also include more slant angles) the satellite instrument measures sun light which is reflected from the surface or scattered in the atmosphere. Such measurements are sensitive not only for stratospheric trace gases, but also for trace gases close to the surface. *Middle*: in limb mode scattered sunlight is observed at several almost horizontal viewing angles. Such observations are very sensitive to stratospheric trace gas profiles. *Bottom*: in occultation geometry, the direct light of sun, moon, or stars is observed. Occultation observations can also be performed during night. In contrast to limb and nadir observations, the light path is well defined.

range (237–793 nm, see Fig. 2.3, having a spectral resolution between 0.2 and 0.4 nm. The spectral range is sampled at a total of 4,096 wavelengths arranged in four "channels". Its standard ground pixel size is 320 × 40 km^2 (East-West × North-South) enabling global coverage to be achieved each 3 days.

In contrast to the few spectral intervals measured by BUV/SBUV and TOMS instruments, GOME spectra yield a plethora of spectral information. In the reflectance spectra not only are the broad band signatures of atmospheric scattering and surface reflectivity observed, but also the detailed features of some strong atmospheric absorbers (e.g. O_3, H_2O, and O_2) can be seen directly (Fig. 2.3). By applying the DOAS method, it has been possible to analyse the atmospheric absorptions of a

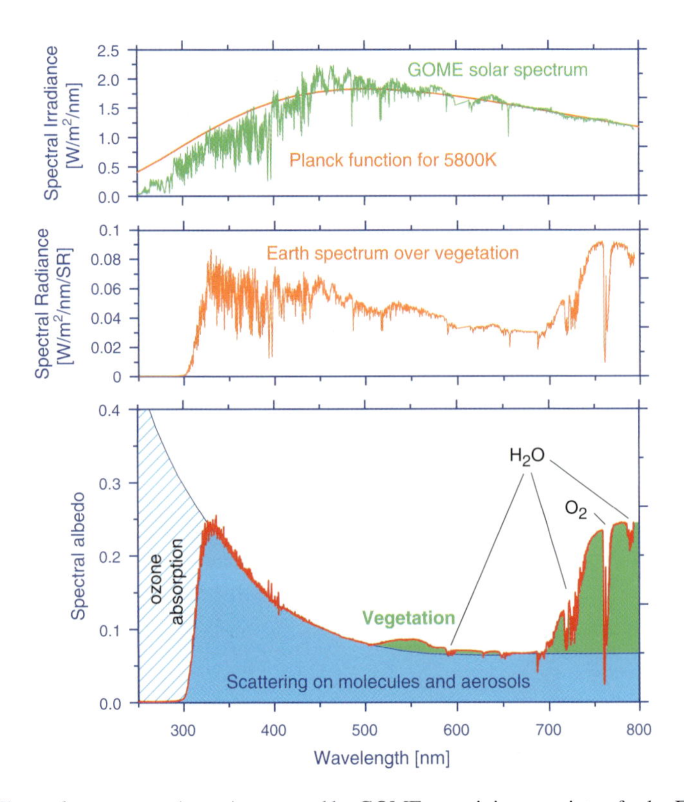

Fig. 2.3 *Top*: solar spectrum (*green*) measured by GOME containing a variety of solar Fraunhofer lines. For comparison, the black-body radiation for 5,800 K is also shown (note that the sun is not a perfect black body; also the radiation in certain parts of the spectrum originates from layers at different temperatures). *Middle*: "earth shine" spectrum of solar radiation reflected and scattered by the Earth's surface and atmosphere (clear sky over vegetation). *Bottom*: the spectral reflectance (*red curve*) calculated from both spectra. The broad band spectral dependence reflects the effects of scattering on air molecules (and aerosols). Atmospheric trace gas absorptions of O_3, O_2, and H_2O are also evident. Enhanced values in the *green* and in the near IR spectral range are due to increased reflectance over vegetation.

large variety of weak atmospheric absorbers such as NO_2, BrO, chlorine dioxide, OClO, formaldehyde, HCHO, glyoxal, CHOCHO, iodine monoxide, IO, the oxygen dimer, O_4, and SO_2, many of them located close to the Earth's surface. In addition, other quantities like aerosol extinction, cloud properties, surface albedo, vegetation properties, ocean colour, and indices characterising the solar cycle are retrieved from the GOME measurements.

In March 2002 the SCanning Imaging Absorption SpectroMeter for Atmospheric CHartographY (SCIAMACHY) (Burrows et al. 1988; Burrows et al. 1995; Bovensmann et al. 1999) was launched on board the European research satellite ENVISAT. Compared to GOME, its spectra cover a wider wavelength range (214–2380 nm), facilitating measurements of the absorption of several greenhouse gases (CO_2, CH_4, and nitrous oxide, N_2O) and the pollutant carbon monoxide,

CO, in the Near-IR spectral region. In addition to nadir viewing, SCIAMACHY also makes observations in limb and solar and lunar occultation geometry, which yield profiles of stratospheric and mesospheric trace constituents (see Fig. 2.2). Another advantage is that the ground pixel size for the nadir viewing mode is appreciably reduced to 30 × 60 km^2 (and in a special mode to 15 × 30 km^2). For the observation of tropospheric trace gases this is most important, because their concentrations can exhibit strong spatial gradients. Smaller ground pixels are also less affected by clouds (Krijger et al. 2007).

In 2004, the Ozone Monitoring Instrument (OMI) was launched (Levelt and Noordhoek 2002). Compared to GOME and SCIAMACHY, it has a limited spectral range (270–500 nm), but a higher spatial resolution (up to 13 × 24 km^2) in nadir viewing mode. Together with its daily global coverage, it provides unprecedented detail of the spatio-temporal variability of tropospheric trace gases. In October 2006, the first of three instruments of a second generation of GOME instruments was launched on MetOp 1. The two other GOME-2 instruments are scheduled for launch at time intervals of 5 years for the MetOp 2 and MetOp 3 platforms, thus extending the time series of GOME, SCIAMACHY, and OMI into the next two decades (EUMETSAT 2005). Compared to GOME-1, the GOME-2 instruments have an almost daily global coverage and much improved spatial resolution (80 × 40 km^2).

2.1.2 *Spectral Retrieval and Radiative Transfer Modelling*

As mentioned above, one important advantage for the retrieval of data products from atmospheric observations in the UV/vis/NIR spectral range is that emission terms can usually be neglected. The measured spectral signatures can be directly related to the absorption spectra of different atmospheric trace gases within the chosen spectral range. In practice, the data analysis usually comprises two steps. First, the spectra are analysed using, for example, the DOAS method: based on Beer-Lambert's law, the spectral analysis yields the trace gas concentration along the atmospheric light path or more precisely the average over all contributing light paths (see Fig. 2.5). For historical reasons the result is nevertheless called slant column density, SCD. In the second step, by taking into account the viewing geometry and scattering, absorption and reflection within the atmosphere and at the Earth's surface, the trace gas SCD is converted into a more useful quantity, that is the VCD, or the trace gas concentration. Usually, for this conversion, numerical atmospheric radiative transfer simulations have to be performed.

The separation of spectral retrieval and radiative transfer modelling has the advantage that it is conceptually simple, and possible retrieval errors can be easily identified and corrected. Nevertheless, it should be noted that this separation is strictly valid only for cases with well-defined atmospheric light paths, such as satellite observations in occultation viewing geometry (Fig. 2.2). For the observation of scattered and reflected light (e.g. up welling from the top of the atmosphere), many different atmospheric light paths contribute to the measured spectra. This causes non-linearities between the measured absorptions and the corresponding atmospheric column densities (Van Roozendael et al. 2006). However, except for strong absorbers (such as O_3, H_2O or CO_2), these non-linearities are small and can usually be neglected (see Section 2.2.4). More details on the spectral analysis and atmospheric radiative transfer modelling are provided in the following sections.

2.2 Spectral Retrieval

The retrieval of information on atmospheric trace species from measurements in the UV/vis/NIR utilises Beer–Lambert's law:

$$dI(\lambda) = -I(\lambda)\sigma(\lambda)\rho(s)ds \tag{2.1}$$

This states that for an infinitesimally thin layer, the change in intensity at a given wavelength λ is proportional to the intensity I, the absorption cross-section σ, the absorber number density ρ and the light path element ds.

After reaching the atmosphere, light from the sun is either scattered by molecules, particles or cloud droplets, or is reflected at the surface towards the observing satellite instrument. Along its path through the atmosphere, part of the light is absorbed by trace species and the resulting reduction in intensity can be used to determine the amount of the species present in the atmosphere. In contrast to other spectral regions, the atmosphere is nearly transparent over large parts of the UV/visible/NIR part of the spectrum. As a result, nadir satellite measurements have sensitivity from the top of the atmosphere down to the surface albeit with varying efficiency. Rayleigh scattering is a strong function of wavelength, and is the dominant extinction process in the UV outside the strong O_3 absorption bands. At short wavelengths (below about 300 nm), O_3 absorption in the stratosphere becomes even stronger, eventually blocking the view from space down to the troposphere. Particle scattering from aerosols is efficient in the UV/vis/NIR spectral range, as is scattering from cloud droplets. For simplicity, this type of scattering is often approximated by Mie theory, although only liquid aerosols or small cloud droplets are strictly spherical. Any retrieval of atmospheric concentrations of a trace species in this spectral region has therefore to account for absorption by molecules, Rayleigh scattering, Mie scattering and surface reflection.

In the UV/vis spectral range, molecular spectra are dominated by electronic transitions and their vibrational structures. In contrast to other wavelength regions, the spectra are quasi-continuous, since most molecules (in particular those containing more than two atoms) have a very large number of overlapping rotational lines, and some are also affected by pre-dissociation. Due to the resulting "band structure" of most molecular spectra, high spectral resolution is usually not required to resolve the cross-sections. This also implies that no vertical information is available from the shape of rotational lines, as is the case for example in the microwave

Table 2.1 Overview of the tropospheric trace species retrieved from measurements of the UV/vis/NIR radiance by nadir viewing satellite instruments, and the spectral windows typically used

Trace species	Typical spectral window used
SO_2	315–327 nm
O_3	325–335 nm
BrO	345–359 nm
HCHO	334–348 nm
IO	416–430 nm
NO_2	425–450 nm
CHOCHO	436–457 nm
H_2O	610–730 nm
CO	2,324–2,335 nm
CH_4	1,629–1,671 nm
CO_2	1,558–1,594 nm

region (see Chapters 1 and 4). The number of species which can be observed in the UV/vis is limited but includes a number of key players in atmospheric chemistry, e.g. O_3, NO_2, HCHO and halogen oxides (see Table 2.1). Towards the IR part of the spectrum, some line absorbers such as O_2, H_2O, CO_2, CH_4, CO have spectral features and can also be retrieved.

In recent years satellite instruments covering the UV/vis/NIR spectral range have evolved from having a few discrete channels with low spectral resolution, to instruments with wide and continuous spectral coverage at moderate spectral resolution. While the latter can usually resolve the relevant structures in the UV and visible range, they cannot resolve the rotational line spectra, in particular in the NIR. This leads to some complications in the data analysis which are discussed later in this section.

Many data analysis approaches used in the UV/visible spectral range separate the retrieval of the amounts of atmospheric constituents into two steps as mentioned before: first, the trace gas concentration $\rho(s)$ of the absorber integrated along the light path s is determined applying Beer-Lambert's law to the spectra. The resulting slant column density, SCD,

$$SCD = \int \rho(s)ds \tag{2.2}$$

is then converted to a vertical column density, VCD, integrated from the surface to the top of atmosphere, TOA,

$$VCD = \int_0^{TOA} \rho(z)dz \tag{2.3}$$

by simulating the effective light paths with a radiative transfer model. This is usually achieved by invoking airmass factors (AMFs), which are discussed in detail in Section 2.3.2. The underlying assumption is that the light path is determined by geometry, scattering from air molecules, aerosols and clouds, as well as surface reflection, but is independent of the amount and vertical distribution of the absorber itself. This is valid for small absorptions (less than a few percent), but for strong absorbers, the retrieval has to be modified as will be discussed in Section 2.2.4.

2.2.1 Discrete Wavelength Techniques

When applying Beer–Lambert's law to atmospheric measurements, three problems need to be addressed:

- What is the initial intensity I_0 at the top of atmosphere?
- How can the effects of different absorbers, scattering and surface reflectance be separated?
- What is the (effective) length of the light path?

a Initial Intensity I_0

Typically satellite instruments make measurements of I_0 by observing the sun without the atmosphere, using a diffuser to reduce light intensity. Therefore, the initial intensity I_0 can be assumed to be known. In practice, instrument degradation related to the diffusers used can be a problem (Hilsenrath et al. 1995), but this will not be discussed here.

A complication is introduced for measurements using scattered light. In this case, the initial intensity does not only depend on I_0 but also on the scattering efficiency in the atmosphere, and, for nadir measurements, on the surface reflectivity. Neither quantity is usually well known. As will be described later, this problem can be solved by using several wavelengths simultaneously and assuming that the change of scattering efficiency with wavelength is small or well known.

b Separating Different Effects

The second point for separating effects (different absorbers, scattering and surface reflectance) can be achieved by using measurements at several wavelengths. To illustrate this, a simplified view of radiative transfer through the atmosphere is considered.

Assuming that the only effect of significance is the absorption by one gas, the intensity measured at the satellite follows from the integrated form of the Beer–Lambert's law (Eq. 2.1):

$$I(\lambda) = cI_0(\lambda)\exp\left\{-\int \sigma(\lambda)\rho(s)ds\right\} \qquad (2.4)$$

where c is an efficiency factor accounting for the fact that scattered light is being measured. This factor c depends not only on the surface reflectance but also on the scattering in the atmosphere and the presence of clouds. If in addition, extinction by Rayleigh and Mie scattering is considered and the absorption cross-section is assumed not to vary along the light path, we obtain:

$$I(\lambda) = cI_0(\lambda)\exp\left\{-\sigma(\lambda)\int \rho(s)ds - \sigma_{Ray}(\lambda)\int \rho_{Ray}(s)ds - \sigma_{Mie}(\lambda)\int \rho_{Mie}(s)ds\right\} \tag{2.5}$$

where σ_{Ray} is the Rayleigh extinction cross-section, σ_{Mie} is the Mie extinction cross-section and ρ_{Ray} and ρ_{Mie} are the number densities of Rayleigh and Mie scatterers, respectively.

Substituting the definition of the slant column density *SCD* (Eq. 2.2), the expression for the received intensity (Eq. 2.5) reads:

$$I(\lambda) = cI_0(\lambda)\exp\{-\sigma(\lambda)SCD - \sigma_{Ray}(\lambda)SCD_{Ray} - \sigma_{Mie}(\lambda)SCD_{Mie}\}. \tag{2.6}$$

We now assume that measurements are made at two selected wavelengths: one, where the absorber of interest has a strong absorption, and the other where the absorber has no absorption or only weak absorption. If one can model the wavelength dependence of scattering with reasonable accuracy, then these two measurements can be used to determine the *SCD* for the absorber of interest. This is basically the approach used for the retrieval of the total column density of O_3 from the measurements by the TOMS instrument.

When several absorbers are present, we have to extend formula in Eq. 2.6 to:

$$I(\lambda) = cI_0(\lambda)\exp\left\{-\sum_i \sigma_i(\lambda)SCD_i - \sigma_{Ray}(\lambda)SCD_{Ray} - \sigma_{Mie}(\lambda)SCD_{Mie}\right\} \tag{2.7}$$

where i is the index of the different absorbers. Measurements at more wavelengths are needed to separate the signals. With this technique, not only O_3 but also SO_2 column densities can be retrieved from measurements of the TOMS instrument, provided the loading of SO_2 is large enough, for example in volcanic plumes.

c The Light Path

The third aspect, the length of the light path, or more precisely the average length of the ensemble of all possible light paths, is determined by using radiative transfer calculations which are discussed in Section 2.3.

The use of discrete wavelengths for measurements of atmospheric O_3 column densities has been applied successfully since the launch of the BUV satellite and extensively using the data from the TOMS instruments (Heath et al. 1978). This technique has the advantage of a relatively simple instrument with good signal to noise ratio, and builds on the heritage of the ground-based Dobson spectrophotometer and its network. As the number of discrete wavelengths used is small, the retrieval algorithms are fast and relatively simple, in particular if pre-calculated values are used for the weighting factors accounting for different atmospheric scenarios and the treatment of surface effects and clouds. By refining these lookup tables based on extensive validation, excellent agreement with ground-based measurements was achieved for the total column densities of O_3. The approaches used to derive the tropospheric O_3 column density from the TOMS measurements is described in Section 2.4. The disadvantage of these measurements is the small number of spectral measurement points. If another signal interferes, there is simply not enough information to separate the effects. This limits the application to strong absorbers with little interference from other effects. Also, for trace gases other than O_3 and SO_2, the exact wavelength positions of the TOMS spectral channels are not suitable. As a result, the method cannot be applied to other trace gases with smaller absorption such as BrO, NO_2 or HCHO.

2.2.2 DOAS Type Retrievals

In order to extend the measurements to weaker signals and more species, instruments have been developed, which cover a large and quasi-continuous spectral range at moderate resolution, e.g. GOME and SCIAMACHY. Such measurements are usually analysed using the Differential Optical Absorption Method (DOAS) which was initially developed for ground-based measurements (Noxon 1975; Platt et al. 1979; Solomon et al. 1987). The basic idea of DOAS is to separate the wavelength dependent absorption signal into two components, the low frequency part and the high frequency part. The high frequency part of the spectrum is used for the retrieval of atmospheric absorptions, while the low frequency part is treated as a closure term and is approximated by a low order polynomial or other smooth function.

In practice, the absorption cross-sections of the absorbers are separated into the slowly varying part $\sigma^*(\lambda)$ and the differential part $\sigma'(\lambda)$: $\sigma(\lambda) = \sigma^*(\lambda) + \sigma'(\lambda)$:

$$I(\lambda) = cI_0(\lambda) \exp\left\{ -\sum_i \sigma'_i(\lambda) SCD_i - \sum_i \sigma^*_i(\lambda) SCD_i \right. \\ \left. - \sigma_{Ray}(\lambda) SCD_{Ray} - \sigma_{Mie}(\lambda) SCD_{Mie} \right\} \quad (2.8)$$

Both the Rayleigh and Mie scattering cross-sections can be approximated by a polynomial in wavelength so that the three slowly varying contributions to the signal are combined into one closure polynomial, which in addition can compensate, in part, for any wavelength dependent changes in the throughput of the instrument:

$$I(\lambda) = cI_0(\lambda)\exp\left\{-\sum_i \sigma_i'(\lambda)SCD_i - \sum_p a_p\lambda^p\right\} \tag{2.9}$$

This equation is often written as a function of optical depth ln (I_0/I)

$$\ln\frac{I_0(\lambda)}{I(\lambda)} = \sum_i \sigma_i'(\lambda)SCD_i + \sum_p c_p\lambda^p. \tag{2.10}$$

This links the measurement signal (the optical depth) in a linear equation with the quantities of interest (the slant column densities SCD_i) and a closure polynomial, which also accounts for the factor c, assuming that the logarithm of the scattering efficiency can be approximated by a polynomial. One such equation can be written for each measurement wavelength, and by solving the resulting system of linear equations in a least squares procedure, the best set of slant column densities and polynomial coefficients result. In addition to its simplicity, the DOAS retrieval has the benefit of cancelling all multiplicative effects in I and I_0, which are measured by the same instrument in a short period of time using near identical experimental configuration. This relaxes the requirements for the absolute radiometric calibration of the measured intensities. Other remaining instrumental issues (e.g. the etalon resulting from multi-beam interference in optical components or ice layers on the detectors) and effects such as spectral changes in surface reflectance are at least partly compensated for by the closure polynomial, making this technique very valuable for the detection of small signals.

In some applications, a second closure polynomial is added to the measured intensities to account for instrumental stray light and the effect of inelastic scattering. This correction is non-linear in the optical densities. If only a constant offset c is used, it can be approximated by including another spectrum c/I in the spectral fitting process (Noxon et al. 1979).

One complication arises from the fact that the measured values of the intensity I and I_0 are usually not fully resolved spectrally but are taken at instrument resolution. Therefore, the quantity measured is the intensity convoluted with the instrument slit function. If this is accounted for in Eq. 2.9, the logarithm can no longer simply be taken on both sides of the equation as it cannot be exchanged with the convolution. For most practical applications it is sufficient to apply the convolution with the instrument function to the cross-sections and then to proceed to (Eq. 2.10). However, for strong and structured absorption, a correction has to be applied which is called I_0-correction (Aliwell et al. 2002). For line absorbers the problem is more complex and will be discussed in Section 2.2.4.

DOAS retrievals using data from GOME, SCIAMACHY, OMI, and GOME-2 have been used to determine tropospheric column densities of many trace species, including SO_2, BrO, IO, HCHO, glyoxal, NO_2, and H_2O. The advantage of using continuous spectra is the higher sensitivity that can be achieved, the ability to separate different absorbers, the intrinsic wavelength calibration on the well known Fraunhofer lines in the solar spectrum and the flexibility in the choice of wavelength regions. The disadvantages are the more complex instrumentation, higher data rates resulting in slower retrievals, the need for input data at high spectral resolution and some complicating effects which become more relevant at higher spectral resolution as discussed below. An overview over more effects relevant for the accuracy of scattered light DOAS measurements can be found in Platt et al. (1997) and Platt and Stutz (2008).

2.2.3 Some Considerations for DOAS Retrievals

a Fraunhofer Spectrum

An ideal light source for an atmospheric DOAS measurement would be bright, constant, continuous and without spectral features. While the sun provides sufficient intensity at visible wavelengths and at least over a day can be considered to be a constant light source, it does have many spectral features. These so called Fraunhofer lines are created by absorption by molecules and ions in the solar atmosphere and result in three important demands on the retrievals from spaceborne observations: the need for accurate wavelength alignment or spectral calibration, the need for spectral oversampling and the need to correct for the so called Ring effect which is discussed in the next section. In addition the requirements on instrument stability (e.g. with respect to the spectral resolution) are especially high.

As shown previously in Fig. 2.3, the depth of the Fraunhofer lines is of the order of 10–50%, much larger than most absorptions in the Earth's atmosphere, which are often on the per mill level. As a result, even a very small wavelength shift between I and I_0 will result in large structures when the ratio is taken in the DOAS equation, rendering the retrievals of small absorptions impossible. To reduce this effect, two approaches are taken. Firstly, instruments are temperature stabilised as far as possible to reduce spectral shifts induced by temperature changes of the spectrometer. Secondly, the spectra I and I_0 are numerically aligned and interpolated, prior to the fit or, iteratively as part of the retrieval, to minimize any spectral misalignment.

Although satellite instruments are usually well stabilised, the need for spectral alignment arises in some instruments from a difference in observation mode between solar irradiance measurement and nadir observations. For GOME and SCIAMACHY solar observations, the measurements are taken in the flight direction. This results in a small Doppler shift of the spectrum, which has to be compensated for. While the exact magnitude of the effect can be calculated, a full correction is only possible if the sampling of the measurements is several times as good as the spectral resolution of the instrument (Chance et al. 2005; Roscoe et al. 1996). If that is not the case (as for example in the GOME instrument), the spectra are “under-sampled”, and artificial structures are introduced in the interpolated spectra.

b The Ring Effect

The combination of the highly structured solar spectrum and inelastic scattering processes can lead to an additional effect which is named the Ring effect after one of its discoverers (Grainger and Ring 1962). Elastic scattering (Rayleigh scattering, Mie scattering) does not change the high frequency spectral shape (i.e. intensity variations over wavelength regions of a few nanometres), which is used by the DOAS approach. Therefore, the Fraunhofer lines in I and I_0 are expected to have the same depth and to cancel when the ratio is taken in the DOAS equation. However, observations have shown that the depth of Fraunhofer lines in scattered light is smaller than that in direct solar light, resulting in a highly structured signal in the ratio (Grainger and Ring 1962). The explanation for this effect is inelastic rotational Raman scattering, which distributes the intensity of the scattered photons over several nanometres (Joiner et al. 1995; Kattawar et al. 1981). As the loss of intensity through Raman scattering is proportional to the local intensity, while the gain in intensity is proportional to the intensity at neighbouring wavelengths, the effect is a filling-in of deep Fraunhofer lines in the scattered light spectrum as illustrated in Fig. 2.4. The amount of filling-in depends on the relative amounts of inelastic and elastic scattering and can be simulated with radiative transfer models (RTMs). In the retrieval, it is accounted for by inclusion of a pseudo-absorber (usually referred to as Ring spectrum) computed by an RTM. It should be noted that the spectral

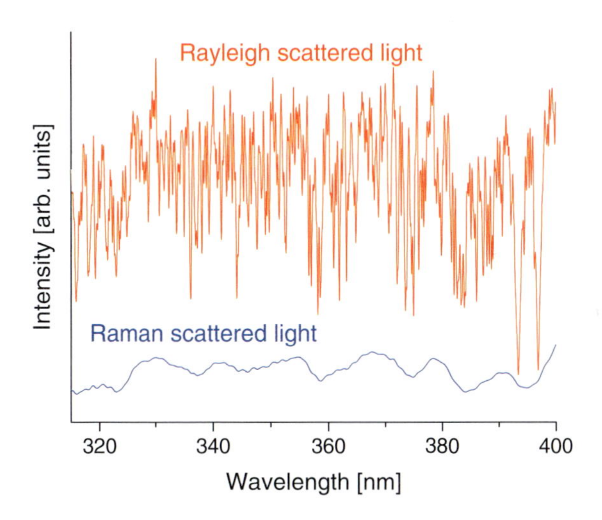

Fig. 2.4 Illustration of the Ring effect. The *upper panel* shows a modelled spectrum of elastically scattered sun light (single Rayleigh scattering) at a spectral resolution of about 0.2 nm. The lower panel shows the modelled spectrum of Raman scattered light. Note that the vertical scales differ: the intensity of the Raman scattered light is only about 4% of that for the elastically scattered light.

shape of inelastically scattered light is relatively smooth and the high frequency component is only introduced by taking the ratio of I and I_0. The strong dependence of the Ring effect on the light path can actually be used to derive information on cloud cover and cloud top height (Park et al. 1986; Joiner and Bhartia 1995; Joiner et al. 1995; deBeek et al. 2001; Joiner and Vasilkov 2006).

A more subtle effect of inelastic scattering is the filling-in of absorption lines of structured absorbers (if the absorption has occurred before the Raman scattering event) (Fish and Jones 1995; Vountas et al. 1998). The effect is similar to that for Fraunhofer lines and can be interpreted as a loss of absorption signal if a photon is scattered inelastically. Therefore, absorption signatures in atmospheric measurements are slightly smaller than expected for an atmosphere without inelastic scattering.

In contrast to rotational Raman scattering, vibrational Raman scattering is a small and negligible effect in the atmosphere. However, in water it can be significant if the light path is long enough. As the energy changes are much larger for vibrational than for rotational transitions, the spectral signature of the two effects is slightly different (Vountas et al. 2003).

c Choice of Fitting Window

When selecting an appropriate spectral region for the retrieval of a specific trace gas, several considerations have to be taken into account. At first glance, one would assume that the optimum retrieval would use all available information and thus all the spectral points. However, in practice, smaller fitting windows are used for several reasons. The most fundamental argument is that the separation of spectral retrieval and radiative transfer calculation is not valid over large spectral ranges. As result of increasing Rayleigh scattering, light paths change towards the UV and therefore the slant column densities at, for example, 350 nm are not identical to those at 450 nm. This effect leads to a mismatch of measured absorption and fitted absorption cross-sections, resulting in poor fits if not accounted for, as discussed in Section 2.2.4 for strong absorbers. Experience also shows that small inaccuracies in cross-sections of interfering species can lead to large errors for weak absorbers, and by using smaller wavelength regions, such errors can partly be compensated for by the scaling factors and the closure polynomial. Moreover, extending the spectral range always increases the noise, if only relatively weak spectral structures of the desired species are added. Therefore the signal to noise ratio might actually deteriorate upon extending the spectral range. Further considerations sometimes suggest the exclusion of spectral regions where spectra of different trace gases or strong Fraunhofer lines overlap.

Appropriate fitting windows are therefore selected by trying to maximise the differential absorption structures of the molecule of interest and, at the same time, minimizing the impact from interfering signals. This can often be done on simulated data, although the results might not be fully representative for real measurements if absorption cross-sections are inaccurate or not all atmospheric effects are included in the model. Some examples for fitting windows used in the literature are given in the Table 2.1.

d Effects of Spectral Surface Reflectivity

In recent years, it became evident that, in addition to trace gas absorptions, the Earth's reflectivity for certain surface types exhibits narrow spectral variations. Examples are absorption features over vegetation in the blue and red spectral range, as well as water absorption structures in the blue spectral range. As mentioned in the previous section, spectral features are caused by vibrational Raman scattering in water (mainly in the UV) and biological activity in the oceans (in the blue and red spectral range) (Vasilkov et al. 2002; Vountas et al. 2003; Vountas et al. 2007; Wagner et al. 2007a; Bracher et al. 2008). Although the amplitude of these spectral structures is usually rather small (in the range of a few percent), they can interfere with trace gas absorption features in the same spectral windows and, if not accounted for, result in large errors in the trace gas results (Vountas et al. 2003; Wagner et al. 2007a). Including the relevant surface spectral features as reference spectra minimizes the interferences. In this way, DOAS analyses of satellite observations can also yield important information on surface properties such as, for example, the primary productivity in sea water (Vountas et al. 2007; Bracher et al. 2008).

2.2.4 Advanced DOAS Concepts

While the DOAS retrieval concept described above is fast and has successfully been used for the analysis of many trace gases, there are situations where the assumptions made are not fulfilled. Modifications to the approach are then needed to achieve accurate results. This is particularly the case in situations where absorption becomes large or the spectral resolution of the instrument is not high enough to resolve the spectrum of line absorbers.

If absorption by a molecule becomes larger than a few percent, as is the case for atmospheric absorption by O_3 in the Huggins bands, or by SO_2 during volcanic eruptions, the average light path is no longer independent of the absorber amount and its vertical distribution. As a result, the separation of spectral retrieval and light path determination, used in standard DOAS retrieval is no longer applicable. One way of visualising the effect is to compare average light paths at two selected wavelengths, one having strong O_3 absorption and one having weak absorption. Photons experiencing weaker absorption will, on average, penetrate deeper into the atmosphere before they are scattered back to the satellite and therefore have a longer light path than photons at O_3 absorption peaks. This leads to a distortion of the absorption structure observed in the measured optical depth as compared to the simple absorption cross-section, which needs to be taken into account. On the other hand, this distortion also provides information on the vertical distribution of the observer and is used for example in the BUV retrievals of O_3 and SO_2 profiles (Bhartia et al. 1996; Hoogen et al. 1999; Liu et al. 2005; Yang et al. 2009).

For line absorbers, especially in the NIR, the spectral resolution of the instruments used is often not sufficient to resolve fully the line structure. The measured signal $I_{meas}(\lambda)$ then depends strongly on the instrument transfer function or slit function $F(\lambda)$ of the instrument:

$$I_{meas}(\lambda) = F(\lambda)^* I(\lambda) \tag{2.11}$$

Here, * denotes the convolution of the true intensity with the instrument slit function.

This has implications for the dependence of the measured optical depth on the absorber amount. While for each monochromatic wavelength the Beer–Lambert law applies, the convoluted intensity will no longer be a simple exponential function of the atmospheric trace gas amount in the presence of non-resolved line absorbers. This is the result of the non-interchangeability of exponential function and convolution. As an additional complication, the spectral shape of the measured optical depth changes with light path, absorber amount and temperature.

One solution to these problems is to replace the slant column density *SCD* by the slant optical depth *SOD* (Noël et al. 2004; Richter et al. 1999). Starting from Eq. 2.5:

$$I(\lambda) = cI_0(\lambda)\exp\left\{-\int \sigma(\lambda)\rho(s)ds - \sigma_{Ray}(\lambda)\int \rho_{Ray}(s)ds - \sigma_{Mie}(\lambda)\int \rho_{Mie}(s)ds\right\}$$

one can replace the first term by the slant optical depth:

$$SOD(\lambda) = \int \sigma(\lambda)\rho(s)ds \tag{2.12}$$

and following the same derivation as for the standard DOAS equation, one obtains:

$$\ln\frac{I_0(\lambda)}{I(\lambda)} = \sum_i SOD_i(\lambda) + \sum_p c_p\lambda^p . \tag{2.13}$$

The $SOD_i(\lambda)$ have to be computed using a radiative transfer model for an *a priori* atmosphere and subsequently be convoluted with the instrument's slit function. To a first order approximation, a simple linear scaling of the *a priori* SOD^0 can be used in place of the exact SOD:

$$\ln \frac{I_0(\lambda)}{I(\lambda)} = \sum_i r_i SOD_i^0(\lambda) + \sum_p c_p \lambda^p \tag{2.14}$$

where SOD^0_i is the slant optical depth for absorber *i* computed for the *a priori* atmosphere and r_i are the fit parameters and give the ratio of the absorber amount in the measurement relative to the amount in the *a priori* atmosphere.

This approach yields accurate results for O_3, SO_2 and H_2O even at large absorption if the *a priori* assumptions are sufficiently close to the real atmospheric situation. The effect of line absorbers is properly taken into account by the forward calculations at high spectral resolution which are then convoluted to the resolution of the instrument. The requirement is to select appropriate *a priori* assumptions using climatologies or alternatively to iterate those using the column densities retrieved in the first retrieval step and the residuals obtained from the fit.

Another, in some aspects similar but more general, formulation is obtained by introducing weighting functions *WF* (Buchwitz et al. 2000; Rozanov et al. 1998). The weighting function is here defined as the change in the logarithm of the intensity for a change in one of the parameters d_i from their value in the *a priori* assumptions, d_i^0:

$$WF_i(\lambda) = \left. \frac{\partial \ln I(\lambda)}{\partial d_i} \right|_{d_i^0} \tag{2.15}$$

The d_i can be the column density of absorbers but also other parameters such as surface albedo, temperature or aerosols. The logarithm of the intensity can then be developed in a first order Taylor expansion around a modelled intensity $I_0^{\mathrm{mod}}$for the *a priori* atmosphere

$$\ln I(\lambda) = \ln I_0^{\mathrm{mod}}(\lambda) + \sum_i WF_i(\lambda)(d_i - d_i^0) + \sum_p c_p \lambda^p \tag{2.16}$$

The fit parameters are $(d_i - d_i^0)$ and the polynomial coefficients for the broadband structures. As for the SODs, the weighting functions are calculated using radiative transfer models and *a priori* assumptions on the atmosphere, surface albedo etc. They can account for the effects of spectral averaging on line absorbers, for strong absorptions and for other parameters such as surface albedo. Weighting functions have been used for the retrieval of O_3, CO, CO_2, and CH_4. As already mentioned for *SOD*s, the main difficulty lies with the choice of proper *a priori* assumptions as the accuracy of the retrieval is best for small values of $(d_i - d_i^0)$. A further alternative was introduced by Frankenberg et al. (2005) who developed a weighting function algorithm (Iterative Maximum *a priori*, IMAP), which adjusts the *a priori* assumptions iteratively during the analysis.

2.3 Interpretation of the Observations Using Radiative Transfer Modelling

The sensitivity of UV/vis/NIR satellite observations depends on the distribution of the paths of the observed radiation through the atmosphere. Since it is only in nadir viewing geometry that a significant fraction of the observed UV, visible and NIR light penetrates into the lowest atmospheric layers, only nadir observations provide sufficient sensitivity for the observation of tropospheric species. This section therefore concentrates on satellite observations in nadir viewing geometry. In the following, the term nadir viewing geometry is defined to also include viewing directions approximately $\pm 60^{\circ}$ around nadir. In contrast, observations in limb or occultation viewing geometry provide information from the upper troposphere and above (see Fig. 2.2). The concepts and methods introduced in this section are not restricted to nadir data but can be and are applied to satellite observations in limb and occultation geometry as well.

2.3.1 Relevant Interaction Processes Between Radiation and Matter

A nadir looking satellite instrument in the UV/vis/NIR spectral range measures sun light backscattered or reflected to the instrument by the Earth's atmosphere or the surface. On its way between entering and leaving the top of the atmosphere, the observed light has been subject to various interactions between radiation and matter. The most relevant processes include scattering on air molecules and particles (aerosols or cloud particles), reflection at the Earth's surface or interaction with the upper layer of the ocean (see Fig. 2.5). In contrast to limb or occultation observations, atmospheric refraction can often be neglected for nadir viewing geometry.

a Molecular Scattering

The most important molecular scattering process is elastic scattering, usually referred to as Rayleigh scattering. Rayleigh scattering is strongly wavelength dependent. The probability that a photon is scattered by a molecule increases as a function of the fourth power of its frequency. This results in the blue colour of the cloud-free sky. For this Rayleigh scattering, the probability distribution of the new directions of the scattered photons (defined as the phase function) depends on the degree of polarisation of the incident light as shown in Fig. 2.6. For unpolarised light such as that from the sun, scattering in forward or backward direction is twice as probable as scattering at 90°. For linearly polarised electromagnetic radiation the probability of being scattered in directions parallel to the electric field vector is zero

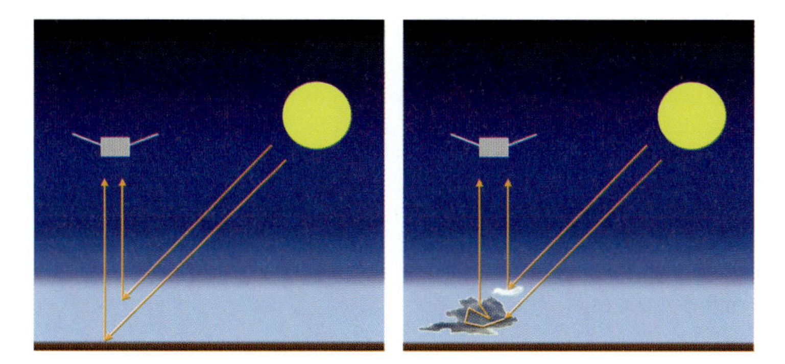

Fig. 2.5 The light received by a satellite instrument has undergone different interactions. In cloud-free scenes it is scattered by air molecules or reflected at the surface (*left*). In the presence of clouds (or aerosols) the light paths become further modified: scattering at high clouds can reduce the light paths, multiple scattering inside thick clouds can even enhance the light paths. Over bright surfaces, multiple scattering in the boundary layer can also be relevant under cloud free conditions (see Fig. 2.10).

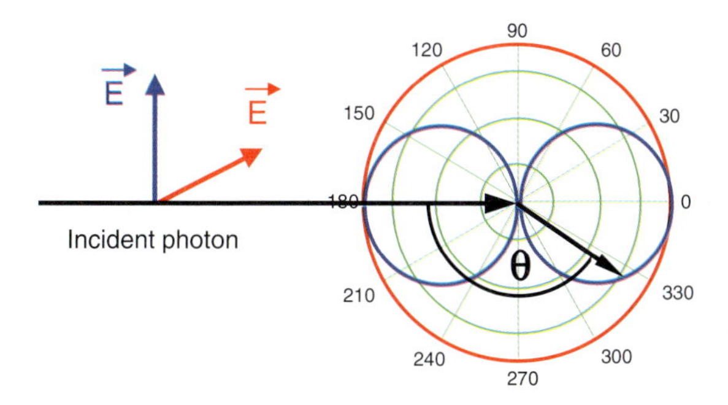

Fig. 2.6 Phase function for Rayleigh scattering. The vector of the electric field indicates the polarisation of the incident photon (*blue*: e-vector in the image plane; *red*: e-vector perpendicular to the image plane). The polar diagram indicates the probability distribution of the new photon direction after the Rayleigh scattering event (the angle θ describes scattering in the image plane). The length of the *arrow* indicates the scattering probability. For electromagnetic radiation with the e-vector perpendicular to the paper plane the scattering probability is independent of the scattering angle θ.

(see Fig. 2.6). Thus Rayleigh scattered light which is scattered at 90° is completely (linearly) polarised; this effect can be observed on clear days during sunset or sunrise, when looking at the zenith through a polariser. The phase function $P(\theta)$ for Rayleigh scattering of unpolarised light is described by:

$$P(\theta) = \frac{3}{4}[1 + \cos^2(\theta)]. \tag{2.17}$$

A small fraction (a few percent) of molecular scattering processes takes place as inelastic rotational Raman scattering. The associated wavelength shifts are in the order of only a few nm, causing the Ring effect (see Section 2.2.3). Also the effect of vibrational Raman scattering on air molecules has been considered, but was found to be negligible at the accuracy level of current measurements. In contrast to Rayleigh scattering, the phase function for rotational Raman scattering depends only weakly on the scattering angle and the state of polarisation (Kattawar et al. 1981).

b Particle Scattering

How atmospheric particles (aerosols or clouds) scatter electromagnetic radiation depends on their size, shape and composition. For spherical particles (e.g. liquid cloud droplets), their scattering properties are well described by Mie theory (Mie 1908; van de Hulst 1981). The scattering properties depend on the real and imaginary parts of the refractive index and on the particle size. With increasing particle size (usually expressed as size parameter $\alpha = 2\pi r/\lambda$), the asymmetry between the probability for backward and forward scattering increases (see Fig. 2.7). The bulk scattering properties of typical size distributions of aerosol and cloud droplets can be approximated by simplified functions, for example, the Henyey-Greenstein function (Henyey and Greenstein 1941). In contrast to spherical particles, the scattering properties of non-spherical particles (e.g. soot or dust particles or crystals in ice clouds) and particles having complex composition cannot be described by a comprehensive theory based on just a few input parameters as is the case for Mie theory. However, in many cases, the bulk scattering properties of such particle populations can be approximated by Mie theory.

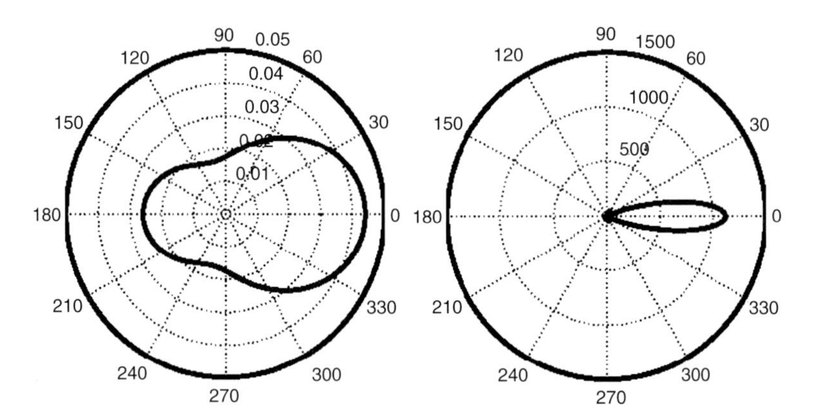

Fig. 2.7 Phase functions for Mie scattering (for refractive index of pure water). *Left*: for small droplet size ($\alpha = 1$); right for large droplet size ($\alpha = 5$). [adapted from (Sanghavi 2003)].

c Reflection and Absorption at the Surface

In addition to scattering from atmospheric constituents, reflection and absorption processes at the Earth's surface play an important role for satellite observations in nadir viewing geometry. The ratio of the reflected diffuse light to the incident irradiance defines the albedo, which usually depends on wavelength. The albedo depends on the characteristic of the incident light: the so called white and black sky albedo is defined for diffuse or directional incident light, respectively. The albedo ranges from zero (all photons are absorbed) to unity (all photons are reflected). The wavelength dependence of the surface albedo differs strongly over the globe (see Fig. 2.8). For most surface types (except snow or ice), the Earth's albedo in the UV/vis spectral range is quite low (between 2 and 30%). Towards the NIR the albedo over the continents increases for most surface types, whereas it decreases to very low values over the oceans. The highest surface albedo is found over snow and ice covered areas; increased reflectivity is also found in the case of sun glint over the ocean. The directional characteristic of surface reflection is described by the bi-directional reflection distribution function (BRDF); the reflected intensity depends on both the angles of the incident and reflected light. Often the BRDF is approximated by a Lambertian reflection characteristic (isotropic distribution of the reflected light, independent of the direction of the incident light). While some surface types (e.g. sand, snow, ice) are well described by this approximation, for many others (e.g. over vegetation, or for sun glint conditions over the ocean) more complex reflection functions are needed (see Fig. 2.9).

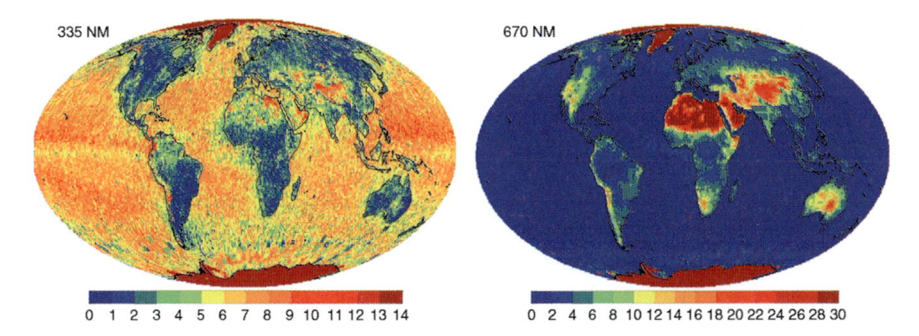

Fig. 2.8 Albedo of the Earth's surface at two wavelengths (*left* 335 nm, *right* 670 nm) [from Koelemeijer et al. (2003)].

d Interactions at the Ocean Surface

Part of the incident light is reflected directly at the ocean surface (by Fresnel reflection) causing sun glint, the intensity of which depends mainly on wind speed (surface roughness) and observation geometry (Cox and Munk 1954; 1956). The remaining part of the incident light can penetrate deeper into the water body. The

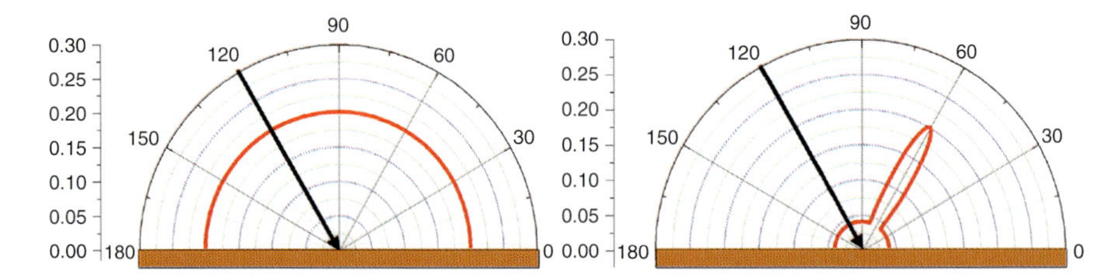

Fig. 2.9 Two examples for the angular dependency of the surface reflection. *Left*: for Lambertian reflection the probability of scattered light in any direction is constant (and independent on the angle of the incident light). *Right*: especially over the ocean, part of the light is reflected under the same angle as the incident ray (specular reflection). This effect is referred to as sun glitter or sun glint.

penetration depth depends strongly on the concentration of particles, in particular of phytoplankton which absorbs and scatters the light. In the presence of high phytoplankton concentrations (e.g. during algae blooms), the ocean albedo can be strongly enhanced in parts of the spectrum. Towards the near IR spectral range, the water absorption increases strongly and the penetration depth into the water body decreases; in these spectral regions the albedo of the oceans is especially low and strongly angle and surface state dependent, being 1% and less.

The incident photons are also scattered by water molecules. Besides elastic scattering, inelastic scattering, mainly due to vibrational Raman scattering, can also occur (see Section 2.2.3). The strongest signal of vibrational Raman scattering is typically found for oceanic regions with the lowest concentrations of particles, for which the penetration depth of the incident sun light is largest.

e Molecular Absorption Processes

Satellite observations of atmospheric trace gases in the UV/vis and NIR spectral range are based on the spectral signatures of the absorption of molecules. Molecular absorption typically contains both broad and narrow spectral structures (within the sub-nm range, see Section 2.1.2). The latter can be easily identified and analysed in spectral retrieval processes (see Section 2.2.2).

For the correct interpretation of satellite observations, all the relevant processes discussed above have to be considered and included in the radiative transfer simulations.

Depending on wavelength, measurement conditions, and targeted quantity, the radiative transfer simulations might be simplified by omitting some of the processes discussed. For example, Raman scattering processes can often be neglected because of their small contribution to the total received light. In the NIR, the probability for Rayleigh scattering becomes so low that it can often be neglected.

2.3.2 *Quantities Used for the Characterisation of the Measurement Sensitivity*

a The Total AMF

For most UV/vis/NIR retrievals, the basic result of the spectral analysis is the slant column density (*SCD*) (Eq. 2.2), the trace gas concentration integrated along the atmospheric light paths. Typically the *SCD* is converted into more universal quantities like the vertical trace gas column density, which is defined as the vertically integrated trace gas concentration (Eq. 2.3). The ratio between the *SCD* and *VCD* is referred to as the air mass factor (*AMF*):

$$AMF = \frac{SCD}{VCD}. \tag{2.18}$$

Only for a few viewing geometries, the *AMF* can be simply determined by geometric considerations. This is possible, for example, for nadir satellite observations made in the near IR where the probability of Rayleigh scattering is very small. For cloud free conditions, the observed light is then almost exclusively reflected by the surface and the elongation of the light path (with respect to the vertical path) can be approximated by:

$$AMF = \frac{1}{\cos(LOS)} + \frac{1}{\cos(SZA)}. \tag{2.19}$$

Here *LOS* denotes the line of sight angle of the instrument (defined with respect to the nadir direction), and *SZA* the solar zenith angle (defined with respect to the zenith direction). At solar zenith angles smaller than 70°, the geometric approximation can also be applied for nadir observations of stratospheric trace gases, even if they are made at visible and UV wavelengths. Since the air density in the stratosphere is low, almost all scattering events take place below the stratosphere, and the light paths through the stratospheric trace gas layers can be adequately described by simple geometry until the curvature of the atmosphere becomes relevant at low sun.

In contrast, for nadir satellite observations of tropospheric trace gases made at visible or UV wavelengths, due to Rayleigh scattering on air molecules the light paths in the tropospheric trace gas layers can become very complex. In such cases numerical radiative transfer simulations have to be performed for determining the *AMF*. The most complex light paths occur in the presence of clouds or strong aerosol loads. In such cases, even for observations in the near IR, radiative transfer simulations have to be performed.

Besides the dependence of the *AMF* on the *SZA* and *LOS* (see Eq. 2.19) the *AMF* usually depends on various other parameters, like the surface albedo, the wavelength, the properties of aerosols and clouds, and the vertical distribution of the trace gas. For optically thick absorbers (the vertical optical thickness $\tau >> 0$) they also become dependent on the trace gas *VCD* itself.

The *AMF* can be derived from radiative transfer simulations in various ways. One universal and often used procedure is based on the simulation of the radiance observed by the detector with or without the atmospheric trace gas of interest. According to the Beer-Lambert law the *SCD* can be expressed as:

$$SCD = \frac{1}{\sigma(\lambda)} \ln \frac{I'(\lambda)}{I(\lambda)} \tag{2.20}$$

here, I and I' denote the modelled radiances with and without the trace gas, respectively and $\sigma(\lambda)$ is the absorption cross-section. Using the definition of the *AMF*, Eq. 2.20 becomes:

$$AMF = \frac{\ln(I') - \ln(I)}{\sigma(\lambda) \cdot VCD} \tag{2.21}$$

VCD is the vertically integrated concentration of the trace gas as it was used as input of the radiative transfer simulations.

It is interesting to note that for weak absorbers the total *AMF* can be approximated by the "intensity weighted approximation" (Solomon et al. 1987; Slusser et al. 1996):

$$AMF = \frac{\sum_i AMF_i \cdot I_i}{\sum_i I_i} \tag{2.22}$$

Here the sum is taken over all light paths contributing to the measured signal. The AMF_i denote the geometrical path length weighted by the vertical profile of the trace gas, and I_i the radiance observed from light path i, respectively. The *AMF* determined using the intensity weighted approximation does not depend on the absorption of the target trace gas, but can depend on other atmospheric absorbers (e.g. an *AMF* determined for a NO_2 profile can depend on the optical depth of the atmospheric O_3 absorption).

The *AMF* as defined in Eqs. 2.18, 2.21, and 2.22, is usually referred to as "total" *AMF*, since it describes the sensitivity of the measurement for the complete atmospheric column density of the trace gas.

b Box-AMF and Weighting Functions

In order to quantify the vertical dependence of the measurement sensitivity, it is useful to define a height-resolved partial *AMF* for specific atmospheric layers, the so called Box-*AMF* (or Block-*AMF*):

$$BAMF_i = \frac{\partial SCD}{\partial VCD_i} \tag{2.23}$$

Here $BAMF_i$ denotes the Box-*AMF* for the atmospheric layer i, and VCD_i the partial vertical column density for layer i. If the absorber concentration ρ_i in layer i is assumed to be constant with altitude, Eq. 2.23 can be rewritten as:

$$BAMF_i = \frac{\partial SCD}{h_i \cdot \partial \rho_i} \tag{2.24}$$

where h_i denotes the vertical extension of layer i.

The Box-*AMF* definition can also be related to the measured radiance. Using the relations of Eq. 2.20, it follows:

$$BAMF_i = \frac{1}{\sigma \cdot h_i} \cdot \frac{\partial(\ln I/I')}{\partial \rho_i} \tag{2.25}$$

Note that in some retrieval algorithms also the derivative of the intensity is calculated (see also Section 2.2.4). The weighting functions, *WF*, are defined as:

$$WF_i = \frac{\partial I}{\sigma \cdot h_i \cdot \partial \rho_i} \tag{2.26}$$

As can be seen, it is closely related to the Box-*AMF* in that it only describes changes in intensity and not in slant column density.

An example for the vertical variation of Box-*AMF*s for satellite observations in nadir viewing geometry is shown in Fig. 2.10. For long wavelengths the height dependence is weak, because most photons are reflected at the surface. For shorter wavelengths the influence of Rayleigh scattering becomes more important; for the troposphere, this leads to a decreased sensitivity at low albedo and increased sensitivity for high albedo (due to multiple scattering).

In Fig. 2.11 Box-*AMF*s for satellite nadir observations are shown for situations when clouds and aerosols are present.

Eqs. 2.23–2.25 are valid for all atmospheric situations, also including observations of optically thick absorbers ($\tau >> 0$) like O_3 in the UV spectral range. However, with increasing optical thickness, the respective Box-*AMF*s also become dependent on the absolute value of the trace gas concentration in the different layers. In general, with increasing optical thickness, the Box-*AMF*s become smaller, indicating decreasing measurement sensitivity.

Fortunately, for most atmospheric species the optical density in the UV/vis/NIR spectral range is small ($\tau << 1$). For these weak absorbers the Box-*AMF*s become essentially independent of the absolute value of the trace gas concentration, and several simplifications can be made.

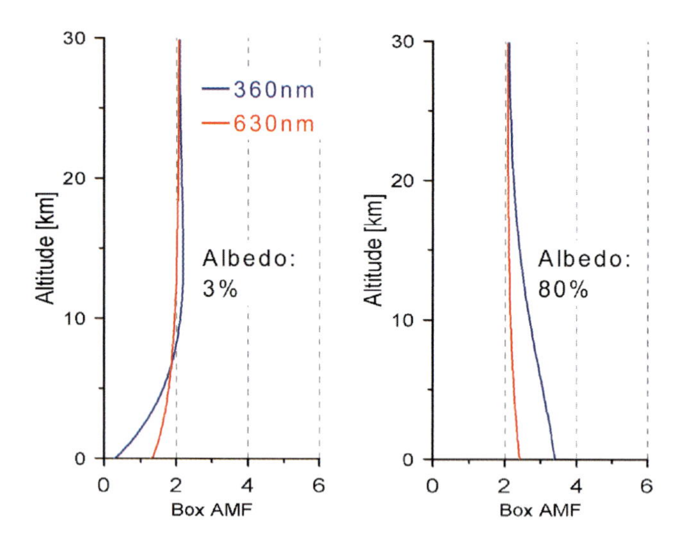

Fig. 2.10 Height-dependence of the sensitivity of satellite observations in nadir viewing geometry, expressed as Box-AMFs. For large wavelengths, only a weak height dependence is found. For shorter wavelengths, the Box-AMFs decrease towards the surface for low albedo (*left panel*) and increase for high albedo (*right panel*). SZA was chosen as 0°.

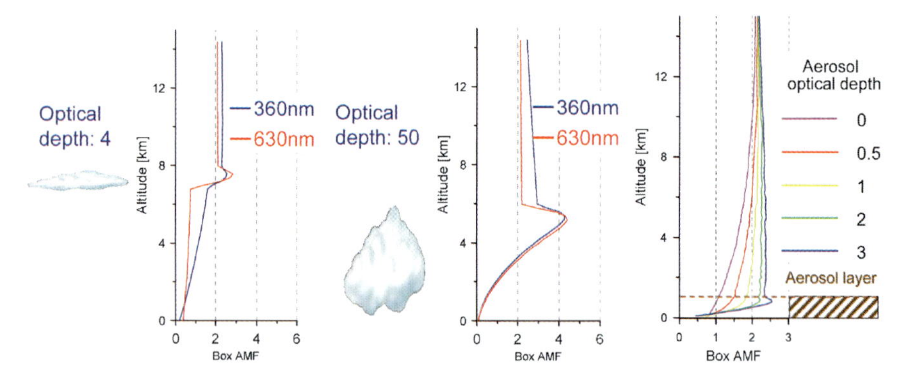

Fig. 2.11 Height-dependence of the sensitivity of satellite observations in nadir viewing geometry in the presence of clouds and aerosols (expressed as Box-AMFs). High, thin clouds mainly shield trace gases below the cloud (*left*). Thick clouds increase the sensitivity in the upper layers of the clouds and decrease the sensitivity towards the surface (*centre*). The effect of near surface aerosols depends on their optical thickness, phase function, and their single scattering albedo. Non-absorbing aerosols can substantially increase the sensitivity (*right*). Surface albedo and SZA were chosen to be 5% and 0°, respectively.

First, the Box-*AMF* can be simply determined by averaging the geometrical path lengths of the individual light paths through layer i, weighted by their intensities (similar to Eq. 2.22 for the total *AMF*):

$$BAMF_i = \frac{\sum_j l_{i,j} \cdot I_{i,j}}{h \cdot \sum_j I_{i,j}} \tag{2.27}$$

Here $l_{i,j}$ and $I_{i,j}$ describe the length and intensity of light path j through layer i, and h the vertical extension of the layer. In addition, for optically thin absorbers for which the light path does not depend on absorber concentration, the Box-*AMF* can simply be expressed as the ratio of the partial SCD_i and the partial VCD_i of an atmospheric layer i:

$$BAMF_i = \frac{SCD_i}{VCD_i} \tag{2.28}$$

From this follows the most important simplification, which is that the total *AMF* (as defined in Eqs. 2.28, 2.21 and 2.22) can be directly determined by combining the height profiles of Box-*AMF*s and the (relative) trace gas concentration:

$$AMF = \frac{SCD}{VCD} = \frac{\sum_{i=0}^{TOA} SCD_i}{VCD} = \frac{\sum_{i=0}^{TOA} BAMF_i \cdot VCD_i}{VCD} = \sum_{i=0}^{TOA} BAMF_i \cdot \frac{VCD_i}{VCD} \tag{2.29}$$

Here VCD_i/VCD describes the relative trace gas concentration profile. Sometimes it is also referred to as shape factor (Palmer et al. 2001).

With this formulation, it becomes possible to separate completely the radiative transfer modelling from the assumptions made for the vertical trace gas distribution. Accordingly, different sets of Box-*AMF*s (e.g. for various aerosol and cloud profiles) can be pre-calculated and stored in look-up tables. By applying Eq. 2.29, the total *AMF* for arbitrary profile shapes can be easily calculated. This procedure is particularly convenient if the assumptions on the relative profile shape change (e.g. after new measurements or model results on the vertical trace gas distributions have become available) or if a large number of different airmass factors have to be calculated as is the case for satellite observations.

Eq. 2.29 is strictly valid only for weak absorbers. Nevertheless, if it is applied for moderate absorbers, the resulting errors are usually small, and can in most cases be neglected compared to other uncertainties (see Section 2.5).

c Averaging Kernels

The knowledge of the height dependence of the measurement sensitivity is crucial for the correct interpretation of individual satellite measurements. In extreme cases, for example, in the presence of thick clouds, the satellite becomes almost blind for trace gases located below the cloud (see Figs. 2.5 and 2.11), and a low value of the retrieved trace gas *SCD* does not necessarily indicate a low trace gas concentration in the atmosphere. The height dependent measurement sensitivity has in particular to be considered if satellite observations are compared to other data sets (e.g. other measurements or model results, in the following referred to as "collocated data sets"). There are two basic ways to consider the height dependent sensitivity in such comparisons.

(a) The first possibility is to simulate the trace gas *SCD* as measured by the satellite based on the collocated data set. This can be achieved by calculating the *AMF* according to the trace gas profile of the collocated data set. Applying Eq. 2.18 then yields the *SCD* which can be compared to the *SCD* retrieved from the satellite measurement. For weak absorbers, instead of Eq. 2.28, the Box-*AMF* concept can be applied (Eq. 2.29):

$$SCD = \sum_{i=0}^{TOA} SCD_i = \sum_{i=0}^{TOA} BAMF_i \cdot VCD_i \tag{2.30}$$

The advantage of this procedure is that the Box-*AMF* can be calculated completely independently from the collocated data set.

(b) The second possibility for the comparison to a collocated data set is to convert the *SCD* obtained from the satellite observation first into a trace gas *VCD* using an (total) *AMF* (according to Eq. 2.18). For the simulation of the *AMF* a (relative) vertical profile shape of the absorber has to be assumed.

By dividing the Box-*AMF*s by the total *AMF*, so called column averaging kernels are determined (Eskes and Boersma 2003) (see Fig. 2.12):

$$AK_i = \frac{BAMF_i}{AMF} \tag{2.31}$$

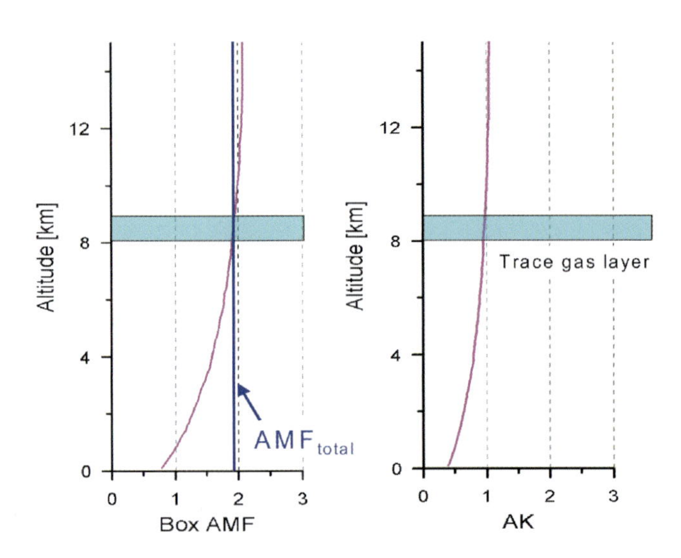

Fig. 2.12 Column averaging kernels (*right*) can be easily derived by dividing the box-AMFs (*left*) by the total AMF for an assumed profile. In this case the total AMF is 1.9 according to a trace gas layer located between 8 and 9 km altitude. Surface albedo and SZA were chosen to be 5% and 0°.

The column averaging kernels provide information about how the retrieved *VCD* changes for a given change of the absorber concentration at a certain altitude range. It can be used to convert the trace gas profile of a collocated data set to the VCD^{instr} defined as the vertical column density which the retrieval should have produced for the collocated profile:

$$VCD^{instr} = \sum_{i=0}^{TOA} AK_i \cdot VCD_i \tag{2.32}$$

This quantity VCD^{instr} can be directly compared to that determined from the satellite observation. Here it is interesting to note that the *VCD* (both measured and simulated) might differ from the true atmospheric *VCD*, if the assumed (relative) profile differs from the true profile. Nevertheless, the comparison between measured and simulated trace gas *VCD* is not affected by this because for both quantities the same (potentially wrong) assumptions are made.

Mathematically, both ways to compare the satellite data to collocated data sets are equivalent. The choice between the two possibilities depends on the technical capabilities for the processing of the involved data sets and the objectives of the study. The comparison of *VCD*s (instead of *SCD*s) has the advantage that the trace gas fields are expressed in a more unified and intuitive way. Thus errors might be more easily identified.

In general, it should be noted that if disagreement between the simulated quantity (either *SCD* or *VCD*) and measured quantity is found, a straight forward interpretation is possible: one can then simply conclude that the data sets do not agree. In the case of agreement (at least within certain limits), the actual agreement of both data sets is only one (usually very probable) possibility. However, since various trace gas profiles can result in the same *SCD* (or *VCD*), also a wrong profile of the collocated data set might yield the same *SCD* (or *VCD*) as measured by the satellite.

d 2-D and 3-D Box-AMF

The Box-*AMF* concept introduced in Eq. 2.23 can be directly extended to two or three dimensions. For that purpose, Eq. 2.25 has to be modified:

$$BAMF_{i,j,k} = \frac{1}{\sigma \cdot h_i} \cdot \frac{\partial(\ln I/I')}{\partial \rho_{i,j,k}} = \frac{\partial SCD}{\partial VCD_{i,j,k}} \tag{2.33}$$

Here, like Eq. 2.25, the index i describes the vertical dimension. In addition, the height layers are also separated in distinct boxes in the two horizontal dimensions; they are described by the indices j and k. The three-dimensional Box-*AMF*s describe the dependence of the measured total SCD on changes in the concentration (or partial *VCD*) of box i, j, k.

Like the one-dimensional Box-*AMF* concept, also the multi-dimensional Box-*AMF*s can be used for the comparison of the satellite observations to collocated data sets.

The *SCD* measured by the satellite can be expressed by a collocated 3-dimensional data set:

$$SCD = \sum_{i,j,k} BAMF_{i,j,k} \cdot VCD_{i,j,k} = \sum_{i,j,k} BAMF_{i,j,k} \cdot h_i \cdot \rho_{i,j,k} \tag{2.34}$$

In this way, the effect of 3-dimensional gradients of trace gases (and aerosols or clouds) inside a satellite ground pixel can also be investigated. This could be relevant around pollution hot-spots or in coastal regions with large spatial variability.

2.3.3 Important Input Data

Depending on the trace gas of interest, information about the various input data is required for the radiative transfer simulations in addition to the settings on viewing geometry and solar position. In contrast to stratospheric absorbers, the radiative transfer simulations for tropospheric trace gases are especially sensitive to the influence of the surface albedo, surface elevation, tropospheric aerosols, clouds, and the vertical trace gas profile. Furthermore, information on the profiles of temperature and pressure is required, but their influence on the radiative transfer simulations is quite weak.

Because of the large ground pixel size of many satellite instruments, horizontal gradients of trace gases, aerosols, clouds and the surface properties can also become important.

Information on several of these parameters is in general difficult to obtain, in particular at the high spatial resolution needed for recent satellite sensors. Most critical are input data on the vertical profiles of the trace gas concentration as well as the aerosol and cloud properties. As input data, independent information, e.g. from simultaneous observations of the cloud and aerosol properties, is often used. Information on the trace gas vertical profile can be obtained from collocated measurements (e.g. from aircraft) or model simulations of the atmospheric chemistry and transport. Also, standardised average profiles or proxy-information, such as the boundary layer height are sometimes used.

Considering their high relevance and limited availability, the largest sources of uncertainty for tropospheric radiative transfer simulations are caused by the influence of the vertical profiles of the trace gas concentration as well as cloud and aerosol properties.

Information on the various input data is summarised in Table 2.2.

As mentioned above, the impact of the *a priori* data on the derived quantities is sometimes large.

Table 2.2 Input information used for radiative transfer simulations of tropospheric trace gases

Input data	Importance	Typically used	Source of information
Trace gas profile	High	Yes	Independent information, e.g., from atmospheric models or general assumptions
Cloud properties	High	Yes	Simultaneous measurements, e.g., cloud fraction and cloud top height
Aerosol properties	High	No	General assumptions, atmospheric models, potentially also simultaneous measurements
Surface albedo	High	Yes	Climatologies, e.g., created from the same satellite instrument, sometimes also retrieved from the measurements
Surface elevation	Medium	No	Databases
Pressure and temperature profile	Low	Yes	Standard profiles
Solar zenith angle	Medium	Yes	Measurement description
Line of sight angle	Medium	No	Measurement description
Relative azimuth angle	Medium	No	Measurement description
Horizontal gradients of clouds, aerosols, trace gases, surface albedo and elevation	Potentially high	No	Data bases, atmospheric models, general assumptions

As discussed in Section 2.3.2, in comparisons with model data, this problem of dependence on *a priori* data can be addressed by either applying the averaging kernels of the measurements to the model data or by converting model profiles to slant column densities.

2.3.4 Overview of Existing Radiative Transfer Models

The direct output of radiative transfer models is usually the radiance measured by a specified detector under defined atmospheric properties and for a given solar position. From the modelled radiance, also other quantities, e.g. air mass factors, can be obtained subsequently (see Eqs. 2.20–2.22, 2.25, 2.26 and 2.33). To describe the observed radiance at a detector, the radiative transfer equation (see also Chapter 1) has to be solved.

$$\frac{dI(\lambda,\varphi,\vartheta)}{ds} = -[\sigma_S(\lambda)+\sigma_A(\lambda)]\cdot\rho\cdot I(\lambda,\varphi,\vartheta)+\sigma_A(\lambda)\cdot\rho\cdot B(\lambda,T) + \int_{4\pi}\frac{d\sigma_s}{d\Omega}(\lambda,\varphi,\vartheta,\varphi',\vartheta')\cdot\rho\cdot I(\lambda,\varphi',\vartheta')d\Omega \quad (2.35)$$

Here $I(\lambda, \varphi, \vartheta)$ describes the intensity along the line of sight (denoted by the two angles φ, ϑ); σ_S and σ_A denote the scattering and absorption cross-section, respectively. ρ is the number density of the molecule and $B(\lambda, T)$ describes the black body radiation. The last term contains the light which is scattered from any direction into the line of sight. The probability for scattering from the original direction φ', ϑ' into the line of sight is described by the differential scattering cross-section $d\sigma_s/d\Omega\ (\lambda, \varphi, \vartheta, \varphi', \vartheta')$, also referred to as a phase function, see Section 2.3.1. In the real atmosphere, this equation has to be expanded to include all species present with their respective densities and absorption and scattering cross-sections. For observations in the UV, visible and near-IR spectral range, the thermal emission term can usually be neglected as it is very small at atmospheric temperatures.

For the purposes of UV/vis/NIR retrievals, the radiative transfer equation can be solved using different methods with specific advantages and disadvantages. Early models often used so called two stream approaches, which assume horizontally homogenous conditions and simulate the integrated upward and downward radiation fluxes (Meador and Weaver 1980). However, such models are usually applied to plane parallel atmospheres and are not able to reproduce the angular dependence of the radiation field. Today, usually more sophisticated models are used, which overcome these limitations (Wendisch and Yang 2010; Perliski and Solomon 1993; Sarkissian et al. 1995; Hendrick et al. 2006; Wagner et al. 2007b).

Analytical models usually describe the direct, single scattered and multiply scattered contributions separately. Such models are in general very fast, but analytical solutions can only be found for single scattering or for plane parallel geometry. To describe adequately measurements at large solar zenith angles, or almost horizontal viewing directions, full multiple scattering, approximations and iterative approaches have to be applied.

Another widely used method is based on the Monte-Carlo technique, which simulates individual photon paths contributing to the measured signal at the detector. From a modelled photon path ensemble, the observed radiance, and also statistical quantities (like the average scattering frequency, etc.) can be determined. To achieve high accuracy, however, a large number of photon paths have to be modelled, which can become very time-consuming. Monte-Carlo models offer the most realistic description of the relevant processes and the boundary conditions like the Earth's curvature and topography. An overview on various methods to simulate the atmospheric radiative transfer can be found in Wendisch and Yang (2010).

Depending on the complexity of the considered situation and on the required accuracy, several simplifications with respect to the full range of interaction processes can be made. These simplifications usually lead to higher computational speed. They include the following aspects.

(a) Most radiative transfer models do not include a correct treatment of polarisation. While the incoming solar light is unpolarised, scattering and reflection processes can cause the observed light to be polarised (depending on the viewing geometry). For most atmospheric applications, only linear polarisation has to be considered. If the atmospheric radiative transfer is described by scalar models, in specific cases, systematic errors can occur. For most situations, these errors are below a few percent (Mishchenko et al. 1994).

(b) Many models only describe first order scattering processes especially in stratospheric applications. Since the air density in the stratosphere is low, the probability for multiple scattering is usually rather small and neglecting higher scattering orders leads to small errors (up to a few percent). Single scattering models are very efficient and were mainly used in early studies (Solomon et al. 1987; Perliski and Solomon 1993). Today, due to the increased computer power, for stratospheric applications, multiple scattering models are usually also used. For the simulation of satellite observations of tropospheric trace gases, multiple scattering has to be included due to the increased air density and the effects of clouds and aerosols.

(c) For many applications, the Earth's curvature (and also refraction) is neglected, since plane parallel models can be operated much more efficiently. However for situations with large SZA ($>85°$) and/or almost horizontal viewing direction (e.g. in limb viewing geometry) neglecting the Earth's curvature leads to considerable errors (Wagner et al. 2007b).

(d) Most models do not consider Raman scattering. Since the probability for Raman scattering is low (e.g. a few percent of molecular scattering is rotational Raman scattering), the resulting error for the modelled radiance is usually negligible. However, if the atmospheric Ring effect is simulated explicitly, Raman scattering has to be included. Such model simulations are usually very time-consuming because the scattering contributions from many wavelengths have to be considered simultaneously (Joiner et al. 1995; Joiner and Bhartia 1995).

2.4 Separation of Tropospheric and Stratospheric Signals

Measurements in the UV/visible spectral range usually do not contain intrinsic information on the vertical distribution of the absorber of interest. This is in contrast to data taken in the IR and microwave regions (see Chapters 3 and 4), where line broadening and temperature dependence can be used to invert vertical atmospheric profiles. While some absorbers have the potential for direct retrieval of altitude information, as will be discussed at the end of this section, current algorithms do not use these possibilities. Instead, they rely on the use of assumptions and external information to extract the tropospheric part of the measured signal.

The signal received at the satellite results from scattering in the atmosphere by molecules, cloud and aerosol particles, and reflection on the surface. Therefore, both absorptions in the troposphere and in the stratosphere contribute to the slant column density *SCD* derived from the measurements. For some species such as HCHO the stratospheric concentrations are so small, that one can safely assume that the observed signal is purely tropospheric. However, for most other species, notably

O_3, NO_2, and BrO, the stratospheric column densities are appreciable and in many cases larger than the tropospheric column densities. This is further aggravated by the fact that the light path and thus the total absorption usually are larger in the stratosphere than in the troposphere. Therefore, an accurate method is needed to estimate the stratospheric column density before the tropospheric part can be determined (Fig. 2.13).

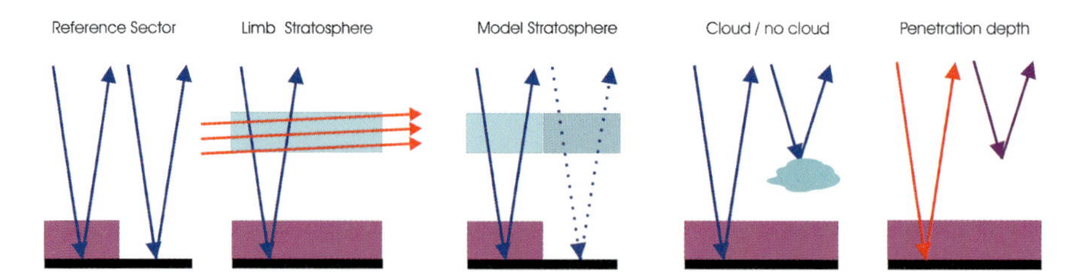

Fig. 2.13 Schematic overview over some methods to separate tropospheric and stratospheric signals. From *left* to *right*: using a measurement over clean regions, independent measurements of the stratospheric column density, (assimilated) model values for the stratosphere, cloud slicing or measurements at different wavelengths having different penetration depths.

2.4.1 Stratospheric Measurement Methods

For O_3, this problem has been addressed by using measurements from two different instruments – TOMS, from which total atmospheric column densities of O_3 can be retrieved, and SAGE and SBUV, which provide stratospheric profiles (Fishman and Larsen 1987). By subtracting the stratospheric column density derived from the profile measurements from the total column density, the tropospheric part can be obtained. If this method is applied to long-term averages, climatological tropospheric O_3 fields can be determined which are in good agreement with independent measurements. The approach relies on several assumptions, in particular that temporal variations in the tropospheric and stratospheric fields are not correlated and therefore independent averaging can be applied. Also, the weighting of the tropospheric signals in the total column densities has to be considered as the sensitivity of the measurements decreases towards the surface (see Section 2.3.2). The method also requires very accurate measurements as only about 10% of the total O_3 column density is located in the troposphere, and therefore a 5% error on the stratospheric column density will result in an error of approximately 45% on the tropospheric amount. The method has been extended to other instruments, e.g. OMI and the Microwave Limb Sounder (MLS) on Aura (Schoeberl et al. 2007), which have nearly coincident measurements and therefore, in combination with trajectory calculations, can be used to retrieve daily tropospheric O_3 maps.

Another variant of the idea to use integrated stratospheric profiles to extract the tropospheric amount from total column information is the limb-nadir matching which can be applied with SCIAMACHY measurements (Bovensmann et al. 1999; Sierk et al. 2006; Sioris et al. 2004; Beirle et al. 2010). The latter instrument is unique in that it takes two measurements of the same air mass within several minutes – first a limb scan providing the stratospheric profile and 7 min later a nadir measurement of the total column density. These data provide optimum coincidence in space and time and, in principle, can be used not only for O_3 but also BrO and NO_2. A complication of the limb-nadir measurements is the reduced sensitivity of limb observations to the lower stratosphere, in particular in mid and high latitudes which results in larger uncertainties of the stratospheric column densities.

2.4.2 *Residual Methods*

Another approach often used to estimate the stratospheric contribution to the measurements is based on assumptions on the spatial distribution of the stratospheric species. Stratospheric O_3 in the tropics, for example, can be assumed to vary mainly with latitude and much less with longitude. Assuming that the tropospheric ozone burden in a clean region, e.g. over the Pacific, is small, then one can compute the tropospheric excess column density as a function of latitude for all longitudes relative to the “clean air sector”. For O_3, this is only applicable in the tropics and has to be refined further by accounting for stratospheric wave patterns (Hudson and Thompson 1998). For NO_2, the method can even be extended to mid-latitudes with good results, at least for instruments in sun-synchronous orbits where the diurnal photochemistry does not come into play (Richter and Burrows 2002; Martin et al. 2002). Because of the short lifetime of NO_2, the assumption of low tropospheric column densities over the Pacific is usually fulfilled but some errors are introduced from dynamical variations in NO_2, in particular those linked to the polar vortex.

An extension of the residual method for NO_2 has been developed for the GOME and OMI instruments, where a 2-dimensional smooth field is fitted through the global total column density measurements after masking out regions with known tropospheric sources (Leue et al. 2001; Bucsela et al. 2006). With this approach, dynamical features are also compensated for, but some large scale tropospheric signals could be potentially removed.

2.4.3 *Model Method*

As global 3D-CTMs simulate stratospheric fields of many absorbers with good accuracy, one can also use model results to remove the stratospheric component of the signal. To compensate biases in both model and measurements, the model results are usually scaled to the measurements over the reference region (Richter et al. 2005) or assimilation techniques are applied (Boersma et al. 2007). For example, the satellite-derived total column densities can be assimilated directly applying specific weighting to reduce the impact of tropospheric contamination (Boersma et al. 2007). These methods further improve the accuracy of the stratospheric correction in particular in the vicinity of the polar vortex. However, as one cannot separate model measurement offsets resulting from inaccuracies in model or measurement, and those resulting from true tropospheric signals, this approach also runs the risk of removing large scale tropospheric signals in the assimilation process.

2.4.4 Cloud Slicing method

In regions where clouds have varying cloud top altitudes sometimes reaching the tropopause, a method called cloud slicing can be used to separate troposphere and stratosphere and in addition to derive vertical profiles within the troposphere (Ziemke et al. 1998; 2001). The method relies on the shielding effect of optically thick clouds (see Section 2.3.2) and assumes that the trace gas is distributed homogeneously, either within a scene with different cloud top heights or over time if measurements from different days are combined. This approach has been used for O_3 data mainly limited to the tropics. Radiative transfer effects within the clouds can have an impact on the results, as well as changes in absorber concentration by vertical redistribution in updrafts, changes in photolysis frequencies, wet scavenging or lightning production introducing uncertainty in the retrieved tropospheric column densities. A method based on a similar idea compares measurements over high mountains with those from low altitude areas in the vicinity (Jiang and Yung 1996).

2.4.5 Other Possible Approaches

Ideally, the stratospheric and tropospheric contributions should be separated solely by using information from the individual measurements. For O_3 in the stratosphere, combination of measurements at different wavelengths provides information on the vertical distribution as the penetration depth of photons into the atmosphere decreases with increasing O_3 absorption towards shorter wavelengths. In principle, use of the Chappuis bands for a total column density with good weighting of the tropospheric contribution in combination with a UV retrieval should provide some information on the tropospheric part (Chance et al. 1997). However, this has not so far been realised, mainly because of interference of chlorophyll and surface absorptions in the Chappuis window (see Section 2.3.3). Using UV measurements only, a separation of troposphere and stratosphere has been achieved for O_3 (Hoogen et al.

1999; Liu et al. 2005) and even more vertical information could be retrieved for strong volcanic SO_2 signals (Yang et al. 2009).

A similar approach can be applied to NO_2 using the variation of vertical sensitivity from 360 to 500 nm which is mainly determined by Rayleigh scattering and surface albedo changes. While this approach has been demonstrated in principle (Richter and Burrows 2000) it has relatively large uncertainties from surface albedo and aerosol and cloud effects. Another source of information on the vertical distribution could be polarisation (Hasekamp and Landgraf 2002) but this needs polarised measurements with high spectral resolution which are not available so far. Potentially, the temperature dependence of absorption cross-sections can also be used to derive vertical information by taking advantage of the temperature contrast between troposphere and stratosphere. The temperature dependence of the absorption is already used in stratospheric O_3 profile retrieval but the resulting signal is small for the troposphere and has not yet been used.

A completely different approach relies on the use of neural networks which are trained on large sets of satellite data in combination with independent measurements. Besides their high efficiency in processing large amounts of data, the main advantages of neural networks are that they can be applied even in cases with insufficient knowledge or too high complexity of the measurement process or instrument performance. However, it should be noted that neural networks usually fail if the measurement conditions (or instrument performance) are outside the range which was used for the training (e.g. due to instrument degradation). For O_3, good consistency of the tropospheric column densities extracted with this method has been demonstrated (Müller et al. 2003).

2.5 Uncertainties in UV/vis/NIR Satellite Measurements

There are several sources of error which need to be taken into account when interpreting satellite measurements of tropospheric composition. Some of the errors are random and can be reduced by averaging, while others are systematic and more difficult to quantify.

The fundamental uncertainty of any remote sensing measurement is linked to the signal strength or the number of photons collected in a single measurement and the associated probability distribution. For the ideal situation where there is no instrument noise, the photon shot noise is the ultimate limit for the noise of the radiance signal. The ratio of this noise to the signal determines the minimum observable optical thickness and thus the minimum detectable *SCD*. For real data, retrievals have to deal with additional error sources, including instrument noise, uncertainties in spectroscopic parameters, imperfect knowledge of the atmospheric light path and errors in the *a priori* assumptions used. Many of the uncertainties have been discussed in the literature (for NO_2 in (Boersma et al. 2004; Martin et al. 2002; van Noije et al. 2006) and references therein). The most relevant are summarised in Table 2.3 and will be briefly discussed below.

Table 2.3 Error sources in some quantities in UV/vis retrievals of tropospheric species

Radiance	Slant column density	Tropospheric slant column density	Tropospheric vertical column density
Photon shot noise	Cross-sections	Stratospheric correction	Surface albedo
Detector noise	Temperature dependence		Cloud height and fraction
Stray-light	Spectral interference		Vertical absorber distribution
Instrument throughput			Aerosol loading
Polarisation sensitivity			

2.5.1 Instrument Noise and Stray Light

The noise of the measured radiance for real instruments comprises the photon shot noise, the noise from the detector dark signal and the readout noise. For measurements over bright scenes, e.g. clouds, ice, or desert, the signal detected by a nadir looking instrument like GOME or SCIAMACHY is usually large and the total noise is dominated by photon shot noise for cooled detectors. However, over dark surfaces, or at short integration times (good spatial resolution), the instrument noise can become a significant contribution to the total noise. This is particularly true in the NIR part of the spectrum, where molecular scattering is less efficient and very low signals are measured, for example, over the ocean. In addition, detectors still suffer from much larger dark signals than in the UV/vis. For most instruments, measurement noise is strongly enhanced in the region of the Southern Atlantic Anomaly where an anomaly in the Earth's magnetic field leads to increased influx of radiation and particles on the instrument detectors and their electronics.

At very short integration times, the sequential readout of the detectors can no longer be neglected causing the spatial footprint of the measurements to vary with wavelength. Over strongly inhomogeneous scenes, e.g. broken clouds, this can lead to aliasing noise in the spectra as changes in intensity related to spatial patterns are misinterpreted as changes in intensity with wavelength. Future instruments could remove this source of error by using improved readout electronics.

In particular for the UV part of the spectrum, stray light generated within the instrument is a potential problem because the light intensity reaching the instrument is many orders of magnitude higher at longer wavelengths than in the UV. Even a small fraction of visible photons scattered to the UV part of the detector can be significant at the low signals found there, leading to distortion of the spectra and underestimation of absorption. Similarly, fluctuations in the dark signal of the detector or the electronic offset added prior to readout can be relevant at low signals if not accounted for. For this reason all instruments measuring in the UV attempt to minimize the stray-light within the instrument.

Photon noise, instrument noise and aliasing noise are statistical error sources which are reduced by averaging of measurements in space or time and therefore affect the precision but not the accuracy of the measurements. Instrumental stray-light in contrast is a systematic error, which reduces the absorption signals and cannot be offset by averaging.

2.5.2 Spectroscopic Uncertainties and Instrument Slit Width

The retrieval of the amounts and distributions of atmospheric trace gases in the UV/visible spectral range is based on molecular absorption spectroscopy. The uncertainty on the absorption cross-section of the trace gas of interest therefore directly translates to an uncertainty of the tropospheric amounts retrieved for the species: this is usually in the range of several percent. In addition to the knowledge of the absolute trace-gas cross section, any uncertainty in the knowledge of the detailed wavelength dependence of the absorption cross-section is relevant for DOAS or equivalent retrieval algorithms. As typical satellite instruments observing in the UV do not resolve the full spectral details caused by atomic and molecular absorption within the sun and the Earth's atmosphere, the slit function of the instrument must be well characterised as the reference spectra must be convolved to the instrument resolution. Any inaccuracy in this convolution results in additional error sources. Similarly for measurements by instruments at discrete wavelengths, the exact band pass of each measurement point and its stability over time are key aspects.

Many of the absorption cross-sections exhibit a temperature dependence which has to be characterised in the laboratory. However, even if the temperature dependence of the cross-section is perfectly known, the atmospheric temperature of the absorbing species is often not known to sufficiently high accuracy, introducing a further potential source of systematic error.

Spectroscopic errors are non-random relative errors and cannot be reduced by averaging. They are best assessed by validation of retrieval results with independent measurements.

2.5.3 Spectral Interference

Arguably the most complex source of uncertainties for the DOAS and related retrieval techniques arises from spectral interference. This is where spectral structures are erroneously attributed to absorption of one absorber although they originate with another absorber, with the instrumental features or with the wavelength dependence of surface reflectance. Ideally in a given spectral window the spectra of the absorbing and scattering features are all orthogonal to one another and have no correlation. However, as result of the spectral correlation between the signatures of different effects, spectral interference occurs. This is observed even for synthetic data with perfect spectra, but is obviously a larger source of uncertainty in real measurements. This increased sensitivity to correlation is the result of errors in the absorption cross-sections used, the slit function and the limited spectral resolution.

In most spectral windows used for retrieval, more than one species contribute to the observed absorption signal and, for weak absorbers, the interfering signals are often more than an order of magnitude larger than the quantity of interest. In addition to other atmospheric absorbers, inelastic scattering in the atmosphere (Ring effect), surface spectral reflectance, and wavelength dependent instrument transmission changes can also interfere with the retrieval of absorption signals. The usual approach, taken to minimise the false identification of molecular absorption, is to select spectral regions where the correlation between the absorption cross-section of the target molecule and that of other molecules is a minimum and the absorptions of the interfering species are small. The remaining impact is then accounted for by including the absorbers in the least squares fit. The same approach is taken for instrumental effects, which are not fully compensated through calibration by introducing appropriate pseudo absorbers in the retrieval.

For instruments observing light at discrete wavelengths, the problem of spectral interference is more severe as usually insufficient information is retrieved from the measurements to separate the effect of different absorbers.

The effect of spectral interference on the retrieved tropospheric products must be quantified for each spectral window selected and the particular instrument, having its unique instrumental characteristics. The best way to assess the impact of spectral interferences is to perform sensitivity studies by varying the analysis properties such as the wavelength range. In most cases, the errors are systematic and cannot be fully removed by averaging.

2.5.4 Light Path Uncertainties

Besides errors associated with the spectroscopic part of the retrieval, the imperfect knowledge of the light path through the atmosphere also introduces uncertainties in the determination of total *VCD*s, tropospheric *VCD*s or vertical profiles of trace gases. Radiative transfer models, when provided with accurate input for the atmospheric and surface conditions, can simulate the radiation field and the *AMF* with high accuracy. However, some of the important input parameters are usually not known perfectly (see Section 2.3.3). This can lead to large uncertainties in individual measurements.

As the measurements of UV/vis radiance beyond ~320 nm contain little or no information on the vertical distribution of the absorbers, this information has to be taken from other measurements, climatologies or atmospheric models. In combination with the altitude dependence of the measurement sensitivity, this is an intrinsic uncertainty for the determination of *VCD*s and tropospheric *VCD*s. All the parameters influencing the altitude dependence of the sensitivity have to be accounted for in the radiative transfer model, including surface spectral reflectance, aerosol load and distribution and clouds. All of these parameters vary in space and time, for example as vegetation changes, ice melts or sand storms develop and terminate.

The presence of clouds has a particularly important potential to reduce the sensitivity of space-borne UV/vis measurements to absorptions in the boundary layer to very low values by their shielding properties. As they are bright compared to the surface, even relatively small cloud fractions within a ground scene can have a significant impact on the measurements. This is because the absorption is weighted by the photon flux from the different parts of the scene.

Current approaches to account for cloud effects rely mostly on the retrieval of cloud properties from the measurements themselves by analysing the signals of absorbers with well known vertical distributions such as O_2 (Kuze and Chance 1994; Koelemeijer et al. 2001; Rozanov and Kokhanovsky 2004) or O_4 (Acarreta et al. 2004) or by using the inelastic scattering signature (Joiner and Bhartia 1995). Measurements which have more than a certain amount of cloud are rejected and the others are corrected using the assumed vertical distribution of the absorber and modelled light path distribution.

The large area observed by current space-borne instruments in each individual measurement adds to the uncertainties in light path. Usually, a ground scene covered by one measurement is not homogeneous but varies in surface altitude, aerosol loading, surface albedo, absorber profile and concentration and, in particular, cloud cover. As a result, the *a priori* assumptions made cannot be correct for all parts of the scene and the retrieval results are a weighted average of the true absorber column density.

As the uncertainties in the light path are mainly the result of insufficient knowledge of the input parameters, they have both a systematic and a random component. Systematic errors are introduced for example by the use of the wrong vertical profile over an industrial area where emissions in the model are not up to date or if the albedo data base is incorrect. Random errors result from statistical variations in aerosol loading or cloud distributions. Consequently, averaging will reduce the effect of light path uncertainties but systematic errors will remain. It must also be realised that the use of *a priori* information from atmospheric models introduces a dependency of the retrieved data products on the model assumptions which is often undesirable.

2.5.5 Uncertainty of Separation Between Stratosphere and Troposphere

As discussed in Section 2.4, the separation of the tropospheric and stratospheric contributions to the atmospheric *VCD* utilizes external information or assumptions, which therefore introduce their associated uncertainty into the results. For example, how well the residual methods work depends on how homogeneous the stratospheric field is as a function of longitude. Overall in the stratosphere constituents are relatively homogeneously distributed in the tropics as a result of the stable tropopause and strong winds, disturbed in mid-latitudes from tropospheric frontal systems, and strongly inhomogeneous in high latitudes in spring in the presence of the polar vortex. Systematic errors in the tropospheric VCDs are reduced by using model results or data assimilation to simulate the longitudinal dependence of the stratospheric column density but a residual source of uncertainty remains. Data smoothing and assimilation techniques run the risk of including large scale tropospheric pollution in the stratospheric field and so underestimate the tropospheric column densities for these situations. In general, it is difficult, if not impossible, to separate signals from the lower stratosphere and upper troposphere as one needs very good vertical resolution and accurate knowledge of the tropopause to do so. This is of significance for the retrieval of O_3 and BrO which have relatively large concentrations in the lower stratosphere.

Errors introduced from incomplete separation of stratospheric and tropospheric signals are systematic where they are linked to large scale dynamical structures but more random where they are linked to high and low pressure systems. The resulting errors are mainly additive and impact most strongly on retrievals for the remote unpolluted tropospheric conditions, and are less significant for regions with large tropospheric column densities.

2.6 Synopsis of the Historic, and Existing, Instruments and Data Products

In this section, a brief review of the most significant past and present generations of satellite instruments for tropospheric observations in the UV/vis/NIR is given. The objective of the first generation of satellite instruments for atmospheric sounding was the retrieval of stratospheric constituents and thus only a few representative missions are included here. Data from some of these instruments were, however, used in combination with information from other instruments to extract the tropospheric *VCD*s from total *VCD* observations. An overview of the satellite instruments is given in Appendix A; for a more detailed discussion see Burrows (1999).

During the last 30 years, the satellite-borne instrumentation has evolved with improvements of basic components: for example the collection/entrance optics, the dispersive element and the detector. Early instruments either had a static field of view (SBUV) or used a scanning mirror (TOMS) which moved orthogonally to the flight direction (the "whiskbroom" configuration). The light was dispersed by a prism and/ or a diffractive grating and detected in distinct wavelength intervals by photo multiplier tubes or photodiode detectors (the slit position being moved to measure at different wavelengths). GOME, SCIAMACHY and GOME-2 all use a similar arrangement of the components on the optical bench with the dispersive element comprising a more complex configuration including a pre-dispersing prism building an intermediate spectrum within the instrument and splitting the wavelength range first into distinct parts. The spectral sub-windows are individually dispersed by diffractive gratings coupled with focusing optics forming an image on one-dimensional photodiode arrays (Burrows et al. 1995; Burrows et al. 1999). For the SCIAMACHY instrument, photodiode arrays for the near-IR spectral range between 1.0 and 2.4 μm were developed and used in space for the first time. A different instrumental design was used for the OMI instrument: instead of employing a scanning mirror to cover a swath perpendicular to the flight direction, a two-dimensional CCD detector is used (Levelt and Noordhoek 2002). One dimension of the detector covers the spectral information, while the other dimension covers the viewing direction perpendicular to the flight direction. This so called "pushbroom" configuration yields two-dimensional horizontal information without the necessity of having moving parts, but results in non-homogenous ground pixel sizes, which depend on the viewing angle.

2.7 Example of the Retrieval Process

In the following section, the retrieval process from UV/visible/NIR nadir measurements is illustrated with an example of tropospheric NO_2 derived from GOME-2 measurements made in March 2008. While details of the analysis vary for other trace gases, the overall approach is similar.

The tropospheric trace gas analysis for an individual satellite ground pixel starts with the measured intensities from the nadir observation and the direct sun measurement (Fig. 2.14a). First, the logarithm of the ratio is taken to determine the optical thickness (Fig. 2.14b) which is then converted into a differential optical thickness by subtraction of a suitable polynomial (Fig. 2.14b–c). In the next step, all known absorbers as well as a correction for the Ring effect are fitted to the optical depth to determine the slant column density of NO_2 (Fig. 2.14d–g). The remaining unexplained part of the signal, the residual, (Fig. 2.14h) is a measure of the quality of the fit. When this process is repeated for all measurements for a day or a month, a global map of NO_2 slant column densities is created (Fig. 2.15a). The slant column densities comprise the absorptions of stratospheric and tropospheric NO_2, and the effects of different light paths at different solar position and viewing geometry are not yet corrected. The slant column density over the Pacific region is assumed to represent the stratospheric column density and is subtracted at all longitudes from the total *SCD* (see box in Fig. 2.15a) yielding the tropospheric slant column density (Fig. 2.15b). As discussed in the previous sections, this is the simplest approach to stratospheric correction and more sophisticated methods can be used to reduce uncertainties. In this step, simultaneously a cloud filter of 0.2 (i.e. maximum cloud coverage of 20%) has been applied to remove those pixels which are strongly influenced by clouds. In the last step of the analysis, the height dependence of the measurement sensitivity needs to be accounted for by applying appropriate tropospheric *AMF*. As these depend on factors such as surface albedo, NO_2 profile and aerosol load, they are a function of time and location (see Fig. 2.15c). The final results of the analysis are the vertical tropospheric NO_2 column densities which are shown in Fig. 2.15d.

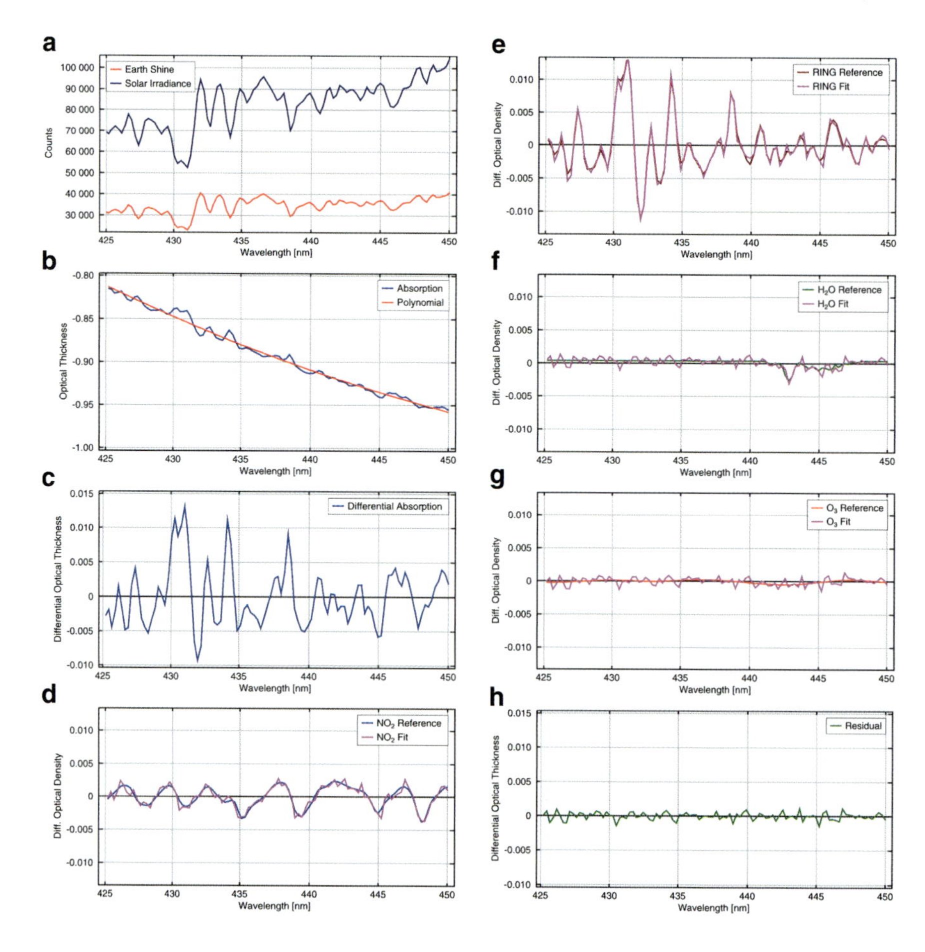

Fig. 2.14 Example of the spectral retrieval of NO_2 from a GOME-2 spectrum. See text for details.

In Fig. 2.15d, a number of interesting features can be discerned. High NO_2 column densities are observed over the industrialised parts of the world, in particular over China where rapid economic development has led to a strong recent increase in NO_x emissions. Enhanced tropospheric NO_2 is also observed over regions of intense biomass burning in Africa and parts of South America. The effect of lightning produced NO_x is not directly discernable but contributes to the high values in Africa. Another striking example of anthropogenic emissions is the thin line of NO_2 visible east of the tip of India, which is the result of ship emissions. There also are some regions with very low or even negative NO_2 column densities indicating that the assumptions made in the reference sector method for stratospheric correction are not adequate. For these regions more sophisticated analyses are necessary. A more detailed discussion of measurement results of tropospheric composition from space can be found in Chapter 8.

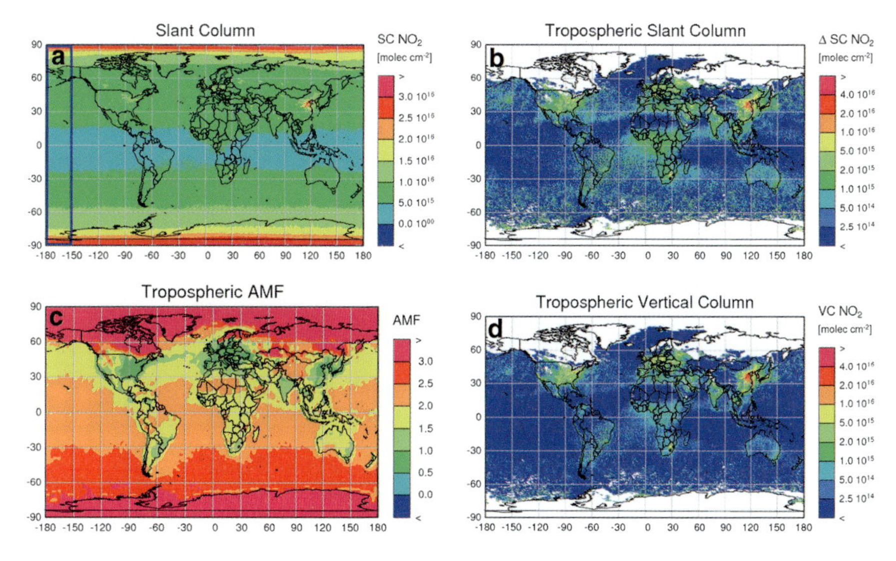

Fig. 2.15 Example of the NO_2 data analysis steps from the initial slant column density to the tropospheric vertical column density. The initial quantities are slant column densities (**a**) which are converted to tropospheric slant column densities (**b**) by subtracting the values over the reference sector (*blue box* in **a**). Application of the tropospheric AMF (**c**) yields the tropospheric vertical column densities (**d**). See text for details.

2.8 Future Developments

Although today's UV/vis/NIR satellite instruments already provide a wealth of useful data for atmospheric chemistry, numerical weather prediction and climate applications, it is interesting to consider, which components of the measurement and analysis system require further development to meet the growing needs of the scientific and operational communities. These are achieved by technical innovation with respect to optical design of the satellite instruments, innovative approaches for the retrieval algorithms and the data analysis and by the synergistic combination of data from complementary sensors.

2.8.1 Technical Design

In general it would be desirable to increase further the spatial and temporal resolution as well as the coverage of the measurements. For polar orbiting satellites, however, these properties are eventually limited by the available flux of reflected and backscattered solar photons. To achieve a sufficient signal to noise ratio, a compromise between global coverage and spatial resolution has to be made. Current UV/vis instruments like OMI and GOME-2 are close to the achievable limits: combining (almost) daily global coverage with spatial resolution (ground pixel size) of the order of several hundred square kilometres. Some improvements can still be expected if larger detector areas and/or reduced spectral resolution are used, which would improve the signal to noise ratio. In contrast to the detectors used in the UV and visible spectral range, the quantum efficiency in the near IR might still be substantially improved, leading to improved signals facilitating higher spatial resolution (Fig. 2.16).

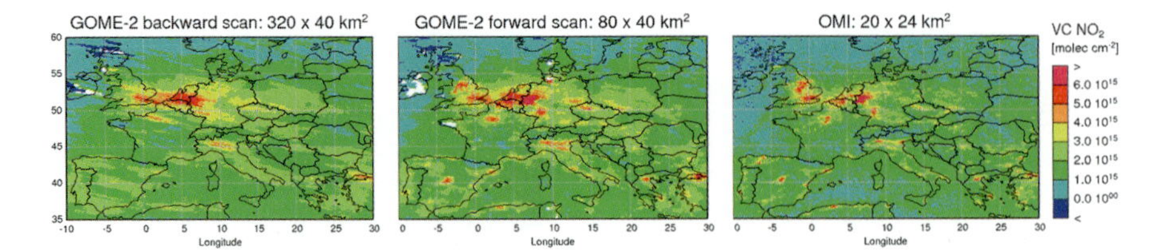

Fig. 2.16 Example of the effect of improved spatial resolution on the observed tropospheric NO_2 field over Europe. From *left* to *right*, spatial resolution is $320 \times 40\,km^2$ (GOME-2 data, only backscan is used), $80 \times 40\,km^2$ (GOME-2 data, only forward scans are used), and $20 \times 40\,km^2$ (OMI data, only inner part of scan is used). With improving spatial resolution, more and more local hot-spots can be identified while background values are slightly reduced. Note that OMI measurements are taken around noon while GOME-2 data are at 9:30 LT which introduces additional differences in the data, which do not result from spatial resolution.

In addition to polar orbiting satellites, it would be very desirable to operate geostationary satellites. These will allow much improved temporal resolution capturing, for example, the whole diurnal cycle (as long as daylight is available). Improvements in the spatial resolution using larger detectors and longer integration times than for low earth orbit instruments would be important for air quality applications, as well as improving the number of cloud free scenes. However, it should be realised that such satellites could only observe part of the globe and again a compromise between the spatial and temporal resolution and the spatial coverage has to be made.

Additional technical modifications might include the construction of instruments with well defined polarisation sensitivities. Such observations could yield improved information on the atmospheric radiative transfer and potentially add tropospheric profile information for O_3 (Hasekamp and Landgraf 2002). Also, the same ground scene could be observed under a variety of viewing angles. Such observations would allow an improved discrimination of aerosol scattering and surface reflection (Delderfield et al. 1986). In addition, for specific cases (with strong gradients and strong signals) tomographic inversion techniques could be applied. Such techniques have already been used for aerosol retrievals (Diner et al. 1998) but could be extended to trace gas observations. Finally, since the effect of aerosol scattering varies with viewing angle, additional profile information for trace gases might be extracted from the combined analysis of multiple-viewing observations.

Besides passive satellite instruments, active instruments might also be constructed. The extension of a lidar instrument (like CALIOP on CALIPSO) to multi-wavelength observations (DIAL) might allow tropospheric trace gas observations with unprecedented vertical resolution albeit with reduced spatial coverage.

2.8.2 Data Analysis

Improvements of the data analysis might be achieved both for the spectral retrieval and the radiative transfer simulations. For the spectral retrieval, the available absorption cross-sections are still not of sufficient quality in many cases. Further laboratory measurements would lead to improved fitting results. Also the treatment of the effects of spectral surface reflectance and of Raman scattering (in the atmosphere and in the ocean) in the spectral fitting process still has limitations and should be improved.

For the radiative transfer simulations, improvements can be expected in the treatment of Raman scattering and polarisation. Improved input data should be sought, especially for the treatment of surface effects (albedo, topography, bi-directional reflection function) and information on aerosol and cloud properties. Also, the consideration of three dimensional gradients of trace gas concentrations, as well as better cloud and aerosol properties, will improve the interpretation of the satellite observations.

Finally, new methods could be used for the comparison between satellite observations and model results. Usually the retrieved trace gas results (or aerosol and cloud properties) are compared to corresponding data from atmospheric model simulations. Instead, the observational quantities (e.g. the spectral radiance or trace gas optical depths measured from satellite) could be calculated directly from the atmospheric model and then compared to the satellite observation. Such methods are especially interesting for data assimilation techniques.

2.8.3 Synergistic Use of Complementary Satellite Observations

From the combined use of information from different observations, many tropospheric data products could be improved substantially. Several combinations are possible: information from different wavelength ranges could yield improved information on the vertical profile. From the combination of observations at different overpass times, information on the diurnal cycle can be obtained as has been already demonstrated for OMI and SCIAMACHY (Boersma et al. 2008). Synergistic use of simultaneous measurements, for example with aerosols and NO_2 from the same platform, could improve the accuracy of the retrieved column densities. Another possible direction is the combination of instruments with high spatial resolution and low spectral resolution, with instruments optimised for trace gas observations. Such synergistic use would combine the benefits of both instruments.

References

Acarreta, J.R., De Haan, J.F., and Stammes, P., 2004, Cloud pressure retrieval using the O_2–O_2 absorption band at 477 nm, *J. Geophys. Res.*, doi:10.1029/2003JD003915.

Aliwell, S.R., M. Van Roozendael, P. V. Johnston, A. Richter, T. Wagner, D. W. Arlander, J. P. Burrows, D. J. Fish, R. L. Jones, K. K. Tornkvist, J.-C. Lambert, K. Pfeilsticker, and I. Pundt, 2002, Analysis for BrO in zenith-sky spectra; An intercomparison exercise for analysis improvement, *J. Geophys. Res.*, doi:10.1029/2001JD000329.

Beirle, S., Kühl, S., Puķīte, J., and Wagner, T., 2010, Retrieval of tropospheric column densities of NO2 from combined SCIAMACHY nadir/limb measurements, Atmos. Meas. Tech., **3**, 283–299, doi:10.5194/amt-3-283-2010.

Bhartia, P.K., R.D. McPeters, C.L. Mateer, L.E. Flynn, and C. Wellemeyer, 1996, Algorithm for the estimation of vertical ozone profiles from the backscattered ultraviolet technique, *J. Geophys. Res.*, **101**, 18793–18806.

Boersma, K. F., Eskes, H. J., Brinksma, E.J., 2004, Error analysis for tropospheric NO_2 retrieval from space, *J. Geophys. Res., Atmospheres*, doi:10.1029/2003JD003962.

Boersma, K. F., Eskes, H. J., Veefkind, J. P., Brinksma, E. J., van der A, R. J., Sneep, M., van den Oord, G. H. J., Levelt, P. F., Stammes, P. , Gleason, J. F., Bucsela, E. J., 2007, Near-real time retrieval of tropospheric NO_2 from OMI, *Atmos. Chem. Phys.*, **7**, 2103–2118.

Boersma, K.F., D.J. Jacob, H.J. Eskes, R.W. Pinder, J. Wang, and R.J. van der A, 2008, Intercomparison of SCIAMACHY and OMI tropospheric NO_2 columns: Observing the diurnal evolution of chemistry and emissions from space, *J. Geophys. Res.*, **113(D16)**, 14.

Bogumil, K., J. Orphal, T. Homann, S. Voigt, P.Spietz, O.C. Fleischmann, A. Vogel, M. Hartmann, H. Bovensmann, J. Frerik, and J.P. Burrows, 2003, Measurements of molecular absorption spectra with the SCIAMACHY pre-flight model: instrument characterization and reference data for atmospheric remote-sensing in the 230–2380 nm Region, *J. Photochem. Photobiol. A*, **157**, 167–184.

Bovensmann, H., J.P. Burrows, M. Buchwitz, J. Frerick, S. Noël, V.V. Rozanov, K. V. Chance, and A.H.P. Goede, 1999, SCIAMACHY - Mission objectives and measurement modes, *J. Atmos. Sci.*, **56(2)**, 127–150.

Bracher, A., M. Vountas, T. Dinter, J. P. Burrows, R. Röttgers, and I. Peeken, 2008, Quantitative observation of cyanobacteria and diatoms from space using PhytoDOAS on SCIAMACHY data, *Biogeosciences Discuss.*, **5**, 4559–4590.

Brewer A.C., C.T. McElroy, and J.B. Kerr, 1973, Nitrogen dioxide concentrations in the atmosphere, *Nature*, **246**, 129–133.

Buchwitz, M., Rozaonv, V.V., Burrows, J. P., 2000, A near-infrared optimized DOAS method for the fast global retrieval of atmospheric CH_4, CO, CO_2, H_2O, and N_2O total column amounts from SCIAMACHY ENVISAT-1 nadir radiances, *J. Geophys. Res. - Atmospheres*, **105(D12)**, 15231–15245.

Burrows, J.P., K.V. Chance, P.J. Crutzen, H. van Dop, J.C. Geary, T.S. Johnson, G.W. Harris, I.S. A. Isaksen, G.K. Moortgat, C. Muller, D. Perner, U. Platt, J.-P. Pommereau and H. Rohde, 1988, SCIAMACHY – A European proposal for atmospheric remote sensing from the ESA polar platform, *Max-Planck-Institute for Chemistry, Mainz, Germany.*

Burrows, J.P., E. Hölzle, A.P.H. Goede, H. Visser, W. Fricke, 1995, SCIAMACHY- Scanning imaging absorption spectrometer for atmospheric chartography, *Acta Astronautica*, **35(7)**, 445.

Burrows, J.P., 1999, Current and future passive remote sensing techniques used to determine atmospheric constituents, in A. F. Bouwman (editor), *Approaches to scaling trace gas fluxes in ecosystems*, Elsevier Science B. V., Amsterdam.

Burrows, J.P., Weber, M., Buchwitz, M., Rozanov, V., Ladstätter-Weißenmayer, A., Richter, A., DeBeek, R., Hoogen, R., Bramstedt, K., Eichmann, K.-U., Eisinger, M., Perner, D., 1999, The global ozone monitoring experiment (GOME): mission concept and first scientific results, *J. Atmos. Sci.*, **56**, 151–175.

Bucsela, E. J., Celarier, E.A., Wenig, M.O., Gleason, J.F., Veefkind, J.P., Boersma K.F., Brinksma, E.J., 2006, Algorithm for NO_2 vertical column retrieval from the ozone monitoring instrument, *IEEE Transactions on Geoscience and Remote Sensing*, **44(5)**, 1245–1258.

Cox, C., Munk, W., 1954, Measurements of the roughness of the sea surface from photographs of the sun's glitter, *J. Opt. Soc. Am.*, **44**, 838–850.

Cox, C., Munk, W., 1956, Slopes of the sea surface deduced from photographs of sun glitter, *Bull. Scripps Inst. Oceanography,* **6(9)**, 401–488.

Chance, K. V., Burrows, J.P., Perner, D., Schneider, W., 1997, Satellite measurements of atmospheric ozone profiles, including tropospheric ozone, from ultraviolet/visible measurements in the nadir geometry: A potential method to retrieve tropospheric ozone, *Journal of Quantitative Spectroscopy & Radiative Transfer*, **57(4)**, 467–476.

Chance, K., Kurosu, T.P., Sioris, C.E., 2005, Undersampling correction for array detector-based satellite spectrometers, *Applied Optics*, **44(7)**, 1296–1304.

Delderfield, J., Llewellyn-Jones, D. T., Bernard, R., de Javel, Y., Williamson, E. J., Mason, I., Pick, D. R. and Barton, I. J., 1986, The along track scanning radiometer (ATSR) for ERS-1, *Proc. SPIE*, **589**, 114–120.

DeBeek, R., M. Vountas, V.V. Rozanov, A. Richter, and J.P. Burrows, 2001, The Ring effect in the cloudy atmosphere, *Geophys. Res. Lett.*, **28**, 721–772.

Diner, D.J., J.C. Beckert, T.H. Reilly, C.J. Bruegge, J.E. Conel, R. Kahn, J.V. Martonchik, T.P. Ackerman, R. Davies, S.A.W. Gerstl, H.R. Gordon, J-P. Muller, R. Myneni, R.J. Sellers, B. Pinty, and M.M. Verstraete, 1998, Multi-angle Imaging SpectroRadiometer (MISR) description and experiment overview, *IEEE Trans. Geosci. Rem. Sens.*, **36 (4)**, 1072–1087.

ESA Publication Division (SP-1182), 1995, GOME, Global ozone monitoring experiment, users manual, edited by F. Bednarz, European Space Research and Technology Centre (ESTEC), Frascati, Italy.

Eskes, H.J., and K.F. Boersma, 2003, Averaging kernels for DOAS total-column satellite retrievals, *Atmos. Chem. Phys.*, **3**, 1285–1291.

EUMETSAT, GOME-2 Products Guide, 2005, http://www.eumetsat.int/en/area4/eps/product_guides/GOME-2/GOME2-PG.pdf.

Farman, J.C., B.G. Gardiner, and J.D. Shanklin, 1985, Large losses of total ozone in Antarctica reveal seasonal ClOx/NOx interaction, *Nature*, **315**, 207–210.

Fish, D.J., and R.L. Jones, 1995, Rotational Raman-Scattering and the Ring Effect in Zenith-Sky Spectra, *Geophys. Res. Lett.*, **22(7)**, 811–814.

Fishman, J., and J.C. Larsen, 1987, Distribution of total ozone and stratospheric ozone in the tropics - implications for the distribution of tropospheric ozone, *J. Geophys. Res., Atmospheres*, **92(D6)**, 6627–6634.

Frankenberg, C., U. Platt, and T. Wagner, 2005, Iterative maximum *a posteriori* (IMAP)-DOAS for retrieval of strongly absorbing trace gases: model studies for CH_4 and CO_2 retrieval from near infrared spectra of SCIAMACHY onboard ENVISAT, *Atmos. Chem. Phys.*, **5**, 9–22.

Frederick, J.E, R. P. Cebula, and D. F. Heath, 1986, Instrument characterization for the detection of long-term changes in stratospheric ozone: an analysis of the SBUV/2 radiometer, *J. Atmos. Oceanic Technol.*, **3**, 472–480.

Grainger, J.F., and J. Ring, 1962, Anomalous Fraunhofer Line Profiles, *Nature,* **193(4817)**, 762.

Hasekamp, O.P., and J. Landgraf, 2002, Tropospheric ozone information from satellite-based polarization measurements, *J. Geophys. Res., Atmospheres*, doi:10.1029/2001/JD001346.

Heath, D. F., Mateer, C. L., and Kreuger, A.J., 1973, The Nimbus-4 BUV atmospheric Ozone experiment- Two Year's operation, *Pure Appl Geophys*, **106–108**, 1238–1253.

Heath, D.F., A.J. Krueger, H.R. Roeder, and B.D. Henderson, 1975, The solar backscatter ultraviolet and total ozone mapping spectrometer (SBUV/TOMS) for Nimbus G, *Opt. Eng.*, **14**, 323–331.

Heath, D.F., Krueger, A.J., Park, H., 1978, The solar backscatter ultraviolet (SBUV) and total ozone mapping spectrometer (TOMS) experiment, in *The Nimbus 7 User's Guide*, NASA Goddard Space Flight Centre, MD, 175–211.

Hendrick, F., M. Van Roozendael, A. Kylling, A. Petritoli, A. Rozanov, S. Sanghavi, R. Schofield, C. von Friedeburg, T. Wagner, F. Wittrock, D. Fonteyn, and M. De Mazière, 2006, Intercomparison exercise between different radiative transfer models used for the interpretation of groundbased zenith-sky and multi-axis DOAS observations, *Atmos. Chem. Phys.*, **6**, 93–108.

Henyey, L., and J. Greenstein, 1941, Diffuse radiation in the galaxy, *Astrophys J.*, **93**, 70–83.

Hilsenrath, E., R.P. Cebula, M.T. Deland, K. Laamann, S. Taylor, C. Wellemeyer, and P.K. Bhartia, 1995, Calibration of the NOAA-11 Solar Backscatter Ultraviolet (SBUV/2) Ozone Data Set from 1989 to 1993 using In-Flight Calibration Data and SSBUV, *J. Geophys. Res.*, **100**, 1351–1366.

Hoogen, R., V.V. Rozanov, J.P. Burrows, 1999, Ozone profiles from GOME satellite data: algorithm description and first validation, *J. Geophys. Res.*, **104**, 8263–8280.

Hudson, R. D., and A. M. Thompson, 1998, Tropical tropospheric ozone from total ozone mapping spectrometer by a modified residual method, *J. Geophys. Res., Atmospheres*, **103(D17)**, 22129–22145.

Jiang, Y. B., and Y. L. Yung, 1996, Concentrations of tropospheric ozone from 1979 to 1992 over tropical Pacific South America from TOMS data, *Science*, **272(5262)**, 714–716.

Joiner, J., and P. K. Bhartia, 1995, The determination of cloud pressures from rotational Raman-scattering in satellite backscatter ultraviolet measurements, *J. Geophys. Res. - Atmospheres*, **100(D11)**, 23019–23026.

Joiner, J., Bhartia, P.K., Cebula, R.P., Hilsenrath, E., McPeters, R.D., Park, H., 1995, Rotational Raman-Scattering (Ring Effect) in Satellite Backscatter ultraviolet measurements, *Applied Optics*, **34(21)**, 4513–4525.

Joiner, J., and A.P. Vasilkov, 2006, First results from the OMI rotational Raman scattering cloud pressure algorithm, *Geoscience and Remote Sensing, IEEE Transactions*, **44**, 5, 1272–1282.

Kattawar, G.W., Young, A.T., Humphreys, T.J., 1981, Inelastic-scattering in planetary-atmospheres. 1. The ring effect, without Aerosols, *Astrophys. J.*, **243(3)**, 1049–1057.

Kirchhoff, G.R., 1859, Über die Fraunhoferschen Linien, Berichte der Königlichen Preußischen Akademie der Wissenschaften, **59**, 662.

Koelemeijer, R.B.A., J.F. de Haan, and P. Stammes, 2003, A database of spectral surface reflectivity in the range 335–772 nm derived from 5.5 years of GOME observations, *J. Geophys. Res.*, doi:10.1029/2002JD002429.

Koelemeijer, R.B.A., Stammes P., Hovenier J.W. and F.J. de Haan, 2010, A fast method for retrieval of cloud parameters using oxygen A-band measurements from the Global Ozone Monitoring Instrument, *J. Geophys. Res.*, D, **106**, 3475–3490.

Komhyr, W. D., Grass, R. D., and Leonhard, R. K., 1989, Dobson spectrophotometer 83: A standard for total ozone measurements, 1962–1987, *J. Geophys. Res.*, **94**, 9847–9861.

Krijger, J.M., M. van Weele, I. Aben, R. Frey, 2007, The effects of sensor resolution on the number of cloud-free observations from space, *Atmos. Chem. Phys.*, **7**, 2881–2891.

Kuze, A., and K.V. Chance, 1994, Analysis of cloud top height and cloud coverage from satellites using the O_2 A and B bands, *J. Geophys. Res.*, **99(D7)**, 14481–14491.

Leue, C., M. Wenig, T. Wagner, U. Platt, and B. Jähne, 2001, Quantitative analysis of NO_x emissions from GOME satellite image sequences, *J. Geophys. Res.*,**106**, 5493–5505.

Levelt P.F., and R. Noordhoek, 2002, OMI Algorithm Theoretical Basis Document Volume I: OMI Instrument, Level 0-1b Processor, Calibration & Operations, *Tech. Rep. ATBD-OMI-01*, Version 1.1.

Liu, X., K. Chance, C.E. Sioris, R.J.D. Spurr, T.P. Kurosu, R.V. Martin, and M.J. Newchurch, 2005, Ozone profile and tropospheric ozone retrievals from the global ozone monitoring experiment: algorithm description and validation, *J. Geophys. Res.*, doi:10.1029/2005JD006240.

Martin, R. V., K. Chance, D. J. Jacob, T. P. Kurosu, R. J. D. Spurr, E. Bucsela, J. F. Gleason, P. I. Palmer, I. Bey, A. M. Fiore, Q. Li, R. M. Yantosca, and R. B. A. Koelemeijer, 2002, An improved retrieval of tropospheric nitrogen dioxide from GOME. *J. Geophys. Res.*, **107(D20)**, 4437, doi:10.1029/2001JD001027.

McPeters, R.D., S.M. Hollandsworth, L.E. Flynn, J.R. Hermans, and C.J. Seftor, 1996, Long-term ozone trends derived from the 16 year combined NIMBUS 7/Meteor 3 TOMS version 7 record, *Geophys. Res. Lett.,* **23**, 3699–3702.

Meador, W.E., and Weaver, W.R., 1980, Two-stream approximations to radiative transfer in planetary atmospheres: A unified description of existing methods and new improvements, *J. Atmos. Sci.* **37**, 630–643.

Mie G, 1908, Beiträge zur Optik trüber Medien, speziell kolloidaler Metallösungen. Annalen der Physik, *Vierte Folge*, Band **25**, No. 3, 377–445.

Mishchenko, M.I., A.A., Lacis, L.D. Travis, 1994, Errors introduced by the neglect of polarization in radiative transfer calculations for Rayleigh scattering atmospheres, J. Quant. Spectrosc. Radiat. Transf., **51**, 491–510.

Müller, M. D., Kaifel, A.K., Weber, M., Tellmann, S., Burrows, J.P., Loyola, D., 2003, Ozone profile retrieval from Global Ozone Monitoring Experiment (GOME) data using a neural network approach (Neural Network Ozone Retrieval System (NNORSY)), *J. Geophys. Res., Atmospheres*, doi:10.1029/2002JD002784.

Noël, S., Buchwitz, M., Burrows, J.P., 2004, First retrieval of global water vapour column amounts from SCIAMACHY measurements, *Atmos Chem Phys*, **4**, 111–125.

Noxon, J. F., 1975, Nitrogen dioxide in the stratosphere and troposphere measured by ground-based absorption spectroscopy, *Science,* **189**, 547.

Noxon, J.F., Whipple Jr., E.C., and R.S. Hyde, 1979, Stratospheric NO 2.1. Observational method and behaviour at mid-latitude, *J. Geophys. Res.*, **84**, 5047–5065.

Palmer, P.I., D.J. Jacob, K. Chance, R.V. Martin, R.J.D. Spurr, T. Kurosu, I. Bey, R. Yantosca, A. Fiore, and Q. Li, 2001, Air-mass factor formulation for differential optical absorption spectroscropy measurements from satellites and application to formaldehyde retrievals from GOME, *J. Geophys. Res.*, 106, 17, 147–17,160.

Park, H., D.F. Heath, and C.L. Mateer, 1986, Possible application of the Fraunhofer line filling in effect to cloud height measurements, *Meteorological Optics, OSA Technical Digest Series*, 70–81, Opt. Soc. Am., Washington, D. C.

Perliski, L.M., and S. Solomon, 1993, On the evaluation of air mass factors for atmospheric near-ultraviolet and visible absorption spectroscopy, *J. Geophys. Res.,* **98**, 10363–10374.

Perner, D., D.H. Ehalt, H.W. Pätz, U. Platt, E.P. Röth, and A. Volz, 1976, OH-radicals in the lower troposphere, *Geophys. Res. Lett.,* **3**, 466–468.

Perner, D. and U. Platt, 1979, Detection of nitrous acid in the atmosphere by differential optical absorption, *Geophys. Res. Lett.*, **7**, 1053–1056.

Platt, U., D. Perner, and W. Pätz, 1979, Simultaneous measurements of atmospheric CH_2O, O_3 and NO_2 by differential optical absorption, *J. Geophys. Res.,* **84**, 6329–6335.

Platt, U., 1994, Differential optical absorption spectroscopy (DOAS), in M. W. Sigrist (Ed.), *Air Monitoring by Spectroscopic Techniques. Chemical Analysis Series,* John Wiley, New York, 127.

Platt, U., Marquard, L., Wagner, T., Perner, D., 1997, Corrections for zenith scattered light DOAS, *Geophys, Res. Lett.*, **24(14)**, 1759–1762.

Platt, U, and Stutz, J., 2008, Differential optical absorption spectroscopy: principles and applications, Springer, Heidelberg.

Richter, A., Eisinger, E., Ladstätter-Weißenmayer, A., Burrows, J.P., 1999, DOAS Zenith sky observations: 2. Seasonal variation of BrO over Bremen (53°N) 1994–1995, *J. Atmos. Chem.*, **32(1)**, 83–99.

Richter, A., and J. P. Burrows, 2000, A multi wavelength approach for the retrieval of tropospheric NO_2 from GOME measurements, in *ERS-ENVISAT symposium, ESA publication SP-461*, edited, Gothenburg.

Richter, A., and J. P. Burrows, 2002, Tropospheric NO_2 from GOME measurements, *Adv Space Res* **29(11)**, 1673–1683.

Richter, A., Burrows, J. P., Nüß, H., Granier, C, Niemeier, U., 2005, Increase in tropospheric nitrogen dioxide over China observed from space, *Nature*, **437(7055)**, 129–132.

Roscoe, H. K., Fish, D.J., Jones, R.L., 1996, Interpolation errors in UV-visible spectroscopy for stratospheric sensing: Implications for sensitivity, spectral resolution, and spectral range, *Applied Optics*, **35(3)**, 427–432.

Rozanov, V.V., Kurosu, T., Burrows, J.P., 1998, Retrieval of atmospheric constituents in the UV-visible: A new quasi-analytical approach for the calculation of weighting functions, *J Quant Spectrosc Radiat Transf*, **60(2)**, 277–299.

Rozanov, V.V., A.A. Kokhanovsky, 2004, Semianalytical cloud retrieval algorithm as applied to the cloud top altitude and the cloud geometrical thickness determination from top of atmosphere reflectance measurements in the oxygen absorption bands, *J. Geophys. Res.*, doi: 10.1029/2003JD004104.

Sanghavi, S., 2003, An efficient Mie theory implementation to investigate the influence of aerosols on radiative transfer, Diploma thesis, University of Heidelberg, Germany.

Sarkissian, A., Roscoe, H. K., Fish, D., Van Roozendael, M., Gil, M., Chen, H. B., Wang, P., Pommereau, J.-P., and Lenoble, J., 1995, Ozone and NO_2 air-mass factors for zenith-sky spectrometers: Intercomparison of calculations with different radiative transfer models, *Geophys. Res. Lett.*, **22(9)**, 1113–1116.

Schoeberl, M.R., Ziemke, J.R., Bojkov, B., Livesey, N., Duncan, B., Strahan, S., Froidevaux, L., Kulawik, S., Bhartia, P.K., Chandra, S., Levelt, P.F., Witte, J. C., Thompson, A.M., Cuevas, E., Redondas, A., Tarasick, D.W., Davies, J., Bodeker, G., Hansen, G., Johnson, B.J., Oltmans, S. J., Voemel, H., Allaart, M., Kelder, H., Newchurch, M., Godin-Beekmann, S., Ancellet, G., Claude, H., Andersen, S.B., Kyrö, E., Parrondos, M., Yela, M., Zablocki, G., Moore, D., Dier, H., von der Gathen, P., Viatte, P., Stübi, R., Calpini, B., Skrivankova, P., Dorokhov, V., De Backer, H., Schmidlin, F.J., Coetzee, G., Fujiwara, M., Thouret, V., Posny, F., Morris, G., Merrill, J., Leong, C.P., König-Langlo, G., Joseph, E., 2007, A trajectory-based estimate of the tropospheric ozone column using the residual method, *J. Geophys. Res. - Atmospheres*, **112 (D24)**, 21.

Sierk, B., Richter, A., Rozanov, A., V. Savigny, C. Schmoltner, A.M. Buchwitz, M., Bovensmann, H., and J. P. Burrows, 2006, Retrieval and monitoring of atmospheric trace gas concentrations in nadir and limb geometry using the space-borne SCIAMACHY instrument, *Environ Monit Assess*, **120(1–3)**, 65–77.

Sioris, C. E., Kurosu, T.P., Martin, R.V., Chance, K., 2004, Stratospheric and tropospheric NO_2 observed by SCIAMACHY: first results, *Trace Constituents in the Troposphere and Lower Stratosphere*, **34(4)**, 780–785.

Solomon, S., A. L. Schmeltekopf, and R. W. Sanders, 1987, On the interpretation of zenith sky absorption measurements, *J. Geophys. Res.*, **92**, 8311–8319.

Slusser, J. R., K. Hammond, A. Kylling, K. Stamnes, L. Perliski, A. Dahlback, D. E. Anderson, and R. DeMajistre, 1996, Comparison of air mass computations. *J. Geophys. Res.*, **101**, 9315–9321.

van Noije, T. P. C., Eskes, H.J., Dentener, F.J., Stevenson, D.S., Ellingsen, K., Schultz, M.G., Wild, O., Amann, M., Atherton, C.S., Bergmann, D.J., Bey, I., Boersma, K.F., Butler, T., Cofala, J., Drevet, J., Fiore, A.M., Gauss, M., Hauglustaine, D.A., Horowitz, L.W., Isaksen, I.S.A., Krol, M.C., Lamarque, J.-F., Lawrence, M.G., Martin, R.V., Montanaro, V., Müller, J.-F., Pitari, G., Prather, M.J., Pyle, J.A., Richter, A., Rodriguez, J.M., Savage, N.H., Strahan, S.E., Sudo, K., Szopa, S., van Roozendael, M., 2006, Multi-model ensemble simulations of tropospheric NO_2 compared with GOME retrievals for the year 2000, *Atmos Chem Phys*, **6**, 2943–2979.

van de Hulst, H.C., 1981, Light scattering by small particles., New York, Dover, ISBN 0486642283.

Van Roozendael, M., Loyola D., Spurr R., Balis D., Lambert J-C., Livschitz Y., Valks P., Ruppert T., Kenter P., Fayt C., Zehner C., 2006, Ten years of GOME/ERS-2 total ozone data – The new GOME data processor (GDP) version 4: 1. Algorithm description, *J. Geophys. Res.*, **111**, D14311, doi:10.1029/2005JD006375.

Vasilkov, A. P., Joiner, J., Gleason, J., Bhartia, P.K., 2002, Ocean Raman scattering in satellite backscatter UV measurements, *Geophys. Res. Lett.*, doi:10.1029/2002GL014955.

Vountas, M., V.V. Rozanov, and J.P. Burrows, 1998, Ring effect: Impact of rotational Raman scattering on radiative transfer in Earth's Atmosphere, *J. Quant. Spec. Rad. Trans.*, **60(6)**, 943–961.

Vountas, M., A. Richter, F. Wittrock, and J.P. Burrows, 2003, Inelastic scattering in ocean water and its impact on trace gas retrievals from satellite data, *Atmos. Chem. Phys.*, **3**, 1365–1375.

Vountas, M., T. Dinter, A. Bracher, J. P. Burrows, and B. Sierk, 2007, Spectral studies of ocean water with space-borne sensor SCIAMACHY using Differential Optical Absorption Spectroscopy (DOAS), *Ocean Sci.*, **3**, 429–440.

Wagner, T., S. Beirle, M. Grzegorski, and U. Platt, 2007a, Satellite monitoring of different vegetation types by differential optical absorption spectroscopy (DOAS) in the red spectral range, *Atmos. Chem. Phys.*, **7**, 69–79.

Wagner, T., J. P. Burrows, T. Deutschmann, B. Dix, C. von Friedeburg, U. Frieß, F. Hendrick, K.-P. Heue, H. Irie, H. Iwabuchi, Y. Kanaya, J. Keller, C. A. McLinden, H. Oetjen, E. Palazzi, A. Petritoli, U. Platt, O. Postylyakov, J. Pukite, A. Richter, M. van Roozendael, A. Rozanov, V. Rozanov, R. Sinreich, S. Sanghavi, F. Wittrock, 2007b, Comparison of Box-Air-Mass-Factors and radiances for multiple-axis differential optical absorption spectroscopy (MAX-DOAS) Geometries calculated from different UV/visible radiative transfer models, *Atmos. Chem. Phys.*, **7**, 1809–1833.

Wendisch, M., and P. Yang, 2010, A Concise Introduction to Atmospheric Radiative Transfer, Wiley & Sons., ISBN: 978-3-527-40836-8.

Yang, K., X. Liu, N.A. Krotkov, A.J. Krueger, and S. A. Carn, 2009, Estimating the altitude of volcanic sulfur dioxide plumes from space borne hyper-spectral UV measurements, *Geophys. Res. Lett.*, doi:10.1029/2009GL038025.

Ziemke, J.R., Chandra, S., Bhartia, P.K., 1998, Two new methods for deriving tropospheric column ozone from TOMS measurements: Assimilated UARS MLS/HALOE and convective-cloud differential techniques, *J. Geophys. Res. - Atmospheres*, **103(D17)**, 22115–22127.

Ziemke, J. R., Chandra, S., Bhartia, P.K.,, 2001, Cloud slicing: A new technique to derive upper tropospheric ozone from satellite measurements, *J. Geophys. Res. – Atmospheres*, **106(D9)**, 9853–9867.

Chapter 3
Using Thermal Infrared Absorption and Emission to Determine Trace Gases

Cathy Clerbaux, James R. Drummond, Jean-Marie Flaud and Johannes Orphal

3.1 Physical Principles

Thermal infrared radiation is more commonly known as radiative heat. The thermal infrared region is the wavelength range of the electromagnetic spectrum which is characteristic of the thermal or heat radiation from the Earth's surface and from the atmosphere (McCartney 1983). In thermal equilibrium, the emission of radiation in the infrared is governed by Planck's law, which describes the spectral distribution of the energy emitted by "black" bodies (i.e. having unit emissivity at all wavelengths) as a function of temperature (see Chapter 1). Both the Earth's surface and the atmosphere emit in the so-called "thermal" infrared, having spectral distributions described by the Planck function corresponding to the local temperature, i.e. the surface temperature of the Earth and the temperature as function of altitude. However, it is important to note that neither the Earth's surface nor the atmosphere are perfect black bodies, so that a correction factor – the "emissivity" which is wavelength-dependent – needs to be used. As the temperature of the Earth's surface is about 300 K (see Fig. 3.1), from Planck's law the maximum of the thermal emission is predicted at a wavelength of around 10 μm (wavenumber: 1,000 cm^{-1}).

Thermal infrared corresponds to the wavelength region from about 3 μm to 100 μm (about 100–3,000 cm^{-1}), but there is actually no strict limit, in particular at longer wavelengths. The Earth's thermal emission is present during day and night which is important for atmospheric remote-sensing applications. Thermal infrared

C. Clerbaux (✉)
UPMC Univ. Paris 06; CNRS/INSU, LATMOS-IPSL, Paris, France

J.R. Drummond
Department of Physics and Atmospheric Science, Dalhousie University, Halifax, Canada

J.-M. Flaud
Université Créteil Paris 12, CNRS UMR 7583, LISA-IPSL, Paris, France

J. Orphal
Institute for Meteorology and Climate Research (IMK), KIT, Karlsruhe, Germany

J.P. Burrows et al. (eds.), *The Remote Sensing of Tropospheric Composition from Space*, Physics of Earth and Space Environments, DOI 10.1007/978-3-642-14791-3_3,

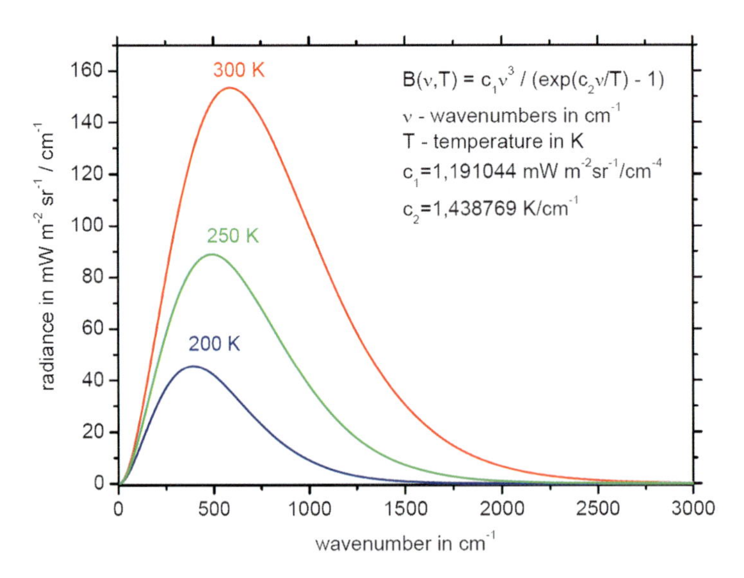

Fig. 3.1 Thermal emission in the infrared as function of temperature, described by Planck's law. Curves for typical atmospheric temperatures are shown. Note the increase of the emitted energy as function of temperature that scales with T^4 (Stefan-Boltzmann law) and the shift of the maximum of the emission to longer wavelengths when the temperature decreases (Wien's law).

is also a part of the solar spectrum and can thus be exploited for satellite measurements that operate in solar occultation mode.

Most atmospheric constituents absorb and re-emit infrared photons, but the strongest spectral features are due to rotational-vibrational transitions allowed by the quantum mechanical selection rules. In particular transitions that lead to a change of the molecular dipole moment upon absorption or emission of a photon are allowed. Homo-nuclear diatomic molecules like N_2 or O_2 therefore do not show strong spectral features in the thermal infrared which is an advantage for remote-sensing since the atmosphere is then largely transparent in this spectral region. However hetero-nuclear diatomic molecules (such as CO, NO, HF, HCl, etc.) and also most polyatomic molecules (like H_2O, CO_2, N_2O, CH_4, NO_2, HNO_3, OCS, SO_2, etc.) show strong absorption and emission in the thermal infrared. In addition, higher-order effects exist such as pressure-induced absorption or emission (that are especially important for N_2, O_2, and CO_2) and radiative effects beyond the wings of spectral lines (the so-called "continua" of N_2, O_2, and H_2O), that have to be included in numerical models of the radiative transfer in the Earth's atmosphere. Thus, the emission spectrum of the Earth's atmosphere contains many spectral features due to the different molecules that absorb and emit radiation (see Fig. 3.2); furthermore, due to the highly variable vertical temperature profile in the atmosphere, the observed spectrum is a superposition of all these processes at different altitudes, so that radiative transfer models are required to calculate (and analyse) infrared atmospheric spectra (see also Chapters 1 and 4).

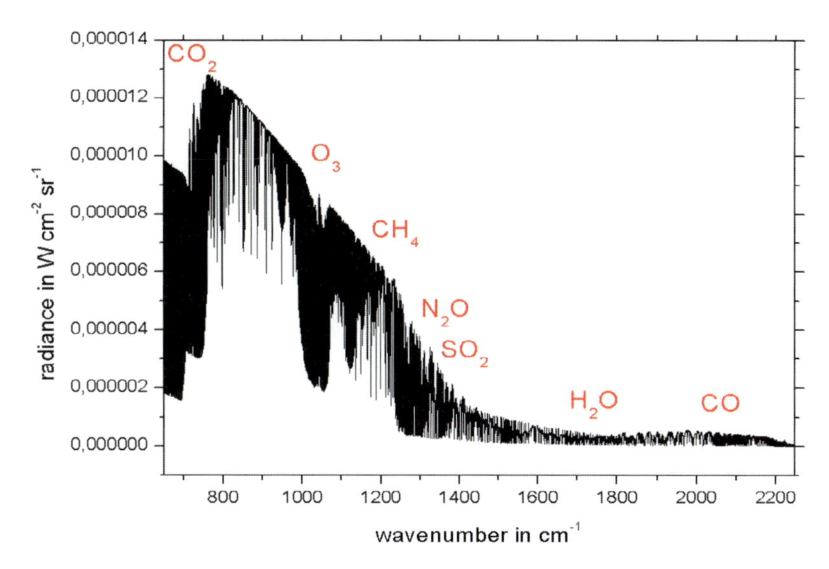

Fig. 3.2 The thermal emission of the Earth's atmosphere in the infrared (nadir geometry) for standard conditions, calculated at high spectral resolution with a radiative transfer model (Orphal et al. 2005). The envelope is given by the Planck function corresponding to the surface temperature (about 300 K) but one can see the strong absorptions due to water vapour, CO_2 and O_3. The radiance below the strongest bands of O_3 (around 1,000 cm^{-1}) and CO_2 (around 650 cm^{-1}) does not fall to zero since these molecules also re-emit radiation corresponding to the Planck curves of their respective temperatures (see Fig. 3.1).

Each molecule possesses a characteristic absorption (and emission) spectrum in the thermal infrared that needs to be measured first in the laboratory. Furthermore, in order to predict the variation of the molecular spectra as a function of temperature, it is essential to know the exact energy of the rotational-vibrational energy levels that are involved in the infrared transition (absorption or emission). For this purpose, quantum mechanical calculations are performed that provide these energy levels as a function of the vibrational and rotational quantum numbers, using theoretical models that take into account different rotation-vibration interactions for each molecule: diatomic or linear molecules (including degenerate modes), symmetric, asymmetric or spherical tops, spin-rotation interactions for radicals, and also large amplitude motions for "floppy" molecules like H_2O or NH_3. It is important to stress that the measurement and theoretical analysis of molecular spectra are a considerable ongoing effort, since they are essential for the understanding and numerical modelling of atmospheric spectra.

An important effect – only visible at relatively high spectral resolution – is the broadening of the individual molecular lines when the total pressure increases; in the Earth's atmosphere this effect leads to a Lorentz profile with molecular linewidths of about 0.2–0.4 cm^{-1} (full width at half maximum, at one atmosphere total pressure). The exact magnitude of this effect depends on the molecule and rotational quantum numbers, and again, laboratory measurements and theoretical models are needed to determine the pressure broadening parameters. The intrinsic line broadening due to the thermal Doppler effect is usually small in the mid-IR spectral region (Chapter 1). Although the "standard" line profile in the thermal infrared is a Voigt profile (a convolution of the Lorentz profile due to the pressure broadening with the Gaussian profile due to the thermal Doppler effect), small deviations from the Voigt profile (so-called "collisional narrowing" effects) have been observed in the laboratory and in atmospheric spectra, and need to be taken into account in the analysis of high-resolution atmospheric spectra. More interestingly, the pressure broadening effect can provide further, independent information for the vertical concentration profiles of atmospheric trace gases, in addition to the observational geometry (see below) and to the variation of the molecular spectra as a function of temperature. An even smaller effect – that is however also important at high spectral resolution – is the shift of the line centre as a function of total pressure (typically a few 0.001 cm^{-1} per atmosphere).

Molecular line parameters (line positions, line intensities, lower state's energies, pressure broadening and pressure shift parameters) are available from spectroscopic databases such as HITRAN (Rothman et al. 2005; 2009) or GEISA (Jacquinet-Husson et al. 2008) that are updated on a regular basis. These databases also contain some parameters for aerosols (refractive indices) and absorption cross-sections for molecules with very dense spectra that are not easily resolved into individual lines (e.g. the CFCs, N_2O_5 and $ClONO_2$). Note that for some remote-sensing experiments like ATMOS, MIPAS or IASI (see below), dedicated spectroscopic databases have been prepared containing reference data that cover only the spectral region and molecules of interest.

For numeric modelling of thermal infrared radiation, many computer codes are available but they always need to be adapted to the detailed purposes of the remote-sensing application. In particular, it is important to stress that the radiative transfer modelling of atmospheric spectra in the thermal infrared is a very challenging task, that can take up to several minutes – even on very modern computers – for a broad spectral region (e.g. some 100 cm^{-1}) at high spectral resolution (e.g. 0.001 cm^{-1}). This is especially important for the inverse problem where concentration profiles of atmospheric trace gases are determined from spectra in the thermal infrared (see below) and many radiative transfer calculations are required for each retrieved profile. It is also important to note that vertical profiles of temperature and pressure are always required for radiative transfer modeling in the thermal infrared, and some other auxiliary parameters, such as the surface temperature and emissivity, the aerosol content, and cloud parameters (e.g. cloud coverage, cloud height, cloud temperature) are also needed. Some information on these parameters can, in principle, also be extracted from the infrared spectra themselves, but this may lead to significant correlations between the retrieved atmospheric parameters.

Finally, there are several intrinsic advantages when using the thermal infrared region for atmospheric remote sensing.

- Many molecules that are observed in the thermal infrared cannot be measured in other spectral regions.
- Since the processes of absorption and emission of infrared photons are governed by Planck's law, they are very sensitive to temperature, as are the molecular line

strengths (due to the temperature-dependent population of the molecular energy levels), and this leads to vertical information that can be extracted from atmospheric spectra in the thermal infrared – even when using nadir (down-looking) geometry from space.

- The pressure-dependence of the molecular line-widths provides additional vertical information, although a high spectral resolution is required to observe this effect.
- Observations in the thermal infrared can be made both day and night, so that – in contrast to remote-sensing measurements in the ultraviolet-visible spectral region – diurnal cycles can be observed, as can species that are only present in significant concentrations during the night.

3.2 Thermal Infrared Instruments: Techniques, History, Specificity

3.2.1 Techniques

Remote sensors using the thermal infrared spectral range use different instrumental design, depending on their primary mission goal. Past and current instruments are based on evolving technology such as cell correlation radiometry, Fourier transform spectroscopy, or grating spectrometry.

a Cell Correlation Radiometry

A correlation cell radiometer (also called gas correlation sensor forms an optical filter specific to the gas being monitored by passing the optical signal through a cell of the same gas within the instrument. The amount of gas (i.e. its density) is varied by varying length or pressure within the cell. The change in signal is greatest near the spectral lines of the gas being investigated, and between the spectral lines both the cell absorption and the change in absorption with the quantity is minimal. The instrument has good selectivity (effective spectral resolution) due to the close association of the change of signal with the spectral lines themselves, good signal-to-noise ratio because many lines are monitored simultaneously, and a modest data rate because only one optical signal is detected.

b Fourier Transform Spectroscopy

Fourier transform infrared spectroscopy (FTS) is a measurement technique for collecting infrared spectra using a Michelson-type interferometer. The thermal infrared radiation emitted from the Earth-Atmosphere system (i.e. the source of photons) is split into two beams by a half-transparent mirror called a beam-splitter, one is reflected from a fixed mirror and one from a moving mirror, which introduces a time delay, or optical path difference. The two beams are then allowed to interfere, and the overall intensity of the light is measured at different time delay settings. By making measurements of the signal at many discrete positions of the moving mirror (the so-called "interferograms"), the source spectrum can finally be reconstructed using the inverse Fourier transform. In order to avoid oscillations around spectral lines that are due to the finite length of the interferograms, the latter are usually multiplied by a numerical function that decreases to zero at maximum delay; this operation is called "apodisation" and slightly reduces the spectral resolution.

c Grating Spectrometry

A grating spectrometer disperses light incident on it, i.e. a spectrum of radiation is separated in space by wavelength, using a prism or a grating. The dispersed light is then recorded by a focal plane detector array (an arrangement of many small detectors in a line, for example, on a semiconductor chip). Multi-aperture array grating spectrometers are used in space and provide high spectral resolution with wide spectral coverage, using advanced imaging design, with wide spectral coverage. Note that both interferometers and grating spectrometers make use of the interference of electromagnetic radiation.

Upwelling radiance enters the system via the cross-track scan mirror, where it is directed into a telescope. The collimated energy exiting the telescope is incident on the spectrometer entrance slit plane containing individual apertures. Ultimately, these slits are imaged onto the focal plane, where each slit image contains the energy from one selected grating order.

3.2.2 History

As satellite technology improved in the late twentieth century, it became possible to expand the range of wavelengths measured from space and to increase the spectral resolution available. Both of these are essential to successful measurements of atmospheric composition. The first instruments were targeted at the middle atmosphere to measure stratospheric ozone and related species (e.g. the UARS mission launched in 1991) and the tropospheric capability was an incidental ability. However substantial progress was made with a number of instruments.

The Atmospheric Trace Molecule Spectroscopy (ATMOS) Experiment (Gunson et al. 1996) was flown by NASA on the space shuttle on four occasions in 1985, 1992, 1993 and 1994. It consisted of a Fourier transform spectrometer with a sun-tracker and measured the atmosphere twice per orbit, by solar occultation at sunrise and sunset. Its spectral resolution was 0.01 cm^{-1} (unapodized) in the wavelength range 600–4,800 cm^{-1}, although this range was accomplished in a number of separated spectral intervals (channels), rather than a continuous scan.

The Measurements of Atmospheric Pollution from Satellites (MAPS) instrument (Reichle et al. 1999) was also a shuttle instrument, but was targeted specifically at CO in the troposphere. It flew on the shuttle in 1981, 1984 and 1994. The instrument was a fixed-cell correlation radiometer operating in the 4.7 μm region of the fundamental band of CO. The instrument viewed downwards – nadir– and therefore had good horizontal resolution, but poor vertical resolution. MAPS produced the first directly measured maps of a tropospheric minor constituent (apart from water vapour). MAPS showed the importance of tropospheric monitoring from space by demonstrating the significant temporal and spatial variation in constituents. It also illustrated the problems of cloud cover for such instruments and the issues of coverage during a limited-duration mission.

The Interferometric Monitor for Greenhouse Gases (IMG) (Kobayashi et al. 1999) was the first satellite-borne instrument launched to use the thermal infrared spectral range to sound the troposphere. It was carried on the ADEOS-1 platform, and provided 10 months of data, from August 1996 to June 1997, until the failure of the platform solar array. IMG was a nadir-looking Fourier transform spectrometer which recorded the thermal emission of Earth between 600 and 3,030 cm^{-1}, with a spectral resolution of ~0.1 cm^{-1}. Owing to the polar orbit of the ADEOS satellite, IMG allowed the simultaneous measurement and global distributions from space for a series of trace gases relevant for climate and chemistry studies: H_2O, CO_2, N_2O, CH_4, O_3, CO, CFCs, and HNO_3 (Clerbaux et al. 2003). For some species, the high spectral resolution allowed vertical profiles to be derived.

3.2.3 *Specificity*

When using the thermal infrared spectral range to sound the atmosphere, a remote sensor on board a satellite records the light passing through different atmospheric layers. From the radiance signal recorded, one can extract information on the vertical concentration of each atmospheric constituent absorbing at a given altitude. The source can be either the thermal emission of the Earth-atmosphere system (nadir and limb viewing, Section 3.1 and Chapter 1), or a section of the solar emission spectra (solar occultation). Solar occultation or limb-viewing instruments provides extremely high sensitivity to trace constituents due to the long atmospheric paths and the strength of the solar source (equivalent to a blackbody at about 5,800– 6,000 K). It also provides excellent vertical resolution, but poor horizontal resolution, each due to the path geometry in the atmosphere. Solar occultation works well in the middle and upper atmosphere where general transparency is high, but runs into difficulty when clouds and heavy aerosol layers, or strong absorbers such as water vapour, are present in the path, as occurs in the troposphere. Therefore as the tangent height – the height of the lowest point on the light path – decreases, the frequency of useful measurements declines.

For cloud free situations, nadir-looking thermal infrared instruments can measure the atmospheric radiation down to the ground and some vertical information can be derived from the shape of the absorption lines, provided the spectral resolution is high enough. The thermal contrast between the surface and the boundary layer determines to what extent one can detect species near the surface. The Earth's surface either heats up or cools down faster than the atmosphere and, therefore, the diurnal variation is larger, and hence thermal contrast more pronounced during day than night. The capability of an infrared nadir sounder to probe the lower atmospheric layers, where local pollution occurs, therefore strongly depends on location, the temperature, the type of surface (emissivity) and the time of the day. Thermal contrasts are generally highest over land during the day and lowest over water at night. This, along with the instrumental characteristics (spectral resolution and radiometric noise) determines the amount of vertical information that can be retrieved for a given species in the nadir geometry, and the uneven sensitivity to the atmospheric layers (averaging kernels, see Section c below).

a Retrieval Algorithms/Inversions

Inversion of geophysical parameters from remotely sensed observations is known to be an ill-posed problem (Rodgers 1976). A variety of methods exist for the retrieval of atmospheric profiles from the spectra measured by remote sounders. A large body of literature is available on the subject, and the most widely used approaches in atmospheric remote sensing are described, for example, in Rodgers (2000) and Tarantola (2005). The accuracy of the retrieved quantities and the ability to retrieve low-resolution vertical profiles from the data provided by an instrument partly rely on the efficiency of the inversion procedure.

b Forward Radiative Transfer

Given the general radiative transfer equation, the measurement $\mathbf{y}$ can be expressed as the vector of measured quantities (radiances):

$$\mathbf{y} = \mathbf{F}(\mathbf{x}, \mathbf{b}) + \boldsymbol{\varepsilon} \tag{3.1}$$

where $\mathbf{F}$ is the forward radiative transfer function, $\mathbf{x}$ denotes the vector of atmospheric states, for example, the trace gas concentration at different altitudes, $\mathbf{b}$ represents model parameters affecting the measurement and $\boldsymbol{\varepsilon}$ is the measurement noise. As described in Section 3.1, a synthetic spectrum can be computed using the line parameters (positions, intensities, broadening and shifting parameters, including their dependence on temperature), and absorption cross sections for the heavier molecules, as collected in spectroscopic databases such as HITRAN and GEISA (Rothman et al. 2005; Rothman et al. 2009; Jacquinet-Husson et al. 2008).

The water vapour, CO_2, O_2 and N_2 continua also have to be included to represent the atmosphere correctly. The resulting spectrum should then be processed to take into account the Instrumental Line Shape (ILS).

Useful variables are the derivatives of the radiance with respect to the parameters to retrieve, the Jacobians $\mathbf{K} = \partial \mathbf{y}/\partial \mathbf{x}$ (x includes the vertical abundances of the target species), as well as to the model parameters $K_b = \partial \mathbf{y}/\partial \mathbf{b}$.

c The Optimal Estimation (OE) Formalism

i Finding an Optimal Solution

Starting from relevant *a priori* information, composed of a mean state $\mathbf{x}_a$, and an *a priori* covariance matrix, $\mathbf{S}_a$, which represents the best statistical knowledge of the state prior to the measurements, the retrieved state can then be found using the Optimal Estimation Method. Assuming a linear problem, the optimal vertical profile can be written as (Rodgers 2000):

$$\hat{\mathbf{x}} = \left(\mathbf{K}^T \mathbf{S}_\varepsilon^{-1} \mathbf{K} + \mathbf{S}_a^{-1}\right)^{-1} \left(\mathbf{K}^T \mathbf{S}_\varepsilon^{-1} \mathbf{y} + \mathbf{S}_a^{-1} \mathbf{x}_a\right) \tag{3.2}$$

where $\mathbf{S}_\varepsilon$ is the measurement covariance matrix

Introducing the gain and averaging kernels matrices **G** and **A**,

$$\mathbf{G} = \frac{\partial \hat{\mathbf{x}}}{\partial \mathbf{y}} = \left(\mathbf{K}^T \mathbf{S}_\varepsilon^{-1} \mathbf{K} + \mathbf{S}_a^{-1}\right)^{-1} \mathbf{K}^T \mathbf{S}_\varepsilon^{-1} \tag{3.3}$$

$$\mathbf{A} = \frac{\partial \hat{\mathbf{x}}}{\partial \mathbf{x}} = \mathbf{GK} \tag{3.4}$$

Eq. 3.2 can also be rewritten as:

$$\hat{\mathbf{x}} = \mathbf{x}_a + \mathbf{A}(\mathbf{x} - \mathbf{x}_a) + \mathbf{G}(\boldsymbol{\varepsilon} + \mathbf{K}_b(\mathbf{b} - \hat{\mathbf{b}})) \tag{3.5}$$

ii Information Content

The element **A**(i,j) of the averaging kernel matrix **A** is the relative contribution of the element **x**(j) of the true state to the element $\hat{\mathbf{x}}(i)$ of the retrieved state. The vertical resolution of the retrieved profile can be defined as the full width at half maximum of the rows of the averaging kernel matrix. The number of independent elements of information contained in the measurement can also be estimated as the Degrees Of Freedom for Signal (DOFS) which is defined as the trace of the averaging kernel matrix (Rodgers 2000). Examples of typical averaging kernel functions for nadir looking thermal infrared instruments are provided in Fig. 3.3.

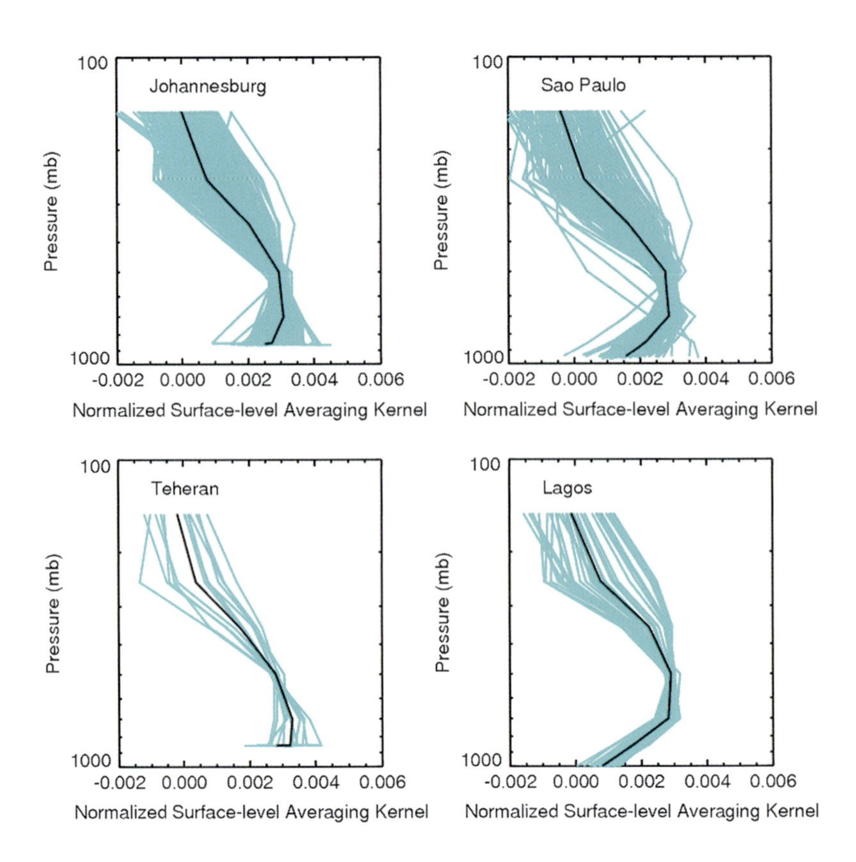

Fig. 3.3 Surface level averaging kernels above Johannesburg (South Africa), Teheran (Iran), Sao Paulo (Brazil) and Lagos (Nigeria) as observed by MOPITT in 2004. The *blue lines* correspond to single observations and the *black line* is an average over all observations (adapted from Clerbaux et al. (2008a)).

iii Error Budget

In the linear approximation, the total error is computed from the linear retrieval equation (Eq. 3.5) as the difference between the true state and the retrieved state:

$$\hat{\mathbf{x}} - \mathbf{x} = (\mathbf{A} - \mathbf{I})(\mathbf{x} - \mathbf{x}_a) + \mathbf{GK}(\mathbf{b} - \hat{\mathbf{b}}) + \mathbf{G}\boldsymbol{\varepsilon} \tag{3.6}$$

where I is the identity matrix. This equation illustrates the three principal sources of error.

1. The smoothing error, $(\mathbf{A} - \mathbf{I})(\mathbf{x} - \mathbf{x}_a)$, which accounts for the smoothing of the true state by the averaging kernels.

2. The model parameters error, $\mathbf{GK}(\mathbf{b} - \hat{\mathbf{b}})$ which accounts for the imperfect knowledge of the direct model parameters.
3. The measurement error, $\mathbf{G}\varepsilon$, associated with the radiometric noise.

d The Tikhonov–Philips Regularization

A modification of the optimal estimation (OE) formalism is the so-called Tikhonov–Philips (TP) regularization. Here, instead of the *a priori* covariance matrix, $\mathbf{S}_a$, which has to represent the best statistical knowledge of the state prior to the measurements, a regularization matrix $\mathbf{R}$ is used to constrain the solution. The method is used, for example, when no precise *a priori* covariance matrix is available, or when the constraint of the retrieval needs to be optimized for a particular vertical region, or for the vertical smoothness of the solution, e.g. for retrievals from limb geometry (Steck 2002). Whereas the strength of the constraint in the "classical" regularization method (Tikhonov 1963; Phillips 2003) is the same for all the altitudes, it may be dependent of the altitude for particular atmospheric retrieval methods (Doicu et al. 2004; Kulawik et al. 2006).

Mathematically, this means that the inverse of matrix $\mathbf{S}_a$ in Eq. 3.3 is replaced by the matrix $\mathbf{R}$ so that the entire retrieval formalism and software codes are the same for both the OE and TP methods. In particular, diagnostic variables like DOFS, are also used for the TP regularization method, and the calculation of the error budget and vertical resolution (using the averaging kernel matrix $\mathbf{A}$) is also essentially the same. The TP method has been applied to the analysis of spectra obtained with several limb and nadir looking sounders (Fischer et al. 2008; Keim et al. 2008; Kulawik et al. 2006; Bowman et al. 2006; Worden et al. 2007a; Eremenko et al. 2008).

The question of which method (OE or TP) is to be used for a particular retrieval problem is difficult to answer since there is no general recipe for defining the "best" covariance ($\mathbf{S}_\mathbf{a}$) or regularization ($\mathbf{R}$) matrices. The OE method will always provide the "optimum" solution that is statistically the most probable (based on the existing data that are represented by the *a priori* covariance matrix $\mathbf{S}_\mathbf{a}$), while the TP method may be interesting for situations where such a matrix is difficult to define, here great care has to be taken to construct regularization matrices that provide physically meaningful results, in particular to avoid excessive or too weak constraints.

e Neural Networks

Statistical methods using neural networks (Fig. 3.4) offer interesting possibilities for solving problems involving complex transfer functions in a time efficient manner. The first step building an efficient neural network is to define its architecture. In the literature, multilayer networks are shown to provide good performance

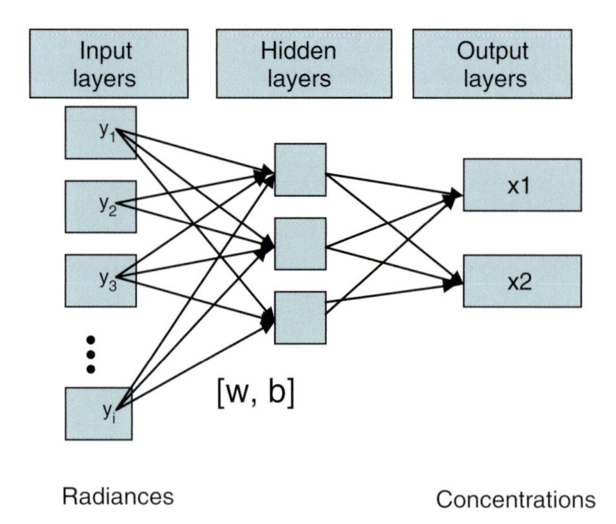

Fig. 3.4 Schematic representation of a neural network, including inputs (recorded radiances), hidden layers with mathematical functions, and outputs (concentrations of trace gases).

in solving problems with geophysical variables. To characterize the architecture of this type of neural network, the number of layers and of neurons. Each neuron is a mathematical function in each layer, and the topology of their connections, must be carefully chosen, as well as the transfer functions used. This is done on the basis of empirical considerations which depend on the complexity of the function associated with the physical problem to solve.

Once the architecture is fixed, the training begins by fitting the parameters of the neural model: weights of the connections and bias associated with neurons. For trace gas retrieval, this training phase consists in associating the input radiance measurement vector (**y**) with the desired column or profile concentration output (**x**). This is done by modifying the parameter weights and bias values, using for example a gradient back-propagation algorithm. In order to build a comprehensive database to train the network, atmospheric concentration profiles, either provided by chemical-transport models, or from independent observations, are used as input to a line-by-line radiative transfer code, and then the spectra are convolved with the instrumental function. The weights and bias are first initialized to random values. Then the square of the differences between the desired and the calculated outputs (trace gas concentration) for the whole the training dataset is minimized through an iterative gradient descent procedure, and the parameters matrix is modified accordingly. This requires a long computation time because of the minimization process. Once the network is trained, the weights and biases are held fixed and the network is ready to operate. This retrieval technique is particularly useful in case of very high data rates and/or time constraints (e.g. if operational retrievals are needed) as it is very efficient in time (Turquety et al. 2004).

3.3 Thermal Infrared: Missions and Products

Information on all the satellite-borne missions using thermal infrared that flew or are still operating now is included in Appendix A.

In the thermal infrared spectral range, most of the key atmospheric species exhibit significant absorptions provided they are present in the atmosphere with sufficient concentrations, which is directly related to their source strength and atmospheric lifetime. Table 3.1 summarises the spectroscopic absorptions of the common species. Figs. 3.5 and 3.6 are illustrations of atmospheric spectra recorded by MIPAS/ENVISAT (limb sounding geometry) and IASI/METOP (nadir looking geometry), together with the identification of the main absorbing molecules, as a function of wavenumber.

3.4 Examples

This section provides some highlights of specific results obtained from thermal infrared remote sensors, both in the limb and nadir-looking modes.

3.4.1 Limb and Solar Occultation Instruments

a ACE-FTS

The Atmospheric Chemistry Experiment mission (ACE, see Appendix A) was launched by CSA (Canada) on SCISAT in August 2003. ACE-FTS is a high resolution (0.02 cm^{-1}) infrared Fourier transform spectrometer operating from 2 to 13 μm (750–4,100 cm^{-1}) that measures the vertical distribution of trace gases and temperature using solar occultation (Bernath et al. 2005). The principal purpose of the ACE-FTS mission is to investigate the chemical and dynamical processes that control the distribution of ozone in the stratosphere and upper troposphere (10–50 km altitude range) with a particular focus on the Arctic winter stratosphere. The ACE concept is similar in some respects to the NASA ATMOS experiment that flew four times on the space shuttle between 1985 and 1994.

A 74°-inclined circular orbit with an altitude of 650 km was chosen for ACE to achieve both global and high latitude coverage. While in orbit, SCISAT observes 15 sunrises and 15 sunsets per day. During sunrise and sunset, the FTS measures infrared absorption signals that contain information about different atmospheric layers; it thus provides vertical profiles of atmospheric constituents. The vertical resolution is about 3–4 km from the cloud tops up to about 100 km. ACE measures O_3 and determines budgets for the H, N, Cl and F families of molecules. Its

Table 3.1 Molecules absorbing in the thermal infrared spectral range, with the location of absorption bands and the associated vibrational modes

Molecules	Band center (cm^{-1})	Absorption band (cm^{-1})	Vibrational mode
O_3	710	550–900	ν_2
	1,043	919–1,243	ν_3
	1,070	940–1,280	ν_1
	2,105	1,880–2,320	$2\nu_1$, $2\nu_3$, $\nu_1 + \nu_3$
	2,800	2,680–2,820	$\nu_1 + \nu_2 + \nu_3$
CO	2,100	2,000–2,260	1–0
SO_2	1,151	1,080–1,260	ν_1
	1,361	1,310–1,400	ν_3
	2,499	2,440–2,530	$\nu_1 + \nu_3$
H_2CO	1,746	1,650 to >1,830	ν_2
	2,780–2,874	2,700 to >3,000	ν_1, ν_5
NO_2	648	650–880	ν_2
	1,621	1,550–1,760	ν_3
	2,910	2,850–2,940	$\nu_1 + \nu_3$
PAN	–	750–1,900	Cross sections
H_2O	1,595	<600 to >3,000	ν_2
CO_2	618.1		$2\nu_2(l_0)$
	667.3	<600–850	$\nu_2(l_1)$
	720.5		ν_1
	1,886	1,870–1,990	$4\nu_2(l_0)$
	2,094	2,000–2,150	$\nu_1 + 2\nu_2(l_0)$
	2,137	2,000–2,150	$2\ \nu_1$
	2,349	2,000–2,700	ν_3
CH_4	1,306.2	900–1,970	ν_4
	3,020.3	2,000 to >3,000	ν_3
CFC-11	850	810–880	ν_4
	1,082	1,050–1,120	ν_1
CFC-12	922	850–950	ν_6
	1,160	1,050–1,200	ν_8
N_2O	1,285	1,210–1,340	ν_1
	2,222	2,120–2,270	ν_3
HNO_3	648.8	615–678	ν_6
	763.2	722–810	ν_8
	879.11	816–960	ν_5
	896.85	816–960	ν_9
	1,205.7	1,165–1,233	$\nu_8 + \nu_9$
	1,303.5	1,098–1,388	ν_4
	1,325.7	1,098–1,388	ν_3
	1,709.57	1,650–1,770	ν_2
NH_3	931(s)–967(a)	750–1,200	ν_3
C_2H_4	945.45	815–1,170	ν_7 (ν_4,ν_{10})
CH_3OH	1,033	966–1,185	ν_8
HCOOH	1,105	960–1,235	ν_6

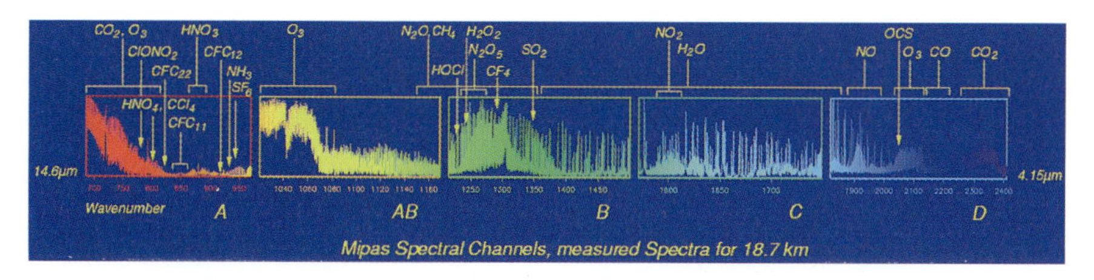

Fig. 3.5 The various spectral bands recorded by MIPAS/ENVISAT at a tangent altitude of about 19 km. The emission domains of a number of minor atmospheric constituents are indicated (Flaud and Oelhaf 2004).

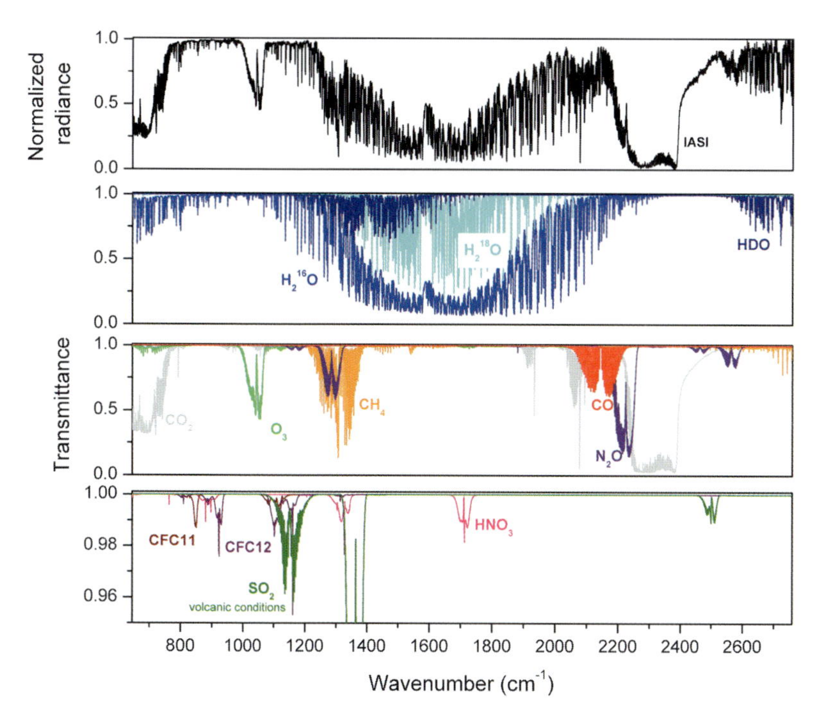

Fig. 3.6 *Top panel*: radiance atmospheric spectrum (in normalized units) recorded by IASI/MetOp, west of Australia, on 20th December 20, 2006. *Middle panels*: Nadir-looking radiative transfer transmittance simulations to identify of the main absorbing gases; *Lower panel*: and the weaker absorbers (Clerbaux et al. 2009).

capability to sound the upper-troposphere/lower stratosphere was exploited by several authors (Rinsland et al. 2006; Dufour et al. 2007; Coheur et al. 2007), see Fig. 3.7 for an illustration of the seasonal variation of CO at 16.5 km.

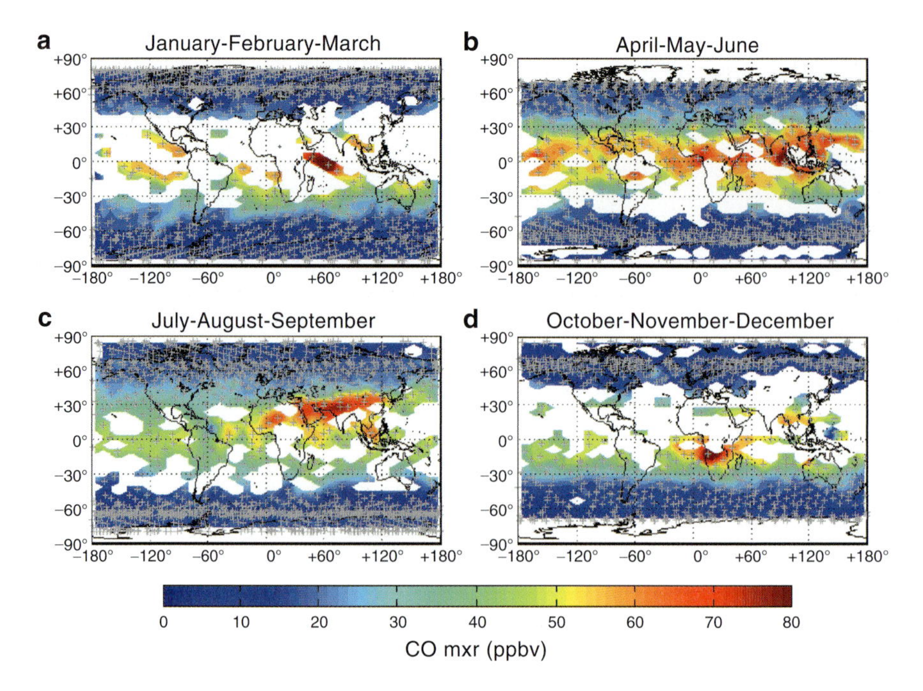

Fig. 3.7 ACE-FTS CO seasonal measurements in 2005 at 16.5 km. The data are interpolated to a 4° latitude × 8° longitude grid. The grey crosses indicate the ACE-FTS measurement locations. Note that the tropical latitudes are not well covered in Jan-Feb-March and Oct-Nov-Dec as the satellite orbit was optimized to study the polar regions in winter (Clerbaux et al. 2008b).

b MIPAS

The Michelson Interferometer for Passive Atmospheric Sounding instrument (MIPAS, see Appendix A) was launched by ESA on ENVISAT into a polar sun-synchronous orbit in March 2002. MIPAS is a Fourier transform infrared spectrometer (Fischer et al. 2008) for the detection of limb emission spectra in the middle and upper atmosphere. It observes a wide spectral range (4.15–14.6 μm, 685–2,410 cm^{-1}) with high spectral resolution (0.025 cm^{-1} unapodized). The primary geophysical parameters of interest are vertical profiles of atmospheric pressure, temperature, and volume mixing ratios of at least 25 trace constituents.

MIPAS observes the emitted radiance from the atmosphere in the limb, i.e. it is most sensitive to the atmospheric signal emitted from the tangent altitude layer and, when a limb sequence with a discrete set of different tangent altitudes is acquired, and it allows the determination of the vertical profiles of atmospheric parameters. MIPAS can measure atmospheric parameters in the altitude range from 5 to 160 km with vertical steps size between 1 and 8 km, respectively. A detailed description of the calibration and characterization of the instrument is given by Kleinert et al. (2007).

The operational MIPAS data processing by ESA generates global distributions of temperature and six key species (O_3, H_2O, CH_4, N_2O, NO_2, HNO_3) while scientific data processing has already proven that many more trace gases can be derived from the mid-infrared spectra (NO, N_2O_5, HNO_4, $ClONO_2$, chlorine monoxide, ClO, HOCl, $BrONO_2$, H_2CO, CO, CFCs, NH_3, OCS, SO_2, SF_6, PAN, HCN, C_2H_6, C_2H_2, H_2O_2, HDO, and O_3 iso-topologues). Furthermore, the MIPAS broadband spectra have been used for the measurement of aerosols and cloud particles (PSCs, cirrus).

Many interesting results have already been achieved from MIPAS measurements (Fig. 3.8). They include contributions to a better understanding of atmospheric processes such as pollution of the upper troposphere, troposphere-stratosphere exchange, chemistry and dynamics of the stratosphere, stratospheric ozone depletion, down-ward transport from the mesosphere into the stratosphere, interaction between varying solar radiation and atmospheric composition, and non-LTE (Local Thermal Equilibrium) effects.

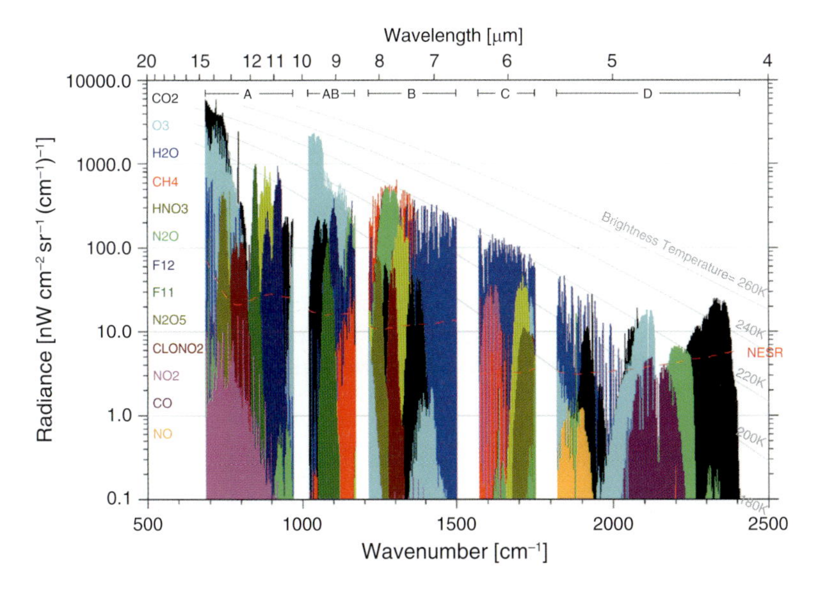

Fig. 3.8 Overview of the spectral coverage and most significant molecules covered by the five channels (A, AB, B, C, and D) of the MIPAS instrument on board ENVISAT (limb geometry, synthetic spectrum calculated for a tangent height of 12 km). Also shown are the emission curves (*grey curves*) corresponding to blackbody radiation at temperatures in the 180–260 K range. The *dashed red line* indicates the noise level (Noise-Equivalent Spectral Radiance NESR) of MIPAS for one single spectrum (Fischer et al. 2008).

Due to problems with the constant movement of the retro-reflectors of the interferometer, the measurements of MIPAS have been performed with reduced spectral resolution and a reduced duty cycle since the beginning of 2005. The spectral resolution has been reduced to 40% which in turn allows for a higher spatial resolution (vertically and horizontally). The duty cycle was further reduced later to about 35%. Nevertheless, MIPAS has detected in this mode more than 500 profiles of various atmospheric parameters, on average, every day. More recently it was found that the problems with the moving retro-reflectors are decreasing, and therefore the duty cycle has been increased step-by-step to 100% since December 2007.

c HIRDLS

The High Resolution Dynamics Limb Sounder (HIRDLS, see Appendix A) is one of the instruments on the Aura spacecraft, which was launched into a near polar, sun-synchronous orbit with a period of approximately 100 min. HIRDLS is a multi-channel, infrared radiometer designed to measure radiated thermal emissions from the atmospheric limb at various spectral intervals in the range 6–17 μm, chosen to correspond to specific gases and atmospheric "windows". The final output is a set of global 3-D fields of atmospheric temperature, several minor constituents and geostrophic winds. The instrument provides measurements of temperature, trace constituents and aerosols from the middle troposphere to the mesosphere, with a key attribute of high vertical resolution. HIRDLS also yields measurements of atmospheric aerosols and cirrus clouds, as well as unique measurements of sub-visible cirrus.

After launch, activation of the HIRDLS instrument revealed that the optical path was blocked so that only 20% of the aperture could view the Earth's atmosphere. Engineering studies suggest that a piece of thermal blanketing material ruptured from the back of the instrument during the explosive decompression of launch. Attempts to remove this material failed. However, even with the 80% blockage, measurements at high vertical resolution can be made at one scan angle. An example of HIRDLS measurements is provided in Fig. 3.9. The HIRDLS chopper stopped on 17th March 2008 and attempts to restart it have been unsuccessful.

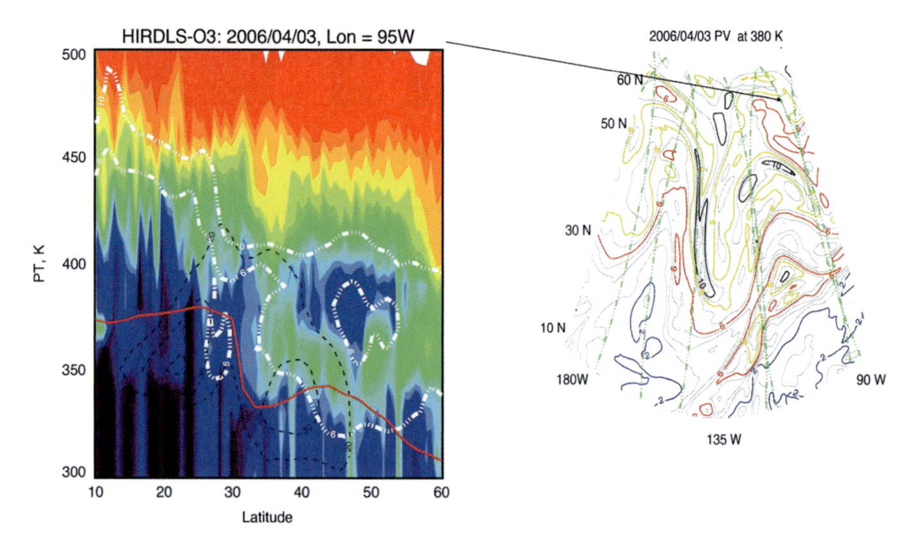

Fig. 3.9 Ozone tropospheric intrusion event as detected from HIRDLS/ Aura on 2nd, 3rd and 4th April 2006 (Courtesy to the HIRDLS Science team, available from http://www.eos.ucar.edu/hirdls/).

3.4.2 Nadir Looking Instruments

a IMG

As mentioned in Section 3.2.2, IMG/ADEOS (see Appendix A) used a nadir-viewing Fourier transform interferometer that recorded the thermal emission of the Earth-atmosphere system between 600 and 3,030 cm^{-1}, with a maximum optical path difference of 10 cm (i.e. 0.1 cm^{-1} unapodized spectral resolution) to sound the atmosphere (Kobayashi et al. 1999).

ADEOS was a sun-synchronous (equator local crossing time at descending node at 10:30 a.m.), ground track repeat, polar-orbiting satellite. The instrument performed a global coverage of the Earth, making 14 1/4 orbits per day with a series of six successive measurements separated by 86 km (every 10 s) along the track. The footprint on the ground was 8 $\times$ 8 km, in three spectral bands, corresponding to three different detectors and three geographically adjacent footprints. Due to the large data rate, the operational mode of IMG was set to 4 days operation/10 days halt alternation, except for one specific period from 1st–10th April 1997 for which 10 consecutive days were available.

b MOPITT

The MOPITT instrument (see Appendix A), Measurements Of Pollution In The Troposphere, is a correlation radiometer instrument with channels at around 4.7 μm and 2.2 μm. The original mission was targeted at CO and CH_4, but only CO measurements have been produced. However the mission, launched in December 1999 has produced, at the time of writing, a ten year record of CO over the planet. MOPITT has a 22 $\times$ 22 km pixel and a scanning swath of ~650 km (29 pixels). It produces continuous coverage within that swath as the satellite flies in a sun-synchronous orbit at 705 km and 98.4° inclination. The planet is mostly covered in about 4 days and the ground coverage has a 16-day repeat cycle. A typical plot of MOPITT L2 "total column" amount of CO over the globe is shown in Fig. 3.10. It is a composite of data taken between 28th November and 6th August 2000 gridded to 1° $\times$ 1°. The gaps in the data are mainly the result of persistent cloud cover in those regions. Biomass burning events in South America and Africa can clearly be seen and a plume of CO extends around the planet eventually fetching up against the Andes.

The MOPITT data products released are from the CO 4.7 μm channels and represent vertical profiles of CO mixing ratios. On the global and seasonal scales the climatology of the CO distribution has been obtained using MOPITT CO data depicting major sources of CO emission on the planet (Edwards et al. 2004; 2006a). Edwards et al. (2006b) showed also that inter-annual components of the CO variation over marine continents and northern Australia are well correlated with El Niño events because dry conditions favour forest fires. On the synoptic scale

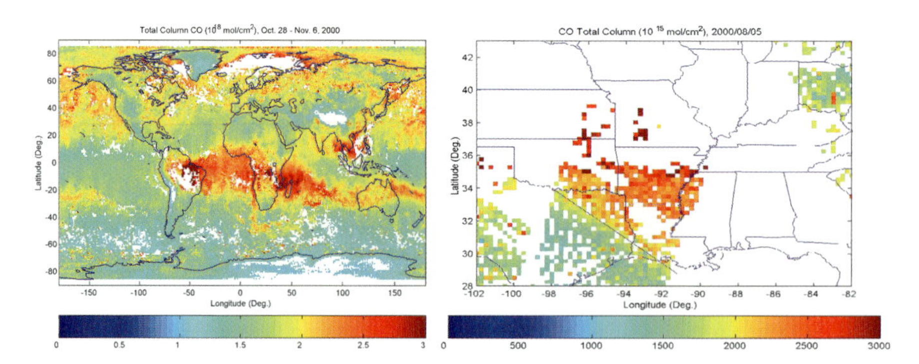

Fig. 3.10 *Left*: Global distribution of MOPITT L2 CO total column amount (8 months average). *Right*: frontal system over Texas on 5th August 2000. (Liu et al. 2006).

large horizontal gradients of CO identified in the MOPITT data have been associated with vertical and horizontal transport of air with different CO concentrations under distinctive meteorological conditions (Liu et al. 2006). Fig. 3.10 shows a frontal system over Texas in May 2000. The transition from a relatively polluted air mass to the south and the cleaner mass to the north is easily seen. The capability of observing vertical structure of tropospheric CO by MOPITT has been demonstrated in a number of papers. There were examples of strong enhancement of upper tropospheric CO in the Asian summer monsoon region due to deep convective transport (Kar et al. 2004), and enhanced CO over the Zagros mountains in Iran generated in a process of mountain venting (Kar et al. 2006). MOPITT data have also been used to isolate plumes of CO in several cities and urban areas where the CO emissions are mainly anthropological (Clerbaux et al. 2008a). A recent study using MOPITT 2.3 μm channels has reported encouraging results in the retrieval of the CO total column in reflected sunlight (Deeter et al. 2009).

c AIRS

Launched into Earth-orbit in May 2002, the Atmospheric Infrared Sounder, AIRS is one of six instruments on board the Aqua satellite, part of the NASA Earth Observing System (see Appendix A). It observes the global water and energy cycles, climate variation and trends, and the response of the climate system to increased greenhouse gases.

AIRS is designed to create three dimensional maps of air and surface temperature, water vapour, and cloud properties. AIRS has 2378 spectral channels in the range 3.74–15.4 μm and good spectral resolution ($\lambda/\Delta\lambda$ ~1,200). It provides accurate information on the vertical profiles of atmospheric temperature and moisture. AIRS can also measure trace greenhouse gases such as O_3, CO, and to some extent CO_2 (see Fig. 3.11), and CH_4.

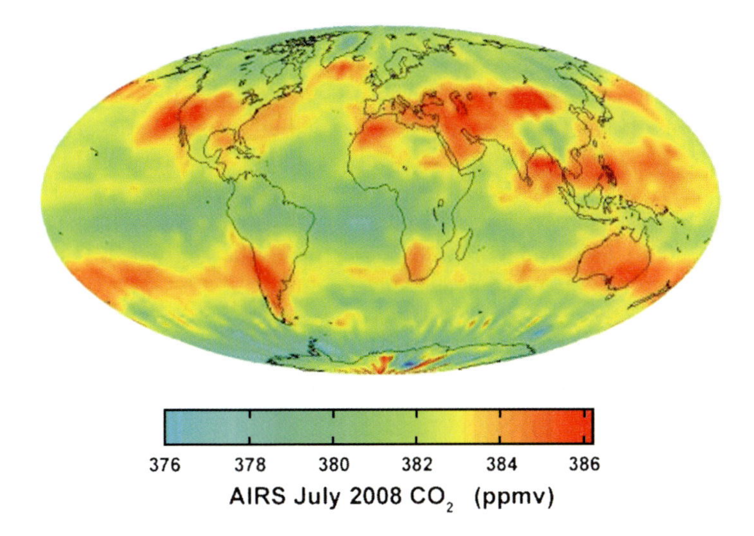

Fig. 3.11 An image created with data acquired by AIRS during July 2008. It shows large scale patterns of CO_2 concentrations that are transported around the Earth by the general circulation of the atmosphere. The effect of the northern hemisphere mid-latitude jet stream is to set the northern limit of enhanced CO_2. The zonal flow of the southern hemisphere mid-latitude jet stream results in a belt of enhanced CO_2 girdling the globe, fed by biogenesis activity in South America, forest fires in both South America and central Africa, the clusters of gasification plants in South Africa and power generation in southeastern Australia. (Image courtesy of NASA/JPL, http://www.nasa.gov/topics/earth/features/airs-20081009.html).

d TES

The Tropospheric Emission Spectrometer (TES, see Appendix A) was launched on the Aura satellite in July 2004 into a sun-synchronous orbit with an equator crossing time of 13:43. TES is a FTS with a resolution of 0.015 cm^{-1} (minimum, unapodized) and a spectral range of 650–2,250 cm^{-1} in a number of bands. It is designed to operate in both limb viewing and nadir viewing modes, although most data have been collected in the nadir viewing mode with a resolution of 0.06 cm^{-1}(unapodized). The entire instrument is cooled with a set of active coolers. The detectors consist of 16×1 arrays aligned along the direction of motion in nadir mode and vertically in limb mode. The 16 pixels on the ground in nadir view an area of 26×42 km^2. The instrument is capable of measuring a large range of molecules including O_3, CO, CO_2, CH_4, NO, N_2O, NO_2, SO_2, NH_3, HNO_3, CFCs etc. The instrument is also capable of making measurements in a number of modes that offer different combinations of coverage and measurement density including a "global survey" mode consisting of measurements about 5° apart along the orbit track and a step/stare mode that increases the measurement density to ~0.4° apart or about 6 s between observations.

The cooled spectrometer has a sufficient signal to noise ratio and spectral resolution to allow the separation of the tropospheric component of gases from the stratospheric component and also to provide some vertical resolution of the concentrations. It has thus been possible to measure tropospheric ozone directly (Jourdain et al. 2007) with some vertical resolution, as shown in Fig. 3.12. In addition it has been possible to detect and measure isotopes of water (Worden et al. 2007b) and also very low concentrations of gases such as NH_3 and CH_3OH (Beer et al. 2008). For several of these measurements the step/stare observing mode was used to enhance the geographical coverage.

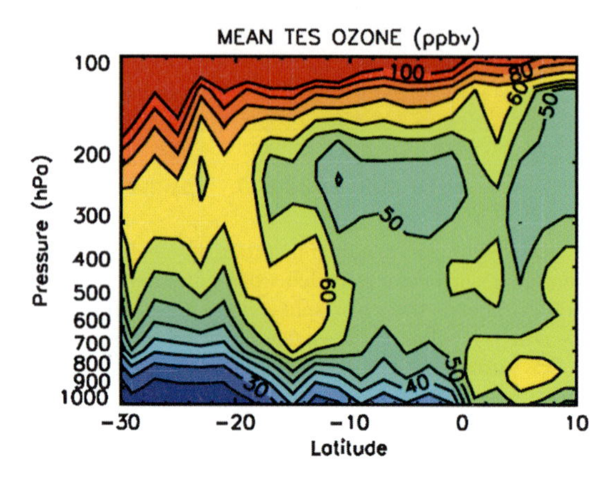

Fig. 3.12 Cross-section of O_3 mixing ratios in ppbv for longitudes over the tropical Atlantic Ocean for 22nd to 25th January 2005. DOFS for the troposphere for these measurements is ~1.5 (Jourdain et al. 2007).

e IASI

IASI, the Infrared Atmospheric Sounding Interferometer (see Appendix A), is a series of tropospheric remote sensors to be carried for a period of 14 years on the Metop-A, B, and C weather satellites deployed as part of the European Polar System which started in 2006. It is a joint undertaking of the French Space agency CNES (Centre National d'Etudes Spatiales) and EUMETSAT, the European organization for the exploitation of meteorological satellites, with CNES managing the instrumental development part and EUMETSAT operating the instrument in orbit.

The instrument comprises a FTS and an imaging system, designed to measure the infrared spectrum emitted by the Earth in the thermal infrared using a nadir geometry. The instrument provides spectra of high radiometric quality at 0.5 cm^{-1} spectral resolution, from 645 to 2,760 cm^{-1}. IASI has a square field of view sampled by a matrix of 2×2 circular pixels of 12 km each and is providing measurements each 25 km at nadir with a good horizontal coverage due to its ability to scan across track with a swath width of $\pm$1,026 km.

IASI provides improved infrared soundings of the temperature profiles in the troposphere and lower stratosphere, moisture profiles in the troposphere, as well as some of the chemical components playing a key role in climate monitoring, global change and atmospheric chemistry (Clerbaux et al. 2009). From the geophysical products that can be derived from the IASI spectra four classes of compounds can be distinguished (see Fig. 3.6):

1. Absorbers with long lifetimes (>50 years) and stable atmospheric concentrations, which require accurate retrievals in order to provide useful information on their global or temporal variability (accuracy lower than a few percent is required). These are principally the strongly absorbing climate gases, namely, CO_2 (Crevoisier et al. 2009a) and N_2O (Ricaud et al. 2009). The retrieval of their concentration from IASI observations requires specific methods and averaging over time and/or space. CFC-11, CFC-12 and HCFC-22, the most abundant substitute of chlorofluorocarbons, are also detectable. Additionally the long-lived OCS is identified in the spectra (Shephard et al. 2009).
2. Strong to medium absorbers which exhibit a significant atmospheric variability (>5%) because of their reactivity (lifetimes from a few weeks to a few years), which can be observed in each individual IASI observation. These species, which contribute to tropospheric and/or stratospheric chemistry, are water vapour (H_2O and isotopologues HDO and $H_2{}^{18}O$) (Herbin et al. 2009), CH_4 (Crevoisier et al. 2009b; Razavi et al. 2009), O_3 (Eremenko et al. 2008; Boynard et al. 2009; Keim et al. 2009), CO (George et al. 2009; Turquety et al. 2009; Fortems-Cheinet et al. 2009) and HNO_3 (Wespes et al. 2009).
3. Weak absorbers that can only be detected above emission sources or in concentrated plumes owing to the good radiometric performance of IASI. These include SO_2 from volcanoes (Clarisse et al. 2008), see Fig. 3.13, NH_3 from biomass burning and intensive land use (Clarisse et al. 2009) and volatile organic compounds such as HCOOH, CH_3OH, C_2H_4 and PAN (peroxyacetyl nitrate) from biomass burning (Coheur et al. 2009).
4. Higher altitude aerosols, such as those resulting from sandstorms, volcanic eruptions or cirrus formation. They mainly manifest themselves in the 700–1,300 cm^{-1} window as broad absorption features.

3.5 Future Plans for Tropospheric Sounders

Retrieval development has been very rapid and is probably approaching the limits of the information contained in the signals detected. Improvements can be seen in vertical resolution and in error treatments and these will benefit from higher performance instrumentation. More complex, or "global" retrievals, whether of multiple species, locations or times, will be used more as computational techniques and computer power improve. Finally, data assimilation and the use of models in retrievals will probably become more routine. These techniques currently

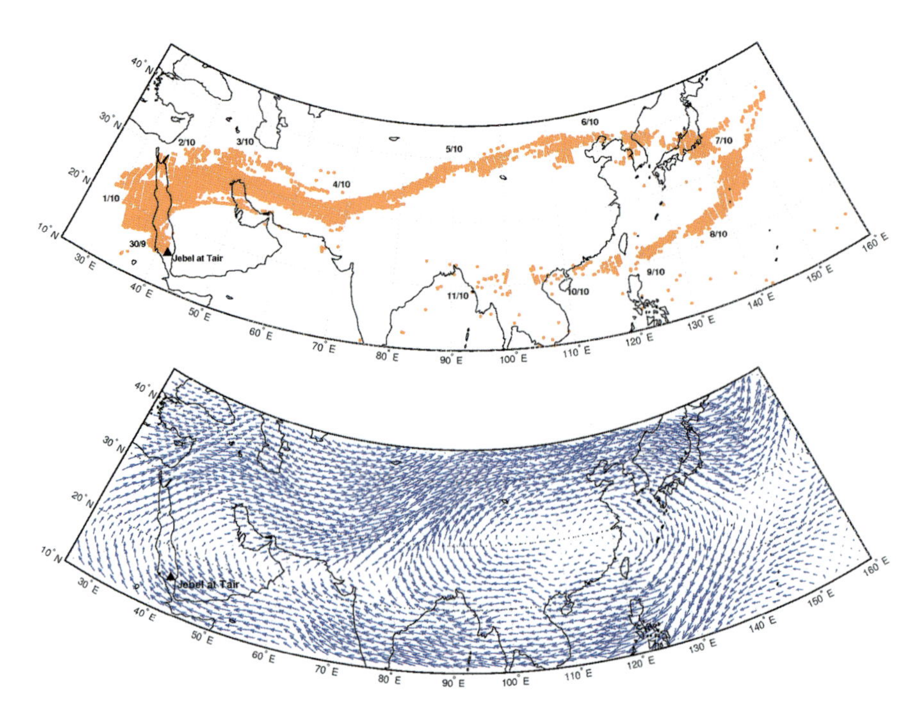

Fig. 3.13 Top: Integrated view of SO_2 as measured by IASI after the Jebel al Tair (Red Sea) eruption, on 30th September 2007. Bottom: ECMWF winds at 100 hPa (Clarisse et al. 2008).

assimilate profile data (level 2 data), but in the near future will switch to the assimilation of radiances (level 1 data) or, in some cases, raw instrument output (level 0 data).

As the population of the planet continues to increase, and with it the industrial activity, the importance of protecting air quality increases. The potential for harm and the consequences become steadily more serious. Thus the forecasting of chemical weather is likely to become main-stream in the next decade and this will require considerable efforts in both the modelling and instrumentation areas to produce timely and accurate forecasts of conditions around major population centres. Already many nations and supra-national organizations are regulating air quality, and compliance with the regulations becomes a significant societal and economic issue.

In the future, infrared remote sounding of the atmosphere will continue to develop. There are strong drivers from the scientific and regulatory areas for higher temporal coverage while maintaining good spatial coverage. There is also the transition from research instrumentation to operational instruments.

All the instruments launched so far have been on polar orbiting satellites (with the major exception of shuttle and shuttle-launched instrumentation). The majority of these satellites have also been sun synchronous. However the time resolution of these instruments is at best two measurements per location per day, and frequently lower.

In order to increase the spatial sampling, two approaches are possible: satellite arrays and higher orbits. Satellite arrays are technologically possible, requiring nothing more than duplication of equipment, costs will need to be examined, but the major issue will become cross-calibration of the array since all of these instruments are extremely sensitive to small changes in the local spacecraft environment. The most frequently discussed higher orbit it the geosynchronous/geostationary orbit which permits continuous measurements over about 25% of the globe from any one satellite. In this configuration hourly measurements at equatorial and mid-latitudes are a possibility. However in order to capture the major populated and industrial areas of the planet, at least three satellites are required and perhaps, more realistically, four. Plans to launch thermal infrared instruments on a geostationary orbit are progressing and are now at different maturity levels in the EU, USA and in Asia[1].

The polar regions are inaccessible from a geostationary orbit but well served by the instrumentation on LEO platforms. In order to measure at high latitudes either polar orbiting satellites will be required or perhaps a more exotic orbit such as the Molniya orbit, but for air quality sensing alone the costs may be prohibitive. There is however a strong motivation for such measurements as the solar zenith angle at the poles is always large and the shorter-wavelength techniques that rely on reflected solar energy are less effective. Thus longer wavelength infrared techniques become essential.

On the technological front, detector performance and array size can continue to grow, the former only modestly because of the fundamental limits of physics: when the quantum efficiency of the detection process approaches unity and one electron is produced for each photon input and so there is no further room for improvement. Array size can continue to increase and cooling requirements can be reduced through detectors that operate at higher temperatures and reduced thermal leakage in the structure. Spectral resolution can be raised in concert with higher performance detectors and more efficient optics.

It is worth noting that there are surprisingly few atmospheric chemistry missions projected for the future, both in the nadir and limb/occultation geometry, compared to those available now.

References

Beer, R., Shephard, M.W., Kulawik, S.S., Clough, S.A., Eldering, A., Bowman, K.W., Sander, S.P., Fisher, B.M., Payne, V.H., Luo, M., Osterman, G.B., and Worden, J.R., 2008, First satellite observations of lower tropospheric ammonia and methanol. *Geophys. Res. Lett.* doi:10.1029/2008GL033642.

[1]For US plans see NASA's NFS decadal study (http://decadal.gsfc.nasa.gov/missions.html)
For EU plans see the Eumetsat MTG satellite concept (http://www.eumetsat.int/Home/Main/What_We_Do/Satellites/Future_Satellites/)

Bernath, P.F., McElroy, C.T., Abrams, M.C., Boone, C.D., Butler, M., Camy-Peyret, C., Carleer, M., Clerbaux, C., Coheur, P.F., Colin, R., DeCola, P., Maziere, M.D., Drummond, J.R., Dufour, D., Evans, W.F.J., Fast, H., Fussen, D., Gilbert, K., Jennings, D.E., Llewellyn, E. J., Lowe, R.P., Mahieu, E., McConnell, J.C., McHugh, M., McLeod, S.D., Michaud, R., Midwinter, C., Nassar, R., Nichitiu, F., Nowlan, C., Rinsland, C.P., Rochon, Y.J., Rowlands, N., Semeniuk, K., Simon, P., Skelton, R., Sloan, J.J., Soucy, M.-A., Strong, K., Tremblay, P., Turnbull, D., Walker, K.A., Walkty, I., Wardle, D.A., Wehrle, V., Zander, R., and Zou, J., 2005, Atmospheric Chemistry Experiment (ACE): mission overview. *Geophys. Res. Lett.* doi:10.1029/2005GL022386.

Bowman, K.W., C.D. Rodgers, S.S. Kulawik, J. Worden, E. Sarkissian, G. Osterman, T. Steck, Ming Lou, A. Eldering, M. Shephard, H. Worden, M. Lampel, S. Clough, P. Brown, C. Rinsland, M. Gunson and R. Beer, 2006, Tropospheric Emission Spectrometer: Retrieval Method and Error Analysis. *IEEE Trans. Geosci. Remote Sens.*, **44**, 1297–1307.

Boynard, A., C. Clerbaux, P.-F. Coheur, D. Hurtmans, S. Turquety, M. George, J. Hadji-Lazaro, C. Keim and J. Meyer-Arnek, 2009, Measurements of total and tropospheric ozone from IASI: comparison with correlative satellite, ground-based and ozonesonde observations. *Atmos.. Chem. Phys.*, **9**, 6255–6271.

Clarisse, L., P.F. Coheur, A.J. Prata, D. Hurtmans, A. Razavi, T. Phulpin, J. Hadji-Lazaro and C. Clerbaux, 2008, Tracking and quantifying volcanic SO_2 with IASI, the September 2007 eruption at Jebel at Tair. *Atmos. Chem. Phys.*, **8**, 7723–7734.

Clarisse, L., Clerbaux, C., Dentener, F., Hurtmans, D., and Coheur, P.-F., 2009, Global ammonia distribution derived from infrared satellite observations. *Nature Geoscience,* doi:10.1038/ngeo551.

Clerbaux, C., J. Hadji-Lazaro, S. Turquety, G. Mégie and P.-F. Coheur, 2003, Trace gas measurements from infrared satellite for chemistry and climate applications. *Atmos. Chem. Phys.,* **3**, 1495–1508.

Clerbaux, C., Edwards, D.P., Deeter, M., Emmons, L., Lamarque, J-F., Tie, X.X., Massie, S.T., and Gille, J., 2008a, Carbon monoxide pollution from cities and urban areas observed by the Terra/MOPITT mission. *Geophys. Res. Lett.* doi:10.1029/2007GL032300.

Clerbaux, C., M. George, S. Turquety, K.A. Walker, B. Barret, P. Bernath, C. Boone, T. Borsdorff, J.P. Cammas, V. Catoire, M. Coffey, P.-F. Coheur, M. Deeter, M. De Mazière, J. Drummond, P. Duchatelet, E. Dupuy, R. de Zafra, F. Eddounia, D.P. Edwards, L. Emmons, B. Funke, J. Gille, D.W.T. Griffith, J. Hannigan, F. Hase, M. Höpfner, N. Jones, A. Kagawa, Y. Kasai, I. Kramer, E. Le Flochmoën, N.J. Livesey, M. López-Puertas, M. Luo, E. Mahieu, D. Murtagh, P. Nédélec, A. Pazmino, H. Pumphrey, P. Ricaud, C.P. Rinsland, C. Robert, M. Schneider, C. Senten, G. Stiller, A. Strandberg, K. Strong, R. Sussmann, V. Thouret, J. Urban and A. Wiacek, 2008b, CO measurements from the ACE-FTS satellite instrument: data analysis and validation using ground-based, airborne and spaceborne observations. *Atmos. Chem. Phys.*, **8**, 2569–2594.

Clerbaux, C., A. Boynard, L. Clarisse, M. George, J. Hadji-Lazaro, H. Herbin, D. Hurtmans, M. Pommier, A. Razavi, S. Turquety, C. Wespes and P.-F. Coheur, 2009, Monitoring of atmospheric composition using the thermal infrared IASI/MetOp sounder. *Atmos. Chem. Phys.*, **9**, 6041–6054.

Coheur, P.-F., H. Herbin, C. Clerbaux, D. Hurtmans, W. Wespes, M. Carleer, S. Turquety, C.P. Rinsland, J. Remedios, D. Hauglustaine, C.D. Boone and P.F. Bernath, 2007, ACE-FTS observation of a young biomass burning plume: first reported measurements of C_2H_4, C_3H_6O, H_2CO and PAN by infrared occultation from space. *Atmos. Chem. Phys.*, **7**, 5437–5446.

Coheur, P.-F., L. Clarisse, S. Turquety, D. Hurtmans and C. Clerbaux, 2009, IASI measurements of reactive trace species in biomass burning plumes. *Atmos. Chem. Phys.,* **9**, 5655–5667.

Crevoisier, C., A. Chédin, H. Matsueda, T. Machida, R. Armante and N. A. Scott, 2009a, First year of upper tropospheric integrated content of CO_2 from IASI hyperspectral infrared observations *Atmos. Chem. Phys.*, **9**, 4797–4810.

Crevoisier, C., D. Nobileau, A. M. Fiore, R. Armante, A. Chédin and N. A. Scott, 2009b, Tropospheric methane in the tropics – first year from IASI hyperspectral infrared observations. *Atmos. Chem. Phys.*, **9**, 6337–6350.

Deeter, M.N., Edwards, D.P., Gille, J.C., and Drummond, J.R., 2009, CO retrievals based on MOPITT near-infrared observations. J. Geophys. Res. doi:10.1029/ 2008JD010872.

Doicu, A., Schreier, F., and Hess, M., 2004, Iterative regularization methods for atmospheric remote sensing. *J. Quant. Spectrosc. Radiat. Transfer*. doi:10.1016/S0022-4073(02)00292–3.

Dufour, G., S. Szopa, D.A. Hauglustaine, C.D. Boone, C.P. Rinsland and P.F. Bernath, 2007, The influence of biogenic emissions on upper-tropospheric methanol as revealed from space. *Atmos. Chem. Phys.*, **7**, 6119–6129.

Edwards, D.P, Emmons, L.K., Hauglustaine, D.A., Chu, A., Gille, J.C., Kaufman, Y.J., Petron, G., Yurganov, L.N., Giglio, L., Deeter, M.N., Yudin, V., Ziskin, D.C., Warner, J., Lamarque, D.F., Francis, G.L., Ho, S.P., Mao, D., Chen, J., Grechko, E.I., and Drummond, J. R., 2004, Observations of Carbon Monoxide and Aerosol From the Terra Satellite: Northern Hemisphere Variability. *J. Geophys. Res.* doi:10.1029/2004JD0047272004.

Edwards, D.P., Pétron, G., Novelli, P.C., Emmons, L.K., Gille, J.C., and Drummond, J.R., 2006a, Southern Hemisphere carbon monoxide interannual variability observed by Terra/Measurement of Pollution in the Troposphere (MOPITT). *J. Geophys. Res.* doi:10.1029/ 2006JD007079.

Edwards, D.P., Emmons, L.K., Gille, J.C., Chu, A., Attie, J.-L., Giglio, L., Wood, S.W., Haywood, J., Deeter, M.N., Massie, S.T., Ziskin, D.C., and Drummond, J. R., 2006b, Satellite-observed pollution from Southern Hemisphere biomass burning. *J. Geophys. Res.* doi:10.1029/ 2005JD006655.

Eremenko, M., Dufour, G., Foret, G., Keim, C., Orphal, J., Beekmann, M., Bergametti, G., and Flaud, J.-M., 2008, Tropospheric ozone distributions over Europe during the heat wave in July 2007 observed from infrared Nadir spectra measured by IASI. *Geophys. Res. Lett.* doi:10.1029/ 2008GL034803.

Fischer, H., M. Birk, C. Blom, B. Carli, M. Carlotti, T. von Clarmann, L. Delbouille, A. Dudhia, D. Ehhalt, M. Endemann, J.M. Flaud, R. Gessner, A. Kleinert, R. Koopmann, J. Langen, M. López-Puertas, P. Mosner, H. Nett, H. Oelhaf, G. Perron, J. Remedios, M. Ridolfi, G. Stiller and R. Zander, 2008, MIPAS: an instrument for atmospheric and climate research. *Atmos. Chem. Phys.*, **8**, 2151–2188.

Flaud, J.-M. and H. Oelhaf, 2004, Infrared spectroscopy and the terrestrial atmosphere. *C. R. Physique* 5, 259–271.

Fortems-Cheinet, A., F. Chevallier, I. Pison, P. Bousquet, C. Carouge, C. Clerbaux, P.F. Coheur, M. George, D. Hurtmans and S. Szopa, 2009, On the capability of IASI measurements to inform about CO surface emissions. *Atmos. Chem. Phys.* Discuss., **9**, 7505–7529.

George, M., C. Clerbaux, D. Hurtmans, S. Turquety, P.-F. Coheur, M. Pommier, J. Hadji-Lazaro, D.P. Edwards, H. Worden, M. Luo, C. Rinsland and W. McMillan, 2009, Carbon monoxide distributions from the IASI/METOP mission: evaluation with other space-borne remote sensors. *Atmos. Chem. Phys. Discuss.*, **9**, 9793–9822.

Gunson, M.R., M.M. Abbas, M.C. Abrams, M. Allen, L.R. Brown, T.L. Brown, A.Y. Chang, A. Goldman, F.W. Irion, L.L. Lowes, E. Mahieu, G.L. Manney, H.A. Michelsen, M.J. Newchurch, C.P. Rinsland, R.J. Salawitch, G.P. Stiller, G.C. Toon, Y.L. Yung and R. Zander, 1996, The Atmospheric Trace Molecule Spectroscopy (ATMOS) experiment: Deployment on the ATLAS Space Shuttle missions. *Geophys. Res. Lett.*, **23**, 2333–2336.

Herbin, H., D. Hurtmans, C. Clerbaux, L. Clarisse and P.-F. Coheur, 2009, $H_2{}^{16}O$ and HDO measurements with IASI/MetOp. *Atmos. Chem. Phys. Discuss.*, **9**, 9267–9290.

Jacquinet-Husson, N., Scott, N.A., Chédin, A., Crépeau, L., Armante, R., Capelle, V., Orphal, J., Coustenis, A., Boonne, C., Poulet-Crovisier, N., Barbe, A., Birk, M., Brown, L.R., Camy-Peyret, C., Claveau, C., Chance, K., Christidis, N., Clerbaux, C., Coheur, P.F., Dana, V., Daumont, L., De Backer-Barilly, M.R., Di Lonardo, G., Flaud, J.M., Goldman, A., Hamdouni, A., Hess, M., Hurley, M.D., Jacquemart, D., Kleiner, I., Köpke, P., Mandin, J.Y., Massie, S.,

Mikhailenko, S., Nemtchinov, V., Nikitin, A., Newnham, D., Perrin, A., Perevalov, V.I., Pinnock, S., Régalia-Jarlot, L., Rinsland, C. P., Rublev, A., Schreier, F., Schult, L., Smith, K.M., Tashkun, S.A., Teffo, J.L., Toth, R.A., Tyuterev, V.I.G., Vander Auwera, J., Varanasi, P., Wagner, G, 2008, The GEISA spectroscopic database: Current and future archive for Earth and Planetary atmosphere studies. *J. Quant. Spectrosc. Rad. Transfer.* doi:10.1016/j.jqsrt.2007.12.015.

Jourdain, L., Worden, H.M., Worden, J.R., Bowman, K, Li, Q., Eldering, A., Kulawik, S.S., Osterman, G., Boersma, K.F., Fisher, B., Rinsland, C.P., Beer, R., and Gunson, M., 2007, Tropospheric vertical distribution of tropical Atlantic ozone observed by TES during the northern African biomass burning season. *Geophys. Res. Lett.* doi:10.1029/2006GL028284.

Kar, J., Bremer, H., Drummond, J.R., Rochon, Y.J., Jones, D.B.A., Nichitiu, F., Zou, J., Liu, J., Gille, J.C., Edwards, D.P., Deeter, M.N., Francis, G., Ziskin, D., and Warner, J., 2004, Evidence of Vertical Transport of Carbon Monoxide From Measurements of Pollution in the Troposphere (MOPITT). *Geophys. Res. Lett.* doi:10.1029/2004GL021128.

Kar, J., Drummond, J.R., Jones, D.B.A., Liu, J., Nichitiu, F., Zou, J., Gille, J.C., Edwards, D.P., and Deeter, M.N., 2006, Carbon monoxide (CO) maximum over the Zagros mountains in the Middle East: Signature of mountain venting? *Geophys. Res. Lett.* doi:10.1029/2006GL026231.

Keim, C., G.Y. Liu, C.E. Blom, H. Fischer, T. Gulde, M. Höpfner, C. Piesch, F. Ravegnani, A. Roiger, H. Schlager and N. Sitnikov, 2008, Vertical profile of peroxyacetyl nitrate (PAN) from MIPAS-STR measurements over Brazil in February 2005 and its contribution to tropical UT NO_y partitioning. *Atmos. Chem. Phys.*, **8**, 4891–4902.

Keim, C., M. Eremenko, J. Orphal, G. Dufour, J.-M. Flaud, M. Höpfner, A. Boynard, C. Clerbaux, S. Payan, P.-F. Coheur, D. Hurtmans, H. Claude, H. Dier, B. Johnson, H. Kelder, R. Kivi, T. Koide, M. López Bartolomé, K. Lambkin, D. Moore, F. J. Schmidlin and R. Stübi, 2009, Tropospheric ozone from IASI: comparison of different inversion algorithms and validation with ozone sondes in the northern middle latitudes. *Atmos. Chem. Phys. Discuss.*, **9**, 11441–11479.

Kleinert, A., G. Aubertin, G. Perron, M. Birk, G. Wagner, F. Hase, H. Nett and R. Poulin, 2007, MIPAS Level 1B algorithms overview: operational processing and characterization. *Atmos. Chem. Phys.*, **7**, 1395–1406.

Kobayashi, H., A. Shimota, C. Yoshigahara, I. Yoshida, Y. Uehara and K. Kondo, Satellite-borne high-resolution FTIR for lower atmosphere sounding and its evaluation. IEEE Trans. *Geosci. Remote Sens.*, **37**, 1496–1507, 1999.

Kulawik, S.S., G. Osterman and D. Jones, 2006, Calculation of altitude-dependent Tikhonov constraints for TES nadir retrievals. *IEEE Trans. Geosci. Remote Sens.*, **44**, 1334–1342.

Liu, J., Drummond, J.R., Jones, D.B.A., Cao, Z., Bremer, H., Kar, J., Zou, J., Nichitiu, F., and Gille, J.C., 2006, Large horizontal gradients in atmospheric CO at the synoptic scale as seen by spaceborne Measurements of Pollution in the Troposphere. *J. Geophys. Res.* doi:10.1029/2005JD006076.

McCartney, E.J., 1983, Absorption and Emission by Atmospheric Gases: The Physical Processes. J. Wiley & Sons, New York.

Orphal, J., G. Bergametti, B. Beghin, J.-P. Hebert, T. Steck and J.-M. Flaud, 2005, Monitoring tropospheric pollution using infrared spectroscopy from geostationary orbit. *C. R. Physique,* **6**, 888–896.

Phillips, C., 2003, A technique for the numerical solution of certain integral equations of the first kind. *J. Assoc. Comput. Math.*, **9**, 84–97.

Razavi, A., C. Clerbaux, C. Wespes, L. Clarisse, D. Hurtmans, S. Payan, C. Camy-Peyret and P. F. Coheur, 2009, Characterization of methane retrievals from the IASI space-borne sounder. *Atmos. Chem. Phys. Discuss.*, **9**, 7615–7643.

Reichle, H.G., B.E. Anderson, V. Connors, T.C. Denkins, D.A. Forbes, B.B. Gormsen, R.L. Langenfelds, D.O. Neil, S.R. Nolf, P.C. Novelli, N.S. Pougatchev, M.M. Roell and L.P. Steele, 1999, Space shuttle based global CO measurements during April and October 1994, MAPS instrument, data reduction, and data validation, *J. Geophys. Res.*, **104**(17), 21,443–21,454.

Ricaud, P., J.-L. Attié, H. Teyssèdre, L. El Amraoui, V.-H. Peuch, M. Matricardi and P. Schluessel, 2009, Equatorial total column of nitrous oxide as measured by IASI on MetOp-A: implications for transport processes. *Atmos. Chem. Phys.*, **9**, 3947–3956.

Rinsland, C.P., Boone, C.D., Bernath, P.F., Mahieu, E., Zander, R., Dufour, G., Clerbaux, C., Turquety, S., Chiou, L., McConnell, J.C., Neary, L., and Kaminski, J.W., 2006, First space-based observations of formic acid (HCOOH): Atmospheric Chemistry Experiment austral spring 2004 and 2005 Southern Hemisphere tropical-mid-latitude upper tropospheric measurements. *Geophys. Res. Lett.* doi:10.1029/2006GL027128.

Rothman, L. S., Jacquemart, D., Barbe, A., Chris Benner, D., Birk, M., Brown, L.R., Carleer, M. R., Chackerian, C., Chance, K., Coudert, L.H., Dana, V., Devi, V.M., Flaud, J.-M., Gamache, R.R., Goldman, A., Hartmann, J.-M., Jucks, K.W., Maki, A.G., Mandin, J.-Y., Massie, S.T., Orphal, J., Perrin, A., Rinsland, C.P., Smith, M.A.H., Tennyson, J., Tolchenov, R.N., Toth, R. A., Vander Auwera, J., Varanasi, P., and Wagner, G, 2005, The HITRAN 2004 molecular spectroscopic database. *J. Quant. Spectrosc. Rad. Transfer* doi:10.1016/j.jqsrt.2004.10.008.

Rothman, L.S., Gordon, I.E., Barbe, A., Benner, D. Chris, Bernath, P.F., Birk, M., Boudon, V., Brown, L.R., Campargue, A., Champion, J.-P., Chance, K., Coudert, L.H., Dana, V., Devi, V.M., Fally, S., Flaud, J.-M., Gamache, R.R., Goldman, A., Jacquemart, D., Kleiner, I., Lacome, N., Lafferty, W.J., Mandin, J.-Y., Massie, S.T., Mikhailenko, S. N., Miller, C.E., Moazzen-Ahmadi, N., Naumenko, O.V., Nikitin, A.V., Orphal, J., Perevalov, V.I., Perrin, A., Predoi-Cross, A., Rinsland, C.P., Rotger, M., Šimečková., M., Smith, M.A.H., Sung, K., Tashkun, S.A., Tennyson, J., Toth, R.A., Vandaele, A.C., Vander Auwera, J, 2009, The HITRAN 2008 molecular spectroscopic database. *J. Quant. Spectrosc. Rad. Transfer* doi:10.1016/j.jqsrt.2009.02.013.

Rodgers, C. D., 1976, Retrieval of atmospheric temperature and composition from remote measurements of thermal radiation. *Rev. Geophys.*, **14**(4), 609–624.

Rodgers, C.D., 2000, Inverse methods for atmospheric sounding: Theory and Practice. Amosph. Oceanic Planet. Phys., **2**, World Sci., River Edge, N.J.

Shephard, M.W., S.A. Clough, V.H. Payne, W.L. Smith, S. Kireev and K.E. Cady-Pereira, 2009, Performance of the line-by-line radiative transfer model (LBLRTM) for temperature and species retrievals: IASI case studies from JAIVEx. *Atmos. Chem. Phys. Discuss.*, **9**, 9313–9366.

Steck, T., 2002, Methods for determining regularization for atmospheric retrieval problems, *Appl. Opt.*, **41**, 1788–1797.

Tarantola, A., 2005, Inverse Problem Theory, siam, http://www.ipgp.jussieu.fr/~tarantola/Files/Professional/Books/InverseProblemTheory.pdf

Tikhonov, A., 1963, On the solution of incorrectly stated problems and a method of regularization. *Dokl. Acad. Nauk SSSR*, **151**, 501–504.

Turquety, S., Hadji-Lazaro, J., Clerbaux, C., Hauglustaine, D.A., Clough, S.A., Cassé, V., Schlüssel, P., and Mégie, G., 2004, Operational trace gas retrieval algorithm for the Infrared Atmospheric Sounding Interferometer. *J. Geophys. Res.* doi:10.1029/2004JD004821.

Turquety, S., D. Hurtmans, J. Hadji-Lazaro, P.-F. Coheur, C. Clerbaux, D. Josset and C. Tsamalis, 2009, Tracking the emission and transport of pollution from wildfires using the IASI CO retrievals: analysis of the summer 2007 Greek fires. *Atmos. Chem. Phys.*, **9**, 4897–4913.

Wespes, C., D. Hurtmans, C. Clerbaux, M.L. Santee, R.V. Martin and P.F. Coheur, 2009, Global distributions of nitric acid from IASI/MetOP measurements. *Atmos. Chem. Phys. Discuss.*, **9**, 8035–8069.

Worden, H.M., Logan, J.A., Worden, J.R., Beer, R., Bowman, K., Clough, S.A., Eldering, A., Fisher, B.M., Gunson, M.R., Herman, R.L., Kulawik, S.S., Lampel, M.C., Luo, M., Megretskaia, I.A., Osterman, G.B., Shephard, M.W., 2007a, Comparisons of Tropospheric Emission Spectrometer (TES) ozone profiles to ozonesondes: Methods and initial results. *J. Geophys. Res.* doi:10.1029/2006JD007258.

Worden, J., Noone, D., Bowman, K., Beer, R., Eldering, A., Fisher, B., Gunson, M., Goldman, A., Herman, R., Kulawik, S.S., Lampel, M., Osterman, G., Rinsland, C., Rodgers, C., Sander, S., Shephard, M., Webster, C.R., and Worden, H., 2007b, Importance of rain evaporation and continental convection in the tropical water cycle. *Nature* doi:10.1038/nature05508.

Chapter 4
Microwave Absorption, Emission and Scattering: Trace Gases and Meteorological Parameters

Klaus Kunzi, Peter Bauer, Reima Eresmaa, Patrick Eriksson, Sean B. Healy, Alberto Mugnai, Nathaniel Livesey, Catherine Prigent, Eric A. Smith and Graeme Stephens

4.1 Introduction

Space-borne remote sensing techniques are widely used today to investigate the atmosphere, both by operational and experimental instruments on a large number of satellites. Sensors operating in the microwave range, defined as being wavelengths from 10 to 0.1 cm, frequency 3–300 GHz *(microwaves also comprise sub-millimetre waves or frequencies up to 3,000 GHz)* of the electromagnetic spectrum were among the first instruments used for this purpose from the ground and on board air- and space-borne platforms. Those instruments measured the thermal emission from a molecular resonance or used the absorption and scattering properties of water droplets or ice crystals to obtain information on atmospheric parameters and composition.

K. Kunzi (✉)
University of Bremen, Institute of Environmental Physics, Bremen, Germany

P. Bauer, R. Eresmaa and S.B. Healy
European Centre for Medium-Range Weather Forecasts (ECMWF), Reading, UK

P. Eriksson
Department of Earth and Space Sciences, Chalmers University of Technology, Gothenburg, Sweden

A. Mugnai
Istituto di Scienza dell'Atmosfera e del Clima, CNR, Roma, Italy

N. Livesey
Microwave Atmospheric Science Team, Jet Propulsion Laboratory, Pasadena, CA, USA

C. Prigent
CNRS, Observatoire de Paris, Paris, France

E.A. Smith
NASA/Goddard Space Flight Center, Greenbelt, MD, USA

G. Stephens
Colorado State University, Fort Collins, CO, USA

J.P. Burrows et al. (eds.), *The Remote Sensing of Tropospheric Composition from Space*, Physics of Earth and Space Environments, DOI 10.1007/978-3-642-14791-3_4,

Measuring the atmospheric temperature profile using space-borne sensors observing thermal emission from molecular oxygen, O_2, was first proposed by Meeks (1961) and applied for the first time to data collected by the Nimbus-E Microwave Sensor (NEMS) on Nimbus-5 (Waters et al. 1975). The total water vapour content was first measured with combined 15 and 22 GHz radiometers on the Mariner-2 probe on Venus (Barath et al. 1964).

In the nineteen seventies, the Nimbus-E Microwave Sounder (NEMS) and the Scanning Microwave Sounder (SCAMS), flown on the Nimbus-5 and 6 satellites respectively, demonstrated the advantage of this type of instrumentation to retrieve parameters such as the atmospheric temperature profile, amounts of water vapour and liquid water. The frequency bands and observing geometries selected for these early instruments are still used today by operational sensors on meteorological satellites in polar orbits.

Microwave limb sensors, designed mainly to measure stratospheric composition, have become prominent since the launch of the Upper Atmosphere Research Satellite UARS by NASA. Such instruments also provide significant information on water vapour and minor constituents in the middle and upper troposphere with good vertical resolution.

Recently active sensors (radar) were used successfully to obtain information about the distribution of hydrometeors (suspended or falling liquid and solid particles). More recently also the signals from satellites of the Global Positioning System (GPS) are being used to retrieve temperature profiles and amounts of water vapour in the atmosphere.

Overall sensors operating in the microwave spectral range have evolved in the past decades to be an extremely successful atmospheric sounding technique, and these types of sensors are expected to remain cornerstones for remote sensing applications in atmospheric research, climatology and meteorology.

4.2 Atmospheric Remote Sensing in the Microwave range

4.2.1 Vector and Scalar Radiative Transfer

Three processes influence the intensity of electromagnetic radiation propagating through the troposphere: absorption, emission and scattering. The assumption of local thermodynamic equilibrium, as explained by Goody and Yung (1989), is valid for microwave radiation. As a consequence the atmospheric emission depends only on the local temperature and the local absorption. Gaseous absorption, and the associated emission, has generally no polarization dependency, one exception being the emission by the oxygen molecule, O_2.

The interaction of the O_2 with the radiation field is due to the magnetic dipole moment of two unpaired electrons; Zeeman splitting in the Earth magnetic field therefore affects the O_2 lines. The effect has to be considered when the lines get very narrow in the upper stratosphere and above, making the thermal radiation dependent on the terrestrial magnetic field strength and orientation with respect to the polarization of the receiving antenna.

On the other hand, the interaction between radiation and particles, as well as the Earth's surface, normally causes the intensity to vary with the observed polarization. Different formalisms to describe the complete polarization state have been developed. In this work the Stokes vector (Bohren and Huffman 1998) will be used. In this case a real vector describes the intensity $\vec{I}$ and the general expression for the radiative transfer equation (Chandrasekhar 1960) can be written as:

$$\frac{d\vec{I}(m)}{ds} = -\bar{K}\vec{I} + \vec{a}B + \int_{4\pi} \bar{P}(m, m')\vec{I}(m')dm' \tag{4.1}$$

where m is the coordinate along the line of propagation, s is the distance along this direction, $\bar{K}$ is a matrix describing the extinction, i.e. the sum of absorption and scattering, $\vec{a}$ is the vector describing absorption, B is the Planck function, and the matrix $\bar{P}$ describes the scattering between the m' and m directions, normally denoted as the phase matrix.

The radiation measured by microwave radiometers is expressed as an equivalent black body temperature or brightness temperature T_B. The brightness temperature is defined as the physical temperature T_P of a black body emitting the same radiative power as that received from the observed target. For temperatures encountered in the Earth atmosphere and for frequencies below 200 GHz, the Planck law is adequately approximated for many applications by the Rayleigh-Jeans law, and the relationship between T_B and T_P becomes linear.

The three terms on the right-hand side of Eq. 4.1 describe respectively extinction, emission and scattering into line-of-sight. It is assumed that particles scattering the electromagnetic radiation are randomly and sparsely distributed and that the scattering by different particles is incoherent. The matrices $\bar{P}$ and $\bar{K}$ can then be set to represent ensemble properties. More detailed descriptions of the vector radiative transfer equation are found in Mishchenko et al. (2002) and Liou (2002).

The differential Eq. 4.1 is simplified for atmospheric conditions where particle scattering can be neglected. A scalar representation then suffices and an analytical solution is given by the following (Ulaby et al. 1981; Liou 2002):

$$I = I_l \exp\left(-\int_0^l k \cdot ds\right) + \int_0^l kB \exp\left(-\int_0^s k \cdot ds'\right) ds \tag{4.2}$$

where k is the absorption coefficient and l is the length of the propagation path. The first term on the right-hand side of Eq. 4.2 describes the intensity of the radiation at the start of the propagation path, I_l. This "background term" is the cosmic background radiation for limb observations, while for downward measurements it is the sum of the surface emission and the reflected down-welling radiation. The second term on the right-hand side of Eq. 4.2 integrates the emission along the propagation path, weighted by the transmission between the observation and emission points.

Eq. 4.1 and Eq. 4.2 describe monochromatic radiation propagating along an infinitely narrow beam. In contrast, the response of an instrument has both a frequency and an angular extension. As a result radiative transfer calculations must normally be performed for a set of frequencies and propagation directions in order to correctly incorporate the frequency and angular responses of a particular instrument.

4.2.2 Gas Absorption in the Microwave Region

For microwave wavelengths H_2O, O_2 and N_2 are the dominating absorbing gases in the troposphere. O_2 dominates the absorption below 10–20 GHz (depending on altitude), by a line complex around 60 GHz and by individual transitions starting at 118.75 GHz. Water vapour is the main absorber close to the surface. Absorption by N_2, which lacks resonant transitions, becomes significant at higher altitudes, especially for frequencies at some distance from H_2O and O_2 resonances. More detailed calculations need to consider the absorption of ozone, O_3, and possibly also other minor atmospheric constituents. The calculated zenith absorption is shown in Fig. 4.1.

The position and strength of molecular transitions in the microwave region are known with a relatively high accuracy. In contrast pressure broadening and non-resonant absorption are less well known. Line broadening resulting from thermal motion (Doppler broadening) becomes important at altitudes around the stratopause and can be neglected for the troposphere. Pressure broadening causes a line shape of van Vleck and Weisskopf type, which, for higher altitudes and frequencies can be simplified to the Lorentz line shape (Rosenkranz 1993). The HITRAN database (Rothman et al. 2005) provides a complete set of parameters with the exception of O_2 around 60 GHz where line mixing must be considered. The Jet Propulsion Laboratory (JPL) spectral catalogue (Pickett et al. 2001) does not treat pressure broadening and covers only the microwave range, but it is considered to contain more accurate data for line position and strength.

Several mechanisms cause additional absorption not covered by summing up the resonant absorption of individual molecular transitions (Rosenkranz 1993). The additional non-resonant absorption is frequently denoted as continuum absorption, and must be considered for all the three main absorbers (H_2O, O_2 and N_2). For this reason, a number of absorption models were created, which provide a combination of data for most important transitions and empirically determined terms to incorporate continuum absorption (Liebe et al. 1993; Rosenkranz 1993; 1998). A detailed review of this topic is given by Kuhn (2003).

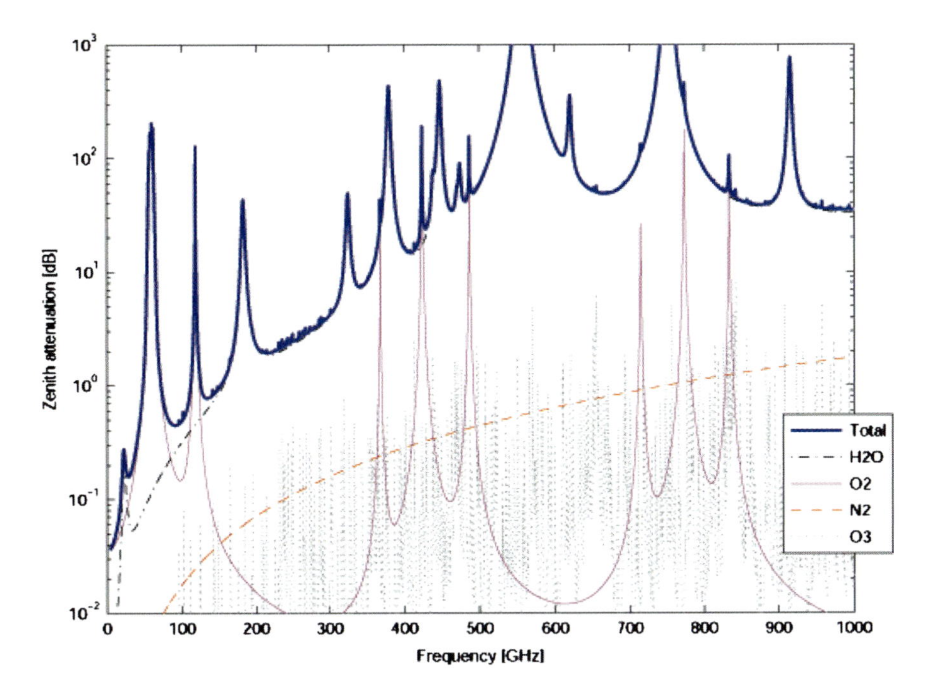

Fig. 4.1 The zenith attenuation for a mid-latitude winter scenario given in dB (Note that dB is a logarithmic measure for power ratios, 10 dB correspond to a factor of 10). The thick solid line gives the total attenuation, while the other lines show the contribution of each gas component. The water vapour profile has a liquid equivalent of 7 mm (*ca.* 2.3×10^{23} molecule/cm^2). Absorption was calculated by the Atmospheric Radiative Transfer Simulator ARTS-1.x (Eriksson and Buehler 2008). The absorption models of Rosenkranz (1998), Rosenkranz (1993) and Liebe et al. (1993) were used for H_2O, O_2 and N_2, respectively. Spectroscopic data for ozone were taken from the Jet Propulsion Laboratory (JPL) and High resolution Transmission Model (HITRAN) catalogues.

4.2.3 Particle Extinction in the Microwave Region

The scattering cross-section of any particle depends strongly on the ratio of its size and the wavelength of the electromagnetic radiation. Aerosols, with typical radii below 1 μm, do not scatter significantly because they are small compared to microwave wavelengths. For the same reason molecular scattering can safely be neglected. Also atmospheric aerosol densities are such that microwave absorption is negligible.

Ice and liquid water particles found in fog, clouds and precipitation are commonly denoted as hydrometeors, and the extinction of such particles can be very high. This extinction can be dominated by either absorption or scattering, depending on the particle size ranging from ~1 μm to ~1 cm, and the complex refractive index n, see Fig. 4.2.

The calculation of the complete phase matrix $\bar{P}$ is a challenge, Eq. 4.1, except for some special particle shapes. Spherical particles are treated by the Lorentz-Mie theory (Liou 2002), while the so-called T-matrix approach addresses any rotationally symmetric particle such as spheroids and cylinders (Mishchenko et al. 2002). An example of a general calculation method is the Discrete Dipole Approximation (DDA) (Draine 2000), but its high demands on computational resources and memory strongly reduce the practical usefulness of the method. As it is so demanding to calculate the absorption and scattering properties for arbitrarily shaped particles and since the particle shapes are not well known, hydrometeors are frequently treated as spherical or have some shape handled by the T-matrix method. This is in general acceptable for liquid clouds and rain droplets, but is problematic for the treatment of ice particles that are highly variable in shape. Extinction and scattering cross-sections for spherical liquid and ice particles are shown in Fig. 4.3.

4.2.4 Simulation Software

A variety of computer codes with the purpose of simulating atmospheric radiative transfer are available, however the emphasis is placed on the general software developed especially for the microwave region, which is publicly available. A complete software package is required to enable the instrument responses to be

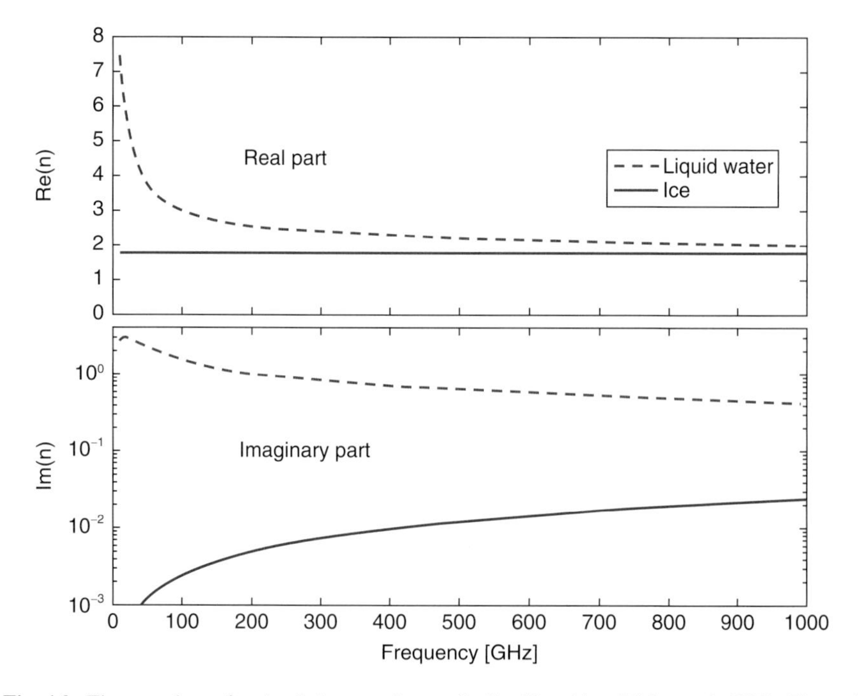

Fig. 4.2 The complex refractive index, n, of water for liquid and ice (Liebe et al. 1993). The real part determines the propagation speed, and the wavelength in the medium, while the imaginary part determines the medium's absorptivity.

incorporated. The term forward model is here used for such codes rather than the more restricted notation, such as radiative transfer code.

Most forward models dedicated to microwave applications describe only scalar radiative transfer without scattering and can thus be denoted as clear-sky models. Eight such forward models are presented and their consistency has been compared by Melsheimer et al. (2005). Most clear-sky forward models are developed for a specific instrument and are not distributed freely. The exceptions include the Atmospheric Transmission at Microwaves (ATM) (Pardo et al. 2001) and the Atmospheric Radiative Transfer Simulator, version 1 (ARTS-1) (Buehler et al. 2005). The retrieval methodologies normally applied for non-scattering measurements require the Jacobian matrix. That is, that the partial derivatives of the measured spectrum, with respect to the quantities to be retrieved, can be calculated. Expressions for Jacobian calculations are found in Buehler et al. (2005) and Read et al. (2006). The inclusion of instrument properties is given special attention in Eriksson et al. (2006). Clear-sky forward models assume in most cases a spherically symmetric one dimensional (1-D) atmosphere. Forward models, which allow atmospheric variables to vary in the horizontal dimension include the two dimensional (2-D) model by Read et al. (2006) and the three dimensional (3-D) development version of ARTS (v1.x) (Eriksson and Buehler 2008).

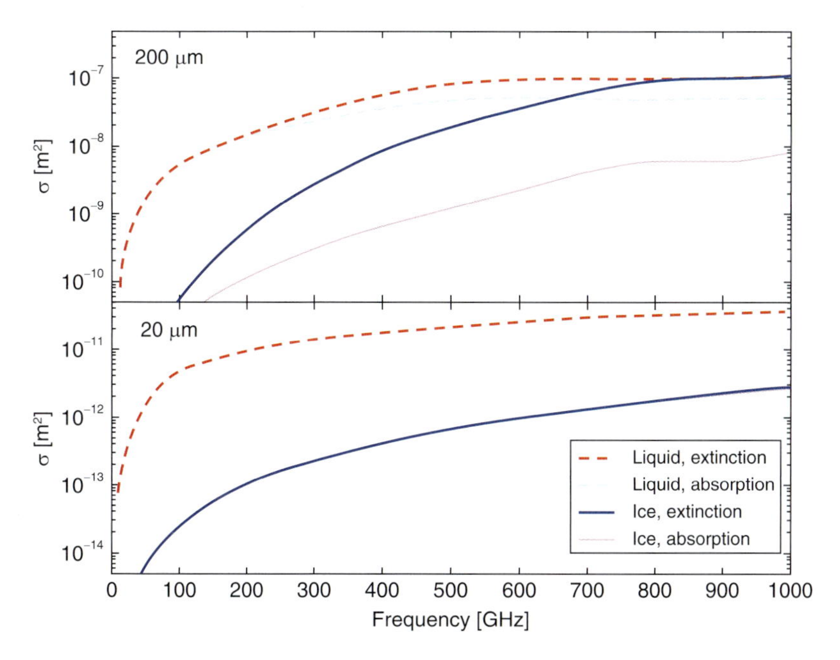

Fig. 4.3 Extinction and absorption cross-section σ of spherical liquid and ice water cloud particles with diameters of 200 μm (top) or 20 μm (bottom). The extinction of 20 μm particles is almost purely due to absorption. Refractive indexes as in Fig. 4.2 are assumed, and cross-sections are calculated with the Lorentz-Mie implementation by Mätzler (2002).

Scattering forward models differ in complexity with respect to radiative transfer implementation (scalar or vector), atmospheric dimensionality (1–3-D), allowed particle shapes, geometrical restrictions and surface scattering properties. The radiative transfer equation has no analytical solution in the case of scattering and different iterative solution methods have been developed. The selection of a solution method largely determines the calculation speed and main applications of a particular forward model. A review of 12 scattering codes is provided by Sreerekha et al. (2006), where, with the exception of ARTS-1.x (Eriksson and Buehler 2008), a flat Earth is assumed. The latter forward model has been developed to deal with vector radiative transfer and 3-D atmospheres. This is required for the accurate simulation of limb sounding observations. ARTS-1.x includes two scattering modules, Discrete Ordinate Iterative method (DOIT) (Emde et al. 2004) and reversed Monte Carlo (MC) (Davis et al. 2005). The latter and other MC algorithms provide comparably high calculation speed for 3-D calculations where the intensities for only a few directions are required (Battaglia et al. 2007). Other methods are favoured if larger portions of the complete radiation field are required. Another possible choice is then Spherical Harmonic Discrete Ordinate Method (SHDOM) (Evans 1998), which is a versatile and well-established scalar 3-D radiative transfer code, which has a rapid 1-D version called SHDOMPP. Another 1-D forward model frequently used for microwave measurements is the MicroWave MODel (MWMOD) (Czekala and Simmer 1998), which performs vector radiative transfer for non-spherical particles with non-random orientations.

As a result of the restrictions imposed by computer power, full radiative transfer calculations are not suitable for numerical weather prediction and fast regression models have therefore been developed. For example the fast Radiative Transfer (RTTOV) for the Television and Infrared Observation Satellite (TIROS) Operational Vertical Sounder (TOVS) (Eyre 1991) provides rapid simulations of radiances for satellite IR and microwave radiometers given an atmospheric profile of temperature, variable gas concentrations, cloud and surface properties (see Section 4.5.3). Later versions of RTTOV contain the forward, tangent linear, adjoint and full Jacobian matrices for the operational sensors. Clouds (liquid and ice) and rain can be considered. In the microwave, RTTOV computes the scattering effects using the delta-Eddington approximation.

The forward model for radar observations describes the particle back scattering as a function of distance to the transmitter assuming single scattering. The forward models and software discussed above describe the scattering of thermal emission. Multiple scattering of radar pulses requires dedicated software, such as Battaglia et al. (2006). Up to the present simulation of radar, which includes multiple scattering, are based on Monte Carlo methods.

4.2.5 The Inverse Problem

In most cases, remote sensing measurements form an ill posed problem, that is, a large, or infinite, number of geophysical states exist, which are consistent with the measurement. One example of what could be considered as a unique measurement is the distance information provided by cloud radars, assuming that multiple scattering can be ignored. The task of extracting geophysical information from atmospheric remote sensing data requires special care because of the problem of non-uniqueness. The data extraction step using remotely sensed data is known as retrieval or inversion. A standard text for atmospheric data retrieval has been published by Rodgers (2000). It contains, for example, an updated presentation of the general method used for characterizing atmospheric inversion problems (Rodgers 1990). Additional information or constraints are needed to determine the best possible solution from a non-unique set of measurements. Several methodologies are used to solve ill posed problems, but for atmospheric sounding, statistical approaches are popular.

A theoretical starting point for statistically oriented retrievals is Bayes' theorem. This theorem is used to determine the probability distribution of all states after the measurement is performed, the *a posteriori* probability density function (PDF). A pure Bayesian method makes direct use of this PDF when choosing a solution. Common solutions take the most likely state (maximum of the PDF) or the expected value (PDF weighted state). The statistical constraint is provided as the *a priori* PDF of the state to be retrieved. This constrains the solution to a distribution of expected or known solutions. The *a posteriori* PDF can be determined in an analytical manner for linear, and in practice also moderately non-linear, inversion cases where both measurement uncertainties and *a priori* information follow Gaussian statistics. This solution is frequently referred to as the *optimal estimation method* (Rodgers 1976), while Rodgers (2000) prefers the general nomenclature of the *maximum a posteriori* solution. The most likely and expected states are identical and the *a posteriori* PDF is symmetric for the assumed conditions. The solution is here found by inverting the radiative transfer model, and the forward model must be able to provide the Jacobian matrix (Section 4.2.4).

Methods have been developed to determine *a posteriori* PDF for highly non-linear or non-Gaussian retrieval cases as well. These methods are, however, normally too computationally demanding to be practical. More efficient and potentially rapid processing is often achieved by obtaining the solution through a pre-calculated retrieval database. The basic challenge for such approaches is to generate an ensemble of atmospheric states that cover all relevant variables and mimic the true atmospheric variability with sufficient accuracy. The states in the retrieval database are distributed according to the *a priori* (multi-dimensional) PDF. The retrieval database is completed by calculating, using a radiative transfer tool, the measurement vectors corresponding to all the considered states of the atmosphere. The inversion is performed by finding a weighted average of the database states that are based on the similarity between the measurement and the simulated radiance. The weights for the database entries are calculated explicitly in the Bayesian Monte Carlo integration approach of Evans et al. (2002).

An implicit weighting is performed when the database is used for creating a regression model, including the training of neural networks (NNs). The application of NNs is growing. For example, early microwave satellite water vapour retrievals used simple regression models (Staelin et al. 1976), while more recent algorithms tend to use NNs (Cabrera-Mercader and Staelin 1995; Jiménez et al. 2005). A general review of NNs for geophysical inversions is provided by Krasnopolsky (2007). To what extent a special implementation of these later methods can be viewed as Bayesian depends on how well the retrieval database matches the true atmospheric statistics and the optimisation constraint when determining the regression model parameters.

The estimation of the atmospheric state at a given time for the initialisation of numerical weather prediction (NWP) models involves complex data assimilation techniques. Assimilation is frequently performed inside a Kalman filtering framework that can be seen as an extension of the Bayesian method also covering time evolving states. For further details, readers are referred to Rodgers (2000). A brief summary of data assimilation and the contribution of microwave remote sensing to current NWP are given in Section 4.5.4.

4.2.6 Observing Technique

Microwave radiometers measure the thermal emission from a target. These types of sensors, which use radiation present in the system, are termed passive; in contrast instruments which emit radiation and then measure the backscattered signal are called radar, and will be considered in Section 4.7. A more general treatment of microwave radiometers can be found in the chapter by Tiuri (1966) in the book by Vowinkel (1988) (pp 236–293) and in the book edited by Janssen (1993).

Fig. 4.4 shows schematically the building blocks of a radiometer. Initially the antenna receives the thermal emission from the target. The spatial resolution of the instrument is given by the refraction limited beam width of the antenna, which is proportional to λ/D, where D is the antenna diameter and λ, the observed wavelength. The next element in the instrument chain is the calibration unit. The very high amplification (*ca.* 100 dB) needed for microwave radiometers require a careful and frequent calibration to minimise unavoidable gain drifts and fluctuations. Such an instrument is called a Total-Power-Radiometer (TPR) and instruments of this type are generally used today. A TPR is calibrated by switching black bodies at well-known temperatures in front of the radiometer. The "hot" black body is typically a microwave absorber at a well-monitored temperature. The cold calibration element is, for space-borne sensors, the cold background radiation of approximately 2.7 K. The frequency for the calibration process is determined by measuring the Allen-Variance (Allan 1966; Rau et al. 1984; Ossenkopf 2008), which determines the period for which gain variations of the instrument are negligible.

The calibration is followed by the amplifier. Up to frequencies in the order of 100 GHz (wavelength 3 mm) direct amplification is feasible by solid-state devices. As a result of the lack of amplifiers for higher frequencies, down-conversion of the original signal to a much lower frequency by using a heterodyne receiver is

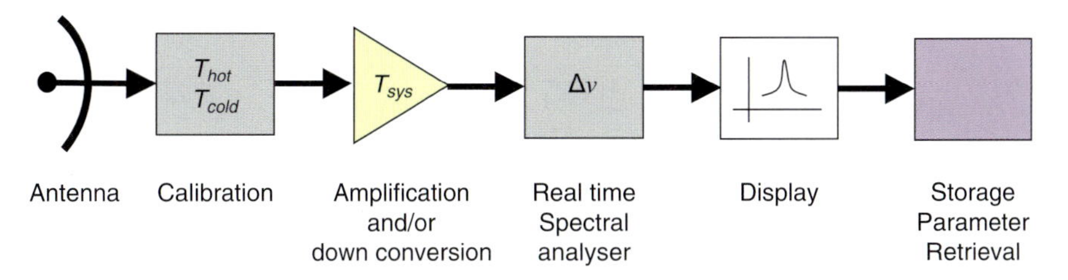

Fig. 4.4 Building blocks of a microwave radiometer.

preferred. This is achieved by using a non-linear element, typically a diode, and a local oscillator. The down-converted signal at the intermediate frequency is then amplified by standard components. The radiometer is characterized by its system noise temperature T_{sys} that is defined as the temperature of a black body at the input of a totally lossless and noiseless instrument, producing the same output power as found for the real system.

The next element in the detection chain is the real time spectral analyser, which is used for resolving the line shape of an emission line. Depending on the application, the width of individual channels range from more than 100 MHz down to less than 0.1 MHz. For applications where no high frequency resolution is needed the bandwidth of a microwave radiometer can be as large as many 1,000 MHz.

The last steps in the detector chain is the data display, storage and processing, providing the desired geophysical parameter.

The quantity, which determines the sensitivity of a radiometer, is the thermal noise generated in the various electronic elements of the receiver and also unavoidable losses in passive components such as transmission lines. The sensitivity of the radiometer is characterised by its system noise temperature, T_{sys}. Typical values for T_{sys} are shown in Fig. 4.5 for receiver types presently used. The quantity characterizing the sensitivity of a radiometer is the minimum detectable temperature difference Δt_{min} when observing a given target, Δt_{min} can be calculated from the following expression

$$\Delta t_{\min} = \frac{T_{sys}}{\sqrt{\Delta\nu \cdot \tau}} \tag{4.3}$$

with $\Delta\nu$ the selected spectral resolution and τ the integration time for one single measurement.

For real time spectrum analysis various techniques are available with Filterbanks, Acousto-Optical-Spectrometers (AOS), being frequently used. Recently broadband Fast Fourier Transform Spectrometer (FFTS) became available. This approach is described in Klein et al. (2006), who also provide a brief review of the other types of real time spectrum analysers. All types offer total bandwidth in excess of 1,000 MHz and can provide a spectral resolution to resolve easily the exact shape of atmospheric emission lines.

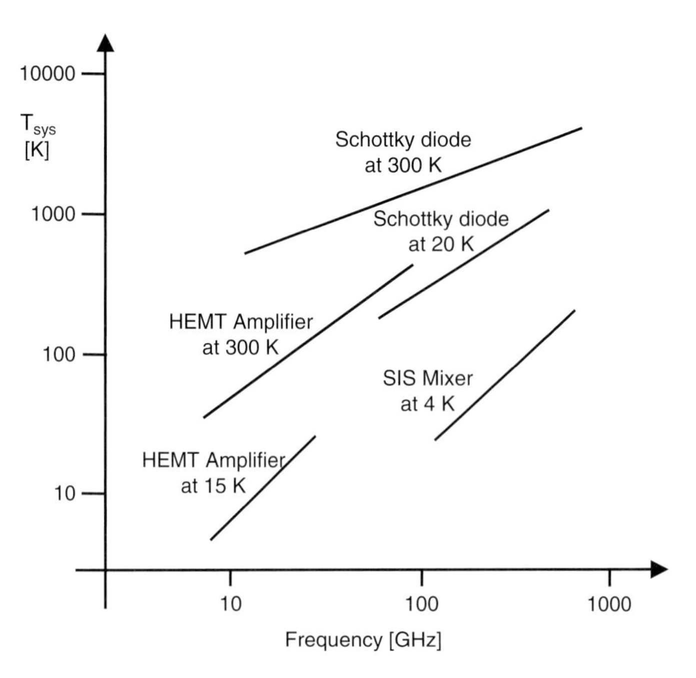

Fig. 4.5 Typical system noise temperature T_{sys} limits of various radiometric receivers. A substantial improvement in noise temperature is possible by cooling; examples are the Schottky diode mixers and the High Electron Mobility Transistor (HEMT) amplifiers. The Superconducter-Insolator-Superconducter (SIS) device is by far the most sensitive receiver but requires cooling to 4 K.

4.3 Temperature and Water Vapour Profiles

4.3.1 Introduction

Similar to the methods first developed for IR observations, the retrieval of atmospheric profiles from microwave observations is based on the different absorption of atmospheric gases at different frequencies around an absorption line. Since in the troposphere, line width is proportional to pressure, the absorption decreases rapidly with altitude at a given frequency off the line centre, for a molecule with a constant mixing ratio such as O_2. If the absorption at a particular frequency off the line centre is high, making the total atmosphere opaque or nearly opaque, it can be seen that the radiation received by a downward looking sensor on a spacecraft originates mainly from a layer at a particular altitude h_{peak}, since for higher altitudes the absorption and therefore also the emission reduces rapidly due to the narrowing of the emission line. On the other hand for altitudes considerably lower than h_{peak}, the radiation will be absorbed by the strong absorption of higher layers. This explains qualitatively the concept of weighting functions as shown in Fig. 4.6.

If the mixing ratio of the particular molecule is constant, as assumed above, the received radiation at various frequencies off line centre, which corresponds to various pressure altitudes, will allow the retrieval of the atmospheric temperature profile. Emission from O_2, which is uniformly mixed up to nearly 100 km, is well suited to estimate the atmospheric temperature profile. On the other hand, if the temperature profile is known, the only unknown is the concentration of the particular molecule. The temperature profile in the atmosphere is not highly variable with time at a given location, and the response of the measured signal to changes in temperature is nearly linear (Rayleigh-Jeans approximation), which is in contrast to thermal IR where a highly non-linear relation exists near the peak of the Planck function. This fact allows for the retrieval of atmospheric species from microwave radiances with good accuracy even with a poorly known temperature.

Fig. 4.6 presents the temperature and water vapour weighting functions for O_2 lines at 57.290 GHz and 118.750 GHz and for the 183.310 GHz H_2O line. The H_2O line around 22 GHz is very weak and not suitable for profile retrievals (the weighting functions for this line are very broad). Nevertheless observations near this line can provide the integrated water vapour in the atmosphere over ocean, where the water vapour appears in emission in front of a cold background due to the low emissivity of the ocean surface. The 50–60 GHz O_2 band and both the 22 GHz and the 183 GHz H_2O lines are the spectral regions currently used by operational meteorological satellites for temperature and water vapour measurements.

Due to the increasing opacity of the atmosphere and the perturbing effect of clouds and precipitation with increasing frequency, O_2 and H_2O lines at higher frequencies are not currently used for atmospheric profiling in a downward looking geometry. However since the satial resolution of microwave sensors is limited by the size of the antenna (Section 4.2.6), it has been suggested that higher frequencies for instruments on geostationary satellites be used, for example the Geostationary Observatory for Microwave Atmospheric Sounding (GOMAS) proposed to the European Space Agency (ESA). From Fig. 4.6 it can be seen that weighting functions at 57.290 and 118.750 GHz are very similar under clear sky conditions. Note that the spatial resolution near 118 GHz is better by nearly a factor of two compared with frequencies near 60 GHz, for a given antenna size. The performances and limitations of sensors operating at higher frequencies are investigated by Klein and Gasiewski (2000) and Prigent et al. (2006).

The microwave spectral regions used to sound the lower atmosphere must be sufficiently transparent; examples are the 22 GHz water vapour line, or the O_2 lines with weighting functions intersecting the surface. However in this case surface emission needs to be treated explicitly. As pointed out above, the ocean emissivity is low, ~0.5, and its dependence on the sea and instrument characteristics, such as temperature, surface wind speed, salinity and the observing geometry, can be modelled. However, land surface emissivity is large, typically being ~0.95, and depends on a large number of poorly known parameters, such as soil moisture, vegetation type, snow cover and type, parameters having high spatial and temporal

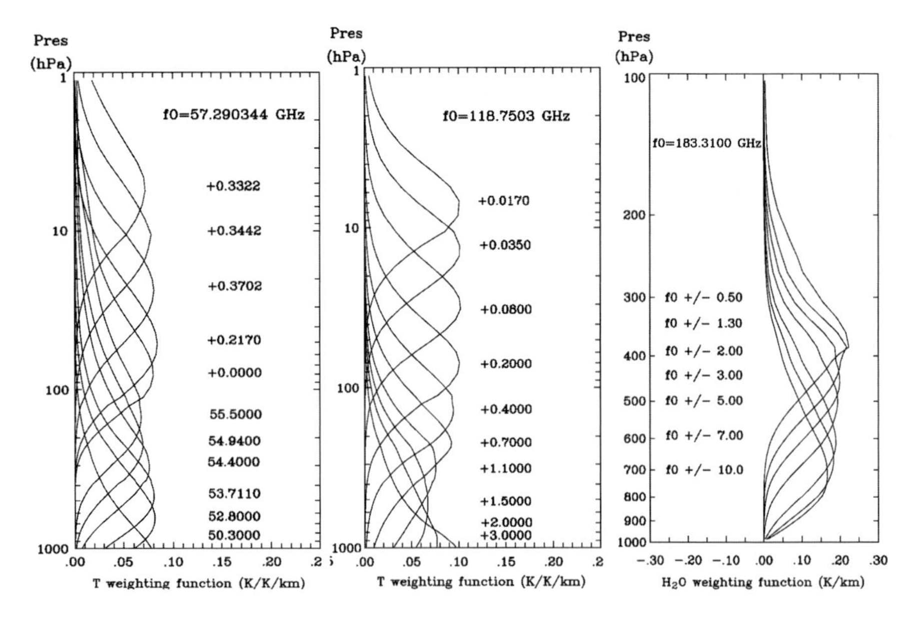

Fig. 4.6 Monochromatic weighting functions for selected frequencies near the 57 GHz O_2 band (left), the 118 GHz O_2 line (centre) and the 183 H_2O line (right), calculated for nadir sounding over land (a surface emissivity of 1 is assumed) for a mean standard atmosphere. Weighting functions show the sensitivity as a function of altitude (Prigent et al. 2005).

variability. As a result of the high emissivity, the brightness temperature of the land surface is close to the physical temperature of the surface. As the temperature of the lower atmospheric layers is very similar to the surface brightness temperature, the contrast between the surface emission and that from the lower atmospheric signal is small, reducing the sensitivity to the lower troposphere and the retrievals of atmospheric parameters requires an adequate knowledge of the surface emission. For this reason, surface-sensitive channels are not used over land to retrieve data products in the operational centres, but attempts have been made in a research mode (Karbou et al. 2005).

4.3.2 Examples

After a few tests in the Earth atmosphere with instruments on the Russian Cosmos satellites (Akvilonova et al. 1973), the NEMS instrument on the Nimbus-5 meteorological satellite provided the first global maps of integrated water vapour column amount over the ocean. The retrieval technique comprised a multi-dimensional regression analysis, trained on simulated radiances computed using data from radiosondes at different latitudes over a 2-year period (Staelin et al. 1976). The accuracy of the retrieved integrated water vapour column was claimed to be 0.2 g/cm^2. Schaerer and Wilheit (1979) described water vapour profiling using the strong H_2O line at 183 GHz, which became available after 1991 with the launch of the Special Sensor Microwave/Temperature-2 (SSM/T-2) instrument (Manning and Wang 2003).

Most operational atmospheric microwave sounders observe with a cross-track scanning geometry at incidence angles between 0° and 55°. Examples are the SSM/T-1 and 2, the Microwave Sounding Unit (MSU), the Advance Microwave Sounding Unit (AMSU-A and B) and the Humidity Sensor for Brasil (HSB). An exception is the Special Sensor Microwave Imager (SSM/I), which includes both sounding and imaging frequencies, using a conical scanning geometry.

Rosenkranz et al. (1997) compared the performances of the two scanning geometries, cross-track and conical, for profiling applications. So-called window channels are included between the molecular resonances to measure surface properties, such as the wind speed over ocean, soil parameters and to obtain additional information on hydrometeors.

4.4 Remote Sensing of Clouds and precipitation

4.4.1 Introduction

Hydrometeors require the radiative transfer to include the effects of scattering. In contrast to visible and IR instruments, which are only sensitive to radiation scattered or emitted from the top of the cloud, microwave radiation can penetrate clouds and precipitation to some degree. The penetration depth depends strongly on the ratio between the hydrometeor size and the wavelength, and on the phase of the hydrometeors, ice, or liquid. As a result of the large imaginary part of the refractive index of liquid water in the microwave range (Fig. 4.2), liquid particles significantly absorb and emit microwave radiation. Ice particles on the other hand, with a small imaginary refractive index, mainly scatter microwave radiation. At frequencies below 50 GHz, the microwave signal is essentially dominated by emission and absorption by liquid clouds and rain and is little affected by the presence of ice clouds. At higher frequencies, scattering effects on frozen particles increase, eventually masking the signal from the liquid cloud and rain.

The remote sensing of clouds and precipitation and the retrieval of data products relies on an accurate understanding of the interaction between the radiation and the hydrometeors. For each frequency the radiation measured by the radiometer is determined by the phase, size and shape distribution, and vertical profile of the hydrometeors.

The microwave radiative transfer in scattering atmospheres has been recently reviewed in depth by Battaglia et al. (2006). A comparison of several multiple scattering radiative transfer models (two-stream, multiple stream and Monte Carlo) concluded that differences between the codes were small (Smith et al. 2002). The main difficulties in simulating the microwave response of clouds and precipitation stem from the lack of information about the particle spatial distribution and their physical properties.

Cloud resolving models, for example Meso-NH (Lafore et al. 1998) or the Goddard Cumulus Ensemble model (Tao and Simpson 1993), can provide high-resolution 3-D structures of the hydrometeor content, along with some indication about the particle size distribution, shape, and phase (ice or water). For instance, Meso-NH calculates the distribution of five parameters (cloud droplet, pristine ice, rain, graupel, and snow) having horizontal and vertical resolutions of the order of 1 km and 100 m respectively. Aircraft campaigns can also give information on particle properties. However, large uncertainties in the description of the physical properties of frozen particles result in significant errors in the interpretation of microwave observations, especially in convective situations at frequencies above 50 GHz.

Fig. 4.7 presents simulations with and without hydrometeors for the ATM Model (Pardo et al. 2001) with a mean standard atmosphere, over ocean and land. The water content of clouds is expressed in cloud liquid water (CLW) which is given by the column in g/m^2 of liquid water integrated over the cloud layer along the line of sight of the instrument. Cloud ice water content is similarly defined.

During the emission/absorption process, liquid particles cause brightness temperatures to increase over a radiatively cold background such as the ocean; the induced warming of the signal increases with frequency up to a saturation level (Fig. 4.7 left, middle). Retrieval of the cloud liquid water path over ocean is currently based on the signals measured between 19 and 85 GHz. Over land, signals from clouds and rain are much weaker (Fig. 4.7 right, middle). However, when frozen particles are large enough with respect to the wavelength, scattering reduces the amount of radiation measured by the sensor, due to scattering of the cold space background. The scattering signal can be used to estimate the cloud ice content and precipitation over ocean and land (Fig. 4.7 bottom). However, the emission originating in the liquid lower part of cloud and rain layers is affected by the scattering process in the upper part of the cloud, especially at higher frequencies.

Satellite observations from the Tropical Rainfall Measuring Mission (TRMM) Microwave Imager (TMI) (TRMM, radar is discussed in Section 4.7) between 10 and 85 GHz are shown in Fig. 4.8 for Hurricane Bret on 22nd August 1999 in the Gulf of Mexico. The top and middle rows show the vertical and horizontal polarizations respectively for the five frequencies. For frequencies below 40 GHz, the cloud results in a warming of the brightness temperatures over ocean and a small cooling over land for both polarizations. At 85 GHz a significant decrease of the brightness temperatures is observed in the spiral arms of the hurricane, where large

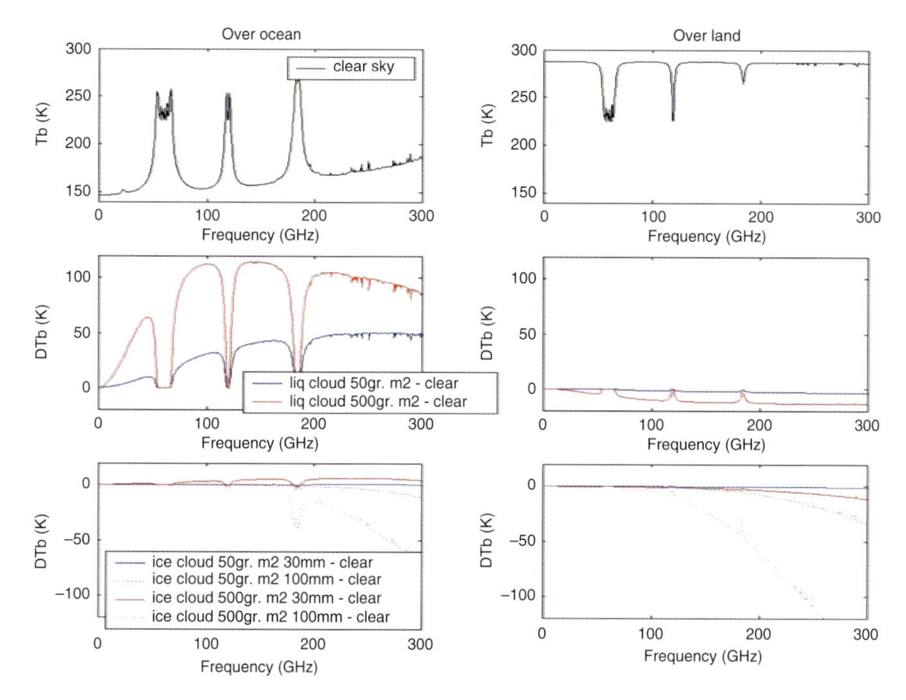

Fig. 4.7 Simulations with the ATM (Pardo et al. 2001), for a mean standard atmosphere, over ocean (left) and land (right). The brightness temperatures Tb under clear sky conditions are presented (top two figures), for an incidence angle of 53°. The surface emissivity is assumed to be 0.5 over ocean and 1 over land. The brightness temperatures difference DTb due a cloud is calculated as DTb = Tb(cloudy) – Tb(clear), for two clouds extending from 2 to 3 km altitude (middle two figures) with a CLW of 50 g/m^2 and 500 g/m^2 and correspondingly for two ice clouds extending from 7 to 8 km altitude (bottom two figures) with a CIW of 50 g/m^2 and 500 g/m^2. Single size spherical particles are considered, water particles have a diameter of 20 μm and ice particles have diameters D of 60 or 200 μm.

quantities of frozen particles scatter the signal. The lower the frequency, the more transparent the clouds become. Large polarisation differences have been observed over ocean at 10 GHz (lower row) even in the presence of clouds because of the almost transparent atmosphere and the strong dependence of ocean surface reflectivity on polarization. At 85 GHz no contrast is observed between ocean and land, because the atmosphere is nearly opaque at this frequency. Note also the increase in spatial resolution from 10 to 85 GHz (from 37·63 km^2 to 5·7 km^2), because the instrument uses the same antenna for all frequencies.

Several studies have characterized the convective activity based on passive microwave scattering observations (Mohr et al. 1999; Hong et al. 2005a; 2005b) or correlate the scattering signal to the radar reflectivity or to the electrical activity (Nesbitt et al. 2000; Cecil et al. 2005; Prigent et al. 2005). They all show that scattering by ice above 80 GHz is a very sensitive indicator of the convective strength, and provides information on the cloud microphysics.

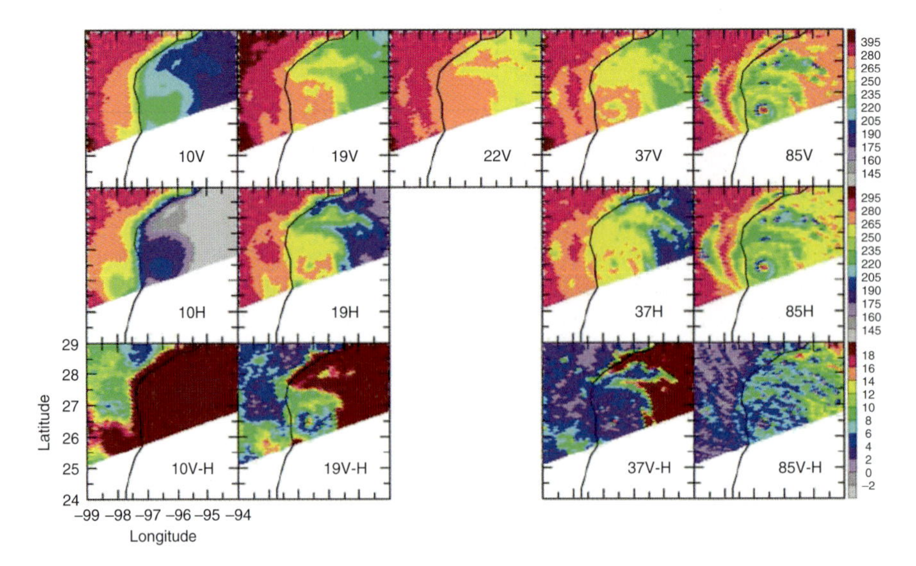

Fig. 4.8 Satellite observations of the brightness temperature from TMI between 10 and 85 GHz (see frequencies and polarization in plots) are presented for Hurricane Bret on 22nd August 1999, in the Gulf of Mexico. The top and middle rows show the vertical and horizontal polarizations for the five frequencies, whereas the polarization difference is displayed in the bottom row. The coastline is indicated by a solid black line. Fig. from Wiedner et al. (2004).

4.4.2 Retrieval of Cloud Liquid Water

Estimates of CLW are used for climate model validation and to study the impact of clouds on the radiative budget as described in Stephens and Greenwald (1991) and Borg and Bennartz (2007). The retrieval of CLW requires a high accuracy of the measurement (L'Ecuyer and Stephens 2003; Turner et al. 2007). An advantage is that microwave measurements yield consistent cloud data during the day and night and are quite insensitive to the cloud particle size distribution, and are not impacted by the presence of thin ice clouds (cirrus) in contrast to passive remote sounding in the visible.

As pointed out earlier, small liquid droplets with a radius below 50 μm, typical for non-precipitating clouds, absorb and emit microwave radiation but do not scatter significantly. The absorption from such liquid particles is proportional to the CLW [Rayleigh approximation, (Ulaby et al. 1981)]. CLW over ocean is routinely estimated from the emission measured between 19 and 85 GHz by imagers such as SSM/I, TMI, or the Advanced Microwave Scanning Radiometer (AMSR).

Over ocean: Different types of algorithms have been developed to retrieve CLW over the ocean from microwave observations, but all of them suffer from the lack of validation measurements. Estimates of CLW from ground-based microwave measurements are considered as the most direct and most reliable data for validation.

Contrary to the satellite case, the ground-based estimates are not contaminated by surface contributions, and are therefore more accurate. Alishouse et al. (1990) developed a statistical algorithm to estimate CLW from a few coincident satellite observations with ground-based radiometric estimates of CLW.

The majority of CLW retrieval algorithms are derived from statistical regressions between simulated radiances and cloud water content, for typical atmospheric and cloud conditions. The difficulty is to find a set of representative situations due to the large spatial and temporal natural variability. Pioneer work was conducted with a two-channel algorithm using data collected at 22 and 31 GHz from observations with the NEMS instrument on Nimbus-5 (Staelin et al. 1976). Subsequent developments tend to use more channels (Chang and Wilheit 1979) or suggest the use of different sets of frequencies, depending on the liquid water range (Weng and Grody 1994), with the low frequency (below 30 GHz) less likely to saturate for high CLW, and the higher frequencies around 85 GHz more sensitive to low CLW. The complexity of the radiative transfer models varies from simple codes for non-scattering clouds (Greenwald et al. 1993; Weng and Grody 1994) to multi-scattering transport models (Liu and Curry 1993; Bauer and Schluessel 1993). The theoretical error is generally claimed to be in the 10 to 30 g/m^2 range. A climatology of CLW has been established from this type of algorithm (Weng et al. 1997). Collocated visible and IR information can be used to separate clear and cloudy pixels to constrain the problem (Liu and Curry 1993).

So-called physical algorithms minimize iteratively the difference between observed and simulated radiances using the radiative transfer equation and a first guess (Prigent et al. 1994; Wentz 1997). Several geophysical variables are retrieved simultaneously, to ensure consistency (usually water vapour, cloud liquid water, ocean surface wind speed, and rain rate). Direct assimilation of the radiances from microwave imagers into a Numerical Weather Prediction (NWP) model was tested by Phalippou (1996) to retrieve the cloud liquid water along with the water vapour and surface wind speed over ocean. This 1-D variational assimilation is now used routinely at the European Centre for Medium Range Weather Forecast (ECMWF), but more advanced CLW estimates from 4-D assimilation is not yet performed by NWP centres.

Due to the lack of systematic CLW *in situ* measurements, the microwave-derived CLW products have been evaluated by comparisons with estimates from IR/vis observations, such as the products from the International Satellite Cloud Climatology Project (ISCCP) (Lin and Rossow 1994), or the estimates from MODIS (Greenwald et al. 2007). An inter-comparison of four published and well-accepted CLW algorithms based on SSM/I data showed significant differences in the tropics with up to a factor of 4 between the lowest and the largest estimates, even when averaged over a month and over latitude bands (Deblonde and Wagneur 1997). The discrepancies were attributed to the partition between CLW and rain, to the lack of sensitivity to low CLW, to the scarcity of *in situ* measurements for calibration, and to the beam-filling problem. Using comparisons between AMSR-derived CLW and MODIS estimates, Greenwald et al. (2007) confirm the influence of the beam filling issue but also found a dependence of the microwave CLW product on the surface wind speed over ocean.

The combination of multi-sensor observations is suggested to partly overcome these problems. Horváth and Davies (2007) compared CLW estimated from microwave observations (TMI) and optical measurements from MODIS and the Multiangle Imaging Spectro Radiometer (MISR). The agreement is satisfying for warm clouds (correlation of 0.85 and rms difference of ~25 g/m^2) with an overestimation for cloud fractions below 65%; however the errors are much larger for cold clouds.

Over land: The land surface emissivity is usually close to unity, making atmospheric features difficult to identify against such a background because of the small contrast. In addition, the emissivity is variable in space and time and difficult to model since it depends on the topography, vegetation, soil moisture, and snow, among other factors. In order to retrieve CLW over land from microwave measurements, Jones and Vonder Haar (1990) suggested first estimating the land surface microwave emissivity from collocated visible, IR, and microwave observations under clear sky conditions, and then to use the emissivity for determining the CLW. Promising results were obtained over central USA. Prigent and Rossow (1999) and Aires et al. (2001) developed a similar technique to be used globally. Aires et al. (2001) is based on a neural network scheme to retrieve simultaneously the surface skin temperature, the water vapour content, and the cloud liquid water, based on a large simulated database, with realistic cloud parameters derived from ISCCP (Rossow and Schiffer 1999). This approach leads to a theoretical rms error in CLW of 80 g/m^2.

4.4.3 Retrieval of Cloud Ice Water

Global coverage of ice clouds is ~30% on average (Stubenrauch et al. 2006), and shows a very large variability from thin cirrus to thick anvils (Heymsfield et al. 2004) with CIW varying from 5 to more than 1,000 g/m^2 and particle equivalent sphere diameter, D_e, from 10 to more than 1,000 μm. No existing satellite instrument is capable alone of observing the full range of CIW and D_e. Due to the wavelength-to-size ratio, visible techniques are very sensitive to small particles. IR techniques become insensitive for particle diameters above ~100 μm. Microwave to sub-millimetre observations can complement the visible and infrared measurements, and provide ice cloud characteristics for larger particles and optically thick clouds.

Before 1987, space borne microwave observations by imagers were limited to frequencies below 37 GHz. At these frequencies, negligible effects due to ice clouds are expected, with the exception of deep convective cores (Spencer et al. 1983). Estimating CIW from microwave observations was first suggested by Evans and Vivekanandan (1990) by simulating the effect of various ice crystal shapes at 37, 85, and 157 GHz. With the availability after 1987 of the SSM/I channel at 85 GHz, CIW have been estimated from microwave observations, using radiative transfer modelling (Weng and Grody 1994; Bauer and Schluessel 1993). Lin and Rossow (1996) and Minnis et al. (2007) have suggested adding visible and IR to the microwave observations in order to constrain the problem and obtain more accurate CIW estimates. Retrieval of CIW from space-borne microwave observations above 100 GHz has been tested by Liu and Curry (1999), using the relationship between CIW and the brightness temperature depression at 150 GHz of the SSM/T instrument. Based on their experience with the aircraft Millimetre-Wave Imaging Radiometer (MIR), Liu and Curry (2000) estimate CIW from measurements with the 89 and 150 GHz AMSU-B channels, using simple scattering efficiency estimates from cloud models and radiative transfer calculations (Ferraro et al. 2000; Weng and Grody 2000). These algorithms are limited to CIW above 300 g/m^2. Greenwald and Christopher (2002) conducted a near-global analysis of coincident 183 GHz AMSU-B measurements and thermal IR data from the Advanced Very High Resolution Radiometer (AVHRR) in order to estimate the effect of cold clouds at millimetre wavelengths. They confirm that non-precipitating cold clouds have a rather weak impact on the 183 GHz brightness temperature (of the order of 1.4K), whereas precipitating clouds have a much larger effect (around 7K). The brightness temperature depression at 183 GHz is difficult to relate to a quantitative characterization of clouds or rain, because it is a complicated function of hydrometeor type, size, and profile.

In order to understand the relationship between the physical and radiative properties of frozen hydrometeors, many efforts have been undertaken recently, with simulation studies (Liu 2004; Kim 2006), with comparison with satellite data (Meirold-Mautner et al. 2007) and with aircraft observations (Skofronick-Jackson et al. 2008).

Several modelling studies explored the potential of the sub-millimetre domain for ice cloud characterization and concluded that several, widely spaced frequencies can determine CIW and ice particle size distribution (Gasiewski 1992; Evans and Stephens 1995a; 1995b). Evans et al. (1998) extended the previous studies with the objective of deriving an algorithm to estimate the global distribution of ice mass from observations at 150, 220, 340, 500, 630, and 880 GHz. Additional observations near the H_2O line at 183 GHz have been suggested for quantifying the water vapour contribution above the cloud (Evans et al. 2005). Millimetre and sub-millimetre receivers have been developed and flown on board aircraft, such as the Goddard Space Flight Center (GSFC) MIR including channels at 85, 150, 183, 220, and 325 GHz (Racette et al. 1996). This sensor has collected data during several measurement campaigns (Wang et al. 1998; Liu and Curry 1998; Deeter and Evans 2000; Skofronick-Jackson et al. 2003). The Conical Scanning Submillimetre-wave Imaging Radiometer (CoSSIR), which has 12 channels at 183, 220, 380, 487, and 640 GHz was used by Evans et al. (2005) during the CRYSTAL-FACE campaign to test a Bayesian retrieval algorithm of CIW and D_e, in comparison to coincident 94 GHz airborne radar observations. A Fourier Transform Spectrometer (FTS), the Far IR Sensor for Cirrus (FIRSC) for use on an aircraft, has also been developed to explore the whole sub-millimetre region between 300 GHz and 3,000 GHz (Evans et al. 1999; Vaneck et al. 2001). This wide coverage in frequency provides high sensitivity for a large range of particle sizes.

Following these simulation studies and results from aircraft campaigns, satellite instruments have been proposed to the space agencies. In Europe, the Cloud Ice

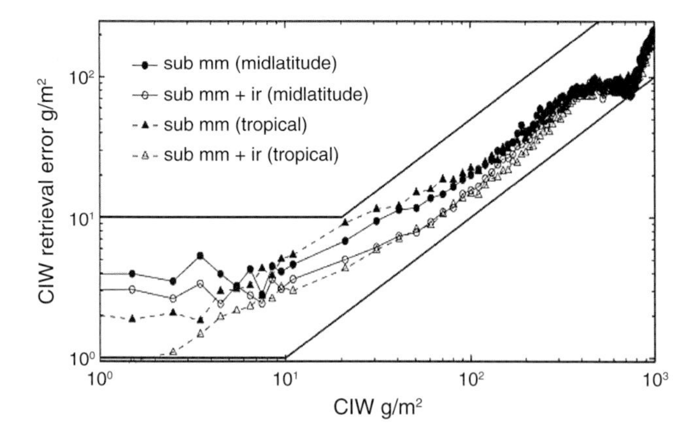

Fig. 4.9 Theoretical errors using the CIWSIR instrument to retrieve the CIW as a function of CIW, for two types of atmospheres (mid-latitude and tropical), with and without IR information (Buehler et al. 2007).

Water Sub-millimetre Imaging Radiometer (CIWSIR) (Kunzi et al. 2001; Buehler et al. 2005; 2007) is a conically scanning instrument to be flown on a polar orbit in tandem with MetOp, with 12 channels around 183, 243, 325, 448 and 664 GHz, and a thermal IR channel. A similar instrument has been proposed in the US by Ackermann et al.

Simulations show that CIWSIR can provide CIW with a detection threshold of 2 g/m^2 and an error of 20%, along with an estimate on the equivalent sphere diameter D_e and a median cloud altitude, with accuracies of 30 μm and 300 m respectively (Jiménez et al. 2007).

Fig. 4.9 shows the CIW retrieval performance as a function of CIW, for tropical and mid-latitude conditions (Buehler et al. 2007).

4.4.4 *Precipitation*

Rain is an intermittent phenomenon, highly variable in intensity, time, and space. Convective rain cells can have rain rates above 50 mm/h lasting a few minutes with characteristic sizes of a few square kilometres, whereas stratiform clouds are typically accompanied by rain rates below 1 mm/h, with life cycles of a few days, covering hundreds of square kilometres.

Precipitation retrievals over the ocean are mostly based on the emission measured at frequencies below 40 GHz (Wilheit et al. 1977; Prabhakara et al. 1992). Over land, the emission-based algorithms are no longer valid since, due to the high surface emissivities, there is no contrast between rain and no rain. However the scattering signal at 85.5 GHz has been used, for example by Spencer et al.

(1989) or Grody (1991), to estimate precipitation indirectly over ocean and land. The emission signal originates in the liquid lower part of the cloud and in the rain layer, whereas the scattering in the upper frozen part of the cloud reduces the emission. As a consequence, the amount of scattering can be an indirect measurement of precipitation, relying on the relationship between the presence of large ice particles in upper layers and the precipitation below. These algorithms are applicable over all surfaces, ocean, land, and mixed.

Channels below 40 GHz are the most sensitive to rain, but are also affected by other parameters such as water vapour, clouds, and surface characteristics. In addition, they suffer from poor spatial resolution. On the other hand, higher frequencies offer spatial resolution in better agreement with the small-scale structure of rain cells, but are mostly sensitive to cloud ice.

The first rain retrieval methods were simple regressions between surface rain rates and their corresponding simulated or measured brightness temperatures (Wilheit et al. 1977). These approaches are still used to generate long-term climatology. Nevertheless the highly non-linear character of the radiative transfer function that links rain and brightness temperatures is a natural limit for such methods. Other algorithms rely on the fact that the space of the possible solutions is limited and can be described by a number of well-documented situations in a database. Retrievals from Kummerow et al. (1996; 2001), Bauer (2001) and Bauer et al. (2001) are all based on probabilistic techniques derived from Bayes theorem.

The performances of these retrievals depend heavily on the simulated database. On the basis of elaborate cloud microphysics obtained from cloud resolving models, radiative transfer calculations can be performed for detailed hydrometeor profiles, taking into account different hydrometeor phases and various size distributions (Smith et al. 1992b; Mugnai et al. 1993; Kummerow and Giglio 1994). The need for consistency between models and measurements has been stressed by Panegrossi et al. (1998) and Tassa et al. (2003). The cloud model and the associated radiative transfer calculations have to be able to explain and reproduce all the observed signatures.

Iterative techniques based on radiative transfer models have also been developed. The simplest one is a profile adjustment using a loop over a radiative transfer model (Kummerow et al. 1989). Unified ocean parameter retrieval algorithms use similar methodology and provide simultaneously near-surface wind speed over ocean, total water vapour, total cloud liquid water, and rain rate, with the advantage of limited correlation between the retrieved parameters (Wentz and Spencer 1998). Most of these algorithms were originally developed for retrieving rain over ocean but some of them also have an over-land module that relies on the scattering at 85 GHz by ice clouds above the rain. Elaborate variational assimilation schemes have also been tested (Moreau et al. 2003) and are currently implemented at ECMWF.

To overcome the sampling problem due to the low-level orbits of the current IR geostationary measurements, they have been combined with microwave observations from low orbits: the microwave rain retrieval is used as a reference and the IR images provide the almost continuous measurement required to follow the evolution of the systems (Sorooshian et al. 2002; Tapiador et al. 2004). The Global Precipitation Climatology Project merges not only microwave and IR observations but also rain gauge data (Adler et al. 2003), to quantify the distribution of precipitation globally from 1979 to the present.

At millimetre and sub-millimetre wavelengths, the measurements are not expected to sense the rain directly, since the cloud opacity in the precipitating area is too large. Correlations between precipitation and brightness temperatures at millimetre waves are expected to arise from the relationship between the cloud, the particular atmospheric profile, and precipitation.

Rain rates have been estimated from AMSU-B observations over ocean and land from a simple differential scattering index in the clouds at 89 and 150 GHz (Ferraro et al. 2000), and an operational product from NOAA is available (Ferraro et al. 2005). These estimates are based on limited cloud model data and *in situ* precipitation measurements. In Staelin and Chen (2000), a neural network scheme is developed to retrieve precipitation rates from AMSU-A and B observations near 54 and 183 GHz, with a training database generated from collocated observations by AMSU and 3 GHz radar. The rms errors, evaluated for different rain-rate categories, are of the order of 50%.

A neural network methodology is also used by Defer et al. (2008), but trained on simulations derived from the coupling of mesoscale cloud models and radiative transfer calculations. The results are tested on AMSU-B observations collocated with radar measurements.

The differential response of the oxygen bands at 50–57 GHz and 118 GHz to the absorption and scattering by hydrometeors has also been suggested for estimating the rain rate (Bauer and Mugnai 2003), but has not yet been tested.

Observations of severe weather events require a high sampling rate. So far, microwave observations are only available from satellites in low orbits, with limited revisiting time. Geostationary orbits provide the adequate sampling but deploying microwave sensors in a geostationary orbit requires the use of millimetre and sub-millimetre radiometers to observe precipitation with an adequate spatial resolution while keeping a reasonable antenna size. The potential of millimetre and sub-millimetre observations for precipitation monitoring has been evaluated by Staelin and Surussavadee (2007), Surussavadee and Staelin (2008a; 2008b) and Defer et al. (2008), showing very good results for rain rates above 1 mm/h over both ocean and land. The GOMAS project was proposed to ESA with channels between 50 and 425 GHz and a 3 m antenna, providing a spatial resolution of 12 km at 380 GHz. Presently simulation studies and aircraft experiments are in progress.

The estimation of snowfall is a challenging problem, with the difficulty first to model accurately the scattering by snow and second to discriminate between the falling snow and the strongly variable emissivity of the snow on the ground. A physical approach has been tried by Kim et al. (2008) to retrieve snowfall rate over land using AMSU-B observations.

4.5 Applications of Microwave Data in Operational Meteorology

4.5.1 Data Assimilation

Data assimilation systems in Numerical Weather Prediction (NWP) provide the technical framework for performing atmospheric analyses that represent the best estimate of the state of the atmosphere at a certain time and that are used to initialise forecast models. Obviously, the quality of the forecast depends on both the accuracy of the model as well as the accuracy of the initial state estimate. The initial state estimate is called the analyses and represents an inversion problem. In general, this inversion is under-determined so that the analysis must employ information from *a priori* data (in NWP usually short-range forecasts initialised with previous analysis) and observations using a mathematical framework to combine optimally the two.

The complexity of the data assimilation system to be used depends on the application and the affordable computational cost and can range from simple interpolation schemes to four-dimensional variational and Ensemble Kalman filter schemes or even non-linear methods (Daley 1991). Global analysis systems are solving the above inversion problem with state vector dimensions of the order of 5×10^7 (the product of the number of grid points, number of levels and number of state variables) and observation vector dimensions of the order of 10^7 (product of number of observation points, number of levels and channels). This fact limits the use of non-linear models and ensemble-based methods, the latter usually being run at much lower resolution and with less detail than deterministic systems.

Most global operational NWP centres are operating efficient incremental four-dimensional variational (4D-Var) data assimilation systems that are based on the assumption that model behaviour is nearly linear in the vicinity of a good short-range forecast of the model state (Bouttier and Courtier 2002). The advantage of 4D-Var methods lies in the fact that they are dynamically consistent because the optimisation is performed over a time window through which the model is integrated, and that computationally efficient adjoint models can be used (Courtier et al. 1993). Systems like this are currently in use at ECMWF, the UK Met Office, Météo-France, the Meteorological Service of Canada (MSC), the Japan Meteorological Agency (JMA) (for the regional model) and in the US soon at the National Center for Environmental Prediction (NCEP).

4.5.2 Microwave Data in Operational Meteorology

Satellite data can be assimilated as level-1 (e.g. calibrated and located radiances), or level-2 (e.g. derived geophysical products) data. The choice depends on various factors, most prominently on the amount of maintenance required in an operational system. Level-2 products often employ a similar inversion framework to retrieve parameters from radiances as is used in assimilation, namely *a priori* information, radiative transfer, error and bias models. If these ingredients are identical to those used in the level-1 data assimilation system, the result between assimilating level-1 or level-2 products should also be identical (Dee and Da Silva 2003).

However, most level-2 products employ different models and *a priori* constraints than used in NWP modelling and their characteristics are often not well defined or difficult to account for in data assimilations systems (Joiner and Dee 2000). The use of level-1 radiance data has advantages because (a) today's operational microwave radiative transfer models are very fast and quite accurate (Saunders et al. 1999), (b) it allows the flexible use of radiometer channels as a function of situation-dependent sensitivity and potential channel corruption, and greatly simplifies error and bias estimation.

In the early days of satellite data assimilation however, retrieved geophysical products were preferred due to less efficient models, more simple observation operators and the uncertain impact of satellite data in general. One of the earliest protagonists of microwave radiometer data in assimilation were Eyre et al. (1993) in Europe, producing retrieved temperature profiles from TOVS including the instruments, High-resolution Infrared Radiation Sounder (HIRS), Microwave Sounding Unit (MSU), and the Stratospheric Sounding Unit (SSU) data. The retrievals were obtained via a 1D-Var algorithm that also used the NWP model forecast as the *a priori* constraint. This approach was later extended to the use of moisture sensitive channels and instruments such as the AMSU-B and the SSM/I (Phalippou 1996). In all cases, the initial retrieval-based systems were replaced by direct radiance assimilation for the above-mentioned reasons (Andersson et al. 1994; Derber and Wu 1998).

The initial concerns over general satellite data impact were mostly overcome by the time 4D-Var data assimilation systems were established, mainly because of the improved treatment of spatial and temporal collocation between data and model trajectory and the interaction of temperature and moisture with model dynamics (Andersson and Thépaut 2008). Since then, data from passive microwave radiometers exploiting the 50–60 GHz oxygen absorption line complex for temperature sounding, and the 183.31 GHz water vapour absorption line for moisture sounding, have proven to be the most important satellite observing system in NWP. At the time of writing, these observations are supplied by AMSU-A, AMSU-B and the Microwave Humidity Sounder (MHS) onboard several NOAA satellites (15–19), Aqua and one METOP spacecraft. (METOP is a series of three satellites to be launched sequentially over 14 years, forming the space segment of EUMETSAT's Polar System (EPS) satellites.) This system is complemented by the so called microwave imagers (e.g. SSM/I, and the follow-on instruments Special Sensor Microwave Imager/Sounder (SSMIS), AMSR-E, TMI) that contribute information on sea-surface temperature, near-surface wind speed, integrated atmospheric moisture, clouds and precipitation.

Fig. 4.10 shows the temporal development of the observing system in the ECMWF 40-year reanalysis project ERA-40 (Uppala et al. 2004). Atmospheric moisture and temperature analyses are largely constrained by MSU, SSU and later AMSU-A and SSM/I data. The strong positive impact of the improved observing system on the Television and IR Observation Satellite (TIROS) with the instruments TIROS Operational Vertical Sounder (TOVS) in 1979 and the Advanced TOVS (ATOVS) in 1998 on the analysis and forecasts is demonstrated by the much improved fit to surface pressure observations at weather stations and buoys (Fig. 4.10).

Apart from clear-sky microwave data, efforts towards cloud-affected data assimilation have been successful in recent years. This was mainly achieved by the greatly improved global model moist physical parameterisations and the enhanced computational capabilities that allow the operational employment of multiple scattering radiative transfer models (Bauer et al. 2006a). The explicit treatment of clouds and precipitation in operational analysis systems is accompanied by a large set of uncertainties. For example, greater model non-linearity, potential dynamic instabilities, large and unknown error structures as well as unknown model biases (Errico et al. 2007). However, several conservative methods have been implemented in operations, for example at ECMWF based on SSM/I data that provide significant and positive impact on both moisture analysis and forecasts (Bauer et al. 2006b; 2006c).

Another recent development in NWP is the use of microwave radiometer observations over land surfaces to constrain the soil moisture analysis. The research has been motivated by the planned ESA Soil Moisture and Ocean Salinity mission (SMOS) (launched in November 2009) that provides moderate resolution at 1.4 GHz from synthetic aperture imagery. Also channels near 6 and 10 GHz provide soil moisture information and is already available from TMI and AMSR-E.

4.5.3 Microwave Radiative Transfer Modelling in Data Assimilation

One of the most relevant issues is the trade-off between the accuracy of the forward radiative transfer model and its computational speed. The latter is optimised by the development of parameterised absorption models that produce regressions as a function of a selection of model predictors (e.g. temperature, humidity etc.) and that are trained with accurate line-by-line absorption models and representative atmospheric profile datasets (Matricardi et al. 2004). Their accuracy is usually better than radiometer noise for both IR and microwave.

Models such as Radiative Transfer for TOVS (RTTOV) are continuously maintained and improved by the NWP community, for example in the framework of RTTOV (by Numerical Weather Predication – Satellite Application Facility, NWP-SAF) or Community Radiative Transfer Model (CRTM) (by the Joint Center for Satellite Data Assimilation, JCSDA) models. On current super-computers, RTTOV produces about 40 forward calculations per millisecond.

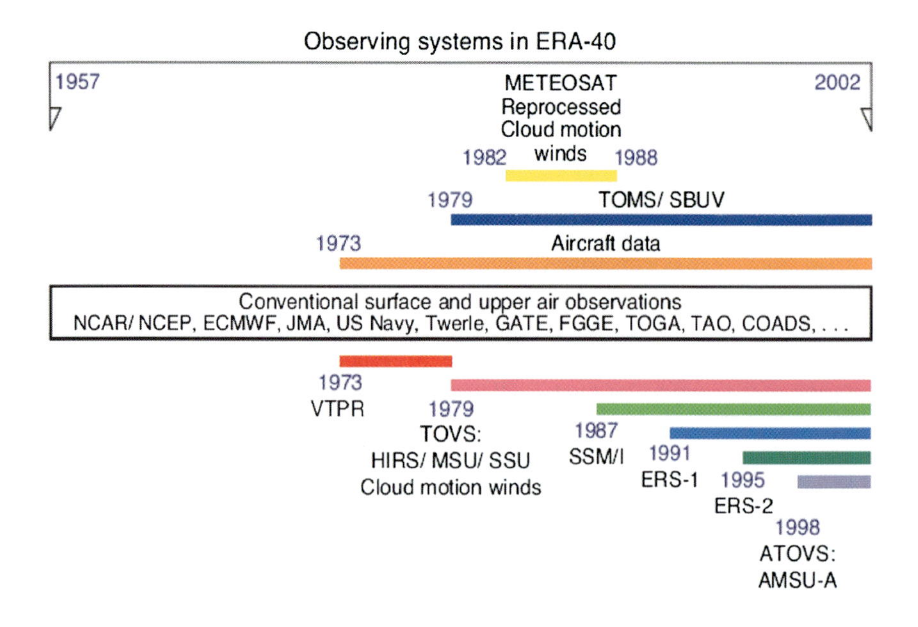

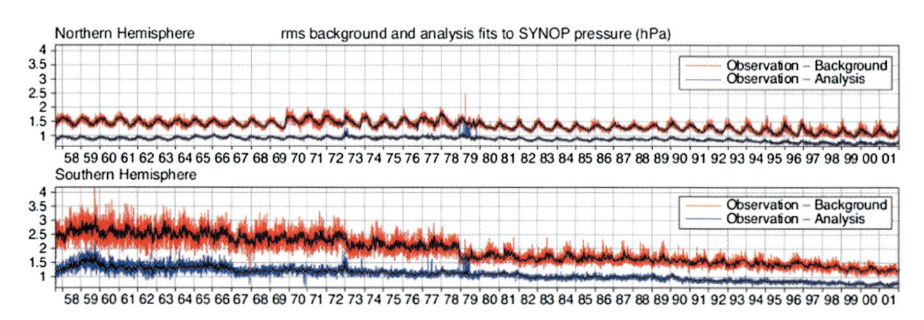

Fig. 4.10 Top: Development of the ERA-40 observing system between 1957 and 2002 with significant contributions from microwave data since 1979 (SSU, MSU), continued by AMSU-A since 1998 and SSM/I since 1987. Bottom: The rms of the background and analysis fits for ERA-40, background (daily (red) and 15-day moving average (black)) and analysis (daily (blue) and 15-day moving average (black)) fits to 00 UTC SYNOP (surface stations) and SHIP (ship observations) surface pressure observations (**a**) over the extratropical northern (upper) and (**b**) southern hemispheres (Uppala et al. 2004).

Note in this context that the temperature information used from AMSU-A observations originates from model-minus-observation departures of 0.1–0.3 K, which sets the target for modelling accuracy requirements. These requirements must be seen in the frame of a data assimilation system in which the short-range forecast of temperature, moisture and wind is already very accurate and where the information from the observations is significant only in relation to the short-range forecast accuracy. The precision is also a function of so called bias-correction schemes (Auligné et al. 2007) that remove the systematic differences between model and observations.

The most apparent problems related to microwave radiative transfer modelling that are relevant for data assimilation in NWP are the simulation of surface emissivity, in particular over land surfaces, the modelling in the presence of clouds and precipitation, and more subtle phenomena such as the impact of the Zeeman line-splitting for channels sensitive to stratospheric layers and Faraday-rotation near 1.4 GHz. A good summary of the status of microwave radiative transfer modelling and all related issues is given by Mätzler (2006).

4.5.4 Impact of Remote Sensing Data on NWP

Data impact in NWP systems can be quantified in various ways by assessing the impact on both analysis and forecast. The latter assumes that better analysis will provide better initial conditions for forecasts. The most prominent impact assessment tool is the Observing System Experiment (OSE) in which new data is added to an existing system and the relative difference to a control system is evaluated. Similarly, individual data sources can be withdrawn from a full system (Andersson et al. 2004).

More sophisticated methods involve the model operators that are used in the data assimilation system. Based on forecast error estimates from the difference between forecasts and verifying analysis, the model and observation operator adjoints can be used to deduce the dependence of this forecast error on individual observation types (Tan et al. 2007) that were used in the initialising analysis (Zhu and Gelaro 2008). An alternative is the use of ensemble-based analysis and forecasting techniques that evaluate forecast impact as a function of ensemble spread with or without specific observation types. Lastly, the Observing System Simulation Experiments (OSSE) provides a framework for observations that do not yet exist and therefore require an observation simulation from independent NWP models (Arnold and Dey 1986).

In clear skies, the main objective for using microwave data in NWP is to constrain temperature and moisture distributions in the analysis. A major OSE impact study was conducted in 2006–2007 to evaluate the impact of the satellite observing system in global NWP at ECMWF (Kelly and Thépaut 2007). The experiments were performed with state-of-the-art modelling and data assimilation system of the time. Fig. 4.10 shows the comparative impact of the Aqua Advanced IR Sounder (AIRS) and the NOAA-16 AMSU-A/B instruments with reference to a baseline (all conventional observations), control (full operational observing system), and the baseline to which only Atmospheric Motion Vectors (AMV) were added. The plot shows the rms error of the forecast for the southern hemisphere (Fig. 4.11a) at 500 hPa geopotential height, a parameter, which is related to large-scale dynamic structures.

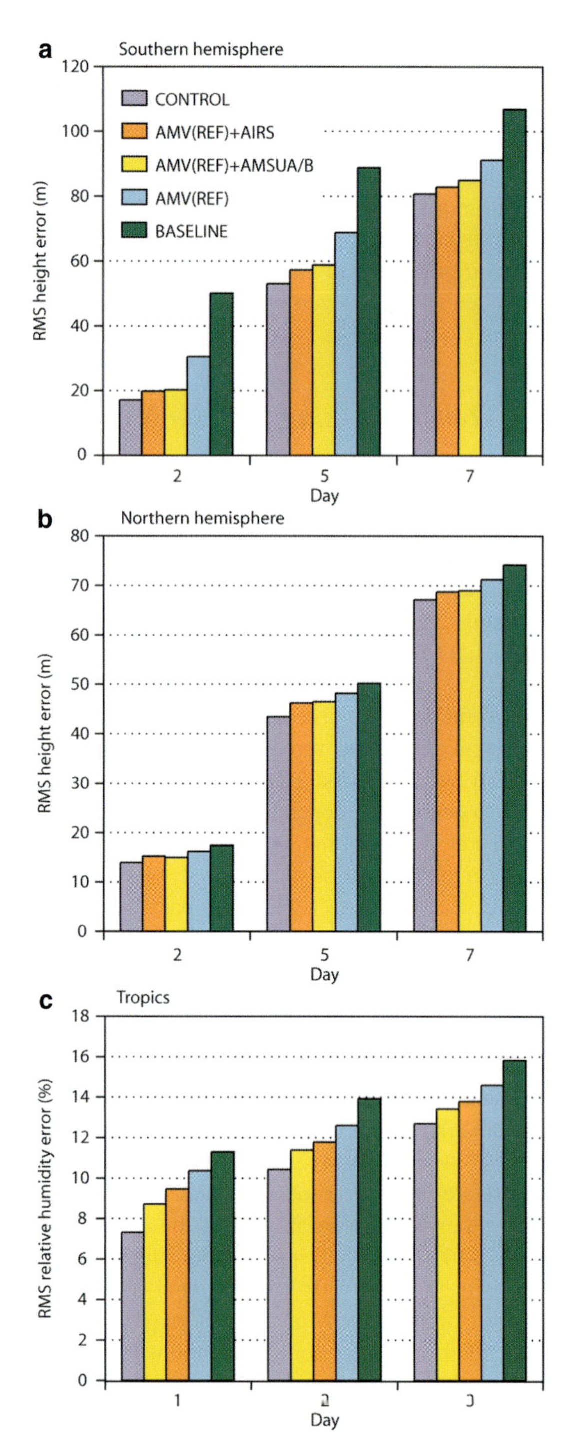

Fig. 4.11 The rms error in the forecast (forecast error) for 500 hPa geopotential height in the southern (**a**) and northern hemisphere (**b**) as well as relative humidity at 500 hPa in the tropics (**c**) from observing system experiments performed at ECMWF (see text for details).

The plot demonstrates that: (a) over the southern hemisphere with few conventional observations even AMV observations produce a significant improvement over the poor baseline configuration despite their limited accuracy and coverage, and (b) that the AMSU-A/B instrument combination produces a very similar relative impact compared to one advanced IR sounder well into the medium range.

The results are similar for the northern hemisphere (Fig. 4.11b) but with a smaller dynamic range due to the stronger constraint from more conventional observations obtained over the continents. Interestingly, the AMSU-A/B combination produces a stronger positive impact for relative humidity than AIRS in the tropics (Fig. 4.11c), mainly due to the weaker sensitivity to clouds. The latter is a significant driver of the valuable contribution of microwave instruments in global NWP.

Fig. 4.12 shows the results from similar OSEs that also include SSM/I observations both in clear and cloud/rain-affected situations. Here, the control and baseline experiments were set up as for Fig. 4.11 but SSM/I observations from two Defense Meteorological Satellite Program (DMSP) satellites (F-13 and 14) were added to a baseline that also contained observations from one AMSU-A instrument. This was necessary for adjusting the large-scale dynamic structures before constraining moisture fields with SSM/I observations.

The SSM/I OSEs tested the separate and combined impact of clear-sky and cloud affected data, both in addition to the baseline and by withdrawing the SSM/I data from the control experiment that contains all observations. The results clearly demonstrate that adding individual systems to a poor baseline always produces a much stronger impact than withdrawing them from a full observing system. This is the consequence of the complementary, and often redundant information provided by various instruments and instrument types. The impact of clear-sky SSM/I data is about as strong as that of cloud-affected data, a remarkable success given the difficulties associated with the assimilation of cloud-affected data mentioned earlier.

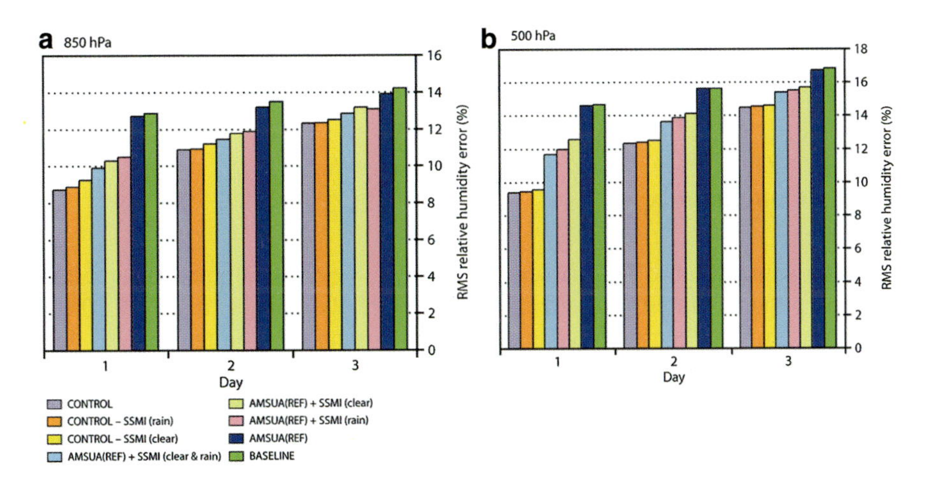

Fig. 4.12 The rms forecast errors for relative humidity at 850 hPa (**a**) and 500 hPa (**b**) in the tropics from SSM/I observing system experiments performed at ECMWF (see text for details).

The results from Kelly and Thépaut (2007) also confirm previous investigations that were dedicated to the assessment of those observing systems that contribute most to the humidity analysis. Andersson et al. (2007) concluded that SSM/I data has the strongest impact in the lower troposphere over oceans complemented by AMSU-B data in the mid and upper troposphere.

Fig. 4.13 shows a different measure of global analysis and forecast impact from selected observation types. Fig. 4.13a shows an example of the information content of rain-affected SSM/I observations in the ECMWF analysis (Cardinali et al. 2004).

The information content shows that a specific observation type contributes to the model analysis. Numbers are between 0 (the observation has no impact on the analysis) and 1 (only the observation determines the analysis). Large values of observation impact are produced along the ITCZ, SPCZ and other areas of significant precipitation occurrence in tropical and mid-latitude weather systems. This impact on the analysis is manifested in the forecast by a significant reduction of forecast errors (in terms of dry energy norm reduction (Zhu and Gelaro 2008), i.e. negative numbers/blue areas) with a similar geographical distribution (Fig. 4.13b).

The same metric of forecast error reduction was produced for selected observation types and for winter 2006/2007 (Fig. 4.13c). They indicate that AMSU-A represents the strongest system, i.e. the largest sensitivity of forecast error reduction (negative numbers) is obtained. This is from the combined effect of the impact of AMSU-A per observation and the large number of observations maintained in the system at that time (AMSU-A onboard NOAA-16/17/18, Aqua and METOP). Microwave observations related to humidity receive much less weight (SSM/I, AMSU-B), which is, to a large extent, explained by the fact that, in the current formulation, the forecast error term does not include the error contributions from atmospheric moisture. Fig. 4.13c shows that SSM/I data in both clear and cloud/rain-affected areas contribute to forecast error reduction. The differences between summer and winter seasons affect both magnitude and relative impact of observing systems (not shown).

4.5.5 Conclusions

Together with advanced IR sounders, microwave sounders and imagers represent the most important satellite observing systems currently available for NWP. These instruments mainly constrain temperature and moisture fields in the analysis, but are increasingly used in cloud and precipitation-affected areas in which IR observations only provide information on the atmosphere above clouds. These satellite observations are mostly assimilated as radiances due to the accuracy obtained with radiative transfer models – this even applies to a large degree to multiple scattering calculations. This conclusion is mainly valid for global modelling systems, but will increasingly be true for regional systems.

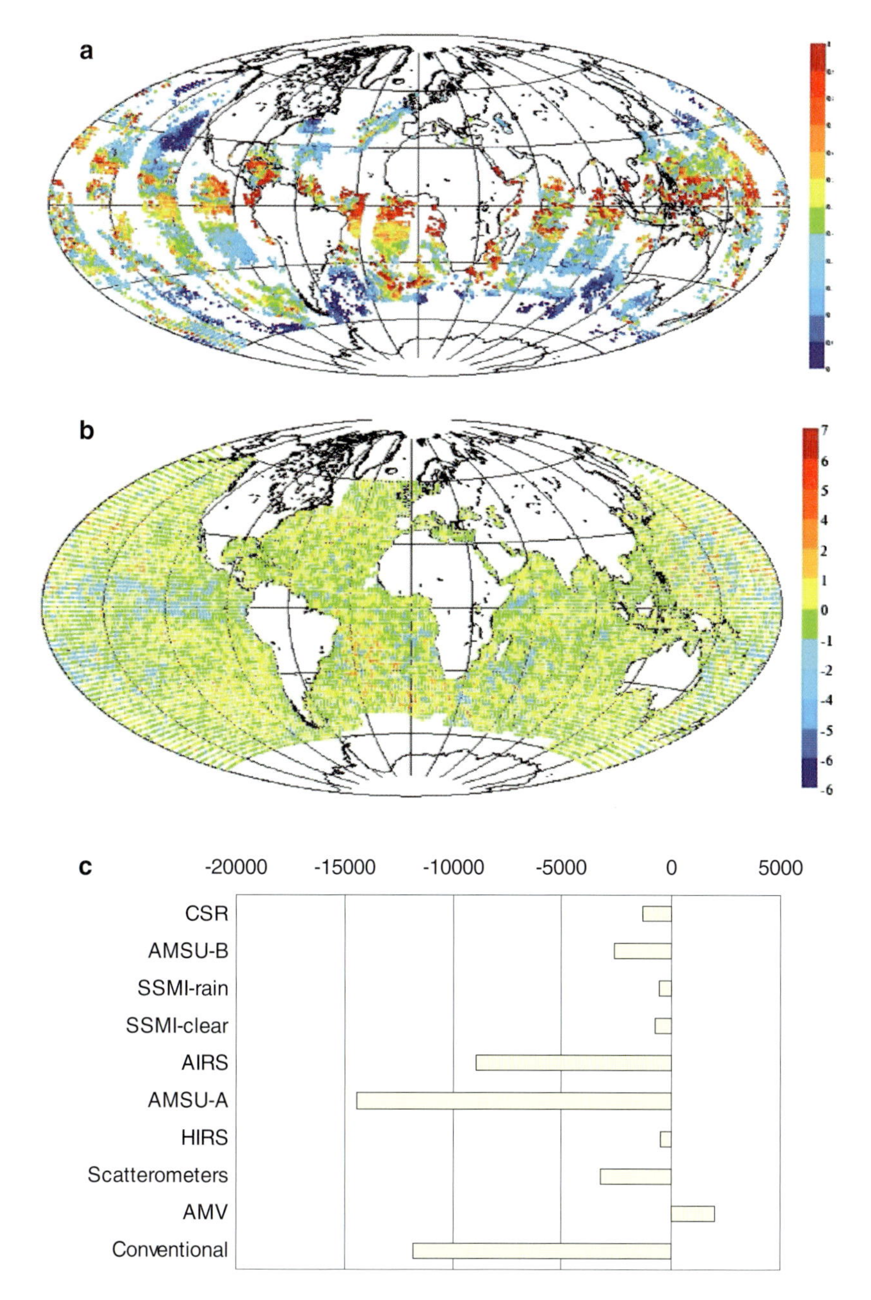

Fig. 4.13 (**a**) Information content of rain-affected SSM/I observations in ECMWF analysis; (**b**) 24-h forecast error sensitivity (dry energy norm in J/kg) to rain-affected SSM/I observations; and (**c**) accumulated sensitivity to all observation types for winter 2006/2007. Courtesy Carla Cardinali).

In the future, active microwave observations will deserve attention: currently Cloudsat radar and precipitation radar (PR) (e.g. TRMM PR see Section 4.7), serve mainly for validation but are expected to become part of the assimilated satellite data suite (Benedetti et al. 2005).

Currently, soil moisture is only analysed from near-surface temperature and humidity, but microwave observations have the potential to improve greatly on this situation through their sensitivity to top-soil water content as a function of land surface type, vegetation, soil moisture itself and observation frequency (Seuffert et al. 2003). Initial results using TMI data show both improved moisture analysis as well as impact on lower tropospheric moisture and clouds (Drusch 2007). Similar studies (Scipal et al. 2008) investigate the impact to be expected from soil moisture missions using existing scatterometers like METOP's Advanced SCATterometer (ASCAT).

4.6 Microwave Limb Sounding of the Troposphere

4.6.1 Background to Microwave Limb Sounding of the Troposphere

The use of limb sounding geometries for space-borne observations of atmospheric composition dates back to 1975 with the Limb Radiance Inversion Radiometer (LRIR) on NASA's Nimbus-6 and in 1978 with the Limb Infrared Monitor of the Stratosphere (LIMS) and Stratospheric and Mesospheric Sounder (SAMS) on Nimbus-7. Limb sounding – viewing the atmosphere "edge on" – offers significant advantages over nadir or near nadir sounding, particularly for atmospheric composition studies. Scanning an instrument's field of view across the Earth's limb brings a wealth of information on the vertical structure of the atmosphere, enabling observations with a higher resolution than is typical of nadir techniques. In addition, the longer viewing path lengths associated with limb sounding (200–500 km compared to *ca.* 20 km for nadir geometries) enhances the signature of trace gases in the observed spectra. However, these advantages typically come at the cost of poorer horizontal resolution than nadir-swath viewing instruments. In addition, limb sounding instruments typically require more precise optics and moving parts, and hence complexity, than nadir sounders using the same wavelength regions.

The microwave region of the electromagnetic spectrum is particularly attractive for limb sounding of atmospheric composition. This is because a wide variety of important atmospheric species are readily observable in the microwave where spectral lines are typically associated with rotational transitions, so that any molecule having a dipole moment is, in principle, observable. In addition microwave signals are unaffected by thin cirrus clouds or atmospheric aerosols that significantly hamper observations at shorter wavelengths, particularly when viewed in limb geometry where the long path length reduces the probability of observing sufficiently cloud-free scenes. Thermal emission by the dry and moist air continua limits microwave limb sounding observations to altitudes above approximately 6 km, except in the extremely dry and cold conditions occasionally found in polar regions.

The millimetre-scale wavelength of microwave radiation observed has important implications for instrument design. In particular, the spatial resolution of microwave instruments is set by diffraction effects rather than being determined mainly by geometrical optics as at shorter wavelengths. The vertical width of a microwave limb sounder field of view is defined by the aperture size and wavelength. Increasing the aperture size proportionally decreases the width of the field-of-view (thus the total power received is invariant). For example, with the Aura Microwave Limb Sounder, the primary antenna length of 1.6 m gives a vertical field-of-view width at the limb of ~4.5 km at 190 GHz and ~1.5 km at 640 GHz.

Thermal emission and absorption lines are subject to both Doppler and pressure broadening, with pressure broadening dominating for microwave lines in the troposphere and stratosphere. Most spectral lines broaden by ~3 MHz/hPa, giving a typical linewidth of ~300 MHz to 1.5 GHz for signatures in the upper troposphere (100–500 hPa). This broadening, combined with increasing interest in remote sounding of tropospheric composition, has driven the development of wide bandwidth microwave observing systems and receivers for atmospheric science applications.

4.6.2 Previous, Existing and Planned Microwave Limb Sounding Instruments

The Microwave Limb Sounder (MLS) on NASA's Aura satellite is a successor to MLS on the Upper Atmosphere Research Satellite (UARS) (Barath et al. 1993), making observations in five spectral regions from 118 GHz to 2.5 THz (Waters et al. 2006), measuring more than 14 atmospheric species. Aura MLS's receiver and spectrometer bandwidths are appreciably increased over those of UARS MLS, allowing composition observations to extend into the upper troposphere (notably for O_3, CO and HNO_3), in addition to the established water vapour and cloud-ice observations.

In addition to the UARS and Aura MLS instruments, three other space-borne microwave limb sounding instruments have flown or are nearing launch. The Millimeterwave Atmospheric Sounder (MAS) (Croskey et al. 1992) was a component of the ATmospheric Laboratory for Application and Science (ATLAS) I, II and III missions flown by the NASA space shuttle in March 1992, March/April 1993 and November 1994. The Odin Sub-Millimeter Radiometer (ODIN-SMR) (Murtagh et al. 2002) is a four-band microwave instrument designed for limb sounding of stratospheric and mesospheric composition, and also measures water vapour and cloud ice in the upper troposphere. The Odin mission is notable in that atmospheric viewing is time-shared with astronomical observations. The Japanese Experiment Module Superconducting Submillimeter-Wave Limb-Emission Sounder (JEM-SMILES) launched successfully in September 2009, and attached to the international space station, uses, for the first time in space, low-noise superconducting receivers to make high precision observations of stratospheric chemistry in tropical and mid-latitude regions.

4.6.3 Applications of Microwave Limb Sounding of the Troposphere

The upper troposphere is an important and somewhat poorly observed region of the atmosphere, with microwave limb sounding (notably from Aura MLS) providing one of the few high vertical resolution daily global satellite datasets. The upper troposphere is where both water vapour and ozone – two strong greenhouse gases – have their largest impact on Earth's radiative balance (Held and Soden 2000; Forster and Shine 1997). The processes that control stratospheric humidity, an important issue for polar stratospheric ozone loss, act in the tropopause transition layer (TTL), essentially the tropical upper troposphere and lowermost stratosphere. Intercontinental transport of air pollution in the upper troposphere, where winds are generally fastest, is an important and currently poorly modelled process affecting global and regional air quality (Stohl et al. 2002; Wang et al. 2006).

Most microwave limb sounding observations are based on spectrally-resolved observations of an individual line or line cluster. However, geophysical information can also be derived from measurements of absolute radiance. In particular, absolute radiance values in the 204 GHz "window" region of the spectrum, used in UARS MLS to measure stratospheric ClO, convey information on upper tropospheric humidity due to strong emission from the water vapour continuum. This information was used to provide the first daily global vertically-resolved observations of upper tropospheric humidity (UTH) from 465 to 147 hPa (Read et al. 1995; 2001). Later this information was combined with the 183 GHz observations of stratospheric water vapour (from 68 to 0.0046 hPa) (Pumphrey et al. 1998) to produce a merged dataset covering the entire TTL (Read et al. 2004a).

UARS MLS UTH data have been used in studies of climate-related phenomena such as El Niño (Chandra et al. 1998; Waters et al. 1999) and the Madden-Julian oscillation (MJO) (Madden and Julian 1971; 1994), see, for example, Stone et al. (1996), Mote et al. (2000) and Ziemke and Chandra (2003). The UARS MLS UTH observations also give needed new insight into the processes in the TTL that regulate the humidity of the stratosphere (Read et al. 2004b) and give rise to the so-called tape-recorder variation in stratospheric humidity (and other species as illustrated in Fig. 4.14, itself discovered in UARS MLS stratospheric H_2O data (Mote et al. 1995; 1996)).

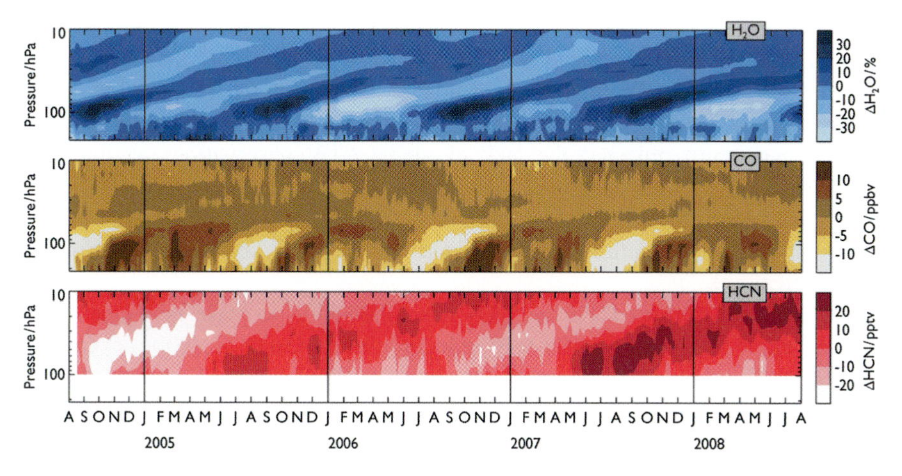

Fig. 4.14 Aura MLS observations of (top) water vapour, (middle) CO, and (bottom) HCN anomalies (differences from long-term mean). All these species show "tape recorder" signatures, with rising bands of alternately enhanced and reduced abundances. In the case of water vapour, as tropical air slowly rises from 12 km (where the bulk of convective outflow occurs) into the stratosphere it passes the "cold point" where it is freeze dried. The humidity of an air parcel thus reflects the coldest temperature it was exposed to. The annual cycle in the cold point temperature is thus recorded in the humidity of air, which subsequently rises in the tropical lower stratosphere. CO is a product of combustion both industrial and biomass related, the CO tape recorder signature reflects the annual cycle in CO emissions due mainly to biomass burning. The shorter chemical lifetime of CO in the stratosphere, compared to water vapour, leads to more rapid dissipation of the tape recorder signal in the vertical. HCN is mainly produced in forest fires. In the record shown, the HCN tape recorder seems to exhibit a two-yearly cycle rather than an annual cycle (Pumphrey et al. 2008), which probably reflects inter-annual variability in forest fires.

The Aura MLS instrument, having a broader measurement bandwidth than UARS MLS, is able to resolve spectrally the 183 GHz water vapour line with more than 3 GHz of instantaneous bandwidth. This enables Aura MLS observations of upper tropospheric humidity to be based on spectral contrast rather than absolute radiance measurement, providing contiguous data from one spectral region from the upper troposphere to the mesosphere (Read et al. 2007). The combination of these data with simultaneous MLS observations of atmospheric composition and cloud ice has provided new insights into important physical processes in the upper troposphere and lower stratosphere, as discussed below.

The UARS and Aura MLS instruments provided the first daily global observations of vertically resolved CIW in the upper troposphere. Scattering from large-particle ice clouds affects microwave limb radiances leading to radiance enhancement or suppression depending on the limb view angle. Initial observations from UARS MLS (Wu et al. 2005) indicated that deep convection was the dominant source of observed variability in relative humidity at 100 hPa. More recent CIW observations from Aura MLS (Wu et al. 2008) offer improved sensitivity and vertical registration, and some particle size information through observations covering a large frequency range.

Cloud processes, while critical to many aspects of weather and climate, are generally poorly captured in global models. Prior to the Aura MLS CIW observations, state-of-the art climate models exhibited differences by factors of *ca.* 20 in their estimates of CIW (compared to ~20% for parameters such as precipitation). This disagreement reflected the lack of global CIW observations available to constrain the model estimates. MLS CIW observations were used by ECMWF to justify changes to their cloud microphysics parameterisations, leading to improvements in their representation of tropical deep convection (Bougeault 2006, personal communication).

MLS observations of UTH and CIW have enabled us to make new quantifications of cloud and water vapour feedbacks on climate. Su et al. (2006) examined the relationship between sea surface temperature (SST) and MLS observations of UTH and upper tropospheric cloud ice (used as a measure of the strength of deep convection) overhead. The authors found a dramatic increase in the positive correlation of both UTH and cloud ice with SST for SSTs above 300K (see Fig. 4.15), and estimated that this accounted for ~65% of the previously-observed "super greenhouse effect".

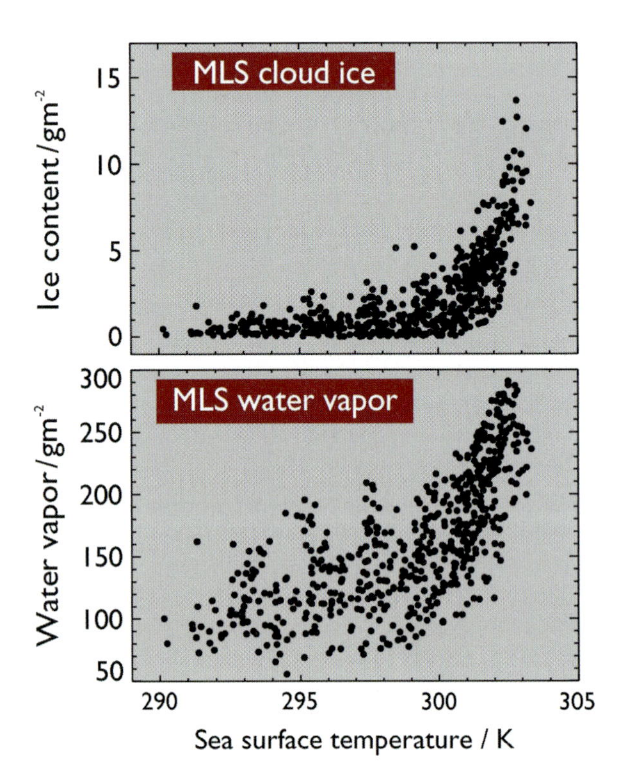

Fig. 4.15 Aura MLS observations of 215 hPa cloud ice (an indicator of deep convection) and water vapour correlated with underlying sea surface temperature (SST). Adapted from Su et al. (2006).

In addition to these data from Aura, similar humidity and cloud-ice datasets have recently been produced from Odin SMR (Eriksson et al. 2007; Ekström et al. 2007). Comparisons of the humidity observations with those from Aura MLS and the airborne measurements of ozone and water vapour by MOZAIC show agreement within 10% – better than the estimated accuracies of the instruments involved (Ekström et al. 2008). Comparisons of SMR cloud ice observations with CloudSat and Aura MLS indicate a common accuracy of 70% – sufficient for directing improvements to model parameterisations (Eriksson et al. 2008b).

4.6.4 Upper Tropospheric Composition and Chemistry

Although Odin-SMR stratospheric composition observations have been used in studies of tropospheric phenomena (Ricaud et al. 2007), the only direct observations of upper tropospheric composition for species other than water vapour, have been observations of O_3, CO and nitric acid, HNO_3 from Aura MLS (Filipiak et al. 2005; Livesey et al. 2008; Santee et al. 2007).

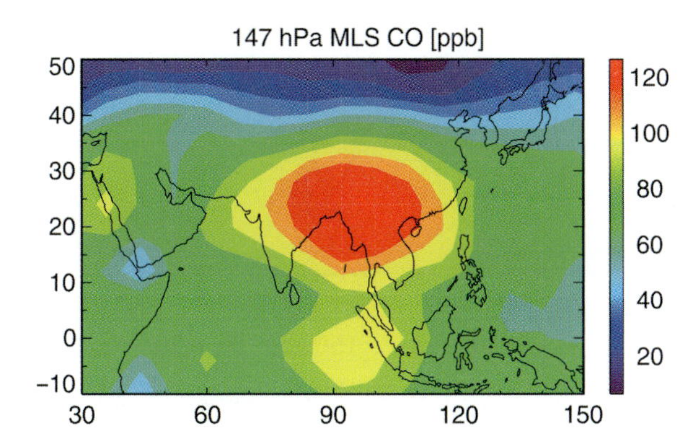

Fig. 4.16 Trapping of pollution in the upper troposphere by the anticyclone over the Asian monsoon region, as seen in 147 hPa observations of CO from the Aura Microwave Limb Sounder. Adapted from Li et al. (2005).

CO is a well established tracer of atmospheric pollution both from biomass burning and industrial sources (Stohl et al. 2002) and MLS upper tropospheric CO data show clear evidence of the trapping of pollution in the upper tropospheric anticyclone over the Asian monsoon region (Li et al. 2005) (see Fig. 4.16). MLS data also show a strong influence of convection on upper tropospheric CO abundance and long range transport (Jiang et al. 2007) (see Fig. 4.17 for an example). MLS CO observations in the upper troposphere and lower stratosphere have shown a strong "tape recorder" signature, as seen in water vapour (Schoeberl et al. 2006).

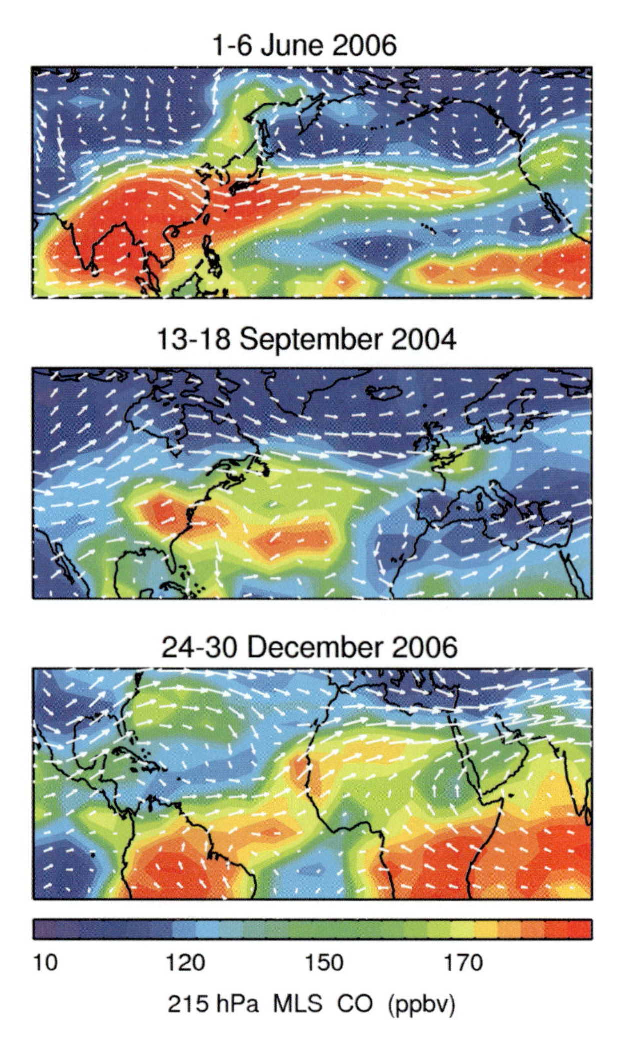

Fig. 4.17 Sample 5 day averages of Aura-MLS CO observations at 215 hPa, showing clear examples of long-range pollution transport in the upper troposphere. The arrows show winds from the GEOS-4 analysis.

The MLS observations of CO and H_2O together with other measurements have shown interesting insights into the processes that act in the TTL (Liu et al. 2007; Read et al. 2008).

MLS observations of O_3, CO and HNO_3 in the upper troposphere and lower stratosphere offer needed new insights into processes such as stratosphere/troposphere exchange that influence the abundance of tropospheric ozone and hence climate. Sample MLS observations of these species are shown in Fig. 4.18 at latitudes with ozone-poor air in the tropical upper troposphere. The "wave one"

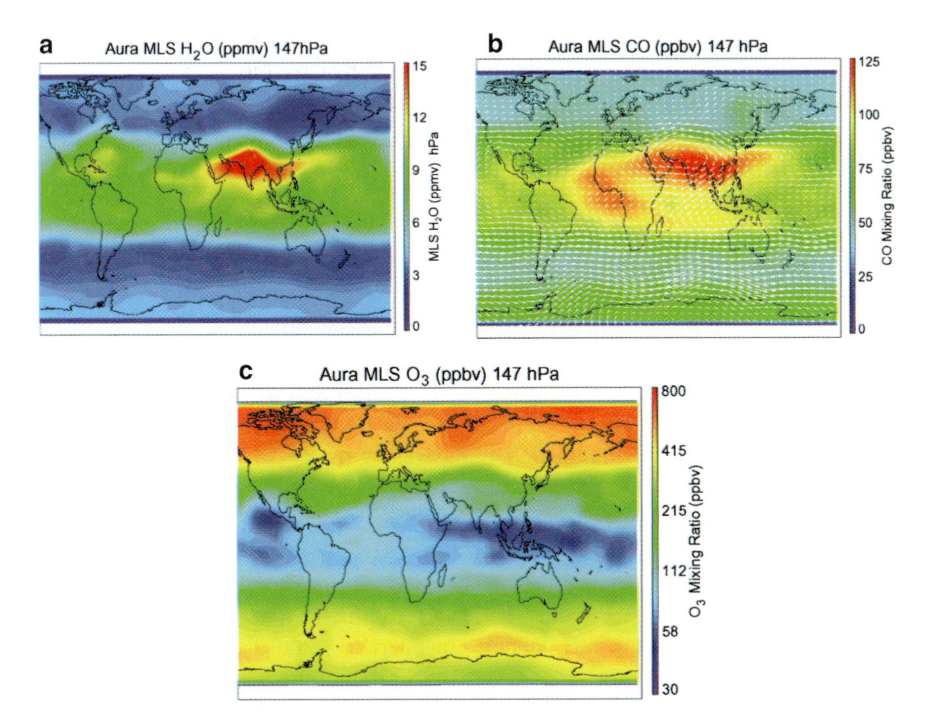

Fig. 4.18 Aura MLS observations of atmospheric composition at 147 hPa (~12 km). This pressure level is in the upper troposphere in the tropics and in the lower stratosphere at higher latitudes. (**a**) Water vapour, showing a clear distinction between the moist tropical upper troposphere and the arid lower stratosphere at higher latitudes. Meridional gradients follow the morphology of the tropopause (not shown). The signature of moist air from strong convection over the Indian subcontinent is clear. (**b**) CO, showing enhanced CO from convective vertical transport of polluted air over India, largely trapped by the anti-cyclonic circulations in this region, with some outflow over the Pacific. There are also indications of the transport of polluted air from the south east coast of the USA into the Atlantic Ocean. (**c**) O_3, showing the clear distinction between ozone-rich lower stratospheric air at mid- and high latitudes with ozone-poor air in the tropical upper troposphere. The "wave one" pattern in the tropics with low O_3 over the Pacific Ocean is a recurrent feature due, in part, to the convective transport of ozone-poor lower tropospheric air to these altitudes in this region.

pattern in the tropics with low O_3 over the Pacific Ocean is a recurrent feature, due in part to the convective transport of ozone-poor lower tropospheric air to these altitudes in this region.

4.6.5 Conclusions

The ability of microwave limb sounding techniques to retrieve tropospheric composition in all but the most cloudy of situations is an important tool for quantifying critical processes in the upper troposphere. These processes include the long-range transport of air pollution, the influence of fast processes such as convection on tropospheric composition and hence radiative forcing, the influx of ozone-rich air from the stratosphere, and the transport of ozone depleting substances from the troposphere to the stratosphere. Observations from Aura MLS and Odin-SMR are providing needed new insights into these processes, however they lack the spatial and temporal resolution needed for full quantitative understanding.

The Stratosphere Troposphere Exchange And climate Monitor Radiometer (STEAM-R) is an instrument concept under development (led by the Odin-SMR team in Sweden) to make measurements in the upper troposphere and lower stratosphere with the improvements in vertical resolution needed to better quantify the processes listed above. STEAM-R uses a multi-beam array of fourteen 310– 360 GHz receivers simultaneously measuring from 5 to 28 km altitude with 1–2 km vertical resolution. Along-track spacing will be 30–50 km. STEAM-R measurement goals include O_3, CO, H_2O, nitric acid, HNO_3, nitrogen dioxide, N_2O, hydrogen cyanide, HCN, acetonitrile, CH_3CN, methyl chloride, CH_3Cl, chlorine monoxide, ClO and HDO. STEAM-R is a component of the Process Exploration through Measurements of IR and millimetre-wave Emitted Radiation (PREMIER) mission currently under consideration by the European Space Agency, and is intended to fly in a sun-synchronous low earth orbit, ideally in formation with meteorological missions such as METOP.

One of the most revolutionary advances in microwave techniques in recent years has been the development of Superconductor-Insulator-Superconductor (SIS) receivers. These offer a factor of *ca.* 30 improvement in signal-to-noise ratio (equivalent to a factor of about 1000 reduction in required integration time for a given signal to noise ratio). SIS technology has been deployed in ground-based and airborne astronomical and atmospheric applications for over 20 years, but it is only the recent development of space flight qualified coolers, capable of achieving the 4K temperatures required, that enable their use in orbit. The Japanese SMILES presently in orbit makes use of SIS technology.

A new SIS-based concept under development is the Scanning Microwave Limb Sounder (SMLS), which adds a horizontal scanning capability enabled by the dramatic reduction in needed integration time. SMLS can measure a ~65°-wide swath, equivalent to ~5,500 km width from an 830 km altitude orbit. SMLS observations would be spaced at ~50×50 km^2 in the horizontal. Such a swath observes 44% of the globe on every orbit. By choosing the appropriate orbital inclination, an SMLS instrument can make multiple measurements per day over large regions of the earth (see Fig. 4.19).

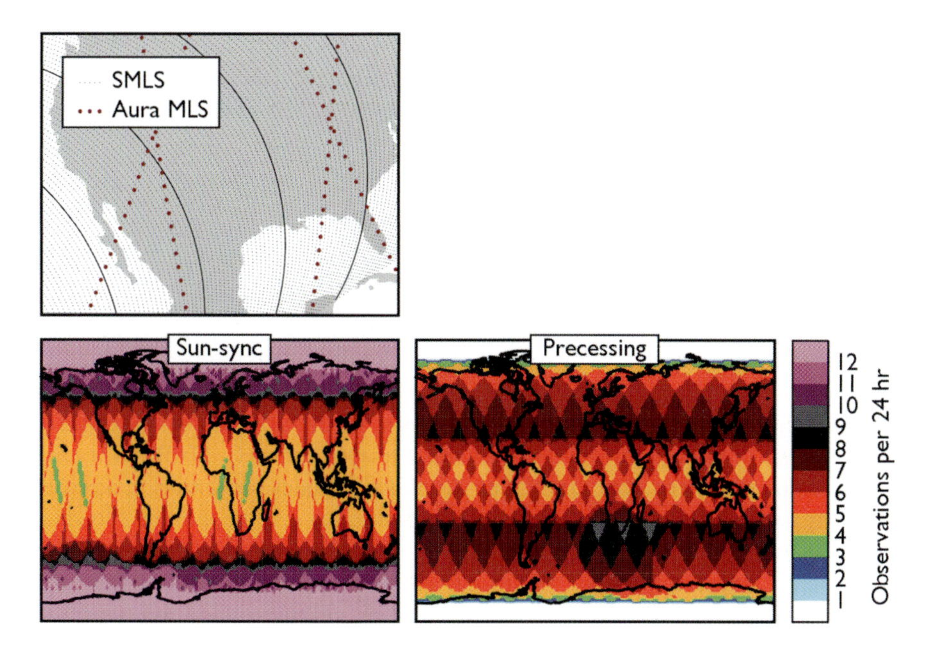

Fig. 4.19 (Top) SMLS observations (black dots) at 50 × 50 km^2 spacing for a single orbit, compared to Aura MLS observations for 24 h. Points in every 20th SMLS scan are connected for illustration. (Lower left and right) number of SMLS observations in 24 h for sun-synchronous and 52° inclined precessing orbits.

4.7 Active Techniques

4.7.1 Introduction

Two radar systems are currently flying in space. One is the Precipitation Radar (PR), which is part of TRMM, (see also Section 4.7.5). The second is the Cloud Profiling Radar (CPR) of the CloudSat mission, which observes both clouds and precipitation. These two radars are short-pulsed profiling radar systems that primarily measure the power backscattered by atmospheric targets (hydrometeors, such as ice crystals and cloud droplets) and any other target intercepted by the antenna beam (e.g. the Earth's surface). The target range for these systems is simply determined from the out-and-back time-of-travel of the pulse and, therefore, ranging by nadir viewing space-borne radars gives the altitude of the targets. The narrow antenna beam and the motion of the spacecraft resolve targets in the along-track direction.

The TRMM PR, launched in November 1997, transmits at 14 GHz and scans approximately 200 km across track. The CloudSat CPR, launched in April 2006, is the first space-borne 94 GHz radar and is non-scanning, staring to Earth slightly offset from the nadir (0.16° off- geodetic nadir). The ESA EarthCare mission is to include a similar radar to CloudSat but with additional capabilities for providing Doppler motion measurements.

4.7.2 The CloudSat Radar

Cloud particles of a size typical of that measured in non-precipitating water and ice clouds are weak scatterers of microwave radiation. Data collected from cloud radars over many years reveal how the reflectivity of clouds varies over several orders of magnitude. The reflectivity factor ranges from below −30 dBZ[1], around the edges of the upper ice layers and low-level water clouds, to approximately 20 dBZ in heavier precipitation (Stephens et al. 2002). Low-level water clouds, in particular, are very dim targets and represent one of the main challenges for the current CloudSat mission. The reflectivity of the lower underlying Earth surface is much larger than that of clouds, typically 40 dBZ or greater, and varies as a function of surface type and condition, and can be influenced over land by vegetation, soil moisture, and snow depth for example, and surface wind speed over oceans. CloudSat is beginning to provide a wealth of new information about surface reflectivity.

The two-way attenuation of the radar pulse as it propagates through the atmosphere is not negligible at 94 GHz. This attenuation results from absorption by gases (chiefly water vapour), liquid water droplets, and precipitation sized particles (see Section 4.2). The dominant attenuation is from precipitation, which, if heavy enough, attenuates the CPR signal completely. Because of its magnitude, this attenuation by precipitation can be exploited to detect and estimate precipitation intensity.

Multiple scattering of radar pulses becomes an issue when space-borne radar footprints begin to exceed a kilometre, that is, become similar in size to the mean free path of microwave photons in hydrometeor suspensions. Through modelling and analysis, multiple scattering has been shown to affect interpretation of ranging of the CPR observations primarily for precipitation that exceeds about 5 mm/h (Haynes et al. 2008).

4.7.3 The CloudSat Mission

The CloudSat Mission was jointly developed by NASA, the JPL, the CSA, CSU, and the US Air Force. For CloudSat's CPR see Im et al. (2005). It is the first space-borne 94-GHz radar and provides unique information about the vertical cloud

[1]The different echo intensities (reflectivity) are measured in dBZ (decibels of Z, the amount of transmitted power returned to the radar receiver). The dBZ values increase as the strength of the signal returned to the radar increases.

profiles over the globe. The CPR instrument on CloudSat began operation in 2006 (Stephens et al. 2008). Since that time, CPR has been acquiring the first-ever continuous global time series of vertical cloud structures and vertical profiles of cloud liquid water and ice content, and precipitation incidence. The vertical resolution is 485 m and the spatial resolution is defined in terms of the antenna 3 dB footprint being 1.4 km. In order to take full advantage of observations by other types of space-borne atmospheric remote sensing instruments, the CloudSat spacecraft flies in formation as part of the Afternoon Constellation of satellites, the so called A-Train, (Stephens et al. 2002). In particular CloudSat flies in close formation within the Cloud-Aerosol Lidar and Infrared Pathfinder Satellite Observations (CALIPSO) (Winker et al. 2007), which carries a lidar system, so that their respective beams cover the same vertical column within about 15 s.

4.7.4 The Cloud Profiling Radar

The need to detect the weak cloud signals was the over-riding requirement on the CPR stated in terms of a minimum detectable cloud reflectivity Z_{min} approximately −28 dBZ at beginning of life. This requirement then dictated the frequency of operation of the radar and related technologies. This minimum sensitivity represents a sensitivity almost five orders of magnitude greater than the TRMM PR.

The CPR provides profiles of atmospheric hydrometeors. The quantity of immediate relevance to these profiles is the range-resolved radar cross-section per unit volume, η, at a specific range r defined as:

$$\eta = \frac{P_{rec}(4\pi)^3 r^2 L_a}{P_t \lambda^2 G_{rec} G^2 \Omega \Delta} \tag{4.4}$$

where P_{rec} is the output power of the receiver, P_t is the transmitted power, λ is the wavelength, G_{rec} is the receiver gain, G is the antenna gain, r is the range to the atmospheric target, Ω is the integral of the normalized two-way antenna pattern, Δ is the integral of the received waveform shape, and L_a is the two-way atmospheric loss. The quantity η is converted to the equivalent (attenuated) range-resolved reflectivity factor:

$$Z_e = \eta \frac{\lambda^4 10^{18}}{\pi^5 |K_w|^2} \tag{4.5}$$

where $|K_W|$ is set to 0.75, representative for water at 10°C at approximately 90 GHz.

Absolute calibration of the radar requires precise knowledge of the various parameters in Eq. 4.4 above. Pre-launch calibration parameters were obtained either directly from laboratory measurements or by analysis of experimental data. In orbit, the transmit power P_t. and receiver gain G_{rec} are routinely measured via internal calibration channels of the radar. Both have remained remarkably stable since launch. End-to-end system calibration is also evaluated using measured backscatter off the ocean surface. The method relies on measuring the backscatter off the ocean surface at an angle at which the sensitivity to wind is minimal. Over the 12 months between August 2006 and August 2007, the CloudSat spacecraft performed several calibration manoeuvres over pre-selected cloud-free oceanic areas and verified that the absolute calibration of the CPR is well within the required 2 dB requirement (Tanelli et al. 2008).

The in-orbit performance of the radar has also been independently verified by comparison with radar measurements obtained from several airborne programs since launch, matched in volume specifically to the space-borne radar observations. Analysis shows the reflectivity of the space-borne and airborne radars agree within 2 dB except in the region of heavy precipitation where multiple scattering in the CloudSat footprint becomes an issue.

Z_{min} is an important instrument design parameter. This minimum factor is defined as the cloud reflectivity factor Z_e, which, after averaging and noise subtraction, yields a power equal to the noise power standard deviation. Z_{min} is therefore determined by the equivalent noise floor and by the number of transmitted pulses. The noise floor ultimately depends on the radiometric temperature of the observed scene at 94 GHz. Based on the current calibration, the minimum detectable reflectivity ranges from -29.9 dBZ to -30.9 dBZ. Seasonal changes in temperature, land cover, and sea ice affect the distribution of Z_{min}.

The first images from CloudSat (Fig. 4.20) presents the historic first-look CPR image of the vertical structure of a warm front over the North Atlantic observed on 20th May 2006.

This image was acquired immediately after activation of CPR as part of a brief 4-h checkout test. The richness of the CloudSat information for studying these classic weather systems has since been highlighted by Posselt et al. (2008). These initial test data were transmitted in near real time and near-real time data are currently being exploited in a number of operational applications described elsewhere (Mitrescu et al. 2008).

4.7.5 The Tropical Rainfall Measurement Mission

Within the Mission to Planet Earth program (MTPE) (Simpson et al. 1988) TRMM was developed in a collaborative effort between NASA and the Japanese National Space Development Agency (NASDA) - now the Japan Aerospace Exploration Agency (JAXA) (Simpson et al. 1996; Kummerow et al. 2000; http://trmm.gsfc.nasa.gov/). The TRMM research program is dedicated to measuring tropical-subtropical rainfall over a long time period (measurements are now in their 12th year), to collect the first representative and consistent ocean climatology of precipitation and latent heating. The TRMM satellite was initially launched into a

a

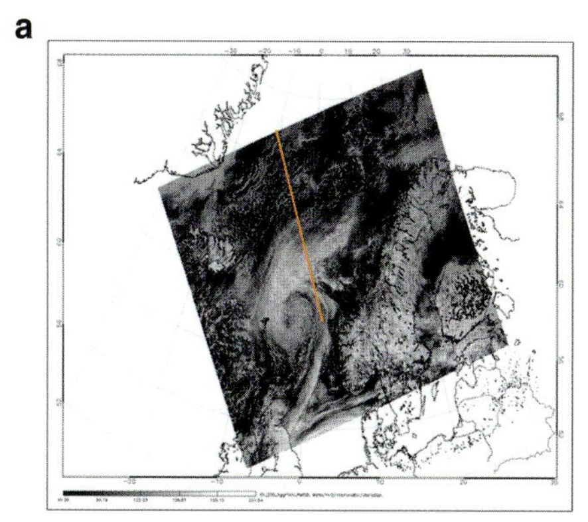

b

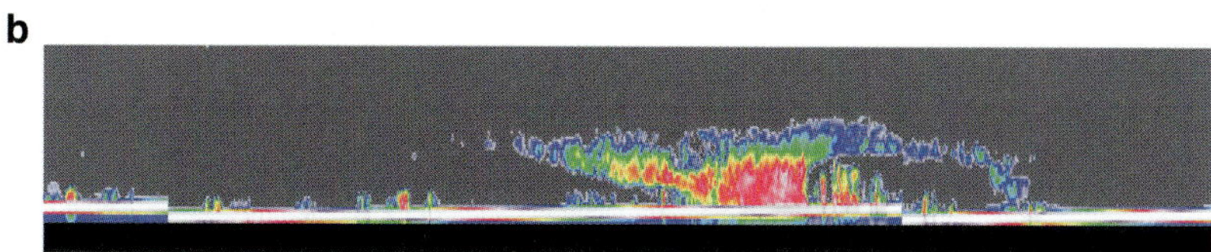

Fig. 4.20 A MODIS image (upper) of a warm frontal system intersected by CloudSat along the orbit track. The first quick-look image of CPR reflectivity gathered for an approximate 1,400 km section of orbit on 20th May 2006 is shown in the lower panel captured immediately after the first turn on of the CPR. The height of the image represents 30 km. (from Stephens et al. 2008).

low-altitude (350 km), non-sun-synchronous orbit inclined 35° to the Earth's equatorial plane, with a nominal mission lifetime of three years. During August 2001, the TRMM orbit altitude was boosted to 400 km in order to lower drag and reduce fuel consumption. The TRMM satellite is expected to remain operational until 2012/2013, at which time it will no longer have sufficient fuel for station keeping. The satellite measures precipitation over tropical latitudes, the coverage extends from 35°S to 35°N, which includes the sub-tropical zones. The tropical-sub-tropical regions are important because more than two thirds of global rainfall occurs there, while precipitation-induced latent heating strongly controls the large-scale general circulation, and various seasonal/intra-seasonal modulated synoptic scale and mesoscale weather disturbances.

TRMM carried a precipitation radar (PR) using a through-nadir scanning, non-coherent, $2 \times 2\ m^2$ slotted wave-guide-phased array antenna providing a ~220 km swath width. The radar system operates at 13.8 GHz, and is capable of generating near real time 3-D pictures of rain rate with a calibration accuracy of ~0.5 dBZ. The radar was developed by NASDA's partner agency in Japan, NiCT. Its design and capabilities have been described in Okamoto et al. (1988), Meneghini and Kozu (1990), Nakamura et al. (1990), Okamoto and Kozu (1993), Kozu et al. (2001), Okamoto (2003) and Okamoto and Shige (2008).

The TRMM satellite also carries the TRMM microwave imager (TMI). This is a conical-scanning radiometer with a swath width of ~760 km, 9 channels at 10.7 GHz, 19 GHZ, 21.3 GHz, 37 GHz and 89 GHz, all channels are dual polarized vertical and horizontal with the exception of the vertical polarized water vapour channel at 21.3 GHz. All channels make use of the same 0.6 m diameter antenna. The TMI has been described by Kummerow et al. (1998; 2000), Smith and Hollis (2003), and Fiorino and Smith (2006). The TMI measurements allow estimates of rain rates to be made because of a well known relationship between the microphysical properties of rain and the up welling microwave radiation. The various frequencies are important for obtaining information at different vertical levels of the atmosphere. The swath width of the TMI is three times greater than the PR (~780 vs. ~220 km) and enables the TMI to measure precipitation over a wide area with a high duty cycle. The PR measures the detailed physics of precipitation along the narrow radar track at the centre of the radiometer track. The combined radar-radiometer algorithms have become the most accurate of the TRMM data analysis (Smith et al. 1995a).

Three additional instruments on the TRMM satellite are used to study precipitation indirectly, these are: (a) the 5-channel Visible and InfraRed Scanner (VIRS), (b) the Lightning Imaging Sensor (LIS), and (c) the prototype Cloud and Earth's Radiant Energy System (CERES), see Kummerow et al. (1998) for details.

4.7.6 Results from TRMM

The TRMM project is very successful in providing high quality geophysical parameters at levels 2 and 3 based on data collected by the different instruments. The L2 algorithm produces instantaneous rain rates at full spatial resolution (i.e. at the resolution of either the radar beam or a mean convolved radiometer beam) while the L3 algorithms produces monthly-averaged rain rates at $5 \times 5°$ spatial resolution.

Fig. 4.21 compares monthly-averaged rain maps over ocean from the most recent algorithm versions. The maps show close agreement between four greatly different retrieval approaches: (a) L2 TMI only, (b) L2 PR only, (c) L2 TMI and PR combined, and (d) L3 TMI only.

Some key characteristics of the algorithms used are given below. The 2a12 TMI-only rain rate profile algorithm combines a mesoscale cloud resolving model (CRM) with a microwave radiative transfer model (Mugnai and Smith 1984; 1988; Mugnai et al. 1990; 1993; Smith and Mugnai 1988; 1989; Smith et al. 1992a; 1992b; Marzano et al. 1999; Di Michele et al. 2003). Kummerow et al. (1996) adopted this strategy making use of a large number of microphysical profiles of the Goddard cumulus ensemble model (GCE) (Tao and Simpson 1993; Simpson and Tao 1993), and the University of Wisconsin nonhydrostatic modelling system (UW-NMS) (Tripoli 1992), to establish a set of profiles, which forms a probability density function of rainfall.

The current 2a12 algorithm uses a CRD which is calculated by linking microwave radiative transfer calculations to a large set of simulated microphysical profiles i.e. pairs of CRM-generated microphysical profiles together with corresponding synthetic top-of-atmosphere microwave brightness temperatures. This approach requires that the CRD of possible cloud-precipitation structures is well

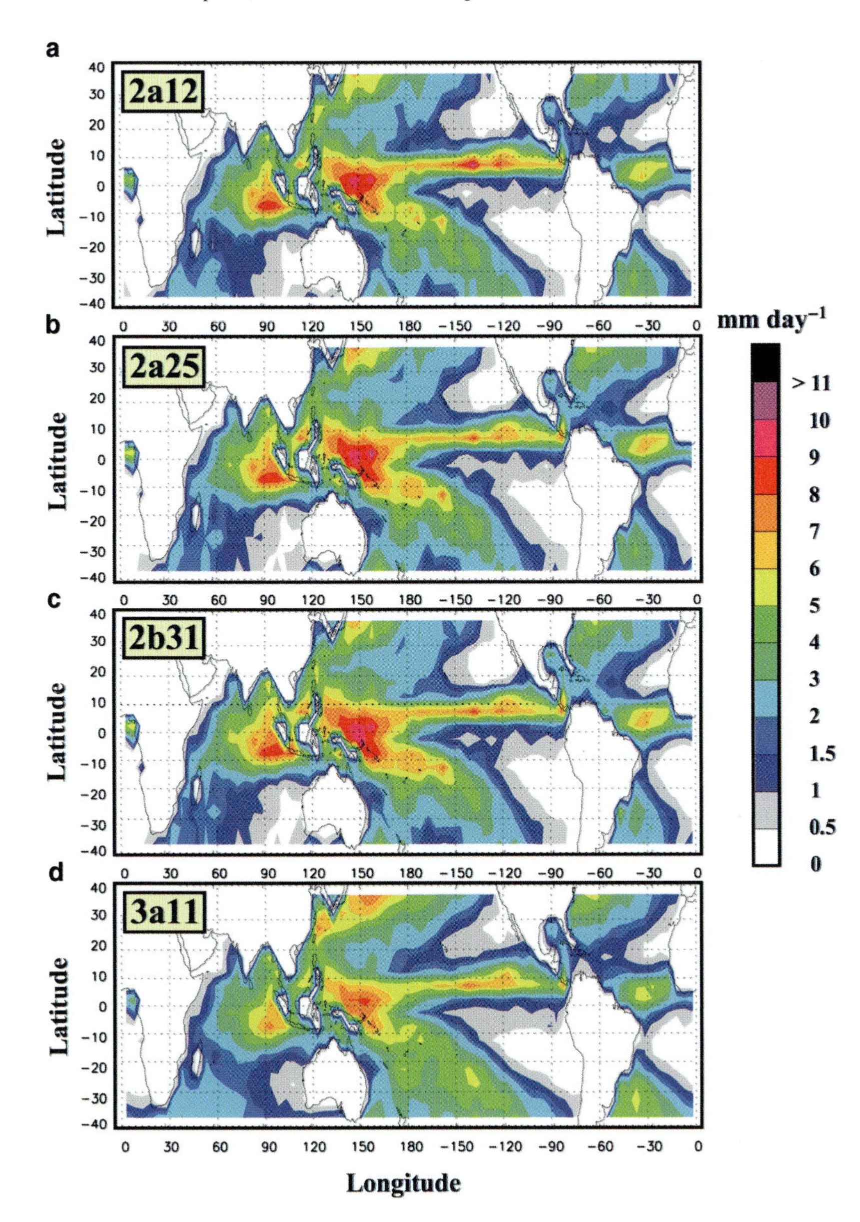

Fig. 4.21 Distributions of monthly rainfall accumulations over tropical oceans for February 1998 produced by most recent versions (V6) of standard TRMM L2 and L3 algorithms. (**a**) shows TMI-only (L2 algorithm 2a12), (Kummerow et al. 1996, 2001; Olson et al. 2001, 2006), (**b**) shows PR-only (L2 algorithm 2a25), (Iguchi et al. 2000; Meneghini et al. 2000; Iguchi 2007), (**c**) shows PR-TMI Combined (L2 algorithm 2b31), (Haddad et al. 1997; Smith et al. 1997), while (**d**) shows TMI-only (L3 algorithm 3a11), (Wilheit et al. 1991a; 1991b; Hong et al. 1997; Tesmer and Wilheit 1998). Colour bar denotes average rain rate in mm/day.

populated, if this is not the case, an iterative variational relaxation technique is more appropriate such as described in Smith et al. (1994a; 1994b; 1995b).

The 2a12 algorithm's greatest strength is that it is an attenuation type solution, i.e. the relationship between radiance (or brightness) temperatures and drop size goes according to D^4 with D the diameter of a spherical droplet instead of the radar backscatter D^6 dependence. This reduces the sensitivity of the retrieved rain rates to the drop size distribution. A weakness is the low spatial resolution of the TMI compared to the PR, which means the retrievals contain greater uncertainty due to beam filling effects, Panegrossi et al. (1998) and Mugnai et al. (2008) have described different ways to mitigate the latter problem.

The 2a25 PR-only L2 rain rate profile algorithm is a hybrid method described by Iguchi et al. (2000), Meneghini et al. (2000) and Iguchi (2007), based on earlier work by Iguchi and Meneghini (1994). The algorithm uses the radar reflectivity vector (Z) to produce a vertical profile of rain. However, in order to get accurate results, it is essential that the path integrated attenuation (PIA) is known. To achieve this, a Hitschfeld-Bordan solution (Hitschfeld and Bordan 1954) is used in a first step to accumulate attenuation in a top-down sequence, followed by application of what is called the surface reference technique (SRT), which provides a correction of the retrieved vertical rain rate vector according to the difference between the Hitschfeld-Bordan PIA and the SRT-derived PIA (Meneghini et al. 1983). For the SRT, the surface return from a rain-filled path is compared to that from a climatological rain-free path created by combining all prior cloud-free pixels over the mission life into a background climatology.

The strength of the 2a25 algorithm is the use of range-gated returns, which results in a well resolved vertical structure. The greatest weakness in using the SRT scheme is the assumption that the differences between two surface returns are caused only by atmospheric path attenuation, and are not due to different surface conditions. In reality, surface reflectivity depends on the roughness of the surface due to wind and internal waves over ocean, and seasonally varying surface properties over land. The greatest error source is the strong dependence of the backscatter on the diameter of the drops, proportional to D^6, and thus is very sensitive to the drop size distribution. Two additional error sources are the sensitivity to residual systematic errors in the PR calibration, and severe attenuation due to high rain rates.

The 2b31 algorithm uses PR and TMI measurements simultaneously. Such a technique is referred to as a "tall vector" algorithm (Farrar 1997); an example is the algorithm to analyse SSM/I data by Smith et al. (1994b) based on hydrometeor profiles from the University of Wisconsin nonhydrostatic modelling system (UW-NMS). Haddad et al. (1997) and Smith et al. (1997) formulate an algorithm based on the design of algorithm 2a25, but use Bayesian probabilities to retrieve the rain rate profile. As with algorithm 2a25, 2b31 first uses the SRT to constrain the Hitschfeld-Bordan solution but then uses the 10.7, 19, and 37 GHz TMI brightness temperatures to estimate the PIA at 13.8 GHz, thus providing a second PIA constraint beyond the SRT. The main weaknesses of this approach are its sensitivity to any residual calibration offset in the PR measurements, and inconsistencies between the radiative transfer model for the passive and active sensors. Farrar (1997) is using a combined Radiative transfer model to overcome this problem. Thus, this type of algorithm shares the strengths of both the 2a25 and 2a12 algorithms.

4.7.7 Conclusions

Since becoming operational, the CloudSat CPR has provided unique, global views of the vertical structure of clouds and precipitation. These new observations have provided the first real estimates of the ice contents of clouds, the direct effects of clouds on radiative heating and how fast and how often clouds produce rainfall (Stephens et al. 2008). The observations have also revealed new knowledge about occurrence and properties of high latitude cloudiness by demonstrating quantitatively the important role of reduced Arctic cloudiness during the 2007 summer to the sea ice loss that occurred (Kay et al. 2008).

The planned aerosol-cloud experiment mission (ACE) will consist of one or more satellites in a LEO, in which the overall science payload is dedicated to obtaining aerosol and cloud profiles for climate and water cycle research. The ACE payload includes a dual-frequency near 13 and 90 GHz (Ka/W-band) Doppler radar designed so that the Ka-band radar scans and the W-band radar acquires multiple beams across-track, enabling vertical profiling of the precipitation rate across a large dynamic range (from very light to heavy), to obtain information on the drop size distribution, to measure the vertical motion, and to determine latent heating (from knowledge of vertical motion profile and thus vertical derivative of horizontal divergence). Finally the optional polarization diversity will give extra information concerning precipitation phase, liquid, frozen, mixed, and melting ice.

The PR/TMI instruments on TRMM have produced new results of vertical structure of precipitation and collected unique data of clouds, convection, frontal zones, precipitating storms, and tropical cyclones. NASA and JAXA are now planning the Global Precipitation Measurement Mission (GPM) (Smith et al. 2007; Mugnai et al. 2007; http://gpm.gsfc.nasa.gov/).

For nearly continuous observations of precipitation at mid- and low-latitudes, the next generation weather radar (NEXRAD) in Space (NIS) consists of a satellite in a GEO carrying a Doppler Ka-band radar used for tropical cyclone monitoring over a ~48° (~5,300 km) diameter great circle disk. The radar design allows for continuous storm pointing enabling rapid-updated storm observations, as frequent as every 15 min at a moderate resolution of ~12 km. The main scientific goal of this mission is to provide a continuous data stream for operational tropical cyclone prediction, as demonstrated through the observing system simulation experiments (OSSE).

4.8 Measuring Atmospheric Parameters Using the Global Positioning System

GPS satellites transmit signals at two microwave frequencies, L1 = 1.57542 GHz and L2 = 1.2276 GHz. The signals are used primarily for precise positioning and navigation, but they can also be used to probe the atmosphere, and this has led to the relatively new field of GPS meteorology. GPS meteorology is concerned with measuring how the Earth's atmosphere affects the propagation of the GPS signals, and then deriving atmospheric state information from these measurements. The atmosphere affects the GPS signal propagation in two ways: firstly the velocity of the signal is reduced because the refractive index of the atmosphere is greater than unity and secondly the ray paths are curved as a result of gradients in the refractive index.

There are two quite distinct GPS measurement types which have very different information content. GPS radio occultation measurements have a satellite-to-satellite, limb geometry and they provide profile information with good vertical resolution properties. In contrast, ground-based GPS measurements have a satellite-to-ground geometry and they primarily give column integrated water-vapour information.

4.8.1 GPS Radio Occultation

The methodology of radio occultation measurements was pioneered by planetary scientists in the 1960s and it formed part of NASA's Mariner 3 and 4 missions to Mars. During the 1980s it was realised that these techniques could be applied to measuring the Earth's atmosphere using the GPS signals. A detailed description of GPS radio occultation (GPSRO) and its error characteristics can be found in Kursinski et al. (1997). The geometry of the measurement is illustrated in Fig. 4.22. A radio signal is transmitted by a GPS satellite, passes through the atmosphere and is measured with a GPS receiver placed on a satellite in LEO. The ray path between the satellites is bent as a result of gradients in the refractive index of the atmosphere, which in turn can be related to gradients in the temperature and humidity.

The bending angle, α, is not measured directly. Given estimates of the satellite locations and velocities, α can be derived from the time derivative of the additional time delay for the radio signal to propagate between the satellites caused by the atmosphere. The motion of the LEO satellite enables the variation of α as a function of the impact parameter, a, to be determined. Bending angles are derived for both the L1 and L2 GPS signals, and this enables the ionospheric bending contribution to be corrected because the ionosphere is dispersive at these frequencies. The ionosphere-corrected bending angle profile can then be inverted with an Abel transform to provide a vertical profile of refractive index. Temperature and humidity information can be derived from the bending angle or refractive index profiles.

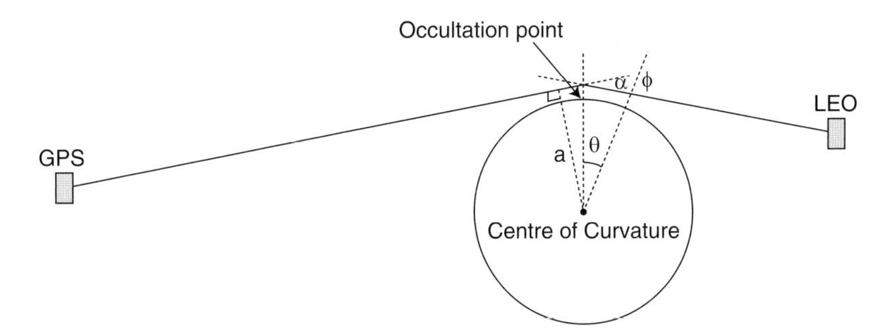

Fig. 4.22 The geometry of the GPSRO measurement technique.

The proof-of-concept was demonstrated in 1995 with the GPS/MET experiment. This showed that the temperature profiles derived from GPSRO measurements were of sub-Kelvin accuracy for heights between 7 km and 25 km (Kursinski et al. 1996; Rocken et al. 1997), despite using a sub-optimal temperature retrieval technique.

4.8.2 Data Availability and Impact

The success of the GPS/MET mission led to a number of missions of opportunity, most notably the Challenging Mini-satellite Payload (CHAMP) (Wickert et al. 2001). This was followed by the Constellation Observing System for Meteorology Ionosphere and Climate (COSMIC) of six receivers (Anthes et al. 2008) and then the Global Navigation Satellite System Receiver for Atmospheric Sounding (GRAS) on METOP-A, which is the first fully operational GPSRO mission (Luntama et al. 2008).

Currently, the COSMIC constellation and GRAS provide around 2000 and 650 profiles per day respectively, distributed globally. Each profile contains between 250 and 300 bending angles, spanning the vertical interval from the surface up to 60 km. To put these numbers in context, ECMWF currently assimilates around 10 million observations per 12 h assimilation window, of which ~90% are radiance measurements. The number of COSMIC and GRAS bending angles actively assimilated account for less than 3% of the total.

Despite the relatively low data numbers, GPSRO measurements are valuable in NWP because their information content complements that provided by satellite radiance measurements. GPSRO measurements have good vertical resolution, an all-weather capability and can be assimilated over both land and sea. Furthermore, they can be assimilated without bias correction. This measurement characteristic is particularly important because it means that the data help to distinguish between model and observation biases, and they provide anchor-points to prevent model drift in adaptive bias correction schemes (Dee 2005). Assimilation without bias correction is possible because the fundamental GPSRO measurement is a time-delay with an atomic clock, and the forward problem is relatively simple to model when compared to the radiative transfer problem.

The GPSRO errors are smallest in a fractional sense in the height interval from 8 to 30 km, and this is where their information content is greatest (Collard and Healy 2003). Theoretical information content studies show that GPSRO measurements contain primarily humidity information in the lower troposphere and temperature information in the upper troposphere and stratosphere. In addition GPSRO measurements potentially provide surface pressure information. This arises because the measurements are assimilated as a function of a height coordinate-ordinate, meaning that the hydrostatic integration is part of the observation operator.

Most operational NWP centres currently assimilate either vertical profiles of bending angle or refractivity, using one-dimensional observation operators. These approaches ignore the two-dimensional, limb geometry of the observation which introduces additional forward model errors, but research into more advanced operators is ongoing (Sokolovskiy et al. 2005; Healy et al. 2007). Observing system experiments at a number of NWP centres have demonstrated that GPSRO measurements provide extremely good temperature information in the upper troposphere and lower stratosphere, confirming the theoretical information content studies (Healy and Thépaut 2006; Cucurull et al. 2007; Aparicio and Deblonde 2008; Poli et al. 2009).

In most cases, the measurements have clearly improved stratospheric analysis and forecast biases with respect to radio-sonde temperature measurements. The measurements have also corrected systematic errors that are in the null space of the radiance measurements as a result of the superior vertical resolution. Furthermore, experiments have demonstrated that the GPSRO measurements have an impact on the bias corrections applied to AMSU-A channels 8, 9, 10 and 11 radiances, generally resulting in an improvement in the fit to radio-sonde measurements.

To date, it has proved more difficult to show any significant impact on the lower tropospheric humidity forecasts, and it is clear that the humidity analysis is being driven by other measurements. This may be because the weight given to the GPSRO observations in the lower troposphere is too low, but the observation processing and forward model errors are certainly larger in this region. Some improvement might be expected with the introduction of two-dimensional observation operators, and data processing techniques in the lower troposphere are continually evolving and improving (Jensen et al. 2003).

GPSRO measurements are now being assimilated in reanalyses. Fig. 4.23 shows the impact of introducing COSMIC measurements in the ECMWF ERA-Interim project on the analyses and short-range forecasts departure statistics for radio-sonde temperature measurements at 100 hPa in the northern and southern hemispheres. The COSMIC measurements are assimilated from 12th December 2006, and there is a clear improvement in the biases from this period with a shift in the bias of order 0.1 K and 0.2 K in the northern and southern hemispheres, respectively.

One outstanding issue that must be addressed in the context of planning future GPSRO missions is the optimum size (number) of the constellation of LEO receivers. There is no indication that the current number of receivers is anywhere near the saturation point, where adding more receivers provide no additional benefit, so the current number probably represents a lower limit. However, estimating the number of receivers where saturation may occur is a challenging problem,

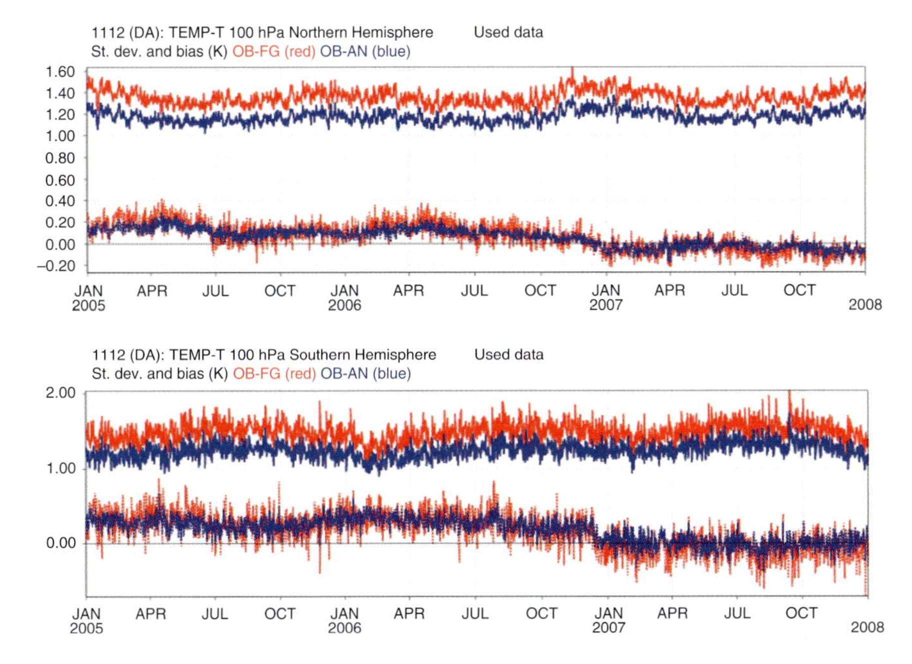

Fig. 4.23 The mean (dotted) and standard deviation (solid) of the ERA-Interim (observation minus background) and (observation minus analysis) temperature departures at 100 hPa in the northern and southern hemispheres. The background departures are in red and the analysis departures in blue. The period is from 1st January 2005 to 31st December 2007.

although the ensemble data assimilation techniques outlined by Tan et al. (2007) may be applicable.

4.8.3 Ground-Based GPS Observations

Ground-based networks of GPS receivers provide information on atmospheric humidity distribution. Meteorological use of these measurements relies on heavy pre-processing to account for various geophysical and geodetic effects. The quantity to be assimilated in NWP is usually Zenith Total Delay (ZTD); sometimes ZTD is converted further to PW prior to data assimilation. ZTD is one of the parameters that are estimated through a least-squares adjustment to find a geodetically consistent solution from a given set of raw measurement data.

ZTD is a measure of the so-called tropospheric refraction that follows from the fact that the propagation speed of microwave signals is decreased in a medium. In case of the neutral atmosphere, this loss of propagation speed is related to distributions of pressure, temperature and water vapour (Bevis et al. 1992). Forward modelling of ZTD observations from given background profiles makes use of these data in NWP. The accuracy of the background humidity field dominates the error budget of ZTD forward modelling. Consequently, the information that is obtained via the ZTD observation minus background departures is translated primarily to the analysis of humidity and only secondarily to the analyses of pressure and temperature.

The major advantages of the ground-based GPS observing system include the high temporal observing resolution (up to around 5 min), the low financial cost of setting up and maintaining the equipment, and the ability of providing ZTD data with an accuracy and usability that is practically independent of the prevailing weather conditions. Due to these advantages, the GPS data is expected to remain an important complement to other humidity-related observing systems. The data assimilation of ZTD observations has already reached the operational status at the UK Met Office and Météo-France.

The ground-based GPS observing system consists of regional and local networks that, in most cases, have been originally set up for geodetic purposes by governmental institutes, or private agencies. The meteorological use of these networks comes as a by-product of the geodetic processing and it is usually of secondary importance from the maintenance point of view. Therefore, the observation density and the data quality vary considerably from one country to another. Moreover, the complexity of geodetic pre-processing makes it necessary in practice to limit the number of receiver stations to only few hundred at a time. This implies that there is a trade-off between the spatial resolution and coverage of the data that is to be pre-processed. The approach taken in Europe is to combine pre-processed data from a number of analysis centres that deal with the geodetic processing with different subsets of the available raw data. The resulting receiver station network, as on 9th September 2009, is shown in Fig. 4.24. Unfortunately, different analysis centres apply slightly different processing methods and software, which results in inhomogeneities in the ZTD data processed at the continental scale.

The ongoing EUMETNET GPS water vapour programme (E-GVAP) has paid significant attention on homogenizing both the ground-based GPS observing system and the pre-processing methods, but considerable inhomogeneities are still likely to be present in the data to be assimilated in Europe. The receiver networks in North America and Japan are likely to be more internally homogeneous. On a global scale, a homogeneous set of GPS ZTD data (from 420 receiver stations on 9th September 2009) are provided by the International GPS Service (IGS).

Pre-processing using geodetic software is a necessary step in order to have the ZTD (or PW) data available in the first place. Unfortunately, the pre-processing step makes observation error characteristics very complicated. The pre-processing introduces a spatially and temporally correlated component of observation error into the data. In order to maintain the statistical optimality of the data assimilation scheme, a way needs be found to deal with the correlation. The data assimilation experiments that have been performed so far have, almost exclusively, assumed that these correlations are negligible (Poli et al. 2007; Yan et al. 2009). This assumption is sometimes justified by applying horizontal thinning to the data (Macpherson et al. 2008). Another way to avoid dealing with observation error correlation is to decrease the weight of the correlated data by setting an overly large value for observation error standard deviation (Vedel and Sattler 2006).

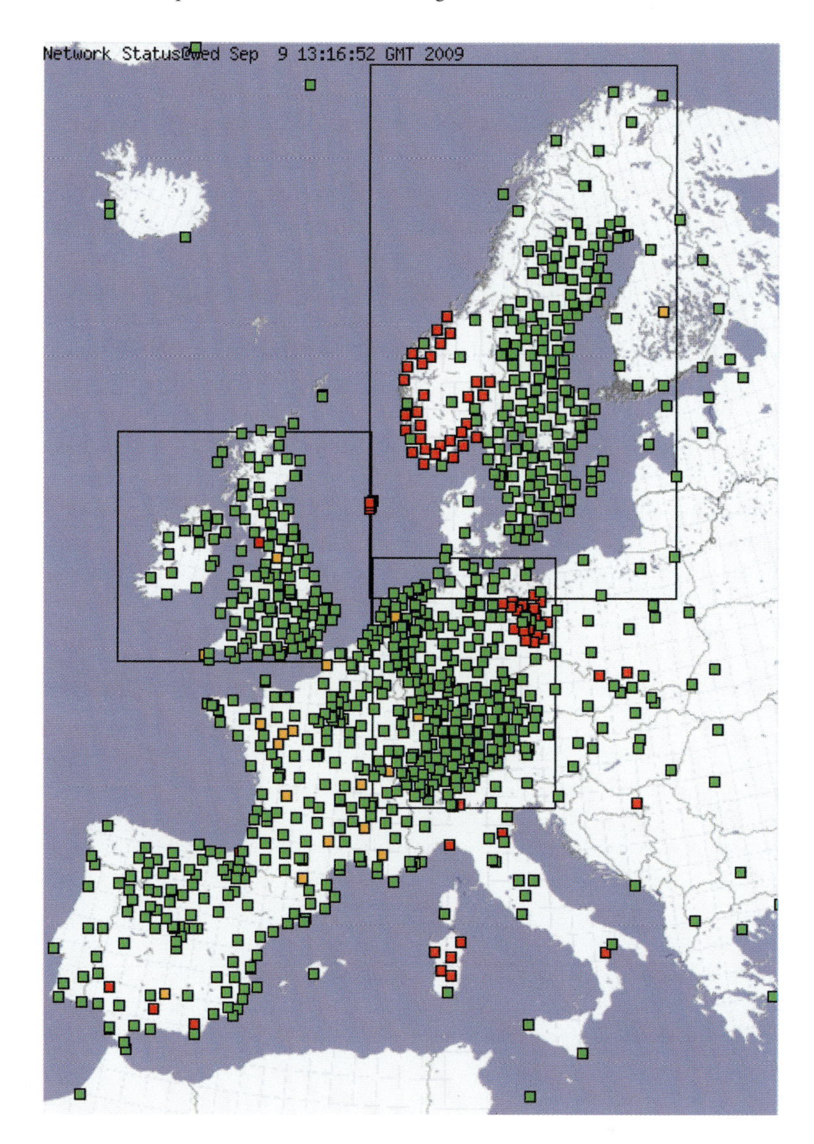

Fig. 4.24 Status of the European ground-based GPS observing network on 9th September 2009, as reported by the E-GVAP programme. Green, yellow and red squares indicate the locations of those receiver stations for which the ZTD observation is as recent as 3, 6, and 54 h, respectively.

Slowly evolving observation biases constitute an even more severe complication to ZTD data assimilation. The origin of the bias is unknown. The fact that the bias seems to be specific to each receiver station makes it difficult to apply sophisticated bias correction schemes in a similar fashion that is applied in, for example, microwave and infrared radiance data assimilation. A common strategy is to derive receiver station-dependent bias corrections from past observation minus background departure statistics.

4.8.4 Impact Studies

Data assimilation experiments using ZTD data have been carried out in both regional 3D- and 4D-Var and global 4D-Var frameworks. The impact of ZTD data assimilation is reported to be slightly positive in terms of reduced forecast error on PW and other humidity-related parameters, surface pressure and geopotential height, and has thus improved the description of the synoptic circulation. The positive impact extends up to the fourth forecast day (Poli et al. 2007). However, the reported positive impacts on long-term verification scores are in most cases below the level of statistical significance. In addition to the standard upper-air meteorological parameters, in most studies a positive impact is found on the categorical forecasts of accumulated precipitation, in particular when relatively high precipitation accumulations are considered (Vedel and Sattler 2006; Macpherson et al. 2008).

More evidence of positive forecast impact is reported in the context of case studies focusing on intense precipitation events (De Pondeca and Zou 2001; Vedel and Sattler 2006; Yan et al. 2009).

Figure 4.25 shows an example of results of such a case study. In this particular case, a convective-scale NWP system Application of Research to Operations in

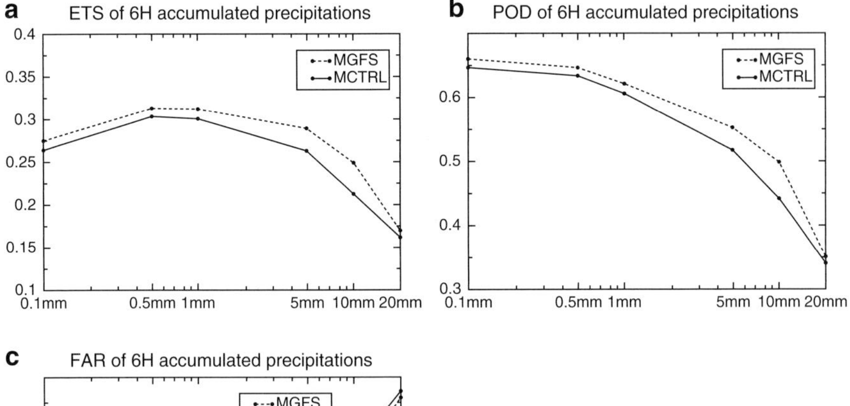

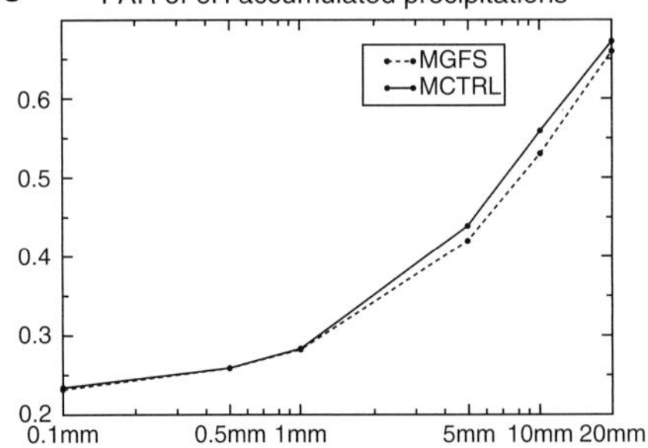

Fig. 4.25 Impact of ZTD data assimilation on (**a**) Equitable Threat Score, (**b**) Probability of Detection, and (**c**) False Alarm Rate of 6-h forecasts of accumulated precipitation in a case study of Mediterranean heavy rainfall. The experiment with ZTD data (dashed lines) shows improved scores with respect to the control run (solid lines) (Yan et al. 2009).

Meso-scale (AROME) has been deployed to investigate the impact of 3D-Var data assimilation of ZTD observations in a case of Mediterranean heavy rainfall. The positive impact of assimilating ZTD data is shown at all thresholds of 6-h accumulated precipitation.

4.9 Outlook

The following section highlights areas where promising work is ongoing or where the authors feel a special effort is needed in the future.

For all applications using microwave radiometry, improvements in microwave spectroscopy will help to interpret the collected data better. For example the question of line pressure broadening is an important topic, and in particular the non-resonant absorption of water vapour needs more work. Also improved modelling of cloud scattering effects, both for liquid and frozen particles, will help to retrieve cloud parameters and enable the observation of atmospheric composition within clouds. Such observations will also help to model chemical transport and climate forecasting better.

Synergetic use of the visible, infrared, and microwave observations has not yet been systematically explored for the retrieval of atmospheric constituents and parameters. Further investigations should be conducted, to benefit from the complementary sensitivities of the different wavelengths to each atmospheric parameter and so help constrain the inversion problem. This also includes the combined analysis of passive and active observations by microwave sensors as has been used successfully for the TRMM project.

While sensors operating in the visible and IR domains are limited to optically thin ice clouds with particle diameters <100 μm, microwave sensors observing below 200 GHz are insensitive to cloud ice water content lower than roughly 300 g/m^2. The sub-millimetre wavelength range appears to be well suited to cover this gap. Space-borne imagers and sounders for meteorological applications are so far limited to frequencies below 190 GHz. New sensors operating over a wide frequency range from 100 to 1,000 GHz to investigate cirrus clouds, would allow for the first time, the collection of global data sets with good horizontal resolution on the ice water content and on particle shapes and size distribution.

The success of the microwave limb sounder made NASA include, in its decadal survey of future earth-science missions, an advanced microwave limb sounder as one of the instruments in a future Global Atmospheric Composition Mission.

For future meteorology instrument scenarios, see for example the US-Europe Joint Polar System, combining the National Polar-Orbiting Operational Environmental Satellite System (NPOESS) and the follow-on to the EUMETSAT's Polar System (EPS), improvements are needed with respect to temperature sounding for the lower troposphere, and more complementary sounding and imaging capabilities to constrain clouds, precipitation and temperature/moisture at the same time. Soil moisture data assimilation also offers significant potential in the future. The potential of soil moisture related observations on NWP forecast accuracy has long been recognized, in particular in situations where the atmospheric hydrological cycle is strongly affected by evaporation rather than oceanic moisture advection.

Future developments require, apart from refined instruments, more advanced data assimilation techniques and improved models. In particular, in areas where current systems do not fully exploit existing data, such as clouds and over land surfaces, better physical parameterisations and more flexible data assimilation techniques will greatly improve forecasting. The great potential of future microwave observations lies in constraining the models' hydrological cycle, namely the diabatic process in the atmosphere and, as mentioned above, soil moisture.

In addition to the GPM Core Satellite instruments, more sophisticated missions are planned, such as the European-GPM (EGPM) forming an important contributor to monitor the Earth's water cycle and climate at mid-to-high latitudes. EGPM consists of a sun-synchronous low Earth orbit satellite carrying two precipitation measurement payloads. A microwave radiometer containing, along with the standard precipitation channels from 18 to 150 GHz, an additional set of 50–55 and 118 GHz O_2 channels that would enable precipitation discrimination and profiling in light-to-moderate rainfall and snowfall, and a radar at Ka-band with a detection threshold of 5 dBZ, to measure light-to-moderate rainfall and snowfall, over both land and oceans, at both mid and high latitudes.

Millimetre and sub-millimetre observations from geostationary satellites are under investigation for near-real time estimates of precipitation, since current visible and IR observations from high orbit do not provide accurate quantification of the rain rates. High sampling intervals of precipitation are particularly crucial in the follow-up to severe weather events.

The recent successful operation of several smaller satellites flying in close formation, the A-Train including active and passive sensors, will certainly lead to similar missions, creating a larger observatory from several smaller spacecraft.

There is potential to extend the use of ground-based GPS measurements towards assimilation of slant delay (SD) observations in convection-resolving NWP systems. SD observations are inherently capable of containing information on azimuthal asymmetry and fine-scaled atmospheric structures, and are therefore of particular interest in forecasting high-impact severe weather phenomena. Unfortunately, the experience gathered to date suggests that the SD data is very noisy and suffers from significant observation error correlation. Operational use of SD data will therefore require a special effort in the further development of geodetic pre-processing, together with improved methods for data assimilation in convective scale NWP systems.

Finally on the hardware side the development of low noise sensors operating up to 1,000 GHz are needed for a number of applications, mainly to investigate clouds or to operate instruments on geostationary orbits. Also extremely low-noise sensors for measurements requiring high spectral resolution are needed, such as the cryogenic sensors presently being used by SMILES and planned for future limb sounders. For microwave sensors in geostationary orbit, as planned for continuous observations in the tropics and mid-latitudes, light-weight large antenna systems have to be developed, offering the required spatial resolution in the order of 10 km.

4.10 Tables of Microwave Sensors

Space borne passive microwave sensors in space are listed in Table 4.1 for meteorological observations and in Table 4.2 for limb sounders.

Table 4.1 A summary of operational meteorological microwave sensors presently in space

Instrument	Platform time coverage	Spectral range (GHz)	Scanning type Spatial resolution	Objectives
SSM/I	DMSP Since 1987	19VH, 22V, 37VH, 85VH	Conical scanning (53°) From 60 to 15 km	Integrated H_2O, cloud and rain, surface properties
SSMI/S	DMSP Since 2003	19VH, 22V, 37VH, 50-59H (7 channels), 91V, 150H, 183H (3 channels)	Conical scanning (53°) From 60 to 15 km	H_2O and T profiles, cloud and rain, surface properties
TMI	TRMM Since 1997	10VH, 19VH, 21V, 37VH, 85VH	Conical scanning (53°) From 50 to 6 km	Integrated H_2O, cloud and rain, surface properties
AMSR-E	NASA/Aqua Since 2002	19VH, 22V, 37VH, 85VH	Conical scanning (53°) From 50 to 5 km	Integrated H_2O, cloud and rain, surface properties
AMSU-A	NOAA, NASA/Aqua ESA/MetOp Since 1998	23, 33, 50–57 (12 channels), 89	Cross-track scanner (±50°) ~45 km at nadir	T profiles
AMSU-B	NOAA Since 1998	89, 157, 183 (3 channels)	Cross-track scanner (±50°) ~15 km at nadir	H_2O profiles
MHS	NASA/Aqua ESA/MetOp Since 2003	150, 183 (3 channels)	Cross-track scanner (±50°) ~15 km at nadir	H_2O profiles
MWRI	FY-3 Since May 2008	10VH, 19VH, 24VH, 36VH, 89VH, 150 VH	Conical scanning (53°) From 15 to 80 km	Integrated H_2O, cloud and rain, surface properties
MWTS	FY-3 Since May 2008	50–57 (4 channels)	Cross-track scanner (±53°) ~50 km at nadir	T profiles
MWHS	FY-3 Since May 2008	150, 183 (3 channels)	Cross-track scanner (±53°) ~15 km at nadir	H_2O profiles

Table 4.2 Summary of previous, current and planned microwave limb sounding instruments

Instrument	Mission timespan	Developer	Radiometers	Main Stratospheric products	Main Tropospheric products
MLS	UARS Sep 1991 – Jan 2000	NASA JPL, USA	63 GHz, 205 GHz, 183 GHz	Temperature, O_3, H_2O, ClO	Temperature, H_2O, Cloud Ice
MAS	ATLAS Mar 1992, Mar/Apr 1993, Nov 1994	MPI, Aeronomy, D, Uni-Bern, CH, NRL, USA	63 GHz, 183 GHz, 205 GHz	Temperature, O_3, H_2O, ClO	Temperature, H_2O, Cloud Ice
SMR	Odin Feb 2001 to present	Chalmers Uni., S, Swedish Space Corp.	119 GHz 486–503 GHz, 541–581 GHz	ClO, O_3, N_2O, H_2O, $H_2{}^{17}O$, $H_2{}^{18}O$, CO, HO_2, HNO_3, NO, HDO,	H_2O, Cloud ice
MLS	Aura July 2004 to present	NASA JPL	118 GHz, 190 GHz, 240 GHz, 640 GHz, 2,500 GHz	Temperature, GPH, H_2O, O_3, HNO_3, N_2O, ClO, HCl, HOCl, BrO, CO, HCN, CH_3CN, SO_2, OH, HO_2	Temperature, H_2O, O_3, CO, HNO_3, Cloud Ice
JEM/SMILES	ISS Launched 2009	JAXA, J	640 GHz superconductingreceiver	O_3, HCl, ClO, HO_2, H_2O_2, HOCl, BrO, HNO_3	

References

Adler, R. F., G. J. Huffman, A. Chang, R. Ferraro, P. Xie, J. Janowiak, B. Rudolf, U. Schneider, S. Curtis, D. Bolvin, A. Gruber, J. Susskind, P. Arkin, and E. Nelkin, 2003, The version 2 global precipitation climatology project (gpcp) monthly precipitation analysis (1979-Present). *J. Hydrometeor.*, **4**, 1147–1167.

Aires, F., C. Prigent, W. Rossow, and M. Rothstein, 2001, A new neural network approach including first guess for retrieval of atmospheric water vapor, cloud liquid water path, surface temperature, and emissivities over land from satellite microwave observations. *J. Geophys. Res.*, **106**, D14, 14887–14907.

Akvilonova, A. B., A. E. Basharinov, A. K. Gorodetskii, A. S. Gurvich, M. S. Krilova, B. G. Kutuza, D. T. Matveev, and A. P. Orlov, 1973, Determination of meteorological parameters wit measurements from the satellite Cosmos-384. *Atmos. Ocean. Phys.*, **9**, 187–198.

Alishouse, J. C., S. A. Snyder, E. R. Westwater, C. T. Swift, C. S. Ruf, S. A. Snyder, J. Vonsathom, and R. R. Ferraro, 1990, Determination of cloud liquid water content using the SSM/I. *IEEE Trans. Geosci. Remote Sens.*, **28**, 817–822.

Allan, D. W., 1966, Statistics of Atomic Frequency Standards. *Proc. IEEE.*, **54**, 221–230.

Andersson, E., J. Pailleux, J.-N. Thépaut, J. R. Eyre, A. P. McNally, G. A. Kelly, and P. Courtier, 1994, Use of cloud-cleared radiances. *Q. J. R. Meteorol. Soc.*, **120**, 627–653.

Andersson, E., R. Dumelow, H. Huang, J.-N. Thépaut, and A. Simmons, 2004, Space/terrestrial link – outline study proposal for consideration by EUCOS management (available at EUCOS secretariat).

Andersson, E., E. Holm, P. Bauer, A. Beljaars, G. A. Kelly, A. P. McNally, A. J. Simmons, and J.-N. Thépaut, 2007, Analysis and forecast impact of the main humidity observing system. *Q. J. R. Meteorol. Soc.*, doi: 10.1002/QJ.112.

Andersson, E., and J.-N. Thépaut, 2008, ECMWF's 4D-Var data assimilation system - the genesis and ten years in operation. *ECMWF Newsletter*, **115**, 8–12.

Anthes, R., P. A. Bernhardt, Y. Chen, L. Cucurull, K. F. Dymond, D. Ector, S. B. Healy, S.-P. Ho, Y.-H. Kuo, D. C. Hunt, H. Liu, K. Manning, C. McCormick, T. K. Meehan, W. J. Randel, C. Rocken, W. S. Schreiner, S. V. Sokolovskiy, S. Syndergaard, D. C. Thompson, K. E. Trenberth, T. K. Wee, N. L. Yen, and Z. Zeng, 2008, The COSMIC/FORMOSAT-3 Mission, Early Results. *Bull. Am. Meteor. Soc.*, **89**, 313–333.

Aparicio, J., and G. Deblonde, 2008, Imapct of the assimilation of [CHAMP] refractivity profiles in environment canada global forecasts. *Mon. Wea. Rev.*, **136**, 257–275.

Arnold, C.P. and C.H. Dey, 1986, Observing-system simulation experiments: Past, present and future, *Bull. Amer. Meteorol. Soc.*, **67**, 687–695.

Auligné, T., A.P. McNally, and D.P. Dee, 2007, Adaptive bias correction for satellite data in a numerical weather prediction system, *Q. J. R. Meteorol. Soc.*, **133**, 631–642.

Barath, F.T., A.H. Barrett, J. Copeland, D.E. Jones, and A.E. Lilley, 1964, Mariner 2 microwave radiometer experiments and results, *Astron. J.*, **69**, 49–58.

Barath, F.T., M.C. Chavez, R.E. Cofield, D.A. Flower, M.A. Frerking, M.B. Gram, W.M. Harris, J.R. Holden, R.F. Jarnot, W.G. Kloezeman, G.J. Klose, G.K. Lau, M.S. Loo, B.J. Maddison, R.J. Mattauch, R.P. McKinney, G.E. Peckham, H.M. Pickett, G. Siebes, F.S. Soltis, R.A. Suttie, J.A. Tarsala, J.W. Waters, and W.J. Wilson, 1993, The Upper Atmosphere Research Satellite Microwave Limb Sounder Experiment, *J. Geophys. Res.*, **98**, 10,751–10,762.

Battaglia, A., M.O. Ajewole and C. Simmer, 2006, Evaluation of radar multiple scattering effects from a GPM perspective. Part I: model description and validation, *J. Appl. Meteor. Clim.*, 45, 12, 1634–1647.

Battaglia, A., C.P. Davis, C. Emde and C. Simmer, 2007, Microwave radiative transfer intercomparison study for 3-D dichroic media. *J. Quant. Spectrosc. Radiat. Transfer*, **105**, 55–67.

Bauer, P., and P. Schluessel, 1993, Rainfall, total water, ice water, and water vapor over sea from polarized microwave simulations and Special Sensor Microwave Imager data. *J. Geophys. Res.*, **98(D11)**, 20737–20759.

Bauer P., P. Amayenc, C.D. Kummerow, E.A. Smith, 2001a, Over-ocean rainfall retrieval from multisensor data of the Tropical Rainfall Measuring Mission. Part I: Design and evaluation of inversion databases. *J. Atmos. Oceanic Technol.*, **18**, 1315–1330.

Bauer P. 2001b, Over-ocean rainfall retrieval from multisensor data of the Tropical Rainfall Measuring Mission. Part II: Algorithm implementation. *J. Atmos. Oceanic Technol.*, **18**, 1838–1855.

Bauer P., A. Mugnai, 2003, Precipitation profile retrievals using temperature-sounding microwave observations, *J. Geophys. Res.*, **108(23)**, 4730, doi:10.1029/2003JD003572.

Bauer, P., E. Moreau, F. Chevallier, and U. O'Keefe, 2006a, Multiple-scattering microwave radiative transfer for data assimilation applications. *Q. J. R. Meteorol. Soc.*, **132**, 1259–1281.

Bauer, P., P. Lopez, A. Benedetti, D. Salmond, and E. Moreau, 2006b, Implementation of 1D+4D-Var assimilation of precipitation affected microwave radiances at ECMWF, Part I: 1D-Var, *Q. J. R. Meteorol. Soc.*, **132**, 2277–2306.

Bauer, P., P. Lopez, A. Benedetti, D. Salmond, S. Saarinen and M. Bonazzola, 2006 c, Implementation of 1D+4D-Var assimilation of precipitation affected microwave radiances at ECMWF, Part II: 4D-Var, *Q. J. R. Meteorol. Soc.*, **132**, 2307–2332.

Benedetti, A., P. Lopez, P. Bauer, and E. Moreau, 2005, Experimental use of TRMM Precipitation Radar observations in 1D+4D-Var assimilation. *Q.J.R. Meteorol. Soc.*, **131**, 2473–2495.

Bevis, M., S. Businger, T.A. Herring, C. Rocken, R.A. Anthes, and R.H. Ware, 1992, GPS Meteorology: Remote Sensing of Atmospheric Water Vapor Using the Global Positioning System. *J. Geophys. Res.*, **97(D14)**, 15787–15801.

Bohren, C., and D.R. Huffman, 1998, Absorption and scattering of light by small particles, Wiley Science Paperback Series.

Borg, L.A. and R. Bennartz, 2007, Vertical structure of stratiform marine boundary layer clouds and its impact on cloud albedo, *Geophys. Res. Lett.*, 34, L05807, doi:10.1029/2006GL028713.

Bouttier, F., and P. Courtier, 2002, Data assimilation concepts and methods, ECMWF Training course lectures, available from ECMWF, Shinfield Park, Reading, UK, 59p.

Buehler, S.A., P. Eriksson, T. Kuhn, A. von Engeln, C. Verdes, 2005, ARTS, the atmospheric radiative transfer simulator, *J. Quant. Spectros. Radiat. Transfer,* **91**, 65–93.

Buehler, S.A., C. Jiménez, K.F. Evans, P. Eriksson, B. Rydberg, A.J. Heymsfield, C. Stubenrauch, U. Lohmann, C. Emde, V.O. John, T.R. Sreerekha and C.P. Davis, 2007, A concept for a satellite mission to measure cloud ice water path and ice particle size, *Q. J. R. Meteorol. Soc.*, **133**, 109–128, doi:10.1002/QJ.143.

Cabrera-Mercader, C.R. and D.H. Staelin, 1995, Passive microwave relative humidity retrievals using feedforward neural networks, *IEEE Trans. Geosci. Remote Sensing,* **33**, 1324.

Cardinali, C., S. Pezzulli, and E. Andersson, 2004, Influence-matrix diagnostic of a data assimilation system, *Q. J. R. Meteorol. Soc.*, **130**, 2767–2786.

Cecil, D.J., S.J. Goodman, D.J. Boccippio, E.J. Zipser and S.W. Nesbitt, 2005, Three years of TRMM precipitation features. Part I: Radar, radiometric, and lightning characteristics, *Month. Weather Rev.*, **133**, 543–566.

Chandra, S., J.R. Ziemke, W. Min, and W.G. Read, 1998, Effects of 1997–1998 El Niño on tropospheric ozone and water vapor, *Geophys. Res. Lett.*, **25**, 3867–3870.

Chandrasekhar, S., 1960, Radiative transfer, Dover Publications, New York.

Chang, A.T.C. and T.T. Wilheit, 1979, Remote sensing of atmospheric water vapor, liquid water, and wind speed at the ocean surface by passive microwave techniques from the Nimbus-5 satellite, *Radio Sci.*, **14**, 793–802.

Collard, A., and S. Healy, 2003, The combined impact of future space-based atmospheric sounding instruments on numerical weather prediction analysis fields: A simulation study. *Quart. J. Roy. Meteorol. Soc.*, **129**, 2741–2760.

Courtier, P., J. Derber, R. Errico, J.-F. Luis, and T. Vukicevic, 1993, Important literature on the use of adjoint, variational methods and the Kalman filter in meteorology, *Tellus,* **45A,** 342–357.

Croskey, C.L., N. Kaempfer, R. Belivacqua, G.K. Hartmann, K.F. Kunzi, P. Schwartz, J.J. Olivero, S.E. Puliafito, C. Aellig, G. Umlauft, 1992, The Millimeter Wave Atmospheric

Sounder (MAS): A shuttle-based remote sensing experiment, *IEEE Trans. Microwave Theory and Techniques,* **40,** 1090–1100.

Cucurull, L., J. Derber, R. Treadon, and R. Purser, 2007, Assimilation of Global Positioning System Radio Occultation Observations into NCEP's Global Data Assimilation System. *Mon. Wea. Rev.*, **135**, 3174–3193.

Czekala, H., and C. Simmer, 1998, Microwave radiative transfer with nonspherical precipitating hydrometeors, *J. Quant. Spectrosc. Radiat. Transfer*, **60**, 3, 365–374.

Daley, R., 1991, Atmospheric Data Analysis, Cambridge Atmospheric and Space Science Series, Cambridge University Press, Cambridge, 457pp.

Davis, C., C. Emde and R. Harwood, 2005, A 3D polarized reversed Monte Carlo radiative transfer model for mm and sub-mm passive remote sensing in cloudy atmospheres, *IEEE Trans. Geosci. Remote Sens.*, **43(6)**, 1096–1101.

Deblonde, G., and N. Wagneur, 1997, Evaluation of global numerical weather prediction analysis and forecasts using DMSP special sensor microwave imager retrievals 1. Satellite retrieval algorithm intercomparison study, *J. Geophys. Res.*, **102, D2**, 1833–1850.

Dee, D.P., and A.M. Da Silva, 2003, The choice of variable for atmospheric moisture analysis, *Mon. Wea. Rev.*, **131**, 155–171.

Dee, D., 2005, Bias and data assimilation. Quart. J. Roy. Meteorol. Soc., 131, 3323–3343.

Deeter, M.N., and K.F. Evans, 2000, A novel ice-cloud retrieval algorithm based on the Millimeter-wave Imaging Radiometer (MIR) 150–220 GHz channels, *J. Appl. Meteorol.*, **39**, 623–633.

Defer, E., C. Prigent, F. Aires, J.R. Pardo, J.C. Walden, O.-Z. Zanife, J.-P. Chaboureau, and J.-P. Pinty, 2008, Developments of precipitation retrievals at millimeter and sub-millimeter wavelength for geostationary satellites, *J. Geophys. Res.*, **113**, D08111, doi:10.1029/2007JD008673.

De Pondeca, M. and X. Zou, 2001, A case study of the variational assimilation of GPS zenith delay observations into a mesoscale model. *J. Appl. Meteorol.*, **40**, 1559–1576.

Derber, J.C. and W.-S. Wu, 1998, The use of TOVS cloud-cleared radiances in the NCEP SSI analysis system, *Mon. Wea. Rev.*, **8**, 2287–2302.

Di Michele, S., F.S. Marzano, A. Mugnai, A. Tassa, and J.P.V. Poiares Baptista, 2003, Physically based statistical integration of TRMM microwave measurements for precipitation profiling. *Radio Sci.*, **38 (8072)**, doi:10.1029/2002RS002636, 16 pp.

Draine, B.T. (2000) The Discrete Dipole Approximation for Light Scattering by Irregular Targets, In: M.I. Mishchenko, J.W. Hovenier, and L.D. Travis (eds) Light Scattering by Nonspherical Particles: Theory, Measurements and Applications. Academic Press, London, pp. 131–144.

Drusch, M., 2007, Initializing numerical weather prediction models with satellite-derived surface soil moisture: Data assimilation experiments with ECMWF's integrated forecasting system and the TMI soil moisture data set, *J. Geophys. Res.*, **112**, doi:10.1029/2006JD007478.

Ekström, M., P. Eriksson, D. Rydberg, and D.P. Murtagh, 2007, First Odin sub-mm retrievals in the tropical upper troposphere: humidity and cloud ice signals, *Atmos. Chem. Phys.*, **7**, 459–469.

Ekström, M., P. Eriksson, W.G. Read, M. Milz, and D.P. Murtagh, 2008, Comparison of satellite limb-sounding humidity climatologies of the uppermost tropical troposphere, *Atmos. Chem. Phys.*, **8**, 309–320.

Emde, C., S.A. Buehler, C. Davis, P. Eriksson, T.R. Sreerekha and C. Teichmann, 2004, A polarized discrete ordinate scattering model for simulations of limb and nadir longwave measurements in 1D/3D spherical atmospheres, *J. Geophys. Res.*, **109(24)**, 24207.

Eriksson, P., M. Ekström, S.A. Buehler and C. Melsheimer, 2006, Efficient forward modelling by matrix representation of sensor responses. *Int. J. Remote Sensing*. **27**, 1793–1808.

Eriksson, P., M. Ekström, B. Rydberg, and D.P. Murtagh, 2007, First Odin sub-mm retrievals in the tropical upper troposphere: ice cloud properties, *Atmos. Chem. Phys.*, **7**, 471–483.

Eriksson, P., S.A. Buehler, (eds), 2008, ARTS-1.x user guide (available at http://www.sat.ltu.se/arts/misc/arts2/doc/uguide/uguide.pdf).

Eriksson, P., M. Ekström, B. Rydberg, D.L. Wu, R.T. Austin, and D.P. Murtagh, 2008, Comparison between early Odin-SMR, Aura MLS and CloudSat retrievals of cloud ice mass in the tropical upper troposphere, *Atmos. Chem. Phys.*, **8,** 1937–1948.

Errico, R., P. Bauer, and J.-F. Mahfouf, 2007, Assimilation of cloud and precipitation data: Current issues and future prospects, *J. Atmos. Sci.*, **64,** 3789-3802.

Evans, K.F. and J. Vivekanandan, 1990, Multiparameter radar and microwave radiative transfer modeling of non spherical atmospheric ice particles, *IEEE Trans. Geosci. Remote Sens.*, **28**, 423–437.

Evans, K.F., and G.L. Stephens, 1995a, Microwave radiative transfer through clouds composed of realistically shaped ice crystals, part I, Single scattering properties, *J. Atmos. Sci.*, **52**, 2041–2057.

Evans, K.F., and G.L. Stephens, 1995b, Microwave radiative transfer through clouds composed of realistically shaped ice crystals, part II, Remote sensing of ice clouds, *J. Atmos. Sci.*, **52**, 2058–2072.

Evans, K.F., S.J. Walter, A.J. Heymsfield, and M.N. Deeter, 1998, Modeling of submillimeter passive remote sensing of cirrus clouds, *J. Appl. Meteorol.*, **37**, 184–205.

Evans, K.F., 1998, The spherical harmonic discrete ordinate method for three-dimensional atmospheric radiative transfer. *J. Atmos. Sci.*, **55**, 429–446.

Evans, K.F., A.H. Evans, I.G. Nolt. and B.T. Marshall, 1999, The prospect for remote sensing of cirrus clouds with a submillimeter-wave spectrometer, *J. Appl. Meteorol.*, **3**8, 514–525.

Evans, K.F., S.J. Walter, A.J. Heymsfield and G.M. McFarquhar, 2002, The submillimeter-wave cloud ice radiometer: Simulations of retrieval algorithm performance. *J. Geophys. Res.*, **107**, doi: 10.1029/2001JD000709.

Evans, K.F., J.R. Wang, P. Racette, G. Heymsfield and L. Li, 2005, Ice cloud retrievals and analysis with data from the Compact Scanning Submillimeter Imaging Radiometer and the Cloud Radar System during CRYSTAL-FACE, *J. Appl. Meteor.*, **44**, 839–859.

Eyre, J.R. 1991, A fast radiative transfer model for satellite sounding systems, ECMWF Research Dept. Tech. Memo. **176** (available from the librarian at ECMWF)

Eyre, J., G.A. Kelly, A.P. McNally, E. Andersson, and A. Persson, 1993, Assimilation of TOVS radiance information through one-dimensional variational analysis, *Q. J. R. Meteorol. Soc.*, **119**, 1427–1463.

Farrar, M.R., 1997, Combined Radar-Radiometer Rainfall Retrieval for TRMM Using Structure Function-Based Optimization. Ph.D. Dissertation (E.A. Smith, Major Prof.), Dept. of Meteorology, The Florida State University, Tallahassee, FL, 185 pp.

Ferraro, R.R, F. Weng, N.C. Grody and L. Zhao, 2000, Precipitation characteristics over land from the NOAA-15 AMSU Sensor. *Geophys. Res. Let.*, **27**, 2669–2672.

Ferraro, R.R., F. Weng, N. Grody, L. Zhao, H. Meng, C. Kongoli, P. Pellegrino, S. Qiu and C. Dean, 2005, NOAA operational hydrological products derived from the AMSU, *IEEE Trans. Geo. Rem. Sens.,* **43**, 1036–1049.

Filipiak, M.J., R.S. Harwood, J.H. Jiang, Q. Li, N.J. Livsey, G.L. Manney, W.G. Read, M.J. Schwartz, J.W. Waters, D.L. Wu, 2005, Carbon monoxide measured by the EOS Microwave Limb Sounder on Aura: First results, *Geophys. Res. Lett.*, **32**, L14825, doi:1029/2005GL022765.

Fiorino, S.T., and E.A. Smith, 2006, Critical assessment of microphysical assumptions within TRMM-radiometer rain profile algorithm using satellite, aircraft and surface datasets from KWAJEX. *J. Appl. Meteor. Clim.*, **45**, 754–786.

Forster, P.M.d.F., and K.P. Shine, 1997, Radiative forcing and temperature trends from stratospheric ozone changes, *J. Geophys. Res.*, **102**, 10841–10855.

Gasiewski, A.J., 1992, Numerical sensitivity analysis of passive EHF and SMMW channels to tropospheric water vapor, clouds, and precipitation, *IEEE Trans. Geosci. Remote Sens.*, **30**, 859–870.

Goody, R.M., and Y.L. Yung, 1989, Atmosphere radiation, Theoretical basis (2ed), Oxford University Press.

Greenwald, T.J., G.L. Stephens, T.H.V. Haar and D.L. Jackson, 1993, A Physical Retrieval of Cloud Liquid Water Over the Global Oceans Using Special Sensor Microwave/Imager (SSM/I) Observations, *J. Geophys. Res.*, **98(10)**, 18,471–18,488.

Greenwald T.J. and S.A. Christopher, 2002, Effect of cold clouds on satellite measurements near 183 GHz, *J. Geophys. Res.*, **107**, 4170, doi:10.1029/2000JD000258.

Greenwald, T.J., T.S. L'Ecuyer and S.A. Christopher, 2007, Evaluating specific error characteristics of microwave-derived cloud liquid water products, *Geophys. Res. Lett.*, **34**, L22807, doi:10.1029/2007GL031180.

Grody, N.C., 1991, Classification of snow cover and precipitation using the Special Sensor Microwave/Imager (SSM/I), *J. Geophys. Res.*, **96,** 7423–7435.

Haddad, Z.S., E.A. Smith, C.D. Kummerow, T. Iguchi, M.R. Farrar, S.L. Durden, M. Alves, and W.S. Olson, 1997, The TRMM 'Day-1' radar/radiometer combined rain-profiling algorithm. *J. Meteor. Soc. Japan*, **75**, 799–809.

Haynes, J. L'Ecuyer, G. Stephens, S. Miller, C. Mitrescu and S. Tanelli, 2008, Rainfall retrieval over the ocean with spaceborne W-band radar, *J. Geophys. Res.*, **114**, D00A22, doi:10.1029/2008JD009973.

Healy, S., and J.-N. Thépaut, 2006, Assimilation experiments with CHAMP GPS radio occultation measurements. *Quart. J. Roy. Meteorol. Soc.*, **132**, 605–623.

Healy, S., J. Eyre, M. Hamrud, and J.-N. Thépaut, 2007, Assimilating GPS radio occultation measurements with two-dimensional bending angle observation operators. *Quart. J. Roy. Meteorol. Soc.*, **133**, 1213–1227.

Held, I. M., and B. J. Soden, 2000, Water vapor feedback and global warming, *Ann. Rev. Energy Environ.*, **24**, 441–475.

Heymsfield, A. J., A. Bansemer, C. Schmitt, C. Twohy and M. R. Poellot, 2004, Effective ice particle densities derived from aircraft data, *J. Atmos. Sci.*, 61, 982–1003.

Hitschfeld, W., and J. Bordan, 1954, Errors inherent in the radar measurement of rainfall at attenuating wavelengths. *J. Meteor.*, **11**, 58–67.

Hong G., G. Heygster, and K. Kunzi 2005 a, Intercomparison of deep convective cloud fractions from passive infrared and microwave radiance measurements, IEEE Geosc. Remote Sensing Lett., **2**, 18–24.

Hong, G., G. Heygster, J. Miao and K. Kunzi, 2005b, Detection of tropical deep convective clouds from AMSU-B water vapor channels measurements, *J. Geophys. Res.*, **110**, D05205, doi:10.1029/2004JD004949.

Hong, Y., T.R. Wilheit, and W.R. Russell, 1997, Estimation of monthly rainfall over oceans from truncated rain-rate samples: Applications to SSM/I data. *J. Atmos. Oceanic Technol.*, **14**, 1012–1022.

Horváth Á. and R. Davies, 2007, Comparison of microwave and optical cloud water path estimates from TMI, MODIS, and MISR, J. Geophys. Res., **112**, D01202, doi:10.1029/2006JD007101.

Iguchi, T., and R. Meneghini, 1994, Intercomparison of single-frequency methods for retrieving a vertical rain profile from airborne or spaceborne radar data. *J. Atmos. Oceanic Tech.*, **11**, 1507–1511.

Iguchi, T., T. Kozu, R. Meneghini, J. Awaka, and K. Okamoto, 2000, Rain-profiling algorithm for the TRMM precipitation radar. *J. Appl. Meteor.*, **39**, 2038–2052.

Iguchi, T., 2007, Space-borne radar algorithms. Measuring Precipitation from Space: EURAINSAT and the Future (V. Levizzani, P. Bauer, and F.J. Turk, eds.), Advances in Global Change Research (Vol. 28), Springer, Dordrecht, Netherlands, 199–212.

Im, E., C. Wu and S.L. Durden, 2005, Cloud profiling radar for the CloudSat mission, IEEE Radar Conference, 9–12 May, pp 483–486.

Janssen, M.A., 1993, Atmospheric remote sensing by microwave radiometry, Wiley, ISBN 0-471-62891-3

Jensen, A., M. Lohmann, H.-H. Benzon, and A. Nielsen, 2003, Full Spectrum Inversion of radio occultation signals. Radio Sci., **38,** 1040, doi:10.1029/2002RS002763.

Jiang, J.H., N.J. Livesey, H. Su, L. Neary, J.C. McConnell, and N.A.D. Richards, 2007, Connecting surface emissions, convective uplifting, and long-range transport of carbon monoxide in the upper-troposphere: New observations from the Aura Microwave Limb Sounder, Geophys. Res. Lett., **34**, L18812, doi:10.1029/2007GL030638.

Jiménez, C., P. Eriksson, V.O. John and S.A. Buehler, 2005, A demonstration of AMSU retrieval precision for mean upper tropospheric humidity by a non-linear multi-channel regression method, Atmos. Chem. Phys., **5**, 451–459.

Jiménez, C., S.A. Buehler, B. Rydberg, P. Eriksson and K.F. Evans, 2007, Performance simulations for a submillimetre wave cloud ice satellite instrument, *Q. J. R. Meteorol. Soc.*, **133,** 129–149, doi:10.1002/QJ.134.

Joiner, J. and D.P. Dee, 2000, An error analysis of radiance and suboptimal retrieval assimilation, *Q. J. R. Meteorol. Soc.*, **126**, 1495–1514.

Jones, A. and T. Vonder Haar, 1990, Passive microwave remote sensing of cloud liquid water over land regions, *J. Geophys. Res.*, **95**, 10, 16673–16683.

Karbou, F., Aires, F., Prigent, C., L. Eymard, 2005, Potential of Advanced Microwave Sounding Unit-A (AMSU-A) and AMSU-B measurements for atmospheric temperature and humidity profiling over land, *J. Geophys. Res.*, **110**, D07109 10.1029/2004JD005318.

Kay, J.E., T. L'Ecuyer, G.L. Stephens, A. Gettelman, and C.O'Dell, 2008, The contribution of cloud and radiation anomalies to the 2007 Arctic sea ice extent minimum, *Geophys. Res. Lett.*, 35, L08503, doi:10.1029/2008GL033451.

Kelly, G.A. and J.-N. Thépaut, 2007, Evaluation of the impact of the space component of the Global Observing System through observing system experiments, ECMWF Newsletter, **113**, 16–28.

Kim M.-J., 2006, Single scattering parameters of randomly oriented snow particles at microwave frequencies, J. Geophys. Res., **111,** 14201, doi:10.1029/2005JD006892.

Kim, M., J.A. Weinman, W.S. Olson, D.-E. Chang, G. Skofronick-Jackson and J.R. Wang, 2008, A physical model to estimate snowfall over land using AMSU-B observations, *J. Geophys. Res.*, **113**, doi: 10.1029/2007JD008589.

Klein B., S.D. Philipp, R. Güsten, I. Krämer, D. Samtleben, 2006, A new generation of spectrometers for radio astronomy: Fast Fourier Transform Spectrometer, SPIE-FFTS2006.pdf

Klein, M., and A.J. Gasiewski, 2000, Nadir sensitivity of passive millimeter and submillimeter wave channels to clear air temperature and water vapor variations, *J. Geophys. Res.*, **105**, 17 481–17 511.

Kozu, T., T. Kawanishi, H. Kuroiwa, M. Kojima, K. Oikawa, H. Kumagai, K. Okamoto, M. Okumura, H. Nakatsuka, K. Nishikawa, 2001, Development of Precipitation Radar on-board the Tropical Rainfall Measuring Mission satellite. IEEE Trans. Geosci. Rem. Sens., **39**, 102–116.

Krasnopolsky, V.M., 2007, Neural network emulations for complex multidimensional geophysical mappings: Applications of neural network techniques to atmospheric and oceanic satellite retrievals and numerical modelling, Rev. Geophys., **45**, RG3009, doi:10.1029/2006RG000200.

Kuhn, T., 2003, Atmospheric absorption models for the millimeter range, PhD thesis, University of Bremen.

Kummerow C.D., R.A. Mack and I.M. Hakkarinen, 1989, A self-consistency approach to improve microwave rainfall estimates from space. *J. Appl. Meteor.*, **39**, 1801–1820

Kummerow C.D. and L. Giglio, 1994, A passive microwave technique for estimating rainfall and vertical structure information from space. Part I: algorithm description. *J. Appl. Meteor.*, **33**, 3–18.

Kummerow, C., W.S. Olson and L. Giglio, 1996, A simplified scheme for obtaining precipitation and vertical hydrometeor profiles from passive microwave sensors. IEEE. Trans. Geosci. Remote Sensing, **34**, 1213–1232.

Kummerow, C., W. Barnes, T. Kozu, J. Shiue, and J. Simpson, 1998, The tropical rainfall measuring mission (TRMM) sensor package. *J. Atmos. Oceanic Tech.*, **15**, 809–817.

Kummerow, C., J. Simpson, O. Thiele, W. Barnes, A.T.C. Chang, E. Stocker, R.F. Adler, A. Hou, R. Kakar, F. Wentz, P. Ashcroft, T. Kozu, Y. Hong, K. Okamoto, T. Iguchi, H. Kuroiwa, E. Im, Z. Haddad, G. Huffman, B. Ferrier, W.S. Olson, E. Zipser, E.A. Smith, T.T. Wilheit, G. North, T. Krishnamurti, and K. Nakamura, 2000, The status of the Tropical Rainfall Measuring Mission (TRMM) after two years in orbit. *J. Appl. Meteor.*, **39**, 1965–1982.

Kummerow, C., Y. Hong, W.S. Olson, S. Yang, R.F. Adler, J. McCollum, R. Ferraro, G. Petty, D.-B. Shin, and T.T. Wilheit, 2001, The evolution of the Goddard Profiling Algorithm (GPROF) for rainfall estimation from passive microwave sensors. *J. Appl. Meteor.*, **40**, 1801–1820.

Kunzi, K., G. Heygster and J. Miao, 2001, Ice cloud and background water vapour by sub-mm and very-high frequency MW radiometer, CLOUDS, A cloud, aerosol, radiation and precipitation explorer, proposal submitted to the ESA Call for Ideas for the Next Earth Explorer Core Missions.

Kursinski, E., G. Hajj, W. Bertiger, S. Leroy, T. Meehan, L. Romans, J. Schofield, D. McCleese, W. Melbourne, C. Thornton, T. Yunck, J. Eyre, and R. Nagatani, 1996, Initial results of radio occultation observations of earth's atmosphere using the Global Positioning System. Science, **271**, 1107–1110.

Kursinski, E., G. Hajj, J. Schofield, R. Linfield, and K. Hardy, 1997, Observing earth's atmosphere with radio occultation measurements using the Global Positioning System. *J. Geophys. Res.*, **102,** 23429–23465.

Lafore, J.-P.J. Stein, N. Asencio, P. Bougeault, V. Ducrocq, J. Duron, C. Fischer, P. Héreil, P. Mascart, V. Masson, J.-P. Pinty, J.-L. Redelsperger, E. Richard, J. Vilà-Guerau de Arellano, 1998, The Meso–NH Atmospheric Simulation System. Part I: adiabatic formulation and control simulations Scientific objectives and experimental design, *Ann. Geophys.*, **16**, 90–109.

L'Ecuyer, T.S. and G.L. Stephens, 2003, The tropical oceanic energy budget from the TRMM perspective, Part I: Algorithm and uncertainties, *J. Clim.*, 16, 1967–1985.

Li, Q. B., J.H. Jiang, W.G. Read, N.J. Livesey, J.W. Waters, Y. Zhang, B. Wang, M.J. Filipiak, C.P. Davis, S. Turquety, S. Wu, R.J. Park, R.M. Yantosca, D.J. Jacob, 2005, Convective outflow of South Asian pollution: A global CTM simulation compared with EOS MLS observations, *Geophys. Res. Lett.*, **32**, L14826, doi:10.1029/2005GL022762.

Liebe, H.J., G.A. Hufford and M.G. Cotton, 1993, Propagation modeling of moist air and suspended water/ice particles at frequencies below 1000 GHz., in AGARD 52nd Specialists Meeting of the Electromagnetic Wave Propagation Panel, Palma de Mallorca, Spain, pp. 3.1–3.10.

Lin, B. and W. Rossow, 1994, Observations of cloud liquid water path over oceans: Optical and microwave remote sensing methods, *J. Geophys. Res.*, **99**, 10, 20907–20927.

Lin, B. and W.B. Rossow, 1996, Seasonal variation of liquid and ice water path in non-precipitating clouds over oceans, *J. Clim.*, **9**, 2890–2902.

Liou, K.N., 2002, An introduction to atmospheric radiation, 2nd ed., Academic Press.

Liu, G. and J. Curry, 1993, Determination of characteristic features of cloud liquid water from satellite microwave measurements, *J. Geophys. Res.*, **98(D3),** 5069–5092.

Liu, G. and J.A. Curry, 1998, Remote sensing of ice water characteristics in tropical clouds using aircraft microwave measurements, *J. Appl. Meteorol.*, **37**, 337-355.

Liu, G. and J.A. Curry, 1999, Tropical ice water amount and its relations to other atmospheric hydrological parameters as inferred from satellite data, *J. Appl. Meteorol.*, **38**, 1182–1194.

Liu, G. and J.A. Curry, 2000, Determination of ice water path and mass median particle size using multichannel microwave measurements, American Meteorological Soc., **39,** 1318–1329.

Liu, G., 2004, Approximation of single scattering properties of ice and snow for high microwave frequencies, *J. Atmos. Sci.*, **61**, 2441–2456.

Liu, C., E. Zipser, T. Garrett, J.H. Jiang, and H. Su, 2007, How do the water vapor and carbon monoxide "tape recorders" start near the tropical tropopause?, *Geophys. Res. Lett.*, **32**, L09804, doi:10.1029/2006GL029234.

Livesey, N.J., M.J. Filipiak, L. Froidevaux, W.G. Read, A. Lambert, M.L. Santee, J.H. Jiang, H.C. Pumphrey, J.W. Waters, R.E. Cofield, D.T. Cuddy, W.H. Daffer, B.J. Drouin, R.A. Fuller,

R.F. Jarnot, Y.B. Yiang, B.W. Knosp, Q.B. Li, V.S. Perun, M.J. Schwartz, W.V. Snyder, P.C. Stek, R.P. Thurstans, P.A. Wagner, M. Avery, E.V. Browell, J.-P Cammas, L.E. Christensen, G.S. Diskin, R.-S. Gao, H.-J-. Jost, M. Loewenstein, J.D. Lopez, P. Nedelec, G.B. Osterman, G.W. Sachse, C.R. Webster, 2008, Validation of Aura Microwave Limb Sounder O_3 and CO observations in the upper troposphere and lower stratosphere, *J. Geophys. Res.*, **113**, D15S02, doi:10.1029/2007JD008805.

Luntama, J.-P., G. Kirchengast, M. Borsche, U. Foelsche, S. Steiner, S. Healy, A. von Engeln, E. O'Clerigh, and C. Marquardt, 2008, Prospects of the EPS GRAS Mission for Operational Atmospheric Applications. *Bull. Amer. Meteor. Soc.*, **89**, 1863–1875.

Macpherson, S.R., G. Deblonde, J.M. Aparicio, and B. Casati, 2008, Impact of NOAA ground-based GPS observations on the Canadian regional analysis and forecast system. *Mon. Wea. Rev.*, **136**, 2727–2746.

Madden, R.A., and P.R. Julian, 1971, Detection of a 40–50 day oscillation in the zonal wind in the tropical Pacific, *J. Atmos. Sci.*, **28**, 702–708.

Madden, R.A., and P.R. Julian, 1994, Observations of the 40–50 day tropical oscillation – A review, *Mon. Weather. Rev.*, **122**, 814–837.

Manning, W., and J.R. Wang, 2003, Retrieval of precipitable water using Special Sensor Microwave/Temperature-2 (SSM/T-2) millimeter-wave radiometric measurements, *Radio Sci.*, 38, 1097, doi:10.1029/2002RS002735.

Marzano, F.S., A. Mugnai, G. Panegrossi, N. Pierdicca, E.A. Smith, and J. Turk, 1999, Bayesian estimation of precipitating cloud parameters from combined measurements of spaceborne microwave radiometer and radar. IEEE Trans. Geosci. Remote Sensing, **37**, 596–613.

Matricardi, M., F. Chevallier, G.A. Kelly, and J.-N. Thépaut, 2004, An improved general fast radiative transfer model for the assimilation of radiance observations, *Q. J. R. Meteorol. Soc.*, **130**, 153–173.

Mätzler, 2002, C., MATLAB functions for Mie scattering and absorption - Version 2, Tech. Rep. 2002–11, Universität Bern.

Mätzler, C. (Ed.), 2006, Thermal microwave radiation: Applications for remote sensing, IET Electromagnetic Waves Series, The Institution of Engineering and Technology, London, ISBN 0-86341-573-3, 555pp.

Meeks, M.L., 1961, Atmospheric emission and opacity at millimeter wavelengths due to oxygen, J. Geophys. Res., **66**, 3749–3757.

Meirold-Mautner, I., C. Prigent, E. Defer, J.-R. Pardo, J.-P. Chaboureau, J.-P. Pinty, M. Mech and S. Crewell, 2007, Radiative transfer simulations using mesoscale cloud model outputs: comparisons with passive microwave and infrared satellite observations for mid-latitudes, *J. Atmos. Sc.*, **64,** 1550–1568.

Melsheimer, C., C. Verdes, S.A. Buehler, C. Emde, P. Eriksson, D.G. Feist, S. Ichizawa, V.O. John, Y. Kasai, G. Kopp, N. Koulev, T. Kuhn, O. Lemke, S. Ochiai, F. Schreier, T.R. Sreerekha, M. Suzuki, C. Takahashi, S. Tsujimaru and J. Urban, 2005, Intercomparison of General Purpose Clear Sky Atmospheric Radiative Transfer Models for the Millimeter/Submillimeter Spectral Range, Radio Sci., 40, RS**1007**, doi:10.1029/2004RS003110.

Meneghini, R.J., J. Eckerman, and D. Atlas, 1983, Determination of rain rate from a spaceborne radar technique. IEEE Trans. Geosci. Rem Sens., **21**, 34–43.

Meneghini, R., and T. Kozu, 1990, Spaceborne Weather Radar. Artech House, 199pp.

Meneghini, R., T. Iguchi, T. Kozu, L. Liao, K. Okamoto, J. Jones, and J. Kwiatkowski, 2000, Use of the surface reference technique for path attenuation estimates from the TRMM Precipitation Radar. J. Appl. Meteor., **39**, 2053–2070.

Minnis P., J. Huang, B. Lin, Y. Yi, R.F. Arduini, T.-F. Fan, J.K. Ayers, G.G. Mace, 2007, Ice cloud properties in ice-over-water cloud systems using Tropical Rainfall Measuring Mission (TRMM) visible and infrared scanner and TRMM Microwave Imager data, J. Geophys. Res., **112**, D06206, doi:10.1029/2006JD007626.

Mishchenko, M.I., L.D. Travis, and A.A. Lacis, 2002, Scattering, Absorption and Emission of Light by Small Particles, Cambridge University Press, Cambridge. ISBN 0-521-78252

Mitrescu,C., S. Miller, G Hawkins, T. L'Ecuyer, J. Turk, P. Partain and G. Stephens, 2008, Near real time applications of CloudSat data, J. App. Met., **47**, 1982–1994.

Mohr, K.I., J.S. Famiglietti and E.J. Zipser, 1999, The contribution to tropical rainfall with respect to convective system type, size, and intensity estimated from the 85-GHz ice-scattering signature, J. Appl. Meteorol., **38**, 596–606.

Moreau E., P. Bauer and F. Chevallier, 2003, Variational retrieval of rain profiles from spaceborne passive microwave radiance observations, J. Geophys. Res., **108**(D16), 4521, doi:10.1029/2002JD003315.

Mote, P.W., K.H. Rosenlof, J.R. Holton, R.S. Harwood, and J.W. Waters, 1995, Seasonal variation of water vapour in the tropical lower stratosphere, Geophys. Res. Lett., **22**, 1093–1096.

Mote, P. W., K.H. Rosenlof, M.E. McIntyre, E.S. Carr, J.C. Gille, J.R. Holten, J.S. Kinnersley, H.C. Pumphrey, J.M. Russell III, J.W. Waters, 1996, An atmospheric tape recorder: the imprint of tropical tropopause temperatures on stratospheric water vapor, J. Geophys. Res., **101**, 3989–4006.

Mote, P.W., H.L. Clark, T.J. Dunkerton, R.S. Harwood, and H.C. Pumphrey, 2000, Intraseasonal variations of water vapor in the tropical upper troposphere and tropopause region, J. Geophys. Res., **105**, 17457–17470.

Mugnai, A., and E.A. Smith, 1984, Passive microwave radiation transfer in an evolving cloud medium. IRS-84: Current Problems in Atmospheric Radiation, Proceedings of the IAMAP International Radiation Symposium (G. Fiocco, editor), ISBN 0-937194-08-5, A. Deepak Publishing, Hampton, VA, 297–300.

Mugnai, A., and E.A. Smith, 1988: Radiative transfer to space through a precipitating cloud at multiple microwave frequencies. Part I: Model description. *J. Appl. Meteor.*, **27**, 1055–1073.

Mugnai, A., H.J. Cooper, E.A. Smith and G.J. Tripoli, 1990, Simulation of microwave brightness temperatures of an evolving hail storm at the SSM/I frequencies. Bull. Amer. Meteor. Soc., **71**, 2–13.

Mugnai, A., E.A. Smith and G.J. Tripoli, 1993, Foundation for statistical-physical precipitation retrieval from passive microwave satellite measurements. Part 1: Emission source and generalized weighting function properties of a time dependent cloud-radiation model, J. Appl. Meteorol., **32**, 17–39.

Mugnai, A., S. Di Michele, E.A. Smith, F. Baordo, P. Bauer, B. Bizzarri, P. Joe, C. Kidd, F.S. Marzano, A. Tassa, J. Testud, and G.J. Tripoli, 2007, Snowfall measurements by proposed European GPM mission. Measuring Precipitation from Space: EURAINSAT and the Future (V. Levizzani, P. Bauer, and F.J. Turk, eds.), Advances in Global Change Research (Vol. 28), Springer, Dordrecht, Netherlands, 655–674.

Mugnai, A., E.A. Smith, G.J. Tripoli, S. Dietrich, V. Kotroni, K. Lagouvardos, and C.M. Medaglia, 2008, Explaining discrepancies in passive microwave cloud-radiation databases in microphysical context from two different cloud-resolving models. Meteor. Atmos. Phys., **101**, 127–145, doi 10.1007/s00703-007-0281-4, 19 pp.

Murtagh, D., U. Frisk, F. Merino, M. Ridal, A. Jonsson, J. Stegman, G. Witt, P. Eriksson, C. Jiménez, G. Megie, J. de la Noö, P. Ricaud, P. Baron, J.R. Pardo, A. Hauchcorne, E.J. Llewellyn, D.A. Degenstein, R.L. Gattinger, N.D. Lloyd, W.F.J. Evans, I.C. McDade, C.S. Haley, C. Sioris, C. von Savigny, B.H. Solheim, J.C. McConnell, K. Strong, E.H. Richardson, G.W. Leppelmeier, E. Kyrölä, H. Auvinen, L. Oikarinenh, 2002, An overview of the Odin atmospheric mission, Can. J. Phys., **80**, 309–319.

Nakamura, K., K. Okamoto, T. Ihara, J. Awaka, and T. Kozu, 1990, Conceptual design of rain radar for the Tropical Rainfall Measuring Mission. Int. J. Sat. Comm., **8,** 257–268.

Nesbitt, S.W., E.J. Zipser and D.J. Cecil, 2000, A census of precipitation features in the tropics using TRMM: radar, ice scattering, and lightning observations, J. Clim., **13**, 4087–4106.

Okamoto, K., J. Awaka, and T. Kozu, 1988, A feasibility study of rain radar for the Tropical Rainfall Measuring Mission 6: A case study of rain radar system. J. Comm. Research Lab., **35**, 183–208.

Okamoto, K., and T. Kozu, 1993, TRMM Precipitation Radar algorithms. Proc. IGARSS'93, IEEE Trans Geosci Remote Sens, **31**, 426–428.

Okamoto, K., 2003, A short history of the TRMM Precipitation Radar. AMS Meteorological Monographs: Cloud Systems, Hurricanes, and the Tropical Rainfall Measuring Mission (TRMM) - A Tribute to Dr. Joanne Simpson, **29**, 187–195.

Okamoto, K., and S. Shige, 2008, TRMM precipitation radar and its observation results (in Japanese). IEICE Trans. Commun., **J91-B**, 723–733.

Olson, W.S., Y. Hong, C.D. Kummerow, and J. Turk, 2001, A texture-polarization method for estimating convective - stratiform precipitation area coverage from passive microwave radiometer data. *J. Appl. Meteorol.*, **40**, 1577–1591.

Olson, W.S., C.D. Kummerow, S. Yang, G.W. Petty, W.-K. Tao, T.L. Bell, S.A. Braun, Y. Wang, S.E. Lang, D.E. Johnson, and C. Chiu, 2006, Precipitation and latent heating distributions from satellite passive microwave radiometry. Part I: Improved method and uncertainties. *J. Appl. Meteor. Climatol.*, **45**, 702–720.

Ossenkopf, V., 2008, The stability of spectroscopic instruments: a unified Allan variance computation scheme, Astronony & Astrophysics **479**(3), 915–926, DOI: 10.1051/0004-6361:20079188

Panegrossi G., S. Dietrich, F.S. Marzano, A. Mugnai, E.A. Smith, X. Xiang, G.J. Tripoli, P.K. Wang, J.P.V. Poiares Baptista, 1998, Use of cloud model microphysics for passive microwave-based precipitation retrieval: Significance of consistency between model and measurement manifolds. *J. Atmos. Sci.*, **55**, 1644–1673.

Pardo, J.R., J. Cernicharo, and E. Serabyn, 2001, Atmospheric Transmission at Microwaves (ATM): An improved model for mm/submm applications, IEEE Trans. on Antennas and Propagation, **49(12)**, 1683–1694.

Phalippou, L., 1996, Variational retrieval of humidity profile, wind speed and cloud liquid water path with the SSM/I: Potential for numerical weather prediction, *Q. J. Roy. Meteorol. Soc.*, **122**, 327–355.

Pickett, H.M., R.L. Poynter, E.A. Cohen, M.L. Delitsky, J.C. Pearson, and H.S.P. Müller, 2001, Submillimeter, millimeter, and microwave spectral line catalog, Tech. rep., Jet Propulsion Laboratory.

Poli, P., P. Moll, F. Rabier, G. Desroziers, B. Chapnik, L. Berre, S.B. Healy, E. Andersson, and F.-Z. El Guelai, 2007, Forecast impact studies of zenith total delay data from European near real-time GPS stations in Météo France 4DVAR. *J. Geophys. Res.*, **112**, D06114, doi:10.1029/2006JD007430.

Poli, P., P. Moll, D. Puech, F. Rabier, and S. Healy, 2009, Quality control, error analysis, and impact assessment of {FORMOSAT-3/COSMIC} in numerical weather prediction. Terr. Atmos. Ocean, **20**, 101–113.

Posselt, D., G.L. Stephens, and M. Miller, 2008, CloudSat: Adding a new dimension to a classical view of extratropical cyclones. Bull. Amer. Met. Soc., **89**, 599–609.

Prabhakara, C., G. Dalu, G.L. Liberti, J.J. Nucciarone and R. Suhasini, 1992, Rainfall estimation over oceans from SMMR and SSM/I microwave data, *J. Appl. Meteorol.*, **31**, 532–552.

Prigent, C., A. Sand, C. Klapisz and Y. Lemaitre, 1994, Physical retrieval of liquid water contents in a North Atlantic cyclone using SSM/I data, *Q. J. R. Meteorol. Soc.*, **120**, 1179–1207.

Prigent, C. and W.B. Rossow, 1999, Retrieval of surface and atmospheric parameters over land from SSM/I: Potential and limitations, Q. J. Roy. Meteorol. Soc., **125**, 2379–2400.

Prigent, C., E. Defer, J. Pardo, C. Pearl, W.B. Rossow and J.-P. Pinty, 2005, Relations of polarized scattering signatures observed by the TRMM Microwave Instrument with electrical processes in cloud systems, Geophys. Res. Lett., **32**, L04810, doi: 10.1029/2004GL022225.

Prigent, C., J. Pardo, and B. Rossow, 2006, Comparisons of the millimeter and submillimeter bands for atmospheric temperature and water vapor soundings for clear and cloudy skies, *J. Appli. Meteorol. Climat.*, **45**, 1622–1633.

Pumphrey, H.C., D. Rind, J.M. Russell, III, and J.E. Harries, 1998, A preliminary zonal mean climatology of water vapour in the stratosphere and mesosphere, Adv. Space. Res., **21**, 1417–1420.

Pumphrey, H.C., C. Boone, K.A. Walker, P. Bernath, and N.J. Livesey, 2008, The tropical tape recorder observed in HCN, *Geophys. Res. Lett.* **35**, L05801, doi:10.1029/2007GL032137.

Rau, G., R.Schieder, B.Vowinkel, 1984, Characterization and Measurement of Radiometer Stability, Proc. 14th European Microwave Conf., Sept. 10–13, Liege, Belgium, 248–253

Racette,P., R.F. Adler, J.R. Wang, A.J. Gasiewski, D.M. Jackson and D.S. Zacharias, 1996, An airborne millimeter-wave imaging radiometer for cloud, precipitation, and atmospheric water vapor studies, *J. Appl. Oceanic Technol.*, **13**, 610–619.

Read, W.G., J.W. Waters, D.A. Flower, L. Froidevaux, R.F. Jarnot, D.L. Hartman, R.S. Harwood, and R.B. Rood, 1995, Upper–tropospheric water vapor from UARS MLS, Bull. Amer. Meteorol. Soc., **76**, 2381–2389.

Read, W.G., J.W. Waters, D.L. Wu, E.M. Stone, Z. Shippony, A.C. Smedley, C.C. Smallcomb, S. Oltmans, D. Kley, H.G.J. Smit, J.L. Mergenthaler, M.K. Karki, 2001, UARS Microwave Limb Sounder upper tropospheric humidity measurement: Method and validation, *J. Geophys. Res.*, **106**, 32207–32258.

Read, W.G., D.L. Wu, J.W. Waters, and H.C. Pumphrey, 2004a, A new 147-56 hPa water vapor product from the UARS Microwave Limb Sounder, *J. Geophys. Res.*, **109**, D06111.

Read, W.G., D.L. Wu, J.W. Waters, and H.C. Pumphrey, 2004b,Dehydration in the tropical tropopause layer: Implications from UARS MLS, *J. Geophys. Res.*, **109,** D06110, doi:10.1029/2003JD004056.

Read, W.G., Z. Shippony, M.J. Schwartz, N.J. Livesey and W.V. Snyder, 2006, The clear-sky unpolarized forward model for the EOS Microwave Limb Sounder (MLS), IEEE Trans. Geosci. Remote Sens., **44**(5), 1367–1379.

Read, W.G., A. Lambert, J. Bacmeister, R.E. Cofield, L.E. Christensen, D.T. Cuddy, W.H. Daffer, B.J. Drouin, E. Fetzer, L. Froidevaux, R. Fuller, R. Herman, R.F. Jarnot, J.H. Jiang, Y.B. Jiang, K. Kelly, B.W. Knosp, H.C. Pumphrey, K.H. Rosenlof, X. Sabounchi, M.L. Santee, M.J. Schwartz, W.V. Snyder, P.C. Stek, H. Su, L.L. Takacs, R.P. Thurstans, H. Vomel, P.A. Wagner, J.W. Waters, C.R. Webster, E.M. Weinstock, and D.L. Wu, 2007, EOS Aura Microwave Limb Sounder upper tropospheric and lower stratospheric humidity validation, *J. Geophys. Res.*, **112**, D24S35, doi:10.1029/2007JD008752.

Read, W.G., M.J. Schwartz, A. Lambert, H. Su, N.J. Livesey, W.H. Daffer, and C.D. Boone, 2008, The roles of convection, extratropical mixing, and *in situ* freeze-drying in the tropical tropopause layer, Atmos. Chem. Phys. Discus., **8**, 3961–4000.

Ricaud, P., B. Barret, J.L. Attié, E.L. Flochmoën, H. Teyssèdre, V.H. Peuch, N. Livesey, A. Lambert, and J.P. Pommerau, 2007, Impact of land convection on troposphere-stratosphere exchange in the tropics, Atmos. Chem. Phys., **7**, 5639–5657.

Rocken, C., R. Anthes, M. Exner, D. Hunt, S. Sokolovsky, R. Ware, M. Gorbunov, W. Schreiner, D. Feng, B. Herman, Y.-H. Kuo, and X. Zou, 1997, Analysis and validation of {GPS/MET} data in the neutral atmosphere. J. Geophys. Res., **102**, 29849–29866.

Rodgers, C., 1976, Retrieval of atmospheric temperature and composition from remote measurements of thermal radiation, Rev. Geophys., **14**, 609–624.

Rodgers, C., 1990, Characterization and error analysis of profiles retrieved from remote sounding measurements, J. Geophys. Res., **95**, 5587–5595.

Rodgers, C., 2000, Inverse methods for atmospheric sounding: Theory and Practice, World Scientific.

Rosenkranz, P.W., 1993, Absorption of microwaves by atmospheric gases, in Atmospheric remote sensing by microwave radiometry, edited by M. A. Janssen, pp. 37–90, John Wiley & Sons, Inc.

Rosenkranz, P.W., K.D. Hutchinson, K.R. Hardy, and M.S. Davis, 1997, An assessment of the impact of satellite microwave sounder incidence angle and scan geometry on the accuracy of atmospheric temperature profile retrievals, J. Atmos. Ocean. Tech., **14**, 488–494.

Rosenkranz, P.W., 1998, Water vapor microwave continuum absorption: A comparison of measurements and models, Radio Sci., **33**, 919–928.

Rossow, W.B. and R.A. Schiffer, 1999, Advances in understanding clouds from ISCCP, Bull. Amer. Meteorol. Soc., **80**, 2261–2287.

Rothman, L.S., D. Jacquemart, A. Barbe, D.C. Benner, M. Birk, L.R. Brown, M.R. Carleer, C. Chackerian Jr, K. Chance, L.H. Coudert, V. Dana, V.M. Davi, J.-M. Flaud, R.R. Gamache,

A. Goldman, J.-M. Hartmann, K.W. Jucks, A.G. Maki, J.-Y. Mandin, S.T. Masie, J. Orphal, A. Perrin, C.P. Rinsland, M.A.H. Smith, J. Tennyson, R.N. Tolchenov, R.A. Toth, J.V. Auwera, P. Varanasi, G. Wagner, 2005, The HITRAN 2004 molecular spectroscopic database, J. Quant. Spectrosc. Radiat. Transfer, **96**, 139–204.

Santee, M.L., A. Lambert, W.G. Read, N.J. Livesey, R.E. Cofield, D.T. Cuddy, W.H. Daffer, B.J. Drouin, L. Froidevaux, R.A. Fuller, R.F. Jarnot, B.W. Knosp, G.L. Manney, V.S. Perun, W.V. Snyder, P.C. Stek, R.P. Thurstans, P.A. Wagner, J.W. Waters, G. Muscari, R.L. de Zafra, J.E. Dibb, D.W. Fahey, P.J. Popp, T.P. Marcy, K.W. Jucks, G.C. Toon, R.A. Stachnik, P.F. Bernath, C.D. Boone, K.A. Walker, J. Urban, D. Murtagh, 2007, Validation of the Aura Microwave Limb Sounder HNO_3 measurements, J. Geophys. Res., **112**, D24S40, doi: 10.1029/2007JD008721.

Saunders, R., M. Matricardi, and P. Brunel, 1999, An improved fast radiative transfer model for assimilation of satellite radiance, *Q. J. R. Meteorol. Soc.*, **125**, 1407–1425.

Schaerer, G., and T. T. Wilheit, 1979, A passive microwave technique for profiling of atmospheric water vapor, Radio Sci., **14**, 371–375.

Schoeberl, M.R., B.N. Duncan, A.R. Douglass, J.W. Waters, N.J. Livesey, W.G. Read, and M.J. Filipiak, 2006, The carbon monoxide tape recorder, Geophys. Res. Lett., **33**, L12811, doi: 10.1029/2006GL026178.

Scipal, K., M. Drusch, W. Wagner, 2008, Assimilation of a ERS scatterometer derived soil moisture index in the ECMWF numerical weather prediction system. Adv. Water Resources, **31,** 1101–1112.

Seuffert, G., H. Wilker, P. Viterbo, J.-F. Mahfouf, M. Drusch, and J.-C. Calvet, 2003, Soil moisture analysis combining screen-level parameters and microwave brightness temperatures: A test with field data. Geophys. Res. Lett., **30**, 1498, doi:10.1029/2003GL017128.

Simpson, J., R.F. Adler, and G. North, 1988, A Proposed Tropical Rainfall Measuring Mission (TRMM) satellite. Bull. Amer. Meteor. Soc., **69**, 278–295.

Simpson, J., and W.-K. Tao, 1993, The Goddard Cumulus Ensemble Model. Part II: Applications for studying cloud precipitating processes and for NASA TRMM. Terrestrial, Atmospheric and Oceanic Sciences, **4**, 73–116.

Simpson, J., C. Kummerow, W.-K. Tao, and R.F. Adler, 1996, On the Tropical Rainfall Measuring Mission (TRMM) satellite. *Meteorol. Atmos. Phys.*, **60**, 19–36.

Skofronick-Jackson, G.M., J.R. Wang, G.M. Heymsfield, R. Hood, W. Manning, R. Meneghini and J.A. Weinman, 2003, Combined radiometer-radar microphysical profile estimations with emphasis on high frequency brightness temperature observations, *J. Appl. Meteorol.*, **42**, 476–487.

Skofronick-Jackson G., A. Heymsfield, E. Holthaus, C. Albers, and M.-J. Kim, 2008, Nonspherical and spherical characterization of ice in Hurricane Erin for wideband passive microwave comparisons, *J. Geophys. Res.*, **113,** D06201, doi:10.1029/2007JD008866.

Smith, E.A., and A. Mugnai, 1988, Radiative transfer to space through a precipitating cloud at multiple microwave frequencies. Part II: Results and analysis. *J. Appl. Meteor.*, **27**, 1074–1091.

Smith, E.A., and A. Mugnai, 1989, Radiative transfer to space through a precipitating cloud at multiple microwave frequencies. Part III: Influence of large ice particles. *J. Meteor. Soc. Japan*, **67**, 739–755.

Smith, E.A., X. Xiang, A. Mugnai, and G. Tripoli, 1992a, A cloud radiation model for spaceborne precipitation retrieval. Extended Abstract Vol. of International TRMM Workshop on the Processing and Utilization of the Rainfall Data Measured from Space, Communications Research Laboratory, Tokyo, Japan, 273–283.

Smith, E.A., A. Mugnai, H.J. Cooper, G.J. Tripoli, and X. Xiang, 1992b, Foundations for statistical - physical precipitation retrieval from passive microwave satellite measurements. Part I: Brightness temperature properties of a time dependent cloud - radiation model. *J. Appl. Meteor.*, **31**, 506–531.

Smith, E.A., C. Kummerow, and A. Mugnai, 1994a, The emergence of inversion-type profile algorithms for estimation of precipitation from satellite passive microwave measurements. Remote Sensing Reviews, **11**, 211–242.

Smith, E.A., X. Xiang, A. Mugnai, and G.J. Tripoli, 1994b, Design of an inversion-based precipitation profile retrieval algorithm using an explicit cloud model for initial guess microphysics. Meteorol. Atmos. Phys., **54**, 53–78.

Smith, E.A., Z. Haddad, and C. Kummerow, 1995a, Overview on TRMM combined algorithm development. In: J. Simpson & C. Kummerow (eds) Contribution to TRMM Science Operations Plan, Florida State University, Tallahassee, FL, pp65.

Smith, E.A., A. Mugnai, and G. Tripoli, 1995b, Theoretical foundations and verification of a multispectral, inversion-type microwave precipitation profile retrieval algorithm. In Passive Microwave Remote Sensing of Land-Atmosphere Interactions (599–621), VSP Press, Utrecht - The Netherlands, 685 pp.

Smith, E.A., J. Turk, M. Farrar, A. Mugnai, and X. Xiang, 1997, Estimating 13.8 GHz path integrated attenuation from 10.7 GHz brightness temperatures for TRMM combined PR-TMI precipitation algorithm. *J. Appl. Meteor.*, **36**, 365–388.

Smith, E.A., P. Bauer, F. S. Marzano, C. D. Kummerow, D. McKague, A. Mugnai, G. Panegrossi, 2002, Intercomparison of microwave radiative transfer models for precipitating clouds, IEEE Trans Geosci. Remote Sens., **40**, 541–549.

Smith, E.A., and T.D. Hollis, 2003, Performance evaluation of level 2 TRMM rain profile algorithms by intercomparison and hypothesis testing. AMS Meteorological Monographs: Cloud Systems, Hurricanes, and the Tropical Rainfall Measuring Mission (TRMM) – A Tribute to Dr. Joanne Simpson, **29**, 207–222.

Smith, E.A., G. Asrar, Y. Furuhama, A. Ginati, A. Mugnai, K. Nakamura, R.F. Adler, M.-D. Chou, M. Desbois, J.F. Durning, F. Einaudi, J.K. Entin, R.R. Ferraro, R. Guzzi, P.R. Houser, P.H. Hwang, T. Iguchi, P. Joe, R. Kakar, J.A. Kaye, M. Kojima, C.D. Kummerow, K.-S. Kuo, D.P. Lettenmaier, V. Levizzani, N. Lu, A.V. Mehta, C. Morales, P. Morel, T. Nakazawa, S.P. Neeck, K. Okamoto, R. Oki, G. Raju, J.M. Shepherd, J. Simpson, B.-J. Sohn, E.F. Stocker, W.-K. Tao, J. Testud, G.J. Tripoli, E.F. Wood, S. Yang, and W. Zhang, 2007, International Global Precipitation Measurement (GPM) Program and Mission: An overview. Measuring Precipitation from Space: EURAINSAT and the Future. In: V. Levizzani, P. Bauer, and F.J. Turk, (eds.), Advances in Global Change Research, Vol. 28, Springer, Dordrecht, Netherlands, 611–653.

Sokolovskiy, S., Y.-H. Kuo, and W. Wang, 2005, Assessing the accuracy of a linearized observation operator for assimilation of radio occultation data: Case simulations with a high-resolution weather model. Mon. Wea. Rev., **133**, 2200–2212.

Sorooshian S., X. Gao, K. Hsu, R.A. Maddox, Y. Hong, H.V. Gupta and B. Imam, 2002, Diurnal Variability of Tropical Rainfall Retrieved from Combined GOES and TRMM Satellite Information. J. Clim., **15**, 983–1001.

Spencer, R.W., B.B. Hinton and W.S. Olson, 1983, Nimbus-7 37 GHz radiances correlated with radar rain rates over the Gulf of Mexico, *J. Appl. Meteorol. Climatol.*, **22**, 2095–2099.

Spencer, R.W., H.M. Goodman and R.E. Hood, 1989, Precipitation retrieval over land and ocean with the SSM/I: identification and characteristics of the scattering signal. J. Atmos. Oceanic Tech., **6**, 254–273.

Sreerekha, T.R., C. Emde, N. Courcoux, C. Teichmann, S.A. Buehler, U. Loehnert, M. Mech, S. Crewell, A. Battaglia, P. Eriksson, B. Rydberg, C. Davis, C. Jimenez, S. English and A. Doherty (2006), Development of an RT model for frequencies between 200 and 1000 GHz, Final Report, ESTEC Contract No 17632/03/NL/FF.

Staelin, D.H., K.F. Kunzi, R.L. PettyJohn, R.K.L. Poon and R.W. Wilcox, 1976, Remote sensing of atmospheric water vapor and liquid water with the Nimbus-5 microwave spectrometer, *J. Appl. Meteorol.*, **15**, 1204–1214.

Staelin, D.H. and F. W. Chen, 2000, Precipitation observations near 54 and 183 GHz using the NOAA-15 satellite," IEEE Trans. Geosci. Remote Sens., **38(5)**, 2322–2332.

Staelin D.H., and C. Surussavadee, 2007, Precipitation Retrieval Accuracies for Geo-Microwave Sounders, IEEE Trans. Geosci. Remote Sens., **45**, 3150–3159.

Stephens, G.L. and T.J. Greenwald, 1991, The Earth's radiation budget in relation to atmospheric hydrology: 2 Observations of cloud effects, *J. Geophys. Res.*, **96**, 15,325–15,340.

Stephens, G.L., D.G. Vane, R.J. Boain, G.G. Mace, K. Sassen, Z. Wang, A.J. Illingworth, E. J. O'Connor, W.B. Rossow, S.L. Durden, S.D. Miller, R.T. Austin, A. Benedetti, C. Mitrescu, and the CloudSat Science Team, 2002, The CloudSat mission and the A-TRAIN: A new dimension to space-based observations of clouds and precipitation. Bull. Am. Met. Soc., **83**, 1771–1790.

Stephens, G.L, D.G. Vane, S. Tanelli, E. Im, S. Durden, M. Rokey, D. Reinke, P. Partain, G.G. Mace, R. Austin, T. L'Ecuyer, J. Haynes, M. Lebsock, K. Suzuki, D. Waliser, D. Wu, J. Kay, A. Gettelman, Z. Wang, R. Marchand, 2008, The CloudSat Mission: Performance and early science after the first year of operation, *J. Geophys. Res.*, **113**, D00A18, doi:10.1029/2008JD009982.

Stohl, A., S. Eckhardt, C. Forster, P. James, and N. Spichtinger, 2002, On the pathways and timescales of intercontinental air pollution transport, *J. Geophys. Res.*, **107**, 4684, doi:10.1029/2001JD001396.

Stone, E.M., W.J. Randel, J.L. Stanford, W.G. Read, and J.W. Waters, 1996, Baroclinic wave variations observed in MLS upper tropospheric water vapour, Geophys. Res. Lett., **23**, 2967–2970.

Stubenrauch, C.J., A. Chédin, G. Rädel, N.A. Scott and S. Serrar, 2006, Cloud properties and their seasonal and diurnal variability from TOVS Path-B, J. Clim., **19**, 5531–5553.

Su, H., W.G. Read, J.H. Jiang, J.W. Waters, D.L. Wu, and E.J. Fetzer, 2006, Enhanced positive water vapor feedback associated with tropical deep convection: New evidence from Aura MLS, Geophys. Res. Lett., **33,** 5709, doi:10.1029/2005GL025505.

Surussavadee C. and D.H. Staelin, 2008 a, Global Millimeter-Wave Precipitation Retrievals Trained with a Cloud-Resolving Numerical Weather Prediction Model, Part I: Retrieval Design, IEEE Trans. Geosci. Remote Sens., **46**, 99–108.

Surussavadee, C. and D.H. Staelin, 2008b, Global Millimeter-Wave Precipitation Retrievals Trained with a Cloud-Resolving Numerical Weather Prediction Model, Part II: Performance Evaluation, IEEE Trans. Geosci. Remote Sens., **46**, 109–118.

Tan, D.G.H., E. Andersson, M. Fisher, and I. Isaksen, 2007, Observing-system impact assessment using a data assimilation ensemble technique: application to the ADM-Aeolus wind profiling mission. *Quart. J. Roy. Meteorol. Soc.*, **133**, 381–390.

Tanelli, S., S.L. Durden, E. Im, K. Pak, D. Reinke, P. Partain, J. Haynes, R. Marchand, 2008, CloudSat's Cloud Profiling Radar after 1 year in orbit: performance, external calibration, and processing, IEEE Trans. Geosc. and Rem. Sens., **46,** 3560–3573.

Tao, W.-K., and J. Simpson, 1993, The Goddard Cumulus Ensemble Model. Part I: Model description. Terrestrial, Atmospheric and Oceanic Sciences, **4**, 35–72.

Tapiador F.J., C. Kidd, V. Levizzani, F.S. Marzano, 2004, A neural networks-based fusion technique to estimate half-hourly rainfall estimates at 0.1° resolution from satellite passive microwave and infrared data, J. Appl. Meteor., **43**, 576–594.

Tassa A., S. Di Michele, A. Mugnai, F.S. Marzano, J.P.V. Poiares Baptista, 2003, Cloud model–based Bayesian technique for precipitation profile retrieval from the Tropical Rainfall Measuring Mission Microwave Imager, Radio Sci., **38**, doi:10.1029/2002RS002674.

Tesmer, J.R., and T.T. Wilheit, 1998, An improved microwave radiative transfer model for tropical oceanic precipitation. *J. Atmos. Sci.*, 55, 1674–1689.

Tiuri, M.E., 1966, Radio-Telescope Receivers, pp 236 – 293 in, J.D. Kraus Radio Astronomy, McGraw-Hill Book Company, New York.

Tripoli, G.J., 1992, A nonhydrostatic model designed to simulate scale interaction. Mon. Wea. Rev., **120**, 1342–1359.

Turner D.D., A.M. Vogelmann, R.T. Austin, J.C. Barnard, K. Cady-Pereira, J.C. Chiu, S.A. Clough, C. Flynn, M.M. Khaiyer, J. Liljegren, K. Johnson, B. Lin, C. Long, A. Marshak, S.Y. Matrosov, S.A. McFarlane, M. Miller, Q. Min, P. Minnis, W. O'Hirok, Z. Wang and W. Wiscombe, 2007,

Thin liquid water clouds: Their importance and our challenge, Bull. Am. Meteorol. Soc., **88**, 177–190.

Ulaby F.T., R.K. Moore, A.K. Fung, 1981, Microwave remote sensing, active and passive, Vol. 1, Artech House Publishers.

Uppala, S., P. Kållberg, A. Hernandez, S. Saarinen, M. Fiorino, X. Li, K. Onogi, N. Sokka, U. Andrae and V. Da Costa Bechtold, 2004, ERA-40: ECMWF 45-year reanalysis of the global atmosphere and surface conditions 1957–2002, ECMWF Newsletter, **101**, 2–21.

Vaneck, M.D., I.G. Nolt, N.D. Tappan, P.A.R. Ade, F.C. Gannaway, P.A. Hamilton, C. Lee, J.E. Davis and S. Predko, 2001, Far-infrared sensor for cirrus (FIRSC): an aircraft-based Fourier-transform spectrometer to measure cloud radiance, Appl. Optics, **40**, 2169–2176.

Vedel, H. and K. Sattler, 2006, Comparison of TOUGH impact studies with ground-based GPS observations. TOUGH Project Rep. D49, 18 pp. Available online at http://web.dmi.dk/pub/tough/deliverables/d49-compare-results.pdf.

Vowinkel, B., 1988, Passive Mikrowellenradiometrie, Vieweg, ISBN 3-528-08959-8

Wang, J.R., P. Racette, J.D. Spinhirne, K.F. Evans and W.D. Hart, 1998, Observations of cirrus clouds with airborne MIR, CLS, and MAS during SUCCESS, Geophys. Res. Lett., **25**, 1145–1148.

Wang, Y., Y. Choi, T. Zeng, B. Ridley, N. Blake, D. Blake, and F. Flocke, 2006, Late-spring increase of trans-Pacific pollution transport in the upper troposphere, Geophys. Res. Lett., **33**, L01811, doi:10.1029/2005GL024975.

Waters, J.W., K.F. Kunzi, R.L. PettyJohn, R.K.L. Poon, and D.H. Staelin, 1975, Remote sensing of atmospheric temperature profiles with the Nimbus 5 microwave spectrometer, J. Atmos. Sci., **32**, 1953–1969.

Waters, J.W., W.G. Read, L. Froidevaux, R.F. Jarnot, R.E. Cofield, D.A. Flower, G.K. Lau, H.M. Pickett, M.L. Santee, D.L. Wu, M.A. Boyles, J.R. Burke, R.R. Lay, M.S. Loo, N.J. Livesey, T.A. Lungu, G.L. Manney, L.L. Nakamura, V.S. Perun, B.P. Ridenoure, Z. Shippony, P.H. Siegel, and R.P. Thurstans, R.S. Harwood, H.C. Pumphrey, and M.J. Filipiak, 1999, The UARS and EOS Microwave Limb Sounder (MLS) experiments, J. Atmos. Sci., **56**, 194–217.

Waters, J.W., L. Froidevaux, R.S. Harwood, R.F. Jarnot, H.M. Pickett, W.G. Read, P.H. Siegel, R.E. Cofield, M.J. Filipiak, D.A. Flower, J.R. Holden, G.K. Lau, N.J. Livesey, G.L. Manney, H.C. Pumphrey, M.L. Santee, D.L. Wu, D.T. Cuddy, R.R. Lay, M.S. Loo, V.S. Perun, M.J. Schwartz, P.C. Stek, R.P. Thurstans, M.A. Boyles, K.M. Chandra, M.C. Chavez, C. Gun-Shing, B.V. Chudasama, R. Dodge,R.A. Fuller, M.A. Girard, J.H. Jiang, Y. Jiang, B.W. Knosp, R.C. LaBelle, J.C. Lam, K.A. Lee, D. Miller, J.E. Oswald, N.C. Patel, D.M. Pukala, O. Quintero, D.M. Scaff, W. Van Snyder, M.C. Tope, P.A. Wagner, M.J. Walch, 2006, The Earth Observing System Microwave Limb Sounder (EOS MLS) on the Aura satellite, IEEE Trans. Geosci. Remote Sens., **44**, 1075–1092.

Weng, F. and N.C. Grody, 1994, Retrieval of cloud liquid water using the special sensor microwave imager (SSM/I), *J. Geophys. Res.*, **99(D12)**, 25535–25551.

Weng, F., N.C. Grody, R. Ferraro, A. Basist and D. Forsyth, 1997, Cloud liquid water climatology from the Special Sensor Microwave/Imager, *J. Clim.*, **10**,1086–1098.

Weng, F. J. and N. C. Grody, 2000, Retrieval of ice cloud parameters using a microwave imaging radiometer, *J. Atmos. Sci.*, **57**, 1069–1081.

Wentz, F.J., 1997, A well calibrated ocean algorithm for special sensor microwave/imager, *J. Geophys. Res.*, **102**, 8703–8718.

Wentz, F.J. and R.W. Spencer, 1998, SSM/I rain retrievals within a unified all-weather ocean algorithm. *J. Atmos. Sci.*, **55**, 1613–1627.

Wickert, J., C. Reigber, G. Beyerle, R. König, C. Marquardt, T. Schmidt, L. Grunwaldt, R. Galas, T. Meehan, W. Melbourne, and K. Hocke, 2001, Atmosphere sounding by GPS radio occultation: First results from (CHAMP). Geophys. Res. Lett., **28**, 3263–3266.

Wiedner, M., C. Prigent, J. Pardo, O. Nuissier, J.-P. Chaboureau, J.-P. Pinty and P. Mascart, 2004, Modeling of passive microwave responses in convective situations using outputs from mesoscale models: comparison with TRMM / TMI satellite observations, *J. Geophys. Res.*, **109**, D06214, doi:10.1029/2003/JD004280.

Wilheit, T.T., A.T.C. Chang, M.S.V. Rao, E.B. Rodgers and J.S. Theon, 1977, Satellite technique for quantitatively mapping rainfall rates over oceans, *J. Appl. Meteorol.*, **16**, 551–560.

Wilheit, T.T., A.T.C. Chang, M.S.V. Rao, E.B. Rodgers and J.S. Theon, 1991a, A satellite technique for quantitatively mapping rainfall over oceans, *J. Appl. Oceanic Technol.*, **8**, 118–136.

Wilheit, T.T., A.T.C. Chang, and L.S. Chiu, 1991b, Retrieval of monthly rainfall indices from microwave radiometric measurements using probability distribution functions. *J. Atmos. Oceanic Tech.*, **8**, 118–136.

Winker, D., W. Hunt, and M. McGill, 2007, Initial performance assessment of CALIOP, Geophys. Res. Lett., **34**, L19803, doi:10.1029/2007GL030135.

Wu, D.L., W.G. Read, A.E. Dessler, S.C. Sherwood, and J.H. Jiang, 2005, UARS/MLS cloud ice measurements: Implications for H_2O transport near the tropopause, *J. Atmos. Sci.*, **62**, 518–530.

Wu, D.L., J.H. Jiang, W.G. Read, R.T. Austin, C.P. David, A. Lambert, G.L. Stephens, D.G. Vane, and J.W. Waters, 2008, Validation of Aura MLS cloud Ice Water Content (IWC) measurements, *J. Geophys. Res.*, **113**, D15S10, doi:10.1029/2007JD008931.

Yan, X., V. Ducrocq, P. Poli, M. Hakam, G. Jaubert, and A. Walpersdorf, 2009, Impact of GPS zenith delay assimilation on convective-scale prediction of Mediterranean heavy rainfall. *J. Geophys. Res.*, **114**, D03104, doi:10.1029/2008JD011036.

Zhu, Y. and R. Gelaro, 2008, Observation sensitivity calculations using the adjoint of the gridpoint statistical interpolation (GSI) analysis system, Mon. Wea. Rev., 136, 335–351.

Ziemke, J.R., and S. Chandra, 2003, A Madden-Julian Oscillation in tropospheric zone, Geophys. Res. Lett., **30**, 2182, doi:10.1029/2003GL018523.

Chapter 5
Remote Sensing of Terrestrial Clouds from Space using Backscattering and Thermal Emission Techniques

Alexander A. Kokhanovsky, Steven Platnick and Michael D. King

5.1 Introduction

Clouds play an important role in terrestrial atmospheric dynamics, thermodynamics, chemistry, and radiative transfer and are key elements of the water and energy cycles. Cloud properties can be modified by anthropogenic and natural gaseous and aerosol emissions (i.e. aerosol indirect effect) and are important for understanding climate change. Therefore, it is of a great importance to understand cloud characteristics and their distributions on a global scale. This can only be achieved using satellite observations.

The first picture of cloud fields from space was recorded after the launch of the unmanned V-2 rocket designed by W. von Braun (USA, 1946) followed by TV images from the low Earth orbit Television Infrared Observation Satellites (e.g. TIROS-1, 1st April 1960). The first visual observations of cloud fields from space were reported by the first cosmonaut, Y. A. Gagarin, who orbited the Earth on the Vostok spacecraft (12th April 1961). Photo, video, and hand-held spectrometry of cloud fields from numerous manned Soviet and American spacecraft soon followed.

However, the era of quantitative long-term cloud observations from space began only 30 years ago with the launch of the first TIROS-N satellite (13th October, 1978). This was an experimental satellite developed by NASA and operated by NOAA. It carried a 4-channel Advanced Very High Resolution Radiometer (AVHRR) to provide day and night cloud top and sea surface temperatures, as well as ice and snow conditions; an atmospheric sounding system (TOVS–TIROS Operational Vertical Sounder) to provide profiles of temperature and water vapour from the Earth's surface to the top of the atmosphere. Since then, many imaging

A.A. Kokhanovsky (✉)
Institute of Environmental Physics, University of Bremen, Germany

S. Platnick
NASA Goddard Space Flight Center, Greenbelt, MD, USA

M.D. King
Laboratory for Atmospheric and Space Physics, University of Colorado, Boulder, CO 80309, USA

J.P. Burrows et al. (eds.), *The Remote Sensing of Tropospheric Composition from Space*, Physics of Earth and Space Environments, DOI 10.1007/978-3-642-14791-3_5,

radiometers and spectrometers have been launched. Collectively, they provide a comprehensive global picture for a range of cloud properties and their spatial distributions. The characteristics of selected passive optical instruments currently in operation and their derived cloud products are summarized in Appendix A.

The main cloud products derived from passive optical satellite observations are:

- Cloud cover,
- Cloud thermodynamic phase,
- Cloud optical thickness,
- Cloud droplet/crystal effective radius,
- Cloud liquid/ice water path, and
- Cloud top properties (temperature, pressure/height).

Recent satellite-borne lidar and radar systems reveal the internal structure of cloud systems on a level of detail not possible with passive optical measurements.

The structure of this chapter is as follows: in Section 5.2 we define the main cloud parameters derived from optical satellite measurements. The corresponding algorithms are outlined and results of retrievals are given. The following section has a focus on the description of cloud validation experiments and satellite cloud retrieval uncertainties. In the last section the modern trends in cloud remote sensing and selected planned satellite missions are reviewed.

5.2 Cloud Parameters and Their Retrievals

On average, about 70% of the Earth's surface is covered by clouds. The cloud fraction is a very important parameter, e.g. for the climate studies and also for the retrievals of the vertical columns of trace gases using space-borne instrumentation. It is equal to the ratio of the area of a pixel covered by a cloud to the total area. Other macroscopic characteristics such as the cloud top height, the cloud geometrical thickness, the cloud base height and the number of cloud layers are of interest as well. The cloud top height is important, e.g. for the correction of column trace gas retrieval algorithms in the presence of clouds. The cloud characteristics must be determined with the highest possible accuracy for the creation of reliable bias-free trace gas vertical columns products and databases.

Microphysical parameters, e.g. the cloud particle number density N, phase (liquid or solid) of cloud particles, size/shape distributions of cloud particles and their refractive index are used to calculate cloud local optical characteristics, which are the cloud extinction k_{ext} and absorption k_{abs} coefficients, single scattering albedo $\omega_0 = 1 - k_{abs}/k_{ext}$, and the phase matrix. The extinction and absorption coefficients can be calculated from the following equations: $k_{ext} = N\langle C_{ext}\rangle$, $k_{abs} = N\langle C_{abs}\rangle$. Here $\langle C_{ext}\rangle$ and $\langle C_{sca}\rangle$ are corresponding average extinction and absorption cross sections of scatterers in a cloud (Liou 2002).

For an idealized vertically homogeneous cloud, the cloud optical thickness τ is defined as: $\tau = k_{ext}L$, where L is the cloud geometrical thickness. For oriented crystals, the extinction matrix must also be calculated.

The global cloud characteristics such as the Stokes vector of the reflected, transmitted and internal light fields can be found from the solution of the vector radiative transfer equation (Liou 2002). The first component of the Stokes vector $I^{\uparrow}$ can be used to find the cloud reflectance $R = \pi I^{\uparrow}/\mu_0 F_0$, where μ_0 is the cosine of the solar zenith angle, F_0 is the solar irradiance at the top of atmosphere. Modern satellite instrumentation is capable of measuring $I^{\uparrow}$, F_0 and, therefore, cloud reflectance can easily be derived and used for the interpretation of measurements and the development of retrieval algorithms.

5.2.1 *Cloud Cover*

Cloud cover or cloud fraction, both terms are used, is defined as the fraction of a given scene covered by cloud, and so ranges from zero for clear skies to unity for overcast scenes. While simple in concept, it is inherently an ill-defined quantity that depends on the spectral region being considered, the spatial resolution of the imager, and the intended application (Schreiner et al. 1993; Ackerman et al. 1998; Wylie et al. 2005). Cloud cover is often derived from algorithms that attempt to identify fields of view contaminated by cloud as part of the pre-processing when determining surface and aerosol optical properties. A cloud mask results from such algorithms. Cloud cover determined from masking approaches is influenced by the spatial resolution of the instrument. Provided the signal to noise ratio of the instrument is sufficient, then smaller errors are achieved on data products for instruments with higher spatial resolution. The instrument spectral and signal-to-noise capabilities are also of great significance as they determine the ability to observe the cloud. For example, a thin cirrus cloud, detectable in a sensitive water vapour absorption band, may not be detectable in the visible. Cloud cover results are also sensitive to algorithm approaches.

Both solar reflectance in daytime observations, and thermal emission bands are used to discriminate cloudy from clear-sky scenes. The identification of cloudy scenes by discrimination of the intensity of backscattered solar radiation is often challenging, for example, as a result of bright underlying surfaces such as snow and desert for solar reflectance bands. A global cloud cover product is often aggregated from individual scenes to a global grid, e.g. such as a $1^{\circ} \times 1^{\circ}$ equal-angle grid for Moderate Resolution Imaging Spectroradiometer, MODIS, products from the NASA Earth Observing System Aqua and Terra satellites. In addition to global maps, cloud cover is often shown as a zonal mean (averaged over discrete latitudinal belts), and separately for ice and liquid water clouds. The cloud fraction derived from a monthly aggregation of the MODIS cloud mask product (product identification MYD35 for MODIS Aqua) is shown in Fig. 5.1 for April 2005.

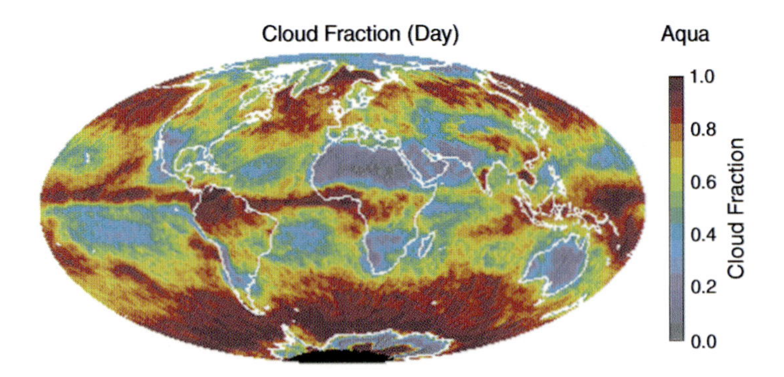

Fig. 5.1 Monthly cloud fraction (daytime observations) derived from the MODIS Aqua cloud mask for April 2005.

Night-time cloud fractions are nearly indistinguishable from those shown in Fig. 5.1. Overall the greatest cloud occurrences are found over oceans, especially in the southern oceans around Antarctica. The mean latitudinal behavior of this cloud fraction is given in Fig. 5.2. The distribution of cloud fraction depends on the underlying surface type.

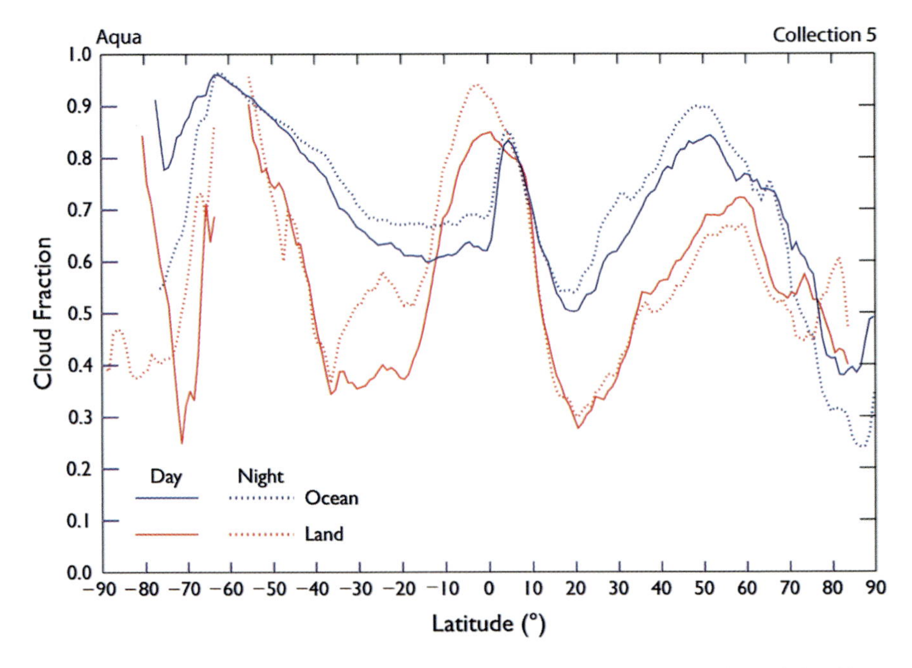

Fig. 5.2 Zonal mean monthly cloud fraction (daytime and night-time observations) derived from the MODIS Aqua cloud mask for April 2005.

5.2.2 *Cloud Phase*

A phase index $P = \mathrm{R}(1550\ \mathrm{nm})/\mathrm{R}(1670\ \mathrm{nm})$ (the numbers refer to the wavelengths), which uses the reflectance ratio within the liquid and ice absorption bands, is applied for the discrimination of liquid water and ice clouds when the spectral resolved measurements are available (Knap et al. 2002; Kokhanovsky et al. 2006). Liquid water and ice have different absorptions at these wavelengths. For liquid water clouds, reflectances at 1550 and 1670 nm are similar but those for ice differ, where $\mathrm{R}(1550\ \mathrm{nm}) \leq P_t\ \mathrm{R}(1670\ \mathrm{nm})$. The threshold value (THV) of the phase index $P_t = 0.7$ is often used to discriminate ice clouds. Calculations and measurements show that P is usually above 0.8 for liquid water clouds.

Mixed phase clouds have intermediate values of P. They can be identified using a P–T diagram, where the cloud brightness temperature at 12 μm (BT_{12}) is plotted along the abscissa and the phase index is plotted along the ordinate axis. The example shown in Fig. 5.3 was created using the SCIAMACHY phase index and AATSR brightness temperature measurements. The collocated measurements of AATSR and SCIAMACHY over Hurricane Isabel (17th September 2003; 30°N, 72°W) were used in the preparation of data shown in Fig. 5.3. Liquid water clouds are separated by the region where BT_{12} is above 273K and the phase index is above 0.8. Values of the phase index between 0.7 and 0.8 are assigned to mixed phase clouds; clouds with $P > 0.8$ and $BT_{12} < 273\mathrm{K}$ correspond to super-cooled water. The FTS instrument on GOSAT, which was launched by JAXA in January 2009, measures simultaneously P and BT_{12} (http://www.jaxa.jp/press/009/02/20090209_ibuki_e.html). Therefore, corresponding dataset is useful for studies of mixed clouds and also for the detection of super-cooled water.

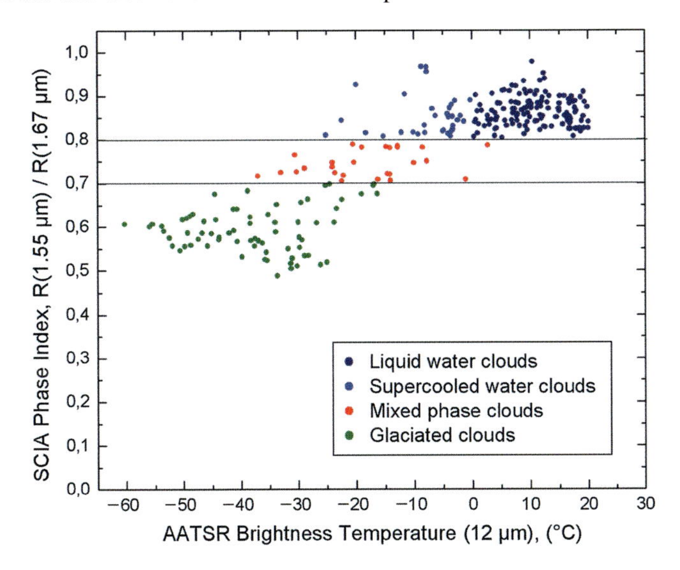

Fig. 5.3 Thermodynamic phase verses brightness temperature (P–T) diagram (Kokhanovsky et al. 2006).

The MODIS instrument measures infrared and solar reflectances, which are used to retrieve the cloud phase (Pilewskie and Twomey 1987; Baum et al. 2000). The IR bi-spectral method relies on a number of THVs for the 8.5 and 11 μm brightness temperatures. In addition, measurements around 1.38, 1.6, and 2.1 μm are used to refine the cloud phase algorithm for use in cloud optical property retrievals. A monthly example of this bi-spectral infrared phase result is shown in Fig. 5.4a. The algorithm distinguishes ice clouds in the Inter-Tropical Convergence Zone (ITCZ) and central Pacific regions, though it probably misclassifies ice clouds and snow/ice on the ground in Antarctica. The cloud retrieval phase algorithm, that also includes near-infrared measurements (Fig. 5.4b), gives somewhat larger ice cloud fractions, especially in the extensive cloud layer surrounding Antarctica (the roaring 40s) and in the continental storm tracks over land in the northern hemisphere.

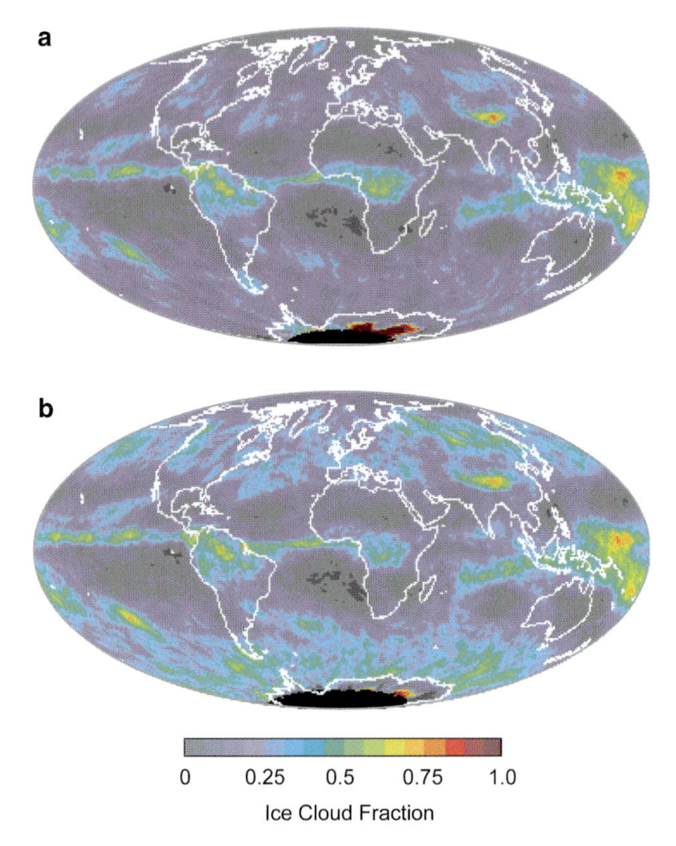

Fig. 5.4 Mean monthly ice cloud fraction for April 2005 derived from the MODIS bi-spectral infrared algorithm (**a**), and the cloud optical property retrieval algorithm (**b**).

Another approach to determine cloud phase is based on using polarization information about the scattered upwelling radiation over a range of scattering angles, where spherical and non-spherical polarised phase functions are distinct

(Goloub et al. 2000). The CNES POLDER instrument (flown on the JAXA ADEOS and ADEOS-II/MIDORI platforms, and currently on PARASOL) provides such capability. Synergistic data products derived from algorithms combining polarized and total radiance/reflectance measurements made by the instruments in the A-Train afternoon constellation of Earth observing satellites, which include POLDER/PARASOL and MODIS/Aqua, are currently being explored.

Multilayer/multiphase cloud scenes are challenging for passive measurements because a single unambiguous phase does not describe the scene well. Approaches for flagging such scenes are being explored by a number of investigators; a scene-level multilayer cloud flag is provided in the Collection 5 MODIS cloud products.

5.2.3 *Cloud Optical Thickness*

Cloud optical thickness, τ, together with the cloud fraction c, has a significant impact on the transfer of solar and infrared radiation through a cloudy atmosphere. τ is defined as the cloud extinction coefficient k_{ext} integrated across the cloud vertical extent. In the framework of the independent pixel approximation, the albedo for a particular scene averaged over all solar incident angles is given as:

$$r = (1 - c)r_{clear} + cr_{cloud}, \tag{5.1}$$

where r_{clear} is the spherical albedo for the clear sky portion and r_{cloud} is the same quantity for the cloudy scene.

For a clear atmosphere over land, r_{clear} in the visible and near-infrared is determined mainly by the surface contribution, which is highly variable with respect to wavelength, season, and surface type and location. Over oceans and outside the sun glint region, the contribution from the atmosphere becomes more important due to the low ocean reflectance.

The value of r_{cloud} not only depends on the cloud optical thickness, but also on the particle asymmetry parameter, g, defined as

$$g = \frac{1}{2}\int_0^{\pi} p(\theta)\sin\theta\cos\theta d\theta, \tag{5.2}$$

where the phase function $p(\theta)$ describes the angular distribution of light scattered by a unit cloud volume. The following approximation is used to estimate the cloud spherical albedo in the visible, where the processes of light absorption by liquid water or ice in clouds can be neglected (Kokhanovsky 2006):

$$r_{cloud} = 1 - t_{cloud}, \; t_{cloud} = \frac{1}{a + b\tau}. \tag{5.3}$$

Here t_{cloud} is the cloud spherical transmittance, $a = 1.07$ and $b = 0.75(1-g)$.

It follows from Eq. 5.3 that larger values of g lead to larger values of light transmission through a cloud, and correspondingly to a smaller reflectance. Ice cloud particles are thought to have values of g around 0.75 or somewhat higher, and liquid water cloud droplets are characterized by $g = 0.85$ across the typical range of effective radii. Therefore, liquid water clouds having the same optical thickness as that of ice clouds are generally less reflective; i.e. there is more transmitted light. Eq. 5.3 can be used for the estimation of the cloud albedo if the value of τ is retrieved from satellite data.

For optically thick clouds, the optical thickness in the visible is estimated from the following equation (Rozenberg et al. 1978; King 1987; Kokhanovsky et al. 2003) for the cloud reflectance:

$$R(\mu, \mu_0, \varphi) = R_{0\infty}(\mu, \mu_0, \varphi) - t_{cloud} K_0(\mu) K_0(\mu_0). \tag{5.4}$$

Here μ is the cosine of the observation zenith angle, μ_0 is the cosine of the solar zenith angle and φ is the relative azimuth. All functions in Eq. 5.4 (except K_0) depend on τ.

Assuming a given cloud model, e.g. spherical particles and a polydisperse distribution with a given effective radius of droplets, a_{ef}, or predefined ice crystals shapes and size distributions, the escape function $K_0(\mu)$, and the reflection function for a semi-infinite non-absorbing layer $R_{0\infty}$ are pre-calculated and stored in look-up-tables (LUTs). Approximate equations for these functions can be used as well. In particular, a good approximation for $K_0(\mu)$ is:

$$K_0(\mu) = \frac{3}{7}(1 + 2\mu). \tag{5.5}$$

This function describes the angular distribution of the radiation escaping a semi-infinite non-absorbing turbid medium with sources located at the infinite depth in the medium. Due to strong multiple scattering effects, its dependence on the microphysical properties of the medium can be neglected.

One derives from Eq. 5.3 and Eq. 5.4:

$$\tau = \frac{1}{b}\left[\frac{K_0(\mu)K_0(\mu_0)}{R_{0\infty}(\mu, \mu_0, \varphi) - R(\mu, \mu_0, \varphi)} - a\right]. \tag{5.6}$$

It follows from this equation that retrievals of τ for very thick clouds ($R \to R_\infty$) are highly uncertain and small errors, e.g. calibration errors, in the measured reflection function will lead to large errors in the retrieved cloud optical thickness. Often, a limiting value of cloud optical thickness is used in the retrieval process, e.g. 100, as most clouds, but certainly not all, have optical thicknesses below 100.

The mean retrieved MODIS Aqua cloud optical thickness (liquid and ice phase) for April 2005 is shown in Fig. 5.5 (aggregated from the MYD06 pixel-level product). One can see that clouds are optically thin in subsidence regions outside the ITCZ along with low cloud fraction (Fig. 5.2). Having lost most of its water vapour to condensation and rain in the upward branch of the circulation, the

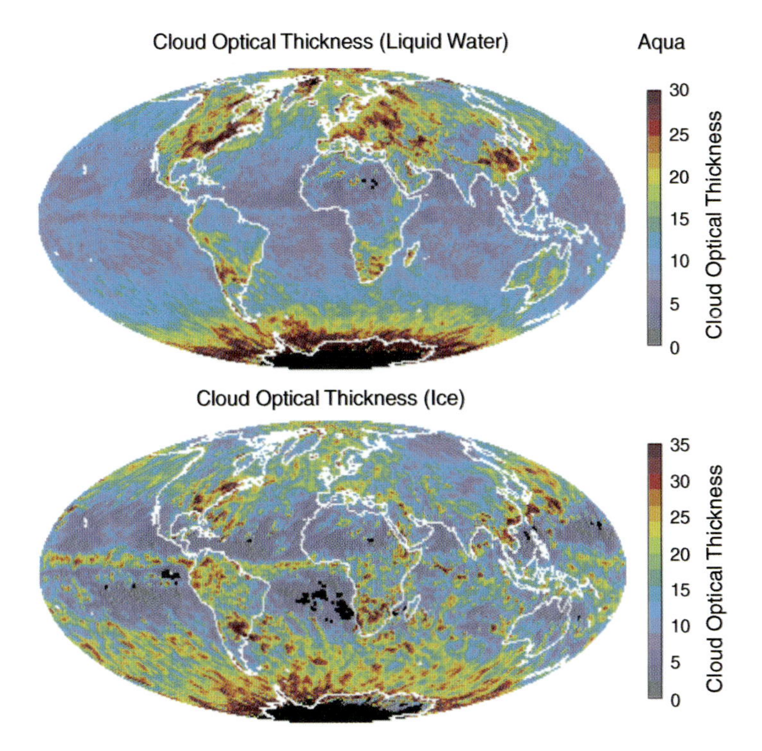

Fig. 5.5 Monthly cloud optical thicknesses derived from the MODIS Aqua cloud optical properties product for April 2005.

descending air is dry in these regions. The corresponding latitudinal variation in the mean cloud optical thickness is shown in Fig. 5.6. One can see that ice clouds are thicker in the ITCZ zone. Thin clouds, defined as having $\tau < 5$ are almost absent in the latitudinal averages shown in Fig. 5.6. There are no latitudinal belts with cloud fraction less than 0.2 (see Fig. 5.2).

It follows from Eq. 5.6 that the retrievals of the product $b\tau$ and, therefore, r_{cloud} is less influenced by the assumptions on the asymmetry parameter, g, which is quite uncertain for ice clouds. Therefore, it is important to report not only cloud optical thickness τ but also the retrieved transport optical thickness $\tau_{\mathrm{tr}} = (1-g)\tau$ in the output of cloud retrieval algorithms.

5.2.4 *Effective Radius*

The effective radius, a_{ef}, for a spherical polydisperse distribution in a cloud is defined as the ratio of the third to the second moment of the size distribution.

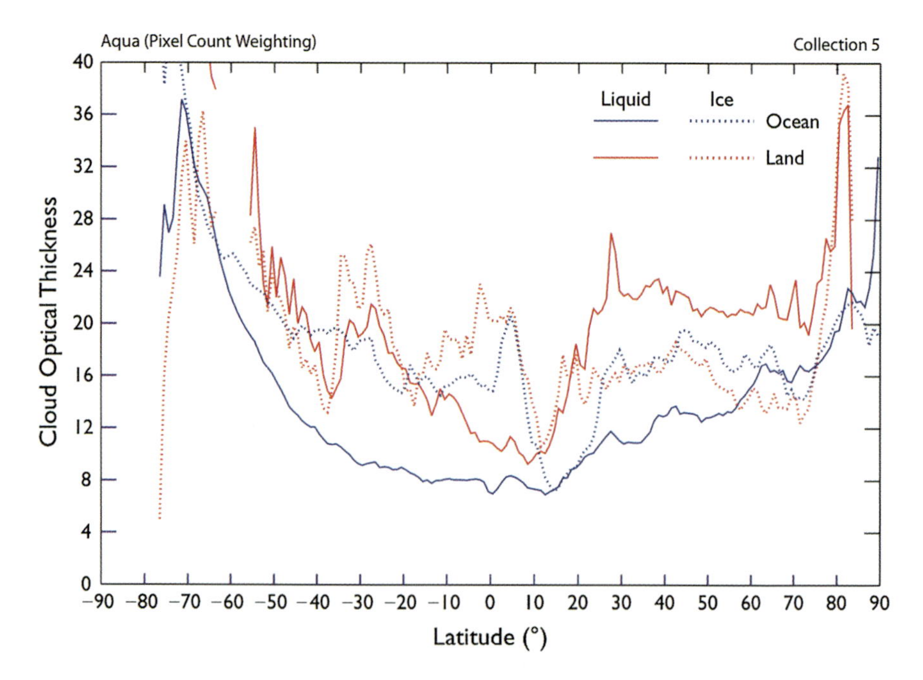

Fig. 5.6 Zonal mean monthly cloud optical thickness (separated by phase, land, and ocean) derived from the MODIS Aqua cloud optical properties product for April 2005.

For non-spherical ice particles, it is defined as the ratio of three times the average volume V to the average surface area S of crystals, i.e.

$$a_{ef} = \frac{3V}{S}, \tag{5.7}$$

which is equivalent to the definition for spherical particles. The determination of a_{ef} from satellite measurements for spherical water droplets is straightforward. LUTs of reflection functions for two channels, e.g. positioned at 670 and 1600 nm, are calculated and then used simultaneously to retrieve both the effective radius and optical thickness that best match the corresponding measurements. It must be emphasised that *a priori* assumptions about the shape of crystals are needed for ice clouds in order to derive their sizes.

The value of the effective radius is determined mostly from light absorption in a near infrared channel (larger particles with larger absorption and smaller reflectance), whereas information about cloud optical thickness comes primarily from the measurements in a non-absorbing visible or shorter wavelength near-infrared channel (Arking and Childs 1985; Twomey and Cocks 1989; Nakajima and King 1990; Nakajima et al. 1991; Han et al. 1994; Platnick et al. 2001). The exclusion is the use of the rainbow feature to extract the value of a_{ef}. Then actually not absorption but scattering processes are used for the effective radius determination. Cloud-bows are more easily observed in the angular patterns of the polarized reflectance. This fact was used by Bréon and Goloub (1998) for the cloud droplet sizing based on POLDER measurements.

Generally, a_{ef} increases towards the bottom of ice clouds. It decreases towards the bottom for liquid water clouds. As electromagnetic radiation of different wavelengths penetrates to different depths inside the cloud, the retrieved value of a_{ef} varies with height and depends on wavelength, λ. Typically measurements in the 1.6 and 2.1 μm bands, which are almost free of gaseous absorption, are used in retrievals of the effective radius of droplets or crystals.

For the case of ice clouds having large crystals, one can also use the reflection function at 1.2 μm for the retrievals of a_{ef}. Then the light absorption is quite small and the simplified version of the asymptotic radiative transfer theory, valid assuming that single scattering albedo $\omega_0 \to 1$, is used in the retrieval procedures (Kokhanovsky et al. 2003; Kokhanovsky 2006).

The reflection function in absorbing near-infrared channels reaches a limiting value with increasing τ relatively quickly, when compared to the non-absorbing visible channels. The following approximation for the reflection function of a semi-infinite absorbing layer is then used (Kokhanovsky et al. 2003):

$$R_\infty(\mu, \mu_0, \varphi) = R_{0\infty}(\mu, \mu_0, \varphi) \exp[-4sQ(\mu, \mu_0, \varphi)], \tag{5.8}$$

where $R_{0\infty}(\mu, \mu_0, \varphi)$ is the reflection function of a non-absorbing semi-infinite layer,

$$Q(\mu, \mu_0, \varphi) = \frac{K_0(\mu)K(\mu)}{R_{0\infty}(\mu, \mu_0, \varphi)}, \tag{5.9}$$

$$s = \sqrt{\frac{1 - \omega_0}{3(1 - \omega_0 g)}}. \tag{5.10}$$

It follows from Eq. 5.8:

$$s = \frac{\ln (R_{0\infty}/R)}{4Q}. \tag{5.11}$$

The parameter s can be also derived from Mie theory for water droplets and from geometrical optics calculations for large ice crystals. It depends on the effective radius because it follows:

$$1 - \omega_0 \sim \kappa a_{ef}, \ s \sim \sqrt{\kappa a_{ef}}, \ \kappa = 4\pi\chi/\lambda \tag{5.12}$$

where χ is the imaginary part of the refractive index of a particle (ice or water). Therefore, a_{ef} can be determined using the measured value of R and also LUTs of Q and $R_{0\infty}$. As mentioned, the determination of a_{ef} for crystalline clouds is not straightforward because the shape (habit) distribution of particles cannot be retrieved from passive satellite measurements (Rolland et al. 2000; King et al. 2004; Ou et al. 2005). *A priori* information on the mixture of shapes in the cloud is needed, and the retrieval results depend on these *a priori* assumptions. The monthly mean effective radius derived from MODIS data is shown in Fig. 5.7. Radii are in the range 5–25 μm for liquid water droplets and they are somewhat larger (up to 30 μm) for ice clouds in the region of the ITCZ. The mean latitudinal distribution of the effective radius is given in Fig. 5.8. It follows from this figure that ice particles have larger sizes compared to water droplets. The retrieved sizes of crystals and droplets characterise the microphysical conditions in the upper portion of the cloud, because solar radiation, which is absorbed in the near-infrared, does not penetrate strongly to the lower levels of an optically thick cloud.

It follows from Fig. 5.7 that particles are generally larger over the ocean. This is explained by the fact that less cloud condensation nuclei are available over ocean as compared with clouds over land, but it might also be partially indicative of different dynamic processes above land and ocean. There is a hemispherical asymmetry in the distribution of a_{ef} (Fig. 5.8). This is related to different areas of land in the northern and southern hemispheres. Droplets in water clouds are generally smaller in the northern latitudes as compared to the southern latitudes. This is consistent with the fact that most industrial activity is in the northern hemisphere, leading to a hemispheric increase in the aerosol particle numbers and, as a consequence, a decrease in droplet sizes and precipitation probabilities. However this hemispheric feature is absent for ice clouds.

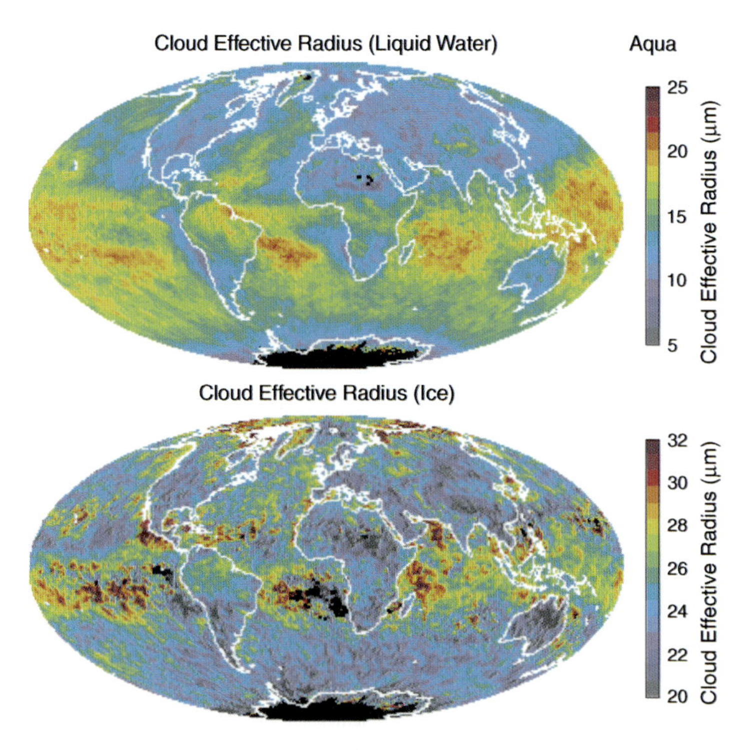

Fig. 5.7 The monthly mean effective radius of water droplets and ice crystals (April 2005, MODIS Aqua).

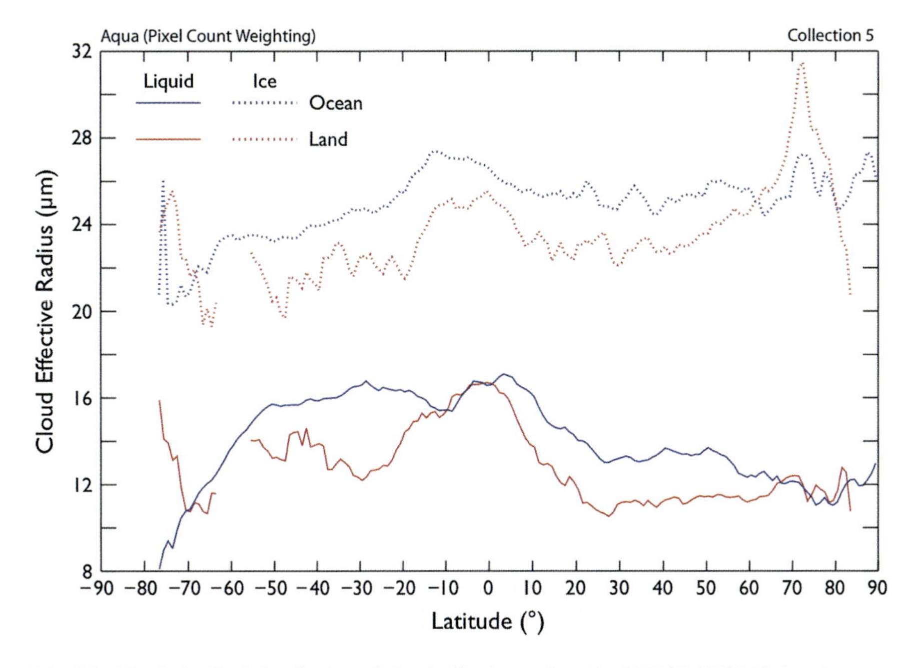

Fig. 5.8 The latitudinal distribution of cloud effective radius (April 2005, MODIS Aqua).

5.2.5 Cloud Liquid Water and Ice Path

As described in the previous chapter on microwave remote sensing, the liquid water path is an important parameter for clouds, which can be determined in different spectral regions from the absorption of liquid water. The amount of liquid water mass in a vertical cloud column of a unit area, or liquid water path, LWP, is calculated from the following equation:

$$\mathrm{LWP} = \rho \int_0^L C_v(z)dz, \tag{5.13}$$

Where ρ is the density of liquid water, L is the geometrical thickness of the cloud and C_v is the dimensionless volumetric concentration of water in the cloud. For a vertically homogeneous cloud, it follows:

$$\mathrm{LWP} = \rho\, C_v L. \tag{5.14}$$

LWP is typically measured in g/m^2 and usually lies in the range 50–200 g/m^2. The value of the LWP is retrieved from the measured values of a_{ef} and τ, explained above. It follows for an idealized vertically homogeneous cloud with large particles ($a_{ef} >> \lambda$):

$$\tau = NSL/2 \tag{5.15}$$

where S is the average surface area of the particles in a cloud and N is the number of particles in a unit volume, which is related to the volumetric concentration *via* the following equation:

$$N = \frac{C_v}{V}. \tag{5.16}$$

It follows from Eqs. 5.15, 5.16 and 5.14:

$$\tau = \frac{3\mathrm{LWP}}{2\rho a_{ef}}, \tag{5.17}$$

and, therefore,

$$\mathrm{LWP} = \frac{2}{3}\rho a_{ef}\tau. \tag{5.18}$$

This equation yields LWP from measurements of a_{ef} and τ. Eq. 5.17 is also used for the determination of the ice water path, IWP, defined via Eq. 5.12 with ρ as the density of ice and C_v as the dimensionless volumetric concentration of ice in the cloud. In this case all parameters in Eq. 5.18 are referred to ice (e.g. ice density ρ and effective size of ice grains a_{ef}).

The LWP cannot be determined from measurements in the visible and near-infrared for thick clouds because the reflection function becomes insensitive to the cloud optical thickness, as the electromagnetic radiation does not penetrate sufficiently into the cloud. For such clouds, the microwave measurements must be used as described in Chapter 4. The global distribution of the LWP as determined from MODIS is shown in Fig. 5.9. Large values of LWP are characteristic for polar regions. Further discussion of MODIS cloud products is provided by Platnick et al. 2003 and references therein.

5.2.6 Cloud Top Height

Cloud altitude and type are associated with the thermodynamic and hydrodynamic structure of the atmosphere and affect the energy budget and the radiative heating profile. Therefore, it is important to monitor cloud top height (CTH) statistics with

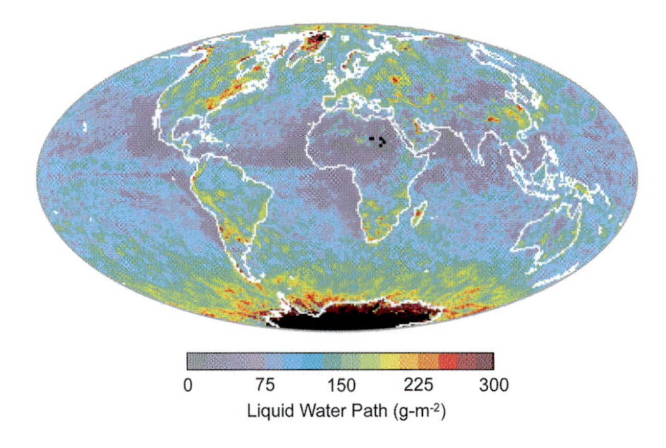

Fig. 5.9 The global distribution of LWP for April 2005 derived from the MODIS Aqua optical property retrievals.

satellite measurements. Several passive techniques have been developed for this purpose, including:

- Stereoscopy,
- Thermal infrared measurements,
- Gaseous absorption measurements.

In the stereoscopic method (Moroney et al. 2002), a cloud is observed from different view angles so enabling the detection of its height from parallax considerations.

The thermal infrared measurements are based either on measurements of brightness temperature at 11 or 12 μm, for example, with colder clouds being higher in atmosphere, and/or CO_2 slicing techniques, which use measurements in the CO_2 absorption bands around 14–15 μm (Strabala et al. 1994; Rossow and Schiffer 1999).

The CO_2 slicing technique assumes that the atmosphere becomes more opaque as the wavelength increases from 13.3 to 15 μm: the radiances obtained from these spectral bands being sensitive to different layers in the atmosphere (Menzel et al. 2008). The cloud top heights (or pressures) data products, determined from MODIS data, use the CO_2 slicing method applied to three channels (13.64, 13.94 and 14.24 μm). As a result of its signal-to-noise range, MODIS CO_2 slicing cloud-top pressures are typically limited to pressures from approximately 700 hPa (about 3 km above sea level) up to the tropopause. Consequently, when low-level clouds are present, the MODIS CTH algorithm defers to an infrared window technique where cloud-top pressure and temperature are determined through comparison of model-calculated and observed 11 μm radiances.

Solar reflectance measurements in the oxygen absorption bands, e.g. A, B, γ and also the absorption band of the molecular complex (O_2–O_2), have often been used to determine cloud top height (Hanel 1961; Yamamoto and Wark 1961;

Saiedy et al. 1965; 1967; Heidinger and Stephens 2000; Koelemeijer et al. 2001; Rozanov and Kokhanovsky 2004). The first CTH satellite measurements using the O_2 A-band were made by astronauts aboard the Gemini satellite in the 1960s (Saiedy et al. 1965; 1967). The technique is similar to that of CO_2 slicing in the sense that the sensitivity to cloud layers depends on absorption of a well-mixed gas (e.g. the O_2 A-band absorption increasing from 758 to 761 nm). Hyper-spectral measurements are needed because the absorption process takes place in a narrow spectral region. The physical principle behind the technique is demonstrated in Fig. 5.10, where calculations of the reflection function at different CTHs in the oxygen A-band are presented. Higher clouds give shallower spectra. The fit of the measured spectra in the oxygen absorption bands (A, B, or γ) enables the CTH to be determined with a_{ef} and τ, and cloud phase determined from other channels almost free of the gaseous absorption.

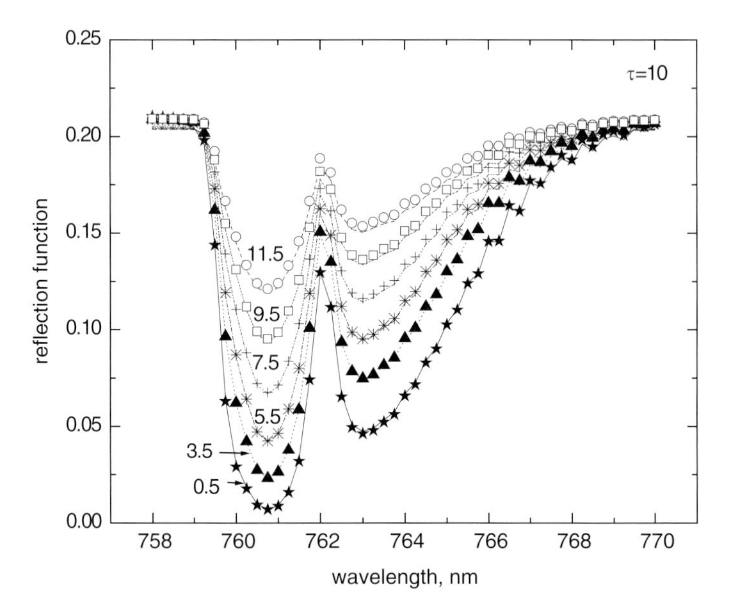

Fig. 5.10 Dependence of the cloud reflection function on the wavelength in the oxygen A-band for cloud top heights equal to 0.5, 3.5, 5.5, 7.5, 9.5, 11.5 km at the cloud optical thickness $\tau = 10$ and $a_{ef} = 6$ μm (the Deirmendjian's Cloud C1 model). The cloud geometrical thickness is equal to 250 m. The solar zenith angle is 60° and the observations are performed along the vertical to the scattering layer. Lines are plotted using the simplified asymptotic radiative transfer theory applied to the calculation in the O_2 A-band (Kokhanovsky, 2006). *Symbols* give results of the exact radiative transfer calculations.

The technique appears to work quite well for low and middle level clouds using relatively poor spectral resolution data. Such data are relatively insensitive to thin cirrus and further development work is being undertaken to improve these retrievals. The observation of large differences between CTH derived from IR and O_2 absorption techniques is used to identify multi-layered clouds. The difference can be explained by the different sensitivities of the emitted thermal IR and the backscattered solar light to CTH. So the techniques are quite complimentary to each other.

For the mapping from cloud pressure (hPa) to cloud top height (km) and back, a climatology of pressure vertical profiles are needed. The profiles will vary to some extent depending on the location and season but, to a good approximation, is given by the following simple expression:

$$p = p_0 \exp(-z/H), \tag{5.18}$$

where H is the scale height, z is the height above the ground, p is the atmospheric pressure at the level z, and p_0 is the pressure at the ground level. The MODIS global mean cloud top pressure for May 2008 is shown in Fig. 5.11.

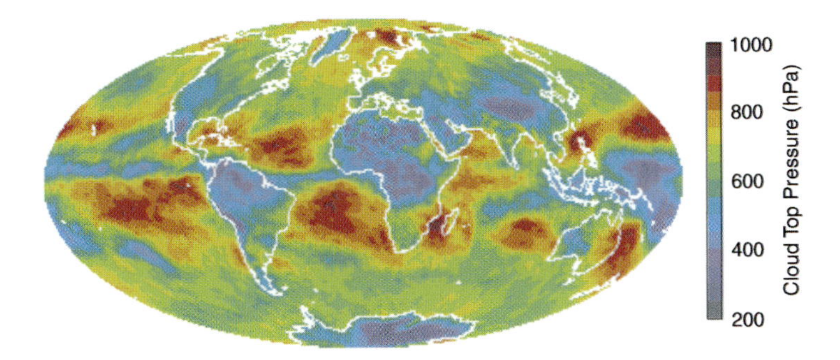

Fig. 5.11 Global mean cloud top pressure derived from MODIS Aqua observations for April 2005.

5.3 Validation of Satellite Cloud Products

The validation of satellite-derived cloud products is essential to establish the accuracy of the data products. Long–term ground and/or airborne measurements are needed in order to have statistically significant results for clouds of different types. Ideally, the most direct measurement of the cloud parameter of interest is desired. This requires diverse instrumentation ranging from particle size spectrometers to lidars and radars. Most importantly, the uncertainties of the validating instruments must be known in order to assess satellite retrieval uncertainties.

As an example, cloud top heights can be determined from ground-based, airborne or satellite lidars and radars (each with different height sensitivities) and compared to CTHs derived from the passive instruments. A difficulty here is that of the averaging scale. Lidars and radars provide information at a given spatial location (ground-based systems) or along thin curtains (CALIPSO and CloudSat). Satellite imagers give information from pixels typically of 1 km^2 in size. Therefore, comparisons are meaningful only when passive and active satellite data can be spatially sampled or averaged in a commensurate manner, and ground-based active systems data are appropriately averaged in time. Fig. 5.12 shows comparisons of

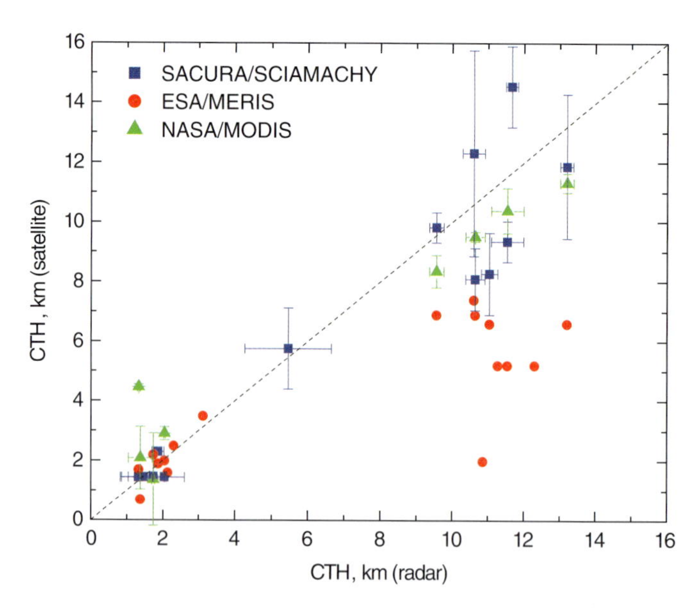

Fig. 5.12 Inter-comparison of ground-radar (35 GHz) and satellite cloud top heights for overcast scenes at the Southern Great Plains (SGP) ARM site (Kokhanovsky et al. 2008).

cloud top heights derived from the 35 GHz MMCR ground radar measurements and official products from NASA (MODIS Collection 5), ESA (MERIS), and the University of Bremen (UB; SCIAMACHY). The oxygen A-band technique is used for the ESA and UB products. The CO_2 slicing (high clouds) and thermal infrared (low clouds) are used in the MODIS CTH product. The comparisons have been performed for the time period 2002–2007. Only those dates when all instruments operated simultaneously and there were extended cloud fields over the site were used in the intercomparison study.

While lidars and radars provide direct information on the cloud top height, their CTHs may differ somewhat due to the different sensitivities of these systems to particles of different sizes and composition. The radar is known to be less sensitive than a lidar to boundaries in low extinction clouds (e.g. lower heights relative to lidar for cirrus clouds, or perhaps missing the cirrus entirely). Further, techniques using different parts of the spectrum have different penetrations into the cloud and thus the definition of cloud top can be ambiguous. With these caveats, Fig. 5.12 shows that low-level cloud heights from MERIS, MODIS and SCIAMACHY UB (SACURA algorithm) are very close to the radar values. However, there are some problems with respect to high clouds where MODIS heights are lower than the radar and MERIS CTHs are even lower. The SCIAMACHY cloud product shows large differences for high clouds although the overall bias relative to the radar is small.

Shortcomings of the current generation of MERIS CTH retrievals are due to the fact that the instrument does not have thermal infrared channels and the measurement in the oxygen A-band is made with one broadband spectral channel. In contrast, SCIAMACHY performs hyperspectral measurements (with the spectral resolution of 0.54 nm) in the oxygen A-band while MODIS has a number of thermal infrared channels that increases the accuracy of the corresponding retrieval algorithms.

Infrared techniques are very sensitive to high cirrus clouds, but this is not the case for oxygen A-band spectrometry. Therefore, the combination of both types of measurements must be used to increase the retrieval information content such as various cloud top definitions, the identification of multi-layered cloud systems, the cloud bottom height determination, etc.

As previously mentioned, the cloud fraction depends on the spatial resolution of the satellite imager. The retrievals of cloud effective radius and cloud phase can be compared with airborne measurements. However, differences between the satellite-derived effective radius and that measured in the cloud may not necessarily be due to algorithm errors but could also include *in situ* sampling errors from an inhomogeneous cloud. The uncertainties in the size distributions obtained from these probes are usually unspecified, especially for ice particles and water droplets in drizzling clouds.

Independent cloud optical thickness and liquid water path measurements, say from ground-based radiometers (τ) and radars (LWP), do not necessarily give unbiased retrievals any more than the satellite retrievals do. For example, radars are less sensitive to small particles and, therefore, the radar-derived LWP can differ from that derived from a satellite optical instrument when size distribution assumptions used in the radar retrieval are incorrect. Ground-based radiometers cannot measure directly transmitted radiation for thick clouds, but rather the transmitted diffuse light from which cloud optical thickness is derived. However, sampling volume problems can once again be important.

As one can conclude, there is an urgent need to develop additional cloud product validation techniques (instrumentation and/or algorithms). Inter-comparisons of products derived from different approaches (whether satellite, ground-based, or aircraft) for the same cloud system must be performed; diverse results serve as an indication of problems in one or more of the techniques.

5.4 Modern Trends in Optical Cloud Remote Sensing from Space

5.4.1 Hyperspectral Remote Sensing

The traditional studies of cloud properties have been performed using the channels positioned at 0.65 (0.865), 1.6, and 12 μm. This enables the determination of cloud liquid water path, the size of droplets/crystals, the cloud optical thickness and also the cloud altitude. However, in recent years, hyper-spectral remote sensing of clouds becomes more and more popular. Unlike traditional methods, where the reflected radiation is measured in quite broad spectral channels (10 nm and even wider), hyper-spectral remote sensing, with a spectral resolution of about a nanometer, offers much more detailed spectral information. SCIAMACHY for example has a resolution of 0.2–1.5 nm. The corresponding retrieval methods are poorly developed at the moment. However, the potential of hyper spectral remote sensing in the vertical profiling of clouds cannot be underestimated. Indeed, the penetration depth of radiation depends on the wavelength. Therefore, the spectral scanning of reflected radiation in the range 0.4–2.2 μm can bring information on the vertical distribution of cloud properties such as the liquid water content and the size of particles. Also positions of cloud boundaries can be found. Corresponding methods are not mature enough. But several important results have been obtained.

The depth of solar Fraunhofer lines in scattered light is less than that observed in the direct sunlight. This is called the Ring effect. The physical mechanism behind this effect is clear: it is largely due to rotational Raman scattering from the wings of the absorption line towards its center. Due to gaseous absorption effects Raman scattering from the centre of the band to the wings is much less pronounced than in the wings→band centre processes, which leads to the filling-in of the gaseous absorption features in the terrestrial atmosphere. De Beek et al. (2001) used Ring effect for the Ca II Ring structure at 393.37 nm to get the cloud top height and also cloud optical thickness (COT) from GOME measurements in the spectral range 392–395 nm. They demonstrated using the software package SCIATRAN that the filling-in decreases with the COT and also with the CTH. The effect of increasing COT and CTH on filling-in of Ca II absorption line is quite obvious: clearly, clouds shield lower atmospheric layers, which reduces molecular scattering events and their Raman scattering contributions. To increase the accuracy of retrievals, the authors also used the measurements in the absorption bands of O_2–O_2 (477 nm) and O_2 (761 nm). Some earlier results in this area have been obtained by Brinkman (1968), Wallace (1972), Price (1977), Park et al. (1986) and Joiner and Bhartia (1995). Comprehensive radiative transfer models of rotational Raman scattering (for forward and inverse modelling) were developed by van Deelen (2007). Joiner and Vasilkov (2006) applied the technique to OMI data. The use of measurements inside gaseous absorption bands (e.g. O_2, CO_2) were used for a long time to get the cloud top altitudes. Saiedy et al. (1965; 1967) reported measurements in O_2 A-band using a hand held spectrometer from a satellite. More recently, the GOME and SCIAMACHY O_2 A-band measurements have been used for the same task by Kuze and Chance (1994), Koelemeijer et al. (2001), Loyola (2004), Rozanov and Kokhanovsky (2004), Grzegorski et al. (2006), Rozanov et al. (2006), Kokhanovsky et al. (2007), van Diedenhoven (2007) and van Diedenhoven et al. (2007). In the last paper the synergetic UV and O_2 A-band measurements have been used to deduce the cloud properties. It is known that clouds screen gaseous atmosphere beneath them, leading to less sharp increase of the top-of-atmosphere reflectance in the UV (Herman et al. 2001). This effect is used for the improvement of the cloud top altitude retrieval algorithm based on O_2 A-band measurements. CO_2 molecular absorption band centered around 14 μm is routinely used for the MODIS CTH determination (Menzel et al. 2008).

SACURA cloud top height retrieval algorithm (Rozanov and Kokhanovsky 2004) is based on the asymptotic radiative transfer theory generalized on the case of gaseous absorption in a vertically inhomogeneous cloud. Therefore, it provides not effective but the true cloud top heights as demonstrated in the previous section (see Fig. 5.12). Other cloud top height retrieval algorithms (e.g. FRESCO (Koelemeijer et al. 2001), ROCINN (Loyola 2004)) are based on the substitution of a cloud by a Lambertian cloud, which brings some biases in the retrieved cloud top heights (usually too low clouds are retrieved). A comprehensive study of various approximations usually applied in cloud top height retrievals using oxygen A-band have been performed by Rozanov and Kokhanovsky (2008).

5.4.2 Lidar Remote Sensing

Active systems for cloud remote sensing from space have quite a short history. There were only three missions up to date:

- Lidar In-space Technology Experiment (LITE, http://www-lite.larc.nasa.gov/, 1994),
- Geoscience Laser Altimeter System (GLAS, http://glas.gsfc.nasa.gov/, 2003–2008), and
- The Cloud-Aerosol Lidar and Infrared Pathfinder Satellite Observation (CALIPSO) satellite mission (http://www-calipso.larc.nasa.gov/, 2006-present) with a lidar system called CALIOP.

The LITE instrument was powered continuously for over 220 h during the mission, with 53 h of lasing at wavelengths 355, 532, and 1064 nm (Winker et al. 1996). LITE unambiguously sensed sub-visible cirrus and has provided a global look at the prevalence and height of very thin clouds that are below the threshold of detection of current passive satellite instruments. Even deep cirrus was generally fully penetrated, so that the vertical structure of the clouds could be observed, and the presence of underlying cloud layers detected. Theoretical results related to LITE were reported by Winker and Poole (1995). Zege et al. (1995) developed an analytical method for the calculation of lidar returns from clouds both from ground and space.

GLAS was successfully launched aboard the ICESat, from Vandenberg Air Force Base, California on 12th January 2003. At an altitude of approximately 600 km, GLAS provides global coverage between 86°N and 86°S. The GLAS laser transmits short pulses (4 nanoseconds) of infrared light at 1064 nm and visible green light at 532 nm 40 times per second. The spatial resolution of the disk illuminated by the laser is 70 m in diameter and spaced at 170 m intervals along the Earth's surface. Many investigations of cloud systems using GLAS have been performed. In particular, Dessler et al. (2006) analyzed cloud-top height data obtained at tropical latitudes between 29th September and 17th November 2003. They found that about 66% of the tropical observations show one or more cloud layers. Of those observations that do show a cloud, about half show two or more cloud layers. Maxima in the cloud-top height distribution occur in the upper troposphere, between 12 and 17 km, and in the lower troposphere, below about 4 km. A less prominent maximum occurs in the mid-troposphere, between 6 and 8 km. The occurrence of cloud layers tends to be consistent with the well known diurnal cycles of continental and oceanic convection, and it was found that cloud layers tend to occur more frequently over land than over ocean, except in the lower troposphere, where the opposite is true. Wang and Dessler (2006) used GLAS measurements to establish the cloud overlap statistics in the tropical area.

The Cloud–Aerosol Lidar and Infrared Pathfinder Satellite Observation satellite mission carries an active lidar (CALIOP with channels at 532 nm (with the depolarization measurements) and 1064 nm), a passive Infrared Imaging Radiometer (IIR), and visible Wide Field Camera. By deriving accurate statistics on cloud height and structure, CALIOP since its launch on 28th April 2006 provided valuable statistical data – especially with respect to thin clouds such as *Ci* and sub-visual *Ci* not seen by passive radiometers. In particular, Sassen et al. (2008) derived valuable information on the global distribution of cirrus clouds. The latitudinal distribution of identified cirrus cloud heights derived for 0.2 km and 2.5° grid intervals is shown in Fig. 5.13. As one may expect both the frequency of *Ci* cloud occurrence and also their altitudes increase towards the equator.

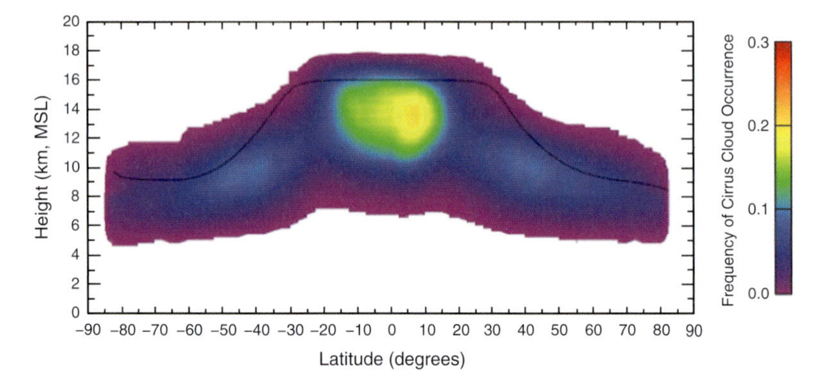

Fig. 5.13 Latitudinal distribution of identified cirrus cloud heights derived for 0.2 km height and grid 2.5° intervals. The line shows the mean tropopause heights averaged over the same one year period, as taken from CALIPSO data files (from Sassen et al. (2008)).

The study of clouds using synergy of lidar and radar (CloudSat, Stephens et al. (2008)) systems in space is also a hot topic in modern satellite cloud remote sensing from space (Barker 2008; Grenier et al. 2009; Haladay and Stephens 2009; Mace et al. 2009; Wu et al. 2009). Clearly, the synergy of multiple satellite systems (including those on a geostationary orbit) will enhance our detailed understanding of the terrestrial cloud system and also on a scale unavailable in the past. This will contribute appreciably to our progress in understanding the terrestrial atmosphere, weather, and climate.

5.4.3 Future Missions

Satellite remote sensing of clouds will be continued through a number of missions planned by space agencies worldwide.

NPP (NASA). The National Polar orbiting Operational Environmental Satellite System Preparatory Project will be launched in 2011.

- CrIS: The Cross-track Infrared Sounder will be combined with the Advanced Technology Microwave Sounder (ATMS). They produce atmospheric temperature, moisture and pressure profiles from space.
- OMPS: Ozone Mapping and Profiler Suite.
- VIIRS: Visible/Infrared Imager/Radiometer Suite collects visible and infrared radiometric data of the Earth's atmosphere, ocean, and land surfaces. Data types include atmospheric aerosols and clouds, land cover and reflectance, land/water and sea surface temperature, ocean color, and low light imagery.
- CERES: Earth's Radiant Energy System.

SENTINEL 3 (ESA). This mission will be launched in 2012.

- A topography system, which includes a dual-band Ku- and C-band altimeter based on technologies used on ESA's Earth Explorer CryoSat mission, a microwave radiometer for atmospheric correction and a DORIS receiver for orbit positioning.
- An Ocean Land Colour Instrument (OLCI), which is based on heritage from ENVISAT's Medium Resolution Imaging Spectrometer MERIS instrument. The OLCI operates across 21 wavelength bands from ultraviolet to near-infrared and uses optimised pointing to reduce the effects of sun glint.
- A surface temperature system called Sea Land Surface Temperature Radiometer (SLSTR), which is based on heritage from ENVISAT's Advanced Along Track Scanning Radiometer (AATSR). The SLSTR uses a dual viewing technique and operates across eight wavelength bands providing better coverage than AATSR because of a wider swath width.

EARTHCARE(ESA). The mission will be launched in 2013.

- Backscatter Lidar (ATLID)
- Cloud Profiling Radar (CPR),
- Multi-Spectral Imager (MSI), 7 channels, 150 km swath, 500 m pixel,
- Broadband Radiometer (BBR) – 2 channels, 3 views (nadir, fore and aft),

GCOM-C (JAXA). This mission will be launched in 2014.

- Second-Generation Global Imager (S-GLI). The spectral coverage is 0.375–12.5 μm in 19 spectral bands with the spatial resolution 0.25–1 km depending on the band.

The instrument set-up is similar in many respects to MODIS and can be used for the droplet to crystal size monitoring and also for the determination of the cloud top height, cloud phase, cloud optical thickness, liquid water path and other relevant cloud parameters. In addition, the instrument has a capability of measuring the polarization state of the reflected light. Measurements in the oxygen A-band are also planned.

5.5 Conclusions

Our need to understand the hydrological cycle is driven by scientific curiosity, the operational requirements for numerical weather prediction and, in the future, climate change. Many instruments and fruitful retrieval techniques have been developed and applied to satellite data to derive cloud data products. The wealth of detailed cloud property information that has been obtained and understood on a global scale would have been impossible before the satellite era. However, to meet the evolving scientific and societal issues related to numerical weather prediction and climate change, many problems still remain to be solved. These address improved missions that are fit for purpose together with algorithm development. They are mostly related to the adequacy of the forward models used in the retrieval algorithms. Until now, all operational cloud retrieval algorithms rely on a homogeneous, single-layered cloud model. In reality, clouds are inhomogeneous objects on all scales in the horizontal and the vertical. 3-D effects are ignored in cloud satellite remote sensing as look-up-tables are calculated using 1-D radiative transfer theory. Therefore, retrievals can be biased for cases where 3-D effects are pronounced, such as scattered cloud fields and extensive vertical convection. Further issues exist for studies of thin clouds, where both cloud inhomogeneity, cloud fraction, and the underlying surface bi-directional reflectance function must be accounted for in the retrieval process. The retrievals of ice clouds rely on the *a priori* assumed models of crystal shapes, which may vary with other cloud microphysical parameters for a given location and time. Again, this leads to biases in the derived products. The retrievals of mixed phase cloud properties such as the ice/water fraction are not yet well developed. However, hyper-spectral measurements show some potential in this respect. The exploration of these important retrieval issues will lead the list of important tasks needed to advance cloud satellite remote sensing in the coming years.

Acknowledgements The authors are grateful to the members of their research groups for the support, advice, and preparation of figures used this work. The support of ACCENT and editors of this book is much appreciated. The use of ESA, NASA and ARM SGP radar data is acknowledged with many thanks.

References

Ackerman, S., K. Strabala, W. Menzel, R. Frey, C. Moeller, and L. Gumley, 1998, Discriminating clear sky from clouds with MODIS. *J. Geophys. Res.*, **103**, 32141–32157.

Arking, A., and J. D. Childs, 1985, Retrieval of cloud cover parameters from multispectral satellite images, *J. Appl. Meteorol.*, **24**, 323–333.

Barker, H. W., 2008, Overlap of fractional cloud for radiation calculations in GCMs: A global analysis using CloudSat and CALIPSO data, *J. Geophys. Res.*, **113**, doi:10.1029/2007JD009677.

Baum, B. A., P. F. Soulen, K. I. Strabala, M. D. King, S. A. Ackerman, W. P. Menzel, and P. Yang, 2000, Remote sensing of cloud properties using MODIS airborne simulator imagery during SUCCESS. II. Cloud thermodynamic phase. *J. Geophys. Res.*, **105**, 11781–11792.

Bréon, F., and P. Goloub, 1998, Cloud droplet effective radius from spaceborne polarization measurements, *Geophys. Res. Lett.*, **25**, 1879–1882.

Brinkman, R. T., 1968, Rotational Raman scattering in planetary atmospheres, *Astrophys. J.*, **15**, 1087–1093.

de Beek, R., M. Vountas, V. V. Rozanov, A. Richter, and J. P. Burrows, 2001, The ring effect in the cloudy atmosphere, *Geophys. Res. Lett.*, **28**, 721–724.

Dessler, A. E., S. P. Palm, and J. D. Spinhirne, 2006, Tropical cloud-top height distributions revealed by the Ice, Cloud, and Land Elevation Satellite (ICESat)/Geoscience Laser Altimeter System (GLAS), *J. Geophys. Res.*, **111**, D12215, doi:10.1029/2005JD006705.

Goloub, P., M. Herman, H. Chepfer, J. Riedi, G. Brogniez, P. Couvert, and G. Sèze, 2000, Cloud thermodynamical phase classification from the POLDER spaceborne instrument, *J. Geophys. Res.*, **105**, 14747–14759.

Grenier, P., J. Blanchet, and R. Muñoz-Alpizar, 2009, Study of polar thin ice clouds and aerosols seen by CloudSat and CALIPSO during midwinter 2007, *J. Geophys. Res.*, **114**, D09201, doi:10.1029/2008JD010927.

Grzegorski, M., M. Wenig, U. Platt, P. Stammes, N. Fournier, and T. Wagner, 2006, The Heidelberg iterative cloud retrieval utilities (HICRU) and its application to GOME data, *Atmos. Chem. Phys.*, **6**, 4461–4476.

Haladay, T., and G. Stephens, 2009, Characteristics of tropical thin cirrus clouds deduced from joint CloudSat and CALIPSO observations, *J. Geophys. Res.*, **114**, doi:10.1029/2008JD010675.

Han, Q., W. B. Rossow, and A. Lacis, 1994, Near-global survey of effective droplet radius in liquid water clouds using ISCCP data, *J. Climate*, 7, 465–497.

Hanel, R. A., 1961, Determination of cloud altitude from a satellite, *J. Geophys. Res.*, **66**, 1300–1300.

Heidinger, A. K., and G. L. Stephens, 2000, Molecular line absorption in a scattering atmosphere. Part II: Application to remote sensing in O_2 A-band, *J. Atmos. Sci.*, **57**, 1615–1634.

Herman, J. R., D. Larko, E. Celarier, and J. Ziemke, 2001, Changes in the Earth's UV reflectivity from the surface, clouds, and aerosols, *J. Geophys. Res.*, **106**, 5353–5368.

Joiner, J., and P. Bhartia, 1995, The determination of cloud pressures from rotational Raman scattering in satellite backscatter ultraviolet measurements, *J. Geophys. Res.*, **100**, 23019–23026.

Joiner, J., and A. P. Vasilkov, 2006, First results from the OMI rotational Raman scattering cloud pressure algorithm, *IEEE Trans. Geosci. Remote Sens.*, **44**, 1272–1282.

King, M. D., 1987, Determination of the scaled optical thickness of clouds from reflected solar radiation measurements. *J. Atmos. Sci.*, **44**, 1734–1751.

King, M. D., S. Platnick, P. Yang, G. T. Arnold, M. A. Gray, J. C. Riedi, S. A. Ackerman, and K. N. Liou, 2004, Remote sensing of liquid water and ice cloud optical thickness, and effective radius in the arctic: Application of airborne multispectral MAS data. *J. Atmos. Oceanic Technol.*, **21**, 857–875.

Koelemeijer, R. B. A., P. Stammes, J. W. Hovenier, and J. F. de Haan, 2001, A fast method for retrieval of cloud parameters using oxygen A band measurements from the Global Ozone Monitoring Experiment, *J. Geophys. Res.*, **106**, 3475–3490.

Knap, W. H., P. Stammes, R. B. A. Koelemeijer, 2002, Cloud thermodynamic-phase determination from near-infrared spectra of reflected sunlight. *J. Atmos. Sci.*, **59**, 83–96.

Kokhanovsky A. A., V. V. Rozanov, E. P. Zege, H. Bovensmann, and J. P. Burrows, 2003, A semianalytical cloud retrieval algorithm using backscattered radiation in 0.4–2.4 μm spectral region, *J. Geophys. Res.*, **108**, doi:10.1029/2001JD001543.

Kokhanovsky, A. A., 2006, *Cloud Optics*, Dordrecht: Springer.

Kokhanovsky, A. A., O. Jourdan, and J. P. Burrows, 2006, The cloud phase discrimination from a satellite, *IEEE Geosci. Rem. Sens. Lett.*, **3**, 103–106.

Kokhanovsky, A. A., M. Vountas, V. V. Rozanov, W. Lotz, H. Bovensmann, and J. P. Burrows, 2007, Global cloud top height and thermodynamic phase distribution as obtained by SCIAMACHY on ENVISAT, *Int. J. Remote Sensing*, **28**, 4499–4507.

Kokhanovsky, A. A., C. M. Naud, and A. Devasthale, 2008, Inter-comparison of ground-based radar and satellite cloud-top height retrievals for overcast single-layered cloud fields, *IEEE Trans. Geosci. Rem. Sens.*, **47**, 1901–1908.

Kuze, A., and K. V. Chance, 1994, Analysis of cloud top height and cloud coverage from satellites using the O_2 *A* and *B* bands, *J. Geophys. Res.*, **99**, 14,481–14,491.

Liou, K.-N., 2002, *Introduction to Atmospheric Radiation*, New York: Academic Press.

Loyola, D.G. R., 2004, Automatic cloud analysis from polar-orbiting satellites using neural network and data fusion techniques, *Proc. Geosci. Remote Sens. Symp.*, **4**, 2530–2533.

Mace, G. G., Q. Zhang, M. Vaughan, R. Marchand, G. Stephens, C. Trepte, and D. Winker, 2009, A description of hydrometeor layer occurrence statistics derived from the first year of merged CloudSat and CALIPSO data, *J. Geophys. Res.*, **114**, D00A26, doi:10.1029/2007JD009755.

Menzel, W. P. , R. A. Frey, H. Zhang, D. P. Wylie, C. C. Moeller, R. E. Holz, B. Maddux, B. A. Baum, K. I. Strabala, and L. E. Gumley, 2008, MODIS global cloud-top pressure and amount estimation: Algorithm description and results. *J. Appl. Meteor. Climatol.*, **47**, 1175–1198.

Moroney, C., R. Davies, and J.-P. Muller, 2002, Operational retrieval of cloud-top heights using MISR data, *IEEE Trans. Geosci. Remote Sens.*, **40**, 1532–1546.

Nakajima, T., and M. D. King, 1990, Determination of the optical thickness and effective particle radius of clouds from reflected solar radiation measurements. Part 1. Theory, *J. Atmos. Sci.*, **47**, 1878–1893.

Nakajima, T., M. D. King, J. D. Spinhirne, and L. F. Radke, 1991, Determination of the optical thickness and effective particle radius of clouds from reflected solar radiation measurements. Part 2. Marine stratocumulus observations, *J. Atmos. Sci.*, **48**, 782–850.

Ou, S.C., K. N. Liou, Y. Takano, and R. L. Slonaker, 2005, Remote sensing of cirrus cloud particle size and optical depth using polarimetric sensor measurements. *J. Atmos. Sci.*, **62**, 4371–4383.

Park H., D. F. Heath, and C. L. Mateer, 1986, Possible application of the Fraunhofer line filling in effect to cloud height measurements, in *Meteorological Optics*, OSA Technical Digest Series, pp. 70–81, Opt. Soc. Am., Washington, D.C.

Pilewskie, P., and S. Twomey, 1987, Discrimination of ice from water in clouds by optical remote sensing. *Atmos. Res.*, **21**, 113–122.

Platnick, S., J. Y. Li, M. D. King, H. Gerber, and P. V. Hobbs, 2001, A solar reflectance method for retrieving the optical thickness and droplet size of liquid water clouds over snow and ice surfaces, *J. Geophys. Res.*, **106**, 15185–15199.

Platnick, S., M. D. King, S. A. Ackerman, W. P. Menzel, B. A. Baum, J. C. Riedi, and R. A. Frey, 2003, The MODIS cloud products: Algorithms and examples from Terra. *IEEE Trans. Geosci. Remote Sens.*, **41**, 459–473.

Price, M. J., 1977, On probing the outer planets with the Raman effect, *Rev. Geophys.*, **15**, 227–234.

Rolland, P., K. N. Liou, M. D. King, S. C. Tsay, and G. M. McFarquhar, 2000, Remote sensing of optical and microphysical properties of cirrus clouds using MODIS channels: methodology and sensitivity to assumptions. *J. Geophys. Res.*, **105**, 11,721–11,738.

Rossow, W. B., and R. A. Schiffer, 1999, Advances in understanding clouds from ISCCP. *Bull. Amer. Meteor. Soc.*, **80**, 2261–2287.

Rozanov, V. V., and A. A. Kokhanovsky, 2004, The semi-analytical cloud retrieval algorithm as applied to the cloud top altitude and the cloud geometrical thickness determination from the top of atmosphere reflectance measurements in the oxygen absorption bands, *J. Geophys. Res.*, **109**, doi:10.1029/2003JD004104.

Rozanov, V. V., Kokhanovsky, A. A., D. Loyola, R. Siddans, B. Latter, A. Stevens, and J. P. Burrows, 2006, Intercomparison of cloud top altitudes as derived using GOME and ATSR-2 instruments onboard ERS-2, *Rem. Sens. Environ.*, **102**, 186–193.

Rozanov V. V. and A. A. Kokhanovsky, 2008, Impact of single- and multi-layered cloudiness on ozone vertical column retrievals using nadir observations of backscattered solar radiation, in: Light scattering reviews, (ed. A. A. Kokhanovsky, vol.3), Berlin: Springer-Praxis, 113–190.

Rozenberg, G. V., M. S. Malkevitch, V. S. Malkova, and V. I. Syachinov, 1978, The determination of optical characteristics of clouds from measurements of the reflected solar radiation using data from the Sputnik "KOSMOS-320", *Izvestiya Acad. Sci. USSR, Phys. Atmos. Okeana*, **10**, 14–24.
Saiedy, F., Hilleary, D. T., and Morgan, W. A., 1965, Cloud-top altitude measurements from satellites, *Appl. Optics*, **4**, 495–500.
Saiedy, F. H., H. Jacobowitz, and D.Q. Wark, 1967, On cloud – top determination from Gemini-5, *J. Atmos. Sci.*, **24**, 63–69.
Sassen, K., Z. Wang, and D. Liu, 2008, Global distribution of cirrus clouds from CloudSat/Cloud-Aerosol Lidar and Infrared Pathfinder Satellite Observations (CALIPSO) measurements, *J. Geophys. Res.*, **113**, doi:10.1029/2008JD009972.
Schreiner, A.J., D. A. Unger, W. P. Menzel, G. P. Ellrod, K. I. Strabala, and J. L. Pellet, 1993, A comparison of ground and satellite observations of cloud cover, *Bull. Amer. Meteor. Soc.*, **74**, 1851–1861.
Stephens, G. L., D. G. Vane, S. Tanelli, E. Im, S. Durden, M. Rokey, D. Reinke, P. Partain, G. G. Mace, R. Austin, T. L'Ecuyer, J. Haynes, M. Lebsock, K. Suzuki, D. Waliser, D. Wu, J. Kay, A. Gettelman, Z. Wang, and R. Marchand, 2008, CloudSat mission: Performance and early science after the first year of operation, *J. Geophys. Res.*, **113**, doi:10.1029/2008JD009982.
Strabala, K. I., S. A. Ackerman, and W. P. Menzel, 1994, Cloud properties inferred from 8–12-μm data. *J. Appl. Meteor.*, **33**, 212–229.
Twomey, S., and T. Cocks, 1989, Remote sensing of cloud parameters from spectral reflectance in the near-infrared, *Contrib. Atmos. Phys.*, **62**, 172–179.
van Deelen, R., 2007, Rotational Raman scattering in The Earth's atmosphere, PhD thesis, Free University of Amsterdam.
van Diedenhoven, B., 2007, Satellite Remote Sensing of Cloud Properties in Support of Trace Gas Retrievals, Ph.D. thesis, Free University, Amsterdam.
van Diedenhoven, B., O. P. Hasekamp and J. Landgraf, 2007, Retrieval of cloud parameters from satellite-based reflectance measurements in the ultraviolet and the oxygen A-band, *J. Geophys. Res.*, **112**, D15208, doi:10.1029/2006JD008155.
Wallace, L., 1972, Rayleigh and Raman scattering by H2 in a planetary atmosphere, *Astrophys. J.*, **176**, 249–257.
Wang, L., and A. E. Dessler, 2006, Instantaneous cloud overlap statistics in the tropical area revealed by ICESat/GLAS data, *Geophys. Res. Lett.*, **33**, L15804, doi:10.1029/2005GL024350, 2006.
Winker D. M. and L. R. Poole, 1995, Monte-Carlo calculations of cloud returns for ground-based and space-based LIDARS , *Appl. Phys. B*, **60**, 341–344.
Winker, D. M., Couch, R. H., and M. P. McCormick, 1996, An overview of LITE: NASA's Lidar In-space Technology Experiment, *Proc. IEEE*, **84**, 164–180.
Wu, D. L., S. A. Ackerman, R. Davies, D. J. Diner, M. J. Garay, B. H. Kahn, B. C. Maddux, C. M. Moroney, G. L. Stephens, J. P. Veefkind, and M. A. Vaughan, 2009, Vertical distributions and relationships of cloud occurrence frequency as observed by MISR, AIRS, MODIS, OMI, CALIPSO, and CloudSat, *Geophys. Res. Lett.*, **36**, doi:10.1029/2009GL037464.
Wylie, D., D. L. Jackson, W. P. Menzel, and J. J. Bates, 2005, Trends in global cloud cover in two decades of HIRS observations, *J. Climate*, **18**, 3021–3031.
Yamamoto, G., and D. Q. Wark, 1961, Discussion of the letter by R. A. Hanel, "Determination of cloud altitude from a satellite", *J. Geophys. Res.*, **66**, 3596–3596.
Zege, E. P., I. L. Katsev, and I. N. Polonsky, 1995, Analytical solution to LIDAR return signals from clouds with regard to multiple scattering, *Appl. Phys. B*, **60**, 345–353.

Chapter 6
Retrieval of Aerosol Properties

Gerrit de Leeuw, Stefan Kinne, Jean-Francois Léon, Jacques Pelon, Daniel Rosenfeld, Martijn Schaap, Pepijn J. Veefkind, Ben Veihelmann, David M. Winker and Wolfgang von Hoyningen-Huene

6.1 Introduction

Atmospheric aerosol is a suspension of liquid and solid particles in air, i.e. the aerosol includes both particles and its surrounding medium; in practice aerosol is usually referred to as the suspended matter, i.e. the particles or the droplets, depending on their aggregation state. Particle and droplet radii vary from a few nm to more than

G. de Leeuw (✉)
Climate Change Unit, Finnish Meteorological Institute, Helsinki, Finland
and
Department of Physics, University of Helsinki, Helsinki, Finland
and
TNO Environment and Geosciences, Utrecht, The Netherlands

S. Kinne
MPI-Meteorology, Hamburg, Germany

J.-F. Léon
LOA, Lille, France

J. Pelon
Université Pierre et Marie Curie, Paris, France

D. Rosenfeld
The Hebrew University of Jerusalem, Jerusalem, Israel

M. Schaap
TNO Environment and Geosciences, Utrecht, The Netherlands

P.J. Veefkind
Royal Netherlands Meteorological Institute (KNMI), De Bilt, The Netherlands

B. Veihelmann
ESA/ESTEC, European Space Agency, Noordwijk, The Netherlands

D.M. Winker
NASA Langley Research Center, Hampton, USA

W. von Hoyningen-Huene
Institute of Environmental Physics, University of Bremen, Bremen, Germany

J.P. Burrows et al. (eds.), *The Remote Sensing of Tropospheric Composition from Space*, Physics of Earth and Space Environments, DOI 10.1007/978-3-642-14791-3_6,

100 μm (Seinfeld and Pandis 1998). Aerosol particles are distinctly different from cloud droplets as regards their physical properties and aggregation state.

The particles can be directly emitted into the atmosphere (e.g. sea spray aerosol, dust, biomass burning aerosol, volcanic ash, primary organic aerosol) or produced from precursor gases (e.g. sulfates, nitrates, ammonium salts, secondary organic aerosol). The total aerosol mass is dominated by particles produced from the surface by natural processes, in particular sea spray aerosol and desert dust. Anthropogenic emission of both primary particles and precursor gases contributes appreciably to the total aerosol load (Andreae and Rosenfeld 2008).

A complete description of aerosol properties requires a multitude of parameters, such as particle size, particle concentration as function of size (particle size distribution), which can vary by roughly ten orders of magnitude, particle shape, and chemical composition. Each of these parameters depends on the type of the sources, chemical reactions in the atmosphere, and removal processes such as wet deposition (induced by precipitation,) and dry deposition (gravitational fallout and turbulence). The chemical composition determines the complex refractive index of the aerosol, which in turn (together with particle shape) determines their optical properties. Chemical composition implicitly includes the amount of water in a particle, which in turn is determined by its hygroscopicity and by the ambient relative humidity (RH). Often aerosol size distributions, which describe the aerosol concentration as function of particle size, are modelled as lognormal size distributions defined by the effective mean radius and standard deviation, whereas the amplitude varies with aerosol concentration. For such conditions the aerosol number concentrations N_i are distributed over a range of aerosol particle radii r_i following an n-mode lognormal distribution:

$$\frac{dN}{d\ln r} = \sum_{i=1}^{n} \frac{N_i}{(2\pi)^{1/2}\ln\sigma_i} \exp\left(-\frac{(\ln r_i - \ln \bar{r}_{gi})^2}{2\ln^2\sigma_i}\right) \tag{6.1}$$

where $\bar{r}_{gi}$ is the geometric mean radius and σ_i is the geometric standard deviation of the i-th lognormal mode. Bimodal size distributions are often assumed. Particles are often hygroscopic; therefore the aerosol radius r can be specified in its dry state (r_d), at a relative humidity RH of 80% (r_{80}), or any other RH.

The complex refractive index m is expressed as $m = n - ik$, where the real part n specifies the refraction in a medium due to the change in the speed of light with respect to that in vacuum, and k determines the absorption; both constants depend on the wavelength of the propagating light beam and on the properties of the medium. Both m and k need to be specified as a function of wavelength to calculate the optical properties. For this purpose, particles are generally assumed to be spherical allowing the application of Mie theory (Mie 1908, Chapter 1) to compute extinction and absorption coefficients and the scattering phase function. The latter describes the angular dependence of the scattering. The sphericity assumption is not appropriate for dust (or other solid) particles, which require more sophisticated methods to determine their optical properties (Dubovik et al. 2006). Aerosol optical properties determine the scattering and absorption of solar radiation, which in turn determine the effects of aerosol on climate and provide a way to observe aerosol properties using electro-optical instruments. Some of these can be used *in situ*, on the ground or in aircraft, whereas others can be used on platforms such as satellites or aircraft and thereby provide the means for the remote sensing of aerosols, which is the topic of this chapter.

Aerosols play an important role in climate and air quality. The impact on climate is induced by the scattering and absorption of incoming solar radiation. The combined effect of scattering and absorption is termed the direct radiative effect of aerosols on climate. Aerosol scattering reduces the amount of incoming radiation reaching the Earth surface and hence constitutes cooling. The net effect of aerosols is cooling which partly offsets the impact of the global warming caused by the absorption by greenhouse gases. Since aerosol particles are often hygroscopic they can act as cloud condensation nuclei (CCN). As the amount of precipitable water available for condensation on CCN is often limited in the atmosphere, an increase in CCN, and thus increased cloud droplet concentrations, results in smaller cloud droplet sizes and enhanced cloud albedo. This has been termed the first indirect effect of aerosols on climate. The smaller cloud droplets precipitate more slowly and evaporate faster, which affects precipitation. This is termed the second indirect effect on climate. Some aerosols contain contaminants, which absorb strongly in the visible in addition to the absorption of liquid water. Overall our understanding of the effects of aerosols on climate is poor (IPCC 2007).

Aerosols have a large impact on our understanding of air quality as they reduce visibility and increase the amount of diffuse radiation. The latter has recently been shown to have impacts on the land carbon sink through its effect on photosynthesis (Mercado et al. 2009). In addition the occurrence of fine particle matter impacts on human health, which is another important aspect of air quality. Concentrations of fine particulate matter are commonly expressed as the mass of dry aerosol particles, PM10 or PM2.5, where 10 (2.5) refers to the maximum aerosol diameter, in μm, for dry particles contributing to the measured mass.

The data from instruments aboard satellites provide a unique method to observe aerosol properties with the same retrieval technique and the same instrument from local to regional to global scales (Kaufman et al. 2002). Instruments used for aerosol measurements from space, see Kokhanovsky and de Leeuw (2009) for an overview, were often not specifically designed for this task, but nevertheless provide valuable information which is complementary to that from instruments dedicated to the remote sensing of aerosols. Algorithms have been developed to determine aerosol properties from satellite observations of back scattered solar radiation.

A brief description of the history of aerosol observations from space is given by Lee et al. (2009). One of the first retrievals of aerosol optical depth from space-borne measurements of the spectral intensity of the reflected solar light was performed using observations from the MSS (Multi Spectral Scanner) on board the Earth Resources Technology Satellite (ERTS-1) (Griggs 1975; Mekler et al. 1977). The first operational aerosol products were generated using data from the AVHRR (Advanced Very High Resolution Radiometer) (Stowe et al. 2002) on board the TIROS-N satellite launched in 19th October 1978. The Nimbus-7, launched in 25th October 1978, carrying the Stratospheric Aerosol Measurement instrument (SAM) (McCormick et al. 1979) and the Total Ozone Mapping Spectrometer (TOMS) (Torres et al. 2002). Thus the retrieval of aerosol properties from satellite-based observations started some 3 decades ago. Initially, retrievals were obtained only for measurements over water; aerosol retrieval results over land have started to become available on a regular basis only in the last decade.

AVHRR and TOMS have provided a long term aerosol product over the ocean spanning a period of roughly 3 decades. A long term aerosol product over land, since 1995, is produced using ATSR-2 (Along Track Scanning Radiometer) (Veefkind et al. 1998) and AATSR (Advanced Along Track Scanning Radiometer) measurements (Grey et al. 2006; Thomas et al. 2007), with the prospect of a longer time series using data from the SLST (Sea and Land Surface Temperature) planned for launch in 2013, as part of Sentinel-3. The long time series over oceans provide information on trends or changes in the last 30 years (Mishchenko et al. 2007). The aerosol data product delivered from TOMS was initially the Absorbing Aerosol Index, a measure for the presence of absorbing aerosol. Subsequently an algorithm was developed to retrieve the aerosol optical depth (AOD, the column-integrated extinction, see Section 6.3) (Torres et al. 1998; 2002).

The launch of lidars in space has added a new dimension to satellite observation of aerosol properties by providing information on the vertical distribution of aerosols and clouds. The first lidar measurements of aerosol back scattering from space were made by the instrument LITE (LIdar in space Technology Experiment) (Winker et al. 1996). LITE is a three-wavelength backscatter lidar which flew on the space shuttle Discovery in September 1994. LITE was operated for 53 h and provided views of multilayer cloud structures and observations of the distribution of desert dust, smoke, and other aerosols. LITE also provided the first global observations of planetary boundary layer height (http://www-lite.larc.nasa.gov/). A second lidar instrument, GLAS (Geoscience Laser Altimeter System) was launched in January 2003 as part of NASA's Earth Observing System (EOS) and was the sole instrument on ICESat (Ice, Cloud, and land Elevation Satellite). One of the secondary objectives of GLAS includes the measurement of cloud and aerosol height profiles (Zwally et al. 2002) (http://icesat.gsfc.nasa.gov/index.php). A dedicated aerosol and cloud lidar, CALIOP (Cloud-Aerosol Lidar with Orthogonal Polarization), was launched on CALIPSO (Cloud Aerosol Lidar and Infrared Pathfinder Satellite Observations) in April 2006. CALIPSO is part of the A-train, a constellation of satellites with instruments designed to study aerosols and clouds: Cloudsat, PARASOL (the follow up of POLDER – POLarization and Directionality of the Earth's Reflectances), MODIS (Moderate Resolution Imaging Spectroradiometer) and OMI (Ozone Monitoring Instrument).

A measure of the effect of aerosols and clouds on the Earth's radiative balance is provided by the direct measurement of the upwelling radiation at the top of the atmosphere. However, the effects of the surface and atmospheric constituents need to be separated. This is accomplished by retrieval algorithms, which provide information on the AOD at the measured wavelengths. More sophisticated research algorithms, see Kokhanovsky and de Leeuw (2009) for a recent review, provide aerosol microphysical properties, such as effective radius, ratio of fine to coarse fraction, single scattering albedo and composition. Unlike ground-based and airborne measurements, instruments aboard satellites provide information over a large area with up to daily global coverage for polar orbiting satellites. Geostationary satellites provide high temporal resolution (15 min) over a large area and hence allow for studies of the diurnal evolution of aerosol properties, as well as the interaction between clouds and aerosols.

The information on aerosols is in the atmospheric reflectance, also referred to as path radiance. Before the aerosol information can be retrieved using algorithms that fit models describing the radiative transfer through the atmosphere with different types of aerosols, the path radiance needs to be separated from the effects due to clouds and surface reflection. Clouds are bright and even small traces of clouds can contaminate the atmospheric signal (Chapter 5). It is therefore important to detect accurately the occurrence of clouds within a ground scene. This is one of the major issues for aerosol retrieval. In particular the presence of sub-visible clouds and high cirrus may contaminate the signal, as well as the distinction between clouds and aerosol near cloud edges (Koren et al. 2007). Another major issue for the retrieval of aerosol data products is the removal of land surface effects from the top of the atmosphere (TOA) radiation. For single-view instruments this is typically dealt with by using assumptions about the reflectance at wavelengths in the near infrared and the spectral dependence of the reflectance by specific surfaces, or by using a surface reflectance data base. When multiple views are available, the surface contribution to the TOA radiation can be eliminated explicitly. The remaining TOA radiation for cloud-free sky, after elimination of the surface contributions, is the atmospheric path radiance, which is compared with the path radiance computed using a radiative transfer model. By minimizing the difference between computed and observed radiances, the AOD can be determined. When more than one wavelength is available, the minimization should be made using all available wavelengths, to determine the most likely aerosol type (or mixture) in the atmospheric column.

Instrumental characteristics required for aerosol retrieval are multiple wavelengths, multiple viewing angles and polarization. PARASOL (Polarization and Anisotropy of Reflectances for Atmospheric Sciences coupled with Observations from a lidar) is an instrument that combines all of these. MISR (Multiangle Imaging SpectroRadiometer) has multiple viewing angles and multiple wavelengths. AATSR has two views and multiple wavelengths. MODIS, MERIS (MEdium Resolution Imaging Spectrometer) and OMI have multiple wavelengths. Therefore the algorithms for each of these instruments are different, and the algorithms for a single instrument may be very different (see Kokhanovsky and de Leeuw (2009) for descriptions of three algorithms used for aerosol retrieval using AATSR data). The algorithms differ in the way they deal with the surface reflectance, which is an important issue over land surfaces. They also differ in the way they screen clouds, and in the aerosol models used in the retrieval. These issues are treated in Sections 6.6 to 6.11. An overview of pertinent characteristics of instruments discussed in this chapter is presented in Appendix A.

The use of aerosol data products derived from instrumentation aboard satellites contributes to our understanding and documenting of regional and global aerosol and cloud properties, including their variations, as well as aerosol-cloud interactions. It provides a 4-D distribution of aerosol and cloud properties on regional and global scales, using data from radiometers (multi-spectral, multi-angle and polarization) which provide the spatial distribution, and lidar data from CALIOP which provide the vertical distribution, all of these as a function of time which allows for following the evolution of the aerosol properties.

These instruments are of experimental nature, and hence have a limited lifetime, in contrast to the operational satellites provided by the METEOSAT (Meteorological satellite – geostationary) and MetOp (Meteorological Operational satellite programme) series of instruments and the GMES (Global Monitoring for Environment and Security) operational system that is developed by ESA (European Space Agency) and EUMETSAT (European Organisation for the Exploitation of Meteorological Satellites). The geostationary instrument SEVIRI (Spinning Enhanced Visible and Infrared Imager – flying on METEOSAT Second Generation, MSG) has the advantage of very high temporal resolution. AATSR flying on ENVISAT (ENVIronmental SATellite) does not provide global coverage in one day, as it has a limited swath, but as indicated above, it is intended to continue the current time series of 15 years (1995–2010) and thus provide more than 20 years of aerosol properties over land.

A description of aerosol retrieval over land using current instrumentation is provided in Kokhanovsky and de Leeuw (2009). This book describes the state of the art about the retrieval of aerosol data products over land, and comprises a series of articles solicited to represent instruments that are currently used for this purpose in a sun synchronous orbit (MODIS, MISR, POLDER, AATSR, MERIS) and SEVIRI which is in a geostationary orbit. For AATSR three different retrieval algorithms, based on different principles, were presented. Tables providing the general characteristics of instruments currently used for aerosol retrieval can also be found in the introduction to the book (de Leeuw and Kokhanovsky 2009).

In this chapter we provide supplementary information about the current state of the art of aerosol retrieval using CALIPSO, POLDER, AATSR, MODIS, OMI and MERIS observations and algorithms that have been developed for these instruments. Brief descriptions of these instruments and their data products are provided in Sections. 6.6 to 6.11. In addition, examples are presented describing applications of satellite data for air quality, climate and aerosol-cloud interaction studies.

6.2 Aerosol Retrieval Algorithms

Aerosol retrieval is based on the comparison of the radiation received by an instrument at TOA with that calculated using a radiative transfer model for the same geometry and atmospheric conditions, for a range of aerosol models. The best fit model is selected to provide the retrieval solution, i.e. AOD and other aerosol properties. Aerosol retrieval algorithms utilize data obtained from observations with instruments mounted on a satellite in a certain orbit, with a certain viewing angle and for a solar zenith angle which varies with the season and the time of day. Aerosol particles scatter light in different directions with an angular distribution that depends on particle size, shape, and chemical composition, and is described by the scattering phase function. The intensity of the scattered light can vary by several orders of magnitude depending on particle size. Any aerosol retrieval algorithm uses the angular dependence of the aerosol scattering and therefore needs to take the specific geometry into account. However, aerosol properties can only be retrieved for cloud-free scenes because scattering by cloud droplets not only overwhelms the aerosol signal, but also has different angular dependence.

Therefore, the first step in any retrieval algorithm is cloud screening. Several criteria may be applied for cloud detection (Ackerman et al. 1998). The retrieval of cloud properties is discussed in Chapter 5. Cloud screening methods are based on a variety of principles, including thresholding of the radiances measured at wavelengths in the near and thermal infrared, the wavelength dependence of the radiances for pairs of channels at wavelengths in the visible to the thermal infrared, the analysis of spatial and temporal patterns, the use of the O_2 A-band, and also the synergistic use of other instruments, including lidars. An example of the application of cloud screening is presented in Section 6.8.

The next problem for clear sky measurements is to account for the land surface contribution to the TOA reflectance. Several different methods are implemented in retrieval algorithms, depending on instrument properties. This is discussed in Sections 6.7 to 6.11 for the individual instruments treated here. After correction for the surface contribution to the TOA radiance, the path radiance, which includes contributions from molecular scattering and absorption, remains. To account for molecular effects for the given sun-satellite geometry properly, a radiative transfer model (RTM) is applied to a set of observation and illumination geometries for a wide range of situations. This is referred to as forward modelling; the retrieval is referred to as inverse modelling.

Radiative transfer calculations are usually time consuming (see Katsev et al. (2009) for an example of the use of a fast radiative transfer code as part of the retrieval algorithm) and are used outside the actual retrieval algorithm to prepare look up tables (LUTs) for a wide variety of situations; these include the viewing geometry expressed by the solar and viewing zenith angles and relative azimuth angle, the wavelengths that are used in the retrieval, a series of reference AOD levels from very low to very high, and other relevant parameters depending on the instrument characteristics used in the retrieval, together with atmospheric information such as surface pressure. The angles and AOD levels are varied in discrete steps. For each combination, forward calculations are made for a series of aerosol models (see Section 6.4) and the results are stored in LUTs. A LUT is thus prepared for each aerosol model and contains parameters such as AOD, single scattering albedo ω, spherical albedo, total and diffuse transmittance along the atmospheric path (downward and upward), total and diffuse downward transmittance of the atmospheric column, surface downward reflectance, and path reflectance. Each of these parameters depends on one or more of the input parameters. The LUTs are used in the retrieval step to speed up the processing. The parameters contained in the LUT are interpolated between the discrete levels to provide for the actual situation, in particular the viewing geometry. Examples of LUT contents are presented in Sections 6.7 and 6.9.

The LUTs are usually prepared using vector radiative transfer calculations for a set of aerosol models which are representative for a certain area (Kaufman et al. 2001; Dubovik et al. 2002; Levy et al. 2007a; 2007b). Ideally, the algorithm must have the ability to select the most appropriate aerosol model or mixture of aerosol models. In many cases, however, aerosol type selection is based on climatology (Levy et al. 2007a; Curier et al. 2008). Such climatologies can be derived from observations (Dubovik et al. 2002; Levy et al. 2007a) or from results from transport models for the area of interest (Curier et al. 2008). In the future, it may be possible to use transport model forecasts (Verver et al. 2002) to constrain the retrieval.

6.3 Aerosol Optical Parameters

Optical properties that are important for the remote sensing of aerosols, and applications for climate and air quality are the extinction and backscatter coefficients, the scattering phase function and the single scattering albedo. These parameters are derived from the aerosol particle size distribution and the refractive index, which is determined by the chemical composition. Moreover, both the aerosol particle size and the chemical composition vary with RH. Chemical and physical properties of aerosols change during their atmospheric lifetime due to a variety of processes (Seinfeld and Pandis 1998), and the RH varies in space and time, so the optical properties change horizontally, vertically, and in time.

When a light beam hits a medium containing aerosol the intensity is reduced due to scattering and absorption by the particles. This is described by the Lambert–Beer law:

$$I(\lambda) = I_o(\lambda)\exp\left(-\int_0^h b_{ext}(\lambda)dz\right), \tag{6.2}$$

where I_o and I are the intensities of the incoming and exiting light beams, respectively, with wavelength λ, z is the position in the medium with thickness h and b_{ext} is the aerosol extinction coefficient, which is given by the sum of the scattering and absorption coefficients, b_{scat} and b_{abs} ($b_{ext} = b_{scat} + b_{abs}$). These coefficients are determined by the product of the particle size distribution $n(r)$ and the extinction, scattering or absorption efficiency, $Q_{ext}(r, m, \lambda)$, $Q_{scat}(r, m, \lambda)$, $Q_{abs}(r, m, \lambda)$, respectively. For scattering:

$$b_{scat}(\lambda) = \int_{r_1}^{r_2} \pi r^2 n(r) Q_{scat}(r, m, \lambda)dr, \tag{6.3}$$

The particle size distribution is a function that describes the concentrations of the aerosol particles as function of radius r (see Eq. 6.1 for a possible formulation). The scattering efficiency for a particle with radius r and complex refractive index $m = n - ik$, at wavelength λ, is given by:

$$Q_{ext}(r, m, \lambda) = C_{ext}(r, m, \lambda)/A(r), \tag{6.4}$$

where $C_{ext}(r, m, \lambda)$ is the extinction cross section and A is the geometric area of that particle. Similar expressions apply to $Q_{scat}(r, m, \lambda)$ and $Q_{abs}(r, m, \lambda)$. The single scattering albedo ω is defined as the ratio of the scattering and extinction efficiencies:

$$\omega(\lambda) = \frac{Q_{scat}(\lambda)}{Q_{ext}(\lambda)}. \tag{6.5}$$

The single scattering albedo is 1 for non-absorbing particles and common values are around 0.97 (0.95–1.0), but much lower values are observed in strongly polluted areas with large amounts of absorbing aerosol (e.g. emitted from forest fires and other combustion processes). Ground-based measurements of absorption and extinction are most reliable in providing such data, although research is needed to improve accuracy and reproducibility. Satellite data of the single scattering albedo are sparse and usually indirectly derived.

The extinction cross section is given by:

$$C_{ext}(r, m, \lambda) = C_{scat}(r, m, \lambda) + C_{abs}(r, m, \lambda) \tag{6.6}$$

and

$$C_{scat}(r, m, \lambda) = I_{scat}(\lambda)/I_o(\lambda) \tag{6.7}$$

with a similar expression for $C_{abs}(r, m, \lambda)$.

Similar equations apply for the aerosol scattering and absorption coefficients and efficiencies, note that $\int_0^h b_{ext}(\lambda)dz = \tau_{aer}(\lambda)$, where $\tau_{aer}(\lambda)$ is the AOD of the layer with depth h, also often called aerosol optical thickness (AOT), which is the primary parameter retrieved by satellites. The wavelength dependence of $\tau_{aer}(\lambda)$ is expressed by the Ångström relationship:

$$\tau_{Aer}(\lambda) = \beta \cdot \lambda^{-\alpha_A} = \tau_{Aer}(\lambda_{ref}) \cdot \left(\frac{\lambda}{\lambda_{ref}}\right)^{-\alpha_A}, \tag{6.8}$$

where β is the AOD at the reference wavelength λ_{ref} (usually taken at 1 μm) and α_A is the Ångström parameter evaluated for the wavelength pair λ_1 and λ_2. Typical values for α_A are in the range 1–2.

For the retrieval of aerosol properties from satellite observations, with a range of illumination and observation angles, information on the angular distribution of the scattering intensity, which is described by the scattering phase function $P(\theta,\alpha,m,\lambda)$, is required:

$$P(\theta, \alpha, m, \lambda) = F(\theta, \alpha, m, \lambda) / \int_0^{\pi} F(\theta, \alpha, m, \lambda) \sin \theta d\theta \tag{6.9}$$

where θ is the scattering angle and α ($=2\pi r/\lambda$) is the size parameter that accounts for the dependence of optical effects of aerosols on their size relative to the wavelength of the incoming light. In all of the above, equations which depend on both r and λ can be replaced by their dependence on α.

Other parameters often encountered are the asymmetry parameter g (see Chapter 5) which provides a measure for the major scattering direction and is given by:

$$g(\lambda) = \frac{1}{2} \int_0^{\pi} \cos \theta P(\theta, \lambda) \sin \theta d\theta, \tag{6.10}$$

For light scattered totally at $\theta = 0^{\circ}$, $g = 1$; $g = -1$ for light scattered totally at $\theta = 180^{\circ}$ and $g = 0$ for isotropically scattered light. The hemispheric backscatter ratio is given by:

$$b(\lambda) = \frac{\int_{\pi/2}^{\pi} P(\theta, \lambda) \sin \theta d\theta}{\int_0^{\pi} P(\theta, \lambda) \sin \theta d\theta}. \tag{6.11}$$

The transmission of a layer of air with thickness h is given by:

$$T(\lambda) = \frac{I(\lambda)}{I_0(\lambda)} = \exp(-\int_0^h b_{ext}(\lambda, z) dz). \tag{6.12}$$

For the calculation of optical properties, it is commonly assumed that the aerosol particles are spherical, which implies that we can use Mie theory (Section 1.9.1). Mie theory is based on the exact solution of the Maxwell equations (Mie 1908) and program codes are readily available. Mie theory needs to be used for spherical particles with sizes on the order of the wavelength of the incident light, i.e. $\alpha \sim 1$. The angular scattering for much smaller particles is symmetric and can be calculated in the Rayleigh limit; the scattering by much larger particles can be calculated using the geometric approximation. Mie theory shows that scattered light has a strong forward lobe, with intensity strongly increasing with increasing particle size. Particles often are hygroscopic and thus absorb water vapour. Liquid particles are spherical for sizes in the optically active range.

For non-spherical particles Mie theory does not apply and approximations about the particle shape, e.g. treatment as spheroids, has to be made to calculate the phase function. This is particularly important for the retrieval of desert dust which has irregular shapes and is non-hygroscopic. For instance, Dubovik et al. (2006) show examples of the application of spheroid models to account for aerosol particle non-sphericity in remote sensing of desert dust. Deviations from Mie theory occur in the scattering phase function for large scattering angles in both the forward and backward directions.

Light scattered by molecules and small aerosol particles is strongly polarized in a plane perpendicular to the scattering plane (the plane defined by the sun, the object being viewed and the observer) while light scattered by surfaces is only weakly polarized. This difference between the polarizing properties of aerosols and molecules as compared to surfaces is used by modern polarimetric remote sensing instruments to determine the amount, size and type of aerosols that are present above the surface. The intensity and polarization of light can be described by the Stokes vector $\boldsymbol{I} = (I, Q, U, V)$ where I is a measure of the intensity of the light, Q and U define the magnitude and orientation of the linearly polarized fraction of the light and V is a measure of the magnitude and helicity of the circular polarization. All four Stokes vector elements have the dimensions of intensity (Wm^{-2}). A detailed discussion of polarization and its use in aerosol remote sensing can be found in Cairns et al. (2009.)

6.4 Databases for Aerosol Properties

Databases for aerosol optical properties are available from analyses of the AERONET sun photometer network (Holben et al. 1998) derived from 8 years of worldwide distributed data for different aerosol types by Dubovik et al. (2002). Established procedures for maintaining and calibrating this global network of radiometers, cloud screening and inversion techniques facilitate the consistent retrieval of the optical properties of aerosols in locations with varying emission sources and conditions. The multi-year, multi-instrument observations show robust differentiation in both the magnitude and spectral dependence of the absorption for desert dust, biomass burning, urban, industrial and marine aerosols. The authors observed significant variability of the absorption for the same aerosol type due to different meteorological and source characteristics, as well as different emission characteristics.

This data base is particularly useful for application in satellite retrieval codes because similar parameters, such as column-integrated aerosol properties, are measured with both, but with much better accuracy and without interference from surface reflectance.

The application of the Dubovik data base to the retrieval of aerosol properties using AATSR observations has provided excellent results in areas with complicated aerosol composition, such as over the Indian Ocean where a transition has been observed from very polluted to very clean aerosol (Robles-Gonzalez et al. 2006), over Africa for biomass burning aerosol (Robles-Gonzalez and de Leeuw 2008) or over the desert for a mixture of fossil fuel and desert dust aerosol (de Leeuw et al.

2005). AERONET data were the basis for the MODIS Collection 5 (C005) aerosol retrieval algorithm (Levy et al. 2007a; 2007b).

Another aerosol data base that provides the parameters describing the aerosol models used in aerosol retrieval algorithms is the Global Aerosol Data Set (GADS). GADS is available from the software package OPAC (Optical Properties of Aerosols and Clouds) (Hess et al. 1998) (http://www.lrz-muenchen.de/~uh234an/www/radaer/opac.html). OPAC provides microphysical and optical properties of ten aerosol components including extinction, scattering and absorption coefficients, the single scattering albedo, the asymmetry parameter and the phase function. The computation of these parameters is based on the microphysical data (size distribution and spectral refractive index), assuming that the particles are spherical. Data are given for up to 61 wavelengths between 0.25 and 40 μm and up to 8 values of the relative humidity. The software package also facilitates the calculation of derived optical properties such as mass extinction coefficients, i.e. the extinction per unit of mass, specific for each aerosol type, and Ångström coefficients.

6.5 Instruments Used for the Retrieval of Aerosol Properties from Space

Data from instruments, which were not explicitly designed for the retrieval of aerosol, have been used to determine aerosol data products. For instance instruments like TOMS, GOME, SCIAMACHY and OMI were designed for the retrieval of trace gas concentrations and the purpose of AVHRR, SeaWiFS, MERIS and ATSR was to measure land/sea surface temperature. GLAS was primarily designed as an altimeter; LITE was set up as a technology experiment. However, data from each of these instruments has also been used for the retrieval of aerosol properties, with varied success. Dedicated instruments for aerosol retrieval are POLDER, MODIS, MISR and CALIOP. Ideally, a sensor should have the capability of observing multiple wavelengths from the UV to the TIR, multiple views, and polarization sensitivity. The combination of spectral polarization and multiple view measurements for a range of wavelengths is only available from the POLDER series of instruments (Deschamps et al. 1994), the latest of which is flying on PARASOL as part of the A-Train. The GLORY mission (Mishchenko et al. 2007) set to launch in February 2011 will carry the Aerosol Polarimetry Sensor (APS) which will collect accurate multi-angle photo-polarimetric measurements of the Earth along the satellite ground track over a wide spectral range extending from the visible to the short-wave infrared. The data from this instrument are expected to provide aerosol retrievals with a higher accuracy than available from current instruments. (A)ATSR and MISR combine two or multiple views, respectively, with multiple wavelengths.

Results from the last decade show that it is possible to obtain a useful set of aerosol parameters even without using the advanced multi-view instruments capable of detecting the polarization state of the reflected solar light. In particular MODIS, which was designed for the measurement of aerosol and cloud properties, has been successful and is the most widely used for aerosol observations from space. The open data policy and accessibility of aerosol products has resulted in numerous publications from the use of MODIS data. Also the global coverage from two MODIS instruments (Terra descending, equator crossing time 10:30, and Aqua ascending, equator crossing time 13:30) ensure a high probability of obtaining useful data.

The parameters retrieved from measurements by instrumentation in space include the AOD at various wavelengths and its wavelength dependence expressed by the Ångström coefficient. Principle component analysis shows which other aerosol parameters could be retrieved using a dedicated aerosol instrument. These could include, for a bimodal aerosol model, the effective radius and effective variance, and the complex refractive index (both real and imaginary parts) for both modes. Examples are presented by Hasekamp and Landgraf (2005) for GOME-2 and Veihelmann et al. (2007) for OMI.

6.6 Retrieval of Aerosol and Cloud Parameters from CALIPSO Observations

CALIPSO is a satellite mission (see Appendix A) developed within the framework of a collaboration between NASA and the French space agency, CNES. CALIPSO provides unique measurements to improve our understanding of the role of aerosols and clouds in the Earth's climate system (Winker et al. 2003; 2009). The CALIPSO payload (see Table 6.1) consists of a two-wavelength polarization-sensitive lidar, and passive imagers operating in the visible and infrared spectral regions. The lidar profiles provide information on the vertical distributions of aerosols and clouds, cloud ice/water phase (via the ratio of signals in two orthogonal polarization

Table 6.1 Characteristics of the CALIPSO instruments

Characteristic	Value
CALIOP	
Wavelengths	532 nm, 1064 nm
Polarization	532 nm, $\parallel$ and $\perp$
Pulse energy	110 mJ each wavelength
Footprint	100 m
Vertical resolution	30–60 m
Horizontal resolution	333 m
WFC	
Wavelength	645 nm
Spectral bandwidths	50 mm
IFOV/swath	125 m/61 km
IIR	
Wavelengths	8.65 μm, 10.6 μm, 12.0 μm
Spectral resolution	0.6 μm–1.0 μm
IFOV/swath	1 km/64 km

channels) and a qualitative classification of aerosol size (via the wavelength dependence of the backscatter). Data from the three instruments are used together to measure the radiative and physical properties of cirrus clouds. CALIPSO is flown in a polar orbit as part of the A-Train constellation which, besides CALIPSO, consists of the Aqua, CloudSat, PARASOL and Aura satellites. The satellites of the constellation fly in a sun-synchronous polar orbit with a nominal ascending node equatorial crossing time of 13:30 local time. The orbit of CALIPSO is maintained to provide space-time coincidence with observations from the other satellites of the constellation. CALIPSO was launched at the end of April 2006, and data have been available from 13th June 2006.

CALIPSO has been designed to provide data to address three major objectives:

- To improve observationally-based estimates of direct and indirect aerosol radiative forcing;
- To improve the characterization of surface radiative fluxes and atmospheric heating rates; and
- To improve model parameterizations of cloud-climate feedbacks.

CALIPSO is also intended to address a number of secondary objectives, which include observing long range transport of pollutants, providing coincident measurements to validate and improve retrievals from other instruments within the A-train, and providing aerosol observations useful for atmospheric chemistry applications.

The Cloud-Aerosol Lidar with Orthogonal Polarization (CALIOP) provides global, vertically-resolved measurements of aerosol spatial distributions (Winker et al. 2007; 2009). It has the ability to perform height-resolved discrimination of aerosol into several types. As seen in Fig. 6.1, CALIOP can observe aerosol over bright surfaces and beneath thin clouds as well as in clear sky conditions. An elevated aerosol layer (yellow and red in the upper panel) between roughly 0.5°S, 12.9°E and 17°S, 8°E overlies a stratus deck (white and red). At the left edge of the plot, two aerosol layers can be seen at altitudes of about 2 km and 5 km. Depolarization signals (lower panel) allow the identification of smoke (depolarization less than 10%) from dust (depolarization greater than 10%). CALIOP also provides vertical profiles of single and multi-layer transmissive clouds. The Imaging Infrared Radiometer (IIR) and Wide Field Camera (WFC) data combined with lidar data are used to retrieve cloud emissivity and effective particle size. Lidar data is incorporated into a split-window retrieval algorithm to provide constraints to improve the retrieval performance.

6.6.1 The CALIPSO Science Payload

The CALIPSO payload consists of three nadir-viewing instruments: CALIOP, IIR and WFC. These instruments are designed to operate autonomously and continuously, although the WFC acquires science data only under daylight conditions. The

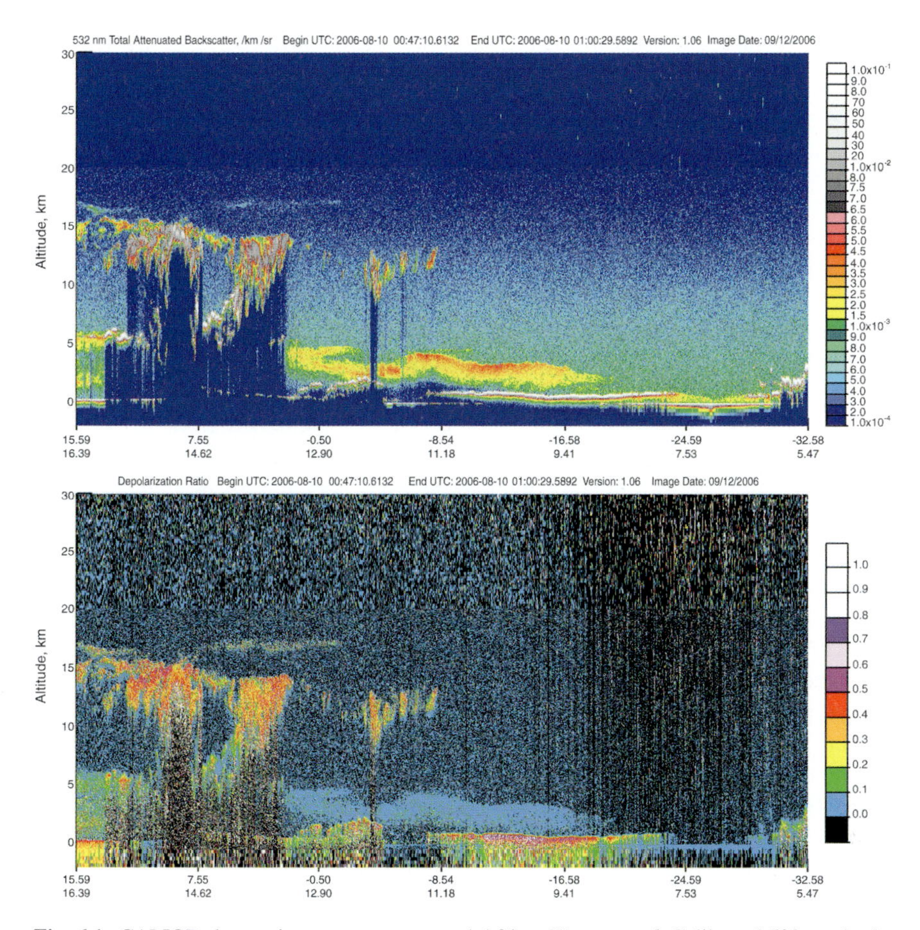

Fig. 6.1 CALIOP observations over west-central Africa. *Upper panel*: Calibrated 532 nm backscatter signals. *Lower panel*: Depolarization profiles from the ratio of the CALIOP 532 nm perpendicular and parallel return signals.

key characteristics of these instruments are listed in Table 6.1. The raw lidar data down-linked to the ground station has the vertical resolution given in Table 6.2.

6.6.2 CALIOP Data Calibration

The calibration algorithms are designed to accomplish two major functions. First, the determination of calibration coefficients for the three lidar channels, and second, the application of these calibration coefficients to produce attenuated backscatter profiles used in Level 2 processing. Determination of the calibration coefficients is basically a three-step process used to derive the Level 1B data products:

Table 6.2 Spatial Resolution of Down-linked Lidar Data

Altitude range	Horizontal resolution	Vertical resolution
30.1 km to 40 km	5.0 km	300 m (532 nm only)
20.2 km to 30.1 km	1.67 km	180 m
8.2 km to 20.2 km	1.0 km	60 m
−0.5 km to 8.2 km	0.33 km	30 m at 532 nm
		60 m at 1064 nm
−2.0 km to −0.5 km	0.33 km	300 m

(a) The calibration coefficient is determined for the 532 nm parallel channel (Powell et al. 2009). For the baseline approach, this is done by comparing the measured 532 nm parallel channel signal from the 30–34 km region to an estimate of the parallel backscatter coefficient computed from a modelled atmospheric density profile. The 30–34 km altitude range is chosen because there is little aerosol in that height range, especially at mid and high latitudes. At low latitudes there is a small bias due to the stratospheric background aerosol. The molecular backscatter coefficients can be estimated well, using knowledge of the molecular number density and theoretically derived estimates of the molecular backscatter cross section (Reagan et al. 2002).
(b) The 532 nm perpendicular channel is then calibrated relative to the calibration obtained for the parallel channel. There is not enough signal to calibrate the perpendicular channel using stratospheric molecular returns, because the depolarization of clear-air 180°-backscatter is only about 0.35%. The calibration is therefore transferred from the parallel to the perpendicular channel using data collected during the Polarization Gain Ratio (PGR) operation (Hunt et al. 2009).
(c) Calibration of the 532 nm parallel and perpendicular channels is then transferred to the 1064 nm channel. As with the 532 nm perpendicular channel, the signal from the 1064 nm channel in the mid-stratosphere is too low to provide a reliable calibration measurement. Transfer of calibration from the 532 nm channels to the 1064 nm channels is accomplished using the backscatter from properly chosen cirrus clouds. Because cirrus cloud particles are large, the ratio of the 532 nm and 1064 nm backscatter coefficients is approximately equal to 1. The 1064 nm calibration coefficient is determined by comparing the 1064 nm backscatter signal with the calibrated 532 nm cirrus backscatter measurements (Reagan et al. 2002).

6.6.3 Description of Available Data Products from CALIOP

The data products generated from the CALIOP measurements are produced according to a protocol which is similar to, but not exactly the same as, that established by NASA's Earth Observing System (EOS).

The data product levels for CALIPSO are reported in Table 6.3. They are defined below.

Table 6.3 CALIPSO product list

Data level	Data products	Production schedule
1b	Calibrated lidar profiles Calibrated IIR radiances Uncalibrated WFC radiances Meteorological profiles Lidar aerosol & cloud browse images	Data produced on 2-day lag following receipt of all required ancillary data (meteorological profiles). Archived and publicly available.
2a	Lidar backscatter profiles Aerosol layer height/thickness Cloud height/thickness	Data produced on 3-day lag thereafter. Archived and publicly available. Last reprocessing Dec. 2007/Jan. 2008.
2b	Aerosol extinction, optical depth Cloud extinction, optical depth Cloud ice/water phase Cloud emissivity (IIR)	First data released in January 2008
2c	Ice particle size (IIR)	First data released in January 2009

- *Level 0*: reconstructed, unprocessed instrument/payload data at full resolution; and with all communications artifacts, e.g. synchronization frames, communications headers, duplicate data removed.
- *Level 1B*: reconstructed, unprocessed instrument data at full resolution which is time-referenced, geo-located, corrected for instrument artifacts, and includes ancillary information processed to sensor units and archived as Level 1 data.
- *Level 2*: geophysical variables derived from Level 1 data, including those derived using measurements from multiple CALIPSO instruments.

The data products are archived upon the completion of the Level 1 processing and include profile products and calibration products.

6.6.4 CALIOP Retrieval Procedure for the Extinction Coefficient

The extinction coefficient determination requires several steps. The first one is the identification of the altitude of the scattering layer using the SIBYL (Selective Iterated Boundary Locator) algorithm (Vaughan et al. 2002; 2005; 2009). SIBYL scans lidar profiles throughout the troposphere and stratosphere, identifies regions of enhanced scattering, and records the location and simple characteristics of these atmospheric features.

Then the SCA (Scene Classification Algorithm (SCA (Liu et al. 2005; 2009), which is actually a set of algorithms, is used to classify these layers by type. It relies on a statistical analysis of observed parameters (Liu et al. 2004; 2009). In addition to being incorporated into the output data products, some of the type classifications performed by the SCA are also required by the hybrid extinction retrieval algorithm (HERA (Young et al. 2005; Young and Vaughan 2009)).

After SIBYL has found a region in a lidar profile, SCA first discriminates between cloud and aerosol and then determines the cloud or aerosol sub-type. Surface, subsurface and totally attenuated regions are also recorded in the Vertical

Feature Mask (VFM). If the region is a feature (cloud or aerosol) SCA then checks to see if the feature is lofted (if the molecular scattering signal is available both above and below the feature for feature layer transmittance retrieval). For a lofted feature, the SCA will derive the lidar ratio using the transmittance-constraint method (Fernald et al. 1972; Young 1995). For both lofted and non-lofted layers, the SCA will conduct a classification of feature types and assign a lidar ratio to the feature corresponding to the extinction processing in HERA. Note that if the feature is lofted and a lidar ratio can be derived using the transmittance method, the computed lidar ratio is selected; if the feature is non-lofted, a lidar ratio is selected based on the model corresponding to the identified feature type. Aerosol models were developed using data from AERONET (Omar et al. 2004; 2009).

For the feature classification, the SCA first determines whether the feature is tropospheric or stratospheric by checking the base altitude of the feature. The tropopause altitude is derived from ancillary data obtained from the Global Modelling and Assimilation Office (GMAO). If the feature base is lower than this altitude, the feature is classified as a tropospheric feature; otherwise, it is classified as a stratospheric feature. If a feature is tropospheric, further classifications (four algorithms) are conducted to sub-type the feature.

The SCA first determines whether a layer is cloud or aerosol, primarily using the layer mean value of the 532 nm attenuated backscatter coefficient, and the attenuated color ratio, which is the ratio of the mean attenuated backscatter coefficients measured at 1064 nm and 532 nm. If the layer is classified as cloud, the SCA will then determine whether it is an ice cloud or water cloud using the measured backscatter intensity and the depolarization ratio profiles, along with ancillary information such as layer height and temperature. The SCA will also use a combination of observed parameters and *a priori* information to select an appropriate extinction-to-backscatter ratio, or lidar ratio (Sa for aerosol layers, Sc for clouds), and multiple scattering function, $\eta(z)$ as defined by Platt (1973), required for retrieving extinction and optical depth. To be consistent, the lidar ratio and multiple scattering function must be based on the same underlying aerosol or cloud particle model. A constant value for the lidar ratio, as well as an array (as a function of range) for the multiple scattering function, are specified for each feature for later use by the optical property retrieval.

If a feature is classified as stratospheric, on the other hand, no further typing is performed. Stratospheric classifications may be included in a future data release. The classification criteria used for features in the stratosphere will differ somewhat from those for features found in the troposphere, though the same general classification approach can be used (Table 6.4).

6.7 Aerosol Remote Sensing from POLDER

The retrieval method used for the retrieval of aerosol properties from POLDER (see Appendix A) data depends on the type of surface below the aerosol layer. We distinguish two cases: land and ocean surfaces. A comprehensive description of

Table 6.4 CALIOP science products and uncertainties

Data Product	Measurement Capabilities and Uncertainties
Aerosols	
Height, thickness	For layers with $\tau > 0.005$
τ, $\sigma(z)$	40%[a]
Clouds	
Height	For layers with $\tau > 0.01$
Thickness	For layers with $\tau < 5$
τ, σ (z)	Within a factor of 2 for $\tau < 5$
Ice/water phase	Layer by layer
Ice cloud emissivity, ε	± 0.05 for $\varepsilon > 0.1$
Ice particle size	$\pm 50\%$ for $\varepsilon > 0.2$

τ – optical depth
σ (z) – profile of extinction cross-section
[a]assumes 30% uncertainty in backscatter-to-extinction ratio

aerosol retrieval from POLDER can be found at the ICARE web site (http://www-icare.univ-lille1.fr/parasol/?rubrique=overview_product).

6.7.1 POLDER Remote Sensing of Aerosols Over Ocean Surfaces

The method is based on a comparison between POLDER measurements and LUTs calculated for a set of aerosol models (size distribution, refractive index, optical thickness) for the POLDER observations. The inversion scheme mainly uses the normalized radiances in the 865 nm channel, where the ocean colour reflectance is zero, and in the 670 nm channel with a constant water reflectance of 0.001. The polarized Stokes parameters at 865 and 670 nm are also used to help to derive the best aerosol model.

The algorithm uses a bimodal aerosol model, which mixes a mode of small particles (S) and a mode of large particles (L) with respective optical thickness τ_S and τ_L, at 865 nm. A mode of small particles (S) consists of a lognormal size distribution of spherical particles with a given refractive index. A mode of large particles (L) consists of a mixture of spherical and non-spherical particles. The spherical particles are lognormally distributed and have a given refractive index. The non-spherical particles are described by the mean model given in Volten et al. (2001). The contributions of large spherical and non-spherical particles to the optical thickness at 865 nm are given by $\tau_{L\text{-}S}$ and $\tau_{L\text{-}NS}$, respectively. The large modes are a combination between spherical large particles and non-spherical particles with a mixture concentration varying from 0 to 1 in steps of 0.25 (Herman et al. 2005). The set of the refractive indices and modal radii of the small and large spherical particles used depends on the viewing conditions. Given a small and a large mode of particles with a total optical thickness $\tau = \tau_S + \tau_L$, the corresponding radiance L is calculated using the approximation of Wang and Gordon (1994). A similar approach can be applied for the normalized Stokes parameters Q and U.

LUTs of the radiances (865, 670 and 565 nm channels) and of the Stokes parameters Q and U (865 nm, 670 nm and blue channels) are calculated for different small modes, large modes of spherical particles and one non-spherical mode, for 11 aerosol optical thicknesses from $\tau = 0$ (molecular case) to $\tau = 2.6$ (extreme turbid atmosphere). These calculations are made for 21 solar angles (3° to 77°), 20 viewing angles (3°–73°) and 37 relative azimuth angles from 0° to 180° (steps of 5°). Computations are performed with a rough ocean surface (Cox and Munk 1954) and a wind speed of 5 m/s. The foam contribution is calculated following Koepke (1984) and a constant value of 0.22 for the foam reflectance. When the aerosol content is low, we only consider a fixed aerosol model for which the aerosol optical thickness is deduced.

The retrieval algorithm follows a two-step procedure. First the concentration of the mode in terms of optical thickness and total optical thickness is adjusted to fit the total radiance for a given combination of the small and the large mode. Then the directional Stokes parameters L, Q and U are interpolated in the LUT in the 865 and 670 nm channels. The difference (rms) between these simulations and the measurements are computed for each couple of modes: the minimum value (best fit) gives the aerosol model (modes and the fine mode concentration) and the corresponding optical thickness at 865 nm.

6.7.2 *POLDER Remote Sensing of Aerosols Over Land Surfaces*

Aerosol remote sensing over land from visible radiance measurements is more difficult than over the ocean because the surface reflectance is generally much greater than that for aerosol, except over dark surfaces (vegetation in the blue channel, lakes in near infrared). Airborne experiments (Deuzé et al. 1993) have shown that the relative contribution of the surface compared to the atmosphere is less important in polarized light than in total light. The aerosol algorithm over land is based on a best fit between polarized POLDER measurements and data simulated for different atmospheres including several aerosol models for different optical thickness, and ground surfaces conditions. The surface contribution depends on the type of surface (Nadal and Bréon 1999). The surface polarized reflectance is multiplied by an exponential factor corresponding to the attenuation through the atmosphere. The atmospheric term contribution to the measured signal is interpolated in LUTs computed using a successive order of scattering code (Lenoble et al. 2007).

Over land, ground based measurements show that the aerosol polarization mainly comes from the small-spherical particles (Vermeulen et al. 2000) with radii less than about 0.5 μm corresponding to the accumulation mode. So, the aerosol models used in the algorithm consist of lognormal size distributions of spherical particles: their characteristics are close to those for the oceanic small mode.

Knowing the super-pixel characteristics (altitude, surface classification, normalized difference vegetation index, NDVI), the surface polarized radiances are computed in the 865 and 670 nm channels, for the given viewing directions. For a given aerosol model, the aerosol optical thickness at 865 nm is adjusted with a root-mean square (rms) method to fit the polarized measurements. Then the rms are compared for the set of aerosol and the best solution corresponds to the minimum value, characterized by its Ångström exponent, and the associated optical thickness and a quality index indicating the confidence degree in the fit.

Fig. 6.2 shows examples of global monthly mean average aerosol products derived from POLDER, for May 2006.

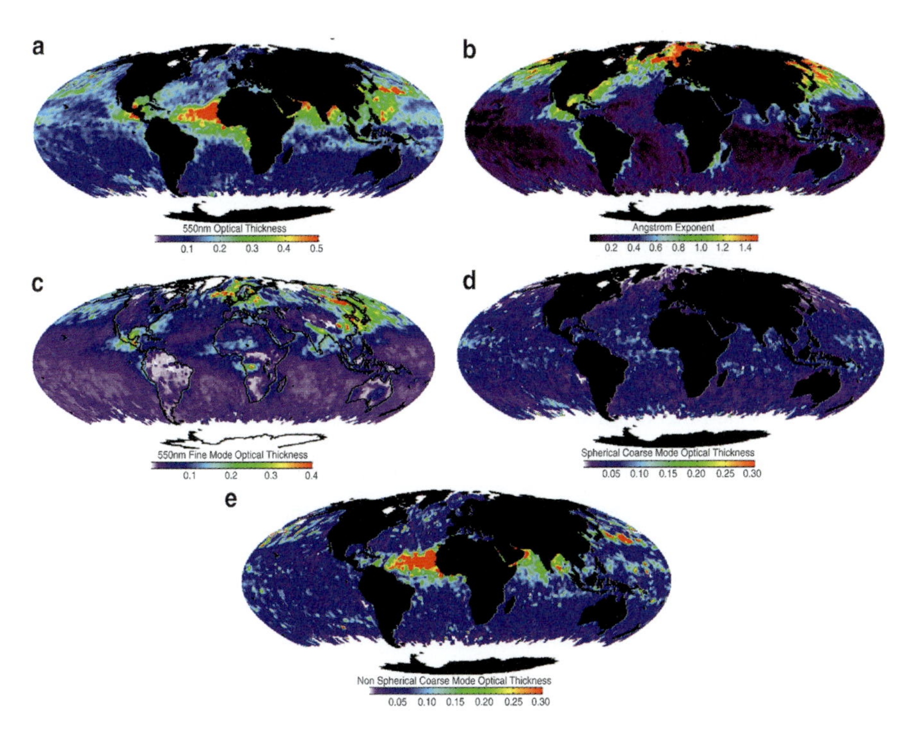

Fig. 6.2 Examples of aerosol information derived from POLDER for May 2006: (**a**) Total aerosol optical thickness; (**b**) Ångström exponent between 670 and 865 nm; (**c**) Fine mode aerosol optical thickness; (**d**) Spherical coarse mode aerosol; (**e**) Non spherical (dust) aerosol optical thickness (Credits: ICARE data center).

6.8 Retrieval of Aerosol Properties Using AATSR

The Advanced Along Track Scanning Radiometer (AATSR) on ENVISAT is the third in a series of ATSR instruments (see Section 6.1 and Appendix A). The data from these instruments have the potential of producing a 20 year aerosol record, starting in 1995. The AATSR dual view algorithm (ADV) has been used to provide aerosol data products over different areas such as the eastern part of the United States (Veefkind et al. 1998), Europe (Robles-Gonzalez et al. 2000; Veefkind et al. 2000), India and the Indian Ocean (Robles-Gonzalez et al. 2006) and Africa (Robles-Gonzalez and de Leeuw 2008). A first step to develop products was made as part of the ESA Data Users Programme project TEMIS (Tropospheric Emission Monitoring Internet Service, www.temis.nl). Other efforts are underway as part of the Globaerosol project (ESA Date Users Element: http://dup.esrin.esa.it/projects/summaryp64.asp) where the algorithm used is that developed by the University of Oxford (Thomas et al. 2007). A third algorithm has been developed by the University of Swansea (Grey et al. 2006) which it is expected to apply to the production of global data sets. The algorithm discussed here (Veefkind et al. 1998; Veefkind and de Leeuw 1998), was developed at TNO (Netherlands Organisation for Applied Scientific Research) and transferred to the University of Helsinki and the Finnish Meteorological Institute (FMI) in 2007, where it is further developed and applied to provide aerosol data products for use in scientific studies.

6.8.1 AATSR Characteristics

AATSR is a dual view imaging spectrometer with seven wavelength bands, four in the visible and NIR (0.555, 0.659, 0.865, and 1.6 μm) and three in the mid- and thermal-infrared (3.7, 11, and 12 μm). The resolution of the instrument is 1×1 km^2 at nadir view and the swath width is 512 km, resulting in a return time of approximately 3 days at mid-latitudes. AATSR has two cameras which provide a nadir view and a forward view at 55° incident angle to the surface. Together these two views allow for near-simultaneous observation of an area on the Earth's surface through two different atmospheric columns within a time interval of about 2 min.

AATSR was primarily designed for the measurement of water temperature but its characteristics render the instrument suitable for aerosol retrieval as well, in particular over land where the dual view is used to eliminate land surface effects on the radiation at the TOA (Veefkind et al. 1998). Over water a single view is used (Veefkind and de Leeuw 1998). Both algorithms include multiple scattering and the bi-directional reflectance of the surface. A drawback is the small swath of 512 km which results in a global coverage at the equator in approximately 5 days.

6.8.2 AATSR Retrieval Algorithm

The upwelling radiances measured at the top of the atmosphere in the visible and NIR channels are used for the retrieval of aerosol properties (AOD, the Ångström parameter and the mixing ratio of dominant aerosol classes). The 0.659 μm, 0.865 μm, and the 11 and 12 μm channels, are additionally used for cloud detection. When clouds are present, their radiance dominates the TOA signal and aerosol properties cannot be retrieved.

To discriminate between cloudy and cloud free areas over land, three tests are applied. These tests are based on the brightness temperature at 11 μm, the reflectance at 0.659 μm and the ratio of the reflectance at 0.865 μm to the reflectance at 0.659 μm. The reflectance of clouds is similar in these channels, whereas over land, the surface reflectance at 0.865 μm is generally higher than at 0.659 μm. Hence, over clouds the ratio should be around 1 and larger over land. Over water the effect is the opposite: a reflectance ratio threshold lower than 1 indicates cloud free pixel over water (Robles Gonzalez 2003).

The core of the algorithm is the derivation of aerosol optical properties for cloud-free pixels, which is accomplished by comparing the measured TOA reflectance to reflectances calculated by a radiative transfer model and stored in LUTs. The difference between the modelled and measured TOA reflectances at each suitable wavelength (0.555 μm, 0.659 μm and 1.600 μm over land) is determined for a range of aerosol mixtures and the error function for all three wavelengths together is minimized to determine the best fit for both the AOD and the aerosol mixing ratio. The radiative transfer model used is DAK (Double Adding KNMI) (de Haan et al. 1987; Stammes 2001) developed at the Royal Netherlands Meteorological Institute (KNMI). A variety of aerosol models and mixtures of these are used as appropriate for the region of interest.

Over dark surfaces, such as over open ocean or dark vegetation, the AOD can be determined directly using a single view (Veefkind and de Leeuw 1998). Over brighter surfaces, the effects of the surface reflection and the atmospheric reflection on the TOA reflectance need to be separated. This is accomplished by taking advantage of the two views provided by AATSR as described in Veefkind et al. (1998).

In the dual-view algorithm, it is assumed that k, the ratio between the surface reflectances in the nadir and the forward views, is independent of the wavelength (Flowerdew and Haigh 1995). Hence k can be determined at 1.6 μm, where the effect of aerosol is minimal and is ignored in the first retrieval step to obtain a first estimate of the AOD. In the next iteration this AOD is used as a first guess and the parameters are adjusted.

6.8.3 AATSR Products

The retrieval is made for single pixels (1 × 1 km at nadir). Results are evaluated by comparison with AERONET Sun Photometer data. Post-processing includes re-gridding to 10 × 10 km^2. In the distribution of AOD values both the highest and the lowest outliers are removed. This procedure is based on the assumption that no large gradients are expected in the aerosol concentrations on a scale of 10 km, unless intensive point sources are present.

Products are the spectral AOD for 0.555 μm, 0.659 μm and 1.6 μm (and at 0.865 μm over water) and the Ångström coefficient. The aerosol mixing ratio, which is the optimum mixture of two aerosol types is, in principle, also available (Robles-Gonzalez et al. 2006; Robles-Gonzalez and de Leeuw 2008). As an example, we show results obtained during UAE2 (the United Arab Emirates Unified

Aerosol Experiment) over the United Arab Emirates and the Persian Gulf (Reid et al. 2005) in Fig. 6.3 (AOD) and the comparison of retrieved AOD with simultaneous AERONET data over water and over land in Fig. 6.4 (de Leeuw et al. 2005).

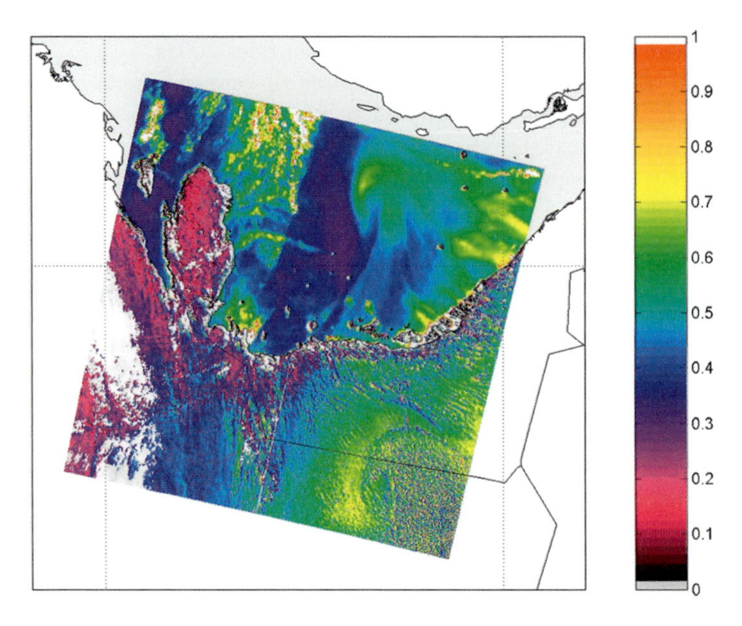

Fig. 6.3 AOD at 0.67 μm over the UAE area retrieved from AATSR data on 7th September 2004 (de Leeuw et al. 2005).

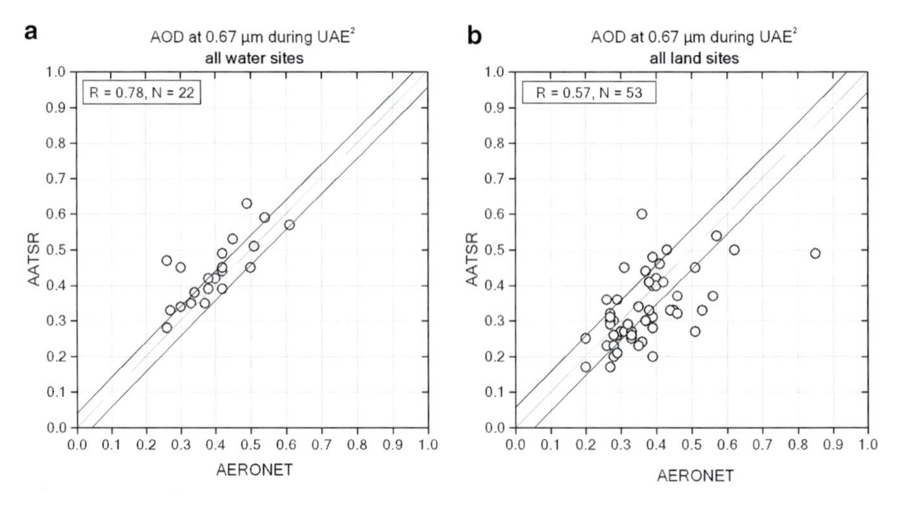

Fig. 6.4 Comparison of AOD derived from AATSR data during UAE2 with AERONET sun photometer data, for all water sites (**a**), and for all land sites (**b**) (de Leeuw et al. 2005).

6.9 Aerosol Remote Sensing from Aqua/MODIS

The first MODIS instrument was launched onboard the EOS-Terra satellite in December 1999 (see Appendix A). In May 2002, a second MODIS instrument was launched on board EOS-Aqua. The MODIS instruments measure sunlight reflected by the Earth's atmosphere and surface and emitted thermal radiation at 36 wavelengths. At least two observations for any place in Europe are obtained per day during daylight hours because the Terra and Aqua satellites cross Europe near 10:30 and 13:30 local solar time, respectively. There are two different algorithms to retrieve aerosol properties over land and over ocean. A comprehensive description of the MODIS aerosol retrieval algorithm can be found at the MODIS atmosphere group web site (http://modis-atmos.gsfc.nasa.gov/MOD04_L2/index.html). The MODIS aerosol products are from Collection 5.

6.9.1 *MODIS Remote Sensing of Aerosols Over Ocean Surfaces*

The first step in the Ocean Algorithm is to organize the reflectance from the six wavelengths used in the procedure (0.55 μm, 0.66 μm, 0.86 μm, 1.24 μm, 1.6 μm and 2.13 μm) into 10 km × 10 km boxes of 20 × 20 pixels at 500 m resolution. The Ocean algorithm requires all 400 pixels in the box to be identified as ocean pixels by the MYD35 mask, which helps minimize problems introduced by shallow water near the coasts. If any land is encountered, the entire box is left for the land algorithm. The major issue with the retrieval of aerosol over ocean is the contamination by bright targets, i.e. either clouds or specular reflection on the water surface. The specular reflection (glint) depends on the geometry of observation. The Ocean Algorithm is designed to retrieve AOT for glint angle only over the dark ocean, away from the glint, i.e. when the glint angle is over 40° and in cloud-free pixels. Moreover, the brightest and darkest 25% of the pixels (reflectance values at 0.87 μm) are discarded to prevent contamination by residual clouds and cloud shadows.

The MODIS aerosol retrieval algorithm is based on LUT inversion. The top-of-the atmosphere radiances for the aerosol channels are computed for several viewing geometries, aerosol optical thicknesses and types. Computations are performed for 15 zenith view angles ($\theta_v = 1.5°$ to 88 by steps of 6°), 15 azimuth angles ($\varphi = 0°$–180° by steps of 12°) and 7 solar zenith angles ($\theta_s = 1.5°$, 12°, 24°, 36°, 48°, 54°, 60°, 66° and 72°). Several values of aerosol total loading are considered for each mode and described by the optical thickness at 0.55 μm. Extreme conditions included in the LUT are pure molecular atmosphere ($\tau = 0.0$) and very turbid atmosphere ($\tau = 2.0$). Three intermediate values are considered ($\tau = 0.2$, 0.5, 1.0), and a linear interpolation between these values is applied. To account for effects of different aerosol types on the radiance at the top of the atmosphere, they are assumed to be an external mixture for which the total radiance can be approximated by the weighted average of the radiances of each individual mode for the same optical thickness (Wang and Gordon 1994). The set of aerosol models is composed of four small modes and five large modes. The goal is to retrieve the ratio η of the small mode optical thickness to the total optical thickness for the set of the small and large modes giving the best fit between observations and measurements. The aerosol optical thickness at 0.550 μm is derived as a by-product. The selection of the aerosol models is performed by minimizing the rms difference between observed and modeled radiances.

The MODIS algorithm is presented by Remer et al. (2005). The flowchart in Fig. 6.5 illustrates the retrieval of aerosol properties over ocean surface.

6.9.2 MODIS Remote Sensing of Aerosols Over Land

Like the ocean algorithm, the land algorithm for MODIS data is an inversion, but it takes only three *nearly* independent observations of spectral reflectance (0.47, 0.66 and 2.1 μm) to retrieve three *nearly* independent pieces of information. These include total AOT at 0.55 μm, fine (model) weighting at 0.55 μm, and the surface reflectance at 2.1 μm. Like the ocean algorithm, the land algorithm is based on an LUT approach, i.e. radiative transfer calculations are pre-computed for a set of aerosol and surface parameters and compared with the observed radiation field. The algorithm assumes that one fine-dominated aerosol model and one coarse dominated aerosol model (each may be comprised of multiple lognormal modes) can be combined with proper weightings to represent the ambient aerosol properties over the target. Spectral reflectance from the LUT is compared with MODIS-measured spectral reflectance to find the best match. This best fit is the solution to the inversion. The processing of radiances can be described by the flowchart in Fig. 6.6 (Remer et al. 2005).

For Collection 5, Levy et al. (2007b) have replaced the surface reflectance assumption, the aerosol models and the LUT. The algorithm performs a simultaneous inversion of two visible (0.47 and 0.66 μm) and one shortwave infrared (2.12 μm) channel, making use of the coarse aerosol information content in the shortwave infrared.

Fig. 6.7 shows examples of global monthly means derived from Aqua/MODIS: AOD and fine mode ratio of the AOD, from Aqua/MODIS at 550 nm.

6.10 Aerosol Properties from OMI

OMI (Appendix A) is an imaging UV-vis solar backscatter spectrometer. It is a Dutch–Finnish instrument onboard the NASA satellite EOS-Aura that was launched in July 2004. Earth radiance spectra are measured simultaneously on a 2,600 km wide swath and global coverage is achieved on a daily basis. The nadir pixel size is

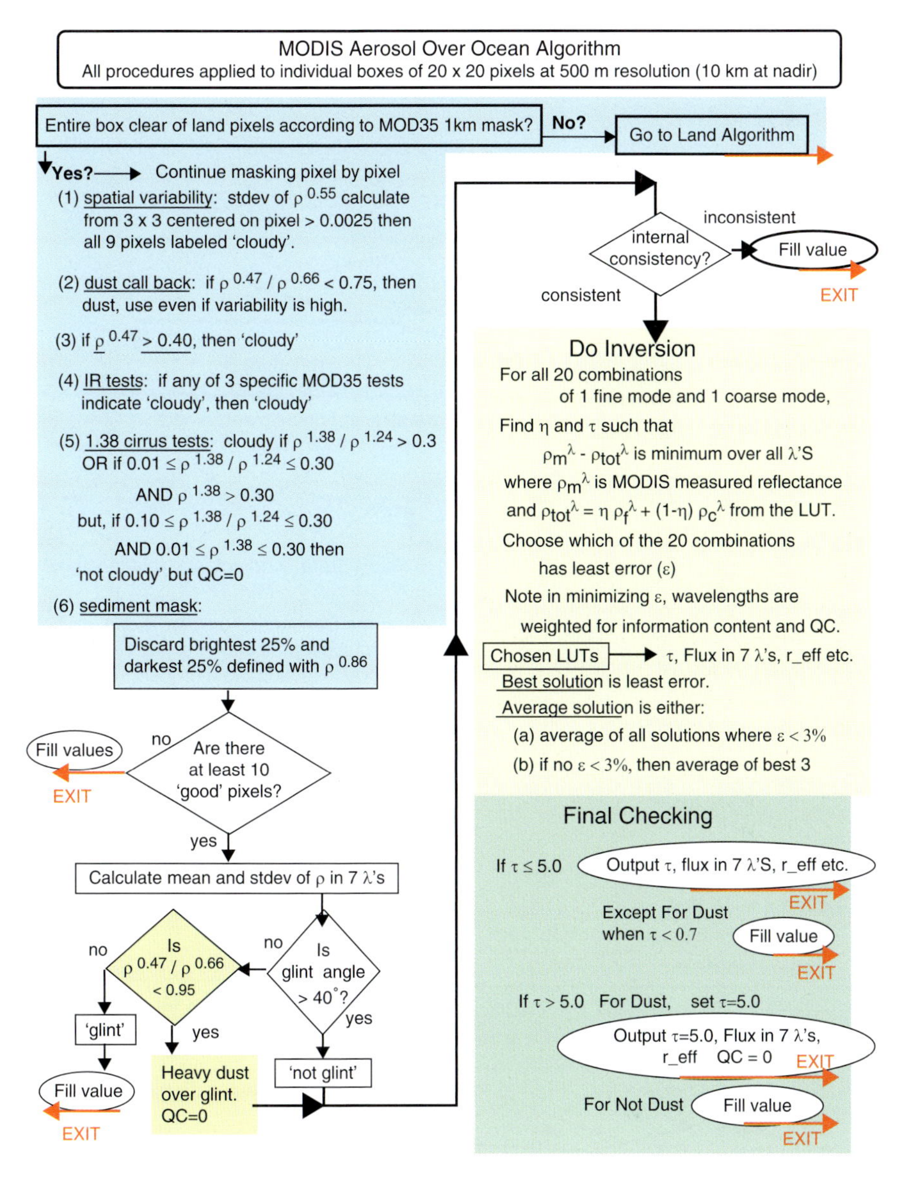

Fig. 6.5 Flowchart illustrating the retrieval of aerosol properties over ocean surfaces. Algorithm for remote sensing of tropospheric aerosol from MODIS Collection 5 (Remer et al. 2005).

13×24 km^2. Two aerosol products are derived from OMI measurements. The OMAERUV product (near-UV algorithm) is based on reflectance measurements at two wavelengths in the near-UV. It provides AOD and AAOD (Absorbing Aerosol Optical Depth) and the aerosol index. AAOD is sensitive to elevated absorbing

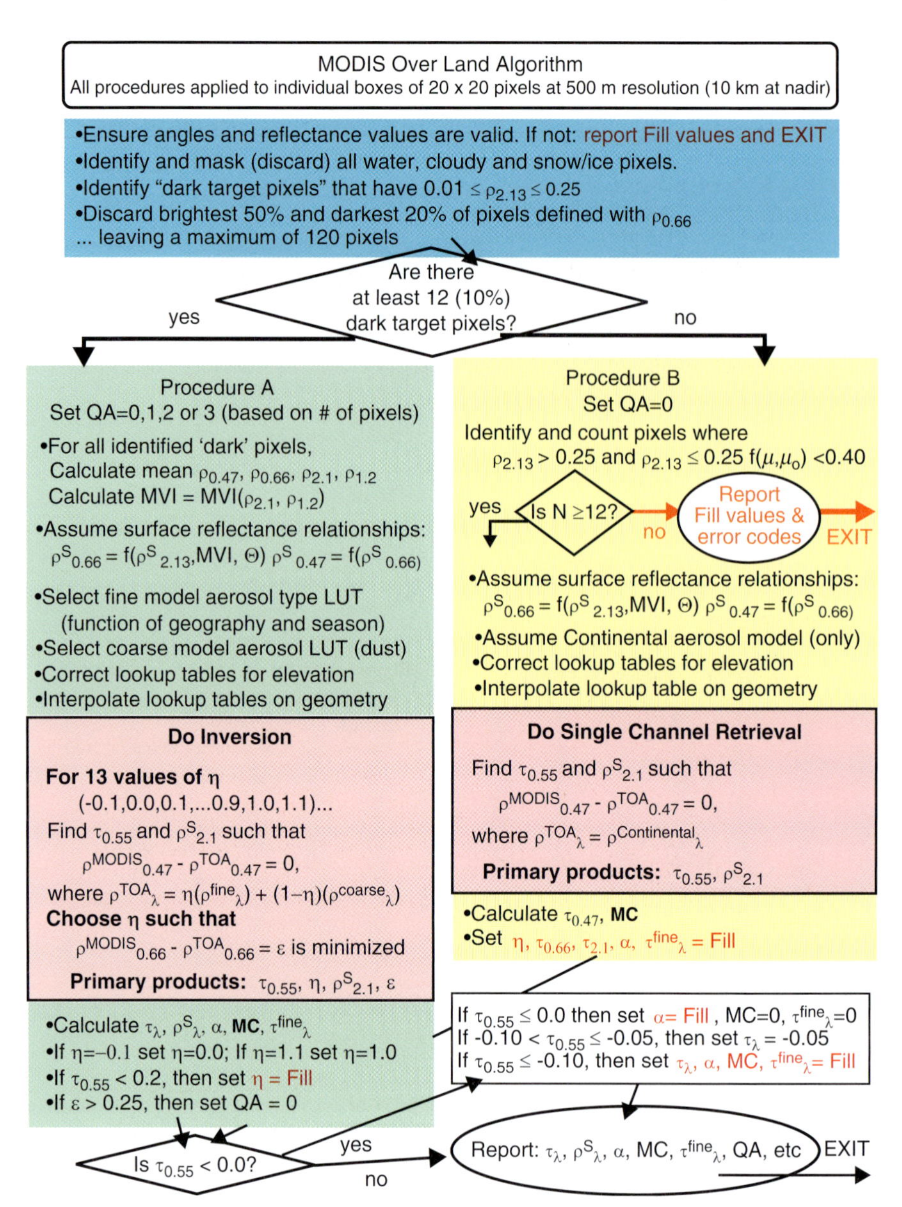

Fig. 6.6 Flowchart illustrating the retrieval of aerosol properties over land. Algorithm for remote sensing of tropospheric aerosol from MODIS Collection 5 (Remer et al. 2005].

aerosols and, in contrast to AOD, can be determined over bright surfaces and clouds. The OMAERO product (multi-wavelength algorithm) is based on measurements in the UV-VIS wavelength region and is explained in the following section.

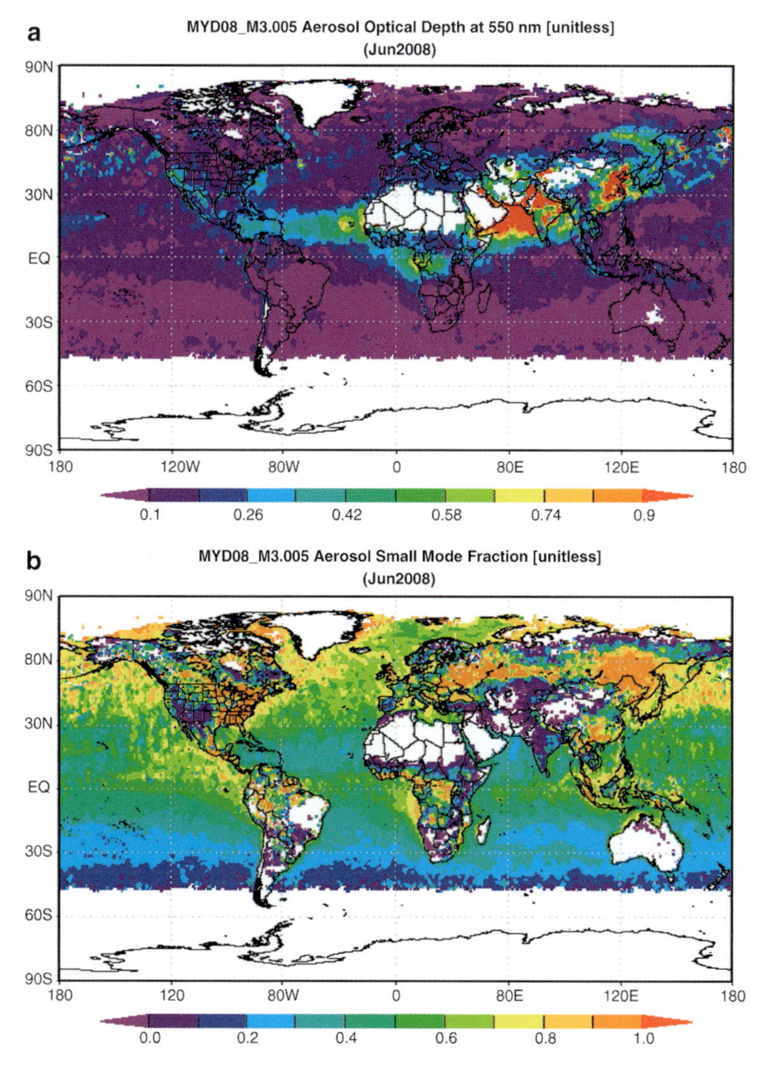

Fig. 6.7 Global distributions of aerosol optical depth (**a**), and small mode fraction of the aerosol optical depth (**b**), retrieved from Aqua/MODIS data at 550 nm, aggregated for the month of June in 2008 (1×1 degree resolution, color scale between 0 and 1) Figure: produced with the Giovanni online data system, developed and maintained by the NASA GES DISC.

6.10.1 Properties from OMI Using the Multi-Wavelength Algorithm

A multi-wavelength aerosol algorithm (OMAERO product) has been developed at KNMI to retrieve aerosol properties from OMI spectral reflectance measurements in the UV-Vis wavelength region (Torres et al. 2007). The AOD is retrieved and a best fitting aerosol type is determined. The single-scattering albedo, the layer height and the size distribution associated with the best fitting aerosol type are also provided. The multi-wavelength algorithm uses the reflectance spectrum in the near UV and the visible wavelength range.

A simulation study on the aerosol information content of OMI spectral reflectance measurements shows that OMI measurements contain two to four degrees of freedom of signal (Veihelmann et al. 2007). Including the near UV enhances the capability of the retrieval to distinguish between weakly absorbing and strongly absorbing aerosol types. Therefore, the OMAERO product can provide additional information about the aerosol type compared with other aerosol products from sensors that do not include the near UV, such as MODIS, MISR or POLDER.

The multi-wavelength algorithm uses a set of aerosol models, including models for desert dust, biomass burning, volcanic and weakly absorbing aerosol. All aerosol types are assumed to be spherical, except desert dust. The non-sphericity of desert dust is taken into account using the spheroidal shape approximation assuming the shape distribution that is used in AERONET retrievals for non-spherical aerosol types. Accounting for particle non-sphericity yields a significant improvement of the retrieved optical thickness when desert dust aerosol is present.

6.10.2 Status of the OMAERO Product

For ocean scenes, global AOD data from the OMAERO product have been compared with other products for June 2006. The comparison with quality assured data (MODIS QA flag = 3) from the MODIS standard product (Fig. 6.8, left) shows excellent agreement between the datasets. Note the use of this data flag excludes many scenes that were not flagged by the OMI cloud screening scheme. A comparison with POLDER data shows good agreement between the datasets (Fig. 6.8, right).

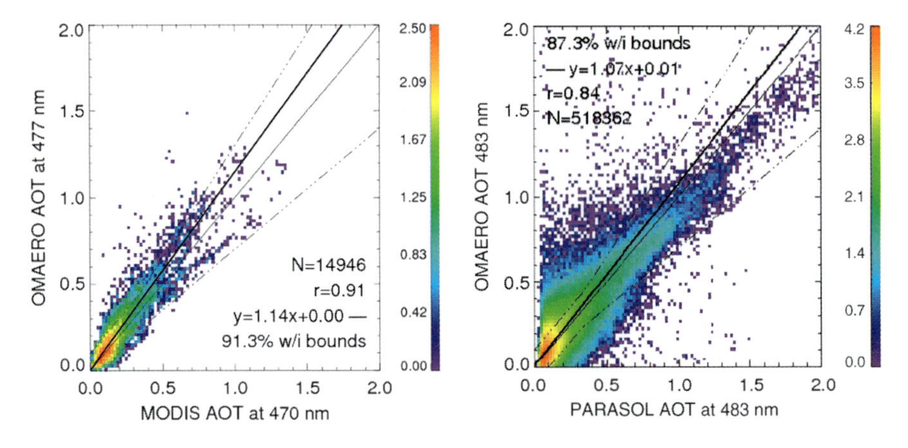

Fig. 6.8 AOD from the OMAERO product compared with quality assured data from the MODIS standard product (*left*) and with quality parameter filtered data from POLDER (*right*).

Curier et al. (2008) compared OMAERO and MODIS data over Europe and adjacent oceans and report correlation coefficients between 0.76 and 0.81 for scenes over ocean and between 0.59 and 0.70 for scenes over land. Current OMAERO data over land may be affected by errors in the surface albedo climatology. Reprocessing with an improved surface albedo climatology is envisaged. The main limitation of OMAERO aerosol data is cloud contamination. Cloud-contaminated scenes cannot be screened out without misclassifying scenes with large aerosol loadings as cloudy. The impact of cloud contamination is being investigated.

6.11 Retrieval of Aerosol Properties Using MERIS

MERIS (Appendix A) on ENVISAT is an instrument designed to measure ocean colour. Other MERIS products include atmospheric properties such as information about clouds and aerosols. The official ESA aerosol product uses the algorithm developed by Santer et al. (1999; 2000). The results have been evaluated by Höller et al. (2007). Non-operational scientific algorithms have been developed; an example is the Bremen AErosol Retrieval (BAER) algorithm discussed below.

In BAER (von Hoyningen-Huene et al. 2003; 2006), the Rayleigh path reflectance is calculated using a radiative transfer model, with the Rayleigh optical thickness for the required wavelengths (Buchholz 1995), the Rayleigh phase function, the illumination and viewing geometry and the actual temperature and pressure conditions at the surface. The barometric height equation and the dry adiabatic lapse rate are used, together with a digital elevation model (GTOPO30), to correct the Rayleigh path reflectance and air mass factors to the actual conditions within the satellite scene.

The correction for surface effects requires the application of a surface model, which can be adapted to the spectral and geometric conditions of the satellite scene. In BAER, a bi-directional reflection function (BRDF), normalized to the nadir position, is used that is based on the Raman-Pinty-Verstraete model (RPV) (Maignan et al 2004). The shape of the BRDF can be described by three parameters which depend on surface type. Currently BAER uses one set of BRDF parameters for the whole scene, ignoring regional variations.

The spectral properties and the magnitude of the surface reflectance are described by a bi-directional scattering distribution function, BSDF. Two basic spectra for "green vegetation" and "bare soil" are used, obtained from averages of the LACE-98 experiment, combined with measurement of the AVIRIS instrument to cover the whole spectral range. The green vegetation and bare soil spectra are linearly mixed, using the vegetation fraction taken from the atmospherically corrected NDVI of the scene, and a scaling factor is used to adapt the spectrum to the radiation conditions in the scene. An initial estimate for the aerosol reflectance is provided by assuming a "black" surface. The accurate estimation of the surface term by the spectral surface model is important, because a deviation of the surface reflectance of 0.01 leads to a change in AOD of about 0.1, depending on aerosol type.

The AOD retrieval is based on a LUT approach. LUTs are derived from radiative transfer modelling for a given BSDF, aerosol phase function, single scattering albedo, for AOD varying between 0 and 2.5, and Rayleigh scattering. The LUT is calculated for each of the MERIS wavelengths and for a range of viewing geometries. The phase function and single scattering albedo are obtained either using an aerosol data base like OPAC (Hess et al. 1998) or from the data base determined from AERONET sun-/sky radiometer measurements (Dubovik et al. 2000; 2002; 2006), or from campaigns such as ACE-2, LACE-98, SAMUM (von Hoyningen-Huene and Posse 1997; von Hoyningen-Huene et al. 1999a; 1999b; 2003; 2008; Silva et al. 2002).

The phase functions are normalized to 1 (in 1/sr). The spectral change in the phase function for wavelengths between 0.412 μm and 0.670 μm is neglected for retrieval over land, because, experimentally, no significant variation was observed. Also the spectral change in single scattering albedo is neglected because it is small within the spectral range. The set of selectable LUTs in BAER is presented in Table 6.5.

The results obtained with LUT No. 6, LACE-98, non-absorbing aerosol, compare well with AERONET AOD values. In cases with strong pollution high AOD tends to be underestimated by up to 20% with this LUT. In such cases a LUT with more absorbing aerosol needs to be selected.

The AOD is determined by minimizing the sum of the deviations between each of the individual estimates $\tau_{Aer}(\lambda_i)$ for channel i and the value $\overline{\tau_{Aer}}(\lambda_i)$ provided by the Ångström power law for the given aerosol type:

$$RMSD = \frac{1}{N}\sqrt{\Sigma_{i=1}^{N}\left(\tau_{Aer}(\lambda_i) - \overline{\tau_{Aer}}(\lambda_i)\right)^2} \tag{6.13}$$

The minimization is achieved by modifying the surface reflectance in an iterative scheme running over k:

$$\rho_{Surf,k}(\lambda) = \rho_{Surf,k-1}(\lambda) \cdot w(\lambda) \cdot \left(1 - \frac{\tau_{Aer,k}(\lambda) - \overline{\tau_{Aer,k}}(\lambda)}{\tau_{Aer,k}(\lambda)}\right) \tag{6.14}$$

until RMSD < 0.005 is reached.

The Ångström parameters are calculated by a least square fit of the retrieved AOD for all used spectral channels, i.e. channels 1–7 over land, instead of the commonly applied two-channel-approaches of Eck et al. (1999). The use of

Table 6.5 Look-up-tables in use and their characteristics

LUT number	Aerosol model	Phase function	Single scattering albedo
1	Clean marine	OPAC, clean marine	1.0
2	Clean continental	OPAC, clean continental	0.975
3	Average continental	OPAC, average continental	0.928
4	ACE-2, marine	ACE-2, experimental	1.0
5	LACE-98, absorbing	LACE-98, experimental	0.98
6	LACE-98, non-abs.	LACE-98, experimental	1.0
7	Desert	SAMUM, experimental	0.97

the Ångström power law ensures a sufficient smoothness of the AOD spectrum. In the first iteration the spectral slope α is determined from the retrieved value of the AOD. It is constrained to $-0.5 \leq \alpha \leq 2.0$. If the retrieved spectral slope is outside this limit, it is set to the climatological average of $\alpha = 1.3$

For the first iteration, the surface reflectance obtained from the BRDF data base is used to determine the AOD. Then the spectral surface reflectance is modified, Eq. 6.14, depending on deviations of the smoothed spectral AOD. Using the modified surface reflectance, an improved AOD is obtained. This procedure is repeated, until RMSD has reached its defined minimum. The retrieval uses 7 wavebands, with different weighing factors. The main steps of BAER described in the previous sections are summarized in Fig. 6.9.

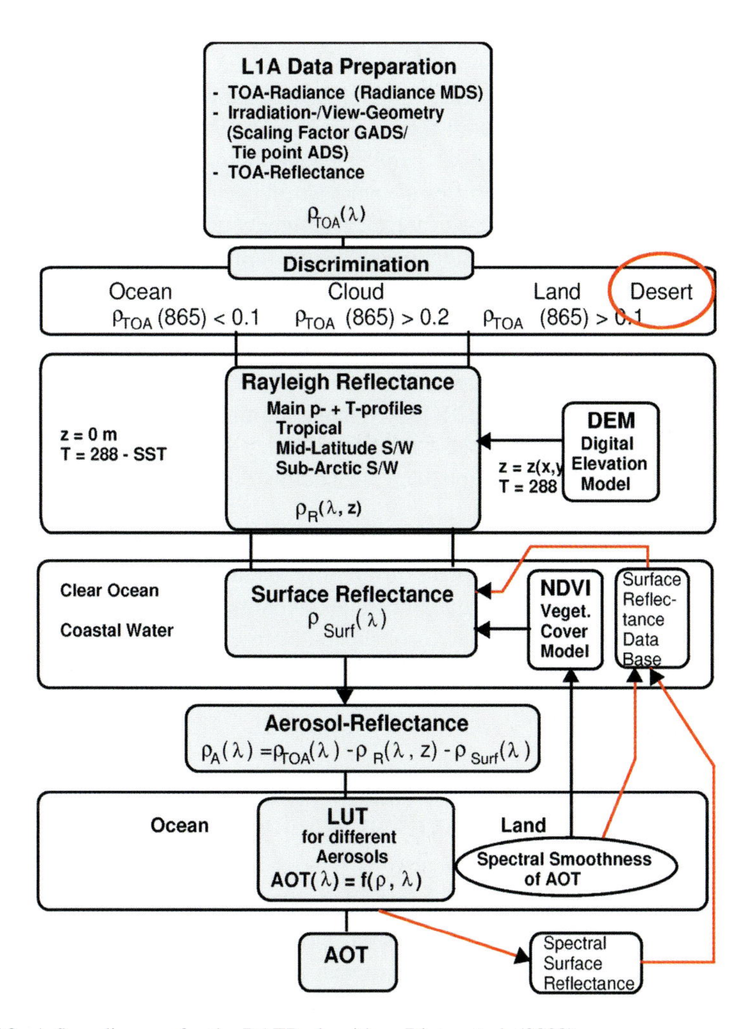

Fig. 6.9 A flow diagram for the BAER algorithm; Dinter et al. (2009).

6.12 Validation

The need for validation is discussed in Chapter 7. For algorithm development it is important to test different schemes using different approaches in order to deal with the surface reflectance, especially over land, and using different aerosol models. This is clearly illustrated by the upgrade of the MODIS algorithm used for Collection 5 (C005) processing where both these aspects were considerably revised. The validation of the treatment of the surface reflectance is a difficult issue and data on the surface reflectance from different instruments often are significantly different due to pixel size and measurement methodology. Usually only the end product can be evaluated, i.e. the AOD which is compared with ground-based measurements available from the AERONET sun photometer network (Holben et al. 1998) with some hundreds of stations over land having similar instruments (CIMEL sun photometers of different types), or lidar networks (see http://www.earlinet.org/) to obtain information on the vertical structure (see Chapter 7).

AERONET was extended with the maritime aerosol network (MAN) in 2008 (Smirnov et al. 2009). MAN utilizes hand held sun photometers (Microtops) which are deployed during research cruises over the world oceans. These are particularly important because they provide surface measurements over the open-ocean where no other data are available for validation and evaluation of satellite data. It is known that a variety of aerosol conditions may occur over the open ocean, for example the transport of dust, transport of biomass burning and pollutants, or a very clean maritime atmosphere with very low AOD close to the measurement uncertainty. Several other sun photometer networks are part of AERONET such as the European PHOTONS and the Canadian AEROCAN, and complimentary networks are maintained such as GAW-PFR (Global Atmosphere Watch – Precision Filter Radiometer) in remote locations. In addition there are a number of national networks making direct sun measurements.

Ground-based *in situ* measurements are performed at many sites world-wide, but they are poorly coordinated and with different procedures and protocols. The EU-funded EUSAAR project (European Supersites for Atmospheric Aerosol Research) aims at harmonizing 20 selected aerosol supersites in Europe measuring chemical, physical and optical properties, and making the data available through a specialized data center. Quality control is an important issue in EUSAAR.

6.13 Air Quality: Using AOD to Monitor PM2.5 in the Netherlands

Satellite measurements provide full spatial coverage of the Earth and are – in principle – consistent for the whole European region. So, although they are less precise than *in situ* observations, satellite measurements may be useful to improve the insight in regional PM distributions and so be complementary to ground-based measurements. The key parameter to derive PM distributions from satellite data is the AOD. Empirical relations between AOD and PM10 or PM2.5 measurements have been reported for different parts of the world (Wang and Christopher 2003; Hutchison 2003; Engel-Cox et al. 2004; 2006; Al-Saadi et al. 2005; Schaap et al. 2008) (and references therein). For example, promising correlations have been found between time-series of AOD and PM2.5 for many stations in eastern and midwest U.S. Other stations, however, particularly in the western US, show hardly any correlation (Wang and Christopher 2003; Hutchison 2003; Engel-Cox et al. 2004). Variations in local meteorological conditions, occurrence of multiple aerosol layers, and variations in aerosol chemical composition are likely to play an important role in determining the strengths of such correlations. To acquire estimates of PM2.5 distributions, one depends critically on an established relation between AOD and ground level PM2.5.

As an example to illustrate the use of satellite data to determine the spatial distribution of PM2.5, a study is described which was aimed at determining an empirical relationship between AOD and PM2.5 for the Netherlands from experimental data, and to explore the ability of mapping PM2.5 over the Netherlands using satellite-retrieved AOD data. The satellite data used are from MODIS because data are available from both the Terra over-flight in the morning (10:30 local time) and the Aqua over-flight in the afternoon (13:30 local time). Thus more data are available than from other instruments with a single daily overpass while the two MODIS over-flights also cover part of the diurnal cycle. In principle any satellite AOD data set could be used as well to provide daily AOD/PM2.5 maps (Kacenelenbogen et al. 2006) using POLDER data.

A field study to establish an empirical PM2.5 – AOD Relationship

To address the relation between AOD and PM2.5, a study was set-up to monitor PM2.5 between 1st August 2006, and 31st May 2007, at the Cabauw experimental site for atmospheric research (CESAR) (51.97°N, 4.93°E). The AOD measurements at Cabauw are made using a CIMEL sun photometer following the AERONET protocol (Holben et al. 2001). Measurements are made every 15 min and transmitted to the AERONET data base in near-real time by satellite. Initial cloud clearing takes place in a first processing step to provide Level 1.5 AOD data which were used in this study. The Level 2 data (pre- and post-field calibration applied, automatically cloud cleared and manually inspected) are updated on an annual basis. Level 2 data were used in this study for a sensitivity analysis. As semi-volatile ammonium nitrate levels are high in the Netherlands (Schaap et al. 2002), PM2.5 was monitored using a Tapered Element Oscillating Microbalance with Filter Dynamics Measurement System (TEOM-FDMS) to avoid losses. In addition to these core instruments, the RIVM aerosol backscatter lidar (Apituley et al. 2000) was extensively used for cloud detection. Three algorithms were used to detect clouds, i.e. a threshold method, a detection method for strong modulations in the lidar signal (Pal et al. 1992) and a method based on retrieval of the backscatter profile from the lidar data (Klett 1985) and setting a threshold at the scattering level of clouds.

6.13.1 Establishing an AOD-PM2.5 Relationship

As a first assessment, all AOD data were plotted against the collocated PM2.5 data in a scatter diagram (Fig. 6.10). At first glance, there seems to be a large variability and no indication for a well defined relation between the variables. Cases were selected with AOD values lower than 1 and PM2.5 concentrations smaller than 100 μg/m^3; some of the points outside these limits are clear outliers. A fit through these data shows that only 13% of the variability in PM2.5 is explained by AOD. Fig. 6.11 shows the complete time series of AOD and PM2.5. PM2.5 is given as a grey line and the AOD data are superimposed as diamonds. The time series for August–September (upper panel) shows that during August the PM2.5 concentrations were relatively low while often AOD was high. The two data sets are virtually uncorrelated during August. This is in contrast to the situation in September and in the spring of 2007 when the AOD and PM2.5 data track each other very well (R^2 ~0.6). The later data illustrate the potential to define situations in which the AOD may be used to estimate PM2.5 levels. However, the statistical analysis presented above (Fig. 6.10) was hampered by the occurrence of cloud contaminated AOD data. The lidar-based cloud screening of the L1.5 data strongly improved the correlation as shown in Fig. 6.11. About 50% of the data points was rejected, possibly because of the occurrence of broken cloud conditions and/or the presence of optically thin high cirrus clouds. The stricter cloud screening substantially improves the correlation between the AOD and PM2.5, resulting in a correlation coefficient $R^2 = 0.41$.

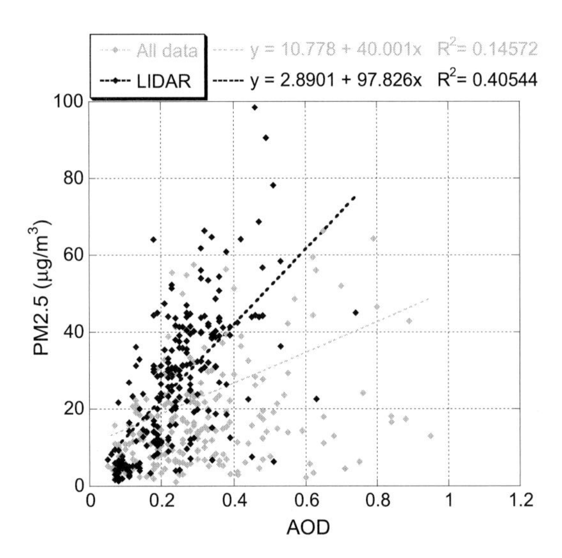

Fig. 6.10 Correlation between PM2.5 and sun photometer AOD at Cabauw before screening (*grey dots*) and after screening (*black dots*) for residual cloud contamination in the Aeronet AOD measurements.

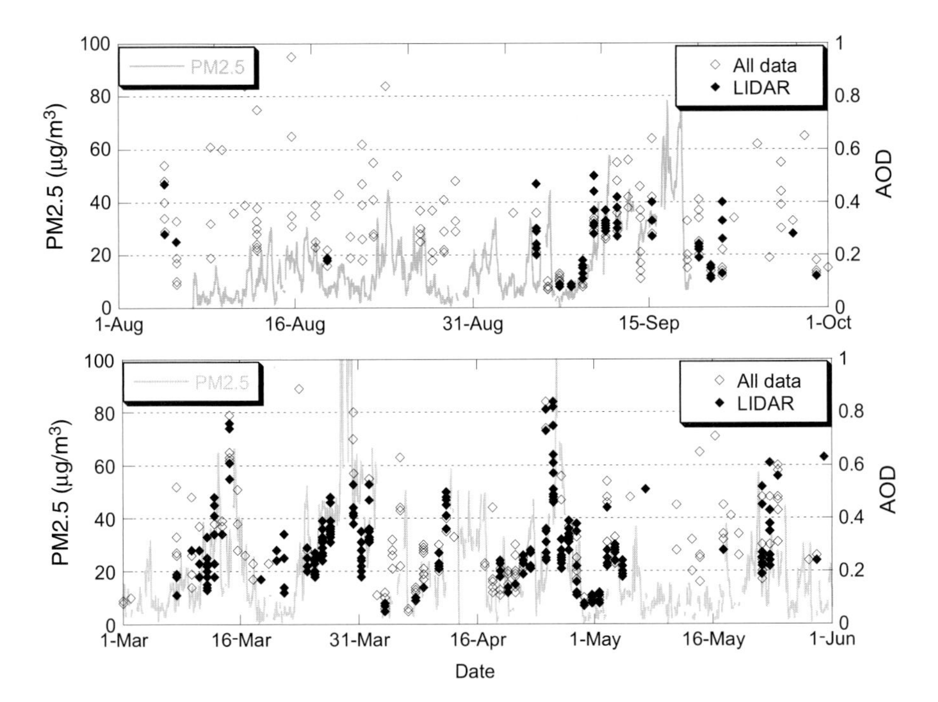

Fig. 6.11 Time series for PM2.5 and AOD for the period August–September (*upper panel*) and March–May (*lower panel*). The AERONET L1.5 AOD data are differentiated between data that did (*filled diamond*) and did not (*open diamonds*) pass our additional cloud screening.

It is noted that the problem of cloud contamination of the AERONET data does not occur when L2 data are used. However, L2 data are not available in NRT as are L1.5 data.

While the ground-based measurements of AOD and PM2.5 are obtained throughout the day, satellite observations of AOD provide "snap-shots" only during their overpass (typically once per day). In order to apply a relation between AOD and PM2.5 to AOD measurements from satellites, an investigation was made to determine whether the AOD-PM2.5 relation changes when the data was limited to the time window in which the MODIS instruments pass over Cabauw. MODIS/TERRA has their overpass in the late morning and MODIS/Aqua in the early afternoon. The effect of constraining the time window is illustrated in Fig. 6.12 for the period between 11:00 and 15:00 UTC. The overall reduction of available data points can be seen. Also, it was observed that a high percentage of the points on the edges of the data cloud are data points associated with early morning or late afternoon measurements. This is especially true for the points with low PM2.5 and moderately high AOD values. Strikingly, the explained variability increases when the time window is centered around midday. For that case, the following relationship between PM2.5 (in μg/m^3) and AOD was derived: PM2.5 = 124.5 AOD – 0.34 with $R^2 = 0.57$.

The relationship between AOD and PM observed at Cabauw may not apply to other areas, because of the spatial variation of aerosol sources and the subsequent

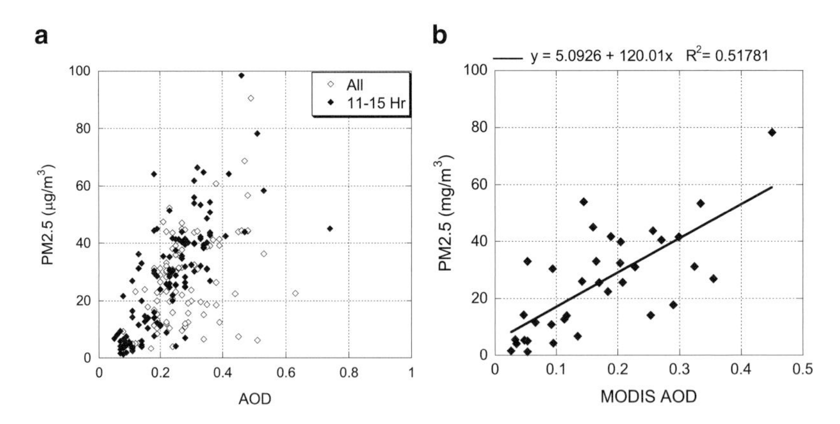

Fig. 6.12 (**a**) The variation of PM2.5 with AOD for all data and those between 11:00 and 15:00 h (Fit: PM2.5 = 124.5 AOD – 0.34); (**b**) The measured PM2.5 concentration as function of MODIS AOD. Note the difference in horizontal scales.

changes in aerosol properties due to chemical and physical processes which also affect the optical properties.

6.13.2 Application of the AOD-PM2.5 Relationship to MODIS Data

MODIS AOD was used to explore the application of the AOD-PM2.5 relation over the Netherlands. Figure 6.12b shows the variability of PM2.5 as function of MODIS AOD from the measurements at Cabauw, which is described by PM2.5 = 120 AOD + 5.1 μg/m^3. This fit explains 52% of the variability in PM2.5. The relation for the MODIS AOD is very similar to that determined with sun photometer data. It is noted that a systematic bias of 0.05 was identified in MODIS AOD compared to the sun photometer data, which explains the cut-off at 5.1 μg/m^3.

To derive a first estimate of the PM2.5 concentration field over the Netherlands based on MODIS data only, the AOD-PM2.5 relation was applied to the annual composite map of MODIS AOD. Results are presented in Fig. 6.13. High AOD values retrieved along the Dutch coast are likely to be an artifact. At land/water boundaries, application of the land algorithm to patches of sea often leads to high AOD values (Chu et al. 2002), while application of the ocean algorithm over coastal waters with suspended sediments (as in the North Sea) often gives rise to high AOD retrievals (Robles-Gonzalez et al. 2000; Chu et al. 2002; Ichoku et al. 2005). Hence, all pixels in which surface water covers more than 10% of the 10 $\times$ 10 km^2 grid were masked. Figure 6.13 shows that MODIS-AOD derived PM2.5 levels over the Netherlands are between 22 and 30 μg/m^3. The lowest PM2.5 concentrations, slightly over 11 μg/m^3, are mapped over the Ardennes and east of the Ruhr area. In the Ruhr area, the resulting PM2.5 levels are between 30 and 42 μg/m^3. Strikingly, the highest PM2.5 levels are mapped over south western Belgium and

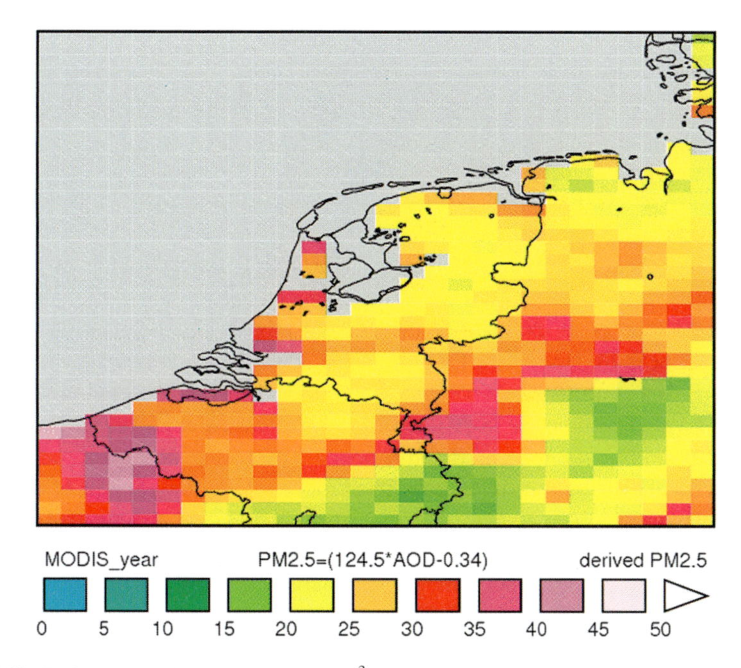

Fig. 6.13 Estimated PM2.5 distribution (μg/m^3) over the Netherlands and its direct surroundings, during situations with predominantly easterly and southerly flow, for which satellite retrievals are available.

northern France, in the region of Lille. Some features of the spatial distribution do not appear to be very realistic: the high values of PM2.5 around Lille and near the northern coast of the Netherlands for example. This might be caused by spatially varying systematic errors in the MODIS AOD data, which could be due to unaccounted variability in surface reflectance such as mixed land/water pixels. Because of the uncertainties in current satellite data of AOD, it is not expected that better PM2.5 maps can be constructed for the Netherlands based on satellite data alone without accounting for atmospheric processes. This conclusion may be specific for the Netherlands while other parts of continental Europe are less affected by the presence of mixed land/water pixels. Furthermore, the atmospheric boundary layer inland is often better mixed than near the coast, resulting in a more homogenous vertical distribution. Satellite measurements of AOD have added-value regarding the *temporal* variation of PM and can be useful in detecting trends and hot spots.

6.14 Application to Climate: Aerosol Direct Radiative Forcing

The aerosol direct effect is estimated in this work using off-line radiative transfer simulations. The necessary model input is based on global data-sets of monthly averages. In this particular approach, measurements and/or measurement-tied data, of sufficient accuracy, are preferred to model simulations. For aerosol, single scattering,

Table 6.6 Annual global averages for aerosol direct forcing

Aerosol forcing in W/m^2	Total (solar + IR)			Total (solar only)			Anthropogenic		
	Clr-sky	All-sky	Cld-eff	Clr-sky	All-sky	Cld-eff	Clr-sky	All-sky	Cld-eff
TOA	−2.7	−1.0	1.7	−3.8	−1.6	2.2	−0.7	−0.2	0.5
Atmosphere	3.4	3.7	0.3	4.3	4.2	−0.1	2.0	1.9	−0.1
Surface	−6.1	−4.7	1.4	−8.1	−5.8	2.3	−2.7	−2.1	0.6

column properties, monthly statistics at AERONET sites (Holben et al 1998) were merged into model median fields of advanced aerosol simulations in global modelling (Kinne et al. 2006). For the surface albedo characterization, MODIS (visible and near-IR land) data (Schaaf et al. 2002) were combined with SSM/I statistics (Basist et al. 1996) on ice and snow cover. Cloud data (required in all-sky simulations) are based on the ISCCP climatology (Rossow et al. 1993). The anthropogenic aerosol fraction (potential anthropogenic dust sources are ignored) and the aerosol vertical distribution are adopted from global modelling (Schulz et al. 2006). Simulated annual global averages for the aerosol direct effect and the aerosol direct anthropogenic forcing at the TOA, within the atmosphere and at the surface separately for cloud-free and all-sky conditions, are summarised in Table 6.6.

In the context of anthropogenic forcing to the entire Earth-Atmosphere-System the relevant aerosol value is defined by the anthropogenic TOA effect at all-sky conditions. Table 6.6 indicates a global annual average of −0.2 W/m^2 at the TOA level. This loss seems negligible, when compared to the 2.6 W/m^2 gain at the TOA level confidently attributed to enhanced greenhouse gas concentrations. Fig. 6.14 (annual map) and Fig. 6.15 (monthly maps) demonstrate, beyond global average values, that anthropogenic aerosol impact displays significant regional diversity. Largely, for a northern hemispheric impact, there are regions with strong cooling (e.g. industrial regions, oceans) or with strong warming (associated with advected pollution over highly reflecting surfaces such as deserts, snow and lower clouds).

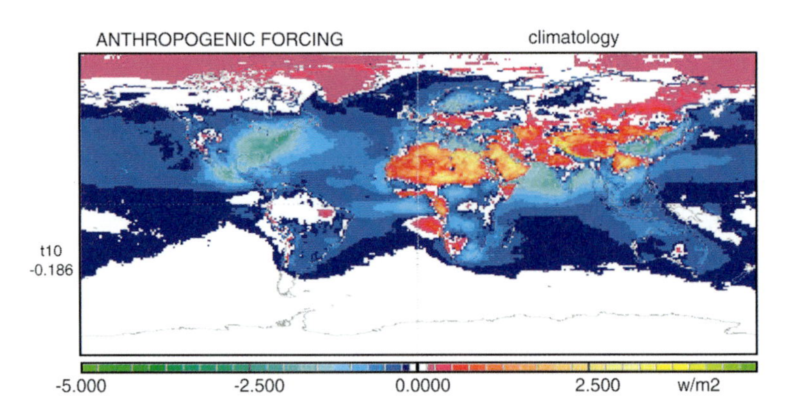

Fig. 6.14 Simulated annual anthropogenic aerosol direct (TOA) forcing at all-sky conditions based on monthly average input data-fields at 1 × 1 degree lat/lon resolution. Aerosol properties (AOD, ω_0 and g (via Ångström)) are tied to AERONET data, sub-spectral surface albedo data are based MODIS, ice and snow cover are obtained from SSM/I data and clouds are prescribed by ISCCP.

The strongest aerosol cooling occurs during the summer months in the northern hemisphere, when solar irradiance is at a maximum and snow cover at a minimum.

The inhomogeneity in Fig. 6.14 and Fig. 6.15 reflects the variability for aerosol and environmental input-fields. Despite a multi-annual data approach and the use of model-median data, the assumed monthly input data fields may contain significant errors, which would affect the overall result. Sensitivity tests were conducted, so that assumptions to individual input parameters could be modified.

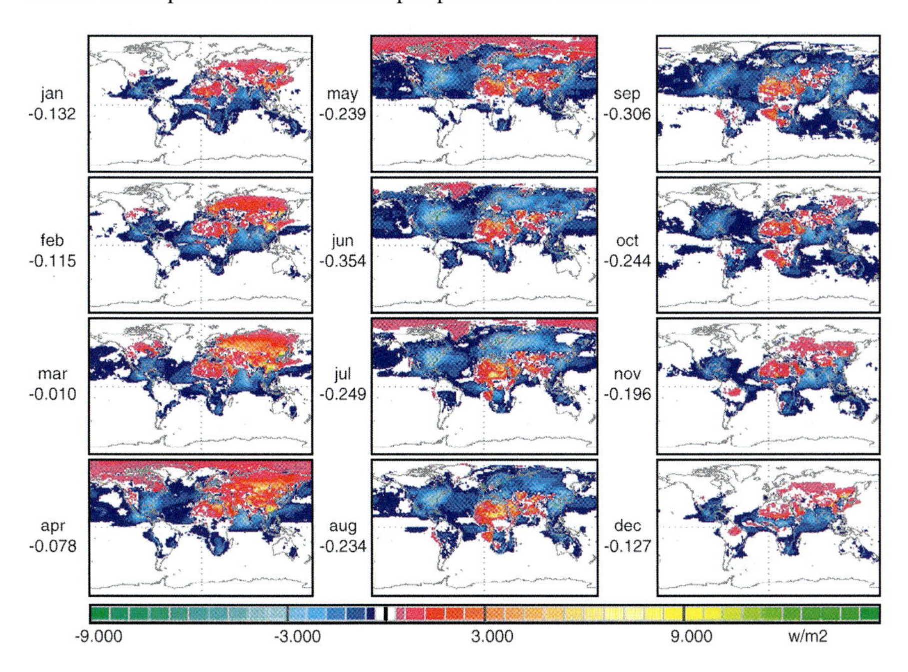

Fig. 6.15 Anthropogenic aerosol direct forcing of Fig. 6.14 at monthly resolution.

6.14.1 Uncertainties in Aerosol Direct Radiative Forcing

Sensitivity studies with modified input data indicate that, in the context of overall aerosol forcing uncertainty, and give the following results.

- The more consistent use of sky-photometer AERONET data for all aerosol properties over the favoured combination of sun and sky-photometer AERONET data (for better coverage and statistics on AOD and Ångström data) has only a small impact (estimated forcing uncertainty -0.01 to $+0.03$ W/m^2 globally). The uncertainty estimates reflect AERONET dataset differences on forcing.
- By using a different Ångström parameter threshold above which all AOD is assigned to the fine mode (here 1.8 instead of 2.1), the fine-mode probability is increased, yielding a 10% larger anthropogenic AOD (estimated forcing uncertainty -0.05 to $+0.01$ W/m^2 globally). The uncertainty estimate takes into account that the reference case assumes a relatively high Ångström threshold.

- A 25% increase in aerosol absorption reduces solar radiation losses to space and results in a significant reduction to aerosol cooling (+0.18 W/m^2 globally). Conversely, a 25% absorption decrease will increase aerosol cooling by a similar amount (estimated forcing uncertainty −0.07 to +0.07 W/m^2 globally). The uncertainty estimate considers that aerosol absorption is known within 10%.
- A 0.05 increase to the solar asymmetry-factor enhances the associated solar backscatter potential of aerosol which results in a reduction to aerosol cooling (+0.10 W/m^2 globally). Conversely, −0.05 decrease of the asymmetry-factor will increase aerosol cooling by a similar amount (estimated forcing uncertainty −0.07 to +0.07 W/m^2 globally). The uncertainty assumes asymmetry-factors to be accurate within 0.03.
- A 5% increase in land albedo strongly decreases the potential of aerosol to cool (+0.22 W/m^2 globally). Conversely, a 5% land albedo decrease will increase aerosol cooling by a similar amount (estimated forcing uncertainty −0.07 to +0.02 W/m^2 globally). The uncertainty estimate assumes the land-albedo on average to be accurate within 1% and the assumed MODIS-SSM/I solar albedo data produces slightly less cooling than with the use of IPCC median solar surface albedo fields.
- A lifting of the all aerosol by 2 km would allow for more "warming" contributions through an increased probability of (absorbing) aerosol above lower lying clouds results in an expected cooling reduction (+0.20 W/m^2 globally) (estimated forcing uncertainty −0.05 to +0.05 W/m^2 globally). The uncertainty estimate assumes that the relative placement between aerosol and clouds to be accurate within 0.5 km.

By combining all individual uncertainties and reducing the sum by 75%, because not all uncertainties are independent, for example, more absorption is associated with lower asymmetry-factors of smaller sizes, a probable range for the global annual aerosol direct forcing is estimated to fall between −0.45 and +0.00 W/m^2 with a high probability of the the most likely value near −0.2 W/m^2.

6.14.2 Comparisons of Aerosol Radiative Forcing with Models

The global annual average near −0.2 W/m^2 agrees well with the average of nine global models participating in the AeroCom forcing exercise (Schulz et al. 2006). The model averages, however, are lacking in regional contrast and in particular the strong warming over desert regions is missing. In that context, it cannot be ruled out that the applied MODIS data, in our hybrid approach, overestimate the surface albedo, especially in the near-infrared spectral region. In that case the most likely value could increase to about −0.25 W/m^2. In contrast, published results of other data-tied methods suggest a significantly more negative aerosol direct forcing of −0.9 W/m^2 (Quaas et al. 2008) or −0.8 W/m^2 (Bellouin et al. 2005). At closer inspection, it became apparent that poor assumptions caused negative biases, such as ignoring snow cover and overestimating anthropogenic AOD in the Bellouin study, or neglecting aerosol above clouds and avoiding desert region in the Quaas study. Once properly corrected, almost all data-tied estimates fall within the expected range (0.45–0.0 W/m^2). Critical issues in data-tied aerosol direct forcing estimates are (1) the definition for aerosol anthropogenic fraction, (2) the aerosol impact in an all-sky environment, for example, absorbing aerosol above lower clouds "warms", (3) the frequent lack of global coverage, and (4) an incorrect representation of the solar surface albedo.

6.14.3 Aerosol Radiative Forcing: Conclusions

The global annual (anthropogenic) aerosol direct radiative forcing is estimated to be near -0.2 W/m^2. From expected uncertainties to aerosol and environmental properties, a likely range (-0.45 to 0.0W/m^2) was derived. This estimate is better than many data tied approaches, suggesting a stronger cooling, because they are biased towards more negative values by poor assumptions or poor data, which lead to too negative estimates for the direct aerosol effect in IPCC-4AR (-1.0 to -0.1 W/m^2 range). The simulated global annual average is in line with estimates from simulations with advanced aerosol modules in global models, but significant differences in the forcing patterns on a regional and seasonal basis need to be resolved. Major uncertainties in simulations of the aerosol direct forcing are the representation of surface albedo properties and also the representation of aerosol properties. With respect to the aerosol properties, comparisons with AERONET data suggest that modelling principally lacks fine mode absorbing aerosol. Larger uncertainties in aerosol direct forcing are associated with the aerosol characterization of absorption and size emphasising a need for regional quality data.

6.15 Use of Satellites for Aerosol-Cloud Interaction Studies

Satellite aerosol products can be used to support the evaluation of impacts of aerosol-cloud interactions on clouds. The relevant aerosol properties are listed below.

- Aerosol optical depth of the fine and coarse modes, which are proxies for small and large CCN, respectively.
- Aerosol absorption, which is relevant to the radiative impacts of the aerosols on clouds, by both "cloud burning" and blocking the surface solar heating.
- Aerosol sphericity, which is an indication of desert dust. Desert dust aerosols typically have relatively small CCN efficiency and high activity as ice nuclei.
- Aerosol index of refraction. This can be useful for restricting the aerosol composition, and/or assessing the amount of water absorbed by the hygroscopic

components of the aerosols. This is important for identification of increase of AOD due to absorption of water and growth into haze particles.
- Vertical distribution of the aerosols. This is important for determining whether the aerosols occur at the same height of the clouds, which is essential for their microphysical interactions.

This discussion shows the wide range of satellite applications and the required resolutions on spatial, vertical and temporal scales. The present satellites represent a trade-off between large coverage in space and time versus high quality data in much smaller domains. Of the parameters listed above, only the spectral AOD is available from most satellite instruments. The data products from several instruments are evaluated below, with an emphasis on the wish list presented above, and for products that are routinely available.

a SEVIRI

The geostationary satellite, SEVIRI, provides imagery for the full disk every 15 min, but its shortest wavelength is 0.6 μm, which restricts the sensitivity to ultra fine aerosols that provide most of the CCN concentrations. The lack of mid-IR channels (2.1 μm) further restricts the usefulness of the MSG aerosols retrievals over land. The MSG has good sensitivity to the large desert dust aerosols, also during night-time due to their emissivity signature that is captured well by the 8.7, 10.8 and 12.0 μm channels of the MSG.

b PARASOL

The POLDER/PARASOL instrument concept allows measurements of the spectral, directional and polarized light reflected by the Earth-Atmosphere system. Over land, where the surface contribution to the total radiance is generally large, the inversion scheme uses the polarized radiances in the 865, 670 nm and blue channels. As the largest particles (radius $>$ 0.5 μm) generate low polarization, only the optical characteristics of the accumulation mode are derived. Over ocean, the characteristics of both aerosol modes (accumulation and coarse) are derived.

POLDER/PARASOL Aerosols products include the following parameters:

- Aerosol Optical Thickness (AOD),
- Ångström coefficient,
- Non-Sphericity Index for the Coarse Mode.

However, inspection of the products shows very sparse coverage and inconsistencies upon transition from land to ocean.

c MODIS

MODIS aerosols are obtained from the polar Terra and Aqua polar orbiting satellites, approximately once a day. The MODIS aerosol products provide AOD over land and ocean with a similar accuracy, except over the bright desert and ice covered areas. MODIS can separate the fine and coarse AOD over ocean, but less accurately over land due to poorer knowledge of the land surface spectral reflectance. The AOD over land for non-polluted situations occasionally provides negative AOD. This underlines the fact that MODIS AOD is not a good proxy for concentrations of small CCN, especially in such conditions.

d OMI

OMI provides UV-Vis measurements with global coverage once per day. The OMAERUV product provides AOD and AAOD. The aerosol index allows the detection of elevated absorbing aerosols also over bright surfaces even over clouds. The OMAERO product provides the AOD and also the single-scattering albedo associated with the best fitting model. Reasonable sensitivity to small aerosols is achieved, although sensitivity is still lost below about 0.1 μm, which is the size that is most important for CCN activity. A limiting factor on the accuracy of the products is the large footprint (13×24 km^2 at nadir), which incurs an appreciable problem of cloud contamination. The influence of absorption by aerosols on the reflectance at the top of the atmosphere is enhanced in the UV as compared to longer wavelengths. Therefore OMI measurements provide complementary information on aerosol absorption when compared with other instruments that do not include the UV.

e CALIPSO

CALIPSO, the space borne lidar, is unique in providing vertical profiles of aerosols, but only at the nadir pixel of the polar orbiting satellite. In addition to vertical profiles, CALIOP can retrieve aerosols in places where other sensors can't – over deserts, ice and snow, and above clouds. Aerosol forcing calculations presented in this chapter use model results for the aerosol profile, but there is a wide diversity in aerosol vertical distribution between different global models. CALIPSO data will be used to evaluate and improve the aerosol profiles predicted by models. Chand et al. (2009) show how CALIPSO data can be used to estimate the radiative forcing of aerosol above cloud and the necessity of considering instantaneous cloud fraction and cloud albedo along with the aerosol properties.

OMI, MODIS, POLDER and CALIPSO are all part of the A-Train and it is best to use them in combination.

6.16 Intercomparison of Aerosol Retrieval Products

Liu and Mishchenko (2008) compared AOD and Ångström parameters retrieved using TOA radiances measured with MISR and MODIS, two instruments that are dedicated to this purpose. They concluded that their analysis cannot be used to determine which algorithm yields a more accurate retrieval in each particular case or which algorithm is better in general. Results from a comparison of MISR and MODIS aerosol products by Kahn et al. (2009) show good correlations between the AOD products (correlation coefficient 0.9 over ocean and 0.7 over land) and the Ångström exponent (correlation coefficient 0.67 over ocean when MISR AOD values > 0.2 are considered). Kahn et al. emphasize the necessity for the proper interpretation of the satellite products. In particular data-quality statements should be followed to ensure proper interpretation and use of the satellite aerosol products. Also other intercomparisons of satellite products reveal significant discrepancies between AOD (order of 0.1) from different instruments, even over ocean where aerosol retrieval should be easiest by virtue of the "dark" surface (Myhre et al. 2004; 2005). A recent comparison over land for a single scene (Kokhanovsky et al. 2007) shows that, even on the scale of a single pixel, there can be large differences in AOD retrieved over land using different retrieval techniques and instruments. However, these differences are not as pronounced for the average AOD over land. For instance, the average AOD at 0.55 μm for the area 7°–12°E, 49°–53°N was equal to 0.14 for MISR, NASA MODIS and POLDER algorithms. It is smaller by 0.01 for the ESA MERIS aerosol product and larger by 0.04 for the MERIS BAER algorithm. AOD derived using AATSR gives on average larger values as compared to all other instruments, while SCIAMACHY retrievals underestimate the aerosol loading. In a second paper, focusing on the AATSR Dual View (ADV) algorithm described in Section 6.8, Kokhanovsky et al. (2009) conclude that the results from the AATSR dual view algorithm compare favourably with the products from orbiting optical instruments dedicated for aerosol retrieval such as MODIS and MISR, which leads to the conclusion that AATSR is well suited for aerosol retrieval over land when the dual view is used together with the ATSR-DV algorithm.

6.17 Conclusions

The use of satellite products has substantially increased over the last decade as a source of data, complimentary to *in situ* measurements and model results, in support of studies on climate, climate change and air quality. Aerosol properties retrieved from satellite instruments provide spatial information on regional to global scales, obtained with the same instrument and the same assumptions and are thus, in this respect, consistent. Based on validation and intercomparison exercises, the uncertainty in the results can be determined. It is emphasized that satellite and *in situ* data are complimentary. Satellite data need to be validated and evaluated versus *in situ* data to understand both their strengths and their limitations. Regular validation is particularly important in order to recognize instrument drift and degradation. One of the strongest merits of satellites is that they provide continuous data with good spatial resolution over areas which are not, or not well, covered by ground-based observations.

An example demonstrating the application in air quality studies was presented in Section 6.13, and an example on the evaluation of the direct radiative effect of aerosols was presented in Section 6.14. In the latter study, sun-photometer data were used as a source for AOD. As an alternative, with better spatial coverage and at the expense of accuracy, satellite data can be used. Kinne (2009b) shows an example on the use of both sun photometer data form AERONET and satellite data from various instruments to provide global AOD maps. The use of satellite data for aerosol-cloud interaction studies is emerging and a preliminary evaluation of satellite products was presented in Section 6.15.

There is an increased demand for satellite aerosol products for model evaluation and AQ assessment. They are also used in scientific studies such as long range transport during Lagrangian experiments and to support observations in, for example, developing countries where satellite data providing spatial coverage are complementary to *in situ* point measurements.

The quality of the aerosol retrieval results varies between different instruments and different retrieval algorithms. Recent comparison exercises are given in Kokhanovsky et al. (2007; 2009). It is particularly important for the correct use of satellite products to consider the proper quality statements when these are available to avoid misinterpretation (Kahn et al. 2009). The use of products with low quality is generally not recommended.

Intercomparison between satellite data shows that no instrument can be singled out as providing the "best" data set. Aerosol retrieval from satellites is still a young science in full development. As discussed above, most instruments used for this purpose were designed with other applications in mind but turn out to provide useful information on aerosol properties. Few dedicated instruments have been launched and the full potential is still being explored. Apart from AOD and derived aerosol properties, aerosol layer height detection using stereo techniques is explored for the detection of aerosol plume properties, in particular for forest fires plumes (Kahn et al. 2007), which require multiple views.

Trends are the synergistic use of different instruments, preferably flying on the same platform. One such approach is the combination of a radiometer and a spectrometer in the SYNAER algorithm (Holzer-Popp et al. 2002). The current version of SYNAER uses AATSR to determine the surface reflectance and the spectral information from SCIAMACHY to retrieve aerosol type information. More recent approaches aim at using the different views provided by AATSR and MERIS to determine aerosol information over ocean (Sogacheva et al. 2009) and over land (North et al. 2008). These initiatives are of particular importance because of the continuation of very similar instrument deployments during future missions such as the sentinels which are prepared by ESA and EUMETSAT.

The GLORY mission (Mishchenko et al. 2007), briefly discussed in Section 6.5, will combine polarimetric measurements with multi-angle viewing, i.e. the instrument characteristics needed to obtain optimum information on aerosol properties. One of the primary objectives of the GLORY mission is to determine the global distribution of aerosol and cloud properties with very high accuracy, and thereby facilitate the quantification of the aerosol direct and indirect effects on climate (Mishchenko et al. 2007).

Efforts to provide dedicated instruments and the exploration of synergies between different instruments will further improve the accuracy and consistency of data products. The expectation is that the use of satellite-retrieved aerosol properties for climate and air quality assessment will continue to increase and contribute to a better understanding of climate change leading to a reduction in the uncertainties. Satellite observations will play a larger role in air quality monitoring and the quantification of the aerosol emissions and their precursor gases.

Acknowledgements The work described in this Chapter was supported by EU-FP6 projects ACCENT, EUCAARI and GEMS, EU-FP7 projects MEGAPOLI and MACC. We gratefully acknowledge the efforts of many global modelling groups contributing to AeroCom exercises and the support of many remote sensing groups, in particular the AERONET group by providing input on site assessments for data quality and regional representation We thank the ICARE thematic centre for providing an easy access to the MODIS and POLDER data and products used in this paper.

References

Al-Saadi, J., J. Szykman, R.B. Pierce, C. Kittaka, D. Neil, D.A. Chu, L. Remer, L. Gumley, E. Prins, L. Weinstock, C. MacDonald, R. Wayland, F. Dimmick and J. Fishman, 2005, Improving national air quality forecasts with satellite aerosol observations. Bulletin of the American Meteorological Society, **86**, 1249–1261.

Ackerman, S.A., K.I. Strabala, W.P. Menzel, R.A. Frey, C.C. Moeller, and L.E. Guclusteringey, 1998, Discriminating clear sky from clouds with MODIS. J. Geophys. Res., **103**, 141 – 157.

Andreae, M.O., and D. Rosenfeld, 2008, Aerosol-cloud-precipitation interactions. Part 1. The nature and sources of could-active aerosols. Earth-Sci Rev, **89**, 13–41.

Apituley, A., A. Van Lammeren and H. Russchenberg, 2000, High time resolution cloud measurements with lidar during CLARA. Phys. Chem. Ear., **25**, 107–113.

Basist, A., D. Garrett, R. Ferraro, N.C. Grody, and K. Mitchell, 1996, A comparison between visible and Microwave snow cover products derived from satellite observations. J. Appl. Meteor., **35**, 163–177.

Bellouin, N., O. Boucher, J. Hayward and M. Reddy, 2005, Global estimate of aerosol direct radiative forcing from satellite measurements, Nature, **438**, 1138–1141.

Buchholz, A., 1995, Rayleigh scattering calculations for the terrestrial atmosphere. Applied Optics **34**, 2765–2773.

Cairns, B., F. Waquet, K. Knobelspiesse, J. Chowdhary and J.-L Deuzé, 2009, Polarimetric remote sensing of aerosols over land surfaces, In: A.A. Kokhanovsky and G. de Leeuw (editors) Satellite Aerosol Remote Sensing Over Land, Springer, Berlin, 295–323.

Chand, D., R. Wood, T.L. Anderson, S.K. Satheesh, and R.J. Charlson, 2009, Satellite-derived direct radiative effect of aerosols dependent on cloud cover, Nature Geoscience, **2**, 181–184, doi:10.1038/NGE0437.

Chu, D.A., Y.J. Kaufman, C. Ichoku, L.A. Remer, D. Tanré, B.N. Holben, 2002, Validation of MODIS aerosol optical thickness retrieval over land. Geophys. Res. Lett., **29**, doi:10.1029/2001GL013205.

Cox, C., and W. Munk, 1954, Measurements of the roughness of the sea surface from photographs of the Sun's glitter. J. Opt. Soc. Am., **44**, 838–850.

Curier R.L., J.P. Veefkind, R. Braak, B. Veihelmann, O. Torres and G. de Leeuw, 2008, Retrieval of aerosol optical properties from OMI radiances using a multiwavelength algorithm: Application to western Europe. J. Geophys. Res., **113**, D17S90, doi:10.1029/2007JD008738.
de Haan J.F., P.B. Bosma, and J.W. Hovenier, 1987, The adding method for multiple scattering calculations of polarized light. Astron. Astrophys., **183**, 371–391.
de Leeuw, G., and A. Kokhanovsky, 2009, Introduction, In: A.A. Kokhanovsky and G. de Leeuw (editors) Satellite Aerosol Remote Sensing Over Land, Springer, Berlin, 1–18.
de Leeuw, G., A.N. de Jong, J. Kusmierczyk-Michulec, R. Schoemaker, M. Moerman, P. Fritz, J. Reid and B. Holben, 2005, Aerosol Retrieval Using Transmission and Multispectral AATSR Data, in: J.S. Reid, S.J. Piketh, R. Kahn, R.T. Bruintjes and B.N. Holben (editors) A Summary of First Year Activities of the United Arab Emirates Unified Aerosol Experiment: UAE2. NRL Report Nr. NRL/MR/7534–05-8899, pp 105–110.
Deschamps, P.Y., F.-M. Breon, M. Leroy, A. Podaire, A. Bricaud, J.C. Buriez, and G. Seze, 1994, The POLDER Mission : Instrument characteristics and scientific objectives. IEEE Trans Geosci Remote Sens., **2**, 598–615.
Deuzé, J.L., F.-M. Breon, P.Y. Dechamps, C. Devaux, M. Herman, A. Podaire and J.L. Roujean, 1993, Analysis of the POLDER (POLarization and Directionnality of Earth's Reflectances) Airborne Instrument Observations over Land Surfaces. Remote Sens. Environ., **45**, 137–154.
Dinter, T., W. von Hoyningen-Huene, J.P. Burrows, A. Kokhanovsky, E. Bierwirth, M. Wendisch, D. Müller, R. Kahn, and M. Diouri, 2009. Retrieval of aerosol optical thickness for desert conditions using MERIS observations during SAMUM campaign. Tellus 61B (2009), 220–237.
Dubovik, O., A. Smirnov, B.N. Holben, M.D. King, Y.J. Kaufman, T.F. Eck, and I. Slutsker, 2000, Accuracy assessments of aerosol optical properties retrieved from AERONET sun and sky-radiance measurements. J. Geophys. Res, **105**, 9791–9806.
Dubovik, O., B.N. Holben, T.F. Eck, A. Smirnov, Y.J. Kaufman, M.D. King, D. Tanré and I. Slutsker, 2002, Variability of Absorption and Optical Properties of Key Aerosol Types Observed in Worldwide Locations. J. Atmos. Sci **59**, 590–698.
Dubovik O., A. Sinyuk, T. Lapyonok, B.N. Holben, M. Mishchenko, P. Yang, T.F. Eck, H. Volten, O. Munoz, B. Veihelmann, W.J. van der Zande, J.-F. Leon, M. Sorokin and I. Slutsker, 2006, Application of spheroid models to account for aerosol particle nonsphericity in remote sensing of desert dust. J. Geophys. Res., **111**, D11208, doi:10.1029/2005JD006619.
Eck, T.F., B.N. Holben, J.S. Reid, O. Dubovik, A. Smirnov, N.T. O'Neill, I. Slutsker, and S. Kinne, 1999, Wavelength dependence of the optical depth of biomass burning, urban, and desert dust aerosol. J. Geophys. Res., 104, **31** 333–31 350.
Engel-Cox, J.A., C.H. Holloman, B.W. Coutant and R.M. Hoff, 2004, Qualitative and quantitative evaluation of MODIS satellite sensor data for regional and urban scale air quality. Atmos. Environ., **38**, 2495–2509.
Engel-Cox, J.A., R.M. Hoff, R. Rogers, F. Dimmick, A.C. Rush, J.J. Szykman, J. Al-Saadi, D.A. Chu, E.R. Zell, 2006, Integrating lidar and satellite optical depth with ambient monitoring for 3-D dimensional particulate characterisation. Atmos. Environ., **40**, 8056–8067.
Fernald, F.G., B.M. Herman and J.A. Reagan, 1972, Determination of aerosol height distributions with lidar. J. Appl. Meteorol., **11**, 482–489.
Flowerdew R.J., and J.D. Haigh, 1995, An approximation to improve accuracy in the derivation of surface reflectances from multi-look satellite radiometers. Geophys. Res. Lett., **23**, 1693–1696.
Grey, W.M.F., P.R.J. North, S.O. Los and R.M. Mitchell, 2006, Aerosol optical depth and land surface reflectance from multi-angle AATSR measurements: Global validation and inter-sensor comparisons. IEEE Trans. Geosci. Remote Sens., **44**, 2184–2197.
Griggs, M., 1975, Measurements of atmospheric aerosol optical thickness over water using ERS-1 data. J. Air Pollut. Conlrol. Assoc., **25**, 622–626.
Hasekamp, O.P., and J. Landgraf, 2005, Retrieval of aerosol properties over the ocean from multispectral single-viewing-angle measurements of intensity and polarization: Retrieval approach, information content, and sensitivity study. J. Geophys. Res., **110**, D20207, doi:10.1029/2005JD006212.

Herman M., J.-L. Deuzé, A. Marchand, B. Roger, P. Lallart (2005), Aerosol remote sensing from POLDER/ADEOS over the ocean: Improved retrieval using a nonspherical particle model, J. Geophys. Res., **110**, D10S02, doi:10.1029/2004JD004798.

Hess, M., P. Koepke and I. Schult, 1998, Optical properties of aerosols and clouds : The software package OPAC. Bull. Am. Met. Soc., **79**, 831–844.

Höller, R., P. Garnesson, C. Nagl, and T. Holzer-Popp, 2007, Using satellite aerosol products for monitoring national and regional air quality in Austria. Proc. 'Envisat Symposium 2007', Montreux, Switzerland, 23–27 April 2007 (ESA SP-636).

Holben, B., T. Eck, I. Slutsker, D. Tanre, J. Buis, E. Vermote, J. Reagan, Y. Kaufman, T. Nakajima, F. Lavenau, I. Jankowiak and A.Smirnov, 1998, AERONET, a federated instrument network and data-archive for aerosol characterization. Rem. Sens. Environ., **66**, 1–66.

Holben, B.N., D. Tanré, A. Smirnov, T.F. Eck, I. Slutsker, N. Abuhassan, W.W. Newcomb, J.S. Schafer, B. Chatenet, F. Lavenu, Y.J. Kaufman, J.V. de Castle, A. Setzer, B. Markham, D. Clark, R. Froin, R. Halthore, A. Karnieli, N.T. O'Neill, C. Pietras, R.T. Pinker, K. Voss and G. Zibordi, 2001, An emerging ground-based aerosol climatology: Aerosol optical depth from AERONET. J. Geophys. Res., **106**, 12.067–12.097.

Holzer-Popp, T., M. Schroedter and G. Gesell, 2002, Retrieving aerosol optical depth and type in the boundary layer over land and ocean from simultaneous GOME spectrometer and ATSR-2 radiometer measurements, 1, Method description. J. Geophys. Res., **107**, 4578, doi:10.1029/2001JD002013.

Hunt W. H., D. M. Winker, M. A. Vaughan, K. A. Powell, P. L. Lucker, C. Weimer, 2009, CALIPSO lidar description and performance assessment. J. Atmos. Oceanic Technol., **26**, 1214–1228, doi: 10.1175/2008JTECHA1221.1.

Hutchison, K.D., 2003, Applications of MODIS satellite data and products for monitoring air quality in the state of Texas. Atmos. Environ., **37**, 2403–2412.

IPCC, 2007: Climate Change 2007: The Physical Science Basis. Contribution of Working Group I to the Fourth Assessment Report of the Intergovernmental Panel on Climate Change, In: S. Solomon, D. Qin, M. Manning, Z. Chen, M. Marquis, K.B. Averyt, M. Tignor and H.L. Miller (editors), Cambridge University Press, Cambridge, United Kingdom and New York, NY, USA, 996 pp.

Ichoku, C., L.A. Remer and T.F. Eck, 2005, Quantitative evaluation and intercomparison of morning and afternoon Moderate Resolution Imaging Spectroradiometer (MODIS) aerosol measurements from Terra and Aqua. J. Geophys. Res., **110**, D10S03, doi:10.1029/2004JD004987.

Kacenelenbogen, M., J.-F. Léon, I. Chiapello, and D. Tanré, 2006, Characterization of aerosol pollution events in France using ground-based and POLDER-2 satellite data. Atmos. Chem. Phys., **6**, 4843–4849.

Kahn, R. A., W.-H. Li, C. Moroney, D. J. Diner, J. V. Martonchik, and E. Fishbein, 2007, Aerosol source plume physical characteristics from space-based multiangle imaging. J. Geophys. Res., **112**, D11205, doi:10.1029/2006JD007647.

Kahn, R.A., D.L. Nelson, M.J. Garay, R.C. Levy, M.A. Bull, D.J. Diner, J.V. Martonchik, S.R. Paradise, E.G. Hansen and L.A. Remer, 2009, MISR Aerosol Product Attributes and Statistical Comparisons With MODIS. IEEE Trans. On Geosc. and Remote Sensing, **47**, 4095–4114.

Katsev, I.L., A.S. Prikhach, E.P. Zege, A.P. Ivanov and A.A. Kokhanovsky, 2009, Iterative procedure for retrieval of spectral optical thickness and surface reflectance from satellite data using fast radiative transfer code and its application to MERIS measurements, in: A.A. Kokhanovsky and G. de Leeuw (editors) Satellite Aerosol Remote Sensing Over Land, Springer, Berlin, 101–132.

Kaufman, Y., A. Smirnov, B. Holben and O. Dubovik, 2001, Baseline maritime aerosol: methodology to derive the optical thickness and scattering properties. Geophys. Res. Letters, **28**, 3251–3254.

Kaufman, Y.J, D. Tanré and O. Boucher, 2002, A satellite view of aerosols in the climate system. Nature **419**, 215–223.

Kinne, S., 2009, Remote Sensing Data Combinations – Superior Global Maps for Aerosol Optical Depth, in: A.A. Kokhanovsky and G. de Leeuw (editors) Satellite Aerosol Remote Sensing Over Land, Springer, Berlin, 361–381.

Kinne, S., M. Schulz, C. Textor, S. Guibert, S. Bauer, T. Berntsen, T. Berglen, O. Boucher, M. Chin, W. Collins, F. Dentener, T. Diehl, R. Easter, J. Feichter, D. Fillmore, S. Ghan, P. Ginoux, S. Gong, A. Grini, J. Hendricks, M. Herzog, L. Horowitz, I. Isaksen, T. Iversen, D. Koch, M. Krol, A. Lauer, J.F. Lamarque, G. Lesins, X. Liu, U. Lohmann, V. Montanaro, G. Myhre, J. Penner, G. Pitari, S. Reddy, O. Seland, P. Stier, T. Takemura and X. Tie, An AeroCom initial assessment – optical properties in aerosol component modules of global models. ACP, **6**, 1–22, 2006.

Klett, J.D., 1985, Lidar inversion with variable backscatter/extinction ratios. Applied Optics, **24**, 1638–1643.

Koepke, P., 1984, Effective reflectance of oceanic whitecaps. Appl. Opt., **23**, 1816–1824.

Kokhanovsky, A.A., and G. de Leeuw, 2009, Satellite Aerosol Remote Sensing Over Land. Springer, Berlin.

Kokhanovsky, A.A., F.-M. Breon, A. Cacciari, E. Carboni, D. Diner, W. Di Nicolantonio, R.G. Grainger, W.M.F. Grey, R. Höller, K.-H. Lee, Z. Li, P.R.J. North, A.M. Sayer, G.E. Thomas and W. von Hoyningen-Huene, 2007, Aerosol remote sensing over land: A comparison of satellite retrievals using different algorithms and instruments. Atmospheric Research, **85**, 372–394.

Kokhanovsky, A.A., R. L. Curier, Y. Bennouna, R. Schoemaker, G. de Leeuw, P.R.J. North, W. M. F. Grey, K.-H. Lee, 2009, The inter-comparison of AATSR dual view aerosol optical thickness retrievals with results from various algorithms and instruments, *Int. J. Remote Sensing*, 30: **17**, 4525–4537, 10.1080/01431160802578012.

Koren, I., L.A. Remer, Y.J. Kaufman, Y. Rudich and J.V. Martins, 2007, On the twilight zone between clouds and aerosols. Geophys. Res. Lett., **34**, L08805, doi:10.1029/2007GL029253, 2007.

Lee, K. H., Z. Li, Y.J. Kim and A.A. Kokhanovsky, 2009, Aerosol monitoring from satellite observations: a history of three decades. In: Y.J. Kim, U. Platt, M.B. Gu and H. Iwahashi (Editors). Atmospheric and Biological Environmental Monitoring, Berlin: Springer, 13–38.

Lenoble, J., M. Herman, J. Deuzé, B. LaFrance, R. Santer and D. Tanré, 2007, A successive order of scattering code for solving the vector equation of transfer in the Earth's atmosphere with aerosols. J. Quant. Spectrosc. Radiat. Transf., **1007**, 479–507.

Levy, R.C., L.A. Remer and O. Dubovik, 2007a, Global aerosol optical properties and application to Moderate Resolution Imaging Spectroradiometer aerosol retrieval over land. J. Geophys. Res., **112**, D13210, doi:10.1029/2006JD007815.

Levy, R.C., L.A. Remer, S. Mattoo, E.F. Vermote and Y.J. Kaufman, 2007b, Second-generation operational algorithm: Retrieval of aerosol properties over land from inversion of Moderate Resolution Imaging Spectroradiometer spectral reflectance. J. Geophys. Res., **112**, D13211, doi:10.1029/2006JD007811.

Liu, L., and M.I. Mishchenko, 2008, Toward unified satellite climatology of aerosol properties: direct comparisons of advanced level 2 aerosol products. J. Quant. Spectrosc. Radiat. Transf., **109**, 2376–2385.

Liu Z., M.A. Vaughan, D.M. Winker, C.A. Hostetler, L.R. Poole, D.L. Hlavka, W.D. Hart and M.J. McGill, 2004, Use of Probability Distribution Functions for Discriminating Between Cloud and Aerosol in Lidar Backscatter Data. J. Geophys. Res., **109**, doi:10.1029/2004JD004732.

Liu Z., A.H. Omar, Y. Hu, M.A. Vaughan, D.M. Winker, L.R. Poole and T.A. Kovacs, 2005, CALIOP Algorithm Theoretical Basis Document Part 4: Scene Classification Algorithms. NASA-CNES document PC-SCI-203.

Liu Z., M. Vaughan, D. Winker, C. Kittaka, B. Getzewitch, R. Kuehn, A. Omar, K. Powell, C. Trepte, and C. Hostetler, 2009, The CALIPSO Lidar Cloud and Aerosol Discrimination: Version 2 Algorithm and Initial Assessment of Performance, J. Atmos. Oceanic Technol., **26**, 1198–1213.

Maignan, F., F.-M. Breòn and R. Lacaze, 2004, Bidirectional reflectance of Earth targets: valuation of analytical models using a large set of spaceborne measurements with emphasis on the Hot Spot. Remote Sens Environ., **90**, 210–220.

McCormick, M.P., P. Hamill, P.J. Pepin, W.P. Chu, T.J. Swissler and L.R. McMaster, 1979, Satellite studies of the Stratospheric aerosol. Bull. American Meteorol. Soc., **60**, 1038–1046.

Mekler, Y., H. Quenzel, G. Ohring and I. Marcus, 1977, Relative atmospheric aerosol content from ERS observations. J. Geophys. Res., **82**, 967–972.

Mercado, L.M., N. Bellouin, S. Sitch, O. Boucher, C. Huntingford, M. Wild and P.M. Cox, 2009, Impact of changes in diffuse radiation on the global land carbon sink. Nature **458**, 1014–1017. doi:10.1038/nature07949

Mie, G., 1908, Beiträge zur Optik trüber Medien, speziell kolloidaler Metallösungen. Ann. Phys., **25**, 377–445.

Mishchenko, M.I., I.V. Geogdzhayev, W.B. Rossow, B. Cairns, B.E. Carlson, A.A. Lacis, L. Liu and L.D. Travis, 2007, Long-term satellite record reveals likely recent aerosol trend. Science 315 (5818), 1543. DOI: 10.1126/science.1136709

Myhre, G., F. Stordal, M. Johnsrud, A. Ignatov, M.I. Mishchenko, I.V. Geogdzhayev, D. Tanré, J.L. Deuzé, P. Goloub, T. Nakajima, A. Higurashi, O. Torres and B.N. Holben, 2004, Intercomparison of satellite retrieved aerosol optical depth over ocean. J. Atmos. Sci., **61**, 499–513.

Myhre, G., F. Stordal, M. Johnsrud, D.J. Diner, I.V. Geogdzhayev, J.M. Haywood, B.N. Holben, T. Holzer-Popp, A. Ignatov, R.A. Kahn, Y.J. Kaufman, N. Loeb, J.V. Martonchik, M.I. Mishchenko, N.R. Nalli, L.A. Remer, M. Schroedter-Homscheidt, D. Tanré, O. Torres, and M. Wang, 2005, Intercomparison of satellite retrieved aerosol optical depth over ocean during the period September 1997 to December 2000. Atmos. Chem. Phys., **5**, 1697–1719.

Nadal, F., and F.-M. Bréon, 1999, Parametrisation of surface polarised reflectance derived from POLDER spaceborne measurements. IEEE Trans. Geosci. Remote Sens., **37**, 1709–1718.

North, P.R.J., C. Brockmann, J. Fischer, L. Gomez-Chova, W. Grey, A. Heckel, J. Moreno, R. Preusker and P. Regner, 2008, MERIS/AATSR synergy algorithms for cloud screening, aerosol retrieval and atmospheric correction. in Proc. 2nd MERIS/AATSR User Workshop, ESRIN, Frascati, 22- 26 September 2008. (CD-ROM), ESA Publications Division, European Space Agency, Noordwijk, The Netherlands.

Omar, A.H., J.-G. Won, S.-C. Yoon, O.D. David, M. Winker and M.P. McCormick, 2004, Development of global aerosol models using cluster analysis of AERONET measurements. J. Geophys. Res., **110**, D10S14, doi:10.1029/2004JD004874.

Omar A. H., D. M. Winker, C. Kittaka, M. A. Vaughan, Z. Liu, Y. Hu, C. T. Trepte, R. R. Rogers, R. A. Ferrare, K.-P. Lee, R. E. Kuehn, and C. A. Hosteler, 2009, The CALIPSO automated aerosol classification and lidar ratio selection algorithm. J. Atmos. Oceanic Technol., **26**, 1994–2014, DOI: 10.1175/2008JTECHA1221.1.

Pal, S.R., W. Steinbrecht and A.I. Carswell, 1992, Automated method for lidar determination of cloud-base height and vertical extent. Applied Optics, 31, 1488–1494.

Platt, C.M.R., 1973, Lidar and radiometer observations of cirrus clouds. J. Atmos. Sci., **30**, 1191–1204.

Powell K. A., C. A. Hostetler, Z. Liu, M. A. Vaughan, R. E. Kuehn, W. H. Hunt, K.P. Lee, C. R. Trepte, R. R. Rogers, S. A. Young and D. M. Winker, 2009, CALIPSO lidar calibration algorithms Part I: Night-time 532-nm parallel channel and 532-nm perpendicular channel. J. Atmos. Oceanic Technol., **26**, 2015–2033, doi: 10.1175/2008JTECHA1221.1.

Quaas J., O. Boucher, N. Bellouin and S. Kinne, 2008, Satellite based estimate of the direct and indirect aerosol climate forcing, J. Geophys. Res., **113**, D05204, doi:10.1029/2007JD008962.

Reagan J. A., X. Wang, and M. J. Osborn, 2002, Spaceborne lidar calibration from cirrus and molecular backscatter returns. IEEE Trans. Geosci. Remote Sens., **40**, 2285–2290.

Reid, J.S., S.J. Piketh, R. Kahn, R.T. Bruintjes and B.N. Holben (Editors), 2005, A Summary of First Year Activities of the United Arab Emirates Unified Aerosol Experiment: UAE2. NRL Report Nr. NRL/MR/7534–05-8899.

Remer, L.A., Y.J. Kaufman, D. Tanré, S. Mattoo, D.A. Chu, J.V. Martins, R.R. Li, C. Ichoku, R.C. Levy, R.G. Kleidman, T.F. Eck, E. Vermote and B.N. Holben, 2005, The MODIS aerosol algorithm, products, and validation. J. Atmos. Sci., **62**, 947–973.

Robles Gonzalez, C., 2003, Retrieval of Aerosol Properties using ATSR-2 Observations and their Interpretation. PhD thesis, University of Utrecht, Utrecht, The Netherlands.
Robles-Gonzalez, C., J.P. Veefkind and G. de Leeuw, 2000, Mean aerosol optical depth over Europe in August 1997 derived from ATSR-2 data. Geophys. Res. Lett. **27**, 955–959.
Robles-Gonzalez, C., G. de Leeuw, R. Decae, J. Kusmierczyk-Michulec, and P. Stammes, 2006, Aerosol properties over the Indian Ocean Experiment (INDOEX) campaign area retrieved from ATSR-2. J. Geophys. Res., **111**, D15205, doi:10.1029/2005JD006184.
Robles-Gonzalez, C., and G. de Leeuw, 2008, Aerosol properties over the SAFARI-2000 area retrieved from ATSR-2. J. Geophys. Res., **113**, D05206, doi:10.1029/2007JD008636.
Rossow, W., A. Walker and C. Garder, 1993, Comparison of ISCCP and other cloud amounts, J. Climate, **6**, 2394–2418.
Santer, R., V. Carrere, P. Dubuisson and J.-C. Roger, 1999, Atmospheric corrections over land for MERIS. Int. J. of Rem. Sens., **20**, 1819–1840.
Santer, R, Carrere, V., Dessailly, D., Dubuisson, P., and Roger, J.-C., 2000: MERIS Algorithm theoretical basis document, ATBD 2.15, Atmospheric corrections over land.
Schaaf, C., F. Gao, A. Strahler, W. Lucht, X. Li, T. Trang, N. Strucknell, X. Zhang, Y. Jin, J.-P. Mueller, P. Lewis, M. Barnsley, P. Hobson, M. Disney, G. Roberts, M. Dunderdale, R. D'Entremont, B. Hu, S. Liang, J. Privette and D. Roy, 2002, First oberservational BRDF, albedo and nadir reflectance from MODIS. Remote Sens. Environ., **83**, 135–148.
Schaap, M., K. Muller and H.M. ten Brink, 2002, Constructing the European aerosol nitrate concentration field from quality analysed data. Atmos. Environ., **36**, 1323–1335.
Schaap, M., R.M.A. Timmermans, R.B.A Koelemeijer, G. de Leeuw and P.J.H. Builtjes, 2008, Evaluation of MODIS aerosol optical thickness over Europe using sun photometer observations. Atmos. Environ., **42**, 2187–2197, doi:10.1016/j.atmosenv.2007.11.044.
Schulz M., C. Textor, S. Kinne, Y. Balkanski, S. Bauer, T. Berntsen, T. Berglen, O. Boucher, F. Dentener, S. Guibert, I.S.A. Isaksen, T. Iversen, D. Koch, A. Kirkevag, X. Liu, V. Montenaro, G. Myhre, J.E. Penner, G. Pitari, S. Reddy, O. Seland, P. Stier and T. Takemura, 2006, Radiative forcing by aerosols as derived from the AeroCom present-day and pre-industrial simulations. ACP, **6**, 5225–5346.
Seinfeld, J.H., and S.N. Pandis, 1998, Atmospheric Chemistry and Physics. Wiley.
Silva, A.M., M.L. Bugalho, M.J. Costa, W.V. Hoyningen-Huene, T. Schmidt, J. Heintzenberg, S. Henning, 2002, Aerosol optical properties from columnar data during the second Aerosol Characterization Experiment an the south coast of Portugal. J. Geophys. Res., **107**, doi: 10.1029/2002JD002196.
Smirnov, A., B.N. Holben, I. Slutsker, D. M. Giles, C. R. McClain, T.F. Eck, S.M. Sakerin, A. Macke, P. Croot, G. Zibordi, P.K. Quinn, J. Sciare, S. Kinne, M. Harvey, T.J. Smyth, S. Piketh, T. Zielinski, A. Proshutinsky, J.I. Goes, N.B. Nelson, P. Larouche, V.F. Radionov, P. Goloub, K. Krishna Moorthy, R. Matarrese, E.J. Robertson, and F. Jourdin, 2009, Maritime Aerosol Network as a component of Aerosol Robotic Network. J. Geophys. Res., **114**, D06204, doi:10.1029/2008JD011257.
Sogacheva, L., P. Kolmonen, L. Curier, G. de Leeuw, A. Kokhanovsky, 2009, Combined AATSR/MERIS algorithm AMARA for aerosol optical depth retrieval over ocean. Proceedings of OceanObs'09, 21–25 September 2009, Venice, Italy.
Stammes, P., 2001, Spectral radiance modelling in the UV-Visible range. in: W.L. Smith and Y.M. Timofeyev (editors), Current problems in Atmospheric Radiation, A. Deepak Publication, Hampton, VA, pp. 385–388.
Stowe, L.L., H. Jacobowitz, G. Ohring, K.R. Knapp and N.R. Nalli, 2002, The Advanced Very High Resolution Radiometer (AVHRR) Pathfinder Atmosphere (PATMOS) climate dataset: Initial analyses and evaluations. J. Clim., **15**, 1243–1260.
Thomas, G.E., C.A. Poulsen, R. L. Curier, G. de Leeuw, S. H. Marsh, E. Carboni, R. G. Grainger and R. Siddans, 2007, Comparison of AATSR and SEVIRI aerosol retrievals over the Northern Adriatic. QJRM, **133**, 85–95, doi: 10.1002/qj.126.

Torres, O., P.K. Bhartia, J.R. Herman and Z. Ahmad, 1998, Derivation of aerosol properties from satellite measurements of backscattered ultraviolet radiation. Theoretical Basis. J. Geophys. Res., **103**, 17099–17110.

Torres, O., P.K. Bhartia, J.R. Herman, A. Sinyuk and B.N. Holben, 2002, A long term record of aerosol optical thickness from TOMS observations and comparison to AERONET measurements. J. Atm. Sci.,**59**, 398–413.

Torres, O., A. Tanskanen, B. Veihelmann, C. Ahn, R. Braak, P.K. Bhartia, J.P. Veefkind, and P.F. Levelt, 2007, Aerosols and Surface UV Products from OMI Observations: An Overview. J. Geophys. Res., **112**, D24S47, doi:10.1029/2007JD008809.

Vaughan M.A., D. M. Winker, and C.A. Hostetler, 2002, SIBYL: a Selective Iterated Boundary Location Algorithm for Finding Cloud and Aerosol Layers in CALIPSO Lidar Data. In: L. R. Bissonnette, G. Roy and G. Vallée (editors), Lidar Remote Sensing in Atmospheric and Earth Sciences, Defence R&D Canada – Valcartier, Québec, Canada, pp. 791–794.

Vaughan M.A., D.M. Winker and K.A. Powell, 2005, CALIOP Algorithm Theoretical Basis Document Part 3: Feature Detection and Layer Properties Algorithms. NASA-CNES document PC-SCI-203.

Vaughan M., K. Powell, R. E. Kuehn, S. Young, D. M. Winker, C. A. Hostetler, W. H. Hunt, Z. Liu, M. J. McGill and B. J. Getzewitch, 2009, Fully automated detection of cloud and aerosol layers in the CALIPSO lidar measurements. J. Atmos. Oceanic Technol., **26**, 2034–2050, DOI: 10.1175/2008JTECHA1221.1.

Veefkind, J.P. and G. de Leeuw, 1998, A new algorithm to determine the spectral aerosol optical depth from satellite radiometer measurements. J. of Aerosol Sciences, **29**, 1237–1248.

Veefkind, J.P., G. de Leeuw and P.A. Durkee, 1998, Retrieval of aerosol optical depth over land using two-angle view satellite radiometry during TARFOX. Geophys. Res. Lett. **25**, 3135–3138.

Veefkind, J.P., G. de Leeuw, P. Stammes and R.B.A. Koelemeijer, 2000, Regional distribution of aerosol over land derived from ATSR-2 and GOME. Remote sens Environ., **74**,377–386.

Veihelmann, B., P.F. Levelt, P. Stammes and J.P. Veefkind, 2007, Aerosol Information Content in OMI Spectral Reflectance Measurements. Atmos. Chem. Phys., **7**, 3115–3127.

Vermeulen, A., C. Devaux and M. Herman, 2000, Retrieval of the scattering and microphysical properties of aerosols from ground-based optical measurements including polarization. I. Method. Appl. Opt., **39**, 6207–6220.

Verver, G.H.L., J.S. Henzing, G. de Leeuw, C. Robles Gonzalez and P.F.J. van Velthoven, 2002, Aerosol retrieval and assimilation (ARIA). Final report Phase 1, NUSP-2, 02-09, KNMI-publicatie: 200.

Volten, H., O. Munoz, E. Rol, J.F. de Haan, W. Vassen, J.W. Hovenier, K. Muinonen and T. Nousiainen, 2001, Scattering matrices of mineral aerosol particles at 441.6 and 632.8 nm. J. Geophys. Res., **106**, 17375–17401.

von Hoyningen-Huene, W., and P. Posse, 1997, Non-sphericity of aerosol particles and their contribution to radiative forcing. J.Quant. Spectr. Rad. Trans. **57**, 651–668.

von Hoyningen-Huene, W., K. Wenzel and S. Schienbein, 1999a, Radiative properties of desert dust and its effect on radiative balance. J. Aeros. Sci., **30**, 489–502.

von Hoyningen-Huene, W., T. Schmidt, S. Schienbein, A.K. Chan and J.T. Lim, 1999b, Climate relevant aerosol parameters of South-East Asian forest fire haze. Atm. Env., **33**, 3183–3190.

von Hoyningen-Huene, W., M. Freitag and J.B. Burrows, 2003, Retrieval of aerosol optical thickness over land surfaces from top-of-atmosphere radiance. J. Geophys. Res., **108**, 4260. doi:10.1029/2001JD002018.

von Hoyningen-Huene, W., A.A. Kokhanovsky, J.B. Burrows, V. Bruniquel-Pinel, P. Regner and F. Baret, 2006, Simultaneous determination of aerosol- and surface characteristics from top-of-atmosphere reflectance using MERIS on board ENVISAT. Adv Space Res., **37**, 2172–2177.

von Hoyningen-Huene, W., A.A. Kokhanovsky and J.P. Burrows, 2008, Retrieval of Particulate Matter from MERIS Observations. In: Y.J. Kim and U. Platt, (editors), Advanced Environmental Monitoring. Springer, Berlin, pp. 190–202.

Wang, M., and H.R. Gordon, 1994, Radiance reflected from the ocean-atmosphere system: synthesis from the individual components of the aerosol size distribution. Appl. Opt., **33**, 7088–7095.

Wang J. and S.A. Christopher, 2003, Intercomparison between satellite-derived aerosol optical thickness and PM25 mass: implications for air quality studies. Geophys. Res. Lett., **30**, 2095, doi:10.1029/2003GL018174.

Winker, D.M., R.H. Couch and M.P. McCormick, 1996, An overview of LITE: NASA's Lidar In-space Technology Experiment. Proc. IEEE, **84**, 164–180.

Winker D.M., J.R. Pelon and M.P. McCormick, 2003, The CALIPSO mission: spaceborne lidar for observation of aerosols and clouds, Proc. of SPIE, **4893**, 1–11.

Winker D.M., W.H. Hunt and M.J. McGill, 2007, Initial performance assessment of CALIOP. Geophys. Res. Lett., 34, L19803, doi:10.1029/2007GL030135.

Winker, D.M., M.A. Vaughan, A.H. Omar, Y. Hu, K.A. Powell, Z. Liu, W.H. Hunt, and S.A. Young, 2009., Overview of the CALIPSO Mission and CALIOP Data Processing Algorithms. J. Atmos. Oceanic Technol., **26**, 2310–2323, doi: 10.1175/2008JTECHA1221.1.

Young, S.A., 1995, Lidar analysis of lidar backscatter profiles in optically thin clouds. Appl. Opt., **34**, 7019–7031.

Young S.A., D.M. Winker, V. Noel, M.A. Vaughan, Y.Hu, R.E. Kuehn, 2005, Algorithm Theoretical Basis Document Part 5: Extinction Retrieval and Particle Property Algorithms, NASA-CNES document PC-SCI-203.

Young S. and M. Vaughan, 2009, The retrieval of profiles of particulate extinction from cloud-aerosol lidar infrared pathfinder satellite observations (CALIPSO) data: algorithm description. J. Atmos. Oceanic Technol., **26**, 1105–1119, DOI: 10.1175/2008JTECHA1221.1.

Zwally, H.J., B. Schutz, W. Abdalati, J. Abshire, C. Bentley, A. Brenner, J. Bufton, J. Dezio, D. Hancock, D. Harding, T. Herring, B. Minster, K. Quinn, S. Palm, J. Spinhirne and R. Thomas, 2002, ICESat's laser measurements of polar ice, atmosphere, ocean, and land. J of Geodynamics **34**, 405–445.

Chapter 7
Data Quality and Validation of Satellite Measurements of Tropospheric Composition

Ankie J.M. Piters, Brigitte Buchmann, Dominik Brunner, Ronald C. Cohen, Jean-Christopher Lambert, Gerrit de Leeuw, Piet Stammes, Michiel van Weele and Folkard Wittrock

7.1 Introduction

When using satellite tropospheric products for atmospheric research and monitoring or for other applications (see Chapters 8 and 9), it is essential to understand their significance. It is therefore important to take into account appropriate estimates of their uncertainties and to understand their capabilities and limitations. Some central questions are listed. How representative are the satellite retrieved products for the actual atmospheric state? Is there a bias and uncertainty with respect to the "truth"? How deep can satellites measure into the boundary layer? How well do they capture temporal variations of atmospheric composition, from daily fluctuations to decadal trends? How well do they capture spatial structures, from local emission sources to global features? It is important to realise that the answers to these questions depend considerably on the atmospheric situation (e.g. cloudy or clear-sky and polluted or clean situations), knowledge of ancillary parameters (e.g. surface elevation and

A.J.M. Piters (✉), P. Stammes and M. van Weele
Royal Netherlands Meteorological Institute (KNMI), De Bilt, The Netherlands

B. Buchmann and D. Brunner
Laboratory for Air Pollution Technology, Empa, Swiss Federal Laboratories for Materials Testing and Research, Dübendorf, Switzerland

R.C. Cohen
Department of Chemistry, University of California, Berkeley, CA, USA

J.-C. Lambert
Belgian Institute for Space Aeronomy(BIRA-IASB), Brussels, Belgium

G. de Leeuw
Climate Change Unit, Finnish Meteorological Institute, Helsinki, Finland
and
Department of Physics, Univeristy of Helsinki, Helsinki, Finland
and
TNO Environment and Geosciences, Utrecht, The Netherlands

F. Wittrock
Institute of Environmental Physics, University of Bremen, Germany

J.P. Burrows et al. (eds.), *The Remote Sensing of Tropospheric Composition from Space*, Physics of Earth and Space Environments, DOI 10.1007/978-3-642-14791-3_7,

reflection, as well as the knowledge of instrumental characteristics and viewing geometries (e.g. high sun elevation versus twilight, and instrument noise which depends on the orbital positions, enhanced noise being typical in the South Atlantic Anomaly).

In part these questions can be answered by "validating" the satellite data, see definitions in Section 7.2.1. A major goal of validation is to describe and to quantify the uncertainty of a satellite product in such a way that it is of direct use for the specific expected research or application areas, in other words to assess its fitness-for-purpose. Another goal is to test and confirm the theoretical error budget of satellite data derived from algorithm sensitivity studies. Table 7.1 lists the estimated current uncertainties of some relevant tropospheric satellite products and the main source of validation measurements.

Usually validation is based on comparisons with independent measurements of the same parameter with known uncertainties. When comparing satellite data to correlative measurements or to modelling results, differences in observed air mass and in observation, and retrieval techniques, have to be taken into account. A summary of the different methods of comparing two data sets, including the use of models is given in Section 7.2. In addition, the methods used for verification and monitoring of quality parameters are also described. The validation of tropospheric satellite products is a relatively young field, and the methods described in Section 7.2 are expected to develop and evolve as tropospheric research with satellite data develops.

Section 7.3 addresses various aspects of quality assurance, which starts before launch and is included in the mission planning and continues during satellite life time.

Validation measurements have, typically, to be performed for a wide range of possible values and for different atmospheric and measurement conditions. Comparison methods have to consider appropriate differences in spatial and temporal resolution and sampling, especially in the presence of significant spatial structures and temporal variations of the species. It is essential that the correlative measurements and validation studies provide the necessary knowledge of all parameters affecting the measurement and the retrieval algorithm (e.g. surface albedo, cloud fraction) and the relevant ranges of the parameters that might have an impact on the uncertainty (see Chapters 2, 3 and 4).

Section 7.4 provides a more comprehensive understanding of the characteristic differences of the various tropospheric species, which have a direct impact on the validation strategy: where, when and how to perform validation measurements and carry out comparisons. The distribution and variability of the tropospheric species are discussed as well as the relevant and significant atmospheric processes. In Section 7.4, we discuss further what measurements are needed to investigate and verify the sensitivities of retrievals, to key parameters such as clouds, albedo and aerosol.

Several measurement techniques are currently used for the validation of tropospheric satellite data products. Section 7.5 details their main characteristics. Most

Table 7.1 Estimated uncertainties of some of the relevant tropospheric satellite products and the main source of validation measurements. Uncertainties will vary for different instruments and situations. Values given here are from the quoted references

Troposp. satellite product	Tropospheric lifetime[a]	Satellite instrument	Dominant validation source	Estimated current best uncertainty	References	Note
NO_2	h to d [b]	GOME SCIAMACHY OMI GOME-2 TES	MAXDOAS/ aircraft	30–50%	Celarier et al. (2008), Brinksma et al. (2008), Bucsela et al. (2008)	
O_3	h to d[b]	GOME SCIAMACHY OMI GOME-2 TES IASI AIRS	O_3 sondes	20–30%	Nassar et al. (2008)	
CO	2 mt	MOPPIT SCIAMACHY AIRS TES IASI	FTIR	20–30%	Dils et al. (2006)	5% in monthly mean
HCHO	h to d[c]	GOME SCIAMACHY OMI GOME-2	MAXDOAS	20–50%	De Smedt et al. (2008)	
CHOCHO	h to d[c]	SCIAMACHY OMI GOME-2	MAXDOAS	30–50%	Wittrock et al. (2006)	In monthly mean
CH_4	12 y	SCIAMACHY AIRS TES IASI	FTIR	1–2%	Dils et al. (2006)	

(*continued*)

Table 7.1 (continued)

Troposp. satellite product	Tropospheric lifetime[a]	Satellite instrument	Dominant validation source	Estimated current best uncertainty	References	Note
SO_2	h	GOME SCIAMACHY OMI GOME-2	Aircraft	1.5 DU	Krotkov et al. (2008)	
BrO	m	GOME SCIAMACHY OMI GOME-2	–	–		
H_2O	–	GOME SCIAMACHY GOME-2 AIRS TES IASI	SSM/I	0.1–0.2 g/cm^2	Noel et al. (2005)	In daily mean
CO_2	100 y	SCIAMACHY GOSAT TES IASI AIRS	FTIR	5%	Dils et al. (2006)	In monthly mean
N_2O	114 y	SCIAMACHY TES IASI	FTIR	10%	Dils et al. (2006)	
Aerosol Optical Depth	up to a week	[d]	Sun-photometer (AERONET)	$\pm 0.05 \pm 0.15\tau$	Remer et al. (2005)	Over land

[a]average life times (or ranges) in the troposphere; *m* minutes, *h* hours, *d* days, *mt* months, *y* years
[b]lower to upper troposphere
[c]depending on latitude (radiation)
[d]Aerosol Optical Depth and many other aerosol properties are retrieved from: POLDER, MODIS, CALIPSO, AVHRR, SeaWiFS, MERIS, MISR, SEVIRI, AATSR, GOME, SCIAMACHY, OMI and GOME-2., see Chapter 6 for details

of the existing instruments contribute to international networks and data centres, of which examples are given.

Recommendations for future validation strategies are given in Section 7.6.

7.2 Methods of Validation

The basis for investigating the quality of satellite data is comparing them to reference data obtained independently and of known quality. The source of the reference data may be from ground-based, air-borne and balloon-borne measurements. In addition, satellite measurements and model output can provide valuable comparisons to help to understand the quality of the satellite data.

The most common methodology for comparing two independent data sets is the comparison of columns and profiles, which are coincident in both time and space. This typically yields an average difference and a spread, which are propagated to estimate for the bias and the uncertainty in the satellite data, provided that those from the reference data are themselves well known.

In the next subsections it is shown that a simple direct comparison as sketched above is often insufficient for the validation of satellite data, and that more sophisticated approaches are needed to characterise the quality of different aspects of the data.

7.2.1 Definitions

The process of assessing the quality of satellite data products involves at least the following aspects: validation, verification, calibration and monitoring.

Validation is defined by the Committee on Earth Observation Satellites (CEOS) as the process of assessing, by independent means, the quality of the data products derived from the system outputs. The ISO guide of metrology vocabulary (VIM) (ISO 2007) further defines *validation* as verification where the specified requirements are adequate for an intended use. Validation therefore addresses the fitness-for-purpose of the data products via comparisons.

Verification is defined as the provision of objective evidence that a given data product fulfils specified requirements. Data products are checked for internal consistency, out-of-bound values, geographical distribution, and statistical behaviour. The comparison of retrieval methods and the comparison of experimental data with that from models are also called verification. Verification identifies errors in the retrieval software or auxiliary data. Validation and verification help to optimise the retrieval algorithms.

Calibration is defined as the process of quantitatively defining the system responses to known, controlled signal inputs (CEOS). It includes routinely checking the quality of the measured reflectance or transmittance with respect to possible changes in instrument behaviour. Applied calibration functions usually take changes of the instrument into account.

Monitoring is the process of routine analysis of specific quality parameters to detect instrumental, processor or auxiliary data problems.

Validation should result in an estimate of the bias and the uncertainty. The ISO guide 99 Vocabulary for International Metrology (VIM) (ISO 2007) defines the *bias* as the systematic error of indication of a measuring system, and the *uncertainty* as the parameter that characterises the dispersion of the values that are being attributed to a measured quantity, based on the information used. The bias therefore is a measure of the total systematic errors, and the uncertainty of the total random errors.

Validation of a data product results in an estimate of the bias and the uncertainty, which may depend on geographical, algorithm and instrumental parameters.

7.2.2 Comparing Data Sets

The comparison of two data sets usually comprises the following: the finding of suitable collocations between data sets, the selection and filtering of data, the treatment of the data, and the analysis of the data values and their differences.

a Finding Collocated Data

The most common way to find collocations is to collect all data within an arbitrary temporal and spatial coincident window spanning, typically, from 200 to 1,000 km and from 1 h to 2 days. The presumed advantage of such a selection window is that the variability caused by differences in air mass is reduced. In Section 7.4.1, the tropospheric processes that underlie this variability are discussed. As expected, this approach works satisfactorily for long-lived species having negligible variability in space and in time, and for which the retrieval has a moderate sensitivity to the vertical structure. When atmospheric variability increases, differences in smoothing and sensitivity result in an increase in the comparison noise. In this context, significant effects, including systematic biases, have been identified for short-lived species, like NO_2 and BrO (e.g. Schaub et al. 2007).

The effective location of a remote sensing measurement can be quite distant from the location of the instrument itself and from the location the instrument is pointing to.

The reason for this is that all remote sensing instruments make use of the absorption, emission or scattering of light by the atmospheric constituent to be measured. Passive instruments measure direct or scattered light from the sun or another light source. Active instruments measure their own scattered light. Absorption, emission or scattering along the path the light has travelled from the light source to the instrument is accumulated. The effective location and extent of these measurements can be calculated using a radiative transfer and atmosphere model.

Therefore, collocation criteria should be applied to the effective locations of the measurements rather than to the locations of the instrument or the location the instrument is pointing to.

The effective horizontal location of profiles measured from a balloon or aircraft is often determined by taking the average location over a relevant altitude range. Emmons et al. (2009), for instance, determine an effective location of aircraft profiles of carbon monoxide, CO, to be compared to MOPITT profiles as the average profile between 500 and 800 hPa, the range where MOPITT has its highest sensitivity.

For long-lived species, the number of collocations can be enhanced using air mass trajectory calculations. In this case, one assumes that the same air mass will be observed at every point along the trajectory.

b Selection and Filtering

The selected data sets for comparison are prepared by filtering the data according to known quality parameters. These quality parameters are usually documented in product description documents or data "disclaimers". Filtering is performed on quality flags, error bars, solar zenith angle values, cloud cover values, temperature values, surface albedo, terrain variability, etc. Historical verification and validation analysis may point to certain low-quality data as well. Nassar et al. (2008), for instance, describe how ARM-SGP ozone-sonde measurements have been critical in identifying erroneous TES retrievals that can sometimes result when the lowest layers of the atmosphere are in emission (Fig. 7.1). This finding led to the inclusion of an "emission layer flag" in a subsequent version of the product.

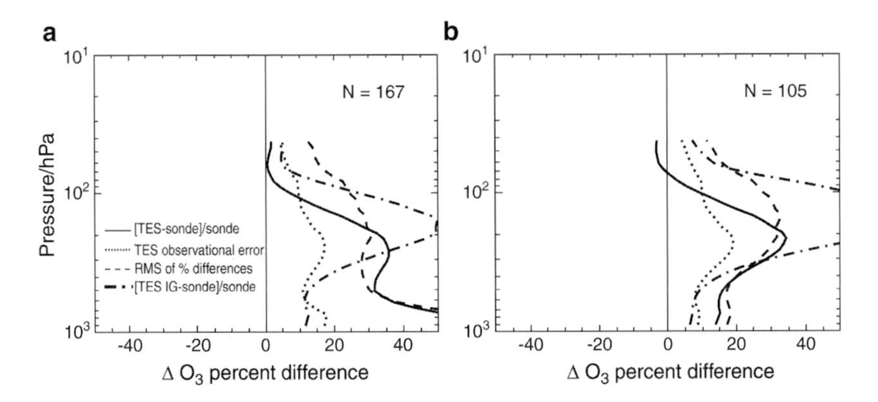

Fig. 7.1 (**a**) Average TES-sonde percent difference and RMS for night observations, screened only by the general data quality flag. Note the large values for both average difference and RMS near the surface. (**b**) Night observations excluding TES scenes with an emission layer identified. IG indicates initial guess. (from Nassar et al. (2008)).

Selection of subsets of data is performed to study the quality of the satellite data for specific situations such as high surface albedo, cloud-free, tropics, or polluted regions. The division in latitude bands or seasons is very common for datasets with several thousands of collocated points. For the validation of TES tropospheric ozone with ozone sondes, for example, Nassar et al. (2008) found ~1,600 collocations in a 2 year data set, so that the comparisons could be performed in 6 latitude bands, with 35–699 collocations in each. The northern mid-latitudes (35°–56°N) data set with 699 collocations could then be subdivided in data sets for four seasons, with 45–409 collocations in each.

c Data Treatment

Vertical Representation

A tropospheric satellite product is often expressed as a column amount below the tropopause or, in the case of profiles, as partial columns in discrete layers or as number densities at discrete altitudes. However, the retrieved values are generally sensitive to variations at other altitudes as well. The *averaging kernel* (Chapter 2) describes the sensitivity of the retrieved values to the actual values at different altitudes. It is a measure for the vertical resolution of the retrieved profiles or, in the case of tropospheric columns, for the sensitivity to, for example, the boundary layer. When comparing satellite products with correlative data, differences in vertical sensitivity between these data sets have to be taken into account, for example by using the averaging kernel information.

Methods for comparing profiles with different degrees of vertical smoothing are described by Rodgers and Connor (2003) and Calisesi et al. (2005). From these methods a bias and uncertainty can be attributed to the satellite retrieved values. These should however not be mistaken as estimates of the deviation from the "true" value, but rather as estimates of the deviation from the expected value based on the "truth" in combination with known retrieval and measurement sensitivities.

An example for the validation of GOME tropospheric nitrogen dioxide, NO_2, is given in Fig. 7.2. Schaub et al (2006) made comparisons with ground-based columns deduced from *in situ* measurements at different altitudes in the Alps, both with and without applying averaging kernels. They found a clear improvement of the comparison under cloudy conditions after multiplying with the averaging kernels, which implies larger errors in the *a priori* NO_2 profiles under cloudy conditions.

Lamsal et al. (2008) constructed ground-level NO_2 (S_O) from OMI tropospheric columns (Ω_O), using the GEOS-CHEM model as interface:

$$S_O = \frac{v S_G \Omega_O}{v \Omega_G + (1 - v)\Omega_G^F} \tag{7.1}$$

where S_G and Ω_G are the ground-level and tropospheric NO_2 from GEOS-CHEM, v is the ratio of the local OMI tropospheric NO_2 column over the mean OMI tropospheric NO_2 column averaged over the GEOS-CHEM grid cell, and Ω_G^F is the modelled free tropospheric NO_2 column (Fig. 7.3).

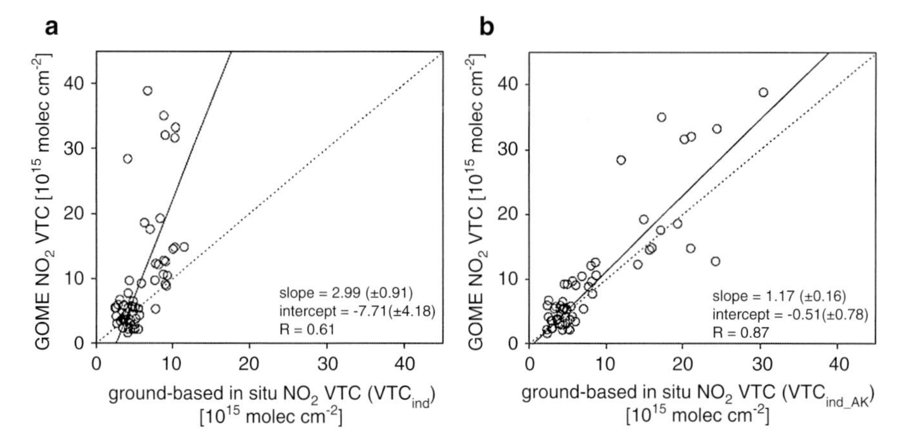

Fig. 7.2 Comparison between GOME tropospheric NO_2 columns and those derived from ground-based *in situ* measurements before (**a**) and after (**b**) multiplying the ground-based profile with the averaging kernel for cloudy conditions. Columns with SCDtrop/SCD > 50% are rejected. (from Schaub et al. (2006)).

Time Differences

For validation of short-lived species like BrO, or NO_2, photochemical corrections can be used to account for time differences between the observations. This has been applied successfully for validation of stratospheric columns and profiles (Theys et al. 2006; Dorf et al. 2006), but is not yet common practice for tropospheric product validation. Brinksma et al. (2008) linearly interpolate ground-based MAX-DOAS NO_2 data to the time of the satellite measurement, which only works if the gaps between the MAX-DOAS measurements are not too large.

Daily averages of FTIR measurements from CO_2, CH_4, N_2O, and CO have been fitted to a third order polynomial in time. The fit-values at the time of satellite measurements have been used to compare with SCIAMACHY values (which are at a given geographic location only available every 6 days) (Sussmann and Buchwitz 2005; Dils et al. 2006), see Fig. 7.4. This only works for rather long-lived and therefore well-mixed species and when the ground-based instrument is far away from variable sources.

Horizontal Representation

Columns measured from a high-altitude station are typically smaller than the total column measured by a satellite, since the average surface elevation in the satellite pixel is lower than the station altitude. To account for these differences the column amounts can be converted to average volume mixing ratios (Dils et al. 2006). For CH_4 and CO, Dils et al. (2006) additionally used a modelled scaling factor to

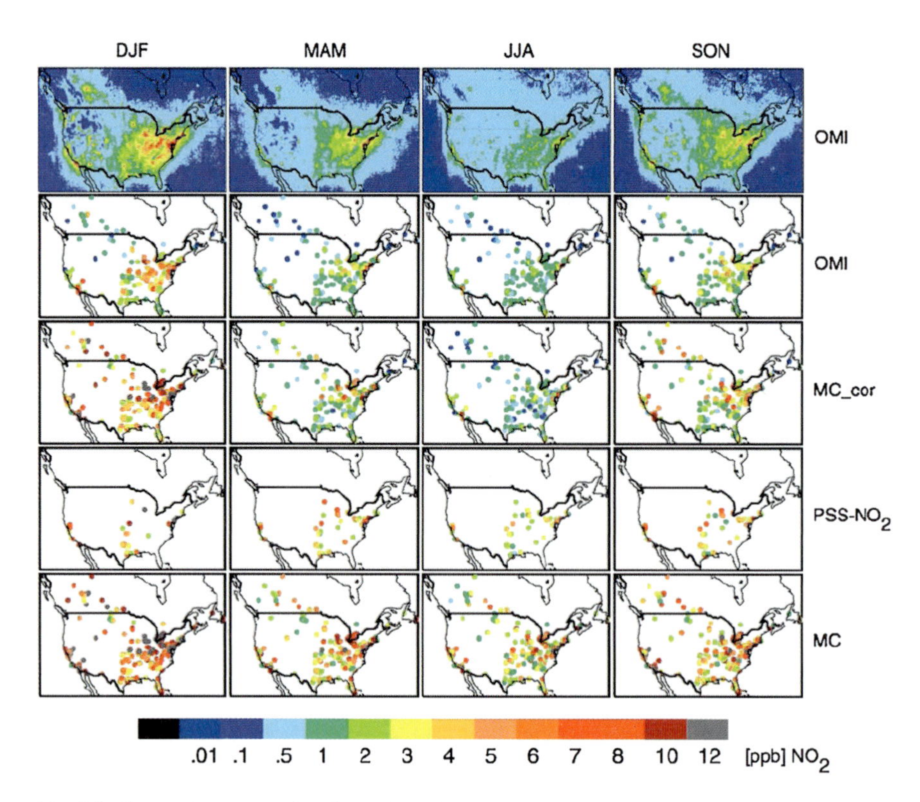

Fig. 7.3 Seasonal average of surface NO_2 mixing ratios for the year 2005. *First row*: A seasonal map of OMI-derived surface NO_2 over North America. *Second row*: The collocated OMI-derived surface NO_2 at the molybdenum converter *in situ* sites. *Third row*: The corrected molybdenum converter *in situ* measurements. *Fourth row:* Alternative photochemical steady-state calculation of surface NO_2. *Fifth row*: The molybdenum converter *in situ* measurements. (from Lamsal et al. (2008)).

account for the fact that the volume mixing ratio is not constant as a function of altitude.

To account for spatial inhomogeneity within the tropospheric NO_2 field, Brinksma et al. (2008) averaged MAX-DOAS measurements in three different directions.

Kramer et al. (2008) constructed representative FOV-weighted *in situ* NO_2 measurements x' for validation of OMI tropospheric NO_2 by weighing the *in situ* NO_2 mixing ratio x_u from an urban station in Leicester and that from a background station x_{bg} with the fraction a of the satellite pixel sampling the Leicester urban area: $x' = ax_u + (1 - a)x_{bg}$ (Fig. 7.5).

Celarier et al. (2008) constructed representative satellite tropospheric NO_2 measurements for comparison to an MFDOAS by integrating the OMI NO_2 field within the MFDOAS Field of View (Fig. 7.6).

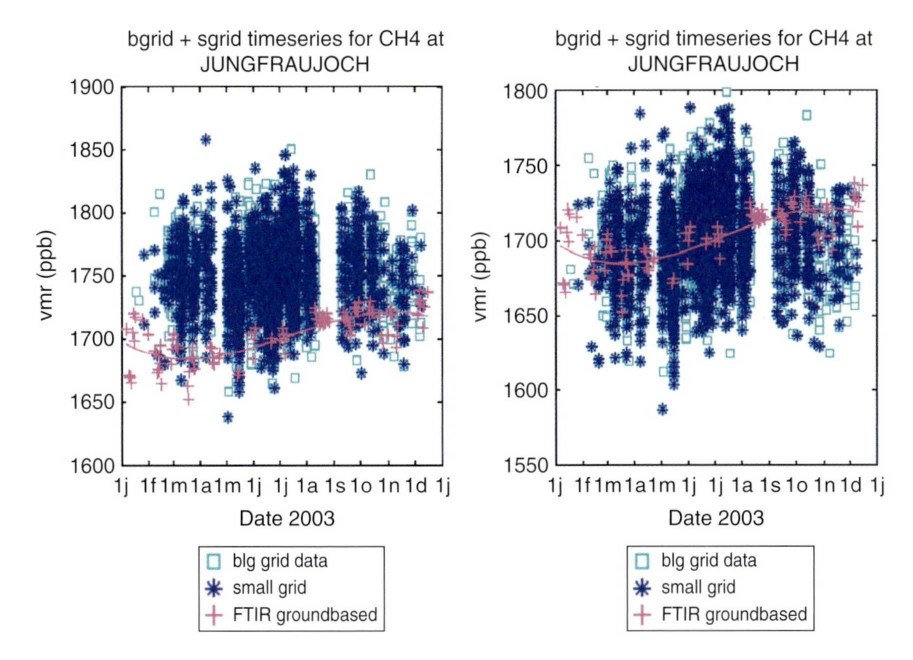

Fig. 7.4 FTIR data relative to their polynomial fit. Time series of CH_4 measurements at Jungfraujoch from ground-based FTIR (+) and SCIAMACHY IMAP-DOAS (*open squares* for large collocation grid; *stars* for small collocation grid). *Left*: original CH_4 mixing ratios (*open squares* and *stars*) and third order polynomial fit through the ground-based FTIR data (*solid line*). *Right*: CH_4 mixing ratios (*open squares* and *stars*) after the application of a correction factor and third order polynomial fit through the FTIR ground-based data (*solid line*). (from Dils et al. (2006)).

Noise Reduction

For satellite products with large retrieval uncertainties, it is often not feasible to compare individual measurements. For these products, (weighted) averages in time and space are compared to correlative data. It is important to realise that estimates for the uncertainty of the satellite data product resulting from such a comparison are valid for the averaged products and not for the individual measurements. An example is given in Buchwitz et al. (2007), where SCIAMACHY and MOPITT data are averaged over specific regions before comparing them (see Fig. 7.7).

d Analysing the Data

The last part of the comparison is the analysis of the data. The data sets can be analysed:

- As function of time: do they have the same temporal behaviour? scatter plot, with correlation coefficient and/or a linear fit.

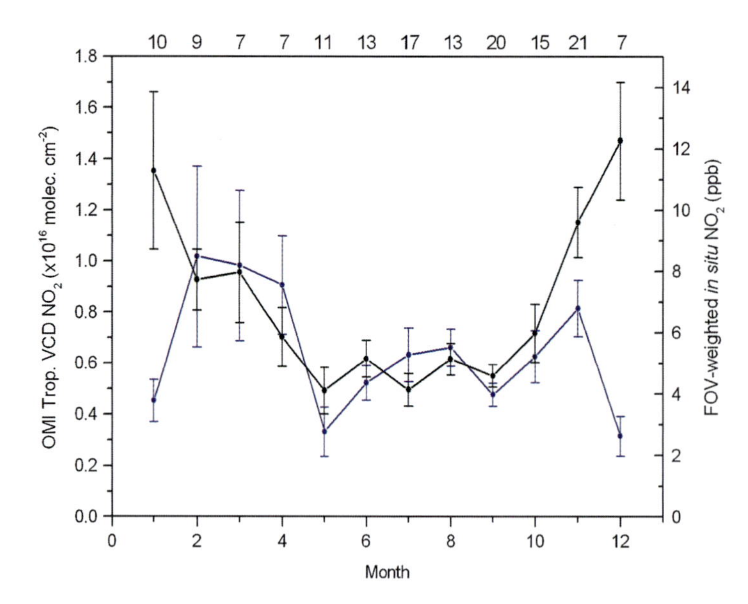

Fig. 7.5 Monthly averages of NO_2 for 150 cloud-free days (cloud fraction less than 20%) between January 2005 and December 2006 for OMI tropospheric columns (*blue*) and coincident mean FOV-weighted *in situ* data from Leicester area (*black*); the error bars show the standard deviation on the mean. The number of cloud-free observations for each month is shown at the top of the plot (from Kramer et al. (2008)).

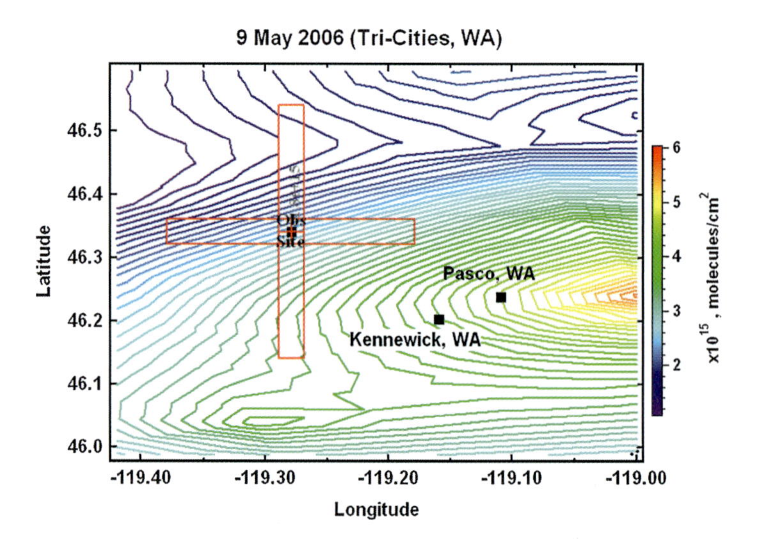

Fig. 7.6 Tropospheric NO_2 vertical column density over the Tri-Cities area of Washington State on the 9th May 2006. The contour map is derived from the individual OMI FOV measurements. The *red rectangles* show the tropospheric region viewed by the MF-DOAS instrument. The centers of population for the cities of Kennewick and Pasco are indicated by *black squares*. (from Celarier et al. (2008)).

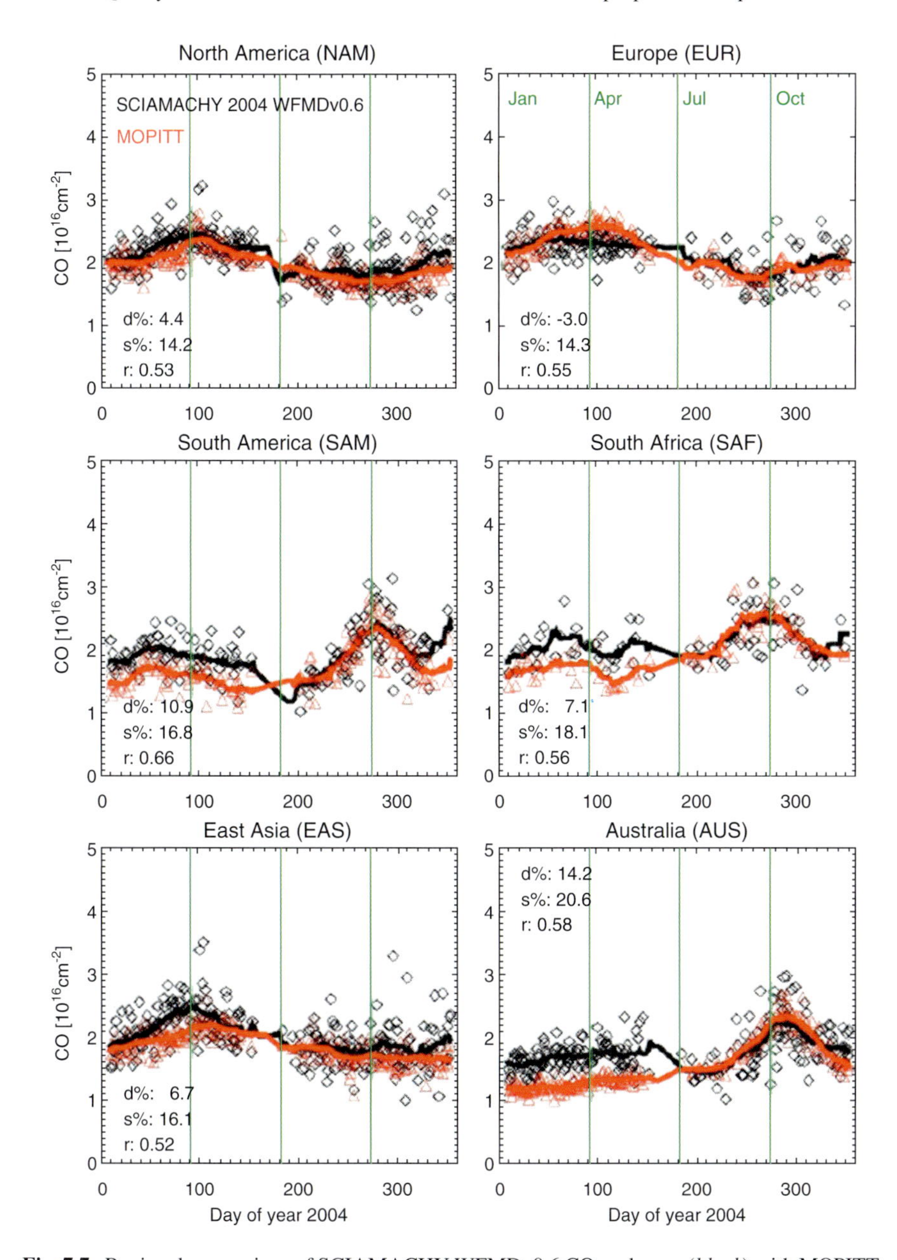

Fig. 7.7 Regional comparison of SCIAMACHY WFMDv0.6 CO columns (*black*) with MOPITT (*red*) for the year 2004. The *symbols* show the daily averages of all coincident grid points. For SCIAMACHY, all measurements have been averaged for which the WFMDv0.6 quality flag indicates a successful measurement. The *solid lines* represent 30 day running averages. For each region the following quantities are shown which have been computed based on the unsmoothed daily averages: d% is the mean difference SCIA–MOPITT in percent, s% denotes the standard deviation of the difference in percent, and r is the correlation coefficient. (from Buchwitz et al. (2007)).

- Difference as a function of instrumental and geophysical parameters.
- Distribution function of the differences, shape, median, mean.
- Geographical patterns: do the patterns look the same?

7.2.3 Use of Models

Models are used in satellite data validation in three ways:

1. model data (e.g. vertical trace gas distributions) are used to invert satellite measurements in order to obtain tropospheric data (e.g. tropospheric columns),
2. satellite data are used to validate models, and
3. model data are used to validate satellite measurements.

Satellite observations of tropospheric constituents have been used widely to validate models (Velders et al. 2001; Blond et al. 2007; Lauer et al. 2002; de Laat et al. 2007) or to improve the spatial and temporal representation of model emissions (Martin et al. 2003; Müller and Stavrakou 2005; Konovalov et al. 2008). In these studies, the satellite observations are considered providing a "truth" within the bounds of the observation errors, thus serving as reference for the models. This implicitly assumes that the measurements are well characterized and more specifically that their uncertainties are well known. However, as demonstrated for instance by van Noije et al. (2006), differences between individual satellite retrievals of tropospheric composition can be larger than expected from available uncertainty estimates indicating an often incomplete understanding of error sources or error correlations.

The reverse approach, using model information to identify inconsistencies in satellite data, has received comparatively little attention so far. Again a realistic characterization of uncertainties is required but, in this case, of uncertainties in the model data. Satellite measurements have frequently been compared with model output to obtain a basic understanding of the quality of the observations (Buchwitz et al. 2006; de Laat et al. 2007; Yudin et al. 2004) but more quantitative evaluations would be desirable. A valuable approach would be to run a model in parallel with a satellite mission and continuously monitor the differences between model fields and observations. This would help identifying a number of possible algorithm or instrument problems such as:

- scanning/viewing or solar zenith angle dependencies,
- data breaks due to changes in satellite operation procedures or instrumental problems,
- slow degradations in data quality and slow instrumental drifts, and
- seasonal biases.

A data assimilation system in which the satellite observations are incorporated into a model framework can be applied in a similar way. Here the statistics of the observation minus forecast (OMF) errors would have to be monitored (Yudin et al.

2004; Eskes et al. 2003). Of particular use would be a system assimilating multiple independent sources as this would allow identifying inconsistencies between different satellite sensors and not only between observations and model.

Finally, models have been demonstrated to be of great use in validation studies by enhancing the comparability between satellite and *in situ* observations. Ordoñez et al. (2006) and Lamsal et al. (2008), for instance, used model profiles to relate *in situ* surface NO_2 observations to vertical tropospheric columns from satellites. Models, in particular in connection with data assimilation, can also serve to establish a link between different satellite observations taken at different times or different locations. Boersma et al. (2008), for instance, employed the GEOS-CHEM model to estimate the photochemical reduction in tropospheric NO_2 columns between the overpass times of SCIAMACHY (10:00 local time) and OMI (13:30). Similarly, Shindell et al. (2005) showed that a three-dimensional global composition model can be used to account for differences in retrieval sensitivity between two different CO sensors and to account for the spatial and temporal separation of the measurements.

In summary, the potential of models for satellite validation and verification and for data quality monitoring has not yet been fully explored. Assimilation of multiple sources of tropospheric constituent observations appears as a promising pathway for the future as it will greatly augment the comparison possibilities between different satellite sensors and potentially even between sensors measuring different species since they are often closely coupled through chemistry or common sources and sinks.

7.2.4 Data Variability

Part of a proper quality assessment is also the description of the variability of a satellite product in space and time, assessing which part is determined by calibration and retrieval uncertainties, and which part is 'real'. The natural variability of a compound is often quite well known from *in situ* and other remotely sensed observations. Even though *in situ* observations do not provide direct information on volume integrated quantities, they can provide important constraints on the expected variability of a tropospheric column or layer. In addition to the simple consideration of variability at a given location, a comprehensive characterisation of the spatio-temporal behaviour of the product is necessary. This includes a description of the seasonal cycle at different locations and a description of meridional gradients and contrasts between polluted (continental) and remote (oceanic) locations.

A larger than expected variability of satellite data or unexpected seasonal or spatial behaviour may be due to various factors:

- limited precision of the satellite measurement,
- an incomplete separation of the stratospheric signal (Section 7.4.2),

- instrumental and calibration problems not included in the uncertainty estimation,
- wrong assumptions or wrong evaluation of surface albedo or terrain elevation,
- wrong assumptions on the presence or effect of clouds, snow, or ice, and
- wrong use of ancillary data in the retrieval.

A comparison between different OMI NO_2 retrievals, for instance, revealed significant differences in the amplitude of the seasonal cycle pointing towards a significant dependence of the seasonal variation on *a priori* assumptions and the specific retrieval algorithm (Richter et al. 2007).

Some products have been found to exhibit unrealistic contrasts between land- and sea-surfaces or, in the case of near-infrared instruments, to show differences over hot and cold surfaces. At high latitudes, the high solar zenith angles and varying snow and ice covers pose additional problems for the retrieval. A close analysis of the geographic variability will thus help identify some of the problems that introduce unrealistic spatial gradients not seen in ground-based measurements.

Since trace-gas concentrations are usually more variable in the planetary boundary layer than in the free troposphere, the sensitivity of the satellite measurement to the near-surface levels needs to be considered when comparing with other observations. Differences in CO levels between SCIAMACHY and MOPITT, for instance, can partly be attributed to the larger sensitivity of the SCIAMACHY measurement to boundary layer CO (Buchwitz et al. 2007). Such differences can also be expected to affect the amplitude of seasonal cycles and contrasts between polluted and unpolluted regions.

Zhou et al. (2009) demonstrated that using a high-resolution topography data set in satellite retrievals of tropospheric NO_2, instead of the global low-resolution one which is usually used, has a significant effect on the accuracy of the data over the Alpine region.

As pointed out in the previous section, data assimilation offers great opportunities for analysing the consistency of a satellite product with other observations, particularly for the consistency in observed seasonal and spatial variations. This was demonstrated for instance by Bergamaschi et al. (2007) who compared the temporal and spatial variation in SCIAMACHY CH_4 observations with inverse model simulations that had been optimized versus high accuracy CH_4 surface measurements from the NOAA ESRL network.

7.3 Quality Assurance

Quality assurance is a prerequisite of all remote sensing products. Apart from a check on operational quality assurance, the derived impact of algorithm improvements, trends of parameters or degradation processes on the product have to be taken in account to provide a product that can provide insights over a long period of time.

7.3.1 Validation and Mission Planning

The satellite mission planning, set-up by the space agencies involved, usually includes several phases which are important to ensure the data quality.

Before launch the main activities to ensure the data quality are the on-ground calibration and the development and testing of the calibration and retrieval software.

The first phase after launch is called the *commissioning phase*, where special measurements are performed to verify the proper performance of all instrument parts.

In the *main validation phase* the satellite instrument is operated optimally for validation measurements. This can be done by measuring over important validation sites or making measurements during campaign periods. Usually several validation campaigns are planned and supported by the agencies for one mission.

Some of the missions also have a *long-term validation phase*. In this phase a few campaigns are planned and routine measurements are collected. This phase is necessary to monitor the data quality over the whole mission duration.

After launch, the agencies perform a day-to-day monitoring of the satellite instrument to detect sudden changes in data quality (Section 7.3.6)

The committee on Earth observation satellites (CEOS) plays an important role in fostering and coordinating interactions between mission scientists and data users. This includes recommendations on network validation sites, development of comprehensive validation methodologies involving ground-based and space-borne assets, and specification of comprehensive and consistent multi-mission validation datasets.

7.3.2 Calibration

Calibration of instruments, whilst not being the main subject of this chapter is inherently related to validation through its impact on the quality of higher level data products.

For all remote sensing instruments three issues have to be calibrated.

a Viewing Geometry

The instrument viewing geometry and spatial location of the observed ground scene can be calibrated with fixed Earth targets or high contrast scenes and with the sun, moon or stars.

b Wavelength

The wavelength registration of the instrument detectors can be calibrated by special lamps with spectral lines at fixed wavelengths, or by the well-known Fraunhofer lines in the solar spectrum.

c Absolute Radiance

The radiometric calibration is, for a large part, performed when the instrument is still on the ground. The behaviour of the detectors is studied in detail and as a function of the angle of the incident light, the detector temperature, etc. In flight, changes in the calibration as a function of orbital position are examined, and degradation is monitored (Section 7.3.5).

7.3.3 Lower-Level Data Products

The accuracy of lower level satellite products such as radiance, irradiance, polarisation, and reflectance directly or indirectly influences the accuracy of the higher level products, and the estimated abundances of tropospheric species. Comparison with correlative data and model output is necessary to quantify instrumental spectral features and reflectance offset. On the other hand, long-term studies of higher level data can point to unresolved problems in lower level data.

Lower-level data accuracy is essential to get good geophysical data products. However, it is often necessary to do a first geophysical data validation to have a good idea of lower-level data quality. Trace gas column retrievals are very sensitive to spectral calibration errors. Relative spectral fitting algorithms, like DOAS, can show spectral calibration errors such as shifts and squeezes of the spectrum or bad pixels. Algorithms including those for ozone profile retrieval, AAI retrieval, cloud retrieval, and aerosol optical thickness retrieval, using the absolute reflectance are sensitive to errors in radiometry.

The most important methods of ensuring the quality of the lower-level data products are listed below.

- Geolocation validation: using coastlines, islands, etc.;
- Radiance validation: usng radiative transfer modelling (RTM) for the reflectance of selected scenes, such as cloud free deserts, snow/ice, ocean, or bright homogeneous clouds; combine reflectance with reference solar reference spectrum to obtain the radiance (Hagolle et al. 1999; Jaross and Warner 2008);
- Irradiance validation: using solar reference spectrum from literature;
- Reflectance validation: using RTM of selected scenes; comparison with other collocated satellite data (Tilstra et al. 2005); and;
- Polarization validation: using physically acceptable values; using RTM for selected scenes; using the sun as an unpolarised source; comparison with other

collocated satellite data (Tilstra and Stammes 2007; Schutgens and Stammes 2003; Schutgens et al. 2004).

7.3.4 Retrieval Algorithm Optimisation

Demonstration of the validity of a retrieval algorithm is often done for a few specific situations. After implementation of a prototype algorithm in an operational environment and processing of large data sets, the resulting products are carefully verified to find internal inconsistencies and erroneous behaviour. Satellite retrieval algorithms use auxiliary data, retrieval assumptions, and simplifications. These can be valid for certain situations and less valid for others. Specific studies of the validity of the retrieval assumptions for well-chosen datasets directly contribute to the improvement of the algorithm.

Simplifications or misinterpretations of auxiliary data used in the retrieval can result in systematic errors in the retrieved quantities that may depend on geophysical, instrumental or algorithmic parameters. It is important to investigate the influence of these parameter-dependent systematic errors on the intended scientific use. For example, global and regional chemical family budgets might be altered by fictitious spatial structures and temporal signals generated by the retrieval algorithms and superimposed on the actual geophysical signals. Therefore, these systematic errors, dependent on retrieval parameters, should be tracked down systematically and characterised in detail.

As a first stage, prior to performing full geophysical validation of a mature data product, validation has often played and still plays a diagnostic role in the improvement of retrieval algorithms. Careful investigation of comparison time series and the use of assimilation tools have been powerful in revealing internal inconsistencies in satellite data, such as gaps, shifts, systematic biases between data acquired at two different viewing angles, drifts, cycles, etc.

Intercomparisons of satellite data retrieved with independent algorithms have suggested possibilities for improvement.

7.3.5 Instrument Degradation

Instruments in space suffer from ageing caused by a variety of processes including contamination of optical surfaces and the impact of cosmic particles. The effect on the quality of the retrieved satellite data products needs to be carefully monitored. This is especially important with respect to the use of the data products to identify changes or trends in the data product. One important effect is degradation of the optical throughput, especially in the UV. This is caused by UV absorbing species depositing on the optics and reducing the signal-to-noise ratio, impacting on the detection limits and information content of the satellite data products. Cosmic particles hit the detectors periodically, causing an increase in dark current and, for certain types of detector, introduce a so-called random telegraph signal (RTS). The first of these effects can, in principle, be corrected if there are enough dark current measurements, the second introduces an extra noise term (Dobber et al. 2008). Decreasing signal, increasing dark current, and increasing noise all influence the uncertainty of the satellite products. Long-term validation is necessary to assess the impact of instrument degradation on the data quality.

7.3.6 Overall Quality Monitoring

During calibration, verification and validation studies it becomes clear which parameters can point to instrument, processor, or auxiliary data problems affecting the quality of the resulting satellite products. These parameters are monitored on a day-to-day basis.

Examples of instrumental problems which can have an impact on the data quality are out-of-bounds detector temperatures, partial field of view blocking, contamination, or degradation. Such problems can occur suddenly or gradually and the impact on the data quality cannot always be predicted. A parameter like detector temperature can easily be monitored as it is part of the house-keeping data, but it is less straightforward to detect other instrumental problems.

Sudden processor problems can occur in various situations, such as a new instrument operation mode, an orbit number passing a certain threshold value, or a new processor operating system.

Auxiliary data problems can occur when, for instance, wrong files are used after a change in the processor environment.

Many problems are identified only by chance, because somebody detects unexpected behaviour in the data. In order to detect most problems early it is important that the data are analysed on a daily basis during the complete mission, looking for sudden changes and trends in the estimated retrieval errors and in the quality flags in the product, for different orbital phases, viewing geometries, instrumental modes, etc. It is also possible to detect problems by looking for sudden or gradual changes in the validation results. For this, it is necessary to have many routine correlative measurements of known accuracy taken during the whole mission and well spread over the globe.

Monitoring the data values, retrieval errors and flags, and even a first-order intercomparison with correlative measurements can be automated, but it might be difficult to describe what kind of anomalies should trigger further study. One way of doing it is to determine for each monitoring parameter the mean, median, standard deviation, minimum and maximum values from a well-behaved subset for different orbital phases, viewing geometries, instrumental modes, etc., and define thresholds based on these statistical values. When a parameter gets out-of-bounds it can trigger further manual analysis.

Quality Assurance (QA) software tools, operated by space agencies, undertake part of this quality monitoring. However, these tools do not usually make routine comparisons with correlative data and lack the more sophisticated algorithms to detect anomalies. They assist the instrument experts to detect anomalies by visual inspection of time series and world maps.

Even in the future when the QA tools become more sophisticated, it will never be possible to detect all problems with such tools, and expert scientific analysis will always remain necessary for the detection of problems with more subtle effects on the data.

7.4 Validation Characteristics of Tropospheric Products

In this section, we detail the differences in characteristics of the various tropospheric species, as listed in Table 7.1, which have a direct impact on the validation strategy.

Validation of tropospheric satellite products helps to detect uncertainties related to instrument characteristics and observational technique in combination with retrieval assumptions. Therefore it is important to plan the validation measurements to minimize the spread due to the real variability of the species. Variations that are "real" are principally determined by controlling atmospheric processes such as chemistry, mixing, and long-range transport or can be attributed to an external factor such as the emission profile of a pollutant.

There exists no single best validation strategy applicable to all satellite-borne products related to tropospheric composition (Section 7.6). The relevant issues depend on the actual trace gas and the relevant processes contributing to the spatial distribution and temporal variation of that trace gas. An important assumption in retrievals typically concerns the vertical distribution in the troposphere. Relevant processes that affect the vertical distribution of trace gases are discussed in Section 7.4.1.

The geographical and temporal range of the validation measurements that are needed may differ between different data products. For example, for tropospheric trace gas products a complete validation measurement programme differentiates between cases that are dominated by boundary layer, free troposphere and stratosphere, respectively. The importance of considering stratospheric variations for tropospheric products is explained in Section 7.4.2. The validation measurements should also comprise a set of regions with different concentration level and that are located in different latitude bands and during different months of the year, in order to capture a sufficient variety of relevant chemical regimes and relevant cycles of variations on, say, seasonal and annual scales.

In an observational data record, there are many possible causes for biases and variations. These may partly be caused by instrument limitations such as the signal-to-noise ratio and calibration issues. To a certain extent, biases and variations are also associated with the physical limitations of the applied observational technique. In general, impacts can be expected from the observation geometry, the actual temperature and humidity profile, cloudiness, land surface characteristics, etc. Retrieval algorithms minimize these impacts where parameterizations and/or corrections are applied. A major impact on trace gas retrievals is expected from the assumptions about clouds, aerosols and the Bidirectional Surface Distribution Function. These are discussed briefly in Section 7.4.3.

7.4.1 Tropospheric Processes Impacting on Trace Gas Distributions

An important time scale with respect to the horizontal distribution of trace gases is the time scale for a plume (e.g. an emission pulse, or a stratospheric intrusion) to dissolve in the background. The time period can be very short, in the case where the plume quickly encounters a strong mixing event or in the case of fast chemical degradation, but typically is of a couple of days up to about 2 weeks.

Most variations on longer-lived pollutants occur in the boundary layer and are related to emission profiles such as rush hours, imperfect mixing and, in the case of ozone, the possibility of rapid ozone formation by fast chemistry under stagnant and warm polluted conditions.

Imperfect knowledge of the trace gas vertical profile in the troposphere is an important contribution to the overall uncertainty for tropospheric trace gas retrievals (Boersma et al. 2004), and therefore also for the validation of these retrievals. For most satellite measurement techniques, the sensitivity of the measurement is a function of altitude in the troposphere. This is true, for example, for measurements of the solar radiation backscattered in the UV-Vis range (observation of NO_2, SO_2, HCHO, O_3) as well as for measurements of the thermal emission (observation of CO, CH_4, O_3). The NO_2 vertical profile in the troposphere is highly variable in time and location and typically determined by a set of interacting processes. NO_x emissions and dry deposition (Ganzeveld and Lelieveld 1995) occur near the surface, while photochemistry and turbulent mixing are important within the boundary layer. In the free troposphere, long-range transport, wet removal, convection and lightning may significantly influence the vertical distribution. Some NO_x is injected at higher altitudes by pyro-convection related to biomass burning, and by aircraft emissions. Finally, the concentration of NO_2 in the atmosphere is in a photochemical steady state with the concentration of NO, N_2O_5, O_3 and ClO, the steady state depending on the ambient temperature, through the Arrhenius equation, and UV light intensity, the actinic flux, having a strong dependence on solar zenith angle and cloudiness. Thus the NO_2 vertical profile will typically undergo rapid changes during a day, some of the changes being recurrent, for example, the diurnal cycle, others being much less predictable such as the effects linked to the presence of clouds and of turbulent transport. Because of the large number of possible processes involved, the validation of NO_2 is challenging. Satellite-to-satellite intercomparison offers some advantages because of the collocation and similarity in spatial representativity. Validation with ground-based remote sensing can offer the advantage of similarities in vertical sensitivity and, for tropospheric data retrieved using residual methods (the tropospheric column is obtained by subtracting from the satellite total column measurement an estimate of the stratospheric column), the possibility of discriminating between the stratospheric and tropospheric contributions to the total column measured by the satellite. Validation with *in situ* observations of NO_2 is very complicated, especially close to emission sources.

Fig. 7.8 shows a satellite-to-satellite comparison of SCIAMACHY and OMI NO_2 tropospheric columns. The purpose of the comparison by Boersma et al. (2008) was to examine the consistency between the two instruments under tropospheric background conditions and the effect of different observation times. The example illustrates the difficulty to distinguish deviations due to "real" atmospheric processes from deviations generated artificially by differences in instrument and/or retrieval characteristics.

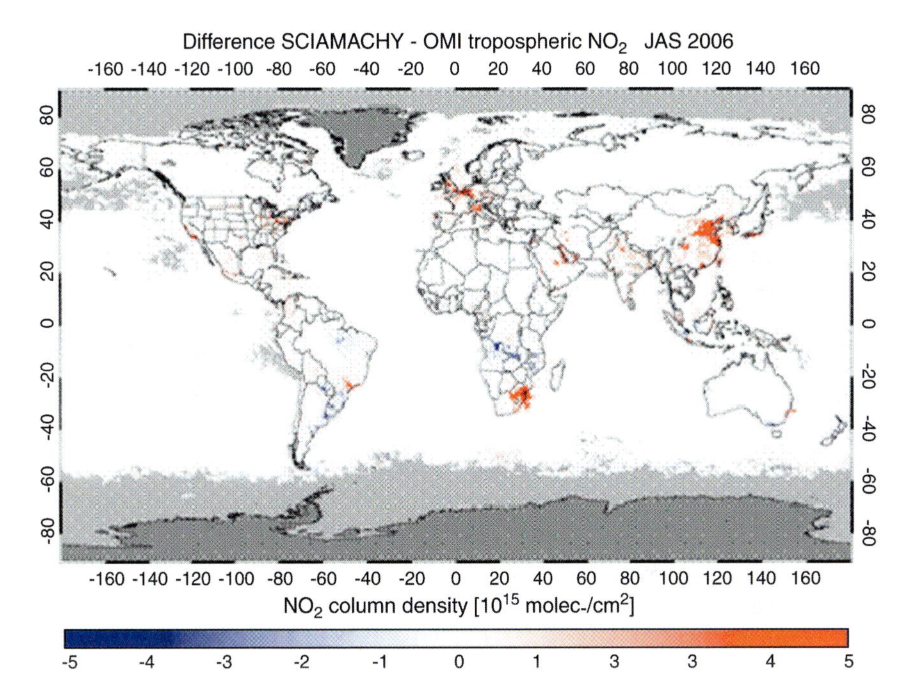

Fig. 7.8 Difference between SCIAMACHY and OMI tropospheric NO_2 columns for the period July-August-September 2006. From a careful analysis of the consistency between SCIAMACHY and OMI tropospheric NO_2 observations in different regions of the world it was concluded that part of the difference could be attributed to the diurnal cycle in emissions and photochemistry. Morning SCIAMACHY NO_2 is higher than afternoon OMI NO_2 over fossil fuel source regions because of photochemical loss in combination with a broad daytime emission profile, and lower than OMI over tropical biomass burning source regions because of a sharp mid-day peak in NO_x emissions as confirmed by fire counts. (from Boersma et al. (2008)).

Although different in detail, similar types of challenges exist for other species. HCHO, CHOCHO, BrO, and SO_2 are short-lived (~hours) and undergo rapid changes during the day in a similar way to NO_2. Water vapour changes in the atmosphere are also rapid, though related to fast processes in the hydrological cycle, including evaporation from the surface, cloud/rain formation, and cloud/rain evaporation. Transport and imperfect mixing in the boundary layer and free troposphere further contribute to water vapour variability on the kilometre scale. Validation of short-lived products is necessarily complex as it should cover a large range of time scales both unpolluted and polluted locations with different chemical regimes, meteorological conditions and emission profiles.

CO, tropospheric O_3 and CH_4 are longer-lived tropospheric gases with chemical lifetimes ranging from days to years. Free-tropospheric variability is mainly related to convection and synoptic scale variability. These are most important for long-range, intercontinental, transport. Stratosphere-troposphere exchange is an important contributor to variations in tropospheric O_3. Stratospheric O_3 intrusions may penetrate deep into the free troposphere before mixing (Roelofs et al. 2003).

It is important for validation that the methods capture the most relevant processes that may contribute to the variation in the product. When different time scales are involved, complex and non-linear relationships may exist between observed variations, instrument and observational technique related variations, and physical/chemical/meteorological controlling processes and factors. Fig. 7.9 shows a comparison of SCIAMACHY CO columns with MOPITT CO columns (de Laat et al. 2010). The explanations of the differences are given in the caption and illustrate the complexities that arise in the comparison of two independent data sets.

7.4.2 Validation Needs for Trace Gases with Stratospheric Contributions

While some species are present primarily in the troposphere, other species may also have a high concentration in the stratosphere, in between the tropospheric target and the satellite. The retrieval of tropospheric columns of O_3, NO_2 and BrO often involves the subtraction of an estimated stratospheric part (Section 2.4). The uncertainty of tropospheric data products will probably be larger above regions where the stratospheric variability is large. Validation can contribute to a better understanding of the error sources of the retrieval methods, especially when it includes verification of the assumed stratospheric contributions and the assumed profile shapes with independent measurements.

a What Causes Stratospheric Variability?

In the stratosphere, and in particular in the lower stratosphere, many trace gases including O_3, CH_4, and the reactive chlorine and nitrogen oxide families, Cl_y and

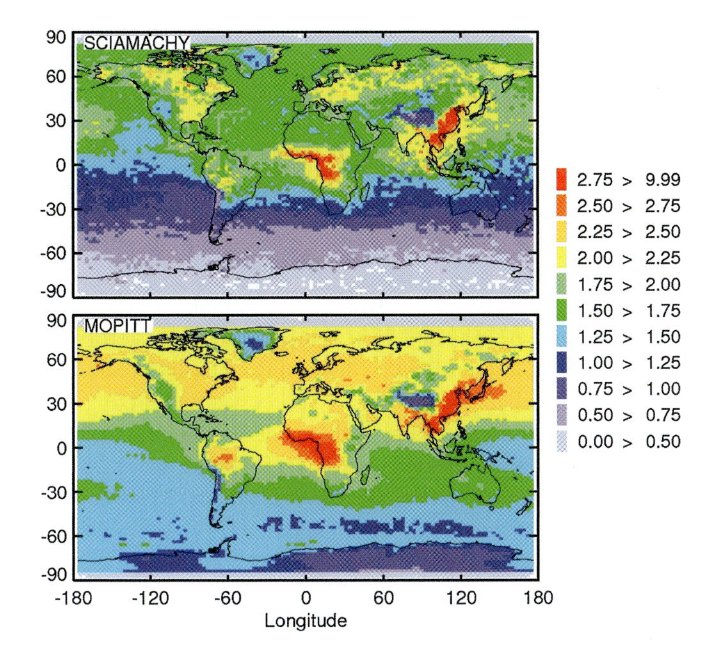

Fig. 7.9 Comparison of CO columns from SCIAMACHY (*top*) and MOPITT (*bottom*). Averages of the period 2004–2005 are shown in 10^{18} molecules/cm^2 on a 3° × 2° resolution. The grey areas indicate regions without observations. Common features are the hemispheric gradient and the emission patterns over tropical biomass burning areas and in eastern Asia. De Laat et al. (2010) showed that there is a bias between the CO columns from both instruments and explain the differences as being related to instrumental precision (SCIAMACHY is somewhat noisier), weighting assumptions (SCIAMACHY is weighted with the noise error), error-covariance and *a priori* assumptions in the retrievals and some other effects, such as the presence of aerosols, differences in spectroscopy of interfering gases such as CH_4 and H_2O and the ice-layer correction to SCIAMACHY. (from de Laat et al. (2010)).

NO_y, have chemical lifetimes considerably longer than the time scales of planetary and synoptic scale wave activity. Their variability is principally dominated by dynamic processes and, consequently, is typically largest at middle and high latitudes. There are exceptions, such as O_3 in polar springtime when its chemical lifetime becomes shorter due to rapid destruction in catalytic reactions with halogen radicals activated on the surfaces of polar stratospheric clouds and aerosols. Here, the variability is caused by a mixture of chemical and dynamical processes. Individual species of a reactive gas family may have much shorter lifetimes and exhibit a pronounced diurnal cycle. A good example is NO_2, which is rapidly photolysed to NO and oxidized by O_3 to NO_3 but is continuously reformed from its products, resulting in a pronounced diurnal cycle of NO_2, whereas the sum NO_x ($NO + NO_2 + NO_3 + 2\ N_2O_5$) is fairly constant throughout the day. Such rapid diurnal variations in the partitioning within a compound family therefore need to be accounted for, as well as the variability induced by dynamic processes.

In the retrieval of NO_2, CH_4 and CO_2, assumptions are made on the vertical distribution of these species (Section 2.3.3). The uncertainty of the tropospheric products might therefore be larger over regions with a deviating or variable vertical distribution.

b What Determines the Vertical Distribution of these Species?

The distribution of trace gases in the atmosphere is generally determined by the distribution of sources and sinks and by transport processes. Vertical exchange is influenced by the temperature profile that determines the (dry) static stability of the profile, as well as by turbulent and convective processes and mixing, and by quasi-horizontally moving air masses crossing in different directions at different altitudes.

Vertical exchange is largely suppressed in regions of high stability, which are therefore regions of enhanced vertical trace gas gradients. The temperature inversion at the top of the boundary layer, for example, separates the (potentially) polluted air in the boundary layer from the cleaner air in the free troposphere. Exchanges between boundary-layer and free troposphere are dominated by orographic effects, the uplift of air masses in warm conveyor belts associated with frontal systems, entrainment and turbulent mixing, cloud processes, and the daily decoupling of the residual layer from the surface as soon as a new inversion appears near the surface at the end of the day.

Another region of high stability is the tropopause inversion which separates tropospheric and stratospheric air masses. In the extratropics, cross-tropopause exchange is dominated by synoptic scale processes and occurs mostly in tropopause folds and cut-off lows generated by baroclinic disturbances in the mean jet stream (Holton et al. 1995). The disturbances are often associated with mesoscale convective complexes and thunderstorms. At tropical latitudes, convectively driven air masses may enter the tropical tropopause layer and subsequently reach the stratosphere by much slower diabatic ascent. The total mass of air entering the stratosphere in the tropics is controlled by the "extra-tropical pump" which drives the Brewer-Dobson circulation (Holton et al. 1995). During exceptionally strong volcanic eruptions aerosols and SO_2 can also reach the stratosphere directly. At higher latitudes, downward motions are dominant and the descending stratospheric air mixes within the troposphere.

The amount of O_3 in the troposphere is an order of magnitude smaller than in the stratosphere. This makes it difficult to retrieve from satellites (Sections 2.4 and 2.5.5), especially in regions where the stratospheric variability is large. Validation analysis of tropospheric ozone should include the distinction between the regions with different stratospheric variability.

The column amount of tropospheric NO_2 above polluted areas is comparable to or larger than the stratospheric part, while above clean areas it is much smaller. Validation analysis should include a progressive distinction from polluted to clean areas.

Also validation analysis for methane should account for stratospheric column variations, which are of the same order as the tropospheric variations caused by emissions and free tropospheric gradients. Total column BrO is typically confined to the stratosphere, with significant tropospheric contributions confined to very specific conditions such as springtime at the edge of first-year sea ice.

7.4.3 Validation Needs Related to Cloud, Albedo and Aerosol Effects

Retrievals of tropospheric gases can be severely affected by the presence of clouds. Clouds shield tropospheric gases from observation, but may also enhance the sensitivity to trace gases above the cloud in the case of measurements with reflected sunlight. Not only clouds, but also aerosols and surface albedo influence the light path in the atmosphere and thereby affect tropospheric trace gas measurements. The actual light path can be measured by using oxygen absorption (using the O_2 A-band and O_2–O_2 lines for example), since oxygen is well mixed. In addition, Raman scattering can be used.

Some retrieval algorithms of tropospheric species, for example, NO_2, only select cloud-free scenes and do not correct for clouds. Other retrieval algorithms correct for the presence of clouds. In the latter case, most algorithms use a simple Lambertian cloud model for the correction in which a surface with high albedo at some pressure level is assumed for the cloud (Koelemeijer et al. 2001; Van Roozendael et al. 2006; Veefkind and de Haan 2001; Boersma et al. 2004). The two parameters in the Lambertian model are the effective cloud fraction and effective cloud pressure (Stammes et al. 2008 and Chapter 5). A scattering cloud model has more parameters: geometric cloud fraction, cloud optical thickness, cloud top and bottom pressures, and the possible presence of multiple cloud layers. Such a model needs more auxiliary data or *a priori* assumptions (van Diedenhoven et al. 2007).

The validation requirements for clouds depend on the trace gas retrieval algorithm. For the algorithms that only need cloud-free scenes, a high quality cloud mask is needed. This is usually derived from a radiance threshold, but can also be derived from high resolution imagery collocated with the trace gas data. Alternatively, the effective cloud fraction can be used because it gives the radiometric weight of clouds in the scene, which is needed in air mass factor calculations. The radiometric weight, w, is defined as the fraction of radiation coming from clouds in the pixel. In the case of the Lambertian cloud model (Stammes et al. 2008):

$$w = cR_c/R_{tot} = c_{eff}A_c/R_{tot} \tag{7.2}$$

Here c is the geometric cloud fraction, c_{eff} is the effective cloud fraction assuming a cloud with albedo A_c, R_c is the reflectance of the cloud and R_{tot} is the measured reflectance of the satellite pixel. In general, it is necessary to validate the product $c_{eff} A_c$.

Cloud properties used in cloud correction algorithms can be validated by comparison with dedicated cloud instruments, such as multi-spectral imagers. In these comparisons, the interpretation of the oxygen absorption cloud properties is important. It has been found that the oxygen cloud pressure, retrieved using a Lambertian cloud retrieval model, does not represent the cloud top pressure, but rather the pressure at a level inside the cloud or between the cloud layers (Sneep et al. 2008; Wang et al. 2008). This is due to the multiple scattering of solar photons inside the cloud. On the other hand, the cloud pressure given by satellite imagers in the thermal IR is the top of the cloud, because IR radiation is emitted from the top of the cloud. Since clouds are strongly absorbing in the IR, IR radiation hardly penetrates through the clouds.

For cloud validation, comparison with other satellite sensors is essential because of the variability of clouds. Most useful are cloud imagers, which have visible, shortwave IR and thermal IR channels for determination of the geometric cloud fraction, optical thickness and cloud top pressure. Cloud radar and lidars from space or ground are important to determine "true values" of cloud top pressure, and cloud mid-level and cloud bottom pressure. Table 7.2 gives an overview.

An important characteristic of clouds is their high variability in time and global variation. Single clouds typically live less than an hour. Statistically, clouds have diurnal cycles depending on the region. Furthermore, cloud types vary over the globe, and detection of clouds depends on the surface albedo. Therefore, it is insufficient to validate at a few specific sites where ground-based instruments are located. The opportunity to validate cloud retrievals with the cloud radar/lidar

Table 7.2 Main cloud parameters for tropospheric trace gas retrievals, their needed accuracy, and sources of validation

Cloud parameter	Symbol	Needed accuracy	Comparison/ validation source	Note
Effective cloud fraction	c_{eff}	0.05	Cloud imager	Compute c_{eff} from c and τ of imagery data
Effective cloud pressure	p_c	100 hPa	Cloud radar	O_2 absorption and Raman scattering give mid-cloud level
Cloud mask	–	5%	Cloud imager	Cloud-free scene selection
Geometric cloud fraction	c	0.1	Cloud imager	Scattering cloud model
Cloud optical thickness	τ	10 %	Cloud imager	Scattering cloud model
Cloud top pressure	p_{top}	100 hPa	Cloud imager Cloud radar/lidar	Scattering cloud model
Cloud bottom pressure	p_{bot}	100 hPa	Cloud radar/lidar	Scattering cloud model

CloudSat/CALIPSO in the A-train is unique for satellite trace gas retrievals (Sneep et al. 2008).

Before validation can be performed, a statistically significant verification of the cloud product is needed. One should verify the quality of cloud retrievals by analysing global cloud retrievals and checking for consistency and absence of jumps depending on the viewing geometry, solar zenith angle, latitude, etc. This verification process can be done easily by using the CAMA tool (Sneep et al. 2006), which is an IDL programme correlating all parameters of a data product with one another. The CAMA tool has been used extensively in the validation of OMI data (Sneep et al. 2008; Stammes et al. 2008; Kroon et al. 2008).

Surface albedo is an important auxiliary parameter needed for trace gas, aerosol and cloud retrievals from satellite. A high surface albedo increases the sensitivity of satellite measurements to tropospheric trace gases, but decreases the sensitivity to (scattering) aerosols. Often a surface albedo climatology is used, for example, the GOME climatology of Koelemeijer et al. (2003) or the OMI climatology of Kleipool et al. (2008). If the assumed surface albedo is not correct, say, due to insufficient spatial resolution, the trace gas amount for the clear sky part of the pixel will be incorrect. However, since the detected cloud fraction will also be incorrect for an incorrect surface albedo, there will be some compensation taking place for the total pixel (cloud-free plus cloudy parts), since cloud fraction will be adjusted to match the reflectance at top of atmosphere. For this compensation it is important that the trace gas and cloud retrievals use the same surface albedo data base. In the validation of trace gases, the cloud information should be taken into account.

The impact of aerosols (Chapter 6) is not explicitly included in trace gas retrievals, but it could be a significant error. In most algorithms aerosols are included implicitly as a type of clouds, which may work for scattering aerosols, but not for absorbing aerosols. This is still a largely unexplored topic, which is relevant for highly polluted regions.

7.4.4 Validation Needs for Aerosols

Chapter 6 describes the details of aerosol retrievals from satellites. Passive measurements yield column integrated aerosol parameters such as AOD, the column integrated aerosol extinction. Active instruments such as a lidar, for example CALIOP on CALIPSO, part of the A-train, provide information on the vertical structure. Instruments with multiple viewing angles (MISR, POLDER, ATSR) can be used to determine the 3-D structure of plumes (Kahn et al. 2007).

Independent data of aerosol physical, chemical, and optical properties are needed to validate the retrieval results. Since the retrieval uses several assumptions on surface and atmosphere parameters, information should also be available on the boundary layer structure and meteorological parameters. These data and additional information can be obtained from dedicated campaigns and networks.

The AERONET network (Holben et al. 1998) was established to obtain information on the aerosol optical properties for satellite validation. Instruments used are CIMEL sun photometers which provide both direct sun measurements and almucantar scans from which information can be derived on the column-integrated aerosol physical and optical properties such as AOD, effective radius and complex refractive index. Lidars provide extinction and backscatter profiles (raman lidars used in the EU-funded EARLINET network) which are used to evaluate CALIOP measurements. Lidar measurements can be integrated to provide the AOD. Lidar measurements provide a very sensitive way to detect cloud reflections, even for sub-visible clouds which can have an adverse effect on aerosol retrieval. A recent comparison of lidar and AERONET data shows that AERONET is very effectively cloud screened (Schaap et al. 2009).

The effect of the surface correction used in the retrieval could in principle be evaluated by comparison with these instruments if the path radiance could be accurately obtained. However, this is usually not the case and the uncertainty in satellite retrieved AOD is larger than that obtained from sun photometers (typically on the order of 0.03 over ocean or 0.05 over land (Robles-Gonzalez et al. 2006), whereas the accuracy of CIMEL sun photometers used in AERONET is between 0.01 and 0.015 (Eck et al. 1999)).

In order to quantify the uncertainties in the aerosol products for clouded scenes it is necessary that the clouded scenes are validated. The cloud mask used in the aerosol retrieval should also be validated and it should be verified that it does not remove aerosol, as might be the case for large AODs from desert dust storms or smoke.

7.5 The Use of Correlative Measurements for Validation

For validation purposes a comprehensive set of correlative data is needed, either acquired during a specially designed short period calibration/validation campaign or used from long term monitoring networks or other special programmes. The correlative data sets ideally comprise collocated ground-based *in situ* and remote sensing of tropospheric and stratospheric profiles and columns, as well as airborne measurements. In this section, the different measurement techniques and their characteristics for validation are discussed. The networks and data centres are summarised in Section 7.5.3 and in Tables 7.3–7.5.

7.5.1 *In Situ Measurements*

Several measurement techniques are currently used for the validation of tropospheric satellite products.

Table 7.3 *In situ* ground based networks and their data centres

Programme	Parameters	Data centre	QA/QC	Web address
EEA/EU	Reactive Gases	Airbase	Yes	http://dataservice.eea.europa.eu/dataservice/metadetails.asp?id=1029
UN-ECE	Reactive gases and chemistry	EMEP/CCC	Yes	http://www.nilu.no/projects/ccc
WMO/GAW	GHGs and reactive gases	WDCGG	Yes	http://gaw.kishou.go.jp/wdcgg
WMO/GAW	Radiation	WRDC	Yes	http://wrdc.mgo.rssi.ru
WMO/GAW	Aerosol and AOD	WDCA	Yes	http://www.gaw-wdca.org
WMO/GAW	Precipitation	WCDPC	Yes	http://www.qasac-americas.org
WMO/GAW	Ozone and UV	WOUDC	Yes	http://www.woudc.org
NOAA	GHGs and reactive gases & aerosols	GMD Data Archive		http://www.esrl.noaa.gov/gmd/dv/ftpdata.html
US national	Climate variables	NCDC		http://www.ncdc.noaa.gov/oa/ncdc.html
NASA/US	Aerosol/Radiation	AERONET	Yes	http://aeronet.gsfc.nasa.gov
PHOTONS	Aerosol/Radiation	PHOTONS	Yes	http://loaphotons.univ-lille1.fr
EUSAAR	Harmonized aerosol data	NILU/IBAS	Yes	http://www.eusaar.net
GEOmon	Trop. and strat. gases and aerosols	NILU/EBAS	Yes	http://www.geomon.eu/data.html

Table 7.4 Remote sensing, balloon and aircraft networks and data centres

Programme	Parameters	Data centre	QA/QC	Web address
NDACC	Remote sensing	NDACC	Yes	http://www.ndacc.org
WMO/GAW	Ozone profiles	WOUDC	Yes	http://www.woudc.org
SHADOZ	Ozone sondes			http://croc.gsfc.nasa.gov/shadoz
CARIBIC	aeroplane	CERA		http://www.caribic-atmospheric.com
EARLINET	Aerosol lidars			http://www.earlinet.org

a *In Situ* Measurements for O_3 and CO

Tropospheric O_3 can be inferred from *in situ* O_3 profiles, as measured by ozone sondes (types: Electrochemical Concentration Cell (ECC), Brewer-Mast, Carbon-Iodine) and by aircraft (UV ozone analyser) during take-off and landing. O_3 soundings are performed between 1 and 3 times a week as part of a continuous program or during special validation campaigns. There are a number of operational ground stations that perform this type of measurements on a regular basis. Most of these submit their data to WOUDC (World Ozone and Ultraviolet radiation Data Centre). This UV/ozone network is part of WMO and is operated by the Meteorological Service of Canada. A smaller and more recent NASA funded network is the SHADOZ (Southern Hemisphere ADditional OZone sondes) network, which focuses on the tropics. The network has been operational since 1998. SHADOZ ozone-sondes operate at different frequencies, all use ECC of various types; twelve active sites participate, and they are also included in the WOUDC network.

In situ measurements are also acquired from UV ozone analysers (Thermo-Electron, Model 49-103) on board the passenger airliners that take part in MOZAIC. This program, which was initiated in 1993, focuses on several species including O_3 and CO. Most of the data (90%) are collected at cruise altitudes of 9–12 km. At mid-latitudes this altitude region covers the tropopause, a region critical for climate change and the exchange between stratosphere and troposphere. The remaining data is acquired during ascents and descents at airports visited by the program.

A less extensive but detailed database has been building up since 2004 in the framework of the CARIBIC project. Measurements of O_3 and CO and a number of other species are made during long distance flights with an Airbus A340-600 from Lufthansa. The ozone analyzer operates with two different measuring principles: a fast O_3 analyzer using fluorescence of an organic dye absorbed on silica gel and a standard O_3 analyzer using UV absorption.

In situ flask measurements are harmonized in the NOAA/ESRL/GMD CCGG cooperative air sampling network. This is a globally distributed network of sites taking regular discrete samples and which includes the 4 NOAA ESRL/GMD baseline observatories, cooperative fixed sites, and commercial ships (http://www.esrl.noaa.gov/gmd/ccgg/flask.html). Air samples are collected on a weekly basis and are analysed in Boulder for CO and a variety of other gases.

Table 7.5 Satellite data centres

Programme	Parameters	Data centre	QA/QC	Web address
NDMC/WMO	Satellite	WDC-RSAT		http://wdc.dlr.de
US national	Climate variables	NCDC		http://www.ncdc.noaa.gov/oa/ncdc.html
CEOS	satellites	CAL/Val portal		http://calvalportal.ceos.org
ESA airborne campaigns	Satellite related	ESA EO campaign		http://earth.esa.int/campaigns
ESA	*In situ*, remote sensing & satellite	EVDC		http://nadir.nilu.no
NASA	Aura	AVDC		http://avdc.gsfc.nasa.gov
NASA	Aerosols	MODIS	Yes	http://modis-atmos.gsfc.nasa.gov
NASA	Aerosols	Giovanni	Yes	http://disc.sci.gsfc.nasa.gov/giovanni/overview
ICARE	Aerosols	ICARE	Yes	http://www.icare.univ-lille1.fr
ESA-DUE	Aerosols, trace gases, clouds	TEMIS		http://www.temis.nl
ESA-GSE	Aerosols & trace gases products	PROMOTE		http://www.gse-promote.org
WMO-GAW	Aerosols	WDC-RSAT		http://wdc.dlr.de

For *in situ* surface measurements of O_3 and CO, the uncertainty contribution resulting from different measurement techniques used in monitoring networks (Klausen et al. 2003; Zellweger et al. 2009) are considered negligible for satellite validation purposes.

b *In Situ* Measurement Techniques for NO_2

For NO_2 the choice of the measurement technique is crucial, since many of the measurement principles used in *in situ* monitoring networks are not specific for the NO_2 detection. Depending on the method and the atmospheric photochemical conditions, interference from non-NO_2 compounds may exceed the NO_2 concentration by up to 250%, because not only NO_2 but also some fraction of the rest of NO_y ($NO_y = NO + NO_2 +$ peroxy nitrates + alkyl and multifunctional nitrates + $HNO_3 + \ldots$) is converted (Steinbacher et al. 2007). For satellite validation, data with low or well defined interference are required.

There are several other specific techniques available to measure NO_2 at atmospheric levels, such as DOAS (Alicke et al. 2002; Platt and Perner 1980), chemiluminescence induced by the reaction with luminol (Kelly et al. 1990), TDLAS (Li et al. 2004), laser induced fluorescence (LIF) (Murphy et al. 2006; Cleary et al. 2002; Day et al. 2003; Matsumoto et al. 2001), or by pulsed cavity ring-down spectroscopy (Kebabian et al. 2005; Osthoff et al. 2006). Recently, more and more *in situ* networks replace non-specific converters with photolytic converters, specific for NO_2. The two most prevalent techniques currently used for validation are LIF detection of NO_2 and conversion of NO_2 to NO followed by detection of NO with chemiluminescence (CL). In LIF, a laser is used to excite NO_2, which then fluoresces. CL detection of NO relies on titrating it with O_3 to produce excited NO_2, which then emits an amount of light proportional to the concentration of NO. Systems employing either LIF or CL are calibrated with gravimetrically prepared standards.

On airborne platforms such as the DLR Falcon, the NASA C130, the NOAA P-3 and others that use CL, NO_2 is converted to NO photolytically, which is a specific conversion. On board NOAA's Ron Brown Research vessel, NO_2 has been measured by NO_2 specific pulsed cavity ring-down spectroscopy (Osthoff et al. 2006).

The first reported comparison of tropospheric NO_2 columns from satellite observations with *in situ* aircraft data was between GOME and the Falcon aircraft and took place over Austria in early May 2001 (Heland et al. 2002). Since then there have been numerous opportunities for comparison between the GOME, SCIAMACHY and OMI satellite instruments and a number of different mobile platforms. In some cases, ground-based measurements have been used to deduce independently an *in situ* column for comparison to the satellites (Schaub et al. 2006).

c Factors Impacting on the Use of *In Situ* Measurements for Satellite NO_2 Data Validation

There are a number of factors concerning both the collection and the subsequent treatment of *in situ* data that must be considered when using observations from mobile or ground-based platforms to generate *in situ* NO_2 columns for comparison with satellite columns. The fundamental problem arises when attempting to compare data taken at a single point (for ground sites) or a point-wise representation of a portion of the column (for airborne platforms) with satellite data, the latter being integrated in both the horizontal and vertical dimensions.

Except for remote locations, the vast majority of the tropospheric NO_2 column is concentrated in the boundary layer, which means that uncertainties and extrapolations of observations within the boundary layer will affect the calculated *in situ* column disproportionately as compared to uncertainties and extrapolations of data in the free troposphere. Aircraft observations require extrapolation from the lowest elevation achieved to ground level while ground-based observations require upward extrapolation. Thus, when using an aircraft platform, sampling within the boundary layer is essential and attempts should be made to sample the column at the lowest possible elevation that safety allows. In addition, in the case of aircraft observations that span several satellite pixels, it is essential to compare only those pixels for which there are *in situ* observations within the boundary layer. Similarly, if ground-based NO_2 observations are to be used, it is essential that they be accompanied by a reliable measurement of the boundary layer depth.

If the widespread existing network of ground-based NO_x sensors is to be used for satellite comparisons, we must assess and quantify accurately the conversion of other NO_y species to NO_2 when using catalytic conversion chemiluminescence. In general, these systems see a positive interference in NO_2 and will thus overestimate boundary layer NO_2, but the magnitude of the interference will vary on a site-to-site basis. The NO_x/NO_y ratio is known to decrease with photochemical aging as NO and NO_2 are converted to higher order NO_y species so, to a first approximation, the interference will be largest at ground sites that are far removed from source regions and which sample well-aged air masses.

7.5.2 *Remote Sensing*

Remote sensing observations from the ground can bridge the gap between satellite columns and surface *in situ* measurements. Using similar techniques as the satellite instruments, they determine the tropospheric columns with good accuracy and sometimes with similar sensitivity to the vertical distribution of the species.

a Multi-Axis Differential Optical Absorption Spectroscopy (MAXDOAS)

Based on the measurement principles pioneered by Dobson and Harrison (1926), DOAS has been used for about three decades to measure amounts of O_3, NO_2, HCHO, BrO, and many other trace gases in the atmosphere (Brewer et al. 1973; Noxon 1975; Platt and Perner 1980; McKenzie and Johnston 1982; Solomon et al. 1987; Pommereau and Goutail 1988a, b). The MAXDOAS technique (Sinreich et al. 2005) to measure aerosol profiles has recently been explored and first results are promising. When using scattered sun light as the light source, a high degree of automation can be obtained and measurements can be taken independently of weather conditions. Originally, DOAS measurements were performed for stratospheric studies and the instrument's telescopes pointed towards the zenith to minimise the impact of tropospheric absorptions. In this viewing geometry, very large sensitivity is obtained for stratospheric absorbers at twilight when the light path in the upper atmosphere is long before the photons are scattered and pass vertically through the lower troposphere before they reach the instrument. In the framework of the international Network for the Detection of Atmospheric Composition Change (NDACC), a contributing network of WMO's Global Atmosphere Watch, over 35 zenith-sky DOAS spectrometers currently perform network operation from the Arctic to the Antarctic. Most of them are located in remote areas characterised by a very clean local troposphere.

Among them, about 50% operate in areas where no significant pollution can be detected at spatial scales of the order of a nadir-viewing satellite pixel. This feature and the enhanced sensitivity to stratospheric absorbers make these spectrometers well suited for validating the satellite column data over clean areas, as well allowing estimates of the stratospheric column to be made by satellite retrieval algorithms based on the residual technique. Most of the NDACC DOAS/UV-visible spectrometers monitor the vertical columns of O_3 and NO_2. The network allows validation studies for NO_2 in the range from 10^{14} molecules/cm^2 during polar winter up to 6.5×10^{15} molecules/cm^2 in polar summer, and under small solar zenith angles to twilight conditions. A few NDACC spectrometers also measure the column abundance of stratospheric BrO and OClO.

In 1993, Sanders et al. (1993) added complementary viewing directions to increase the signal-to-noise ratio of stratospheric absorptions and they noticed that off-axis measurements also increase the sensitivity to tropospheric absorbers. Resulting multi-axis or MAXDOAS measurements (Hönninger and Platt 2002; Hönninger et al. 2004b; Wittrock et al. 2004) are based on the idea that very long light paths in the lower troposphere are achieved when pointing the telescope to the horizon while the stratospheric light path is largely independent of the pointing of the telescope. By combining the measurements from different viewing directions, vertical profile information can be retrieved on the lower troposphere in much the same way as for limb measurements from balloon and satellite instruments. The profile information is valuable for atmospheric chemistry applications and satellite validation but it also improves the accuracy of the primary measurement quantity of

MAXDOAS measurements, the tropospheric columns. Fig. 7.10 shows two MAX-DOAS instruments in operation.

Fig. 7.10 MAXDOAS instruments at the University of Bremen.

In recent years, a number of studies have been performed using MAXDOAS measurements for the investigation of pollution events (Heckel et al. 2005; Leigh et al. 2006), for studies of halogen oxide chemistry in high latitudes (Honninger et al. 2004a; Wagner et al. 2007) and a few for satellite validation (Brinksma et al. 2008; Irie et al. 2008; Wittrock et al. 2006; Fig. 7.11). The main difference between MAXDOAS and satellite measurements is the viewing geometry, which determines the vertical sensitivity of the measurements. As an example, while the sensitivity of a satellite UV/Vis measurement of NO_2 decreases towards the surface over dark

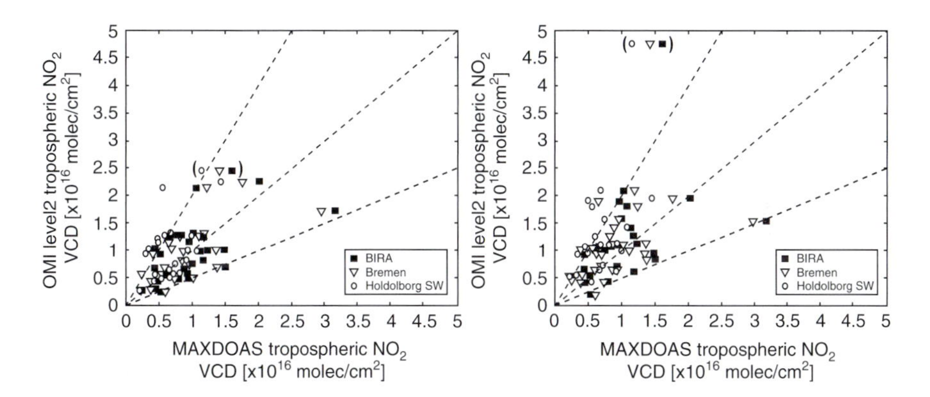

Fig. 7.11 Example of a recent study on satellite validation of tropospheric NO_2. OMI tropospheric NO_2 from NASA, (version 1.0, *left*), and KNMI (DOMINO, *right*) for cloud fractions less than 20%, compared to those from three MAXDOAS instruments at Cabauw during the DANDELIONS campaign (from Brinksma et al. (2008)).

areas, the sensitivity of a MAXDOAS measurement decreases from the surface to the middle troposphere. Consequently, the tropospheric columns determined from the ground have a higher accuracy than the satellite measurements and depend much less on the *a priori* assumptions made in the retrieval. While the vertical integration problem can be solved by using ground-based remote sensing data, the complications from horizontal and temporal variability remain and have to be taken into account when selecting appropriate locations for validation.

b Fourier Transform Infrared Spectroscopy (FTIR)

Many atmospheric trace species, including O_3, NO_2, N_2O, CO, HNO_3, CH_4, exhibit absorption signatures in the infrared range. By the application of FTIR, the vertical column of these species can be detected with high-spectral-resolution measurements of solar spectrum absorption by the atmosphere (Chapter 3). If weather permits, FTIR can measure throughout the day, and observe the diurnal cycle of species such as NO_2 and HCHO. The need to observe the solar disc directly is a limiting feature. This is especially the case for instruments requiring manual intervention, at stations, which are frequently overcast, and at polar stations that experience polar night for several months of the year. In polar regions, the full moon can be used as infrared light source (Notholt et al. 1993). Water vapour (Schmid et al. 1996) and SO_2 (Mellqvist et al. 2005) are found mainly in the troposphere. Their column measurements can be used directly for tropospheric studies (Toon et al. 1989) and, if the comparison method handles appropriately spatial and temporal variability, they can be used directly for the validation of satellite tropospheric data products. For tropospheric species having a non-negligible stratospheric abundance, such as O_3, N_2O, CH_4 and HNO_3, profiling techniques have been developed and validated against independent measurements (Pougatchev et al. 1995; 1996; Nakajima et al. 1997). It is interesting to note that nearly all FTIR stations are affiliated with the NDACC; consequently, the instruments and algorithms have to comply with rigorous validation protocols. For instance at the NDACC station Jungfraujoch in Switzerland, the FTIR column data record of CO, CH_4 and N_2O extends back to the 1950s, from which secular trends have been calculated (Zander et al. 1989a; 1989b; 1994). FTIR measurements have been used successfully for the validation of SCIAMACHY CO, CH_4, CO_2 and N_2O columns (Dils et al. 2006). They have also been used for the validation of upper troposphere/lower stratosphere measurements by ENVISAT MIPAS (Cortesi et al. 2007; Ceccherini et al. 2008) and by SCISAT-1 ACE. Recent developments in ground-based FTIR include the retrieval of CO_2 (Yang et al. 2002), of SO_2 (Mellqvist et al. 2005), and of HCHO (Jones et al. 2009).

c Light Detection and Ranging (lidar)

Information about the vertical distribution of several tropospheric species can be retrieved from lidar measurements. Aerosol profiles can be measured as backscatter profiles, but for the retrieval of extinction, assumptions are needed about the backscatter to extinction ratio, the lidar ratio, as a function of height above the surface. Raman lidars, where Raman-shifted wavelengths for N_2 or O_2 are used, measure the extinction directly. Raman lidars are used in the EARLINET network. Vertical profiles of trace gases can be measured by DIAL systems. The elastic backscatter ratio by air is measured at several wavelengths where there are differences in the absorption by the target molecule (Measures 1992; Schoulepnikoff et al. 1998). Applying a spectral analysis technique similar to DOAS, DIAL measurements yield the vertical distribution of O_3 concentration and water vapour mixing ratio in the free troposphere (3 km to about 12 km at middle latitudes) with a temporal resolution of about 15 min and a vertical resolution of 50–300 m (Vogelmann and Trickl 2008). Unfortunately, the high sensitivity of the DIAL technique to aerosols and other effects limit its sensitivity and accuracy for gases in the PBL and its capacity to reach higher altitudes than those near the tropopause for O_3, and lower for H_2O and NO_2.

d Sun Photometers

Sun photometers are used to measure the column integrated atmospheric extinction, i.e. AOD (or Aerosol Optical Thickness – AOT). The AOD is the primary aerosol parameter retrieved from satellites. When AOD is available at several wavelengths, the Ångström exponent, describing the AOD wavelength dependence, can be derived. These parameters are available from sun photometer networks such as AERONET (Holben et al. 1998) and PHOTONS, which use CIMEL instruments, or the GAW PFR network that uses precision filter radiometers. AERONET provides additional aerosol parameters retrieved from Almucantar scans. AERONET provides data every 15 min for cloud free conditions, in near real time (L1 and L1.5 products). The CIMEL instruments are calibrated each year and after calibration, the data are re-processed to provide L2 products. GAW PFR measures every minute but the products are not available in NRT.

7.5.3 Networks and Data Centres

Many of the essential *in situ* and ground based measurements for air pollution and climate change (Essential Climate Variables, ECVs) contribute to long-term monitoring activities on national and international levels carried out under the auspices of WMO's GAW and GCOS. The use of these datasets gives long-term availability, traceability and known quality of the needed data. They have been acquired and checked following the defined QA/QC procedures of the respective networks and are usually publicly available through databases (Table 7.3), Additionally within EU framework programmes such as GEOmon (http://www.geomon.eu) or EUSAAR (http://www.eusaar.net), harmonized datasets of representative sites are provided on the European scale for selected parameters. An overview of databases and additional

Table 7.6 Web sites listing validation activities and results

Instrument	website
SCIAMACHY	http://www.sciamachy.org/validation
OMI	http://www.knmi.nl/omi/research/validation
TES	http://tes.jpl.nasa.gov/data/validation
GOME-2	http://o3saf.fmi.fi/documents.html
MOPITT	http://mopitt.eos.ucar.edu/Validation

information about sites and measurement programmes are available on http://www.gosic.org or http://gaw.empa.ch/gawsis and data can be accessed through a large variety of data centres (Tables 7.3–7.5).

As geophysical validation of Earth Observation data remains a high priority, ESA has initiated a project to develop a Generic Environment for Calibration/validation Analysis (GECA), which is considered to become the next generation validation data centre. The evolution part of GECA is in the interoperability between various validation data centres through standardisation of metadata, and catalogue and data exchange. Currently data centre interoperability has started with the Aura Validation Data Centre (AVDC), ENVISAT Validation Data Centre (EVDC), EARLINET, GAW, GEOmon and NDACC.

7.5.4 Validation Activities

As described in Section 7.3.1 each satellite instrument undergoes a planned validation phase, often including dedicated measurement campaigns. Table 7.6 lists a number of web sites where validation activities and results on tropospheric satellite products are archived. These web sites also contain additional documentation on validation strategies and requirements.

7.6 Future Validation strategies

7.6.1 Requirements for Future Validation Measurements

For the validation of tropospheric satellite products, a comprehensive set of correlative measurements is needed. These measurements should have ample coverage in space and time and in the range of values that can occur. In addition, they should be performed with a certain quality standard. The following general requirements can be formulated.

- Instruments for correlative measurements should ideally be placed in a global network and perform continuous measurements.
- A quality procedure, including regular intercomparisons, should be implemented in such a network.

- In choosing the sites for this network, it has to be considered that validation should be performed under various conditions (clean/polluted), but that the measurements should be representative for a larger area (i.e. a satellite ground pixel). This means that the surface height and albedo should not be too variable.
- The local variability for the measured species should be properly characterized or measured. This could mean that it is necessary to measure at various locations within a satellite pixel or in various directions (for remote sensing).
- A number of sites should have ample cloud-free conditions.
- Remote sensing measurements should be accompanied by *in situ* surface measurements.

Apart from the tropospheric species of interest, additional measurements should be made of aerosols, clouds, and boundary layer height.

7.6.2 Validation Strategy for Tropospheric O_3

Currently, the main validation source for tropospheric O_3 is the ozone-sonde network. There is a need for more independent measurements to complement the ozone-sondes, preferably from a network of comparable instruments. The network should at least cover the tropics and northern mid-latitudes. Candidate instruments for such a network would be Brewer, MAXDOAS, and DIAL.

The DIAL tropospheric ozone lidar measures an O_3 profile in the free troposphere, but it needs to be complemented with additional boundary layer measurements. More DIAL instruments would be needed and they would need to be operated routinely. A Brewer network is already in place. It would be worthwhile to develop a retrieval algorithm for tropospheric O_3 from Brewers. In addition, MAXDOAS is a candidate for measuring tropospheric O_3. A retrieval algorithm has to be developed and a network has to be set up.

In general, measurements should be done close to the satellite measuring time. Balloon, sonde and aircraft measurements, which often have constraints on the flight times, should be complemented with surface measurements and transport models with actual meteorological information to correct as much as possible for the inevitable time differences. Close to (precursor) emission sources, the collocation in time is even more important.

7.6.3 Validation Strategy for Tropospheric NO_2

The only instruments currently measuring the tropospheric column of NO_2 are remote sensing instruments. *In situ* instruments have also been used in a few aircraft campaigns. The various remote sensing instruments have been developed relatively recently, and they still need thorough characterisation and validation. The current accurate *in situ* instruments are rather heavy and are not suited for use on light aircrafts or small balloons.

A network of remote sensing instruments, complemented by *in situ* profiling techniques would be needed.

Light-weight/portable *in situ* measuring techniques should be developed so that sonde type measurements can be made in a network. The use of existing instruments such as (mini-)MAXDOAS and NO_2 lidar should be enhanced in existing networks such as NDACC. Further studies should assess the accuracy of these instruments.

Tropospheric NO_2 has a strong daily cycle, which makes the timing of validation measurements important. Since tropospheric NO_2 is so variable, the collocation with the satellite is critical. Also NO_2 has a high spatial variability. It is therefore important to place instruments in locations that are representative of the background concentration and not too close to the source. The most extensive campaign so far, focused on tropospheric NO_2 in particular, has been the CINDI campaign, June–July 2009, in Cabauw, The Netherlands (Fig. 7.12). Similar efforts should be encouraged in the future.

Fig. 7.12 One of the more than 20 MAXDOAS-like instruments in use during the Cabauw Intercomparison campaign of nitrogen dioxide measuring instruments (CINDI), June–July 2009, Cabauw, The Netherlands.

7.6.4 Validation Strategy for CO

CO is measured at only a few sites, using flasks or continuous measurement methods (*in situ*; WMO-GAW 2010) and FTIR (remote sensing). FTIR measures total CO columns, given that CO is mostly in the troposphere, this approximates to tropospheric CO.

There is a small quasi-global network of ground-based FTIR instruments, but places with high CO emissions are not well covered, and the location of the FTIR instruments is often not ideal. Some FTIR instruments are situated too high and only measure part of the column, some are too close to the sea, where satellite measurements are inaccurate. The flask measurements from the Cooperative Air Sampling Network cover a wide region, but only give surface CO concentrations.

The FTIR network should be extended to optimize the coverage for CO validation. Large CO variability can be expected near source regions. There is a lack of CO column data over regions with high surface albedo.

The deployment of light instruments for sonde applications is highly recommended.

References

Alicke, B., U. Platt and J. Stutz, 2002, Impact of nitrous acid photolysison the total hydroxyl radical budget during the Limitation of Oxidant Production/Pianura Padana Produzione di Ozone study in Milan, *J. Geophys. Res.*, **107(D22)**, 8196, doi:10.1029/2000JD000075.

Bergamaschi, P., C. Frankenberg, J.F. Meirink, M. Krol, F. Dentener, T. Wagner, U. Platt, J.O. Kaplan, S. Körner, M. Heimann, E.J. Dlugokencky and A. Goede, 2007, Satellite chartography of atmospheric methane from SCIAMACHY on board ENVISAT: 2. Evaluation based on inverse model simulations, *J. Geophys. Res.*, **112**, D02304, doi: 10.1029/2006JD007268.

Blond, N., K.F. Boersma, H.J. Eskes, R.J. van der A, M. Van Rozendael, I. De Smedt, G. Bergametti and R. Vautard, 2007, Intercomparison of SCIAMACHY nitrogen dioxide observations, *in situ* measurements and air quality modeling results over Western Europe, *J. Geophys. Res.*, **112**, doi:10.1029/2006JD007277.

Boersma, K.F., H.J. Eskes and E.J. Brinksma, 2004, Error analysis for tropospheric NO_2 retrieval from space, *J. Geophys. Res.*, **109**, D04311, doi:10.1029/2003JD003962.

Boersma, K.F., D.J. Jacob, H.J. Eskes, R.W. Pinder, J. Wang and R.J. van der A, 2008, Intercomparison of SCIAMACHY and OMI tropospheric NO_2 columns: observing the diurnal evolution of chemistry and emissions from space, *J. Geophys. Res.*, **113**, 16S27, doi:10.1029/2007JD008832.

Brinksma, E., G. Pinardi, R. Braak, H. Volten, A. Richter, A. Schönhardt, M. Van Roozendael, C. Fayt, C. Hermans, R. Dirksen, T. Vlemmix, A.J.C. Berkhout, D.P.J. Swart, H. Oetjen, F. Wittrock, T. Wagner, O.W. Ibrahim, G. de Leeuw, M. Moerman, L. Curier, E.A. Celarier, W.H. Knap, J.P. Veefkind, H.J. Eskes, M. Allaart, R. Rothe, A.J.M. Piters and P. Levelt, 2008, The 2005 and 2006 DANDELIONS NO_2 and Aerosol Intercomparison Campaigns, *J. Geophys. Res.*, **113**, 46, doi:10.1029/2007JD008808.

Brewer, A.W., J.B. Kerr and C.T. McElroy, 1973, Nitrogen dioxide concentrations in the atmosphere, Nature, **246**, 129–133.

Buchwitz, M., R. de Beek, S. Noël, J.P. Burrows, H. Bovensmann, O. Schneising, I. Khlystova, M. Bruns, H. Bremer, P. Bergamaschi, S. Körner and M. Heimann, 2006, Atmospheric carbon gases retrieved from SCIAMACHY by WFM-DOAS: version 0.5 CO and CH_4 and impact of calibration improvements on CO_2 retrieval, *Atmos. Chem. Phys.*, **6**, 2727–2751.

Buchwitz, M., I. Khlystova, H. Bovensmann and J.P. Burrows, 2007, Three years of global carbon monoxide from SCIAMACHY: comparison with MOPITT and first results related to the detection of enhanced CO over cities, *Atmos. Chem. Phys.*, **7**, 2399–2411.

Bucsela, E.J., A.E. Perring, R.C. Cohen, K.F. Boersma, E.A. Celarier, J.F. Gleason, M.O. Wenig, T.H. Bertram, P.J. Wooldridge, R. Dirksen and J.P. Veefkind, 2008, Comparison of tropospheric NO_2 from *in situ* aircraft measurements with near-real-time and standard product data from OMI, *J. Geophys. Res.*, **113**, D16S31, doi:10.1029/2007JD008838.

Calisesi, Y., V.T. Soebijanta and R. van Oss, 2005, Regridding of remote soundings: Formulation and application to ozone profile comparison, *J. Geophys. Res.*, **110**, D23306, doi:10.1029/2005JD006122.

Ceccherini, S., U. Cortesi, P.T. Verronen and E. Kyrölä, 2008, Technical Note: Continuity of MIPAS-ENVISAT operational ozone data quality from full- to reduced-spectral-resolution operation mode, *Atmos. Chem. Phys.*, **8**, 2201–2212.

Celarier, E.A., E.J. Brinksma, J.F. Gleason, J.P. Veefkind, A. Cede, J.R. Herman, D. Ionov, F. Goutail, J.-P. Pommereau, J.-C. Lambert, M. van Roozendael, G. Pinardi, F. Wittrock, A. Schönhardt, A. Richter, O.W. Ibrahim, T. Wagner, B. Bojkov, G. Mount, E. Spinei, C.M. Chen, T.J. Pongetti, S.P. Sander, E.J. Bucsela, M.O. Wenig, D.P.J. Swart, H. Volten, M. Kroon and P.F. Levelt, 2008, Validation of Ozone Monitoring Instrument nitrogen dioxide columns, *J. Geophys. Res.*, **113**, D15S15, doi:10.1029/2007JD008908.

Cleary, P.A., P.J. Wooldridge and R.C. Cohen, 2002, Laser-induced fluorescence detection of atmospheric NO_2 with a commercial diode laser and a supersonic expansion, *Appl. Optics*, **41**, 6950–6956.

Cortesi, U., J.C. Lambert, C. De Clercq, G. Bianchini, T. Blumenstock, A. Bracher, E. Castelli, V. Catoire, K.V. Chance, M. De Mazière, P. Demoulin, S. Godin-Beekmann, N. Jones, K. Jucks, C. Keim, T. Kerzenmacher, H. Kuellmann, J. Kuttippurath, M. Iarlori, G.Y. Liu, Y. Liu, I.S. McDermid, Y.J. Meijer, F. Mencaraglia, S. Mikuteit, H. Oelhaf, C. Piccolo, M. Pirre, P. Raspollini, F. Ravegnani, W.J. Reburn, G. Redaelli, J.J. Remedios, H. Sembhi, D. Smale, T. Steck, A. Taddei, C. Varotsos, C. Vigouroux, A. Waterfall, G. Wetzel and S. Wood, 2007, Geophysical validation of MIPAS-ENVISAT operational ozone data, *Atmos. Chem. Phys.*, **7**, 4807–4867.

Day, D.A., M.B. Dillon, P.J. Wooldridge, J.A. Thornton, R.S. Rosen, E.C. Wood, and R.C. Cohen, 2003, On alkyl nitrates, O_3, and the "missing NO_y", *J. Geophys. Res.*, **108(D16)**, 4501, doi:10.1029/2003JD003685.

de Laat, A.T.J., A.M.S. Gloudemans, I. Aben, M. Krol, J.F. Meirink, G.R. van der Werf and H. Schrijver, 2007, Scanning Imaging Absorption Spectrometer for Atmospheric Chartography carbon monoxide total columns: Statistical evaluation and comparison with chemistry transport model results, *J. Geophys. Res.*, **112**, doi:10.1029/2006JD008256.

de Laat, A. T. J., A. M. S. Gloudemans, I. Aben, and H. Schrijver, 2010, Global evaluation of SCIAMACHY and MOPITT carbon monoxide column differences for 2004–2005, J. Geophys. Res., 115, D06307, doi:10.1029/2009JD012698.

De Smedt, I., J.-F. Müller, T. Stavrakou, R. van der A, H. Eskes and M. Van Roozendael, 2008, Twelve years of global observations of formaldehyde in the troposphere using GOME and SCIAMACHY sensors, *Atmos. Chem. Phys.*, **8**, 4947–4963.

Dils, B., M. De Mazière, J.F. Müller, T. Blumenstock, M. Buchwitz, R. de Beek, P. Demoulin, P. Duchatelet, H. Fast, C. Frankenberg, A. Gloudemans, D. Griffith, N. Jones, T. Kerzenmacher, I. Kramer, E. Mahieu, J. Mellqvist, R.L. Mittermeier, J. Notholt, C.P. Rinsland, H. Schrijver, D. Smale, A. Strandberg, A.G. Straume, W. Stremme, K. Strong, R. Sussmann, J. Taylor, M. van den Broek, V. Velazco, T. Wagner, T. Warneke, A. Wiacek and S. Wood, 2006, Comparisons between SCIAMACHY and ground-based FTIR data for total columns of CO, CH_4, CO_2 and N_2O, *Atmos. Chem. Phys.*, **6**, 1953–1976.

Dobber M., Q. Kleipool, R. Dirksen, P. Levelt, G. Jaross, S. Taylor, T. Kelly, L. Flynn, G. Leppelmeier, N. Rozemeijer, 2008, Validation of Ozone Monitoring Instrument level 1b data products, *J. Geophys. Res.*, **113**, D15S06, doi:10.1029/2007JD008665.

Dobson, G.M.B. and D.N. Harrison, 1926, Measurements of the amount of ozone in the Earth's Atmosphere and its Relation to other Geophysical Conditions, *Proc. R. Soc. London*, **110**, 660–693.

Dorf, M., H. Bösch, A. Butz, C. Camy-Peyret, M.P. Chipperfield, A. Engel, F. Goutail, K. Grunow, F. Hendrick, S. Hrechanyy, B. Naujokat, J.-P. Pommereau, M. Van Roozendael, C. Sioris, F. Stroh, F. Weidner, and K. Pfeilsticker, 2006, Balloon-borne stratospheric BrO measurements: comparison with Envisat/SCIAMACHY BrO limb profiles, *Atmos. Chem. Phys.*, **6**, 2483–2501.

Eck, T.F., B.N. Holben, J.S. Reid, O. Dubovik, A. Smirnov, N.T. O'Neill, I.Slutsker and S. Kinne, 1999, Wavelength dependence of the optical depth of biomass burning, urban and desert dust aerosols, *J. Geophys. Res.*, **104**, 31333–31350.

Emmons, L.K., D.P. Edwards, M.N. Deeter, J.C. Gille, T. Campos, P. Nédélec, P. Novelli and G. Sachse, 2009, Measurements of Pollution In The Troposphere (MOPITT) validation through 2006, *Atmos. Chem. Phys.*, **9**, 1795–1803.

Eskes, H.J., P.F.J. van Velthoven, P.J.M. Valks and H.M. Kelder, 2003, Assimilation of GOME total-ozone satellite observations in a three-dimensional tracer-transport model, *Q. J. R. Met. Soc.*, **129**, 1663–1681, doi: 10.1256/qj.02.14.

Ganzeveld, L. and J. Lelieveld, 1995, Dry deposition parameterization in a chemistry general circulation model and its influence on the distribution of reactive trace gases, *J. Geophys. Res.*, **100(10)**, 20999–21012.

Hagolle, O., Ph. Goloub, P.Y. Deschamps, H. Cosnefroy, X. Briottet, T. Bailleul, J.M. Nicolas, F. Parol, B. Lafrance and M. Herman, 1999, Results of POLDER in-flight calibration, *IEEE Transactions on Geoscience and Remote Sensing*, **37**, 03.

Heckel, A., A. Richter, T. Tarsu, F. Wittrock, C. Hak, I. Pundt, W. Junkermann and J.P. Burrows, 2005, MAX-DOAS measurements of formaldehyde in the Po-Valley, *Atmos. Chem. Phys.*, **5**, 909–918.

Heland, J., H. Schlager, A. Richter and J.P. Burrows, 2002, First comparison of tropospheric NO_2 column densities retrieved from GOME measurements and *in situ* aircraft profile measurements, *Geophys. Res. Lett.*, **29(20)**, 1983.

Holben, B. N., T.F. Eck, I. Slutsker, D. Tanré, J.P. Buis, A. Setzer, E. Vermote, J.A. Reagan, Y.J. Kaufman, T. Nakajima, F. Lavenu, I. Jankowiak and A. Smirnov, 1998, AERONET - A federated instrument network and data archive for aerosol characterization, *Remote Sens. Environ.*, **66(1)**, 1–16.

Holton, J.R., P.H. Haynes, M.E. McIntyre, A.R. Douglas, R.B. Rood and L. Pfister, 1995, Stratosphere-troposphere exchange, *Rev. Geophys.*, **33**, 403 – 439.

Hönninger G. and U. Platt, 2002, The Role of BrO and its Vertical Distribution during Surface Ozone Depletion at Alert, *Atmos. Environ.*, **36**, 2481–2489.

Hönninger, G., H. Leser, O. Sebastián and U. Platt, 2004a, Ground-based measurements of halogen oxides at the Hudson Bay by active longpath DOAS and passive MAX-DOAS, *Geophys. Res. Lett.*, **31**, L04111, doi:10.1029/2003GL018982.

Hönninger, G., C. von Friedeburg and U. Platt, 2004b, Multi axis differential optical absorption spectroscopy (MAX-DOAS), *Atmos. Chem. Phys.*, **4**, 231–254.

Irie, H., Y. Kanaya, H. Akimoto, H. Iwabuchi, A. Shimizu, and K. Aoki, 2008, First retrieval of tropospheric aerosol profiles using MAX-DOAS and comparison with lidar and sky radiometer measurements, *Atmos. Chem. Phys.*, **8(2)**, 341–350.

ISO/IEC Guide 99-12: 2007 International Vocabulary of Metrology – Basic and General Concepts and Associated Terms, VIM.

Jaross G. and J. Warner, 2008, Use of Antarctica for validating reflected solar radiation measured by satellite sensors, *J. Geophys. Res.*, **113**, D16S34, doi:10.1029/2007JD008835.

Jones, N.B., K. Riedel, W. Allan, S. Wood, P.L. Palmer, K. Chance and J. Notholt, 2009, Long-term tropospheric formaldehyde concentrations deduced from ground-based fourier transform solar infrared measurements, *Atmos. Chem. Phys.*, **9**, 7131–7142.

Kahn, R. A., W.-H. Li, C. Moroney, D. J. Diner, J. V. Martonchik and E. Fishbein, 2007, Aerosol source plume physical characteristics from space-based multiangle imaging, *J. Geophys. Res.*, **112**, D11205, doi:10.1029/2006JD007647.

Kebabian, P.L., S.C. Herndon and A. Freedman, 2005, Detection of nitrogen dioxide by cavity attenuated phase shift spectroscopy, *Anal. Chem.*, **77**, 724– 728.

Kelly, T.J., C.W. Spicer and G.F. Ward, 1990, An assessment of the luminol chemiluminescence technique for measurements of NO_2 in ambient air, *Atmos. Environ.*, **24A**, 2397– 2403.

Klausen, J., C. Zellweger, B. Buchmann and P. Hofer, 2003, Uncertainty and bias of surface ozone measurements at selected Global Atmosphere Watch sites, *J. Geophys. Res.*, **108**, 4622, doi:10.1029/2003JD003710.

Kleipool, Q.L., M.R. Dobber, J.F. de Haan, and P.F. Levelt, 2008, Earth surface reflectance climatology from 3 years of OMI data, *J. Geophys. Res.*, **113**, doi:10.1029/2008JD010290.

Koelemeijer, R., P. Stammes, J. Hovenier and J. de Haan, 2001, A fast method for retrieval of cloud parameters using oxygen A band measurements from the Global Ozone Monitoring Experiment, *J. Geophys. Res.*, **106(4)**, 3475–3490.

Koelemeijer, R.B.A., J.F. de Haan and P. Stammes, 2003, A Database of spectral surface reflectivity in the range 335 - 772 nm derived from 5.5 years GOME observations, *J. Geophys. Res.*, **108**, 4070, doi:10.1029/2002JD002429.

Konovalov, I.B., M. Beekman, J.P. Burrows and A. Richter, 2008, Satellite measurement based estimates of decadal changes in European nitrogen oxides emissions, *Atmos. Chem. Phys.*, **8**, 2723–2641.

Kramer, L.J., R.J. Leigh, J.J. Remedios, and P.S. Monks, 2008, Comparison of OMI and ground-based *in situ* and MAX-DOAS measurements of tropospheric nitrogen dioxide in an urban area, *J. Geophys. Res.*, **113**, D16S39, doi:10.1029/2007JD009168.

Kroon M., J.P. Veefkind, M. Sneep, R.D. McPeters, P.K. Bhartia and P.F. Levelt, 2008, Comparing OMI-TOMS and OMI-DOAS total ozone column data, *J. Geophys. Res.*, **113**, D16S28, doi:10.1029/2007JD008798.

Krotkov, N.A., B. McClure, R.R. Dickerson, S.A. Carn, C. Li, P.K. Bhartia, K. Yang, A.J. Krueger, Z. Li, P.F. Levelt, H. Chen, P. Wang and D. Lu, 2008, Validation of SO_2 retrievals from the Ozone Monitoring Instrument over NE China, *J. Geophys. Res.*, **113**, 40, doi:10.1029/2007JD008818.

Lamsal L.N., R.V. Martin, A. van Donkelaar, M. Steinbacher, E.A. Celarier, E. Bucsela, E.J. Dunlea and J.P. Pinto, 2008, Ground-level nitrogen dioxide concentrations inferred from the satellite-borne Ozone Monitoring Instrument, *J. Geophys. Res.*, **113**, D16308, doi:10.1029/2007JD009235.

Lauer, A., M. Dameris, A. Richter, and J.P. Burrows, 2002, Tropospheric NO_2 columns: a comparison between model and retrieved data from GOME measurements, *Atmos. Chem. Phys.*, **2**, 67–78.

Leigh, R.J., G.K. Corlett, U. Friess and P.S. Monks, 2006, Concurrent multiaxis differential optical absorption spectroscopy system for the measurement of tropospheric nitrogen dioxide, *Appl. Optics*, **45(28)**, 7504–7518.

Li, Y.Q., K.L. Demerjian, M.S. Zahniser, D.D. Nelson, J.B. McManus and S.C. Herndon, 2004, Measurement of formaldehyde, nitrogen dioxide, and sulfur dioxide at Whiteface Mountain using a dual tunable diode laser system, *J. Geophys. Res.*, **109**, D16S08, doi:10.1029/2003JD004091.

Martin, R.V., D.J. Jacob, K. Chance, T.P. Kurosu, P.I. Palmer and M.J. Evans, 2003, Global inventory of nitrogen oxide emissions constrained by space-based observations of NO_2 columns, *J. Geophys. Res.*, **108**, doi:10.1029/2003JD003453.

Matsumoto, J. , J. Hirokawa, H. Akimoto and Y. Kajii, 2001, Direct measurement of NO_2 in the marine atmosphere by laser-induced fluorescence technique, *Atmos. Environ,*, **35**, 2803–2814, doi:10.1016/S1352-2310(01)00078-4.

McKenzie, R.L. and P.V. Johnston, 1982, Seasonal variation in stratospheric NO_2 at 45 degrees S, *Geophys. Res. Lett.*, **9**, 1255–1258.

Measures, R.M., 1992, *Laser Remote Sensing. Fundamentals and Applications*, Krieger, New-York, pp 510.

Mellqvist, J., B. Galle, M. Kihlman and M. Burton, 2005, Mobile solar FTIR measurements of SO_2 and halogens in the gas plumes of active volcanoes, *Geophys. Res. Abstracts*, **7**, 08987.

Müller, J.-F. and T. Stavrakou, 2005, Inversion of CO and NO_x emissions using the adjoint of the IMAGES model, *Atmos. Chem. Phys.*, **4**, 1157–1186.

Murphy, J.G., D.A. Day, P.A. Cleary, P.J. Wooldridge and R.C. Cohen, 2006, Observations of the diurnal and seasonal trends in nitrogen oxides in the western Sierra Nevada, *Atmos. Chem. Phys.*, **6**, 5321– 5338.

Nakajima, H., X. Liu, I. Murata, Y. Kondo, F.J. Murcray, M. Koike, Y. Zhao and H. Nakane, 1997, Retrieval of vertical profiles of ozone from high resolution infrared solar spectra at Rikubetsu, Japan, *J. Geophys. Res.*, **102**, 29981–29990.

Nassar, R., J.A. Logan, H.M. Worden, I.A. Megretskaia, K.W. Bowman, G.B. Osterman, A.M. Thompson, D.W. Tarasick, S. Austin, H. Claude, M.K. Dubey, W.K. Hocking, B.J. Johnson, E. Joseph, J. Merrill, G.A. Morris, M. Newchurch, S.J. Oltmans, F. Posny, F.J. Schmidlin, H. Vömel, D.N. Whiteman and J.C. Witte, 2008, Validation of Tropospheric Emission Spectrometer (TES) nadir ozone profiles using ozonesonde measurements, *J. Geophys. Res.*, **113**, D15S17, doi:10.1029/2007JD008819.

Noel, S., M. Buchwitz, H. Bovensmann and J.P. Burrows, 2005, Validation of SCIAMACHY AMC-DOAS water vapour columns, *Atmos. Chem. Phys.*, **5**, 1835–1841.

Notholt, J., R. Neuber, O. Schrems and T. von Clarmann, 1993, Stratospheric trace gas concentrations in the Arctic polar night derived by FTIR-spectroscopy with the moon as IR light source, *Geophys. Res. Lett.*, **20**, 2059–2062.

Noxon, J. F., 1975, Nitrogen Dioxide in the Stratosphere and Troposphere measured by Ground-based Absorption Spectroscopy, *Science*, **189**, 547–549.

Ordoñez, C., A. Richter, M. Steinbacher, C. Zellweger, H. Nüss, J.P. Burrows and A.S.H. Prévôt, 2006, Comparison of 7 years of satellite-borne and ground-based tropospheric NO_2 measurements around Milan, Italy, *J. Geophys. Res.*, **111**, doi:10.1029/2005JD006305.

Osthoff, H.D., S.S. Brown, T.B. Ryerson, T.J. Fortin, B.M. Lerner, E.J. Williams, A. Pettersson, T. Baynard, W.P. Dubé, S.J. Ciciora and A.R. Ravishankara, 2006, Measurement of atmospheric NO_2 by pulsed cavity ring-down spectroscopy, *J. Geophys. Res.*, **111**, D12305, doi:10.1029/2005JD006942.

Platt, U. and D. Perner, 1980, Direct measurements of atmospheric CH_2O, HNO_2, NO_2, and SO_2 by differential optical absorption in the near UV, *J. Geophys. Res. Oceans*, **85**, 7453–7458.

Pommereau, J.P. and F. Goutail, 1988a, O_3 and NO_2 Ground-Based Measurements by Visible Spectrometry during Arctic Winter and Spring 1988, *Geophys. Res. Lett.*, **15**, 891.

Pommereau, J.P. and F. Goutail, 1988b, Stratospheric O_3 and NO_2 Observations at the Southern Polar Circle in Summer and Fall 1988, *Geophys. Res. Lett.*, **15**, 895.

Pougatchev, N.S., B.J. Connor, and C.P. Rinsland, 1995, Infrared measurements of the ozone vertical distribution above Kitt Peak, *J. Geophys. Res.*, **100**, 16689–16698.

Pougatchev, N.S., B.J. Connor, N.B. Jones, C.P. Rinsland, 1996, Validation of ozone profile retrievals from infrared ground-based solar spectra, *Geophys. Res. Lett.*, **23(13)**, 1637–1640.

Remer, L. A., Y. J. Kaufman, D. Tanré, S. Mattoo, D. A. Chu, J. V. Martins, R-R. Li, C. Ichoku, R. C. Levy, R. G. Kleidman, T. F. Eck, E. Vermote and B. N. Holben, 2005, The MODIS aerosol algorithm, products, and validation, *J. Atmos. Sci.*, **62(4)**, 947– 973.

Richter, A., J. Leitão, A. Heckel and J.P. Burrows, 2007, Synergistic use of multiple sensors for tropospheric NO_2 measurements, Presentation at the *ACCENT AT2 workshop "Tropospheric NO_2 measured by satellites"*, 10–12 Sept 2007, KNMI, De Bilt, The Netherlands.

Robles-Gonzalez, C., G. de Leeuw, R. Decae, J. Kusmierczyk-Michulec and P. Stammes, 2006, Aerosol properties over the Indian Ocean Experiment (INDOEX) campaign area retrieved from ATSR-2, *J. Geophys. Res.*, 111, D15205, doi:10.1029/2005JD006184.

Rodgers, C.D. and B.J. Connor, Intercomparison of remote sounding instruments, 2003, *J. Geophys. Res.*, **108 (D3)**, 4116, doi:10.1029/2002JD002299.

Roelofs G. J., A.S. Kentarchos, T. Trickl, A. Stohl, W.J. Collins, R.A. Crowther, D. Hauglustaine, A. Klonecki, K.S. Law, M.G. Lawrence, R. von Kuhlmann and M. van Weele, 2003 Intercomparison of tropospheric ozone models: Ozone transport in a complex tropopause folding event, *J. Geophys. Res.*, **108 (12)**, 8529, doi:10.1029/2003JD003462.

Sanders, R.W., S. Solomon, J.P. Smith, L. Perliski, H.L. Miller,G.H. Mount, J.G. Keys and A.L. Schmeltekopf, 1993,Visible and Near-Ultraviolet Spectroscopy at McMurdo Station Antarctica, 9. Observations of OClO from April to October 1991, *J. Geophys. Res.*, **98(D4)**, 7219–7228.

Schaap, M., A. Apituley, R.M.A. Timmermans, R.B.A. Koelemeijer and G. de Leeuw, 2009, Exploring the relation between aerosol optical depth and PM2.5 at Cabauw, the Netherlands, *Atmos. Chem. Phys.*, **9**, 909–925.

Schaub, D., K.F. Boersma, J.W. Kaiser, A.K. Weiss, D. Folini, H.J. Eskes and B. Buchmann, 2006, Comparison of GOME tropospheric NO_2 columns with NO_2 profiles deduced from ground-based *in situ* measurements, *Atmos. Chem. Phys.*, **6**, 3211–3229.

Schaub, D., D. Brunner, K.F. Boersma, J. Keller, D. Folini, B. Buchmann, H. Berresheim and J. Staehelin, 2007, SCIAMACHY tropospheric NO_2 over Switzerland: estimates of NO_x lifetimes and impact of the complex Alpine topography on the retrieval, *Atmos. Chem. Phys.*, **7**, 5971–5987.

Schmid, B., K.J. Thome, Ph. Demoulin, R. Peter, C. Matzler and J. Sekler, 1996, Comparison of modeled and empirical approaches for retrieving columnar water vapor from solar transmittance measurements in the 0.94 micron region, *J. Geophys. Res.*, **101**, 9345–9358.

Schoulepnikoff, L., H. van den Bergh, V. Mitev and B. Calpini, 1998, Tropospheric air pollution monitoring lidar, in: Meyers, R.A. (editor), *The Encyclopedia of Environmental Analysis and Remediation*, John Wiley and Sons, New York, 4873–4909.

Schutgens, N.A.J. and P. Stammes, 2003, A novel approach to the polarization correction of spaceborne spectrometers, *J. Geophys. Res.*, **108 (D7)**, 4229, doi:10.1029/2002JD002736.

Schutgens, N.A.J., L.J. Tilstra, P. Stammes and F.-M. Bréon, 2004, On the relationship between Stokes parameters Q and U of atmospheric ultraviolet/visible/near-infrared radiation, *J. Geophys. Res.*, **109**, D09205, doi:10.1029/2003JD004081.

Shindell, D.T., G. Faluvegi and L.K. Emmons, 2005, Inferring carbon monoxide pollution changes from space-based observations, *J. Geophys. Res.*, **110**, doi:10.1029/2005JD006132.

Sinreich, R., U. Frieß, T. Wagner and U. Platt, 2005, Multi axis differential optical absorption spectroscopy (MAXDOAS) of gas and aerosol distributions, *Faraday Discuss.*, **130**, 153–164, doi:10.1039/B419274P.

Sneep, M., R. Braak, E. Brinksma and M. Kroon, 2006, Documentation for the 'CAMA' verification and validation toolkit, Version 1.2, MA-OMIE-KNMI-832, KNMI, De Bilt, The Netherlands.

Sneep M., J.F. de Haan, P. Stammes, P. Wang, C. Vanbauce, J. Joiner, A.P. Vasilkov and P.F. Levelt, 2008, Three-way comparison between OMI and PARASOL cloud pressure products, *J. Geophys. Res.*, **113**, D15S23, doi:10.1029/2007JD008694.

Solomon, S., A.L. Schmeltekopf and R.W. Sanders, 1987, On the interpretation of zenith sky absorption measurements, *J. Geophys. Res.*, **92**, 8311–8319.

Stammes P., M. Sneep, J.F. de Haan, J.P. Veefkind, P. Wang, P.F. Levelt, 2008, Effective cloud fractions from the Ozone Monitoring Instrument: Theoretical framework and validation, *J. Geophys. Res.*, **113**, D16S38, doi:10.1029/2007JD008820.

Steinbacher, M., C. Zellweger, B. Schwarzenbach, S. Bugmann, B. Buchmann, C. Ordonez, A.S. H. Prevot and C. Hueglin, 2007, Nitrogen oxide measurements at rural sites in Switzerland: Bias of conventional measurement techniques, *J. Geophys. Res.*, **112**, D11307, doi:10.1029/2006JD007971.

Sussmann, R. and M. Buchwitz, 2005, Initial validation of ENVISAT/SCIAMACHY columnar CO by FTIR profile retrievals at the Ground-Truthing Station Zugspitze, *Atmos. Chem. Phys.*, **5**, 1497–1503.

Theys, N., F. Hendrick, M. Van Roozendael, I. De Smedt, C. Fayt and R. van der A, 2006, Retrieval of BrO Columns from SCIAMACHY and their Validation Using Ground-Based DOAS Measurements, *Proc. of the First Atmospheric Science Conference*, ESRIN, Frascati, Italy, 8 – 12 May 2006, ESA SP-628.

Tilstra L.G., G. van Soest, P. Stammes, 2005, Method for in-flight satellite calibration in the ultraviolet using radiative transfer calculations, with application to Scanning Imaging Absorption Spectrometer for Atmospheric Chartography (SCIAMACHY), *J. Geophys. Res.*, **110**, D18311, doi:10.1029/2005JD005853.

Tilstra L.G. and P. Stammes, 2007, Earth reflectance and polarization intercomparison between SCIAMACHY onboard Envisat and POLDER onboard ADEOS-2, *J. Geophys. Res.*, **112**, D11304, doi:10.1029/2006JD007713.

Toon, G.C., C.B. Farmer, P.W. Schaper, J.-F. Blavier and L.L. Lowes, 1989, Ground-based Infrared Measurements of Tropospheric Source Gases over Antarctica during the 1986 Austral Spring, *J. Geophys. Res.*, **94**, 11613–11624.

van Noije, T.P.C., H J. Eskes, F.J. Dentener, D.S. Stevenson, K. Ellingsen, M.G. Schultz, O. Wild, M. Amann, C.S. Atherton, D.J. Bergmann, I. Bey, K.F. Boersma, T. Butler, J. Cofala, J. Drevet, A.M. Fiore, M. Gauss, D.A. Hauglustaine, L.W. Horowitz, I.S.A. Isaksen, M.C. Krol, J.-F. Lamarque, M.G. Lawrence, R.V. Martin, V. Montanaro, J.-F. Müller, G. Pitari, M.J. Prather, J. A. Pyle, A. Richter, J.M. Rodriguez, N.H. Savage, S.E. Strahan, K. Sudo, S. Szopa and M. Van Roozendael, 2006, Multi-model ensemble simulations of tropospheric NO_2 compared with GOME retrievals for the year 2000, *Atmos. Chem. Phys.*, **6**, 2943–2979.

van Diedenhoven B., O.P. Hasekamp, J. Landgraf, 2007, Retrieval of cloud parameters from satellite-based reflectance measurements in the ultraviolet and the oxygen A-band, *J. Geophys. Res.*, **112**, D15208, doi:10.1029/2006JD008155.

Van Roozendael, M., D. Loyola, R. Spurr, D. Balis, J-C. Lambert, Y. Livschitz, T. Ruppert, P. Valks, P. Kenter, C. Fayt and C. Zehner, 2006, Ten years of GOME/ERS-2 total ozone data – The new GOME Data Processor (GDP) Version 4: I Algorithm Description, *J. Geophys. Res.*, **111**, D14311, doi:10.1029/2005JD006375.

Veefkind, J.P. and J.F. de Haan, 2001, DOAS Total Ozone Algorithm, in Barthia, P.K. (editor), *OMI Algorithm Theoretical Basis Document*, Volume II – Chapter 3, ATBD-OMI-02, Version 1.0

Velders, G.J., C. Granier, R.W. Portmann, K. Pfeilsticker, M. Wenig, T. Wagner, U. Platt, A. Richter, and J.P. Burrows, 2001, Global tropospheric NO_2 column distributions: Comparing three-dimensional model calculations with GOME measurements, *J. Geophys. Res.*, **106**, 12643–12660.

Vogelmann, H. and T. Trickl, 2008, Wide-range sounding of free-tropospheric water vapor with a differential-absorption lidar (DIAL) at a high-altitude station, *Appl. Optics*, **47**, 2116–2132.

Wagner, T., O. Ibrahim, R. Sinreich, U. Frieß, R. von Glasow and U. Platt, 2007, Enhanced tropospheric BrO over Antarctic sea ice in mid winter observed by MAX-DOAS on board the research vessel Polarstern, *Atmos. Chem. Phys.*, **7(12)**, 3129–3142.

Wang, P., P. Stammes, R. van der A, G. Pinardi and M. VanRoozendael, 2008, FRESCO+: an improved O_2 A-band cloud retrieval algorithm for tropospheric trace gas retrievals, *Atmos. Chem. Phys.*, **8**, 6565–6576.

Wittrock, F., H. Oetjen, A. Richter, S. Fietkau, T. Medeke, A. Rozanov and J.P. Burrows, 2004, MAX-DOAS measurements of atmospheric trace gases in Ny-Alesund - Radiative transfer studies and their application, *Atmos. Chem. Phys.*, **4**, 955–966.

Wittrock, F., A. Richter, H. Oetjen, J.P. Burrows, M. Kanakidou, S. Myriokefalitakis, R. Volkamer, S. Beirle, U. Platt and T. Wagner, 2006, Simultaneous global observations of glyoxal and formaldehyde from space, *Geophys. Res. Lett.*, **33**, L16804, doi:10.1029/2006GL026310.

WMO-GAW Report No. 192, 2010, Guidelines for the Measurement of Atmospheric Carbon Monoxide, WMO/TD-No. 1551.

Yang, A., G.C. Toon, J.S. Margolis and P.O. Wennberg, 2002, Atmospheric CO_2 retrieved from ground-based near IR spectra, *Geophys. Res. Lett.*, **29(9)**, doi: 10.1029/2001GL014537.

Yudin, V.A., G. Pétron, J.-F. Lamarque, B.V. Khattatov, P.G. Hess, L.V. Lyjak, J.C. Gille, D.P. Edwards, M.N. Deeter and L.K. Emmons, 2004, Assimilation of the 2000–2001 CO MOPITT retrievals with optimized surface emissions, *Geophys. Res. Lett.*, **31**, doi:10.1029/2004GL021037.

Zander, R., Ph. Demoulin, D.H. Ehhalt, U. Schmidt and C.P. Rinsland, 1989a, Secular increase of the total vertical column abundance of carbon monoxide above central Europe since 1950, *J. Geophys. Res.*, **94**, 11020–11028.

Zander, R., Ph. Demoulin, D.H. Ehhalt and U. Schmidt, 1989b, Secular increase of the vertical column abundance of methane derived from IR solar spectra recorded at the Jungfraujoch station, *J. Geophys. Res.*, **94**, 11029–11039.

Zander, R., D.H. Ehhalt, C.P. Rinsland, U. Schmidt, E. Mahieu, J. Rudolph, P. Demoulin, G. Roland, L. Delbouille and A.J. Sauval, 1994, Secular trend and seasonal variability of N_2O above the Jungfraujoch station determined from IR solar spectra, *J. Geophys.Res.*, **99**, 16745–16756.

Zellweger C., C. Hüglin, J. Klausen, M. Steinbacher, M. Vollmer and B. Buchmann, 2009, Intercomparison of four different carbon monoxide measurement techniques and evaluation of the long-term carbon monoxide time series of Jungfraujoch, *Atmos. Chem. Phys.*, **9**, 3491–3503.

Zhou, Y., D. Brunner, K.F. Boersma, R. Dirksen and P. Wang, 2009, An improved tropospheric NO_2 retrieval for OMI observations in the vicinity of mountainous terrain, *Atmos. Meas. Tech.*, **2(2)**, 401–416

Chapter 8
Applications of Satellite Observations of Tropospheric Composition

Paul S. Monks and Steffen Beirle

8.1 Introduction

The advent of satellite measurements of the troposphere has taken us from a local/regional view of composition previously available from ground-based measurements to a global view. This revolution in tropospheric research has only occurred in the last three decades. Although instruments for surface mapping and meteorological parameters were recognized as remote sensing applications from the start of the space age, the remote sounding of tropospheric constituents by satellite instrumentation, often conceived for stratospheric measurement, has been used initially to give a new global view of tropospheric composition.

The time scales of atmospheric processes range from seconds to decades. For example, a pollution episode may only be apparent for a week or less, but the wider impact of such an episode may last much longer. This reflects the fact that processes such as the emission rate to the planetary boundary layer, chemical production, homogenous (*e.g.* radical–molecule, radical–radical, photolysis) and heterogeneous reactions, including both wet and dry deposition, determine the production and removal rates of species within the atmosphere. Remote sounding of the troposphere from satellites yields measurements of atmospheric composition, which give regional and global views on spatial and temporal scales not available from any other observing system. The number and type of measurements of a species required to provide a true global representation depends on the atmospheric lifetimes of the species involved.

The challenge of remote sensing the troposphere from satellite platforms is substantial and it is only with the current generation of satellite instruments and improvements in retrieval that a view of the troposphere has become available from space. An era is dawning in which long time series measurements of the troposphere

P.S. Monks (✉)
Department of Chemistry, University of Leicester, Leicester, UK

S. Beirle
Max-Planck-Institut für Chemie, Mainz, Germany

J.P. Burrows et al. (eds.), *The Remote Sensing of Tropospheric Composition from Space*, Physics of Earth and Space Environments, DOI 10.1007/978-3-642-14791-3_8,

from space will become available leading to new understanding of regional and global change. One of the challenges currently being tackled is the assimilation and fusion of different data streams to give a more holistic view of chemical and physical processes in the troposphere.

After the launch of a satellite for each instrument there are a number of scientific stages for the application of satellite data from the first-light in space to the calibration and validation of the satellite product which is a continuous process. The next stage is an observational stage, where the scientific discovery comes from an appreciation of the temporal and spatial distribution of the trace compound, the final stage becomes quantitative analysis with a model.

The aim of this chapter is to give an overview about the utility of satellite observations for measuring tropospheric composition. The focus will be on probing the chemical composition of the atmosphere. The retrieval and applications of aerosols are considered separately in Chapter 6 and the methodologies for the retrieval and validation of the trace species are described in Chapters 2, 3 and 7. This chapter summarises the tropospheric chemical species that can be measured from space and looks at the generalised applications of these measurements at the primary observational level. For a full list of satellites that have measured atmospheric composition in the troposphere and stratosphere readers are referred to the table in Appendix A of this book and also to Burrows (1999) and Martin (2008).

8.2 Overview of the Tropospheric Chemical Species Measured from Space

In this section the tropospheric gases that can be measured from space will be reviewed. Table 8.3 lists the tropospheric chemical species that have been measured from space. A recent overview of the main scientific questions and drivers for tropospheric composition can be found in Monks et al. (2009).

8.2.1 Tropospheric Ozone, O_3

Ninety percent of atmospheric O_3 can be found in the stratosphere; on average only about 10% resides in the troposphere. While stratospheric O_3 determines the amount of short wavelength radiation available to initiate photochemistry (Monks 2005), tropospheric O_3 acts as initiator, reactant and product in much of the oxidation chemistry that takes place in the troposphere.

Tropospheric O_3 was one of the first chemical species (other than water) to derived from space-based total O_3 column observations based on the UV radiances measured by the TOMS instrument combined with stratospheric measurements and/or SAGE and SBUV to give tropospheric residuals (Fishman 1991b; Fishman et al. 1990; 1996). This technique utilises back scattered solar radiation and was refined by Hudson, Thompson and co-workers (Hudson et al. 1995; Hudson and Thompson 1998; Kim et al. 1996) to give a tropical tropospheric O_3 product. Ziemke and co-workers have advanced the residual-type retrieval methodologies using cloud slicing techniques as well as synergistic use of TOMS (Ahn et al. 2003; Ziemke et al. 1998; 2000; 2001; 2003) and latterly OMI (Choi et al. 2008; Schoeberl et al. 2007) with stratospheric O_3 measurements from HALOE and MLS. These methods have also been advanced using assimilation techniques (Stajner et al. 2008).

Measurements of tropospheric O_3 were expanded using GOME-1 (Hoogen et al. 1999; Liu et al. 2006b; Munro et al. 1998). A number of groups have gone on to improve the retrievals of tropospheric O_3 using a variety of methodologies (Del Frate et al. 2005; Iapaolo et al. 2007; Liu et al. 2007; Muller et al. 2003; Tellmann et al. 2004; van der A et al. 2002). There has been an extensive study to compare GOME-1 O_3 profiles from nine different algorithms (Meijer et al. 2006).

Recent tropospheric O_3 products have become available from TES (Nassar et al. 2008; Osterman et al. 2008; Richards et al. 2008; Worden et al. 2007a; 2007b) that offers direct measurements of tropospheric O_3 from mid-IR spectra. Using IASI data for example, Eremenko et al. (2008) have shown the ability to map out the O_3 distributions during the European heatwave of 2007.

The general distribution and inter-annual variability of tropospheric O_3 as measured from space has been shown using a range of satellite sensors (Fishman and Brackett 1997; Fishman et al. 2005; Liu et al. 2006b; Ziemke et al. 2006). Satellite derived tropospheric O_3 has been used to investigate the influence of stratospheric air masses on tropospheric vertical O_3 columns over the Pacific (Ladstatter-Weissenmayer et al. 2004). Various tropospheric and stratospheric O_3 satellite data has been combined with O_3 sondes to produce O_3 climatologies (Lamsal et al. 2004). Early work indentified tropospheric pollution episodes (Fishman et al. 1987) which have recently been improved to show regional pollution (Fishman et al. 2003).

In the tropics, there has been a comparison of tropical O_3 columns from GOME with a model (Valks et al. 2003), as well as studies of O_3 over Africa (Meyer-Arnek et al. 2005a). The impact of biomass burning (BB) on tropical Atlantic O_3 has also been assessed (Jourdain et al. 2007). Long terms trends in satellite derived tropospheric O_3 over the Pacific have been derived showing a significant upward trend in the mid-latitudes of both hemispheres but not in the tropics (Ziemke et al. 2005). O_3 has been used in combination with other tracers to investigate lightning (Martin et al. 2007), oxidant budgets over the Indian ocean (Ladstatter-Weissenmayer et al. 2007a), pollution flows from north America (Choi et al. 2008) and the effects of the 2006 El Nino (Logan et al. 2008).

The distribution of O_3 and other trace gases are governed by the complex interaction of dynamical, chemical and radiative processes. Feedbacks within the chemistry climate system, in particular the impact of changing O_3 on the Earth's climate system via radiative forcing in the upper troposphere/lower stratosphere region which is sensitive to such perturbations, has been thought to be of particular importance (Worden et al. 2008). There are number of satellite instruments that give solely upper tropospheric (and lower stratospheric) views of O_3 (and a range of other tracers, see for example Section 8.2.6) (Coheur et al. 2005; Fischer et al. 2008;

Hegglin et al. 2008; Nardi et al. 2008; Raspollini et al. 2006; Rozanov et al. 2007). Studies of upper tropospheric O_3 have included assessing the impact of biomass burning (Clarmann et al. 2007).

8.2.2 *Nitrogen Dioxide,* NO_2

Nitrogen oxides ($NO_x = NO + NO_2$) are released into the troposphere from a variety of biogenic, anthropogenic and physical sources including fossil fuel combustion, biomass burning, microbial activity in soils and lightning discharges. There is still some debate about the exact magnitude of the various sources and sinks for NO_x (Lerdau et al. 2000). According to present estimates, about 30% of the global budget of NO_x comes from fossil fuel combustion with almost 86% of the NO_x emitted in one form or the other into the planetary boundary layer from surface processes. Other major sources are biomass burning ca. 19%, microbial release from soil 32% and lightning 13% (Schumann and Huntrieser 2007). Typical NO/NO_2 ratios in surface air are 0.2–0.5 during the day tending to zero at night. Over the timescales of hours to days NO_x is converted to nitric acid and nitrates, which are subsequently removed by rain and dry deposition.

Satellite measurements of tropospheric NO_2 have found widespread utility. Observational analyses have demonstrated the strong weekly cycles in the observed NO_2 (Beirle et al. 2003), the influence of biomass burning (Burrows et al. 1999; Ladstatter-Weissenmayer and Burrows 1998; Thomas et al. 1998) and continental scale outflow (Leue et al. 2001; Richter and Burrows 2002).

Owing to the importance of nitrogen dioxides in air quality (DEFRA 2003) there has been a focus on comparisons of tropospheric NO_2 measurements with ground sites for regional air quality (Blond et al. 2007). Measurements have been compared to ground based measurements in Kyrgyzstan (Ionov et al. 2006), St Petersburg (Poberovskii et al. 2007), the Moscow region (Timofeev et al. 2000), Switzerland (Schaub et al. 2005; 2007), the Milan area (Ordonez et al. 2006) and in the urban UK (Kramer et al. 2008). Comparisons have also been made with airborne data over the south-eastern USA (Martin et al. 2004b), the Alps/Mediterranean (Heue et al. 2005), over the Atlantic (Bucsela et al. 2008), and over Shanghai (Chen et al. 2009). On a regional scale, there have been comparisons with regional models over North America (Chun'e and Baoning 2008; Kim et al. 2009), Asia (Han et al. 2009; Uno et al. 2007), Africa (Meyer-Arnek et al. 2005a) and Europe (Konovalov et al. 2005). The data over Europe showed varying agreement with bottom-up emission inventories highlighting both apparent over and under-estimates (Konovalov et al. 2006; 2008).

Tropospheric NO_2 data have been used to quantify NO_x emissions from soil (Bertram et al. 2005; Jaegle et al. 2004), shipping (Beirle et al. 2004a; Richter et al. 2004; Franke et al. 2009), power plants (Kim et al. 2006; 2009) and lightning

(Beirle et al. 2004c; 2006; Boersma et al. 2005; Choi et al. 2005; Hild et al. 2002; Martin et al. 2007; Sioris et al. 2007; Thomas et al. 2003).

The length and quality of the space-based NO_2 records has allowed the observation of trends. Trends in tropospheric column densities of NO_2 can highlight the effectiveness of legislative abatement methods as well as the accuracy of emission inventories. There has been particular interest in the NO_2 trend over the developing countries in eastern Asia (Irie et al. 2005) and, in particular, China (Richter et al. 2005; van der A et al. 2006; He et al. 2007; Zhang et al. 2007) as well as the global trend (Richter et al. 2005; van der A et al. 2008; Hayn et al. 2009). Richter et al.

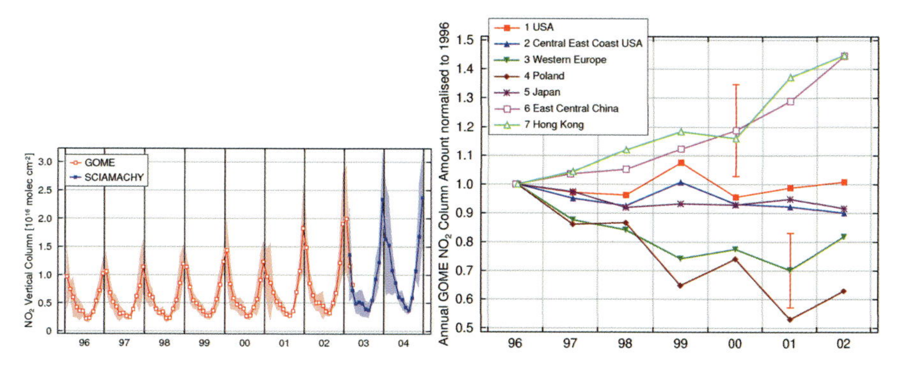

Fig. 8.1 *Left*: monthly averages of tropospheric vertical columns of NO_2 over eastern central China. The temporal evolution of tropospheric NO_2 columns from GOME and SCIAMACHY. *Right*: the mean annual NO_2 column amount normalized to that in 1996 for the geographical regions USA, central east coast USA, western Europe, Poland, Japan, eastern central China, and Hong Kong. The *error bars* represent the estimated uncertainty (s.d.) for an individual year, the values over China being larger as a result of the poorer knowledge and therefore larger uncertainty of the aerosol loading and its change (Richter et al. (2005) (Reprinted by permission of Macmillan Publishers Ltd)).

(2005) showed that there had been a 50% increase in the emissions of NO_2 over China in the period 1996–2004 (see Fig. 8.1 and Section 8.3.3).

Owing to the nature of NO_x emissions there have been many studies that use the satellite data in combination with other data sources and models to constrain emission budgets. For example, Jaegle et al. (2005) used inverse methods to take GOME NO_2 data and partition the emissions between fossil fuel, biomass burning and soil emissions. A number of groups have inverted the space data in order to derive emissions for China (Wang et al. 2007a), Europe (Konovalov et al. 2008) and globally (Jaegle et al. 2005; Konovalov et al. 2006; Muller and Stavrakou 2005; Stavrakou et al. 2008). Stavrakou et al. (2008) inverted the 10 year GOME/ SCIAMACHY record such that the largest emission increases were found over eastern China, and in particular in the Beijing area (growth rate of 9.6% per year), whereas appreciable emission decreases were calculated over the United States (−4.3% per year in the Ohio River Valley), and to a lesser extent over Europe

(−1.4% per year in Germany, −1.0% per year in the Po Basin). They noted that the emission changes result in significant trends in surface O_3, amounting to increases of more than 15% per decade over large parts of China in summertime.

A number of groups have used a combination of satellite data and models to estimate global emissions of NO_2 (Ma et al. 2006; Martin et al. 2003; Toenges-Schuller et al. 2006; Zhang et al. 2007) without the formal inversion of the data using emission inventory (bottom-up) methods. Over China, Ma et al. (2006) have compared the satellite data to a model constrained by differing emission inventories, while Zhang et al. (2007) used a dynamic methodology to fit the observed trends over China to the emission inventories. Non-model based statistical methods have been used to infer global emission budgets from GOME measurements (Leue et al. 2001). The data have been used to assess model performance (Savage et al. 2004). The distribution and budget of tropospheric NO_x over Asia, especially India, are examined using a global 3-D chemistry-meteorology model and GOME NO_2 columns (Kunhikrishnan et al. 2004a). Pollution and the influence of stratospheric input in the Mediterranean (Ladstatter-Weissenmayer et al. 2003; 2007c) region has been explored using multi-tracer satellite observations.

A global comparison of NO_2 data with models has been undertaken showing the utility of these data (Lauer et al. 2002; Velders et al. 2001). The approach has been extended with an extensive multi-model comparison (van Noije et al. 2006) that highlighted a combination of model uncertainty and retrieval bias affected the goodness of fit.

The role of NO_2 in O_3 budgets has been investigated globally (Choi et al. 2008; Edwards et al. 2003), regionally over the Mediterranean (Ladstatter-Weissenmayer et al. 2007c), the tropics (Sauvage et al. 2007) and over China (Zhao et al. 2006).

As recently stated by the HTAP report "Observations from the ground, aircraft and satellites provide a wealth of evidence that ozone (O_3) and fine particle concentrations in the UNECE region and throughout the Northern Hemisphere are influenced by intercontinental and hemispheric transport of pollutants." Tropospheric satellite observations of composition have been key observational indicators and quantitative constraints (Keating and Zuber 2007).

Damoah et al. (2004) have shown that Russian forest fires produced NO_2 and other products as well as CO plumes which can be transported appreciable distances. Spichtinger et al. (2001; 2004) have observed the long-range transport (LRT) of NO_2 from Canadian boreal fires (Fig. 8.2). Stohl et al. (2003) used the space-data to characterise express pathways for LRT over the North Atlantic. Wenig et al. (2003) have tracked the long-range transport of South African power plant emissions across the Pacific. Kunhikrishnan et al. (2004b) have examined the export of NO_2 plumes from Africa and Indonesia over the central Indian Ocean.

Guerova et al. (2006) have used a combination of model and satellite data to estimate the impacts of transatlantic transport episodes on summer O_3 in Europe. More recently Bousserez et al. (2007) used model and satellite data to characterise both the anthropogenic and biomass burning influences in the North Atlantic LRT during the summer of 2004.

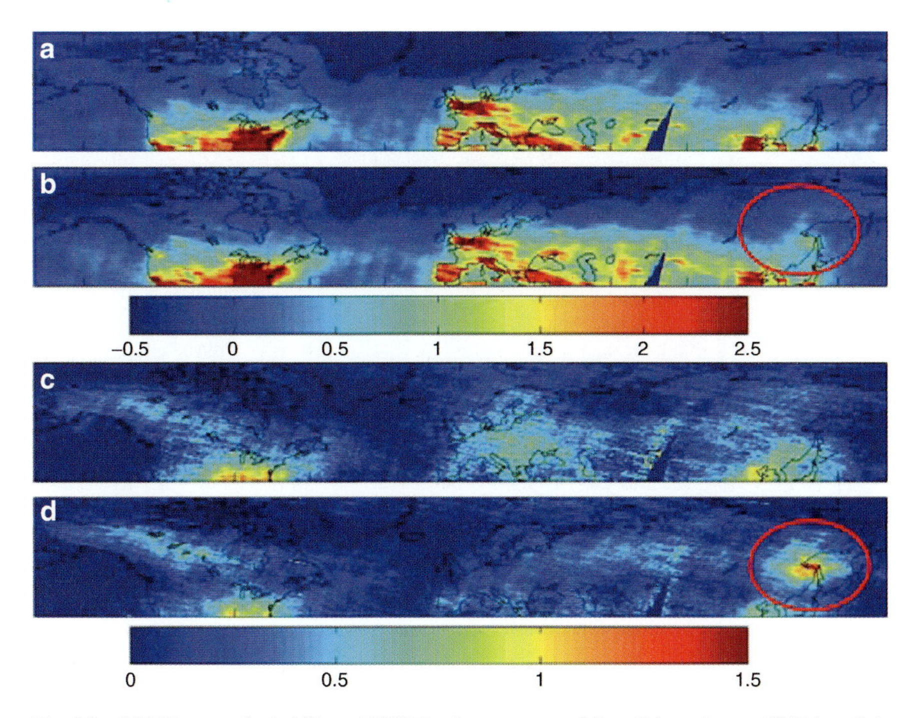

Fig. 8.2 GOME tropospheric NO_2 and HCHO columns averaged from July to August 1997 (**a** and **c**), and 1998 (**b** and **d**). The red circles mark enhanced NO_2 and HCHO columns over eastern Siberia where the strongest fire activity occurred in 1988 (Spichtinger et al. 2004).

8.2.3 *Carbon Monoxide*, CO

CO plays a central role in tropospheric chemistry, being a universal product of hydrocarbon photochemical degradation chemistry as well as a primary pollutant. Owing to its relatively long atmospheric lifetime (in the order of months) it is also a useful tracer for tropospheric dynamical phenomena. In the background atmosphere a combination of CO and CH_4 are the main loss routes for OH. The main sources of CO are the oxidation of CH_4 and other non-methane hydrocarbons (NMHC) with 40–60% of surface CO levels over the continents, slightly less over the oceans, and 30–60% of CO levels in the free troposphere, being estimated to come from NMHC oxidation (Poisson et al. 2000). The other major source of about equal magnitude is the incomplete combustion of either fossil fuels or biomass.

The first global space-based CO measurements were made from the MAPS instrument on board the space shuttle (Connors et al. 1999; Newell et al. 1999). Many of the gross features of global CO e.g. biomass burning, anthropogenic pollution, NH/SH gradients where identified (Connors et al. 1999; Newell et al. 1999). The data were used in early satellite data assimilation experiments (Lamarque et al. 1999). Fig. 8.3 shows CO data taken from the IMG on ADEOS which flew for

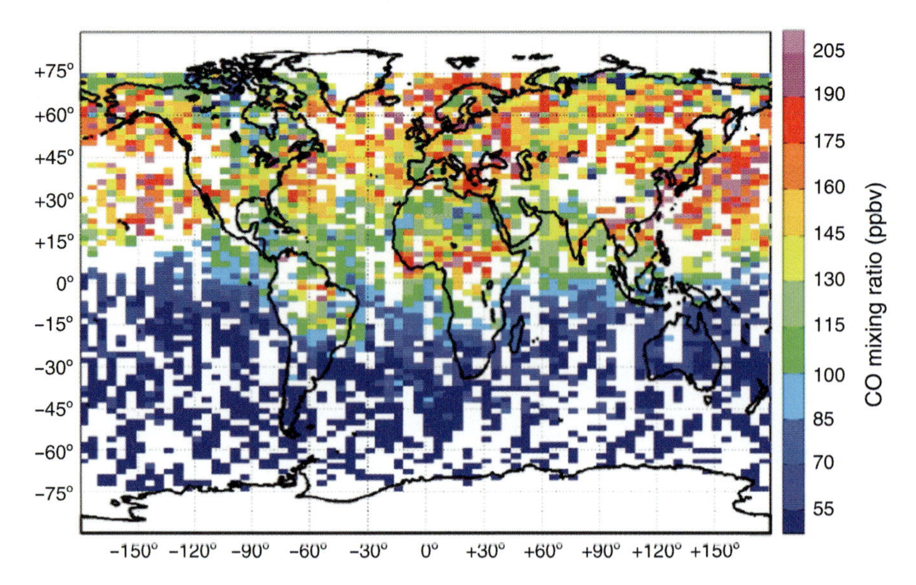

Fig. 8.3 CO volume mixing ratios (ppb) in the lower troposphere (1.2 km) for 1st–10th April 1997 retrieved from the cloud filtered IMG spectra (averages on a 2° × 5° grid) (Barret et al. 2005). Compare this with Fig. 8.13c.

a period in 1996/7 (Barret et al. 2005) giving a glimpse of the global distribution and the role of fires.

With the advent of MOPITT measurements (Deeter et al. 2003; 2004; Drummond and Mand 1996; Edwards et al. 2004; Emmons et al. 2004) global measurements of tropospheric CO have revolutionised our understanding of the natural and man-made tropospheric pollution. As an example, Fig. 8.4 shows average CO mixing ratios at the surface level, as derived from the measurements from March 2000 to June 2007 over China and parts of India and Japan, which are among the most populated areas of the globe (Clerbaux et al. 2008b). Daytime observations over land were used as the highest thermal contrast is expected (Deeter et al. 2007b) giving maximum information content along with an increased sensitivity towards the surface.

These measurements can be taken down to city scale, for example in Mexico City (Massie et al. 2006). Extensive validation of MOPITT has taken place (Emmons et al. 2007). MOPITT has been shown to be able to capture the influence of synoptic processes on the horizontal and vertical distribution of CO (Liu et al. 2006a).

Measurements from SCIAMACHY (Buchwitz et al. 2004; 2005b; 2006; de Laat et al. 2006; Frankenberg et al. 2005b; Gloudemans et al. 2008) which measures in the NIR, is more weighted to the surface than thermal IR measurements and shows enhancements of CO over urban regions (Buchwitz et al. 2007a). A comparison of SCIAMACHY data has been made with MOPITT data (Buchwitz et al. 2006, 2007a; Gloudemans et al. 2005; Straume et al. 2005; Turquety et al. 2008).

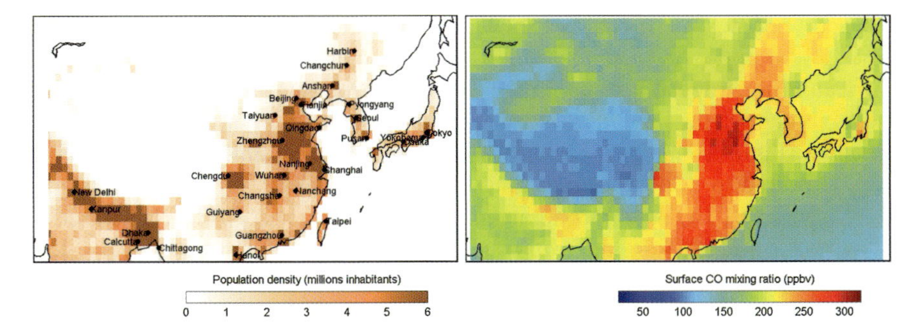

Fig. 8.4 *Left panel*: Population density (source CIESIN, in million inhabitants http://sedac.ciesin.columbia.edu/gpw) over China and surroundings. All cities with more than two million inhabitants are indicated. *Right panel*: MOPITT CO mixing ratios at the surface level (obtained by averaging the MOPITT L2 measurements from March 2000 to June 2007). Adapted from Clerbaux et al. (2008b).

Comparison of SCIAMACHY CO data has been made with models showing that the seasonal variation of the model is very similar to that of the SCIAMACHY measurements. For certain locations, significant differences were found, which are probably related to modelling errors owing to CO emission uncertainties (de Laat et al. 2006; 2007). Validation of the measurements has been undertaken with FTIR (Dils et al. 2006) and ship-borne observations (Warneke et al. 2005).

CO measurements are also available from AIRS (McMillan et al. 2005) and have been used to look at long-range transport of pollution (Stohl et al. 2007; Zhang et al. 2008). Comparisons between AIRS CO and MOPITT and ground-based remote sensing measurements have been made (Warner et al. 2007).

Recently, TES has provided profile measurements of CO (Rinsland et al. 2006b) which have been validated against aircraft measurements (Luo et al. 2007a) and MOPITT (Luo et al. 2007b) observations. Vertically-resolved CO concentration profiles have been retrieved from ACE, extending from the mid-troposphere to the thermosphere (from about 5 to 110 km) (Clerbaux et al. 2005; 2008c; Hegglin et al. 2008). Similarly, there are upper tropospheric CO measurements made from MLS (Filipiak et al. 2005; Livesey et al. 2008; Pumphrey et al. 2007). Observations from MLS have been used to show persistent maxima in CO and minima in O_3 within the anticyclone in the upper troposphere-lower stratosphere (UTLS) throughout summer, and variations in these tracers are closely related to the intensity of underlying deep convection (Park et al. 2007).

A range of new observations of CO are just becoming available from IASI (Clerbaux et al. 2009).

a General Transport Phenomena

Owing to the nature of CO as a marker of anthropogenic pollution and biomass burning, it has been used in a number of studies to assess the importance of transport pathways. The data has shown evidence of vertical transport of CO (Kar et al. 2004) in the Asian summer monsoon. The transport pathways of CO in the African upper troposphere during the West African monsoon have been investigated through the assimilation of CO observations by the Aura Microwave Limb Sounder (Barret et al. 2008). Assimilation of CO into a chemical transport model was also used to evaluate spring time transport over West Africa (Pradier et al. 2006). MOPITT CO data in combination with trajectory analysis has been used to show that transport from the tropics to the extratropics is a comparatively slow process, giving rise to the appearance of "transport barriers" in the subtropics (Bowman 2006).

Tropical climatologies of water vapour, O_3, CO, and nitric acid from a variety of satellite, aircraft, and balloon-based measurement platforms have been used to test convective parameterisations (Folkins et al. 2006a). Satellite and sonde observations have been use to explore the seasonal cycles of O_3, CO and convective outflow at the tropical tropopause (Folkins et al. 2006b). Bhattacharjee et al. (2007) have explored the influence of a dust storm on CO and water vapour over the Indo-Gangetic Plains showing that these can lift surface CO rapidly throughout the troposphere.

An intriguing feature has been observed in daytime measurements of CO over the Middle East during spring and summer. Enhanced CO is observed over the Zagros Mountains of Iran, following the local topography over this region. It has been argued that this feature is formed by the processes of mountain venting by thermal winds caused by strong daytime differential heating (Kar et al. 2006).

b Hemispheric Transport of Air Pollution

Satellite CO data has found great utility in the investigation of the magnitude and impact of the hemispheric transport of air pollution (Keating and Zuber 2007; Singh et al. 2006). For example, Asian outflow and trans-Pacific transport of CO and O_3 pollution has been tracked using an integrated satellite, aircraft and models (Heald et al. 2003). High CO levels measured at surface background stations and by satellite in/over China (Zhao et al. 2007) have been shown to arise from LRT of biomass burning and biofuel burning areas located in the border areas of Pakistan and India.

Satellite observations of CO columns from MOPITT and of aerosol optical depths from MODIS have been useful in mapping North American pollution outflow and the trapping of convectively lifted pollution by upper-level anticyclones (Li et al. 2005). Guerova et al. (2006) have used a combination of models and satellite data to assess the impact of transatlantic transport episodes on summertime O_3 in Europe. Interlaced long-range pollution events of contrasting origin and age the tropical Atlantic Ocean have been delineated with a combination of *in situ* measurements, satellites, trajectories, emission inventories and global models (Gros et al. 2004). A combination of two satellite CO measurements has been used to look at long-range transport of CO over the land and ocean (Warner et al. 2007).

Satellite CO data in combination with *in situ* aircraft data and models has been used to characterize the Asian chemical outflow and relate it quantitatively to its sources as well as determining its chemical evolution during transport (Allen et al. 2004; Jacob et al. 2003). The relationship between satellite and *in situ* observations of CO has been explored in pollutant outflow based on a large-scale feature sampled over central northern Pacific (Crawford et al. 2004). Observations from multiple satellite sources of CO (and other tracers) has been used to investigate emission and export from Asia (Turquety et al. 2008) showing that, compared to observations, there seem be an underestimation of emissions, especially in eastern Asia. Zhang et al. (2008) used an ensemble of aircraft, satellite, sonde, and surface observations to better understand the mechanisms for transpacific O_3 pollution and its implications for North American air quality using CO measurements from TES and AIRS.

c Emission Estimates

CO measurements from space have been used to estimate global emission sources of CO (Arellano et al. 2004; Lin et al. 2007; Petron et al. 2004; Stavrakou and Muller 2006). There has been many attempts to look at regional emission budgets for Europe (Pfister et al. 2004) and Asia (Heald et al. 2004; Peng and Zhao 2007; Tanimoto et al. 2008; Yumimoto and Uno 2006). Emissions in Asia for a range of compounds have been predicted to rise sharply owing to increasing industrialisation (Monks et al. 2009).

For Asia, a combination of satellite and aircraft observations have been used to derive estimates of emissions of 361 Tg y^{-1} for CO (Heald et al. 2004). Tropospheric CO budget analysis suggests that in northern China, surface emission is the largest source of tropospheric CO (Peng and Zhao 2007). Further, model results underestimate CO by 23% in northern China (Peng and Zhao 2007). Arellano et al. (2004) noted that CO emissions in eastern Asia are about a factor of 1.8–2 higher than bottom-up estimates. Adjoint inverse modelling of satellite CO data gives annual anthropogenic (fossil and biofuel combustion) CO emissions over China of 147 Tg (Yumimoto and Uno 2006). Inverse estimates of the CO emissions from China up to 2005 suggested an increase of 16% since 2001, in good agreement with MOPITT satellite observations and the bottom-up estimates up to 2006 (Tanimoto et al. 2008).

Global estimates for the total anthropogenic surface sources of CO (fossil fuel + biofuel + biomass burning) have been derived using inverse modelling of the satellite observations (Petron et al. 2004; Yudin et al. 2004). From this approach sources of CO (all in Tg(CO) per year) were determined to be 509 in Asia, 267 in Africa, 140 in North America, 90 in Europe and 84 in Central and South America (Petron et al. 2004). Emission changes in CO have been inferred from two sets of satellite observations separated by 10 years (Shindell et al. 2005). In a case study using SCIAMACHY CO, it has been possible to estimate the emission of CO on a country scale for the UK (Khlystova et al. 2009).

d Fires (Biomass Burning)

Biomass burning (BB) is a significant process in the earth system which, as a stochastic process, is difficult to model. CO measurements from space have been used for a range of scientific purposes such as the estimation of global biomass burning emission sources of CO (Arellano et al. 2006), the identification of CO plumes from forest fires (Lamarque et al. 2003; Lee et al. 2005) (see Table 8.1) and the impact of biomass burning on regional air quality (Choi and Chang 2006a).

An extensive study has been undertaken by Edwards et al. (2006a) to investigate southern hemisphere BB using a combination of satellite CO and aerosol measurements coupled to models. Further in a separate study, Edwards et al. (2006b) used a 5 year CO data record to examine the inter-annual variability of the southern hemisphere CO loading and show how this relates to climate conditions which determine the intensity of fire sources. The observations showed an annual austral springtime peak in the SH zonal CO loading each year with dry season BB emissions in South America, southern Africa, the maritime continent, and north-western Australia. Although fires in southern Africa and South America typically produce the greatest amount of CO, the most significant interannual variation is due to varying fire activity and emissions from the maritime continent (SE Asia including the Philipines, Indonesia and Malaysia) and northern Australia (Edwards et al. 2006b; Rinsland et al. 2008). In related work (Gloudemans et al. 2006), model results show a large contribution of South American BB CO over Australian BB regions during the 2004 BB season of up to 30–35% and up to 55% further south, with smaller contributions for 2003. BB CO transported from southern Africa contributed up to a similar to 40% in 2003 and around 30% in 2004 (Gloudemans et al. 2006). Elevated SH upper tropospheric CO as well as a range of small molecule organic tracers (C_2H_6, HCN, and C_2H_2) has been detected from a combination of biomass burning emissions and long-range transport (Rinsland et al. 2005). A combination of ship-based FTIR observations and satellite observations has been used to investigate the BB over the South Atlantic (Velazco et al. 2005) with observations of recurring enhancements of CO in the upper troposphere (10–15 km) in the equatorial regions and the South Atlantic.

Table 8.1 Overview of regionalised biomass burning episodes identified using CO from space

Fire Area	Year	Reference
Idaho-Montana Forest	August 2000	Lamarque et al. (2003)
East coast of Korea	April 2000	Choi and Chang (2006b)
N.W. USA Forest	2000	Liu et al. (2005)
SE Asia	2001	Zheng et al. (2004)
Russian Forest	May 2003	Lee et al. (2005)
Russian Forest	2003	Generoso et al. (2007)
Central Asia	2003	Wang et al. (2006)
Alaska	2004	Pfister et al. (2005), Turquety et al. (2007)
Indonesia	2006	Rinsland et al. (2008)
European Arctic	2006	Stohl et al. (2007)

Using MOPITT data and emission inversion techniques Pfister et al. (2005) showed that the Alaskan wildfires of 2004 emitted 30 $\pm$ 5 Tg CO during June to August 2004 which is comparable to the anthropogenic emissions of the continental US. Similarly, Turquety et al. have shown the importance of peat burning and pyroconvective injection (Turquety et al. 2007) for the same fires. Turquety et al. (2007) have made an estimate of North American fire emissions during the summer of 2004 to be 30 Tg CO, that includes 11 Tg from peat burning.

A temporary increase in northern hemispheric tropospheric CO burden in 2002 and 2003 (Yurganov et al. 2005) has been ascribed to boreal fires in Russia. Using CO data from MOPITT and AIRS, Yurganov et al. (2008) have concluded that fires can explain a substantial fraction of the interannual variability of CO_2.

Large differences in CO and O_3 have been measured over Indonesia and the eastern Indian Ocean in October to December 2006 relative to 2005, in 2006 O_3 was higher by 15–30 ppb (30–75%) while CO was higher by up to 80 ppb in October and November, and about 25 ppb in December. These differences were caused by high fire emissions from Indonesia in 2006 associated with the lowest rainfall since 1997, reduced convection during the moderate El Nino, and reduced photochemical loss because of lower H_2O (Logan et al. 2008; Rinsland et al. 2008). Space-based CO measurements and O_3 sondes have been used to look at the spatial and temporal variation of biomass burning over South Africa and South America (Bremer et al. 2004).

The global impact of biomass burning is very much dependent on the injection height and temporal nature of the fires. Satellite CO and surface measurements have been used to examine the injection properties of boreal forest fires (Hyer et al. 2007a) as well as the effects of source temporal resolution on transport simulations of the emissions (Hyer et al. 2007b).

Biomass burning is impacting the lower stratospheric composition. CO in the lower stratosphere (LS) observed with the Aura Microwave Limb Sounder (MLS) shows an annual oscillation in its composition that results from the interaction of an annual oscillation in slow ascent from the TTL to the LS and seasonal variations in sources, including a semi-annual oscillation in CO from biomass burning (Duncan et al. 2007).

e Model Performance

CO measurements retrieved from satellite instruments have been used extensively to assess model performance (Arellano et al. 2007; Bousserez et al. 2007). Twenty six state-of-the-art atmospheric chemistry models have been run to study future air quality and climate change. They were compared to near-global satellite observations from the MOPITT instrument and local surface measurements for CO. The models show large underestimates of northern hemisphere extratropical CO, while typically performing reasonably well elsewhere. The results suggested that year-round emissions, probably from fossil fuel burning in eastern Asia and seasonal biomass burning emissions in southern and central Africa, are greatly underestimated in the current emission inventories. Variability among models was large, resulting primarily from intermodel differences in representations and emissions of nonmethane volatile organic compounds and in hydrologic cycles, which affect OH and soluble hydrocarbon intermediates (Shindell et al. 2006).

A combination of satellite measurements and models have been used to make a comparison of the chemical nature of air pollution in eastern China and the eastern United States (Tie et al. 2006) highlighting the difference in influence of biogenic and anthropogenic hydrocarbons in both regions. More detailed studies in the same vein have looked at Eastern China (Zhao et al. 2006) showing that during summer, local emissions produce about 50–70% of the O_3 concentration in eastern China. Kar et al. (2008) have used multiple satellite measurement to map the air pollution over the Indo-Gangetic basin.

8.2.4 Formaldehyde, HCHO

HCHO is an important intermediate formed in the oxidation of CH_4 and many other hydrocarbons. The two major loss processes for HCHO, namely photolysis and reaction with OH are relatively fast giving an atmospheric lifetime of typically a few hours. Owing to its high water solubility HCHO can also be removed by wet deposition.

Much of the focus of HCHO measurements from space has been in using it as a proxy measure for isoprene emissions (Palmer et al. 2003). Isoprene can be the dominant precursor of HCHO in regions with strong biogenic emissions and its relatively short lifetime (<1 h) means that emissions of the precursor can be localised. The approach has been used to look at seasonal and annual variability in isoprene emissions over the USA (Abbot et al. 2003; Chance et al. 2000; Millet et al. 2008; Palmer et al. 2006), Asia (Fu et al. 2007), Africa (Meyer-Arnek et al. 2005a), Europe (Dufour et al. 2009), tropical regions (Palmer et al. 2007) and globally (Shim et al. 2005). In Asia, the approach was also extended to constrain emissions of alkenes and xylenes as well as the influence of biomass burning (Fu et al. 2007). Work in Europe (Dufour et al. 2009) has demonstrated that the methodology can be extended to regions without "strong" emissions of isoprene.

In combination with other tracers HCHO has been used to trace pollution over the Mediterranean (Ladstatter-Weissenmayer et al. 2003), detect biomass burning over south-eastern Asia (Thomas et al. 1998), quantify the influence of boreal forest fires (Spichtinger et al. 2004), estimate O_3 production sensitivities (Martin et al. 2004a), quantify biogenic and biomass burning budget contributions over Africa (Meyer-Arnek et al. 2005a), constrain tropical tropospheric O_3 (Sauvage et al. 2007) and correlate with ground-based data in the south-eastern USA (Martin et al. 2004b). Space-based HCHO measurements have recently been combined with glyoxal (see Section 8.2.5) measurements (Wittrock et al. 2006). Marbach et al. (2009) to report on enhanced HCHO column density over a ship track in the Indian Ocean. De Smedt et al. (2008) (Fig. 8.5) have derived a 12 year dataset of HCHO from GOME and SCIAMACHY.

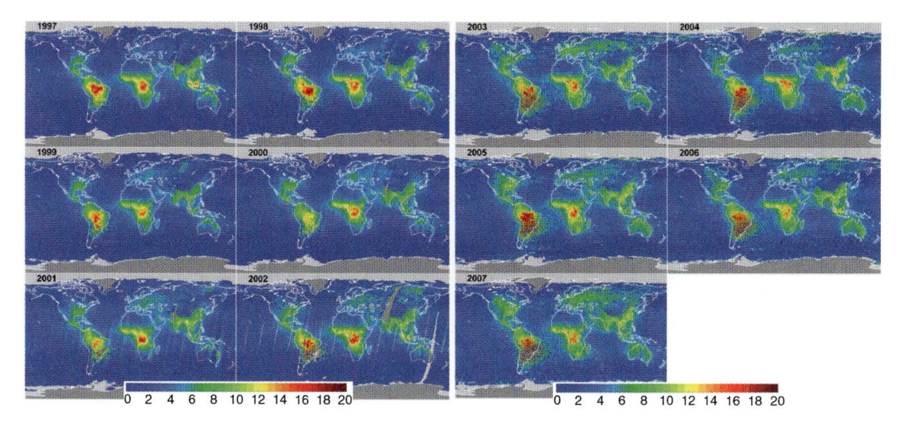

Fig. 8.5 Yearly averaged GOME HCHO columns from 1997 to 2002 (units are 10^{15} molecule/cm^2) (*left*) and SCIAMACHY 2003 to 2007 (*right*) (De Smedt et al. 2008).

HCHO measurements have also been made in the upper troposphere from ACE-FTS and MIPAS (Coheur et al. 2007; Steck et al. 2008). Upper tropospheric HCHO in combination with a range of other trace organics has been used to characterise biomass burning plumes (Coheur et al. 2007).

8.2.5 *Glyoxal*, CHOCHO

CHOCHO is the smallest α-dicarbonyl, an oxidation product of numerous VOC (Volkamer et al. 2001). Direct and time resolved CHOCHO measurements can provide a useful indicator to constrain VOC oxidation processes (Volkamer et al. 2005) owing to the mixture of anthropogenic and biogenic sources. The atmospheric residence time of CHOCHO is limited by rapid photolysis and reaction with OH radicals, and is about 1.3 h for overhead sun conditions (Volkamer et al. 2005). There are some indications that CHOCHO possibly contributes to secondary organic aerosol (SOA) formation (Liggio et al. 2005a; 2005b; Volkamer et al. 2007). However, the atmospheric relevance of CHOCHO uptake on aerosols is presently not clear. Global observations of CHOCHO from space offer the potential of identifying photochemical hot spots in the Earth's atmosphere, and, coupled with observations of HCHO, improve the understanding of biogenic emissions, biomass burning, and urban pollution.

Wittrock et al. (2006) demonstrated the first measurements of CHOCHO from space. The global pattern of CHOCHO columns was found to be similar to that of HCHO, indicating common atmospheric sources, in particular (biogenic) isoprene. The ratio between CHOCHO and HCHO was found to be about 0.05 in source regions such as the tropical rain forests. At some locations, larger ratios are found and this is attributed to unidentified additional sources of CHOCHO. Large

CHOCHO columns are found primarily over areas having strong biogenic emissions in the tropics which appear to be the dominant global source. During strong biomass burning events, CHOCHO was clearly observed from fires in Alaska (Fig. 8.6)(Wittrock et al. 2006).

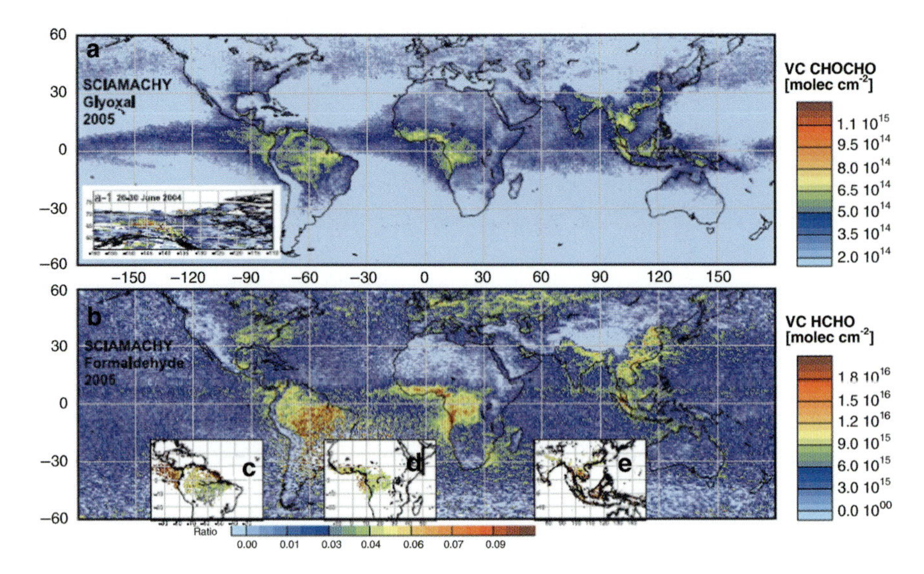

Fig. 8.6 Yearly mean for (**a**) glyoxal and (**b**) formadelyde derived from SCIAMACHY observations in 2005. The sub figures (**c–e**) show the ration between glyoxal and formaldehyde (Wittrock et al. 2006).

Five years worth of data has been compiled (Vrekoussis et al. 2009) and the largest columns are seen in tropical and sub-tropical regions associated with high biological activity and the plumes from vegetation fires. The majority of the identified hot spots are characterized by a well-defined seasonality: the highest values being observed during the warm and dry periods as a result of the enhanced biogenic, primarily isoprene, emissions and/or biomass burning from natural or man-made fires. The regions influenced by anthropogenic pollution also encounter enhanced amounts of CHOCHO. There is growing evidence for a potential marine source of CHOCHO (Fu et al. 2008).

8.2.6 Sulfur Dioxide, SO_2

Sulfur chemistry is an integral part of life, owing to its role in plant and human metabolism. Sulfur compounds have both natural and anthropogenic sources in the atmosphere. In modern times, the atmospheric sulfur budget has been dominated by anthropogenic emissions, particularly from fossil fuel burning. It is estimated that 75% of the total sulfur emission budget (ca. 102 Tg S per year) is dominated by anthropogenic sources with 90% of it occurring in the northern hemisphere. The natural sources include volcanoes, plants, soil and biogenic activity in the oceans.

The oxidation of sulfur compounds in the atmosphere causes a number of different environmental problems such as acidification, climate balance and the formation of a sulfate layer in the stratosphere, the so-called Junge layer. By far the largest sulfur component emitted into the atmosphere is SO_2. Coal and oil combustion contribute up to 80% of the global budget of SO_2 while volcanoes contribute around 10% (Bates et al. 1992).

Satellite measurements of tropospheric SO_2 (and ash) have seen extensive application to identification and quantification of volcanic emissions (Table 8.2). Most of the early measurements (Krueger 1983; Krueger et al. 1995) came from the TOMS instrument whose functionality has been recently extended to OMI (Yang et al. 2007). Authors have demonstrated that the combination of different satellite sensors can give a more holistic chemical and physical picture of volcanic eruptions (Eckhardt et al. 2008; Prata et al. 2007). In particular the use of instruments with sensitivity to the mid-/upper-troposphere have been demonstrated as

Table 8.2 Application of tropospheric satellite SO_2 measurements to volcanic emissions

Volcano	Year(s) of Eruption(s)	Satellite	References
Anatahan	2003	TOMS/ASTER	Wright et al. (2005)
Anatahan	2003–2004	TOMS	Guffanti et al. (2005)
Anatahan	2003	TOMS	Pallister et al. (2005)
Etna	2001	GOME	Zerefos et al. (2006)
Etna	2001 & 2002	GOME	Thomas et al. (2005)
Etna	2002	AIRS	Carn et al. (2005)
El Chichon	1982	TOMS	Seftor et al. (1997)
Galunggung	1982–1983	TOMS	Bluth et al. (1994)
Hekla	1980, 2000	TOMS	Sharma et al. (2004)
Karthala	2005	SEVERI	Prata and Kerkmann, (2007)
Jebel at Tair	2007	IASI	Clarisse et al. (2008)
Jebel at Tair	2007	AIRS/OMI	Eckhardt et al. (2008)
Krafla	1984	TOMS	Sharma et al. (2004)
Manam	2005	OMI/TES	Clerbaux et al. (2008a)
Mauna Loa	1984	TOMS	Sharma et al. (2004)
Miyakejima	2000	ASTER	Urai (2004)
Multiple (20)	1996–2002	GOME	Khokhar et al. (2005)
Nyamuragira	1978–2002	TOMS	Carn and Bluth (2003)
Nyamuragira	1996	GOME	Eisinger and Burrows (1998)
Nyamuragira	2006	OMI; TES	Clerbaux et al. (2008a)
Mt. Spurr	1992	TOMS	Rose et al. (2001)
Mt. St. Helens	1991	TOMS	Bluth et al. (1992)
Pinatubo	1991	TOMS/TOVS	Guo et al. (2004)
Popocatepetl	1996	GOME	Eisinger and Burrows (1998)
Popocatepetl	2000–2001	MODIS	Novak et al. (2008)
Rabual	2006	TES	Clerbaux et al. (2008a)
Redoubt	1989–1990	TOMS	Schnetzler et al. (1994)
Soufriére Hills	2006	OMI/AIRS	Prata et al. (2007)

a way of ascertaining injection height (Ackerman et al. 2008; Prata and Bernardo 2007). The emission of bromine compounds from volcanoes is dealt with in Section 8.2.12. There are practical uses of satellite imagery for protecting international airways from volcanic ash (Tupper et al. 2004).

Owing to the relatively low sensitivity of SO_2 measurements from space there are few reports on anthropogenic sources. Eisinger and Burrows detected widespread SO_2 (Eisinger and Burrows 1998) over south-eastern Europe that they attributed to lignite coal-burning in power plants and these data have been used by Zerefos et al. (2000) to attribute the different sources of SO_2 over Greece. Strong point source emissions such as Peruvian copper smelters (Carn et al. 2007) and a fire at an Iraqi sulfur plant has been identified (Carn et al. 2004). Lee et al. (2008) have used a combination of *in-situ*, ground-based remote sensing and satellite data to follow and attribute the long-range transport of SO_2 from China to Korea, while Krotkov et al. (2008) have used a combination of aircraft and OMI data to validate SO_2 pollution over north-eastern China. Khokhar et al. (2008) have analysed time series of SO_2 from GOME over non-ferrous metal smelters in Peru and Russia.

8.2.7 *Ammonia,* NH_3

The global emission of NH_3 is about 54 Mt Ny^{-1}. The major global sources are excreta from domestic animals and fertilizers, but oceans, biomass burning and crops are also important (Asman et al. 1998; Schlesinger and Hartley 1992). About 60% of the global NH_3 emission is estimated to come from anthropogenic sources. Boundary layer NH_3 concentrations can vary widely as NH_3 is readily absorbed by surfaces, and reacts with OH and acidic aerosols leading to a rather short atmospheric lifetime in the order of a few hours (Dentener and Crutzen 1994).

The first tropospheric NH_3 measurements from space were made by Beer et al. (2008) using data from TES. NH_3 concentrations over China ranged from 5 to almost 25 ppb, while over North America they were consistently less than 5 ppb. Significant spatial variations were observed over China. The authors note that that the next step is to globalise these observations to understand regional and temporal variations. Recent results from IASI (Clarisse et al. 2009) have demonstrated the first global pictures of NH_3 from space (Fig. 8.7). The strong fires that have occurred in the Mediterranean Basin, and particularly Greece in August 2007, and those in southern Siberia and eastern Mongolia in the early spring of 2008 have shown strong NH_3 biomass burning signatures (Coheur et al. 2009).

8.2.8 *Carbon Dioxide,* CO_2

CO_2 is the ultimate form of oxidised carbon in the Earth's atmosphere. Much focus in recent times has been on the anthropogenic driven growth in CO_2 concentration and as a source of carbon for plants. In the modern epoch the concentration of CO_2

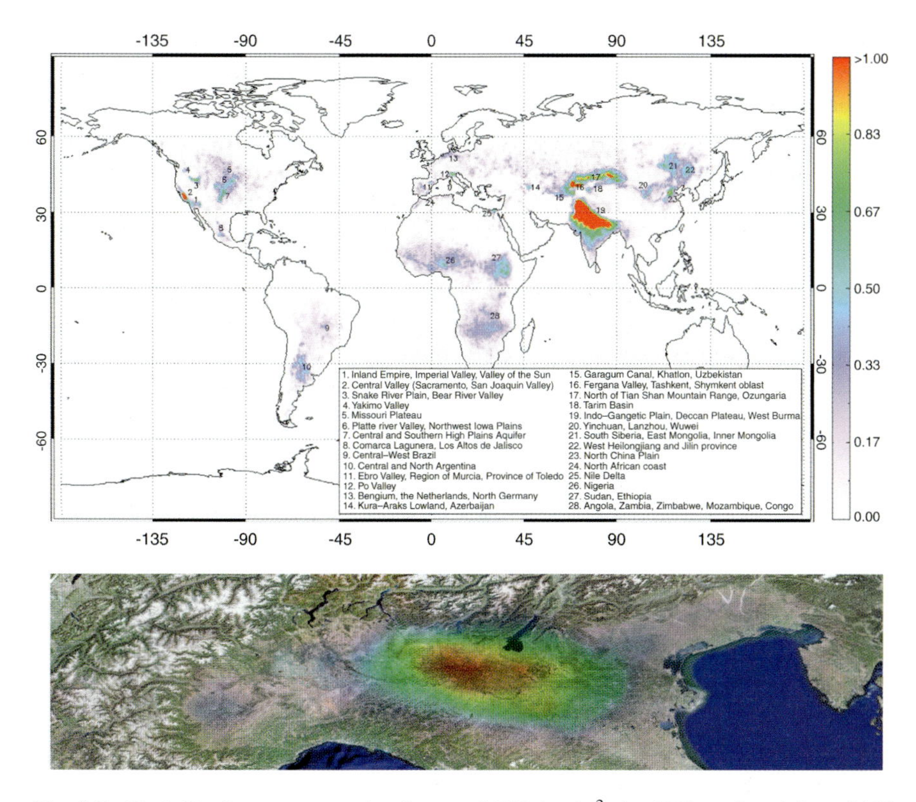

Fig. 8.7 (*Top*) Yearly average total columns of NH_3(mg/m^2) in 2008 retrieved from IASI measurements on a 0.25° by 0.25° grid. (*Bottom*) NH_3 concentrations derived from IASI observations above the Po Valley. The aerial photographs are ©2008 Google – Imagery ©2008 Terrametrics (http://maps.google.com/). Both adapted from Clarisse et al. (2009) (Reprinted by permission of Macmillan Publishers Ltd).

has increased from pre-industrial values of about 280 ppm to values in excess of 380 ppm today. The current concentrations far exceed values inferred from ice cores for the last 650,000 years. The main sinks for atmospheric CO_2 are oceanic and uptake by the terrestrial biosphere.

The first measurements of CO_2 from space (Chedin et al. 2002; 2003; 2005; 2008; Peylin et al. 2007) were mid-tropospheric measurements made in the thermal infrared from TIROS-N/TOVS on NOAA-10. The data were collected for 4 years (1987–1991) and have been used to estimate the influence of tropical biomass burning on mid-tropospheric CO_2 (Chedin et al. 2005; 2008).

AIRS is a thermal infrared spectrometer/radiometer (see Chapters 1 and 3) (Aumann et al. 2003) the data from which a number of authors have used to demonstrate and validate methods for retrieving mid-tropospheric CO_2 (Aumann et al. 2005; Chahine et al. 2005; Crevoisier et al. 2004; Engelen et al. 2004; Engelen and McNally 2005; Maddy et al. 2008; Tiwari et al. 2006).

Buchwitz et al. (2005a; 2005b; 2006) retrieved full tropospheric columns of CO_2 from SCIAMACHY in the short-wave infrared for the first time giving a view of surface CO_2 owing to the enhanced sensitivities at these wavelengths. The data have been used to map out the increasing trend in global CO_2 from space (Buchwitz et al. 2007b; Schneising et al. 2008). Barkley et al. (2006a; 2006b; 2006c; 2007) have extended the Buchwitz methodology looking at assessing validity with a combination of surface, aircraft and model data. Further, they have demonstrated spatial and temporal correlations between SCIAMACHY data more sensitive to surface CO_2 and mid tropospheric data from AIRS over North America (Barkley et al. 2006b) and interesting correlations of CO_2 with surface and vegetation type (Barkley et al. 2006c, 2007) (Fig. 8.8).

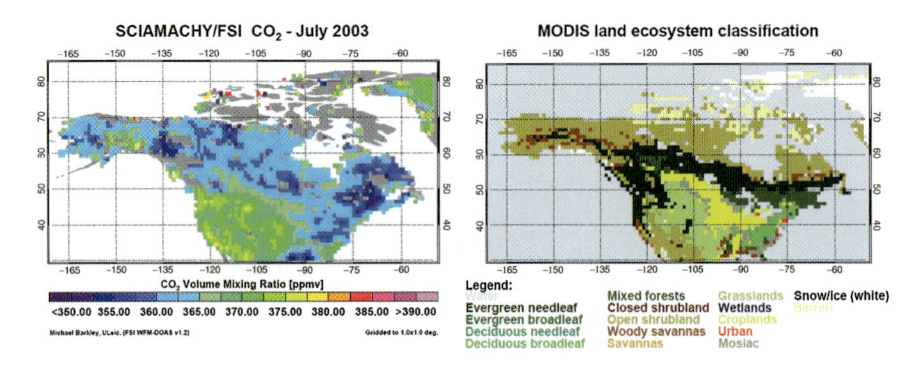

Fig. 8.8 SCIAMACHY CO_2 observations over North America for July 2003 (*left*) with a map of land vegetation cover over this scene (*right*) (Barkley et al. 2006c). The transition from low CO_2 VMRs along the Canadian shield and the eastern coast to the higher values over the mid-western US corresponds to a change in the vegetation type from evergreen needle leaf, mixed and deciduous broadleaf forests to land covered by crops and large grass plains.

Much of the effort in measurement of tropospheric CO_2 from space has been focussed on the precision and accuracy of the measurements, owing to the overwhelming need to use these measurements in an inverse mode to assess emissions. The measurements are being supplemented by new ones from specially designed sensors *i.e.* GOSAT (Hamazaki et al. 2004) and from the second version of OCO (the first instrument was lost in launch) (Crisp et al. 2004).

8.2.9 Methane, CH_4

CH_4 is the second most important anthropogenic greenhouse gas. It also has an indirect effect on climate through chemical feedbacks. More than 50% of present-day global CH_4 emissions are anthropogenic, the largest contributors being fossil fuel production, ruminants, rice cultivation, and waste handling (Bergamaschi et al. 2007). CH_4 levels have almost doubled since preindustrial times. The natural source strength of CH_4, mainly constituted by emission from wetlands, is particularly uncertain, because these emissions vary considerably in time and space. The dominant atmospheric sink for CH_4 is reaction with OH. Available ground-based measurements are sparse, albeit precise, and limitedly representative at larger scales (Walter et al. 2001). Better knowledge of CH_4 distribution and emissions is indispensable for a correct assessment of its impact on global change.

Measurements of CH_4 from SCIAMACHY show the expected latitudinal gradient (NH 10% > SH) with large-scale patterns of anthropogenic and natural CH_4 emissions (Buchwitz et al. 2005a; 2005b; Frankenberg et al. 2005a; 2006; Schneising et al. 2009; Straume et al. 2005).

Frankenberg et al. (2005a; 2006) using SCIAMACHY data observed pronounced enhancements of CH_4 in the Red Basin of China and in large areas of Asia in general, followed by northern parts of South America and central Africa. In particular, high mixing ratios in Columbia and Venezuela are larger than given by the CH_4 inventories. The largest seasonal variations are caused by rice emissions in Asia, which are very intense during a relatively short time period. The measurements indicate that these emissions already start towards the end of July and decline sharply in November, which is earlier than predicted by the model based inventories. Strong deviations between observed and modelled CH_4 abundances in tropical rainforest regions are observed, hinting at underestimated tropical emissions (Frankenberg et al. 2005a; 2006). This observation has been much debated in the context of the occurrence and role of plants in the production of CH_4 (Dueck and van der Werf 2008). Frankenberg et al. (2008) have recently reinvestigated the relevant CH_4 spectroscopy which may have introduced some biases into the initial retrievals. Separate data from Schneising et al. (2009) still show higher CH_4 over the tropics compared to a model. Upper tropospheric CH_4 profiles have been retrieved from a number of satellites including most recently ACE-FTS and MIPAS (De Maziere et al. 2008).

8.2.10 Water, H_2O

H_2O and the hydrological cycle are key elements of the earth system. As well as being key to life on Earth it is a major part of the climate system. More than 99% of the atmospheric H_2O is found and exists in all three phases in the troposphere. H_2O enters the atmosphere by evaporation mostly over the oceans, but also by transpiration from plants over the continents. The evaporation flux is balanced by the return of H_2O to the surface by various forms of precipitation. In the vapour phase, H_2O can be transported several thousands of kilometres before condensing which leads to greater precipitation than evaporation over the continents, the net balance being provided by surface flow of H_2O. The lifetime of a H_2O molecule in the atmosphere is estimated to be about 10 days.

Noel et al. (1999; 2002) have retrieved total H_2O column measurements from GOME-1 using DOAS in the 700 nm region and Maurellis et al. (2000) in the

585–600 nm region. Such column data have been validated against microwave measurements from SSM/I. Wagner et al. (2005; 2006) have used a similar approach to look at trends in total column precipitable water over the period 1996–2003. In general, the trends showed a strong correlation with near surface temperature though not always over land in the northern hemisphere (Wagner et al. 2006). Mieruch et al. (2008) have observed increasing water trends in GOME-1 data for Greenland, eastern Europe, Siberia and Oceania, whereas decreasing trends have been observed for the northwest USA, central America, Amazonia, central Africa and the Arabian Peninsula. Lang et al. (2007) have evaluated the GOME water climatology against independent *in-situ* measurements from the operational WMO radio-sonde network, against high spatial resolution H_2O vapour columns from MERIS and with ERA40 model results.

The GOME-1 H_2O column series from UV/VIS data have been extended to SCIAMACHY (Noel et al. 2004; 2005) and more recently the operational GOME-2 instrument (Noel et al. 2008). Recently, nadir measurements from TES have been used to retrieve water profiles (Shephard et al. 2008). A study of the 2006 El Nino (Logan et al. 2008) showed a movement in H_2O vapour owing to eastward movement of convection during El Nino leading to higher H_2O over eastern Africa and the western Indian Ocean. Similar results have been found by Wagner et al. (2005) for H_2O column anomalies for the El Nino 1997/1998 in GOME data.

Isotopologue measurements of H_2O are becoming available *e.g.* HDO (Herbin et al. 2007; Steinwagner et al. 2007; Zakharov et al. 2004; Frankenberg et al. 2009) (see Section 8.2.16). The use of such measurements are explained in Monks et al. (2009).

Microwave retrievals of H_2O vapour are dealt with in Chapter 4.

8.2.11 Bromine Monoxide, BrO

In comparison to the chemistry taking place in the stratosphere where halogen chemistry is well known and characterised, there has been much debate as to the role of halogen species in the oxidative chemistry of the troposphere (Platt and Hönninger 2003). There is growing experimental evidence about the prevalence of halogen chemistry as part of tropospheric photochemistry (Platt and Hönninger 2003; Read et al. 2008; Saiz-Lopez et al. 2007b; Simpson et al. 2007). Much of the proposed halogen chemistry is propagated through the reactions of a series of halogen atoms and radicals (Monks 2005).

BrO species can be formed in the polar boundary layer (Barrie et al. 1988; Fan and Jacob 1992; Saiz-Lopez et al. 2007b), mid-latitude marine boundary layer (Saiz-Lopez et al. 2004; 2006), salt pans (Hebestreit et al. 1999) as well as in volcanic emissions (Bobrowski et al. 2003). There are claims that BrO is omnipresent in the atmosphere (Hegels et al. 1998).

The major source of gas-phase bromine in the lower troposphere is thought to be the release of species such as IBr, ICl, Br_2 and BrCl from sea-salt aerosol, following the uptake from the gas-phase and subsequent reactions of hypohalous acids (HO_X, where X = Br, Cl, I) (Vogt et al. 1996). The halogen release mechanism is autocatalytic (Fan and Jacob 1992) and has become known as the "bromine explosion" (Platt and Lehrer 1997). The initiation of the BrO chemistry is relatively simple:

$$BrX + h\nu \rightarrow Br + X \tag{8.1}$$

$$Br + O_3 \rightarrow BrO + O_2 \tag{8.2}$$

The chemistry of halogen radicals has been recently reviewed by Monks (Monks 2005) and the impact of BrO chemistry by Simpson et al. (2007) and von Glasow et al. (2004).

The space-based measurements have clearly illustrated the spatial extent of the BrO in the Antarctic and Arctic boundary layer in spring (Richter et al. 1998; Wagner et al. 2001; Wagner and Platt 1998) (See Fig. 8.9). Various authors using the space-based data have investigated trends (Hollwedel et al. 2004), long-range transport of BrO_x (Ridley et al. 2007), the correlation with surface remote sensing measurements (Theys et al. 2007; Wagner et al. 2007), assessment with hydrocarbon-loss methods (Zeng et al. 2006), the role of the marginal ice zone (Jacobi et al. 2006) and total columns over Arrival Heights (Antarctica) during sunrise (Schofield et al. 2006).

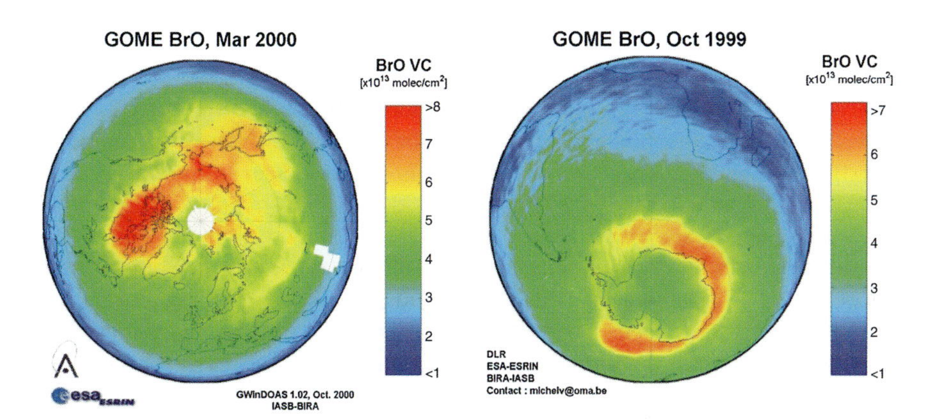

Fig. 8.9 BrO in the Antarctic and Arctic during spring derived from GOME satellite measurements (Van Roozendael et al. 2002).

BrO has been detected, probably in the free troposphere, outside polar regions such as over the Maldives (Ladstatter-Weissenmayer et al. 2007a), mid-latitudes (Van Roozendael et al. 2002) and globally (Hegels et al. 1998). The impact of these observations on, for example, the O_3 budget of the troposphere (von Glasow et al. 2004) remains an open question.

Space based measurements have been used to survey the BrO emissions from volcanoes (Afe et al. 2004). There is some debate as to whether there is an expectation that the BrO emissions should be correlated with SO_2 emissions (Afe et al. 2004). An unambiguous detection of volcanic BrO was possible for the 2008 eruption of the Kasatochi volcano in Alaska (Theys et al. 2009).

8.2.12 Iodine Monoxide, IO

IO has been detected by ground-based measurements both in the marine boundary layer (Alicke et al. 1999; Allan et al. 2000), the Dead sea (Zingler and Platt 2005) and the Antarctic boundary layer (Friess et al. 2001; Saiz-Lopez et al. 2007b). The major sources of iodine into the troposphere are thought to be from macroalgal sources releasing either molecular iodine and/or organoiodine compounds (Carpenter 2003). Photolysis of I_2 and the organo-iodine compounds releases the iodine.

$$\mathrm{RI} + h\nu \rightarrow \mathrm{R} + \mathrm{I} \tag{8.3}$$

$$\mathrm{I} + \mathrm{O_3} \rightarrow \mathrm{IO} + \mathrm{O_2} \tag{8.4}$$

During daylight hours IO exists in a fast photochemical equilibrium with iodine viz:

$$\mathrm{IO} + h\nu \rightarrow \mathrm{I} + \mathrm{O} \tag{8.5}$$

The aerosol "explosion" mechanism, previously described for bromine, acts effectively to recycle the iodine back to the gas-phase (Platt and Hönninger 2003). The potential impact of this chemistry has been demonstrated by Read et al. (2008) who have shown evidence for widespread destruction of tropical O_3 by bromine and iodine monoxides.

Saiz-Lopez et al. (2007a) and Schönhardt et al. (2007) first derived IO from SCIAMACHY measurements. There are significant differences between the two retrievals with Saiz-Lopez et al. observing IO above cloudy regions whereas Schönhardt et al. see negligible IO. Schönhardt et al. (2008) have produced the first global pictures of IO. Figure 8.13 shows the global measurements of IO and Fig. 8.10 shows measurements over Antarctica.

From space, the largest amounts of IO have been detected in springtime over the Antarctic. The seasonal variation of IO in Antarctica showed high values in springtime, slightly less IO in the summer period and again larger amounts in autumn (Schönhardt et al. 2008). In winter, no elevated IO levels were found in the areas accessible to the satellite measurements. The observed satellite seasonal cycle is in good agreement with recent ground-based measurements in Antarctica (Friess et al. 2001; Saiz-Lopez et al. 2007b). Conversely, in the Arctic region, no elevated IO levels were detected in the period analysed. Schönhardt et al. (2008) suggest that this observation is evidence for different conditions with respect to iodine release in the two polar regions.

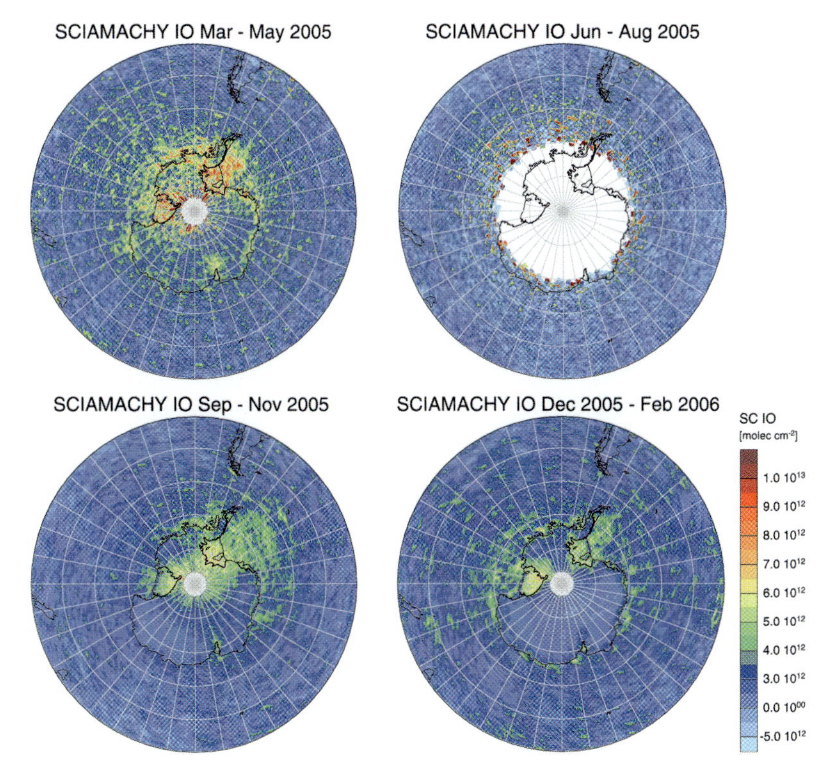

Fig. 8.10 Seasonally averaged slant column amounts of IO above the southern hemisphere from Antarctic autumn to summer (Schönhardt et al. 2008). Maxima of IO columns occur over the Weddell Sea, the Ross Sea and along the coast especially in spring and in autumn with levels remaining positive at some places throughout the summer.

8.2.13 Methanol, CH_3OH

CH_3OH is the most abundant oxygenated hydrocarbon gas in the atmosphere and is therefore a major contributor to non-methane volatile organic compounds, NMVOC (Singh et al. 1995). The primary source of atmospheric CH_3OH is plant growth and decay, the second largest source is atmospheric production with minor sources from biomass burning and anthropogenic emissions (Jacob et al. 2005). There is considerable uncertainty in the atmospheric CH_3OH budget. In the remote troposphere, CH_3OH concentrations are 0.1–1 ppb (Singh et al. 1995) while the concentrations in the continental boundary layer are an order of magnitude larger. The atmospheric lifetime is about 16 days in the free troposphere owing, primarily, to OH oxidation to produce HCHO (Singh et al. 1995).

The first lower troposphere measurements of CH_3OH have been demonstrated by Beer et al. (2008) from TES. Upper tropospheric measurement of CH_3OH have been made by IR occultation from ACE-FTS (see Fig. 8.11) (Dufour et al. 2006; 2007). Dufour have shown the utility of the limb measurements of CH_3OH measurements

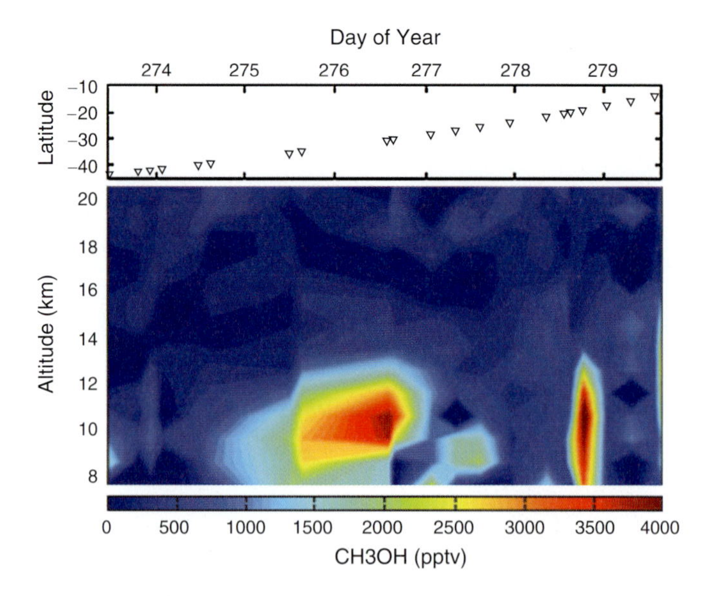

Fig. 8.11 Time series of ACE-FTS measurements of methanol for sunsets between 30th September and 6 October 2004 (Dufour et al. 2006). The latitude and time of each individual measurement are shown at the *top* of the panel.

in the upper troposphere for quantifying surface budgets and the influence of biomass burning (Dufour et al. 2007).

8.2.14 Nitrous Oxide, N_2O

N_2O is the fourth largest single contributor to positive radiative forcing and serves as a long-lived marker of the anthropogenic influence of the N-cycle. Atmospheric concentrations have risen by 16% since pre-industrial times to a value of around 319 ppbV. N_2O is also the main source of NO_x to the stratosphere. N_2O is mainly produced by microbial nitrification and denitrification processes in soils and water. Owing to its long atmospheric lifetime (*ca.* 1 century) the mixing ratio of N_2O shows very little spatial variation, <1%, throughout the troposphere.

There have been relatively limited measurements of full tropospheric column N_2O from space. Estimates in validation exercises suggest precisions in the order of 20% (Piters et al. 2006). Comparisons of early products with FTIR measurements showed agreement within 13% (Dils et al. 2006). Measurements on a ship-borne FTIR campaign showed a strong deviation (*ca.* 30%) between the satellite and ship based data (Warneke et al. 2005). All of these studies make the point that the datasets are not large enough to draw any statistical significance and that there is room for further improvement in the retrievals.

8.2.15 *Nitric Acid,* HNO_3

HNO_3 is the end point of NO_x chemistry in the atmosphere (see also Section 8.2.2) (Monks 2005). The dominant HNO_3 sink is wet removal (contributing to acid deposition) and dry deposition.

First global distributions of HNO_3 in the troposphere and the stratosphere derived from IR satellite measurements have been measured recently from IMG-ADEOS (Wespes et al. 2007) (Fig. 8.12).

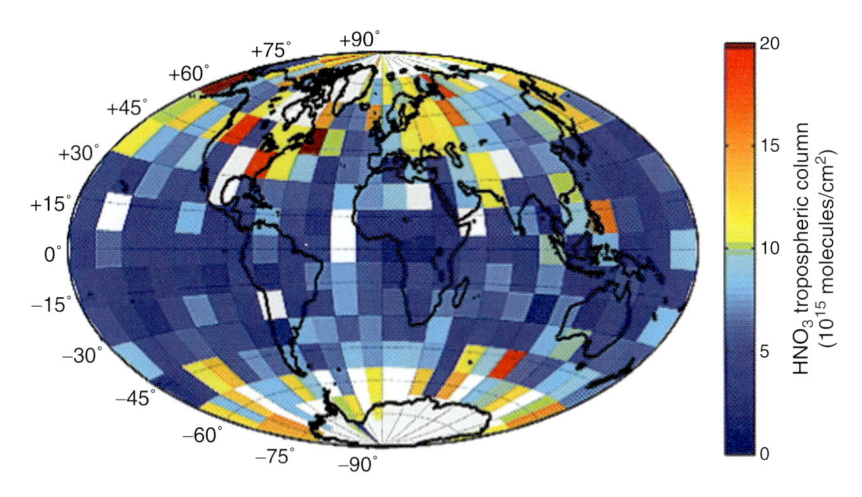

Fig. 8.12 Global distribution of HNO_3 in the troposphere (0–10 km), in 10^{15} molecules/cm^2. Data are averaged on a 15° × 12° grid (Wespes et al. 2007).

There are also several upper tropospheric measurements (Table 8.3). A number of satellite measurements including those of HNO_3 have been used for space-based constraints on the production of NO_2 by lightning (Martin et al. 2007).

8.2.16 *Other Trace Species*

There are a plethora of organic compounds in the atmosphere. The atmospheric budget is controlled by a combination of anthropogenic and natural (frequently biogenic) emissions tensioned against atmospheric chemical loss processes (photolysis or reaction with atmospheric oxidants) or physical loss (mainly heterogeneous removal). Many of these organic compounds are intrinsically linked to the chemistry that controls the global oxidising ability. A range of mainly small lightweight hydrocarbons have been measured in the upper troposphere such as C_2H_2 (Park et al. 2008; Rinsland et al. 2005; 2007b), C_3H_4 (Coheur et al. 2007) and C_2H_6 (Clarmann et al. 2007; Park et al. 2008; Rinsland et al. 2005; 2007b). The oxygenated compounds measured in the upper troposphere have been acetone

Table 8.3 Tropospheric trace gases measured from space. A full list of tropospheric satellite instruments and their characteristics is given in Appendix A

Gas	Instrument	Reference
Ozone	ACE-FTS[a]	Hegglin et al. (2008)
	GOME-1	Del Frate et al. (2005), Hansen et al. (2003), Hoogen et al. (1999), Iapaolo et al. (2007), Ladstatter-Weissenmayer et al. (2004, 2007c), Lamsal et al. (2004), Liu et al. (2006b, 2007), Meijer et al. (2006), Meyer-Arnek et al. (2005b), Muller et al. (2003), Munro et al. (1998), Tellmann et al. (2004), Valks et al. (2003), van der A et al. (2002)
	HIRDLS[a]	Nardi et al. (2008)
	IMG	Coheur et al. (2005)
	OMI	Choi et al. (2008), Martin et al. (2007), Schoeberl et al. (2007), Stajner et al. (2008), Ziemke et al. (2006)
	MIPAS[a]	Clarmann et al. (2007), Fischer et al. (2008), Raspollini et al. (2006)
	MLS[a]	Froidevaux et al. (2008), Ziemke et al. (2006)
	SCIAMACHY	Geer et al. (2006), Rozanov et al. (2007)
	TES	Choi et al. (2008), Eremenko et al. (2008), Jourdain et al. (2007), Logan et al. (2008), Nassar et al. (2008), Osterman et al. (2008), Richards et al. (2008), Worden et al. (2007a, b), Bowman et al. (2009)
	TOMS	Ahn et al. (2003), Browell et al. (1996), Chandra et al. (1998, 1999, 2002, 2003, 2004), Chatfield et al. (2004), Creilson et al. (2003, 2005), Cros et al. (1992), Edwards et al. (2003), Fishman, (1991a, b), Fishman and Balok (1999), Fishman and Brackett, (1997), Fishman et al. (1987, 1990, 1991, 1992, 2002, 2003, 2005), Hudson et al. (1995), Hudson and Thompson, (1998), Kim et al. (1996), Morris et al. (2006), Nganga et al. (1996), Olson et al. (1996), Peters et al. (2002, 2004), Portmann et al. (1997), Richardson et al. (1991), Thompson and Hudson (1999), Thompson et al. (1993, 1996a, 1996b, 2000, 2001, 2003), Tie et al. (2007), Vukovich et al. (1996, 1997), Watson et al. (1990), Ziemke and Chandra, (1999, 2003a, b), Ziemke et al. (1998, 2000, 2001, 2003, 2005)
NO_2	GOME-1	Beirle et al. (2003, 2004a, b, c, 2006), Blond et al. (2007), Boersma et al. (2005), Choi et al. (2005), Chun'e and Baoning (2008), Damoah et al. (2004), Edwards et al. (2003), Franke et al. (2009), Guerova et al. (2006), Han et al. (2009), Hayn et al. (2009), Hild et al. (2002), Ionov et al. (2006), Irie et al. (2005), Jaegle et al. (2004, 2005), Konovalov et al. (2005), Kunhikrishnan et al. (2004a, b), Ladstatter-Weissenmayer et al. (2003, 2007c), Lauer et al. (2002), Leue et al. (2001), Ma et al. (2006), Martin et al. (2002, 2003, 2004b), Meyer-Arnek et al. (2005a), Muller and Stavrakou (2005), Ordonez et al. (2006), Poberovskii et al. (2007), Richter and Burrows, (2002), Richter et al. (2005), Sauvage et al. (2007), Savage et al. (2004),

		Schaub et al. (2005), (2007), Spichtinger et al. (2001, 2004), Stohl et al. (2003), Thomas et al. (1998, 2003), Timofeev et al. (2000), Toenges-Schuller et al. (2006), Uno et al. (2007), van der A et al. (2006, 2008), van Noije et al. (2006), Velders et al. (2001), Wang et al. (2007a), Wenig et al. (2003), Zhang et al. (2007), Zhao et al. (2006)
	MIPAS[a]	Fischer et al. (2008), Raspollini et al. (2006)
	OMI	Boersma et al. (2007, 2008a, 2009), Bucsela et al. (2006, 2008), Celarier et al. (2008), Choi et al. (2008), Kim et al. (2009), Kramer et al. (2008), Lamsal et al. (2008), Mijling et al. (2009), Wang et al. (2007b), Wenig et al. (2008)
	OSIRIS[a]	Sioris et al. (2007)
	SCIAMACHY	Bertram et al. (2005), Boersma et al. (2008b), Bousserez et al. (2007), Franke et al. (2009), He et al. (2007), Heue et al. (2005), Irie et al. (2005), Kaynak et al. (2009), Kim et al. (2006, 2009), Konovalov et al. (2006, 2008), Martin et al. (2007), Richter et al. (2004, 2005), Stavrakou et al. (2008), van der A et al. (2008), Zhang et al. (2007)
	GOME-2	Franke et al. (2009)
N_2O	MIPAS[a]	Fischer et al. (2008), Raspollini et al. (2006)
	SCIAMACHY	Dils et al. (2006), Warneke et al. (2005)
NH_3	MIPAS[a]	Burgess et al. (2006a)
	TES	Beer et al. (2008)
	IASI	Clarisse et al. (2009)
CO	ACE[a]	Clerbaux et al. (2005, 2008c), Folkins et al. (2006a), Hegglin et al. (2008), Rinsland et al. (2005, 2007b, 2008), Turquety et al. (2008)
	AIRS[a]	McMillan et al. (2005), Stohl et al. (2007), Warner et al. (2007), Yurganov et al. (2008), Zhang et al. (2008)
	IASI	Clerbaux et al. (2009)
	IMG	Barret et al. (2005), Turquety et al. (2004)
	MAPS	Connors et al. (1999), Lamarque et al. (1999), Newell et al. (1999)
	MLS[a]	Barret et al. (2008), Duncan et al. (2007), Filipiak et al. (2005), Livesey et al. (2008), Park et al. (2007), Pumphrey et al. (2007)
	MOPITT	Allen et al. (2004), Arellano et al. (2004, 2006, 2007), Bhattacharjee et al. (2007), Bousserez et al. (2007), Bowman (2006), Bremer et al. (2004), Choi and Chang, (2006a, b), Choi et al. (2005), Clerbaux et al. (2008b, c), Crawford et al. (2004), Deeter et al. (2003, 2004, 2007a, b), Edwards et al. (2003, 2004, 2006a, b), Emmons et al. (2004, 2007), Generoso et al. (2007),

(continued)

Table 8.3 (continued)

Gas	Instrument	Reference
		Gros et al. (2004), Guerova et al. (2006), Heald et al. (2003, 2004), Ho et al. (2005), Hyer et al. (2007a, b), Ito et al. (2007), Jacob et al. (2003), Kampe and Sokolik (2007), Kar et al. (2004, 2006, 2008), Kim et al. (2005), Lamarque et al. (2003), Lee et al. (2005), Li et al. (2005), Liang et al. (2007), Lin et al. (2007), Liu et al. (2005, 2006a), Luo et al. (2007b), Massie et al. (2006), Peng and Zhao (2007), Petron et al. (2004), Pfister et al. (2004, 2005), Pradier et al. (2006), Rinsland et al. (2006b), Shindell et al. (2005, 2006), Singh et al. (2006), Stavrakou and Muller (2006), Tanimoto et al. (2008), Tie et al. (2006), Turquety et al. (2007, 2008), Velazco et al. (2005), Wang et al. (2006), Warner et al. (2007), Yudin et al. (2004), Yumimoto and Uno (2006), Yurganov et al. (2005, 2008), Zhao et al. (2006, 2007), Zheng et al. (2004)
	SCIAMACHY	Buchwitz et al. (2004, 2005b, 2006, 2007a), de Laat et al. (2006, 2007), Dils et al. (2006), Frankenberg et al. (2005b), Gloudemans et al. (2005, 2006, 2008), Khlystova et al. (2009), Straume et al. (2005), Turquety et al. (2008), Warneke et al. (2005)
	TES	Bowman et al. (2009), Logan et al. (2008), Luo et al. (2007a, b), Rinsland et al. (2006b, 2008), Warner et al. (2007), Zhang et al. (2008)
CO_2	AIRS[a]	Aumann et al. (2005), Barkley et al. (2006b), Chahine et al. (2005), Crevoisier et al. (2004), Engelen et al. (2004), Engelen and McNally (2005), Maddy et al. (2008)
	SCIAMACHY	Barkley et al. (2006a, b, c, 2007), Bosch et al. (2006), Buchwitz et al. (2005a, b, 2007b), Dils et al. (2006)
	NOAA-10 [TIROS-N, TOVS][a]	Chedin et al. (2002, 2003, 2005, 2008), Peylin et al. (2007)
CH_4	ACE[a]	De Maziere et al. (2008), Rinsland et al. (2007b)
	AIRS[a]	Xiong et al. (2008)
	IMG	Clerbaux et al. (2003), Turquety et al. (2004)
	MIPAS[a]	De Maziere et al. (2008), Fischer et al. (2008), Raspollini et al. (2006)
	SCIAMACHY	Bergamaschi et al. (2007), Buchwitz et al. (2005a, b, 2006), Dils et al. (2006), Frankenberg et al. (2005a, 2006), Gloudemans et al. (2005, 2008), Straume et al. (2005), Bloom et al. (2010)
methanol (CH_3OH)	ACE[a]	Coheur et al. (2007), Dufour et al. (2006, 2007), Rinsland et al. (2007b)
	TES	Beer et al. (2008)

HCHO	ACE[a]	Coheur et al. (2007)
	GOME	Abbot et al. (2003), Barkley et al. (2008b), Chance et al. (2000), De Smedt et al. (2008), Fu et al. (2007), Ladstatter-Weissenmayer et al. (2003, 2007c), Marbach et al. (2009), Martin et al. (2004a, b), Meyer-Arnek et al. (2005b), Millet et al. (2006), Palmer et al. (2001, 2003, 2006, 2007), Sauvage et al. (2007), Shim et al. (2005), Spichtinger et al. (2004), Thomas et al. (1998)
	MIPAS[a]	Steck et al. (2008)
	OMI	Millet et al. (2008)
	SCIAMACHY	De Smedt et al. (2008), Dufour et al. (2009), Wittrock et al. (2006)
glyoxal (CHOCHO)	SCIAMACHY	Fu et al. (2008), Vrekoussis et al. (2009), Wittrock et al. (2006)
H_2O	ACE-FTS[a]	Hegglin et al. (2008)
	GOME-1	Lang et al. (2002, 2007), Maurellis et al. (2000), Mieruch et al. (2008), Noel et al. (1999, 2002), Wagner et al. (2005, 2006)
	GOME-2	Noel et al. (2008)
	MIPAS[a]	Milz et al. (2005)
	MLS[a]	Read et al. (2007)
	SCIAMACHY	Mieruch et al. (2008), Noel et al. (2004, 2005)
	TES	Logan et al. (2008), Shephard et al. (2008)
SO_2	AIRS[a]	Carn et al. (2005), Eckhardt et al. (2008), Prata and Bernardo (2007), Prata et al. (2007)
	ASTER[a]	Urai (2004)
	GOES/HIRS[a]	Ackerman et al. (2008)
	GOME-1	Eisinger and Burrows (1998), Khokhar et al. (2005), Thomas et al. (2005), Zerefos et al. (2000, 2006)
	IASI	Clarisse et al. (2008)
	OMI	Carn et al. (2007), Eckhardt et al. (2008), Krotkov et al. (2006, 2008), Prata et al. (2007), Yang et al. (2007)
	SCIAMACHY	Lee et al. (2008)
	SEVERI	Eckhardt et al. (2008), Prata and Kerkmann (2007)
	TES	Clerbaux et al. (2008a)
	TOMS	Bluth et al. (1992, 1994), Carn and Bluth (2003), Carn et al. (2004), Guffanti et al. (2005), Guo et al. (2004), Gurevich and Krueger (1997), Krotkov et al. (1997), Krueger et al. (1995),

(continued)

Table 8.3 (continued)

Gas	Instrument	Reference
		Massie et al. (2004), Novak et al. (2008), Pallister et al. (2005), Rose et al. (2001), Schnetzler et al. (1994), Seftor et al. (1997), Sharma et al. (2004), Tupper et al. (2004), Wright et al. (2005)
	TOVS	Guo et al. (2004)
BrO	GOME	Afe et al. (2004), Chance (1998), Hegels et al. (1998), Hollwedel et al. (2004), Ladstatter-Weissenmayer et al. (2007b), Richter et al. (1998), Ridley et al. (2007), Schofield et al. (2006), Van Roozendael et al. (2002), Wagner et al. (2001, 2007), Wagner and Platt (1998), Zeng et al. (2006)
	SCIAMACHY	Afe et al. (2004), Jacobi et al. (2006), Theys et al. (2009), Theys et al. (2007)
IO	SCIAMACHY	Saiz-Lopez et al. (2007a), Schönhardt et al. (2008)
OCS	ACE-FTS[a]	Barkley et al. (2008a), Rinsland et al. (2007b)
acetone	ACE-FTS[a]	Coheur et al. (2007)
PAN	ACE-FTS[a]	Coheur et al. (2007)
	MIPAS[a]	Glatthor et al. (2007)
acetylene (C_2H_2)	ACE-FTS[a]	Park et al. (2008), Rinsland et al. (2005, 2007b), Turquety et al. (2008)
ethane (C_2H_6)	ACE-FTS[a]	Coheur et al. (2007), Park et al. (2008), Rinsland et al. (2005, 2007b), Senten et al. (2008), Turquety et al. (2008)
	MIPAS[a]	Clarmann et al. (2007)
ethene (C_2H_4)	ACE-FTS[a]	Coheur et al. (2007), Herbin et al. (2009)
SF_6	ACE-FTS[a]	Rinsland et al. (2007b)
	MIPAS[a]	Burgess et al. (2004, 2006b)
CFC-11	ACE-FTS[a]	Mahieu et al. (2008)
	IMG	Coheur et al. (2003)
	MIPAS[a]	Coheur et al. (2003), Hoffmann et al. (2005, 2008)

CFC-12	ACE-FTS[a]	Mahieu et al. (2008)
	IMG	Coheur et al. (2003)
	MIPAS[a]	Hoffmann et al. (2005)
CFC-113	ACE-FTS[a]	Dufour et al. (2005)
HCFC-142b	ACE-FTS[a]	Dufour et al. (2005)
HCFC-22	MIPAS[a]	Moore and Remedios (2008)
	IMG	Coheur et al. (2003)
methyl chloride (CH_3Cl)	ACE-FTS[a]	Rinsland et al. (2007b)
formic acid (HCOOH)	ACE-FTS[a]	Rinsland et al. (2006a, 2007b)
HDO	IMG	Herbin et al. (2007), Zakharov et al. (2004)
	MIPAS[a]	Steinwagner et al. (2007)
	SCIAMACHY	Frankenberg et al. (2009)
HNO_3	ACE-FTS[a]	Martin et al. (2007), Wolff et al. (2008)
	HIRDLS[a]	Kinnison et al. (2008)
	IMG	Wespes et al. (2007)
	MIPAS[a]	Fischer et al. (2008), Raspollini et al. (2006)
	MLS[a]	Santee et al. (2007)
peroxy nitric acid (HO_2NO_2)	MIPAS[a]	Stiller et al. (2007)
HCN	ACE-FTS[a]	Park et al. (2008), Rinsland et al. (2005, 2007b), Turquety et al. (2008)
HCl	ACE-FTS[a]	Mahieu et al. (2008), Rinsland et al. (2007b), Senten et al. (2008)
H_2O_2	ACE-FTS[a]	Rinsland et al. (2007a)

[a]upper troposphere only

(CH_3COCH_3) (Coheur et al. 2007), formic acid (HCOOH) (Rinsland et al. 2006a; 2007b) and H_2O_2 (Rinsland et al. 2007a). Oxidised and reduced nitrogen compounds measured in the upper troposphere are peroxyacetylnitrate (PAN) (Coheur et al. 2007; Glatthor et al. 2007), HO_2NO_2 (Stiller et al. 2007), HCN (Park et al. 2008; Rinsland et al. 2005; 2007b). Upper tropospheric inorganic compounds such as OCS (Barkley et al. 2008a; Rinsland et al. 2007b), SF_6 (Rinsland et al. 2007b), HCl (Mahieu et al. 2008; Senten et al. 2008) and HDO (Herbin et al. 2007; Steinwagner et al. 2007; Zakharov et al. 2004) have also been measured.

There are a range of CFC and HCFC compounds that have been measured in the upper troposphere, namely, CFC-11 (Coheur et al. 2003; Hoffmann et al. 2005, 2008; Mahieu et al. 2008), CFC-12 (Coheur et al. 2003; Hoffmann et al. 2005; Mahieu et al. 2008), CFC-113 (Coheur et al. 2003; Dufour et al. 2005), HCFC-142b (Dufour et al. 2005) and HCFC-22 (Coheur et al. 2003; Moore and Remedios 2008). Moore and Remedios (2008) used a combination of two satellite systems to derive a mean northern hemisphere mid-latitude (20–50°N) HCFC-22 growth rate between 1994 and 2003 of 5.4 ± 0.7 pptv per year and a mean southern hemisphere growth rate (60 80°S) of 6.0 ± 0.7 pptv per year in the same period.

There have been important applications of upper tropospheric trace organics to mapping the age and composition of biomass burning plumes (Coheur et al. 2007; Dufour et al. 2006; Rinsland et al. 2005). H_2O_2 has been detected in young biomass burning plumes in the tropics (Rinsland et al. 2007a). A HCOOH emission factor relative to CO of 1.99 ± 1.34 g/kg during 2004 in upper tropospheric biomass burning plumes is inferred from a comparison with lower mixing ratios measured during the same time period (Rinsland et al. 2006a). Upper tropospheric CO, C_2H_6, HCN, CH_3Cl, CH_4, C_2H_2, CH_3OH, HCOOH, and OCS measurements show enhancement in biomass burning plumes of up to 185 ppbv for CO, 1.36 ppbv for C_2H_6, 755 pptv for HCN, 1.12 ppbv for CH_3Cl, 0.178 ppbv for C_2H_2, 3.89 ppbv for CH_3OH, 0.843 ppbv for HCOOH, and 0.48 ppbv for OCS in western Canada and Alaska at 50°–68°N latitude between 29 June and 23 July 2004 (Rinsland et al. 2007b).

Enhancements of C_2H_6 and O_3 observed in the southern hemisphere have been attributed to a biomass burning plume, which covers wide parts of the southern hemisphere, from South America, the Atlantic Ocean, Africa, the Indian Ocean to Australia. The chemical composition of the part of the plume-like pollution belt associated with South American fires, where rainforest burning is predominant, appears to differ from the part of the plume associated with southern African savannah burning. In particular, African savannah fires lead to a larger O_3 enhancement than equatorial American fires (Clarmann et al. 2007). At the end of the biomass burning season in South America and South and East Africa, elevated PAN amounts of 200–700 pptv were measured in a large plume extending from Brazil over the South Atlantic, central and South Africa, the southern Indian Ocean as far as Australia at altitudes between 8 and 16 km (Glatthor et al. 2007).

Park et al. (2008) have used speciated data to investigate the role of the Asian monsoon as a transport barrier from the upper troposphere to the lower stratosphere, the range of chemical tracers being able to tag different air masses origins and quantify photochemical lifetimes.

8.3 Satellite Observations of Tropospheric Composition: What Can We Learn?

Satellite observations have for the first time provided simultaneous global measurements of several important tropospheric trace gases. These datasets open new horizons in atmospheric science. Characteristic spatial and temporal patterns allow the identification, location, and quantification of different sources to study phenomena such as transport or trends.

The comparison of satellite measurements with CTMs/GCMs provides a substantive challenge to the models. The measurements help to constrain the models and the comparisons indicate the shortcomings of both measurements and models. The applications involving models are discussed in Chapter 9. However, in the following sections we highlight the potential of satellite based trace gas measurements to investigate several central questions in tropospheric chemistry research.

8.3.1 Column Density Maps as Proxies for Emissions

Global maps of tropospheric trace gases, as shown in Fig. 8.13, impressively illustrate the power of tropospheric satellite measurements. The spatial patterns of column densities are determined by (1) sources of the respective trace gas, (2) transport and (3) the lifetime. Enhanced column densities indicate source regions, where spatial patterns are most distinct for short lifetimes.

Thus, to start with a straightforward application, column densities serve as a first-order proxy of emissions. For example, this is particularly distinct in the case of NO_2, owing to the sensitivity of UV/VIS satellite measurements to the lower troposphere and the short lifetime of NO_x, of only a few hours, in the boundary layer, leading to high spatial gradients (Fig. 8.13b). As a consequence, besides the regions of high population density paired with industrial activity, like the US East coast, western Europe, or eastern China, where column densities are generally high, Megacities like Los Angeles, Mexico City, Moscow, Seoul, or Tokyo, clearly show up as "hot spots". Even emissions from ships can be recognized for the highly frequented route between Sri Lanka and Indonesia (Beirle et al. 2004a; Richter et al. 2004). The emissions from other NO_x sources, *i.e.* biomass burning, soil emissions, and lightning, are less localized and the respective columns are therefore smeared out over large areas in a multi-annual mean.

Similarly, different sources can be identified in the global maps for other trace gases as well. Anthropogenic emissions can also clearly be seen in the maps for CO (Fig. 8.13c) and SO_2 (Fig. 8.13f), particularly over eastern Asia, though the lifetimes of these gases are generally higher than for NO_x, and signals are thus less localized. Nevertheless, maps of mean SO_2 show sharp peaks over copper smelters in Peru (Fig. 8.14). Also recent maps of NH_3 (see Fig. 8.7) show global distributions in relation to strong source regions.

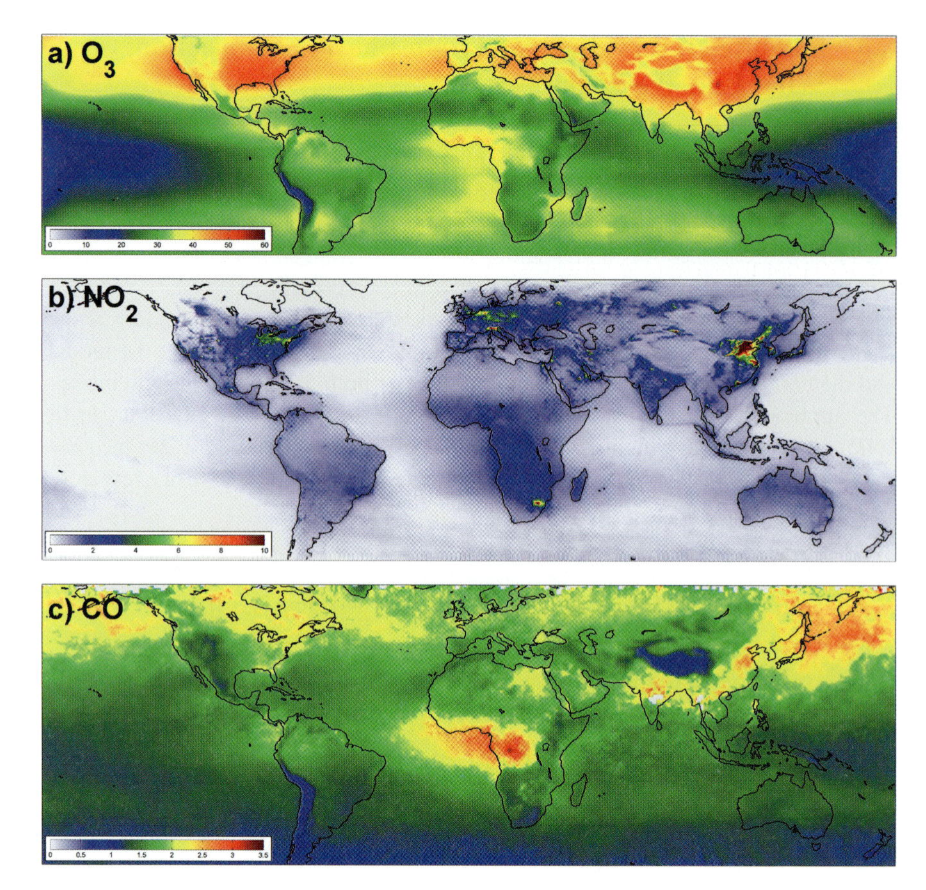

Fig. 8.13 Atlas of tropospheric species observed from space. (**a**) Mean summer column density of tropospheric O_3 (DU) from TOMS measurements 1979–2005. Data provided by Jack Fishman (Fishman et al. 2003). http://asd-www.larc.nasa.gov/TOR/TOR_Data_and_Images.html. (**b**) Mean tropospheric column density of NO_2 (10^{15} molecules/cm^2) derived from SCIAMACHY measurements 2003–2007. Data provided by Steffen Beirle. (**c**) Mean column density of CO (10^{18} molecules/cm^2) derived from nighttime IASI measurements July 2008. Data provided by Matthieu Pommier/Cathy Clerbaux.

There are particularly strong enhancements of SO_2 after volcanic eruptions (Khokhar et al. 2005; Thomas et al. 2005; Yang et al. 2007), as illustrated in Fig. 8.14b (note the change in colour scale compared to a). Satellite measurements can be used to monitor volcanoes remotely (see http://www.gse-promote.org/) and give information on the extent and location of plumes, this information being important for air traffic routing.

Analogous highly sporadic enhancements can be observed for strong biomass burning episodes in a wide range of tracers such as CO, NO_2, HCHO, and CHOCHO (Thomas et al. 1998; Edwards et al. 2006; Wittrock et al. 2006). For example, Fig. 8.15 shows the enhancement of CO from the Alaskan forest fires in July 2004 (see also Section 8.2.3).

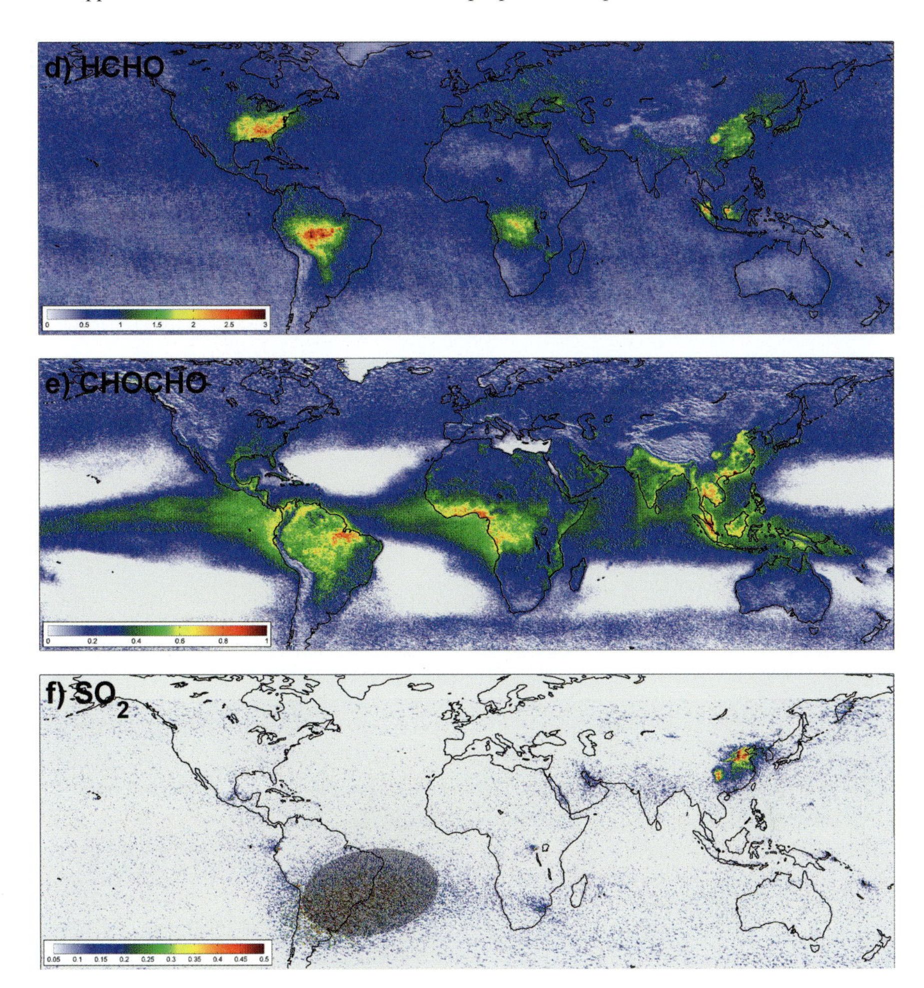

Fig. 8.13 (continued) Atlas of tropospheric species observed from space. (**d**) Mean column density of HCHO (10^{16} molecules/cm^2) derived from OMI August 2006. Data provided by Thomas Kurosu. (**e**) Mean glyoxal (CHOCHO) column density (10^{15} molecules/cm^2) from SCIAMACHY measurements 2003–2007. Data provided by Mihalis Vrekoussis (Vrekoussis et al. 2009). (**f**) Mean column density of SO_2 (DU) from SCIAMACHY measurements 2007. The shaded area over South America/South Atlantic masks improper fit results owing to the South Atlantic Anomaly (SAA). Data provided by Andreas Richter.

The origin, amount and impact of halogen oxides on tropospheric composition is still highly debated (Platt and Hönninger 2003; von Glasow et al. 2004; Monks et al. 2009). The detection of tropospheric BrO (Wagner and Platt 1998; Richter et al. 1998) and IO (Saiz-Lopez et al. 2007a; Schönhardt et al. 2008) from space were milestones for polar tropospheric chemistry research, proving the existence of halogen oxides over extended areas in polar spring (Fig. 8.9) and indicating where and when *in-situ* measurements should be performed for in-depth analysis of the basic chemical mechanisms.

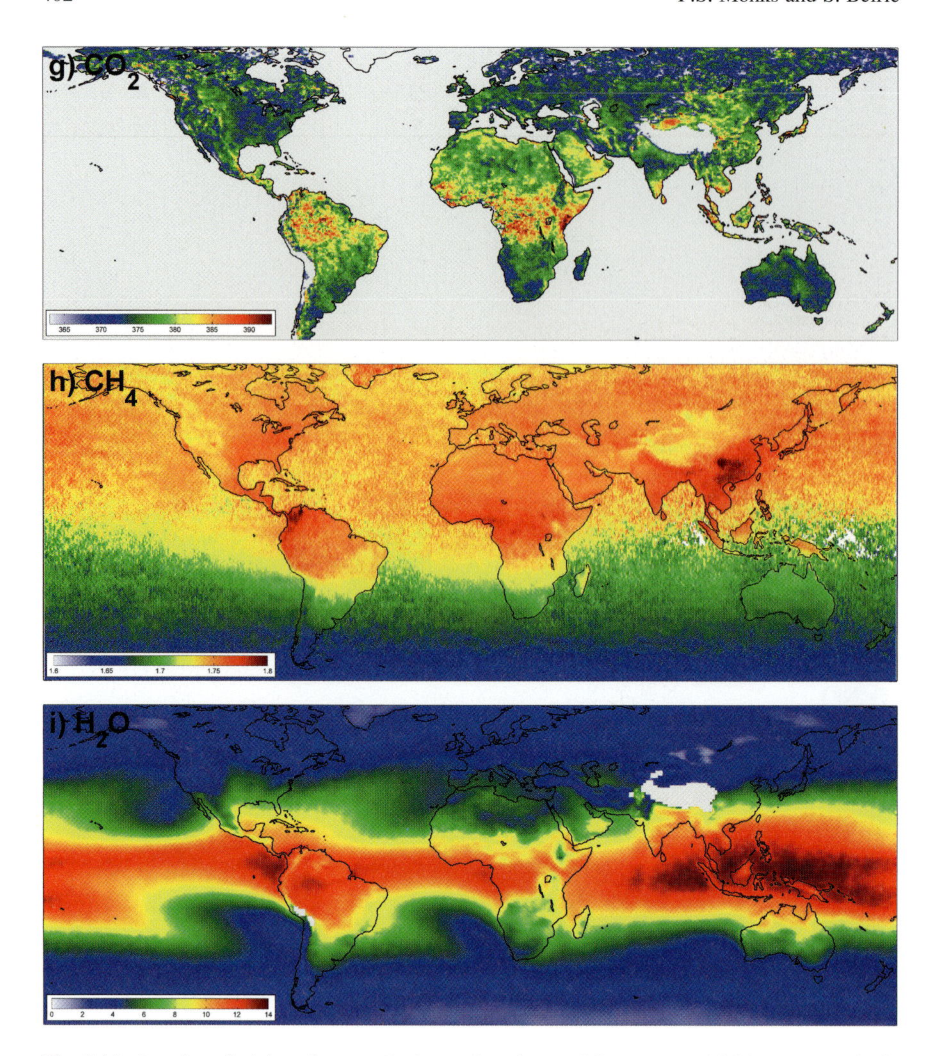

Fig. 8.13 (continued) Atlas of tropospheric species observed from space. (**g**) Mean tropospheric mixing ratio (ppm) of CO_2 derived from SCIAMACHY measurements in May 2003. Data provided by Michael Buchwitz (Buchwitz et al. 2007b). (**h**) Mean tropospheric mixing ratio (ppm) of CH_4 derived from SCIAMACHY measurements 2004. Data provided by Christian Frankenberg (Frankenberg et al. 2008). (**i**) Mean tropospheric column density of H_2O (10^{22} molecules/cm^2) derived from GOME measurements 1996–2004. Data provided by Thomas Wagner (Wagner et al. 2006).

In the cases of some VOCs observable from space, in particular HCHO and CHOCHO, (Fig. 8.13d, e), the secondary production from photochemical degradation exceeds direct emissions. Enhanced column measurements thus serve as proxy for the emissions of precursors and indicators of photochemistry (Millet et al. 2008; Palmer et al. 2003; 2006). For example, enhanced HCHO columns over south-eastern US are used as a quantitative proxy for biogenic isoprene emissions (Palmer et al. 2006).

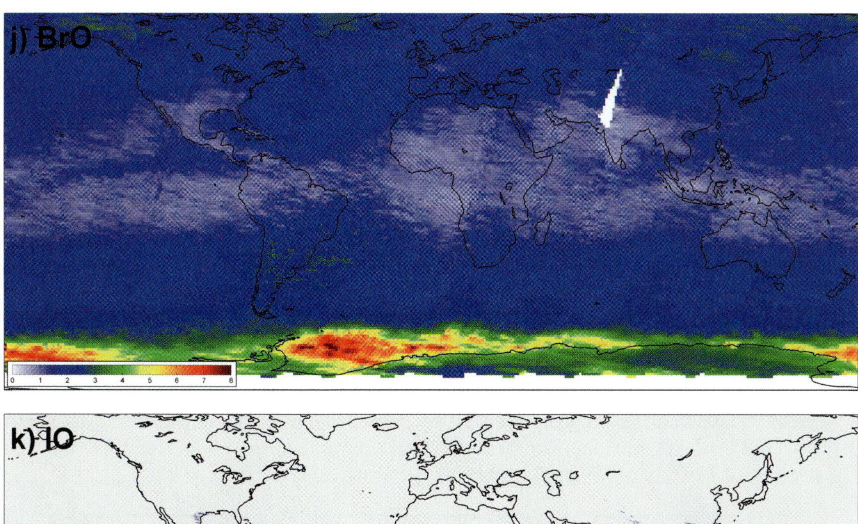

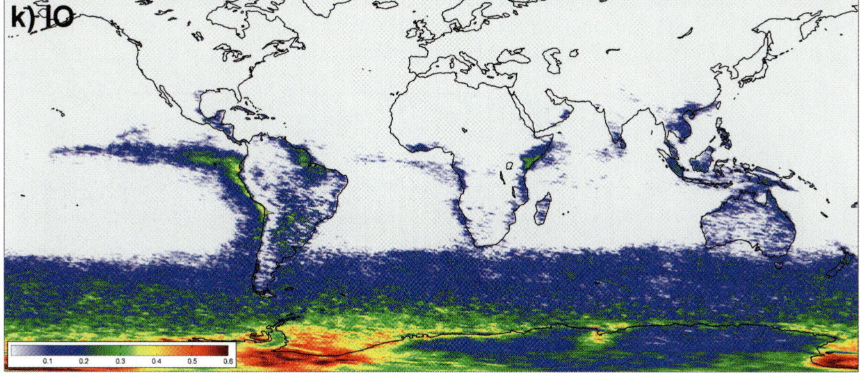

Fig. 8.13 (continued) Atlas of tropospheric species observed from space. (**j**) Mean tropospheric column density of BrO (10^{13} molecules/cm^2) derived from GOME measurements September 1997. Data provided by Nicolas Theys/Michel van Roozendael. (**k**) Mean tropospheric column density of IO (10^{13} molecules/cm^2) derived from SCIAMACHY measurements Sep–Nov 2005. Data provided by Anja Schönhardt (Schönhardt et al. 2008).

The global pattern of tropospheric O_3 shown in Fig. 8.13a reflects the availability of O_3 precursors, *i.e.* NO_x and VOCs. Highest columns being found over eastern Asia and the US East coast. The enhancement over the western Atlantic from Africa is because of the presence of the compounds (see CO, HCHO, CHOCHO), NO_x from biomass burning and NO_x from lightning (Martin et al. 2007).

From the spatial distribution alone as measured from space, on individual days or temporally averaged, it is possible to learn about sources of different trace gases. Consequently, satellite measurements have been used to estimate or constrain emissions in several studies (Arellano et al. 2004 (CO); Bergamaschi et al. 2007 (CH_4); Leue et al. 2001; Martin et al. 2003; Jaegle et al. 2005 (NO_2)) on a global scale, as well as in numerous studies focusing on particular regions and/or source types. For such inversion studies, knowledge of lifetimes and transport is needed, which is provided by chemical transport or general circulation models, to link emissions to columns. The link of satellite observations to chemistry models is

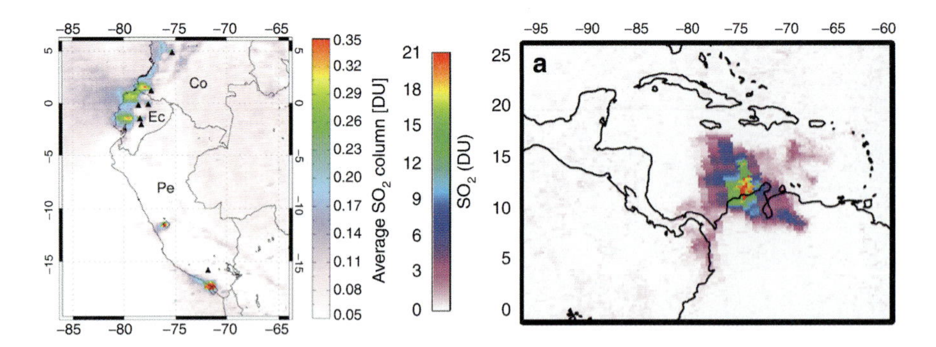

Fig. 8.14 (**a**) Average SO_2 column amounts measured by OMI over southern Colombia (Co), Ecuador (Ec), and Peru (Pe) between 1st September 2004 and 30th June 2005. Triangles mark Volcanoes, while Peruvian copper smelters are indicated by diamonds (Carn et al. 2007). (**b**) Observed OMI SO_2 column over the volcanic plume emitted from Soufriere Hills Volcano (Montserrat; 16.72°N, 62.18°W) on 21st May 2006 (Yang et al. 2007).

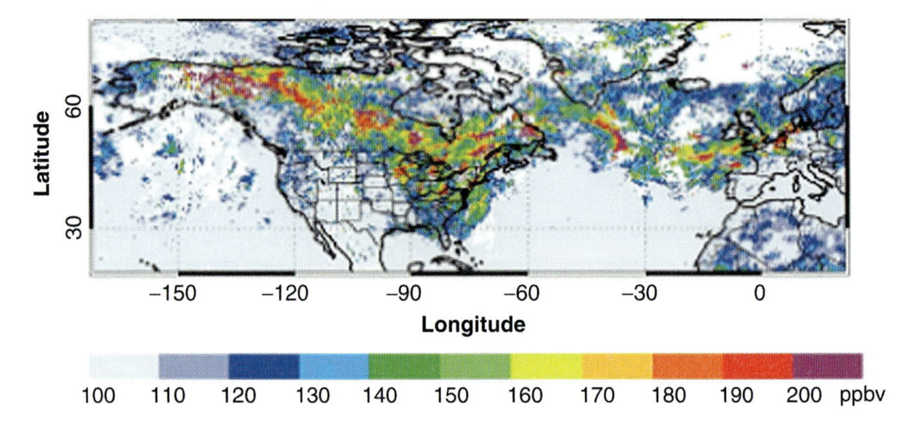

Fig. 8.15 MOPITT 700-hPa CO mixing ratio for 15–23 July 2004 (Fishman et al. 2008).

discussed in Chapter 9. However, independent lifetime information can also be gained from the satellite measurements themselves from spatial patterns downwind the sources (Beirle et al. 2004a).

8.3.2 Monitoring Transport and Circulation

The availability of temporally consecutive global measurements allows an investigation of transport of various trace gases for individual episodes as well as in terms of predominant transport patterns. Satellite observations not only reveal the location and strength of sources, but also the fate of the different trace gases visible from space. In addition, transport patterns (in particular of trace gases with lifetimes of weeks to months) serve as tracers for the validation of transport models. In particular, CO is a good marker for long range transport (LRT) owing to its lifetime of a few months (see Section 8.2.3).

Local emissions from industry or biomass burning can contribute appreciably to levels in remote regions. However, even close to sources, the question arises as to what fraction is due to these local emissions, and how much is due to long-range/intercontinental transport. For instance, Liang et al. (2007) reported that the Asian influence on pollution levels observed in the free troposphere over North America in summertime contributes about 7%. Heald et al. (2003) and Zhang et al. (2008) report on Trans-Pacific transport events of CO (Fig. 8.16), suggesting that similar LRT episodes affect North American levels of O_3. Gloudemans et al. (2006) found evidence for long-range transport of CO from biomass burning in the southern hemisphere using SCIAMACHY measurements, concluding that South American biomass burning emissions contribute up to 30–35% of the CO levels over Australian biomass burning regions.

The transport of SO_2 plumes from volcanic eruptions can be investigated from a time series of satellite maps (Prata et al. 2007) (Fig. 8.27). Such volcanic plumes can travel thousands of km over several days. The heavy eruption of the Kasatochi volcano in Alaska on 7th August 2008 led to the first detection of volcanic BrO in satellite spectra. The BrO plume could be tracked over a time period of about 6 days (Theys et al. 2009) (Fig. 8.17).

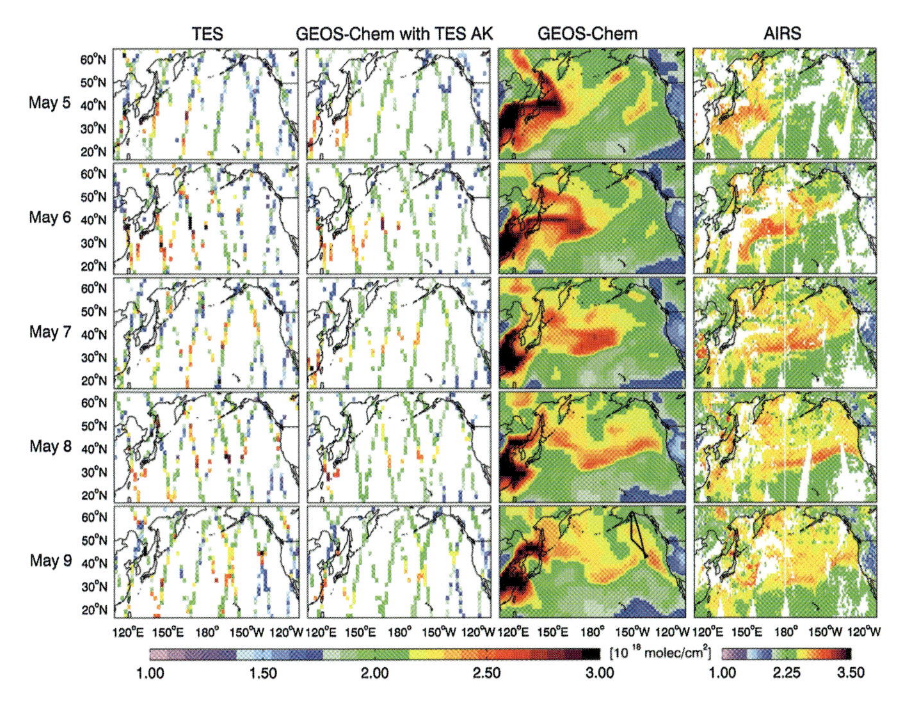

Fig. 8.16 CO columns from AIRS, TES and the GEOS-Chem model during the transpacific Asian pollution event, 5th–9th May, observed by the INTEX-B aircraft (Zhang et al. 2008).

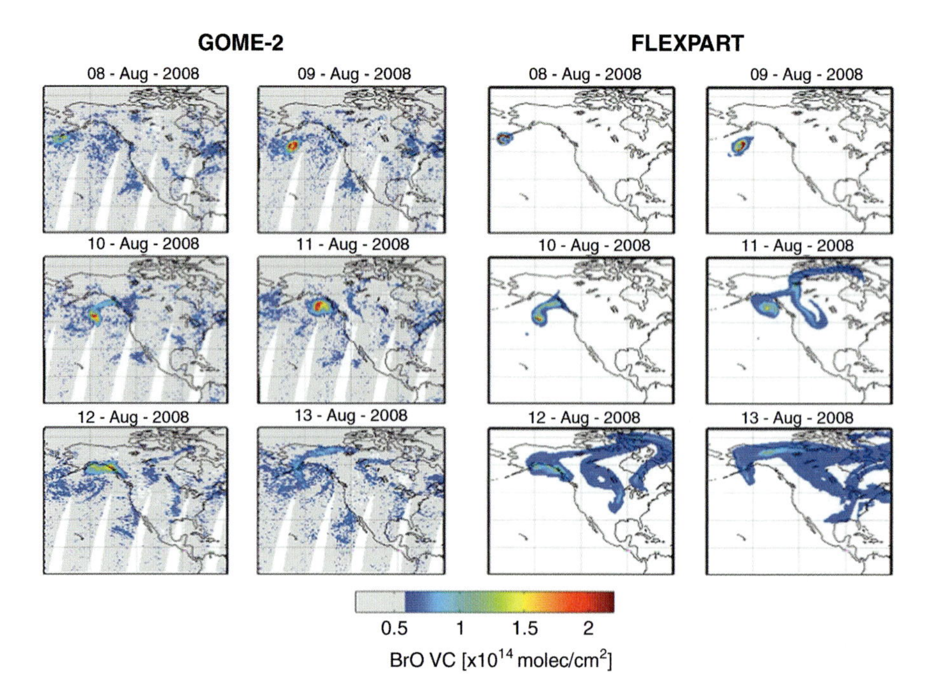

Fig. 8.17 GOME-2 measurements of BrO total columns after the eruption of the Kasatochi volcano and corresponding FLEXPART column simulations (Theys et al. 2009).

In the case of NO_2, LRT is unusual owing to the short lifetime of NO_x in the boundary layer, but it occurs occasionally when the boundary layer pollution is lifted to the upper troposphere, where the NO_x lifetime can reach several days (Wenig et al. 2003). In addition, PAN acts as reservoir for NO_x and plays an important role for the LRT of nitrogen oxides (Singh et al. 1992).

A prominent example of intercontinental transport of an anthropogenic NO_x plume within a meteorological "bomb" from the US East coast to Europe within five days was reported by Stohl et al. (2003) (Fig. 8.18).

A statistical analysis of transport patterns of NO_x is given in Eckhardt et al. (2003) where NO_2 distributions for high and low NAO (North Atlantic Oscillation) index are compared. For high NAO, NO_2 levels over northern Europe are significantly higher than for low NAO in winter, owing to the changes in wind patterns. An increase in NAO as reported by Hurrell (1995) implies an increase in O_3 precursors in the Arctic. Similarly, Creilson et al. (2003) report a positive correlation of NAO indices and tropospheric O_3 columns over Europe due to transport of O_3 precursors from North America to Europe.

Thus, analysis of range, frequency and significance of transport events permits us to verify our knowledge about atmospheric processes, in particular the lifetimes of different trace gases. Satellite measurements are also a powerful tool to investigate the question of how far regional pollution levels are local or impacted by LRT (Keating and Zuber 2007).

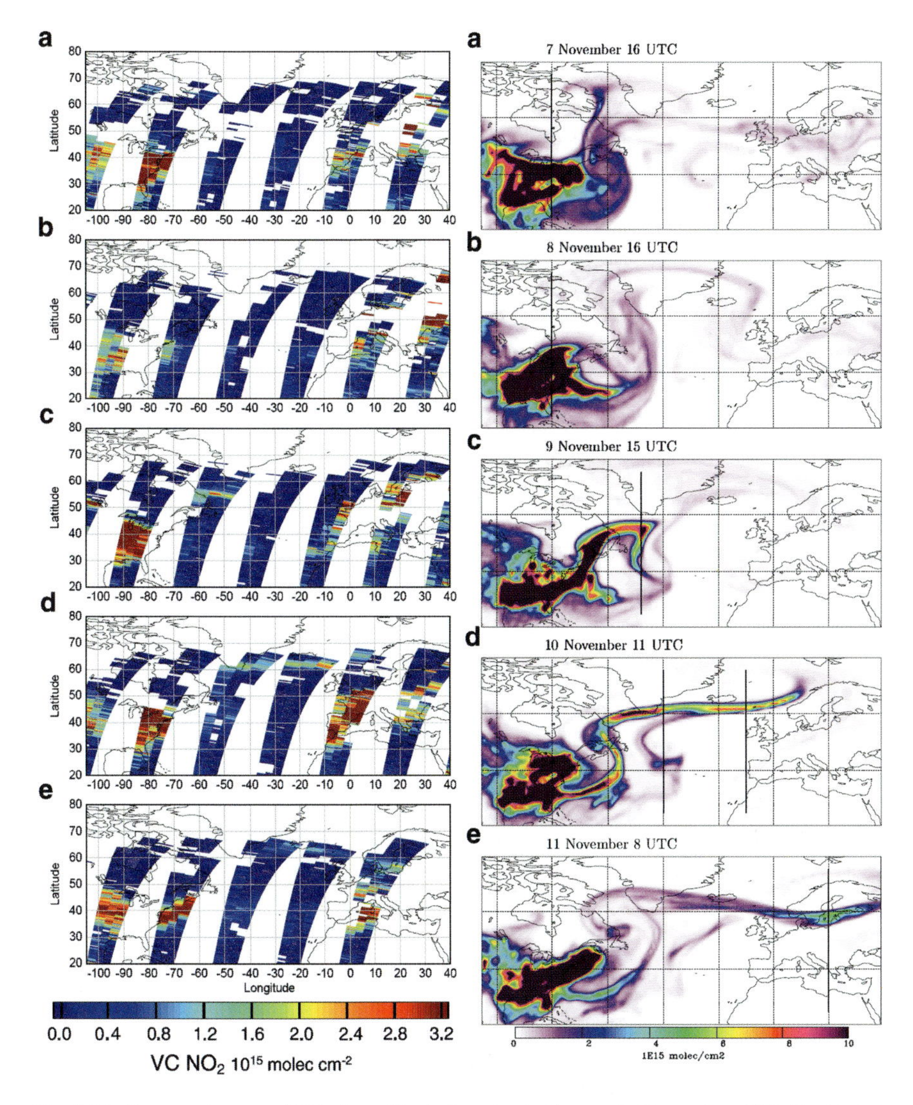

Fig. 8.18 Transport event of NO_2 from North America to Europe in November 2001 (Stohl et al. 2003). *Left*: TVCDs of NO_2 (GOME). *Right*: Total columns of the FLEXPART NO_x tracer.

8.3.3 Trends

There is now more nearly 15 years worth of tropospheric column measurements particularly in the UV/Vis spectral range. Using these data, studies on long-term temporal changes or trends of a range of different trace gases are becoming possible.

Fig. 8.13b shows that China is the region with the highest observed NO_2 columns in the world. There has been a tremendous growth in industrial activity during recent years. The pattern was different some years ago when it was dominated by the U.S. and Europe (Leue et al. 2001). Richter et al. (2005) first studied the trend of NO_2 columns and found a strong increase of 50% over 8 years in the Chinese region, while NO_2 columns over Europe and parts of the U.S. decreased (Fig. 8.19). Similar trend studies have been performed by Irie et al. (2005), Konovalov et al. (2008), Stavrakou et al. (2008), van der A et al. (2008), Hayn et al. (2009) and Kim et al. (2006) focussed on changes in NO_2 over U.S. power plants and found significant reductions owing to the implementation of pollution controls by utility companies in the eastern U.S.

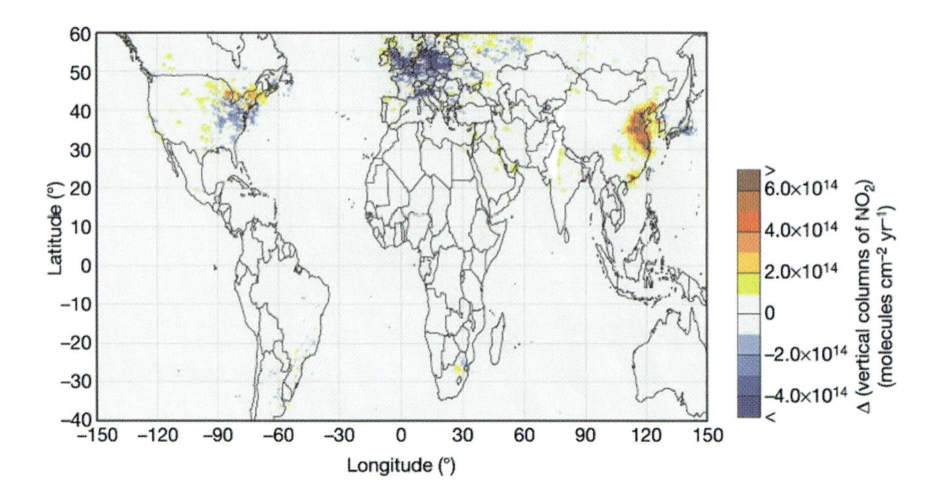

Fig. 8.19 Average annual changes in tropospheric NO_2 as observed by GOME from 1996 to 2002 (Richter et al. 2005 (Reprinted by permission of Macmillan Publishers Ltd)).

The increase of NO_2 columns over China illustrates the need for the updating of the relevant emission inventories on annual basis. Models using outdated emission inventories cannot be expected to simulate atmospheric chemistry realistically.

Similar trend investigations have been performed for CO (Yurganov et al. 2008) and SO_2 (Khokhar et al. 2008). Results are not as clear as with NO_x emissions over China; nevertheless, Yurganov et al. (2008) find an increase of global CO of about 2% per year for the second half of the year from MOPITT measurements between 2000 and 2006, while for the first half of the year, no significant change was detected. Ongoing measurements and an increasing number of CO sensors will allow more detailed regional studies to be performed in the near future. Khokhar et al. (2008) report on a decrease of SO_2 columns of 25% from 1996 to 2002 over the copper smelter Ilo in Peru, derived from GOME measurements.

For O_3, the time-series are available from 1979. However, the determination of trends of tropospheric O_3 is difficult owing to the dominant stratospheric column that also changes. Nevertheless, Ziemke et al. (2005) report an increase of tropospheric O_3 of 2 to 3 DU per decade for mid-latitudes for both hemispheres, while for the tropics, no significant change could be found, as also reported by Thompson and Hudson (1999).

The combustion of fossil fuels has lead to a large increase in atmospheric CO_2 levels, which is overlaid by the seasonal variation imprinted by the uptake and release of CO_2 by plants. This is impressively documented in the long-time series of CO_2 mixing ratios over Mauna Loa, Hawaii (Keeling et al. 1976). Buchwitz et al. (2007b) present a good match for the Mauna Loa time-series with SCIAMACHY northern hemispheric mean columns, and show an appreciable increase in CO_2 even for the relatively short time-series available from SCIAMACHY (Fig. 8.20).

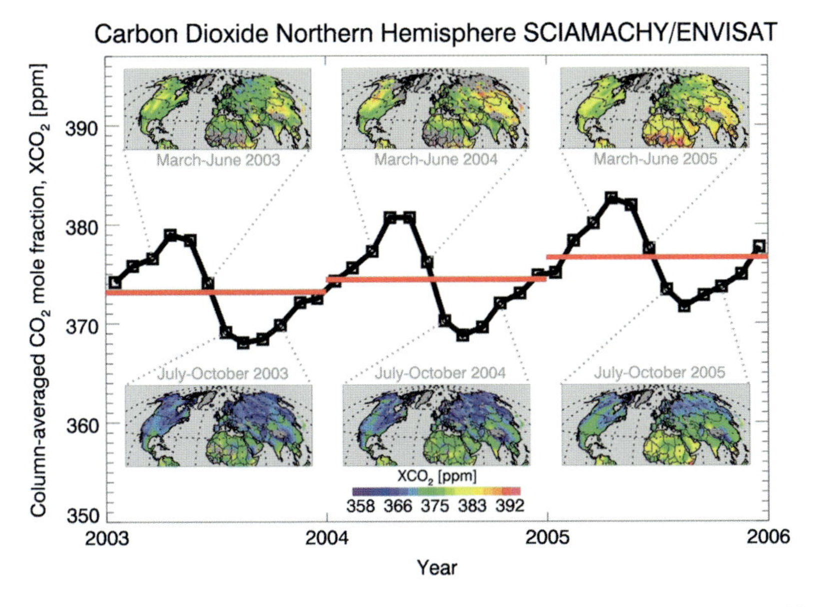

Fig. 8.20 Atmospheric CO_2 over the northern hemisphere from 2003 to 2005 as retrieved from SCIAMACHY satellite measurements (Buchwitz et al. 2007b).

As well as trace gases that are directly affected by anthropogenic emissions, water vapour, a natural greenhouse gas, is subject to changes owing to climate change. Since water vapour is strongly coupled to temperature, global warming should enhance water vapour columns, establishing a positive feedback mechanism. From GOME measurements, an increase of total column precipitable water of about 3% per year is reported by Wagner et al. (2006), in parallel with the increase in surface temperature, and hence indicates a strong positive climate feedback for water vapour. Similar results are found by Mieruch et al. (2008).

A continuous time series of satellite measurements will allow us to monitor future changes of various trace gases, and so to check the efficiency of measures taken to reduce air pollution. Satellite measurements have the potential to verify compliance of international climate protocols such as the Kyoto Treaty.

8.3.4 Periodical Temporal Patterns

As well as longer-term trends, periodical temporal patterns can be analyzed. Annual cycles are the usual dominant patterns within a time series of atmospheric trace gases owing to changes in emissions, chemistry and/or viewing conditions of the same periodicity.

The annual cycle of CO reflects the emissions during tropical biomass burning seasons (Edwards et al. 2006a; 2006b; Frankenberg et al. 2005b). Similar response of biomass burning emissions can be found in measurements of NO_2, HCHO or CHOCHO (Jaegle et al. 2005; Stavrakou et al. 2009; Myriokefalitakis et al. 2008). Palmer et al. (2006) investigated seasonal variability of isoprene emissions over North America using satellite measurements of HCHO. On shorter time scales, the clear coincidence of the onset of precipitation with enhanced NO_2 column densities was used to identify and estimate soil emissions of NO_x (Jaegle et al. 2004; Bertram et al. 2005).

Tropospheric BrO in polar regions shows an annual peak in hemispheric spring (Fig. 8.9), which is an important clue to its origin (Monks 2005). The specific annual cycle indicates that newly-formed sea ice plays an important role in the heterogeneous release of bromine leading to the bromine explosion. This could either be directly or indirectly due to the formation of highly saline frost flowers or aerosols resulting from frost flowers (Kaleschke et al. 2004).

Van der A et al. (2008) determined maps of the month with highest NO_2 column density and, with simple assumptions on yearly cycles of different NO_x sources, derived maps of the dominant NO_x source (Fig. 8.21).

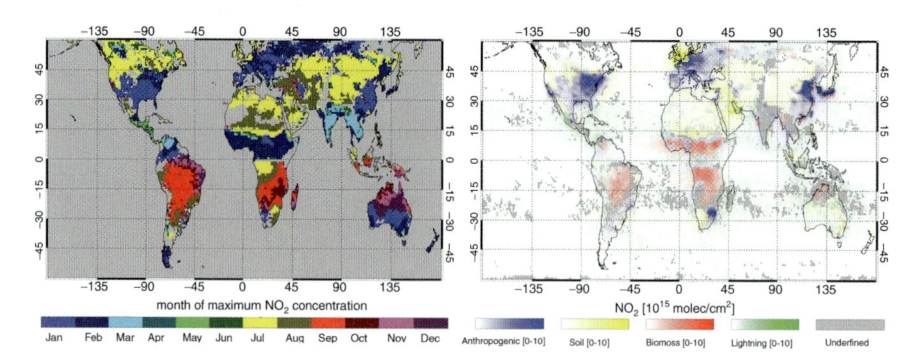

Fig. 8.21 (*Left*) Month of maximum NO_2; (*Right*) Dominant NO_x source identification (van der A et al. 2008).

Anthropogenic emissions in most industrialized countries follow a weekly cycle with emission reductions on Sundays. This is reflected in the weekly pattern of NO_2 column densities (Fig. 8.22): A clear reduction of NO_2 columns on Sundays can be found for the United States, Europe, South Korea and Japan, while columns reach a minimum on Friday in the Middle East and Saturday in Israel. In China, no weekly cycle is observed.

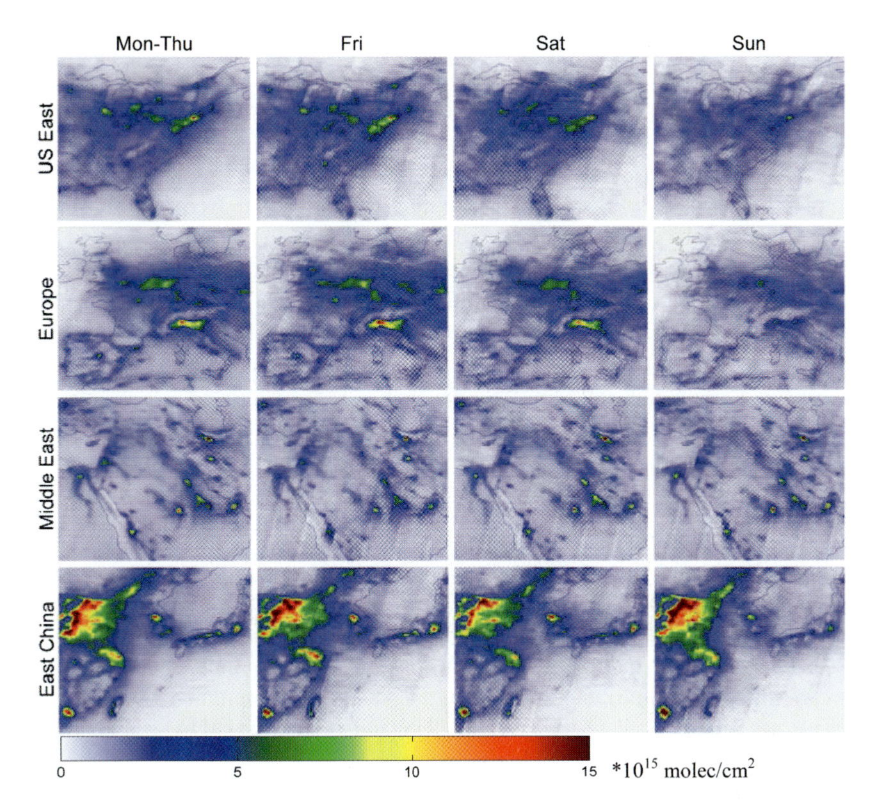

Fig. 8.22 Weekly cycle of NO_2 column densities for the United States, Europe, the Middle East, and eastern Asia (SCIAMACHY measurements 2003–2007).

The observation of a weekly cycle helps to discriminate between different NO_x sources that show different reductions, such as traffic and power generation (Beirle et al. 2003; Kaynak et al. 2009). In addition, analysis of the complete weekly pattern holds information on the NO_x lifetime. For example, Monday levels of NO_2 over Germany are significantly lower than Tuesday levels in winter, since Monday "inherits" comparably clean Sunday air owing to the longer lifetime of about 1 day (Beirle et al. 2003).

8.3.5 Synergistic Use of Different Measurements

Several years of satellite measurements and a growing number of instruments in space have resulted in extensive datasets providing information on many atmospheric trace gases. At the same time, there is considerable additional information on clouds, aerosols, or ground albedo which is of importance for the quantitative interpretation of column measurements from space. Furthermore, satellite measurements of other quantities such as lightning flashes, fires, night-time light pollution or vegetation indices, can serve as proxies for various emission sources such as lightning NO_x, biomass burning emissions of CO, VOC and NO_x, or anthropogenic versus biogenic activity.

The possibilities offered by comparisons between the different datasets and their combined use are manifold, e.g., combining different species from one sensor, the same species from different sensors, different species from different sensors, or trace gases with auxiliary data like lightning etc. The potential of synergistic use of the available information has not yet fully been exploited, but recent studies have begun to demonstrate the insights that can be gained from the combined use of various datasets. Here follows a short discussion on some aspects of different synergistic applications and their potential.

a Improving Retrievals

Additional measurements are required for the retrieval of tropospheric column densities. The sensitivity of satellite measurements for tropospheric trace gases is affected by aerosols and clouds as well as the spectral ground albedo. Information about these parameters can be gained from the spectral measurements themselves (Chapters 5 and 6), with the important advantage that, for example, cloud properties directly match the time and location of the column measurements. In addition, imaging spectrometers like MODIS or MERIS, and space-born LIDARs (CALIPSO), provide detailed information on clouds and aerosols with high spatial resolution.

Satellite observations designed for stratospheric trace gas observations can be used to extract tropospheric columns from total column measurements. This is of particular importance for the retrieval of tropospheric O_3, since the total O_3 column is dominated by the stratosphere. For instance, satellite SBUV measurements have been used to derive tropospheric O_3 from TOMS measurements (Fishman and Balok 1999). For a range of tropospheric gases, for example, NO_2 or BrO, the stratospheric column has to be known to retrieve the residual tropospheric column. SCIAMACHY operates in an alternating limb-nadir mode to allow the direct retrieval of stratospheric columns for the correction of nadir column measurements (Sierk et al. 2006; Beirle et al. 2009).

b Identifying Sources

Trace gas columns from satellite measurements can be compared to independent measurements of different parameters like fire or flash counts, which are proxies for different trace gas sources. For instance, fire counts from the satellite instruments ATSR or MODIS, are indicators of biomass burning and have been compared to CO (Edwards et al. 2006), HCHO and NO_2 (Spichtinger et al. 2004) and CHOCHO (Wittrock et al. 2006). Flash counts from the Lightning Imaging Sensor (LIS) have been compared to NO_2 columns over Australia to estimate NO_x production by lightning (Beirle et al. 2004c) (Fig. 8.23).

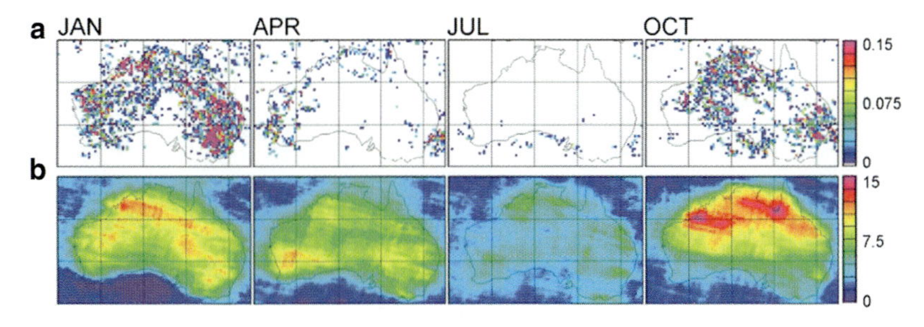

Fig. 8.23 Monthly mean LIS flash counts (upper row, flashes per day and km^2) and GOME tropospheric NO_2 column (lower row, 10^{14} molecules/cm^2) in Australia for several months in1999 (Beirle et al. 2004c).

Toenges-Schuller et al. (2006) used light pollution at night, as measured from DMSP, as proxy for anthropogenic emissions; it shows a good spatial correlation with NO_2 columns.

c Learning About Atmospheric Chemistry

From the UV/vis spectral measurements, several trace gases (*e.g.* NO_2, SO_2, HCHO, CHOCHO, CO, O_3) can be derived simultaneously, giving information on different sources, such as anthropogenic versus biomass burning, and allows insights into their chemistry and lifetimes. For instance, Martin et al. (2003) have differentiated NO_x saturated from NO_x sensitive regions from the observed ratio of HCHO to NO_2 columns (Fig. 8.24).

Results from different sensors for the same trace gas can be used synergistically. Currently, four nadir spectrometers in the UV/vis (GOME, SCIAMACHY, OMI and GOME-2) are in operation simultaneously. Comparisons between different instruments can provide consistency checks. In addition, the different overpass times provide information on the diurnal variations of trace gas columns. Boersma et al. (2008a) compared monthly mean NO_2 columns from SCIAMACHY (overpass time: 10:00) and OMI (13:30) (Fig. 8.25). OMI columns are lower than SCIAMACHY columns over fossil fuel combustion regions, mainly because of higher OH concentrations, and thus shorter lifetimes, in the early afternoon. Over biomass burning regions, OMI columns are higher owing to the diurnal cycle of fire activity, which is not considered in emission inventories.

d Learning About Profiles

Measuring one trace gas at different wavelengths provides additional profile information owing to different altitude sensitivities. For instance, MOPITT (4.7 μm) is more sensitive to CO in the free troposphere while CO columns in the NIR spectra

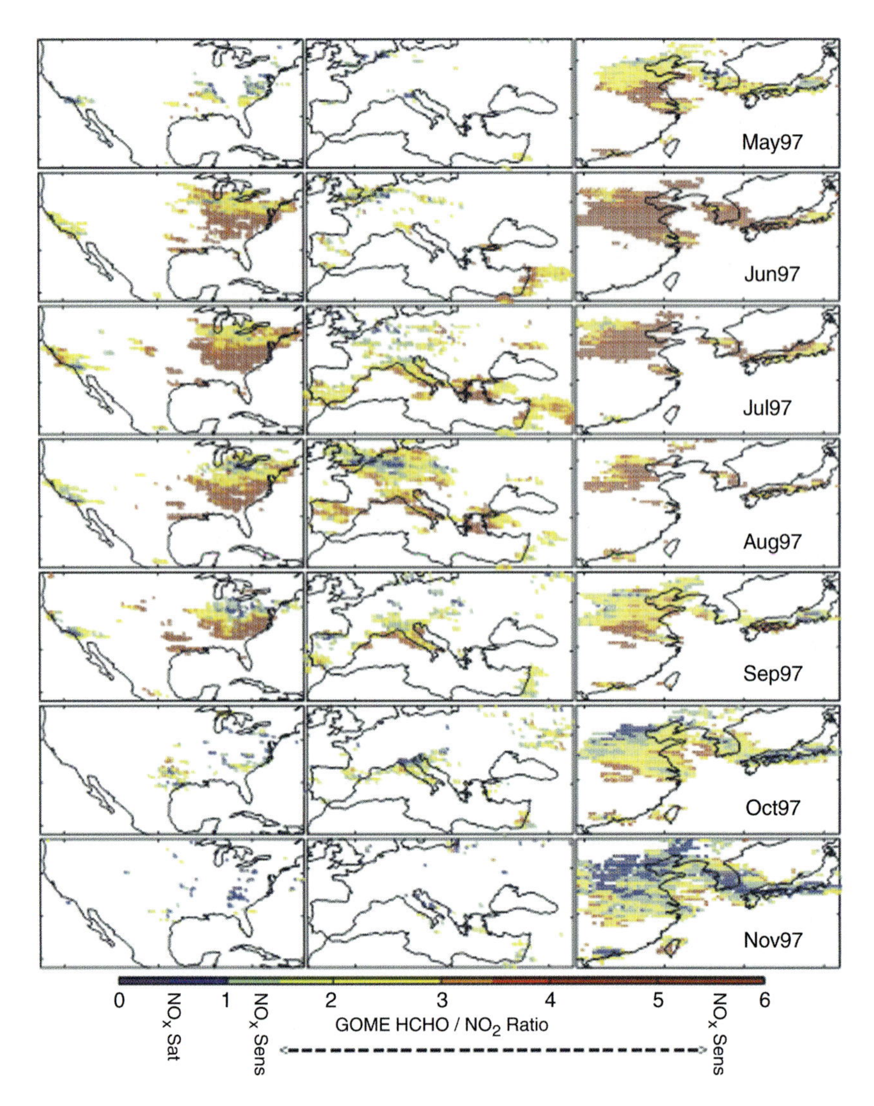

Fig. 8.24 Monthly mean tropospheric $HCHO/NO_2$ column ratio from GOME (Martin et al. 2004a).

from SCIAMACHY (2.3 μm) reach towards the ground; the difference allows the boundary layer concentration of CO to be estimated (Fig. 8.26) (Turquety et al. 2008).

e Multi-Platform Observations

Similar comparisons of trace gas columns from different sensors have been performed and are in progress for many trace gases. For CO, there are several new

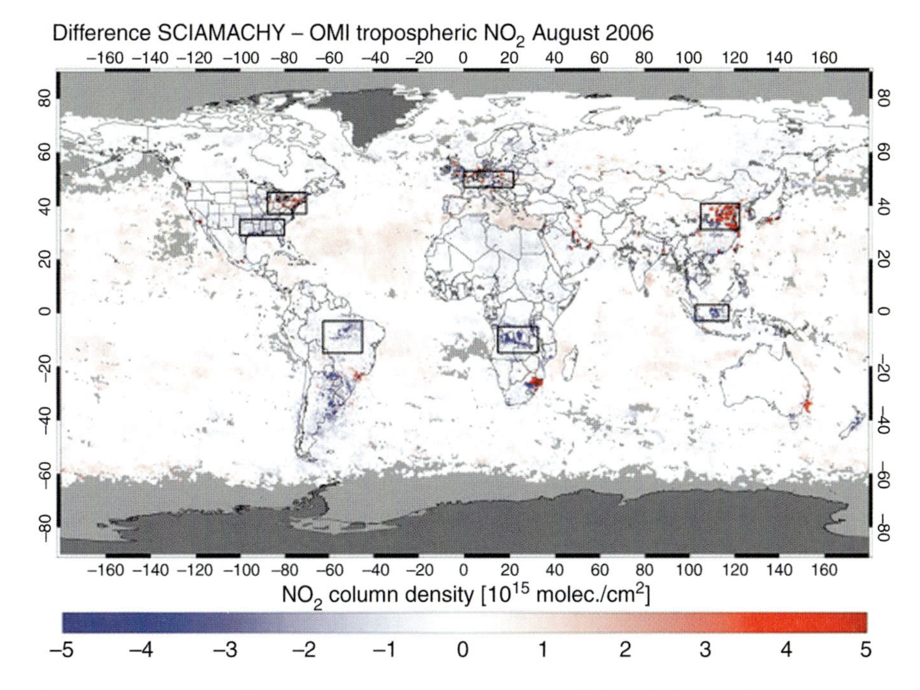

Fig. 8.25 Absolute difference between monthly mean SCIAMACHY and OMI tropospheric columns for August 2006 (Boersma et al. 2008a).

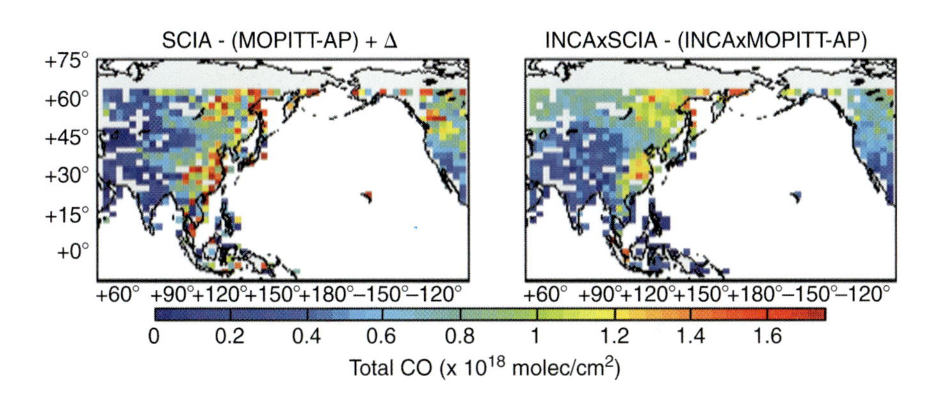

Fig. 8.26 Boundary layer residual of CO derived from the difference of SCIAMACHY and MOPITT columns of CO (Turquety et al. 2008).

instruments, so that now five (N)IR instruments (MOPITT, SCIAMACHY, AIRS, TES, IASI) are available for synergistic use. For instance, Prata et al. (2007) used measurements from AIRS (IR), SEVIRI (IR), MLS (MW, Limb), and OMI (UV/vis), to track the early evolution and long range transport of a volcanic cloud from Soufrière Hills volcano, Montserrat (Fig. 8.27) in order to estimate the total

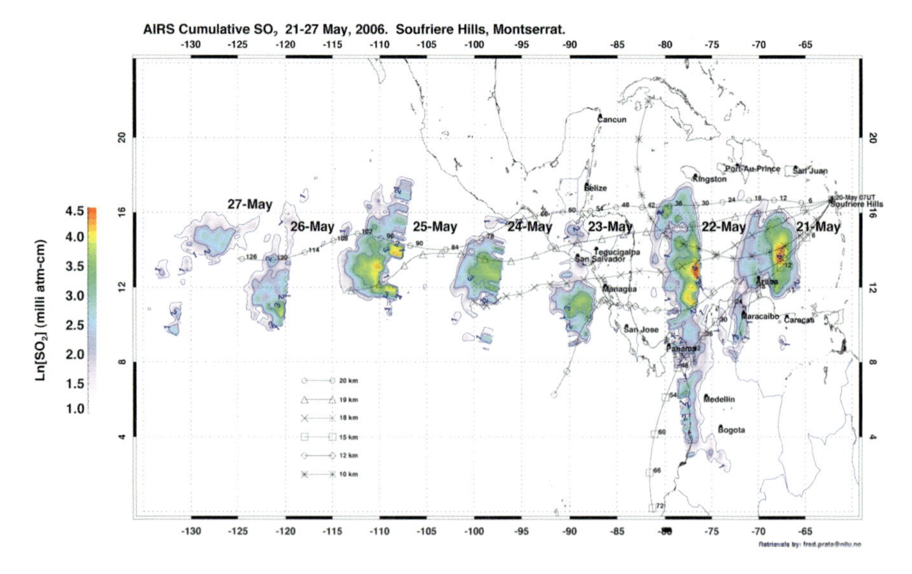

Fig. 8.27 AIRS SO_2 total column retrievals for 21st–26th May 2006 from the Soufriere Hills volcano, Montserrat (Prata et al. 2007).

SO_2 emissions. Such comparisons of the results from several satellite measurements, having different measurement times, wavelengths (and thus averaging kernels) etc. will further improve our understanding of tropospheric composition and chemistry.

8.3.6 *Operational Use*

Most of the tropospheric composition data collected from satellites so far has been from "research" instruments on individual science missions. However, both TOMS-OMI series and GOME-SCIMACHY missions have yielded long term measurements. As has been shown previously in Section 8.3.2 there is much to gain from long-time series of satellite data in mapping change in the earth-system.

However time series and monitoring, requiring consistent long-term measurements delivered in a timely fashion require operational measurements: measurements made with satellite instruments which are intended to deliver the data for the foreseeable future. An example is provided by the satellites used by the meteorological services.

IGACO is a strategy for bringing together ground-based, aircraft and satellite observations of 13 chemical species in the atmosphere. IGACO will be implemented as a strategic element of the GAW programme of the WMO.

Using the current generation of satellites, pilot operational services are being developed, *e.g.* PROMOTE/GMES (http://www.gse-promote.org/), or TEMIS (www.temis.nl) that provide information on air quality, UV and climate gases.

Long-term operational tropospheric chemical measurements from satellites have been initiated in concert with the meteorological community of MetOp (Klaes et al. 2007), providing measurements in the UV/vis (GOME-2) and the IR (IASI). In a parallel area, data fusion of various satellite sensors has been used to generate an operational aerosol prediction (Al-Saadi et al. 2005).

Despite the uncertainties of tropospheric trace gas columns derived from satellite measurements, particularly due to clouds and trace gas profiles, the improving spatial resolution and the growing number of different sensors available will allow us to plan different applications. For instance, future satellite instruments with footprints of some km^2 would allow the monitoring of air quality, in particular NO_x, on urban scale. A number of groups have indentified the utility of geostationary observations, not only for air quality applications where high spatial and temporal measurements are required, but also for operational assimilation into predictive models (Bovensmann et al. 2002, 2004; Burrows et al. 2004; Fishman et al. 2008; Munzenmayer et al. 2008). In particular, the existing satellite instruments on sun-synchronous orbits are not capable of resolving the diurnal chemistry cycle.

There has been some discussion on the application and use of so-called chemical weather (Lawrence et al. 2005), a direct analogy to meteorological weather, and the role satellites would play in delivery of operational data.

In future, satellite measurements could also play an important role in monitoring emission policies by measuring levels and trends of pollutants (NO_x, CO, SO_2), O_3, and greenhouse gases (CO_2 and CH_4).

8.4 Summary and Outlook

Satellite observations of tropospheric trace gases are a new and powerful tool to study tropospheric composition. The spatial and temporal information obtained globally provides unique information on sources, transport, and sinks for a range of gas-phase and particulate species. Tropospheric composition satellite data is beginning to make the transition from observation to quantification.

Looking to the future, a new view will be afforded with the long term continuous measurements from new satellites. There are now a number of programs to make the supply of high quality tropospheric composition data operational. In many senses, the satellites that have given so much insight into the troposphere were designed for research purposes and we await data from a generation of satellites designed for that purpose. A significant challenge still exists to merge the satellite data to the *in situ* observing systems in a meaningful and consistent way.

There is a clear need for improved spatial resolution, spatial cover and increased temporal coverage. Geostationary observations have a lot to offer (Fishman et al. 2008; Munzenmayer et al. 2008) for greater regional insights into tropospheric chemistry from space. There are scientific challenges across the globe, such as the continuing challenge of urbanisation and Megacities (Molina and Molina 2004), and a trans-national observing system for the tropics, that will require a new generation of satellite measurements. Synergistic use of multiple instruments has just started and there is much data still to be analysed; there are many further applications ahead. Tropospheric satellite observations have to be ready to meet the challenges of climate change as we look to the future and this will put great demands on facilities, data, analysis and understanding.

Acknowledgements The authors are particularly grateful to Dr. John Remedios and Prof. Peter Bernath for help and guidance through the worlds of MIPAS and ACE-FTS. We would also like to thank Dr. Andreas Richter, Dr. Pierre Coheur and Dr. Cathy Clerbaux for helpful discussions and suggestions.

References

Abbot, D.S. P.I. Palmer, R.V. Martin, K.V. Chance, D.J. Jacob and A. Guenther, 2003, Seasonal and interannual variability of North American isoprene emissions as determined by formaldehyde column measurements from space, *Geophys. Res. Lett.* **30**, 1886.

Ackerman, S.A. A.J. Schreiner, T.J. Schmit, H.M. Woolf, J. Li and M. Pavolonis, 2008, Using the GOES Sounder to monitor upper level SO_2 from volcanic eruptions, *J. Geophys. Res.-Atmospheres* **113**, D14s11.

Afe, O.T. A. Richter, B. Sierk, F. Wittrock and J.P. Burrows, 2004, BrO emission from volcanoes: A survey using GOME and SCIAMACHY measurements, *Geophys. Res. Lett.* **31**, L24113.

Ahn, C. J.R. Ziemke, S. Chandra and P.K. Bhartia, 2003, Derivation of tropospheric column ozone from the Earth Probe TOMS/GOES co-located data sets using the cloud slicing technique, *J. Atmos. and Solar-Terrestrial Physics* **65**, 1127–1137.

Al-Saadi, J. J. Szykman, R.B. Pierce, C. Kittaka, D. Neil, D.A. Chu, L. Remer, L. Gumley, E. Prins, L. Weinstock, C. MacDonald, R. Wayland, F. Dimmick and J. Fishman, 2005, Improving national air quality forecasts with satellite aerosol observations, *Bull. Amer. Met. Soc.* **86**, 1249.

Alicke, B. K. Hebestreit, J. Stutz and U. Platt, 1999, Iodine oxide in the marine boundary layer, *Nature* **397**, 572–573.

Allan, B.J. G. McFiggans, J.M.C. Plane and H. Coe, 2000, Observations of iodine monoxide in the remote marine boundary layer, *J. Geophys. Res.-Atmospheres* **105**, 14363–14369.

Allen, D. K. Pickering and M. Fox-Rabinovitz, 2004, Evaluation of pollutant outflow and CO sources during TRACE-P using model-calculated, aircraft-based, and Measurements of Pollution in the Troposphere (MOPITT)-derived CO concentrations, *J. Geophys. Res.-Atmospheres* **109**, D15s03.

Arellano, A.F. P.S. Kasibhatla, L. Giglio, G.R. van der Werf and J.T. Randerson, 2004, Top-down estimates of global CO sources using MOPITT measurements, *Geophys. Res. Lett.* **31**, L01104.

Arellano, A.F. P.S. Kasibhatla, L. Giglio, G.R. van der Werf, J.T. Randerson and G.J. Collatz, 2006, Time-dependent inversion estimates of global biomass-burning CO emissions using Measurement of Pollution in the Troposphere (MOPITT) measurements, *J. Geophys. Res.-Atmospheres* **111**, D09303.

Arellano, A.F. K. Raeder, J.L. Anderson, P.G. Hess, L.K. Emmons, D.P. Edwards, G.G. Pfister, T.L. Campos and G.W. Sachse, 2007, Evaluating model performance of an ensemble-based chemical data assimilation system during INTEX-B field mission, *Atmos. Chem. and Phys.* **7**, 5695–5710.

Asman, W.A.H. M.A. Sutton and J.K. Schjorring, 1998, Ammonia: emission, atmospheric transport and deposition, *New Phytologist* **139**, 27–48.

Aumann, H.H. M.T. Chahine, C. Gautier, M.D. Goldberg, E. Kalnay, L.M. McMillin, H. Revercomb, P.W. Rosenkranz, W.L. Smith, D.H. Staelin, L.L. Strow and J. Susskind, 2003,

AIRS/AMSU/HSB on the Aqua mission: Design, science objectives, data products, and processing systems, *IEEE Trans Geosci Remote Sens* **41**, 253–264, doi: 10.1109/tgrs. 2002.808356.

Aumann, H.H. D. Gregorich and S. Gaiser, 2005, AIRS hyper-spectral measurements for climate research: Carbon dioxide and nitrous oxide effects, *Geophys. Res. Lett.* **32**, doi: 10.1029/ 2004gl021784.

Barkley, M.P. U. Friess and P.S. Monks, 2006a, Measuring atmospheric CO_2 from space using full spectral initiation (FSI) WFM-DOAS, *Atmos. Chem. and Phys.* **6**, 3517–3534.

Barkley, M.P. P.S. Monks and R.J. Engelen, 2006b, Comparison of SCIAMACHY and AIRS CO_2 measurements over North America during the summer and autumn of 2003, *Geophys. Res. Lett.* **33**, L20805.

Barkley, M.P. P.S. Monks, U. Friess, R.L. Mittermeier, H. Fast, S. Korner and M. Heimann, 2006c, Comparisons between SCIAMACHY atmospheric CO_2 retrieved using (FSI) WFM-DOAS to ground based FTIR data and the TM3 chemistry transport model, *Atmos. Chem. and Phys.* **6**, 4483–4498.

Barkley, M.P. P.S. Monks, A.J. Hewitt, T. Machida, A. Desai, N. Vinnichenko, T. Nakazawa, M.Y. Arshinov, N. Fedoseev and T. Watai, 2007, Assessing the near surface sensitivity of SCIAMACHY atmospheric CO_2 retrieved using (FSI) WFM-DOAS, *Atmos. Chem. and Phys.* **7**, 3597–3619.

Barkley, M.P. P.I. Palmer, C.D. Boone, P.F. Bernath and P. Suntharalingam, 2008a, Global distributions of carbonyl sulfide in the upper troposphere and stratosphere, *Geophys. Res. Lett.* **35**, doi: 710.1029/2008gl034270.

Barkley, M. P. P. I. Palmer, U. Kuhn, J. Kesselmeier, K. Chance, T. P. Kurosu, R. V. Martin, D. Helmig, and A. Guenther, 2008b, Net ecosystem fluxes of isoprene over tropical South America inferred from GOME observations of HCHO columns, *J. Geophys. Res.* **113**, D20304, doi:10.1029/2008JD009863.

Barret, B. S. Turquety, D. Hurtmans, C. Clerbaux, J. Hadji-Lazaro, I. Bey, M. Auvray and P.F. Coheur, 2005, Global carbon monoxide vertical distributions from spaceborne high-resolution FTIR nadir measurements, *Atmos. Chem. Phys.* **5**, 2901–2914.

Barret, B. P. Ricaud, C. Mari, J.L. Attie, N. Bousserez, B. Josse, E. Le Flochmoen, N.J. Livesey, S. Massart, V.H. Peuch, A. Piacentini, B. Sauvage, V. Thouret and J.P. Cammas, 2008, Transport pathways of CO in the African upper troposphere during the monsoon season: a study based upon the assimilation of spaceborne observations, *Atmos. Chem. and Phys.* **8**, 3231–3246.

Barrie, L.A. J.W. Bottenheim, R.C. Schnell, P.J. Crutzen and R.A. Rasmussen, 1988, Ozone Destruction and Photochemical-Reactions at Polar Sunrise in the Lower Arctic Atmosphere, *Nature* **334**, 138–141.

Bates, T.S. B.K. Lamb, A. Guenther, J. Dignon and R.E. Stoiber, 1992, Sulfur Emissions to the Atmosphere from Natural Sources, *J. Atmos. Chem.* **14**, 315–337.

Beer, R. M.W. Shephard, S.S. Kulawik, S.A. Clough, A. Eldering, K.W. Bowman, S.P. Sander, B.M. Fisher, V.H. Payne, M.Z. Luo, G.B. Osterman and J.R. Worden, 2008, First satellite observations of lower tropospheric ammonia and methanol, *Geophys. Res. Lett.* **35**, L09801.

Beirle, S. U. Platt, M. Wenig and T. Wagner, 2003, Weekly cycle of NO_2 by GOME measurements: a signature of anthropogenic sources, *Atmos. Chem. and Phys.* **3**, 2225–2232.

Beirle, S. U. Platt, R. von Glasow, M. Wenig and T. Wagner, 2004a, Estimate of nitrogen oxide emissions from shipping by satellite remote sensing, *Geophys. Res. Lett.* **31**, L18102.

Beirle, S. U. Platt, M. Wenig and T. Wagner, 2004b, Highly resolved global distribution of tropospheric NO_2 using GOME narrow swath mode data, *Atmos. Chem. and Phys.* **4**, 1913–1924.

Beirle, S. U. Platt, M. Wenig and T. Wagner, 2004c, NOx production by lightning estimated with GOME. *Adv. Space Res.* **34**, 793–797.

Beirle, S. N. Spichtinger, A. Stohl, K.L. Cummins, T. Turner, D. Boccippio, O.R. Cooper, M. Wenig, M. Grzegorski, U. Platt and T. Wagner, 2006, Estimating the NOx produced by lightning from GOME and NLDN data: a case study in the Gulf of Mexico, *Atmos. Chem. and Phys.* **6**, 1075–1089.

Beirle, S. S. Kühl, J. Pukite, and T. Wagner, 2009, Retrieval of tropospheric column densities of NO_2 from combined SCIAMACHY nadir/limb measurements, *Atmos. Meas. Tech. Discuss.* **2**, 2983–3025.

Bergamaschi, P. C. Frankenberg, J.F. Meirink, M. Krol, F. Dentener, T. Wagner, U. Platt, J.O. Kaplan, S. Korner, M. Heimann, E.J. Dlugokencky and A. Goede, 2007, Satellite chartography of atmospheric methane from SCIAMACHY on board ENVISAT: 2. Evaluation based on inverse model simulations, *J. Geophys. Res.-Atmospheres* **112**, D02304.

Bertram, T.H. A. Heckel, A. Richter, J.P. Burrows and R.C. Cohen, 2005, Satellite measurements of daily variations in soil NOx emissions, *Geophys. Res. Lett.* **32**, L24812.

Bhattacharjee, P.S. A.K. Prasad, M. Kafatos and R.P. Singh, 2007, Influence of a dust storm on carbon monoxide and water vapor over the Indo-Gangetic Plains, *J. Geophys. Res.-Atmospheres* **112**, D18203.

Blond, N. K.F. Boersma, H.J. Eskes, R.J. van der A, M. Van Roozendael, I. De Smedt, G. Bergametti and R. Vautard, 2007, Intercomparison of SCIAMACHY nitrogen dioxide observations, *in situ* measurements and air quality modeling results over Western Europe, *J. Geophys. Res.-Atmospheres* **112**, D10311.

Bloom, A.A. P.I. Palmer, A. Fraser, D.S. Reay, C. Frankenberg, 2010, Large-scale controls of methanogenesis inferred from methane and gravity spaceborne data, *Science*, **327**, 322–325. doi: 10.1126/science.1175176.

Bluth, G.J.S. S.D. Doiron, C.C. Schnetzler, A.J. Krueger and L.S. Walter, 1992, Global Tracking of the SO_2 Clouds from the June, 1991 Mount-Pinatubo Eruptions, *Geophys. Res. Lett.* **19**, 151–154.

Bluth, G.J.S. T.J. Casadevall, C.C. Schnetzler, S.D. Doiron, L.S. Walter, A.J. Krueger and M. Badruddin, 1994, Evaluation of Sulfur-Dioxide Emissions from Explosive Volcanism - the 1982-1983 Eruptions of Galunggung, Java, Indonesia, *J. Volcanology and Geothermal Res.* **63**, 243–256.

Bobrowski, N. G. Hönninger, B. Galle and U. Platt, 2003, Detection of bromine monoxide in a volcanic plume, *Nature* **423**, 273–276 10.1038/nature01638.

Boersma, K.F. H.J. Eskes, E.W. Meijer and H.M. Kelder, 2005, Estimates of lightning NOx production from GOME satellite observations, *Atmos. Chem. Phys.* **5**, 2311–2331.

Boersma, K. F. H. J. Eskes, J. P. Veefkind, E. J. Brinksma, R. J. van der A, M. Sneep, G. H. J. van den Oord, P. F. Levelt, P. Stammes, J. F. Gleason, and E. J. Bucsela, 2007, Near-real time retrieval of tropospheric NO_2 from OMI, *Atmos. Chem. Phys.* **7**, 2103–2118.

Boersma, K.F. D.J. Jacob, H.J. Eskes, R.W. Pinder, J. Wang and R.J. van der A, 2008a, Intercomparison of SCIAMACHY and OMI tropospheric NO_2 columns: Observing the diurnal evolution of chemistry and emissions from space, *J. Geophys. Res.-Atmospheres* **113**, D16s26.

Boersma, K. F. D. J. Jacob, E. J. Bucsela, A. E. Perring, R. Dirksen, R. J. van der A, R. M. Yantosca, R. J. Park, M. O. Wenig, T. H. Bertram, and R. C. Cohen, 2008b, Validation of OMI tropospheric NO_2 observations during INTEX-B and application to constrain NO_x emissions over the eastern United States and Mexico, *Atmos. Environ.* **42**(19), 4480–4497, doi:10.1016/j.atmosenv.2008.02.004.

Boersma, K. F. D. J. Jacob, M. Trainic, Y. Rudich, I. DeSmedt, R. Dirksen, and H. J. Eskes, 2009, Validation of urban NO_2 concentrations and their diurnal and seasonal variations observed from space (SCIAMACHY and OMI sensors) using *in situ* measurements in Israeli cities, *Atmos. Chem. Phys.* **9**, 3867–3879, 2009.

Boesch, H. G.C. Toon, B. Sen, R.A. Washenfelder, P.O. Wennberg, M. Buchwitz, R. de Beek, J.P. Burrows, D. Crisp, M. Christi, B.J. Connor, V. Natraj and Y.L. Yung, 2006, Space-based near-infrared CO_2 measurements: Testing the Orbiting Carbon Observatory retrieval algorithm and validation concept using SCIAMACHY observations over Park Falls, Wisconsin, *J. Geophys. Res.-Atmospheres* **111**, D23302.

Bousserez, N. J.L. Attie, V.H. Peuch, M. Michou, G. Pfister, D. Edwards, L. Emmons, C. Mari, B. Barret, S.R. Arnold, A. Heckel, A. Richter, H. Schlager, A. Lewis, M. Avery, G. Sachse, E.V. Browell and J.W. Hair, 2007, Evaluation of the MOCAGE chemistry transport model during the ICARTT/ITOP experiment, *J. Geophys. Res.-Atmospheres* **112**, D10s42.

Bovensmann, H. S. Noel, P. Monks, A.P.H. Goede and J.P. Burrows, 2002, The geostationary scanning imaging absorption spectrometer (GEOSCIA) mission: Requirements and capabilities. *Adv. Space Res.* **29**, 1849–1859.

Bovensmann, H. K.U. Eichmann, S. Noel, J.M. Flaud, J. Orphal, P.S. Monks, G.K. Corlett, A.P. Goede, T. von Clarmann, T. Steck, V. Rozanov and J.P. Burrows, 2004, The geostationary scanning imaging absorption spectrometer (GeoSCIA) as part of the geostationary tropospheric pollution explorer (GeoTROPE) mission: requirements, concepts and capabilities. *Adv. Space Res.* **34**, 694–699.

Bowman, K.P. 2006, Transport of carbon monoxide from the tropics to the extratropics, *J. Geophys. Res.-Atmospheres* **111**, D02107.

Bowman, K. W. D. B. A. Jones, J. A. Logan, H. Worden, F. Boersma, R. Chang, S. Kulawik, G. Osterman, P. Hamer, and J. Worden, 2009, The zonal structure of tropical O_3 and CO as observed by the Tropospheric Emission Spectrometer in November 2004 – Part 2: Impact of surface emissions on O_3 and its precursors, *Atmos. Chem. Phys.* **9**, 3563–3582.

Bremer, H. J. Kar, J.R. Drummond, F. Nichitu, J.S. Zou, J. Liu, J.C. Gille, M.N. Deeter, G. Francis, D. Ziskin and J. Warner, 2004, Spatial and temporal variation of MOPITT CO in Africa and South America: A comparison with SHADOZ ozone and MODIS aerosol, *J. Geophys. Res.-Atmospheres* **109**, D12304.

Browell, E.V. M.A. Fenn, C.F. Butler, W.B. Grant, M.B. Clayton, J. Fishman, A.S. Bachmeier, B.E. Anderson, G.L. Gregory, H.E. Fuelberg, J.D. Bradshaw, S.T. Sandholm, D.R. Blake, B.G. Heikes, G.W. Sachse, H.B. Singh and R.W. Talbot, 1996, Ozone and aerosol distributions and air mass characteristics over the South Atlantic Basin during the burning season, *J. Geophys. Res.-Atmospheres* **101**, 24043–24068.

Buchwitz, M. R. de Beek, K. Bramstedt, S. Noel, H. Bovensmann and J.P. Burrows, 2004, Global carbon monoxide as retrieved from SCIAMACHY by WFM-DOAS, *Atmos. Chem. and Phys.* **4**, 1945–1960.

Buchwitz, M. R. de Beek, J.P. Burrows, H. Bovensmann, T. Warneke, J. Notholt, J.F. Meirink, A.P.H. Goede, P. Bergamaschi, S. Korner, M. Heimann and A. Schulz, 2005a, Atmospheric methane and carbon dioxide from SCIAMACHY satellite data: initial comparison with chemistry and transport models, *Atmos. Chem. and Phys.* **5**, 941–962.

Buchwitz, M. R. de Beek, S. Noel, J.P. Burrows, H. Bovensmann, H. Bremer, P. Bergamaschi, S. Korner and M. Heimann, 2005b, Carbon monoxide, methane and carbon dioxide columns retrieved from SCIAMACHY by WFM-DOAS: year 2003 initial data set, *Atmos. Chem. and Phys.* **5**, 3313–3329.

Buchwitz, M. R. de Beek, S. Noel, J.P. Burrows, H. Bovensmann, O. Schneising, I. Khlystova, M. Bruns, H. Bremer, P. Bergamaschi, S. Korner and M. Heimann, 2006, Atmospheric carbon gases retrieved from SCIAMACHY by WFM-DOAS: version 0.5 CO and CH_4 and impact of calibration improvements on CO_2 retrieval, *Atmos. Chem. Phys.* **6**, 2727–2751.

Buchwitz, M. I. Khlystova, H. Bovensmann and J.P. Burrows, 2007a, Three years of global carbon monoxide from SCIAMACHY: comparison with MOPITT and first results related to the detection of enhanced CO over cities, *Atmos. Chem. Phys.* **7**, 2399–2411.

Buchwitz, M. O. Schneising, J.P. Burrows, H. Bovensmann, M. Reuter and J. Notholt, 2007b, First direct observation of the atmospheric CO_2 year-to-year increase from space, *Atmos. Chem. and Phys.* **7**, 4249–4256.

Bucsela, E.J. E.A. Celarier, M.O. Wenig, J.F. Gleason, J.P. Veefkind, K.F. Boersma and E.J. Brinksma, 2006, Algorithm for NO_2 vertical column retrieval from the ozone monitoring instrument, *IEEE Trans Geosci Remote Sens* **44**, 1245–1258.

Bucsela, E.J. A.E. Perring, R.C. Cohen, K.F. Boersma, E.A. Celarier, J.F. Gleason, M.O. Wenig, T.H. Bertram, P.J. Wooldridge, R. Dirksen and J.P. Veefkind, 2008, Comparison of tropospheric NO_2 from *in situ* aircraft measurements with near-real-time and standard product data from OMI, *J. Geophys. Res.-Atmospheres* **113**, D16s31.

Burgess, A.B. R.G. Grainger, A. Dudhia, V.H. Payne and V.L. Jay, 2004, MIPAS measurement of sulphur hexafluoride (SF_6), *Geophys. Res. Lett.* **31**, doi: 10.1029/2003gl019143.

Burgess, A.B. A. Dudhia, R.G. Grainger and D. Stevenson, 2006a, Progress in tropospheric ammonia retrieval from the MIPAS satellite instrument, Adv. Space Res. **37**, 2218–2221, doi: 10.1016/j.asr.2005.06.073.

Burgess, A.B. R.G. Grainger and A. Dudhia, 2006b, Zonal mean atmospheric distribution of sulphur hexafluoride (SF_6), *Geophys. Res. Lett.* **33**, doi: 10.1029/2005gl025410.

Burrows, J.P. 1999, Current and future passive remote sensing techniques used to determine atmospheric constituents. In: Bouwman, A.F. (Editor), *Developments in Atmospheric Sciences 24: Approaches to Scaling Trace Gas Fluxes in Ecosystems*, Elsevier, Amsterdam, The Netherlands, 315–347.

Burrows, J.P. M. Weber, M. Buchwitz, V. Rozanov, A. Ladstatter-Weissenmayer, A. Richter, R. DeBeek, R. Hoogen, K. Bramstedt, K.U. Eichmann and M. Eisinger, 1999, The global ozone monitoring experiment (GOME): Mission concept and first scientific results, *J. Atmos. Sci.* **56**, 151–175.

Burrows, J.P. H. Bovensmann, G. Bergametti, J.M. Flaud, J. Orphal, S. Noel, P.S. Monks, G.K. Corlett, A.P. Goede, T. von Clarmann, T. Steck, F. Fischer and F. Friedl-Vallon, 2004, The geostationary tropospheric pollution explorer (GeoTROPE) mission: objectives, requirements and mission concept. *Adv Space Res.* **34**, 682–687.

Carn, S.A. and G.J.S. Bluth, 2003, Prodigious sulfur dioxide emissions from Nyamuragira volcano, DR Congo, *Geophys. Res. Lett.* **30**, 2211.

Carn, S.A. A.J. Krueger, N.A. Krotkov and M.A. Gray, 2004, Fire at Iraqi sulfur plant emits SO_2 clouds detected by Earth Probe TOMS, *Geophys. Res. Lett.* **31**, L19105.

Carn, S.A. A.J. Krueger, N.A. Krotkov, K. Yang and P.F. Levelt, 2007, Sulfur dioxide emissions from Peruvian copper smelters detected by the Ozone Monitoring Instrument, *Geophys. Res. Lett.* **34**, L09801.

Carn, S.A. L.L. Strow, S. de Souza-Machado, Y. Edmonds and S. Hannon, 2005, Quantifying tropospheric volcanic emissions with AIRS: The 2002 eruption of Mt. Etna (Italy), *Geophys. Res. Lett.* **32**, doi: 510.1029/2004gl021034.

Carpenter, L.J. 2003, Iodine in the marine boundary layer, *Chem. Revs.* **103**, 4953–4962, doi: 10.1021/cr0206465.

Celarier, E.A. E.J. Brinksma, J.F. Gleason, J.P. Veefkind, A. Cede, J.R. Herman, D. Ionov, F. Goutail, J.P. Pommereau, J.C. Lambert, M. van Roozendael, G. Pinardi, F. Wittrock, A. Schönhardt, A. Richter, O.W. Ibrahim, T. Wagner, B. Bojkov, G. Mount, E. Spinei, C.M. Chen, T.J. Pongetti, S.P. Sander, E.J. Bucsela, M.O. Wenig, D.P.J. Swart, H. Volten, M. Kroon and P.F. Levelt, 2008, Validation of ozone monitoring instrument nitrogen dioxide columns, *J. Geophys. Res.-Atmospheres* **113**, D15s15.

Chahine, M. C. Barnet, E.T. Olsen, L. Chen and E. Maddy, 2005, On the determination of atmospheric minor gases by the method of vanishing partial derivatives with application to CO_2, *Geophys. Res. Lett.* **32**, doi: 510.1029/2005gl024165.

Chance, K. 1998, Analysis of BrO measurements from the Global Ozone Monitoring Experiment, *Geophys. Res. Lett.* **25**, 3335–3338.

Chance, K. P.I. Palmer, R.J.D. Spurr, R.V. Martin, T.P. Kurosu and D.J. Jacob, 2000, Satellite observations of formaldehyde over North America from GOME, *Geophys. Res. Lett.* **27**, 3461–3464.

Chandra, S. J.R. Ziemke, W. Min and W.G. Read, 1998, Effects of 1997–1998 El Nino on tropospheric ozone and water vapor, *Geophys. Res. Lett.* **25**, 3867–3870.

Chandra, S. J.R. Ziemke and R.W. Stewart, 1999, An 11-year solar cycle in tropospheric ozone from TOMS measurements, *Geophys. Res. Lett.* **26**, 185–188.

Chandra, S. J.R. Ziemke, P.K. Bhartia and R.V. Martin, 2002, Tropical tropospheric ozone: Implications for dynamics and biomass burning, *J. Geophys. Res.-Atmospheres* **107**, 4188.

Chandra, S. J.R. Ziemke and R.V. Martin, 2003, Tropospheric ozone at tropical and middle latitudes derived from TOMS/MLS residual: Comparison with a global model, *J. Geophys. Res.-Atmospheres* **108**, 4291.

Chandra, S. J.R. Ziemke, X.X. Tie and G. Brasseur, 2004, Elevated ozone in the troposphere over the Atlantic and Pacific oceans in the Northern Hemisphere, *Geophys. Res. Lett.* **31**, L23102.

Chatfield, R.B. H. Guan, A.M. Thompson and J.C. Witte, 2004, Convective lofting links Indian Ocean air pollution to paradoxical South Atlantic ozone maxima, *Geophys. Res. Lett.* **31**, L06103.

Chedin, A. A. Hollingsworth, N.A. Scott, S. Serrar, C. Crevoisier and R. Armante, 2002, Annual and seasonal variations of atmospheric CO_2, N_2O and CO concentrations retrieved from NOAA/TOVS satellite observations, *Geophys. Res. Lett.* **29**, doi: 10.1029/2001gl014082.

Chedin, A. S. Serrar, N.A. Scott, C. Crevoisier and R. Armante, 2003, First global measurement of midtropospheric CO_2 from NOAA polar satellites: Tropical zone, *J. Geophys. Res.-Atmospheres* **108**, doi: 1310.1029/2003jd003439.

Chedin, A. S. Serrar, N.A. Scott, C. Pierangelo and P. Ciais, 2005, Impact of tropical biomass burning emissions on the diurnal cycle of upper tropospheric CO_2 retrieved from NOAA 10 satellite observations, *J. Geophys. Res.-Atmospheres* **110**, doi: 10.1029/2004jd005540.

Chedin, A. N.A. Scott, R. Armante, C. Pierangelo, C. Crevoisier, O. Fosse and P. Ciais, 2008, A quantitative link between CO_2 emissions from tropical vegetation fires and the daily tropospheric excess (DTE) of CO_2 seen by NOAA-10 (1987-1991), *J. Geophys. Res.-Atmospheres* **113**, doi: 1310.1029/2007jd008576.

Chen, D. B. Zhou, S. Beirle, L.M. Chen and T. Wagner, 2009, Tropospheric NO_2 column densities deduced from zenith-sky DOAS measurements in Shanghai, China, and their application to satellite validation, *Atmos. Chem. Phys.* **9**, 3641–3662.

Choi, Y. Y.H. Wang, T. Zeng, R.V. Martin, T.P. Kurosu and K. Chance, 2005, Evidence of lightning NOx and convective transport of pollutants in satellite observations over North America, *Geophys. Res. Lett.* **32**, L02805.

Choi, S.D. and Y.S. Chang, 2006a, Carbon monoxide monitoring in Northeast Asia using MOPITT: Effects of biomass burning and regional pollution in April 2000, *Atmos. Environ.* **40**, 686–697.

Choi, S.D. and Y.S. Chang, 2006b, Evaluation of carbon uptake and emissions by forests in Korea during the last thirty years (1973–2002), *Environ Monit Assess* **117**, 99–107.

Choi, Y. Y.H. Wang, Q. Yang, D. Cunnold, T. Zeng, C. Shim, M. Luo, A. Eldering, E. Bucsela and J. Gleason, 2008, Spring to summer northward migration of high O_3 over the western North Atlantic, *Geophys. Res. Lett.* **35**, L04818.

Chun'e, S. and Z. Baoning, 2008, Tropospheric NO_2 columns over Northeastern North America: Comparison of CMAQ model simulations with GOME satellite measurements, *Adv. Atmos. Sci.* **25**, 59–71.

Clarisse, L. P.F. Coheur, A.J. Prata, D. Hurtmans, A. Razavi, T. Phulpin, J. Hadji-Lazaro and C. Clerbaux, 2008, Tracking and quantifying volcanic SO_2 with IASI, the September 2007 eruption at Jebel at Tair, *Atmos. Chem. Phys.* **8**, 7723–7734.

Clarisse, L. C. Clerbaux, F. Dentener, D. Hurtmans and P.-F. Coheur, 2009, Global ammonia distribution derived from infrared satellite observations, *Nat. Geosci.* **2**, 479–483

Clarmann, T.V. N. Glatthor, M.E. Koukouli, G.P. Stiller, B. Funke, U. Grabowski, M. Hoepfner, S. Kellmann, A. Linden, M. Milz, T. Steck and H. Fischer, 2007, MIPAS measurements of upper tropospheric C_2H_6 and O_3 during the southern hemispheric biomass burning season in 2003, *Atmos. Chem. and Phys.* **7**, 5861–5872.

Clerbaux, C. J. Hadji-Lazaro, S. Turquety, G. Magie and P.F. Coheur, 2003, Trace gas measurements from infrared satellite for chemistry and climate applications, *Atmos. Chem. Phys.* **3**, 1495–1508.

Clerbaux, C. P.F. Coheur, D. Hurtmans, B. Barret, M. Carleer, R. Colin, K. Semeniuk, J.C. McConnell, C. Boone and P. Bernath, 2005, Carbon monoxide distribution from the ACE-FTS solar occultation measurements, *Geophys. Res. Lett.* **32**, doi: 10.1029/2005gl022394.

Clerbaux, C. P.F. Coheur, L. Clarisse, J. Hadji-Lazaro, D. Hurtmans, S. Turquety, K. Bowman, H. Worden and S.A. Carn, 2008a, Measurements of SO_2 profiles in volcanic plumes from the NASA Tropospheric Emission Spectrometer (TES), *Geophys. Res. Lett.* **35**, doi: 510.1029/2008gl035566.

Clerbaux, C. D.P. Edwards, M. Deeter, L. Emmons, J.F. Lamarque, X.X. Tie, S.T. Massie and J. Gille, 2008b, Carbon monoxide pollution from cities and urban areas observed by the Terra/MOPITT mission, *Geophys. Res. Lett.* **35**, L03817.

Clerbaux, C. M. George, S. Turquety, K.A. Walker, B. Barret, P. Bernath, C. Boone, T. Borsdorff, J.P. Cammas, V. Catoire, M. Coffey, P.F. Coheur, M. Deeter, M. De Maziere, J. Drummond, P. Duchatelet, E. Dupuy, R. de Zafra, F. Eddounia, D.P. Edwards, L. Emmons, B. Funke, J. Gille, D.W.T. Griffith, J. Hannigan, F. Hase, M. Hopfner, N. Jones, A. Kagawa, Y. Kasai, I. Kramer, E. Le Flochmoen, N.J. Livesey, M. Lopez-Puertas, M. Luo, E. Mahieu, D. Murtagh, P. Nedelec, A. Pazmino, H. Pumphrey, P. Ricaud, C.P. Rinsland, C. Robert, M. Schneider, C. Senten, G. Stiller, A. Strandberg, K. Strong, R. Sussmann, V. Thouret, J. Urban and A. Wiacek, 2008c, CO measurements from the ACE-FTS satellite instrument: data analysis and validation using ground-based, airborne and spaceborne observations, *Atmos. Chem. and Phys.* **8**, 2569–2594.

Clerbaux, C. A. Boynard, L. Clarisse, M. George, J. Hadji-Lazaro, H. Herbin, D. Hurtmans, M. Pommier, A. Razavi, S. Turquety, C. Wespes and P.F. Coheur, 2009, Monitoring of atmospheric composition using the thermal infrared IASI/MetOp sounder, *Atmos. Chem. Phys.* **9**, 6041–6054.

Coheur, P.F. C. Clerbaux and R. Colin, 2003, Spectroscopic measurements of halocarbons and hydrohalocarbons by satellite-borne remote sensors, *J. Geophys. Res.-Atmospheres* **108**, doi: 1410.1029/2002jd002649.

Coheur, P.F. B. Barret, S. Turquety, D. Hurtmans, J. Hadji-Lazaro and C. Clerbaux, 2005, Retrieval and characterization of ozone vertical profiles from a thermal infrared nadir sounder, *J. Geophys. Res.-Atmospheres* **110**, D24303.

Coheur, P.F. H. Herbin, C. Clerbaux, D. Hurtmans, C. Wespes, M. Carleer, S. Turquety, C.P. Rinsland, J. Remedios, D. Hauglustaine, C.D. Boone and P.F. Bernath, 2007, ACE-FTS observation of a young biomass burning plume: first reported measurements of C_2H_4, C_3H_6O, H_2CO and PAN by infrared occultation from space, *Atmos. Chem. Phys.* **7**, 5437–5446.

Coheur, P.F. L. Clarisse, S. Turquety, D. Hurtmans and C. Clerbaux, 2009, IASI measurements of reactive trace species in biomass burning plumes, *Atmos. Chem. and Phys.* **9**, 5655–5667.

Connors, V.S. B.B. Gormsen, S. Nolf and H.G. Reichle, 1999, Spaceborne observations of the global distribution of carbon monoxide in the middle troposphere during April and October 1994, *J. Geophys. Res.-Atmospheres* **104**, 21455-21470.

Crawford, J.H. C.L. Heald, H.E. Fuelberg, D.M. Morse, G.W. Sachse, L.K. Emmons, J.C. Gille, D.P. Edward, M.N. Deeter, G. Chen, J.R. Olson, V.S. Connors, C. Kittaka and A.J. Hamlin, 2004, Relationship between Measurements of Pollution in the Troposphere (MOPITT) and *in situ* observations of CO based on a large-scale feature sampled during TRACE-P, *J. Geophys. Res.-Atmospheres* **109**, D15s04.

Creilson, J.K. J. Fishman and A.E. Wozniak, 2003, Intercontinental transport of tropospheric ozone: a study of its seasonal variability across the North Atlantic utilizing tropospheric ozone residuals and its relationship to the North Atlantic Oscillation, *Atmos. Chem. and Phys.* **3**, 2053–2066.

Creilson, J.K. J. Fishman and A.E. Wozniak, 2005, Arctic Oscillation - induced variability in satellite-derived tropospheric ozone, *Geophys. Res. Lett.* **32**, L14822.

Crevoisier, C. S. Heilliette, A. Chedin, S. Serrar, R. Armante and N.A. Scott, 2004, Midtropospheric CO_2 concentration retrieval from AIRS observations in the tropics, *Geophys. Res. Lett.* **31**, doi: 410.1029/2004gl020141.

Crisp, D. R.M. Atlas, F.M. Breon, L.R. Brown, J.P. Burrows, P. Ciais, B.J. Connor, S.C. Doney, I.Y. Fung, D.J. Jacob, C.E. Miller, D. O'Brien, S. Pawson, J.T. Randerson, P. Rayner, R.J. Salawitch, S.P. Sander, B. Sen, G.L. Stephens, P.P. Tans, G.C. Toon, P.O. Wennberg, S.C. Wofsy, Y.L. Yung, Z. Kuang, B. Chudasama, G. Sprague, B. Weiss, R. Pollock, D. Kenyon and S. Schroll, 2004, The orbiting carbon observatory (OCO) mission. *Adv. Space Res*, **34**, 700–709.

Cros, B. D. Nganga, A. Minga, J. Fishman and V. Brackett, 1992, Distribution of Tropospheric Ozone at Brazzaville, Congo, Determined from Ozonesonde Measurements, *J. Geophys. Res. -Atmospheres* **97**, 12869–12875.

Damoah, R. N. Spichtinger, C. Forster, P. James, I. Mattis, U. Wandinger, S. Beirle, T. Wagner and A. Stohl, 2004, Around the world in 17 days - hemispheric-scale transport of forest fire smoke from Russia in May 2003, *Atmos. Chem. and Phys.* **4**, 1311–1321.

de Laat, A.T.J. A.M.S. Gloudemans, H. Schrijver, M.M.P. van den Broek, J.F. Meirink, I. Aben and M. Krol, 2006, Quantitative analysis of SCIAMACHY carbon monoxide total column measurements, *Geophys. Res. Lett.* **33**, doi: 510.1029/2005gl025530.

de Laat, A.T.J. A.M.S. Gloudemans, I. Aben, M. Krol, J.F. Meirink, G.R. van der Werf and H. Schrijver, 2007, Scanning Imaging Absorption Spectrometer for Atmospheric Chartography carbon monoxide total columns: Statistical evaluation and comparison with chemistry transport model results, *J. Geophys. Res.-Atmospheres* **112**, D12310.

De Maziere, M. C. Vigouroux, P.F. Bernath, P. Baron, T. Blumenstock, C. Boone, C. Brogniez, V. Catoire, M. Coffey, P. Duchatelet, D. Griffith, J. Hannigan, Y. Kasai, I. Kramer, N. Jones, E. Mahieu, G.L. Manney, C. Piccolo, C. Randall, C. Robert, C. Senten, K. Strong, J. Taylor, C. Tetard, K.A. Walker and S. Wood, 2008, Validation of ACE-FTS v2.2 methane profiles from the upper troposphere to the lower mesosphere, *Atmos. Chem. Phys.* **8**, 2421–2435.

De Smedt, I. J.F. Müller, T. Stavrakou, R. van der A, H. Eskes and M. Van Roozendael, 2008, Twelve years of global observations of formaldehyde in the troposphere using GOME and SCIAMACHY sensors, *Atmos. Chem. Phys.* **8**, 4947–4963.

Deeter, M.N. L.K. Emmons, G.L. Francis, D.P. Edwards, J.C. Gille, J.X. Warner, B. Khattatov, D. Ziskin, J.F. Lamarque, S.P. Ho, V. Yudin, J.L. Attie, D. Packman, J. Chen, D. Mao and J.R. Drummond, 2003, Operational carbon monoxide retrieval algorithm and selected results for the MOPITT instrument, *J. Geophys. Res.-Atmospheres* **108**, 4399.

Deeter, M.N. L.K. Emmons, D.P. Edwards, J.C. Gille and J.R. Drummond, 2004, Vertical resolution and information content of CO profiles retrieved by MOPITT, *Geophys. Res. Lett.* **31**, L15112.

Deeter, M.N. D.P. Edwards and J.C. Gille, 2007a, Retrievals of carbon monoxide profiles from MOPITT observations using lognormal *a priori* statistics, *J. Geophys. Res.-Atmospheres* **112**, D11311.

Deeter, M.N. D.P. Edwards, J.C. Gille and J.R. Drummond, 2007b, Sensitivity of MOPITT observations to carbon monoxide in the lower troposphere, *J. Geophys. Res.-Atmospheres* **112**, D24306.

DEFRA, 2003, Nitrogen Dioxide in the United Kingdom. The Air Quality Expert Group, Department for Environment, Food and Rural Affairs, The Scottish Executive, The National Assembly for Wales and The Department of the Environment in Northern Ireland.

Del Frate, F. M.F. Iapaolo and S. Casadio, 2005, Intercomparison between GOME ozone profiles retrieved by neural network inversion schemes and ILAS products, *J. Atmos. and Oceanic Tech.* **22**, 1433–1440.

Dentener, F.J. and P.J. Crutzen, 1994, A 3-Dimensional Model of the Global Ammonia Cycle, *J. Atmos. Chem.* **19**, 331–369.

Dils, B. M. De Maziere, J.F. Muller, T. Blumenstock, M. Buchwitz, R. de Beek, P. Demoulin, P. Duchatelet, H. Fast, C. Frankenberg, A. Gloudemans, D. Griffith, N. Jones, T. Kerzenmacher, I. Kramer, E. Mahieu, J. Mellqvist, R.L. Mittermeier, J. Notholt, C.P. Rinsland, H. Schrijver, D. Smale, A. Strandberg, A.G. Straume, W. Stremme, K. Strong, R. Sussmann, J. Taylor, M. van den Broek, V. Velazco, T. Wagner, T. Warneke, A. Wiacek and S. Wood, 2006,

Comparisons between SCIAMACHY and ground-based FTIR data for total columns of CO, CH_4, CO_2 and N_2O, *Atmos. Chem. and Phys.* **6**, 1953–1976.

Drummond, J.R. and G.S. Mand, 1996, The measurements of pollution in the troposphere (MOPITT) instrument: Overall performance and calibration requirements, *J.Atmos. and Oceanic Tech.* **13**, 314–320.

Dueck, T. and A. van der Werf, 2008, Are plants precursors for methane? *New Phytologist* **178**, 693–695.

Dufour, G. C.D. Boone and P.F. Bernath, 2005, First measurements of CFC-113 and HCFC-142b from space using ACE-FTS infrared spectra, *Geophys. Res. Lett.* **32**, doi: 10.1029/2005gl022422.

Dufour, G. C.D. Boone, C.P. Rinsland and P.F. Bernath, 2006, First space-borne measurements of methanol inside aged southern tropical to mid-latitude biomass burning plumes using the ACE-FTS instrument, *Atmos. Chem. and Phys.* **6**, 3463–3470.

Dufour, G. S. Szopa, D.A. Hauglustaine, C.D. Boone, C.P. Rinsland and P.F. Bernath, 2007, The influence of biogenic emissions on upper-tropospheric methanol as revealed from space, *Atmos. Chem. and Phys.* **7**, 6119–6129.

Dufour, G. F. Wittrock, M. Camredon, M. Beekmann, A. Richter, B. Aumont and J.P. Burrows, 2009, SCIAMACHY formaldehyde observations: constraint for isoprene emission estimates over Europe?, *Atmos. Chem. and Phys.* **9**, 1647–1664.

Duncan, B.N. S.E. Strahan, Y. Yoshida, S.D. Steenrod and N. Livesey, 2007, Model study of the cross-tropopause transport of biomass burning pollution, *Atmos. Chem. and Phys.* **7**, 3713–3736.

Eckhardt, S. A. Stohl, S. Beirle, N. Spichtinger, P. James, C. Forster, C. Junker, T. Wagner, U. Platt and S.G. Jennings, 2003, The North Atlantic Oscillation controls air pollution transport to the Arctic, *Atmos. Chem. and Phys.* **3**, 10.

Eckhardt, S. A.J. Prata, P. Seibert, K. Stebel and A. Stohl, 2008, Estimation of the vertical profile of sulfur dioxide injection into the atmosphere by a volcanic eruption using satellite column measurements and inverse transport modeling, *Atmos. Chem. Phys.* **8**, 3881–3897.

Edwards, D.P. J.F. Lamarque, J.L. Attie, L.K. Emmons, A. Richter, J.P. Cammas, J.C. Gille, G.L. Francis, M.N. Deeter, J. Warner, D.C. Ziskin, L.V. Lyjak, J.R. Drummond and J.P. Burrows, 2003, Tropospheric ozone over the tropical Atlantic: A satellite perspective, *J. Geophys. Res.-Atmospheres* **108**, 4237.

Edwards, D.P. L.K. Emmons, D.A. Hauglustaine, D.A. Chu, J.C. Gille, Y.J. Kaufman, G. Petron, L.N. Yurganov, L. Giglio, M.N. Deeter, V. Yudin, D.C. Ziskin, J. Warner, J.F. Lamarque, G.L. Francis, S.P. Ho, D. Mao, J. Chen, E.I. Grechko and J.R. Drummond, 2004, Observations of carbon monoxide and aerosols from the Terra satellite: Northern Hemisphere variability, *J. Geophys. Res.-Atmospheres* **109**, D24202.

Edwards, D.P. L.K. Emmons, J.C. Gille, A. Chu, J.L. Attie, L. Giglio, S.W. Wood, J. Haywood, M.N. Deeter, S.T. Massie, D.C. Ziskin and J.R. Drummond, 2006a, Satellite-observed pollution from Southern Hemisphere biomass burning, *J. Geophys. Res.-Atmospheres* **111**, D14312.

Edwards, D.P. G. Petron, P.C. Novelli, L.K. Emmons, J.C. Gille and J.R. Drummond, 2006b, Southern Hemisphere carbon monoxide interannual variability observed by Terra/Measurement of Pollution in the Troposphere (MOPITT), *J. Geophys. Res.-Atmospheres* **111**, D16303.

Eisinger, M. and J.P. Burrows, 1998, Tropospheric sulfur dioxide observed by the ERS-2 GOME instrument, *Geophys. Res. Lett.* **25**, 4177–4180.

Emmons, L.K. M.N. Deeter, J.C. Gille, D.P. Edwards, J.L. Attie, J. Warner, D. Ziskin, G. Francis, B. Khattatov, V. Yudin, J.F. Lamarque, S.P. Ho, D. Mao, J.S. Chen, J. Drummond, P. Novelli, G. Sachse, M.T. Coffey, J.W. Hannigan, C. Gerbig, S. Kawakami, Y. Kondo, N. Takegawa, H. Schlager, J. Baehr and H. Ziereis, 2004, Validation of Measurements of Pollution in the Troposphere (MOPITT) CO retrievals with aircraft *in situ* profiles, *J. Geophys. Res. -Atmospheres* **109**, D03309.

Emmons, L.K. G.G. Pfister, D.P. Edwards, J.C. Gille, G. Sachse, D. Blake, S. Wofsy, C. Gerbig, D. Matross and P. Nedelec, 2007, Measurements of Pollution in the Troposphere (MOPITT) validation exercises during summer 2004 field campaigns over North America, *J. Geophys. Res.-Atmospheres* **112**, D12s02.

Engelen, R.J. E. Andersson, F. Chevallier, A. Hollingsworth, M. Matricardi, A.P. McNally, J.N. Thepaut and P.D. Watts, 2004, Estimating atmospheric CO_2 from advanced infrared satellite radiances within an operational 4D-Var data assimilation system: Methodology and first results, *J. Geophys. Res.-Atmospheres* **109**, doi: 10.1029/2004jd004777.

Engelen, R.J. and A.P. McNally, 2005, Estimating atmospheric CO_2 from advanced infrared satellite radiances within an operational four-dimensional variational (4D-Var) data assimilation system: Results and validation, *J. Geophys. Res.-Atmospheres* **110**, doi: 10.1029/2005jd005982.

Eremenko, M. G. Dufour, G. Foret, C. Keim, J. Orphal, M. Beekmann, G. Bergametti and J.-M. Flaud, 2008, Tropospheric ozone distributions over Europe during the heat wave in July 2007 observed from infrared nadir spectra recorded by IASI, *Geophys. Res. Lett.* **35**, L18805.

Fan, S.M. and D.J. Jacob, 1992, Surface Ozone Depletion in Arctic Spring Sustained by Bromine Reactions on Aerosols, *Nature* **359**, 522–524.

Filipiak, M.J. R.S. Harwood, J.H. Jiang, Q.B. Li, N.J. Livesey, G.L. Manney, W.G. Read, M.J. Schwartz, J.W. Waters and D.L. Wu, 2005, Carbon monoxide measured by the EOS Microwave Limb Sounder on Aura: First results, *Geophys. Res. Lett.* **32**, doi: 510.1029/2005gl022765.

Fischer, H. M. Birk, C. Blom, B. Carli, M. Carlotti, T. von Clarmann, L. Delbouille, A. Dudhia, D. Ehhalt, M. Endemann, J.M. Flaud, R. Gessner, A. Kleinert, R. Koopman, J. Langen, M. Lapez-Puertas, P. Mosner, H. Nett, H. Oelhaf, G. Perron, J. Remedios, M. Ridolfi, G. Stiller and R. Zander, 2008, MIPAS: an instrument for atmospheric and climate research, *Atmos. Chem. Phys.* **8**, 2151–2188.

Fishman, J. 1991a, The Global Consequences of Increasing Tropospheric Ozone Concentrations, *Chemosphere* **22**, 685–695.

Fishman, J. 1991b, Probing Planetary Pollution from Space, *Environ. Sci. and Tech.* **25**, 612–621.

Fishman, J. F.M. Vukovich, D.R. Cahoon and M.C. Shipham, 1987, The Characterization of an Air-Pollution Episode Using Satellite Total Ozone Measurements, *J. Climate Appl. Meteorol.* **26**, 1638–1654.

Fishman, J. C.E. Watson, J.C. Larsen and J.A. Logan, 1990, Distribution of Tropospheric Ozone Determined from Satellite Data, *J. Geophys. Res -Atmospheres* **95**, 3599–3617.

Fishman, J. K. Fakhruzzaman, B. Cros and D. Nganga, 1991, Identification of Widespread Pollution in the Southern-Hemisphere Deduced from Satellite Analyses, *Science* **252**, 1693–1696.

Fishman, J. V.G. Brackett and K. Fakhruzzaman, 1992, Distribution of Tropospheric Ozone in the Tropics from Satellite and Ozonesonde Measurements, *J. Atmos. and Terr. Phys.* **54**, 589–597.

Fishman, J. V.G. Brackett, E.V. Browell and W.B. Grant, 1996, Tropospheric ozone derived from TOMS/SBUV measurements during TRACE A, *J. Geophys. Res.-Atmospheres* **101**, 24069–24082.

Fishman, J. and V.G. Brackett, 1997, The climatological distribution of tropospheric ozone derived from satellite measurements using version 7 Total Ozone Mapping Spectrometer and Stratospheric Aerosol and Gas Experiment data sets, *J. Geophys. Res.-Atmospheres* **102**, 19275–19278.

Fishman, J. and A.E. Balok, 1999, Calculation of daily tropospheric ozone residuals using TOMS and empirically improved SBUV measurements: Application to an ozone pollution episode over the eastern United States, *J. Geophys. Res.-Atmospheres* **104**, 30319–30340.

Fishman, J. A.E. Balok and F.M. Vukovich, 2002, Observing tropospheric trace gases from space: Recent advances and future capabilities. *Adv. Space Res*, **29**, 1625–1630.

Fishman, J. A.E. Wozniak and J.K. Creilson, 2003, Global distribution of tropospheric ozone from satellite measurements using the empirically corrected tropospheric ozone residual technique: Identification of the regional aspects of air pollution, *Atmos. Chem. and Phys.* **3**, 893–907.

Fishman, J. J.K. Creilson, A.E. Wozniak and P.J. Crutzen, 2005, Interannual variability of stratospheric and tropospheric ozone determined from satellite measurements, *J. Geophys. Res.-Atmospheres* **110**, D20306.

Fishman, J. K.W. Bowman, J.P. Burrows, A. Richter, K.V. Chance, D.P. Edwards, R.V. Martin, G.A. Morris, R.B. Pierce, J.R. Ziemke, J.A. Al-Saadi, J.K. Creilson, T.K. Schaack and A.M. Thompson, 2008, Remote sensing of tropospheric pollution from space, *Bull. Amer. Met. Soc.* **89**, 805–821, doi: 10.1175/2008bams2526.1.

Folkins, I. P. Bernath, C. Boone, L.J. Donner, A. Eldering, G. Lesins, R.V. Martin, B.M. Sinnhuber and K. Walker, 2006a, Testing convective parameterizations with tropical measurements of HNO_3, CO, H_2O, and O_3: Implications for the water vapor budget, *J. Geophys. Res.-Atmospheres* **111**, doi: 1410. 1029/2006jd007325.

Folkins, I. P. Bernath, C. Boone, G. Lesins, N. Livesey, A.M. Thompson, K. Walker and J.C. Witte, 2006b, Seasonal cycles of O_3, CO, and convective outflow at the tropical tropopause, *Geophys. Res. Lett.* **33**, 510. doi: 1029/2006gl026602.

Franke, K. Richter, A. Bovensmann, H. Eyring, V. Jöckel, P. and J. P. Burrows, 2009, Ship emitted NO_2 in the Indian Ocean: comparison of model results with satellite data, *Atmos. Chem. Phys.* **9**, 7289–7301.

Frankenberg, C. J.F. Meirink, M. van Weele, U. Platt and T. Wagner, 2005a, Assessing methane emissions from global space-borne observations, *Science* **308**, 1010–1014.

Frankenberg, C. U. Platt and T. Wagner, 2005b, Retrieval of CO from SCIAMACHY onboard ENVISAT: detection of strongly polluted areas and seasonal patterns in global CO abundances, *Atmos. Chem. and Phys.* **5**, 1639–1644.

Frankenberg, C. J.F. Meirink, P. Bergamaschi, A.P.H. Goede, M. Heimann, S. Korner, U. Platt, M. van Weele and T. Wagner, 2006, Satellite chartography of atmospheric methane from SCIAMACHY on board ENVISAT: Analysis of the years 2003 and 2004, *J. Geophys. Res. -Atmospheres* **111**, D07303.

Frankenberg, C. T. Warneke, A. Butz, I. Aben, F. Hase, P. Spietz and L.R. Brown, 2008, Pressure broadening in the $2\upsilon_3$ band of methane and its implication on atmospheric retrievals, *Atmos. Chem. Phys.* **8**, 5061–5075.

Frankenberg C, Yoshimura K, Warneke T, Aben I, Butz A, Deutscher N, Griffith D, Hase F, Notholt J, Schneider M, Schrijver H, Röckmann T. 2009, Dynamic processes governing lower-tropospheric HDO/H_2O ratios as observed from space and ground, *Science*. **325**(5946):1374–7.

Friess, U. T. Wagner, I. Pundt, K. Pfeilsticker and U. Platt, 2001, Spectroscopic measurements of tropospheric iodine oxide at Neumayer Station, Antarctica, *Geophys. Res. Lett.* **28**, 1941–1944.

Froidevaux, L. Y.B. Jiang, A. Lambert, N.J. Livesey, W.G. Read, J.W. Waters, E.V. Browell, J.W. Hair, M.A. Avery, T.J. McGee, L.W. Twigg, G.K. Sumnicht, K.W. Jucks, J.J. Margitan, B. Sen, R.A. Stachnik, G.C. Toon, P.F. Bernath, C.D. Boone, K.A. Walker, M.J. Filipiak, R.S. Harwood, R.A. Fuller, G.L. Manney, M.J. Schwartz, W.H. Daffer, B.J. Drouin, R.E. Cofield, D.T. Cuddy, R.F. Jarnot, B.W. Knosp, V.S. Perun, W.V. Snyder, P.C. Stek, R.P. Thurstans and P.A. Wagner, 2008, Validation of Aura Microwave Limb Sounder stratospheric ozone measurements, *J. Geophys. Res.-Atmospheres* **113**, doi: 2410.1029/2007jd008771.

Fu, T.M. D.J. Jacob, P.I. Palmer, K. Chance, Y.X.X. Wang, B. Barletta, D.R. Blake, J.C. Stanton and M.J. Pilling, 2007, Space-based formaldehyde measurements as constraints on volatile organic compound emissions in east and south Asia and implications for ozone, *J. Geophys. Res.-Atmospheres* **112**, D06312.

Fu, T.M. D.J. Jacob, F. Wittrock, J.P. Burrows, M. Vrekoussis and D.K. Henze, 2008, Global budgets of atmospheric glyoxal and methylglyoxal, and implications for formation of secondary organic aerosols, *J. Geophys. Res.-Atmospheres* **113**, 15303, doi: 1710.1029/2007jd009505.

Geer, A.J. W.A. Lahoz, S. Bekki, N. Bormann, Q. Errera, H.J. Eskes, D. Fonteyn, D.R. Jackson, M.N. Juckes, S. Massart, V.H. Peuch, S. Rharmili and A. Segers, 2006, The ASSET intercomparison of ozone analyses: method and first results, *Atmos. Chem. and Phys.* **6**, 5445–5474.

Generoso, S. I. Bey, J.L. Attie and F.M. Breon, 2007, A satellite- and model-based assessment of the 2003 Russian fires: Impact on the Arctic region, *J. Geophys. Res.-Atmospheres* **112**, D15302.

Glatthor, N. T. von Clarmann, H. Fischer, B. Funke, U. Grabowski, M. Hopfner, S. Kellmann, A. Linden, M. Milz, T. Steck and G.P. Stiller, 2007, Global peroxyacetyl nitrate (PAN) retrieval in the upper troposphere from limb emission spectra of the Michelson Interferometer for Passive Atmospheric Sounding (MIPAS), *Atmos. Chem. and Phys.* **7**, 2775–2787.

Gloudemans, A.M.S. H. Schrijver, Q. Kleipool, M.M.P. van den Broek, A.G. Straume, G. Lichtenberg, R.M. van Hees, I. Aben and J.F. Meirink, 2005, The impact of SCIAMACHY near-infrared instrument calibration on CH_4 and CO total columns, *Atmos. Chem. and Phys.* **5**, 2369–2383.

Gloudemans, A.M.S. M.C. Krol, J.F. Meirink, A.T.J. de Laat, G.R. van der Werf, H. Schrijver, M.M.P. van den Broek and I. Aben, 2006, Evidence for long-range transport of carbon monoxide in the Southern Hemisphere from SCIAMACHY observations, *Geophys. Res. Lett.* **33**, doi: 510.1029/2006gl026804.

Gloudemans, A.M.S. H. Schrijver, O.P. Hasekamp and I. Aben, 2008, Error analysis for CO and CH_4 total column retrievals from SCIAMACHY 2.3 μm spectra, *Atmos. Chem. Phys.* **8**, 3999–4017.

Gros, V. J. Williams, M.G. Lawrence, R. von Kuhlmann, J. van Aardenne, E. Atlas, A. Chuck, D.P. Edwards, V. Stroud and M. Krol, 2004, Tracing the origin and ages of interlaced atmospheric pollution events over the tropical Atlantic Ocean with *in situ* measurements, satellites, trajectories, emission inventories, and global models, *J. Geophys. Res.-Atmospheres* **109**, D22306.

Guerova, G. I. Bey, J.L. Attie, R.V. Martin, J. Cui and M. Sprenger, 2006, Impact of transatlantic transport episodes on summertime ozone in Europe, *Atmos. Chem. and Phys.* **6**, 2057–2072.

Guffanti, M. J.W. Ewert, G.M. Gallina, G.J.S. Bluth and G.L. Swanson, 2005, Volcanic-ash hazard to aviation during the 2003–2004 eruptive activity of Anatahan volcano, Commonwealth of the Northern Mariana Islands, *J. Volcanology and Geothermal Res.* **146**, 241–255.

Guo, S. G.J.S. Bluth, W.I. Rose, I.M. Watson and A.J. Prata, 2004, Re-evaluation of SO_2 release of the 15 June 1991 Pinatubo eruption using ultraviolet and infrared satellite sensors, *Geochemistry Geophysics Geosystems* **5**, Q04001.

Gurevich, G.S. and A.J. Krueger, 1997, Optimization of TOMS wavelength channels for ozone and sulfur dioxide retrievals, *Geophys. Res. Lett.* **24**, 2187–2190.

Hamazaki, T. A. Kuze and K. Kondo, 2004, Sensor system for greenhouse gas observing satellite (GOSAT). In: Strojnik, M. Editor, *12th Conference on Infrared Spaceborne Remote Sensing*, Spie-Int Soc Optical Engineering, Denver, CO, 275–282.

Han, K.M. C.H. Song, H.J. Ahn, R.S. Park, J.H. Woo, C.K. Lee, A. Richter, J.P. Burrows, J.Y. Kim and J.H. Hong, 2009, Investigation of NOx emissions and NOx- related chemistry in East Asia using CMAQ-predicted and GOME-derived NO_2 columns, *Atmos. Chem. and Phys.* **9**, 1017–1036.

Hansen, G. K. Bramstedt, V. Rozanov, M. Weber and J.P. Burrows, 2003, Validation of GOME ozone profiles by means of the ALOMAR ozone lidar, *Annales Geophysicae* **21**, 1879–1886.

Hayn, M. S. Beirle, F. A. Hamprecht, U. Platt, B. H. Menze and T. Wagner, 2009, Analysing spatio-temporal patterns of the global NO_2-distribution retrieved from GOME satellite observations using a generalized additive model, *Atmos. Chem. Phys.* **9**, 6459–6477.

He, Y. I. Uno, Z. Wang, T. Ohara, N. Sugirnoto, A. Shimizu, A. Richter and J.P. Burrows, 2007, Variations of the increasing trend of tropospheric NO_2 over central east China during the past decade, *Atmos. Environ.* **41**, 4865–4876.

Heald, C.L. D.J. Jacob, A.M. Fiore, L.K. Emmons, J.C. Gille, M.N. Deeter, J. Warner, D.P. Edwards, J.H. Crawford, A.J. Hamlin, G.W. Sachse, E.V. Browell, M.A. Avery, S.A. Vay, D.J. Westberg, D.R. Blake, H.B. Singh, S.T. Sandholm, R.W. Talbot and H.E. Fuelberg, 2003, Asian outflow and trans-Pacific transport of carbon monoxide and ozone pollution: An integrated satellite, aircraft, and model perspective, *J. Geophys. Res.-Atmospheres* **108**, 4804.

Heald, C.L. D.J. Jacob, D.B.A. Jones, P.I. Palmer, J.A. Logan, D.G. Streets, G.W. Sachse, J.C. Gille, R.N. Hoffman and T. Nehrkorn, 2004, Comparative inverse analysis of satellite (MOPITT) and aircraft (TRACE-P) observations to estimate Asian sources of carbon monoxide, *J. Geophys. Res. Atmospheres* **109**, D23306.

Hebestreit, K. J. Stutz, D. Rosen, V. Matveiv, M. Peleg, M. Luria and U. Platt, 1999, DOAS measurements of tropospheric bromine oxide in mid-latitudes, *Science* **283**, 55–57.

Hegels, E. P.J. Crutzen, T. Klupfel, D. Perner and J.P. Burrows, 1998, Global distribution of atmospheric bromine-monoxide from GOME on earth observing satellite ERS-2, *Geophys. Res. Lett.* **25**, 3127–3130.

Hegglin, M.I. C.D. Boone, G.L. Manney, T.G. Shepherd, K.A. Walker, P.F. Bernath, W.H. Daffer, P. Hoor and C. Schiller, 2008, Validation of ACE-FTS satellite data in the upper troposphere/lower stratosphere (UTLS) using non-coincident measurements, *Atmos. Chem. and Phys.* **8**, 1483–1499.

Herbin, H. D. Hurtmans, S. Turquety, C. Wespes, B. Barret, J. Hadji-Lazaro, C. Clerbaux and P.F. Coheur, 2007, Global distributions of water vapour isotopologues retrieved from IMG/ADEOS data, *Atmos. Chem. Phys.* **7**, 3957–3968.

Herbin, H. D. Hurtmans, L. Clarisse, S. Turquety, C. Clerbaux, C.P. Rinsland, C. Boone, P.F. Bernath and P.F. Coheur, 2009, Distributions and seasonal variations of tropospheric ethene (C_2H_4) from Atmospheric Chemistry Experiment (ACE-FTS) solar occultation spectra, *Geophys. Res. Lett.* **36**, doi: 510.1029/2008gl036338.

Heue, K.P. A. Richter, M. Bruns, J.P. Burrows, C. von Friedeburg, U. Platt, I. Pundt, P. Wang and T. Wagner, 2005, Validation of SCIAMACHY tropospheric NO_2-columns with AMAXDOAS measurements, *Atmos. Chem. and Phys.* **5**, 1039–1051.

Hild, L. A. Richter, V. Rozanov and J.P. Burrows, 2002, Air mass factor calculations for GOME measurements of lightning-produced NO_2. *Adv. Space Res.* **29**, 1685–1690.

Ho, S.P. D.P. Edwards, J.C. Gille, J.M. Chen, D. Ziskin, G.L. Francis, M.N. Deeter and J.R. Drummond, 2005, Estimates of 4.7 Mm surface emissivity and their impact on the retrieval of tropospheric carbon monoxide by Measurements of Pollution in the Troposphere (MOPITT), *J. Geophys. Res. Atmospheres* **110**, D21308.

Hoffmann, L. R. Spang, M. Kaufmann and M. Riese, 2005, Retrieval of CFC-11 and CFC-12 from ENVISAT MIPAS observations by means of rapid radiative transfer calculations. *Atmospheric Remote Sensing: Earth's Surface, Troposphere, Stratosphere and Mesosphere – I*, Elsevier Science Ltd, Oxford, 915–921.

Hoffmann, L. M. Kaufmann, R. Spang, R. Muller, J.J. Remedios, D.P. Moore, C.M. Volk, T. von Clarmann and M. Riese, 2008, ENVISAT MIPAS measurements of CFC-11: retrieval, validation, and climatology, *Atmos. Chem. Phys.* **8**, 3671–3688.

Hollwedel, J. M. Wenig, S. Beirle, S. Kraus, S. Kuhl, W. Wilms-Grabe, U. Platt and T. Wagner, 2004, Year-to-year variations of spring time polar tropospheric BrO as seen by GOME. *Adv. Space Res.* **34**, 804–808.

Hoogen, R. V.V. Rozanov and J.P. Burrows, 1999, Ozone profiles from GOME satellite data: Algorithm description and first validation, *J. Geophys. Res.-Atmospheres* **104**, 8263–8280.

Hudson, R.D. J.H. Kim and A.M. Thompson, 1995, On the Derivation of Tropospheric Column Ozone from Radiances Measured by the Total Ozone Mapping Spectrometer, *J. Geophys. Res.-Atmospheres* **100**, 11137–11145.

Hudson, R.D. and A.M. Thompson, 1998, Tropical tropospheric ozone from total ozone mapping spectrometer by a modified residual method, *J. Geophys. Res.-Atmospheres* **103**, 22129–22145.

Hurrell, J.W. 1995, Decadal Trends in the North-Atlantic Oscillation - Regional Temperatures and Precipitation, *Science* **269**, 676–679.

Hyer, E.J. D.J. Allen and E.S. Kasischke, 2007a, Examining injection properties of boreal forest fires using surface and satellite measurements of CO transport, *J. Geophys. Res.-Atmospheres* **112**, D18307.

Hyer, E.J. E.S. Kasischke and D.J. Allen, 2007b, Effects of source temporal resolution on transport simulations of boreal fire emissions, *J. Geophys. Res.-Atmospheres* **112**, D01302.

Iapaolo, M. S. Godin-Beekmann, F. Del Frate, S. Casadio, M. Petitdidier, I.S. McDermid, T. Leblanc, D. Swart, Y. Meijer, G. Hansen and K. Stebel, 2007, Gome ozone profiles retrieved by neural network techniques: A global validation with lidar measurements, *J. Quant. Spectros. and Rad. Trans.* **107**, 105–119.

Ionov, D.V. V.P. Sinyakov and V.K. Semenov, 2006, Validation of GOME (ERS-2) NO_2 vertical column data with ground-based measurements at Issyk-Kul (Kyrgyzstan). *Adv. Sapce Res.* **37**, 2254–2260.

Irie, H. K. Sudo, H. Akimoto, A. Richter, J.P. Burrows, T. Wagner, M. Wenig, S. Beirle, Y. Kondo, V.P. Sinyakov and F. Goutail, 2005, Evaluation of long-term tropospheric NO_2 data obtained by GOME over East Asia in 1996–2002, *Geophys. Res. Lett.* **32**, L11810.

Ito, A. A. Ito, and H. Akimoto, 2007, Seasonal and interannual variations in CO and BC emissions from open biomass burning in Southern Africa during 1998–2005, Global Biogeochem Cycles **21**, Gb2011.

Jacob, D.J. J.H. Crawford, M.M. Kleb, V.S. Connors, R.J. Bendura, J.L. Raper, G.W. Sachse, J.C. Gille, L. Emmons and C.L. Heald, 2003, Transport and Chemical Evolution over the Pacific (TRACE-P) aircraft mission: Design, execution, and first results, *J. Geophys. Res.-Atmospheres* **108**, 1–19.

Jacob, D.J. B.D. Field, Q.B. Li, D.R. Blake, J. de Gouw, C. Warneke, A. Hansel, A. Wisthaler, H.B. Singh and A. Guenther, 2005, Global budget of methanol: Constraints from atmospheric observations, *J. Geophys. Res.-Atmospheres* **110**, D08303.

Jacobi, H.W. L. Kaleschke, A. Richter, A. Rozanov and J.P. Burrows, 2006, Observation of a fast ozone loss in the marginal ice zone of the Arctic Ocean, *J. Geophys. Res.-Atmospheres* **111**, D15309.

Jaegle, L. R.V. Martin, K. Chance, L. Steinberger, T.P. Kurosu, D.J. Jacob, A.I. Modi, V. Yoboue, L. Sigha-Nkamdjou and C. Galy-Lacaux, 2004, Satellite mapping of rain-induced nitric oxide emissions from soils, *J. Geophys. Res.-Atmospheres* **109**, D21310.

Jaegle, L. L. Steinberger, R.V. Martin and K. Chance, 2005, Global partitioning of NO_x sources using satellite observations: Relative roles of fossil fuel combustion, biomass burning and soil emissions, *Faraday Discussions* **130**, 407–423.

Jourdain, L. H.M. Worden, J.R. Worden, K. Bowman, Q. Li, A. Eldering, S.S. Kulawik, G. Osterman, K.F. Boersma, B. Fisher, C.P. Rinsland, R. Beer and M. Gunson, 2007, Tropospheric vertical distribution of tropical Atlantic ozone observed by TES during the northern African biomass burning season, *Geophys. Res. Lett.* **34**, L04810.

Kaleschke, L. A. Richter, J. Burrows, O. Afe, G. Heygster, J. Notholt, A.M. Rankin, H.K. Roscoe, J. Hollwedel, T. Wagner and H.W. Jacobi, 2004, Frost flowers on sea ice as a source of sea salt and their influence on tropospheric halogen chemistry, *Geophys. Res. Lett.* **31**, doi: 410.1029/2004gl020655.

Kampe, T.U. and I.N. Sokolik, 2007, Remote sensing retrievals of fine mode aerosol optical depth and impacts on its correlation with CO from biomass burning, *Geophys. Res. Lett.* **34**, L12806.

Kar, J. H. Bremer, J.R. Drummond, Y.J. Rochon, D.B.A. Jones, F. Nichitiu, J. Zou, J. Liu, J.C. Gille, D.P. Edwards, M.N. Deeter, G. Francis, D. Ziskin and J. Warner, 2004, Evidence of vertical transport of carbon monoxide from Measurements of Pollution in the Troposphere (MOPITT), *Geophys. Res. Lett.* **31**, L23105.

Kar, J. J.R. Drummond, D.B.A. Jones, J. Liu, F. Nichitiu, J. Zou, J.C. Gille, D.P. Edwards and M.N. Deeter, 2006, Carbon monoxide (CO) maximum over the Zagros mountains in the Middle East: Signature of mountain venting?, *Geophys. Res. Lett.* **33**, L15819.

Kar, J. D.B.A. Jones, J.R. Drummond, J.L. Attie, J. Liu, J. Zou, F. Nichitiu, M.D. Seymour, D.P. Edwards, M.N. Deeter, J.C. Gille and A. Richter, 2008, Measurement of low-altitude CO over the Indian subcontinent by MOPITT, *J. Geophys. Res.-Atmospheres* **113**, doi: 1310.1029/2007jd009362.

Kaynak, B. Y. Hu, R.V. Martin, A.G. Russell, and C.E. Sioris, 2009, Comparison of weekly cycle of NO_2 satellite retrievals and NOx emission inventories for the continental U.S. *J. Geophys. Res.* **114**, 05302, doi: 10.1029/2008JD010714.

Keating, T. and A. Zuber, 2007, Hemispheric Transport of Air Pollution 2007. *Air Pollution Studies No 16*, UNECE, Geneva.

Keeling, C.D. R.B. Bacastow, A.E. Bainbridge, C.A. Ekdahl, P.R. Guenther, L.S. Waterman and J.F.S. Chin, 1976, Atmospheric Carbon-Dioxide Variations at Mauna-Loa Observatory, Hawaii, *Tellus* **28**, 538–551.

Khlystova, I. M. Buchwitz, J.P. Burrows, H. Bovensmann and D. Fowler, 2009, Carbon monoxide spatial gradients over source regions as observed by SCIAMACHY: A case study for the United Kingdom, *Adv. Space Res.* **43**, 923-929, doi: 10.1016/j.asr.2008.10.012.

Khokhar, M.F. C. Frankenberg, M. Van Roozendael, S. Beirle, S. Kuhl, A. Richter, U. Platt and T. Wagner, 2005, Satellite observations of atmospheric SO_2 from volcanic eruptions during the time-period of 1996–2002. *Adv Space Res*, **36** 879–887.

Khokhar, M.F. U. Platt and T. Wagner, 2008, Temporal trends of anthropogenic SO_2 emitted by non-ferrous metal smelters in Peru and Russia estimated from Satellite observations, *Atmos. Chem. Phys. Discuss.* **8**, 17393–17422.

Kim, J.H. R.D. Hudson and A.M. Thompson, 1996, A new method of deriving time-averaged tropospheric column ozone over the tropics using total ozone mapping spectrometer (TOMS) radiances: Intercomparison and analysis using TRACE A data, *J. Geophys. Res.-Atmospheres* **101**, 24317–24330.

Kim, J.H. S. Na, M.J. Newchurch and R.V. Martin, 2005, Tropical tropospheric ozone morphology and seasonality seen in satellite and *in situ* measurements and model calculations, *J. Geophys. Res.-Atmospheres* **110**, D02303.

Kim, S.W. A. Heckel, S.A. McKeen, G.J. Frost, E.Y. Hsie, M.K. Trainer, A. Richter, J.P. Burrows, S.E. Peckham and G.A. Grell, 2006, Satellite-observed US power plant NOx emission reductions and their impact on air quality, *Geophys. Res. Lett.* **33**, L22812.

Kim, S.W. A. Heckel, G.J. Frost, A. Richter, J. Gleason, J.P. Burrows, S. McKeen, E.Y. Hsie, C. Granier and M. Trainer, 2009, NO_2 columns in the western United States observed from space and simulated by a regional chemistry model and their implications for NOx emissions, *J. Geophys. Res.-Atmospheres* **114**, doi: 2910.1029/2008jd011343.

Kinnison, D.E. J. Gille, J. Barnett, C. Randall, V.L. Harvey, A. Lambert, R. Khosravi, M.J. Alexander, P.F. Bernath, C.D. Boone, C. Cavanaugh, M. Coffey, C. Craig, V.C. Dean, T. Eden, D. Ellis, D.W. Fahey, G. Francis, C. Halvorson, J. Hannigan, C. Hartsough, C. Hepplewhite, C. Krinsky, H. Lee, B. Mankin, T.P. Marcy, S. Massie, B. Nardi, D. Packman, P.J. Popp, M.L. Santee, V. Yudin and K.A. Walker, 2008, Global observations of HNO_3 from the High Resolution Dynamics Limb Sounder (HIRDLS): First results, *J. Geophys. Res.-Atmospheres* **113**, doi: 2110.1029/2007jd008814.

Klaes, K.D. M. Cohen, Y. Buhler, P. Schlussel, R. Munro, J.P. Luntama, A. von Engelin, E.O. Clerigh, H. Bonekamp, J. Ackermann and J. Schmetz, 2007, An introduction to the EUMETSAT Polar System, *Bull. Amer. Met. Soc.* **88**, 1085, doi: 10.1175/bams-88-7-1085.

Konovalov, I.B. M. Beekmann, R. Vautard, J.P. Burrows, A. Richter, H. Nuss and N. Elansky, 2005, Comparison and evaluation of modelled and GOME measurement derived tropospheric NO_2 columns over Western and Eastern Europe, *Atmos. Chem. and Phys.* **5**, 169–190.

Konovalov, I.B. M. Beekmann, A. Richter and J.P. Burrows, 2006, Inverse modelling of the spatial distribution of NOx emissions on a continental scale using satellite data, *Atmos. Chem. and Phys.* **6**, 1747–1770.

Konovalov, I.B. M. Beekmann, J.P. Burrows and A. Richter, 2008, Satellite measurement based estimates of decadal changes in European nitrogen oxides emissions, *Atmos. Chem. and Phys.* **8**, 2623–2641.

Kramer, L.J. R.J. Leigh, J.J. Remedios and P.S. Monks, 2008, Comparison of OMI and ground based *in situ* and MAX-DOAS measurements of tropospheric nitrogen dioxide in an urban area, *J. Geophys. Res.-Atmospheres* **113**, D16S39.

Krotkov, N.A. A.J. Krueger and P.K. Bhartia, 1997, Ultraviolet optical model of volcanic clouds for remote sensing of ash and sulfur dioxide, *J. Geophys. Res.-Atmospheres* **102**, 21891–21904.

Krotkov, N.A. S.A. Carn, A.J. Krueger, P.K. Bhartia and K. Yang, 2006, Band residual difference algorithm for retrieval of SO_2 from the Aura Ozone Monitoring Instrument (OMI), *Ieee Transactions on Geoscience and Remote Sensing* **44**, 1259–1266.

Krotkov, N.A. B. McClure, R.R. Dickerson, S.A. Carn, C. Li, P.K. Bhartia, K. Yang, A.J. Krueger, Z.Q. Li, P.F. Levelt, H.B. Chen, P.C. Wang and D.R. Lu, 2008, Validation of SO_2 retrievals from the Ozone Monitoring Instrument over NE China, *J. Geophys. Res.-Atmospheres* **113**, doi: 1310.1029/2007jd008818.

Krueger, A.J. 1983, Sighting of El-Chichon Sulfur-Dioxide Clouds with the Nimbus-7 Total Ozone Mapping Spectrometer, *Science* **220**, 1377–1379.

Krueger, A.J. L.S. Walter, P.K. Bhartia, C.C. Schnetzler, N.A. Krotkov, I. Sprod and G.J.S. Bluth, 1995, Volcanic Sulfur-Dioxide Measurements from the Total Ozone Mapping Spectrometer Instruments, *J. Geophys. Res.-Atmospheres* **100**, 14057–14076.

Kunhikrishnan, T. M.G. Lawrence, R. von Kuhlmann, A. Richter, A. Ladstatter-Weissenmayer and J.P. Burrows, 2004a, Analysis of tropospheric NOx over Asia using the model of atmospheric transport and chemistry (MATCH-MPIC) and GOME-satellite observations, *Atmos.Environ.* **38**, 581–596.

Kunhikrishnan, T. M.G. Lawrence, R. von Kuhlmann, A. Richter, A. Ladstatter-Weissenmayer and J.P. Burrows, 2004b, Semiannual NO_2 plumes during the monsoon transition periods over the central Indian Ocean, *Geophys. Res. Lett.* **31**, L08110.

Ladstatter-Weissenmayer, A. and J.P. Burrows, 1998, Biomass burning over Indonesia, *Earth Observation Quarterly* **58**.

Ladstatter-Weissenmayer, A. J. Heland, R. Kormann, R. von Kuhlmann, M.G. Lawrence, J. Meyer-Arnek, A. Richter, F. Wittrock, H. Ziereis and J.P. Burrows, 2003, Transport and build-up of tropospheric trace gases during the MINOS campaign: comparision of GOME, *in situ* aircraft measurements and MATCH-MPIC-data, *Atmos. Chem. and Phys.* **3**, 1887–1902.

Ladstatter-Weissenmayer, A. J. Meyer-Arbek, A. Schlemm and J.P. Burrows, 2004, Influence of stratospheric airmasses on tropospheric vertical O_3 columns based on GOME (Global Ozone Monitoring Experiment) measurements and backtrajectory calculation over the Pacific, *Atmos. Chem. and Phys.* **4**, 903–909.

Ladstatter-Weissenmayer, A. H. Altmeyer, M. Bruns, A. Richter, A. Rozanov, V. Rozanov, F. Wittrock and J.P. Burrows, 2007a, Measurements of O_3, NO_2 and BrO during the INDOEX campaign using ground based DOAS and GOME satellite data, *Atmos. Chem. and Phys.* **7**, 283–291.

Ladstatter-Weissenmayer, A. M. Kanakidou, J. Meyer-Arnek, E.V. Dermitzaki, A. Richter, M. Vrekoussis, F. Wittrock and J.P. Burrows, 2007c, Pollution events over the East Mediterranean: Synergistic use of GOME, ground-based and sonde observations and models, *Atmos. Environ.* **41**, 7262–7273, doi: 10.1016/j.atmosenv.2007.05.031.

Lamarque, J.F. B.V. Khattatov, J.C. Gille and G.P. Brasseur, 1999, Assimilation of Measurement of Air Pollution from Space (MAPS) CO in a global three-dimensional model, *J. Geophys. Res.-Atmospheres* **104**, 26209–26218.

Lamarque, J.F. D.P. Edwards, L.K. Emmons, J.C. Gille, O. Wilhelmi, C. Gerbig, D. Prevedel, M.N. Deeter, J. Warner, D.C. Ziskin, B. Khattatov, G.L. Francis, V. Yudin, S. Ho, D. Mao, J. Chen and J.R. Drummond, 2003, Identification of CO plumes from MOPITT data: Application to the August 2000 Idaho-Montana forest fires, *Geophys. Res. Lett.* **30**, 1688.

Lamsal, L.N. M. Weber, S. Tellmann and J.P. Burrows, 2004, Ozone column classified climatology of ozone and temperature profiles based on ozonesonde and satellite data, *J. Geophys. Res.-Atmospheres* **109**, doi: 1510.1029/2004jd004680.

Lamsal, L.N. R.V. Martin, A. van Donkelaar, M. Steinbacher, E.A. Celarier, E. Bucsela, E.J. Dunlea, and J. Pinto, 2008, Ground-level nitrogen dioxide concentrations inferred from the satellite-borne Ozone Monitoring Instrument, *J. Geophys. Res.* **113**, D16308, doi: 10.1029/2007JD009235.

Lang, R. A.N. Maurellis, W.J. van der Zande, I. Aben, J. Landgraf and W. Ubachs, 2002, Forward modeling and retrieval of water vapor from the Global Ozone Monitoring Experiment: Treatment of narrowband absorption spectra, *J. Geophys. Res.-Atmospheres* **107**, doi: 2310.1029/2001jd001453.

Lang, R. S. Casadio, A.N. Maurellis and M.G. Lawrence, 2007, Evaluation of the GOME water vapor climatology 1995-2002, *J. Geophys. Res.-Atmospheres* **112**, D17110.

Lauer, A. M. Dameris, A. Richter and J.P. Burrows, 2002, Tropospheric NO2 columns: a comparison between model and retrieved data from GOME measurements, *Atmos. Chem. and Phys.* **2**, 67–78.

Lawrence, M.G. O. Hov, M. Beeknann, J. Brandt, H. Elbern, H. Eskes, H. Feichter and M. Takigawa, 2005, The chemical weather, *Environ. Chem.* **2**, 6–8, doi: 10.1071/en05014.

Lee, C. A. Richter, H. Lee, Y.J. Kim, J.P. Burrows, Y.G. Lee and B.C. Choi, 2008, Impact of transport of sulfur dioxide from the Asian continent on the air quality over Korea during May 2005, *Atmos. Environ.* **42**, 1461–1475.

Lee, S. G.H. Choi, H.S. Lim, J.H. Lee, K.H. Lee, Y.J. Kim and J. Kim, 2005, Detection of Russian fires using MOPITT and MODIS data. *On the Convergence of Bio-Information-, Environmental-, Energy-, Space- and Nano-Technologies, Pts 1 and 2*, 816–823.

Lerdau, M.T. L.J. Munger and D.J. Jacob, 2000, Atmospheric chemistry – The NO_2 flux conundrum, *Science* **289**, 2291.

Leue, C. M. Wenig, T. Wagner, O. Klimm, U. Platt and B. Jahne, 2001, Quantitative analysis of NOx emissions from Global Ozone Monitoring Experiment satellite image sequences, *J. Geophys. Res.-Atmospheres* **106**, 5493–5505.

Li, Q.B. D.J. Jacob, R. Park, Y.X. Wang, C.L. Heald, R. Hudman, R.M. Yantosca, R.V. Martin and M. Evans, 2005, North American pollution outflow and the trapping of convectively lifted pollution by upper-level anticyclone, *J. Geophys. Res.-Atmospheres* **110**, D10301.

Liang, Q. L. Jaegle, R.C. Hudman, S. Turquety, D.J. Jacob, M.A. Avery, E.V. Browell, G.W. Sachse, D.R. Blake, W. Brune, X. Ren, R.C. Cohen, J.E. Dibb, A. Fried, H. Fuelberg, M. Porter, B.G. Heikes, G. Huey, H.B. Singh and P.O. Wennberg, 2007, Summertime influence of Asian pollution in the free troposphere over North America, *J. Geophys. Res.-Atmospheres* **112**, D12s11.

Liggio, J. S.M. Li and R. McLaren, 2005a, Heterogeneous reactions of glyoxal on particulate matter: Identification of acetals and sulfate esters, *Environ. Sci. and Tech.* **39**, 1532–1541.

Liggio, J. S.M. Li and R. McLaren, 2005b, Reactive uptake of glyoxal by particulate matter, *J. Geophys. Res.-Atmospheres* **110**, doi: 10.1029/2004jd005113.

Lin, Y.P. C.S. Zhao, L. Peng and Y.Y. Fang, 2007, A new method to calculate monthly CO emissions using MOPITT satellite data, *Chinese Science Bulletin* **52**, 2551–2558.

Liu, J. J.R. Drummond, Q.B. Li, J.C. Gille and D.C. Ziskin, 2005, Satellite mapping of CO emission from forest fires in Northwest America using MOPITT measurements, *Remote Sensing of Environment* **95**, 502–516.

Liu, J. J.R. Drummond, D.B.A. Jones, Z. Cao, H. Bremer, J. Kar, J. Zou, F. Nichitiu and J.C. Gille, 2006a, Large horizontal gradients in atmospheric CO at the synoptic scale as seen by spaceborne Measurements of Pollution in the Troposphere, *J. Geophys. Res.-Atmospheres* **111**, D02306.

Liu, X. K. Chance, C.E. Sioris, T.P. Kurosu, R.J.D. Spurr, R.V. Martin, T.M. Fu, J.A. Logan, D.J. Jacob, P.I. Palmer, M.J. Newchurch, I.A. Megretskaia and R.B. Chatfield, 2006b, First directly retrieved global distribution of tropospheric column ozone from GOME: Comparison with the GEOS-CHEM model, *J. Geophys. Res.-Atmospheres* **111**, doi: 1710.1029/2005jd006564.

Liu, X. K. Chance and T.P. Kurosu, 2007, Improved ozone profile retrievals from GOME data with degradation correction in reflectance, *Atmos. Chem. and Phys.* **7**, 1575–1583.

Livesey, N.J. M.J. Filipiak, L. Froidevaux, W.G. Read, A. Lambert, M.L. Santee, J.H. Jiang, H.C. Pumphrey, J.W. Waters, R.E. Cofield, D.T. Cuddy, W.H. Daffer, B.J. Drouin, R.A. Fuller, R.F. Jarnot, Y.B. Jiang, B.W. Knosp, Q.B. Li, V.S. Perun, M.J. Schwartz, W.V. Snyder, P.C. Stek, R.P. Thurstans, P.A. Wagner, M. Avery, E.V. Browell, J.P. Cammas, L.E. Christensen, G.S. Diskin, R.S. Gao, H.J. Jost, M. Loewenstein, J.D. Lopez, P. Nedelec, G.B. Osterman, G.W. Sachse and C.R. Webster, 2008, Validation of Aura Microwave Limb Sounder O_3 and CO observations in the upper troposphere and lower stratosphere, *J. Geophys. Res.-Atmospheres* **113**, doi: 1510.1029/2007jd008805.

Logan, J.A. I. Megretskaia, R. Nassar, L.T. Murray, L. Zhang, K.W. Bowman, H.M. Worden and M. Luo, 2008, Effects of the 2006 El Nino on tropospheric composition as revealed by data from the Tropospheric Emission Spectrometer (TES), *Geophys. Res. Lett.* **35**, L03816.

Luo, M. C. Rinsland, B. Fisher, G. Sachse, G. Diskin, J. Logan, H. Worden, S. Kulawik, G. Osterman, A. Eldering, R. Herman and M. Shephard, 2007a, TES carbon monoxide validation with DACOM aircraft measurements during INTEX-B 2006, *J. Geophys. Res.-Atmospheres* **112**, D24s48.

Luo, M. C.P. Rinsland, C.D. Rodgers, J.A. Logan, H. Worden, S. Kulawik, A. Eldering, A. Goldman, M.W. Shephard, M. Gunson and M. Lampel, 2007b, Comparison of carbon monoxide measurements by TES and MOPITT: Influence of *a priori* data and instrument characteristics on nadir atmospheric species retrievals, *J. Geophys. Res.-Atmospheres* **112**, D09303.

Ma, J.Z. A. Richter, J.P. Burrows, H. Nuss and J.A. van Aardenne, 2006, Comparison of model-simulated tropospheric NO_2 over China with GOME-satellite data, *Atmospheric Environment* **40**, 593–604.

Maddy, E.S. C.D. Barnet, M. Goldberg, C. Sweeney and X. Liu, 2008, CO_2 retrievals from the Atmospheric Infrared Sounder: Methodology and validation, *J. Geophys. Res.-Atmospheres* **113**, doi: 10.1029/2007jd009402.

Mahieu, E. P. Duchatelet, P. Demoulin, K.A. Walker, E. Dupuy, L. Froidevaux, C. Randall, V. Catoire, K. Strong, C.D. Boone, P.F. Bernath, J.F. Blavier, T. Blumenstock, M. Coffey, M. De Maziere, D. Griffith, J. Hannigan, F. Hase, N. Jones, K.W. Jucks, A. Kagawa, Y. Kasai, Y. Mebarki, S. Mikuteit, R. Nassar, J. Notholt, C.P. Rinsland, C. Robert, O. Schrems, C. Senten, D. Smale, J. Taylor, C. Tactard, G.C. Toon, T. Warneke, S.W. Wood, R. Zander and C. Servais, 2008, Validation of ACE-FTS v2.2 measurements of HCl, HF, CCl_3F and CCl_2F2 using space-, balloon- and ground-based instrument observations, *Atmos. Chem. Phys.* **8**, 6199–6221.

Marbach, T. S. Beirle, U. Platt, P. Hoor, F. Wittrock, A. Richter, M. Vrekoussis, M. Grzegorski, J. P. Burrows, and T. Wagner, 2009, Satellite measurements of formaldehyde from shipping emissions, *Atmos. Chem. Phys.* **9**, 8223–8234.

Martin, R.V. K. Chance, D.J. Jacob, T.P. Kurosu, R.J.D. Spurr, E. Bucsela, J.F. Gleason, P.I. Palmer, I. Bey, A.M. Fiore, Q.B. Li, R.M. Yantosca and R.B.A. Koelemeijer, 2002, An improved retrieval of tropospheric nitrogen dioxide from GOME, *J. Geophys. Res. -Atmospheres* **107**, 4437.

Martin, R.V. D.J. Jacob, K. Chance, T.P. Kurosu, P.I. Palmer and M.J. Evans, 2003, Global inventory of nitrogen oxide emissions constrained by space-based observations of NO_2 columns, *J. Geophys. Res.-Atmospheres* **108**, 4537.

Martin, R.V. A.M. Fiore and A. Van Donkelaar, 2004a, Space-based diagnosis of surface ozone sensitivity to anthropogenic emissions, *Geophys. Res. Lett.* **31**, L06120.

Martin, R.V. D.D. Parrish, T.B. Ryerson, D.K. Nicks, K. Chance, T.P. Kurosu, D.J. Jacob, E.D. Sturges, A. Fried and B.P. Wert, 2004b, Evaluation of GOME satellite measurements of tropospheric NO_2 and HCHO using regional data from aircraft campaigns in the southeastern United States, *J. Geophys. Res.-Atmospheres* **109**, D24307.

Martin, R.V. B. Sauvage, I. Folkins, C.E. Sioris, C. Boone, P. Bernath and J. Ziemke, 2007, Space-based constraints on the production of nitric oxide by lightning, *J. Geophys. Res. -Atmospheres* **112**, D09309.

Martin, R.V. 2008, Satellite remote sensing of surface air quality, *Atmos. Environ.* **42**, 7823–7843, doi: 10.1016/j.atmosenv.2008.07.018.

Massie, S.T. O. Torres and S.J. Smith, 2004, Total Ozone Mapping Spectrometer (TOMS) observations of increases in Asian aerosol in winter from 1979 to 2000, *J. Geophys. Res. -Atmospheres* **109**, D18211.

Massie, S.T. J.C. Gille, D.P. Edwards and S. Nandi, 2006, Satellite observations of aerosol and CO over Mexico city, *Atmos. Environ.* **40**, 6019–6031.

Maurellis, A.N. R. Lang, W.J. van der Zande, I. Aben and W. Ubachs, 2000, Precipitable water column retrieval from GOME data, *Geophys. Res. Lett.* **27**, 903–906.

McMillan, W.W. C. Barnet, L. Strow, M.T. Chahine, M.L. McCourt, J.X. Warner, P.C. Novelli, S. Korontzi, E.S. Maddy and S. Datta, 2005, Daily global maps of carbon monoxide from NASA's Atmospheric Infrared Sounder, *Geophys. Res. Lett.* **32**, doi: 410.1029/2004gl021821.

Meijer, Y.J. D.P.J. Swart, F. Baier, P.K. Bhartia, G.E. Bodeker, S. Casadio, K. Chance, F. Del Frate, T. Erbertseder, M.D. Felder, L.E. Flynn, S. Godin-Beekmann, G. Hansen, O.P. Hasekamp, A. Kaifel, H.M. Kelder, B.J. Kerridge, J.C. Lambert, J. Landgraf, B. Latter, X. Liu, I.S. McDermid, Y. Pachepsky, V. Rozanov, R. Siddans, S. Tellmann, R.J. van der A, R.F. van Oss, M. Weber and C. Zehner, 2006, Evaluation of Global Ozone Monitoring Experiment (GOME) ozone profiles from nine different algorithms, *J. Geophys. Res.-Atmospheres* **111**, D21306.

Meyer-Arnek, J. A. Ladstatter-Weissenmayer, A. Richter, F. Wittrock and J.P. Burrows, 2005a, A study of the trace gas columns O_3, NO_2 and HCHO over Africa in September 1997, *Faraday Discussions* **130**, 387–405.

Mieruch, S. S. Noel, H. Bovensmann and J.P. Burrows, 2008, Analysis of global water vapour trends from satellite measurements in the visible spectral range, *Atmos. Chem. and Phys.* **8**, 491–504.

Mijling, B. R. J. van der A, K. F. Boersma, M. Van Roozendael, I. DeSmedt, and H. M. Kelder, 2009, Reductions of NO_2 Detected from Space During the 2008 Beijing Olympic Games, *Geophys. Res. Lett.* **36**, L13801, doi:10.1029/2009GL038943.

Millet, D.B. D.J. Jacob, S. Turquety, R.C. Hudman, S.L. Wu, A. Fried, J. Walega, B.G. Heikes, D.R. Blake, H.B. Singh, B.E. Anderson and A.D. Clarke, 2006, Formaldehyde distribution over North America: Implications for satellite retrievals of formaldehyde columns and isoprene emission, *J. Geophys. Res.-Atmospheres* **111**, D24s02.

Millet, D.B. D.J. Jacob, K.F. Boersma, T.M. Fu, T.P. Kurosu, K. Chance, C.L. Heald and A. Guenther, 2008, Spatial distribution of isoprene emissions from North America derived from formaldehyde column measurements by the OMI satellite sensor, *J. Geophys. Res. -Atmospheres* **113**, D02307.

Milz, M. T. von Clarmann, H. Fischer, N. Glatthor, U. Grabowski, M. Hopfner, S. Kellmann, M. Kiefer, A. Linden, G.M. Tsidu, T. Steck, G.P. Stiller, B. Funke, M. Lopez-Puertas and M.E. Koukouli, 2005, Water vapor distributions measured with the Michelson Interferometer for Passive Atmospheric Sounding on board Envisat (MIPAS/Envisat), *J. Geophys. Res. -Atmospheres* **110**, doi: 1410.1029/2005jd005973.

Molina, M.J. and L.T. Molina, 2004, Megacities and atmospheric pollution, *J. Air Waste Manage. Assoc.* **54**, 644–680.

Monks, P.S. 2005, Gas-phase radical chemistry in the troposphere, *Chem Soc Rev* **34**, 376–395.

Monks, P.S. C. Granier, S. Fuzzi, A. Stohl, M. Williams, H. Akimoto, M. Amann, A. Baklanov, U. Baltensperger, I. Bey, N. Blake, R.S. Blake, K. Carslaw, O.R. Cooper, F. Dentener, E. Fragkou, G. Frost, S. Generoso, P. Ginoux, V. Grewe, A. Guenther, H.C. Hansson, S. Henne, J. Hjorth, A. Hofzumahaus, H. Huntrieser, M.E. Jenkin, J. Kaiser, M. Kanakidou, Z. Klimont, M. Kulmala, M.G. Lawrence, J.D. Lee, C. Liousse, G. McFiggans, A. Metzger, A. Mieville, N. Moussiopoulos, J.J. Orlando, P.I. Palmer, D. Parrish, A. Petzold, U. Platt, U. Poeschl, A.S.H. Prévôt, C.E. Reeves, S. Reiman, Y. Rudich, K. Sellegri, R. Steinbrecher, D. Simpson, H.T. Brink, J. Theloke, G.V.D. Werf, R. Vautard, V. Vestreng, C. Vlachokostas and R. vonGlasow, 2009, Atmospheric Composition Change – Global and Regional Air Quality, *Atmos. Environ.* **43**, 5268–5350.

Moore, D.P. and J.J. Remedios, 2008, Growth rates of stratospheric HCFC-22, *Atmos. Chem. and Phys.* **8**, 73–82.

Morris, G.A. S. Hersey, A.M. Thompson, S. Pawson, J.E. Nielsen, P.R. Colarco, W.W. McMillan, A. Stohl, S. Turquety, J. Warner, B.J. Johnson, T.L. Kucsera, D.E. Larko, S.J. Oltmans and J.C. Witte, 2006, Alaskan and Canadian forest fires exacerbate ozone pollution over Houston, Texas, on 19 and 20 July 2004, *J. Geophys. Res.-Atmospheres* **111**, D24s03.

Muller, J.F. and T. Stavrakou, 2005, Inversion of CO and NOx emissions using the adjoint of the IMAGES model, *Atmos. Chem. and Phys.* **5**, 1157–1186.

Muller, M.D. A.K. Kaifel, M. Weber, S. Tellmann, J.P. Burrows and D. Loyola, 2003, Ozone profile retrieval from Global Ozone Monitoring Experiment (GOME) data using a neural network approach (Neural Network Ozone Retrieval System (NNORSY)), *J. Geophys. Res.-Atmospheres* **108**, 4497.

Munro, R. R. Siddans, W.J. Reburn and B.J. Kerridge, 1998, Direct measurement of tropospheric ozone distributions from space, *Nature* **392**, 168–171.

Munzenmayer, R. H. Bovensmann, B. Grafmuller, R. Mager and T. Reinert, 2008, Future geostationary earth observation systems, *JBIS-J. Br. Interplanet. Soc.* **61**, 78–90.

Nardi, B. J.C. Gille, J.J. Barnett, C.E. Randall, V.L. Harvey, A. Waterfall, W.J. Reburn, T. Leblanc, T.J. McGee, L.W. Twigg, A.M. Thompson, S. Godin-Beekmann, P.F. Bernath, B.R. Bojkov, C.D. Boone, C. Cavanaugh, M.T. Coffey, J. Craft, C. Craig, V. Dean, T.D. Eden, G. Francis, L. Froidevaux, C. Halvorson, J.W. Hannigan, C.L. Hepplewhite, D.E. Kinnison, R. Khosravi, C. Krinsky, A. Lambert, H. Lee, J. Loh, S.T. Massie, I.S. McDermid, D. Packman, B. Torpy, J. Valverde-Canossa, K.A. Walker, D.N. Whiteman, J.C. Witte and G. Young, 2008, Initial validation of ozone measurements from the High Resolution Dynamics Limb Sounder, *J. Geophys. Res.-Atmospheres* **113**, doi: 1810.1029/2007jd008837.

Nassar, R. J.A. Logan, H.M. Worden, I.A. Megretskaia, K.W. Bowman, G.B. Osterman, A.-M. Thompson, D.W. Tarasick, S. Austin, H. Claude, M.K. Dubey, W.K. Hocking, B.J. Johnson, E. Joseph, J. Merrill, G.A. Morris, M. Newchurch, S.J. Oltmans, F. Posny, F.J. Schmidlin, H. Vomel, D.N. Whiteman and J.C. Witte, 2008, Validation of Tropospheric Emission Spectrometer (TES) nadir ozone profiles using ozonesonde measurements, *J. Geophys. Res.-Atmospheres* **113**, D15s17.

Newell, R.E. Y. Zhu, V.S. Connors, H.G. Reichle, P.C. Novelli and B.B. Gormsen, 1999, Atmospheric processes influencing measured carbon monoxide in the NASA Measurement of Air Pollution From Satellites (MAPS) experiment, *J. Geophys. Res.-Atmospheres* **104**, 21487–21501.

Nganga, D. A. Minga, B. Cros, C.B. Biona, J. Fishman and W.B. Grant, 1996, The vertical distribution of ozone measured at Brazzaville, Congo during TRACE A, *J. Geophys. Res.-Atmospheres* **101**, 24095–24103.

Noel, S. M. Buchwitz, H. Bovensmann, R. Hoogen and J.P. Burrows, 1999, Atmospheric water vapor amounts retrieved from GOME satellite data, *Geophys. Res. Lett.* **26**, 1841–1844.

Noel, S. M. Buchwitz, H. Bovensmann and J.P. Burrows, 2002, Retrieval of total water vapour column amounts from GOME/ERS-2 data. Adv Space Res, **29**, 1697–1702.

Noel, S. M. Buchwitz and J.P. Burrows, 2004, First retrieval of global water vapour column amounts from SCIAMACHY measurements, *Atmos. Chem. and Phys.* **4**, 111–125.

Noel, S. M. Buchwitz, H. Bovensmann and J.P. Burrows, 2005, Validation of SCIAMACHY AMC-DOAS water vapour columns, *Atmos. Chem. and Phys.* **5**, 1835–1841.

Noel, S. S. Mieruch, H. Bovensmann and J.P. Burrows, 2008, Preliminary results of GOME-2 water vapour retrievals and first applications in polar regions, *Atmos. Chem. and Phys.* **8**, 1519–1529.

Novak, M.A.M. I.M. Watson, H. Delgado-Granados, W.I. Rose, L. Cardenas-Gonzalez and V.J. Realmuto, 2008, Volcanic emissions from Popocatepetl volcano, Mexico, quantified using Moderate Resolution Imaging Spectroradiometer (MODIS) infrared data: A case study of the December 2000–January 2001 emissions, *J. Volcanology and Geothermal Res.* **170**, 76–85.

Olson, J.R. J. Fishman, V. Kirchhoff, D. Nganga and B. Cros, 1996, Analysis of the distribution of ozone over the southern Atlantic region, *J. Geophys. Res.-Atmospheres* **101**, 24083–24093.

Ordonez, C. A. Richter, M. Steinbacher, C. Zellweger, H. Nuss, J.P. Burrows and A.S.H. Prevot, 2006, Comparison of 7 years of satellite-borne and ground-based tropospheric NO_2 measurements around Milan, Italy, *J. Geophys. Res.-Atmospheres* **111**, D05310.

Osterman, G.B. S.S. Kulawik, H.M. Worden, N.A.D. Richards, B.M. Fisher, A. Eldering, M.W. Shephard, L. Froidevaux, G. Labow, M. Luo, R.L. Herman, K.W. Bowman and A.M. Thompson, 2008, Validation of Tropospheric Emission Spectrometer (TES) measurements of the total, stratospheric, and tropospheric column abundance of ozone, *J. Geophys. Res.-Atmospheres* **113**, D15s16.

Pallister, J.S. F.A. Trusdell, I.K. Brownfield, D.F. Siems, J.R. Budahn and S.F. Sutley, 2005, The 2003 phreatomagmatic eruptions of Anatahan volcano - textural and petrologic features of deposits at an emergent island volcano, *J. Volcanology and Geothermal Res.* **146**, 208–225.

Palmer, P.I. D.J. Jacob, K. Chance, R.V. Martin, R.J.D. Spurr, T.P. Kurosu, I. Bey, R. Yantosca, A. Fiore and Q.B. Li, 2001, Air mass factor formulation for spectroscopic measurements from satellites: Application to formaldehyde retrievals from the Global Ozone Monitoring Experiment, *J. Geophys. Res.-Atmospheres* **106**, 14539–14550.

Palmer, P.I. D.J. Jacob, A.M. Fiore, R.V. Martin, K. Chance and T.P. Kurosu, 2003, Mapping isoprene emissions over North America using formaldehyde column observations from space, *J. Geophys. Res.-Atmospheres* **108**, 4180.

Palmer, P.I. D.S. Abbot, T.M. Fu, D.J. Jacob, K. Chance, T.P. Kurosu, A. Guenther, C. Wiedinmyer, J.C. Stanton, M.J. Pilling, S.N. Pressley, B. Lamb and A.L. Sumner, 2006, Quantifying the seasonal and interannual variability of North American isoprene emissions using satellite observations of the formaldehyde column, *J. Geophys. Res.-Atmospheres* **111**, D12315.

Palmer, P.I. M.P. Barkley, T.P. Kurosu, A.C. Lewis, J.E. Saxton, K. Chance and L.V. Gatti, 2007, Interpreting satellite column observations of formaldehyde over tropical South America, *Phil. Trans. Roy. Soc. a-Mathematical Physical and Engineering Sciences* **365**, 1741–1751.

Park, M. W.J. Randel, A. Gettelman, S.T. Massie and J.H. Jiang, 2007, Transport above the Asian summer monsoon anticyclone inferred from Aura Microwave Limb Sounder tracers, *J. Geophys. Res.-Atmospheres* **112**, doi: 1310.1029/2006jd008294.

Park, M. W.J. Randel, L.K. Emmons, P.F. Bernath, K.A. Walker and C.D. Boone, 2008, Chemical isolation in the Asian monsoon anticyclone observed in Atmospheric Chemistry Experiment (ACE-FTS) data, *Atmos. Chem. and Phys.* **8**, 757–764.

Peng, L. and C.S. Zhao, 2007, Analysis of carbon monoxide budget in North China, *Chemosphere* **66**, 1383–1389.

Peters, W. M. Krol, F. Dentener, A.M. Thompson and J. Lelieveld, 2002, Chemistry-transport modeling of the satellite observed distribution of tropical tropospheric ozone, *Atmos. Chem. and Phys.* **2**, 103–120.

Peters, W. M.C. Krol, J.P.F. Fortuin, H.M. Kelder, A.M. Thompson, C.R. Becker, J. Lelieveld and P.J. Crutzen, 2004, Tropospheric ozone over a tropical Atlantic station in the Northern hemisphere: Paramaribo, Surinam, *Tellus Series B-Chemical and Physical Meteorology* **56**, 21–34.

Petron, G. C. Granier, B. Khattatov, V. Yudin, J.F. Lamarque, L. Emmons, J. Gille and D.P. Edwards, 2004, Monthly CO surface sources inventory based on the 2000–2001 MOPITT satellite data, *Geophys. Res. Lett.* **31**, L21107.

Peylin, P. F.M. Breon, S. Serrar, Y. Tiwari, A. Chedin, M. Gloor, T. Machida, C. Brenninkmeijer, A. Zahn and P. Ciais, 2007, Evaluation of Television Infrared Observation Satellite (TIROS-N) Operational Vertical Sounder (TOVS) spaceborne CO_2 estimates using model simulations and aircraft data, *J. Geophys. Res.-Atmospheres* **112**, doi: 10.1029/2005jd007018.

Pfister, G. G. Petron, L.K. Emmons, J.C. Gille, D.P. Edwards, J.F. Lamarque, J.L. Attie, C. Granier and P.C. Novelli, 2004, Evaluation of CO simulations and the analysis of the CO budget for Europe, *J. Geophys. Res.-Atmospheres* **109**, D19304.

Pfister, G. P.G. Hess, L.K. Emmons, J.F. Lamarque, C. Wiedinmyer, D.P. Edwards, G. Petron, J.C. Gille and G.W. Sachse, 2005, Quantifying CO emissions from the 2004 Alaskan wildfires using MOPITT CO data, *Geophys. Res. Lett.* **32**, L11809.

Piters, A.J.M. K. Bramstedt, J.C. Lambert and B. Kirchhoff, 2006, Overview of SCIAMACHY validation: 2002-2004, *Atmos. Chem. and Phys.* **6**, 127–148.

Platt, U. and E. Lehrer, 1997, Arctic tropospheric ozone chemistry, ARCTOC, Final Report of EU-Project No. EV5V-CT93-0318, Heidelberg, EU, *Brussels*, **42**.

Platt, U. and G. Hönninger, 2003, The role of halogen species in the troposphere, *Chemosphere* **52**, 325–338.

Poberovskii, A.V. A.V. Shashkin, D.V. Ionov and Y.M. Timofeev, 2007, NO_2 content variations near St. Petersburg as inferred from ground-based and satellite measurements of scattered solar radiation, *Izvestiya Atmospheric and Oceanic Physics* **43**, 505–513.

Poisson, N. M. Kanakidou and P.J. Crutzen, 2000, Impact of non-methane hydrocarbons on tropospheric chemistry and the oxidizing power of the global troposphere: 3-dimensional modelling results, *J. Atmos. Chem.* **36**, 157–230.

Portmann, R.W. S. Solomon, J. Fishman, J.R. Olson, J.T. Kiehl and B. Briegleb, 1997, Radiative forcing of the Earth's climate system due to tropical tropospheric ozone production, *J. Geophys. Res.-Atmospheres* **102**, 9409–9417.

Pradier, S. J.L. Attie, M. Chong, J. Escobar, V.H. Peuch, J.F. Lamarque, B. Khattatov and D. Edwards, 2006, Evaluation of 2001 springtime CO transport over West Africa using MOPITT CO measurements assimilated in a global chemistry transport model, *Tellus Series B-Chemical and Physical Meteorology* **58**, 163–176.

Prata, A.J. and C. Bernardo, 2007, Retrieval of volcanic SO_2 column abundance from atmospheric infrared sounder data, *J. Geophys. Res.-Atmospheres* **112**, D20204.

Prata, A.J. and J. Kerkmann, 2007, Simultaneous retrieval of volcanic ash and SO_2 using MSG-SEVIRI measurements, *Geophys. Res. Lett.* **34**, doi: 610.1029/2006gl028691.

Prata, A.J. S.A. Carn, A. Stohl and J. Kerkmann, 2007, Long range transport and fate of a stratospheric volcanic cloud from Soufriere Hills volcano, Montserrat, *Atmos. Chem. Phys.* **7**, 5093–5103.

Pumphrey, H.C. M.J. Filipiak, N.J. Livesey, M.J. Schwartz, C. Boone, K.A. Walker, P. Bernath, P. Ricaud, B. Barret, C. Clerbaux, R.F. Jarnot, G.L. Manney and J.W. Waters, 2007, Validation of middle-atmosphere carbon monoxide retrievals from the Microwave Limb Sounder on Aura, *J. Geophys. Res.-Atmospheres* **112**, 12, doi: 10.1029/2007jd008723.

Raspollini, P. C. Belotti, A. Burgess, B. Carli, M. Carlotti, S. Ceccherini, B.M. Dinelli, A. Dudhia, J.M. Flaud, B. Funke, M. Hopfner, M. Lopez-Puertas, V. Payne, C. Piccolo, J.J. Remedios, M. Ridolfi and R. Spang, 2006, MIPAS level 2 operational analysis, *Atmos. Chem. and Phys.* **6**, 5605–5630.

Read, W.G. A. Lambert, J. Bacmeister, R.E. Cofield, L.E. Christensen, D.T. Cuddy, W.H. Daffer, B.J. Drouin, E. Fetzer, L. Froidevaux, R. Fuller, R. Herman, R.F. Jarnot, J.H. Jiang, Y.B. Jiang, K. Kelly, B.W. Knosp, L.J. Kovalenko, N.J. Livesey, H.C. Liu, G.L. Manney, H.M. Pickett, H.C. Pumphrey, K.H. Rosenlof, X. Sabounchi, M.L. Santee, M.J. Schwartz, W.V. Snyder, P.C. Stek, H. Su, L.L. Takacs, R.P. Thurstans, H. Vomel, P.A. Wagner, J.W. Waters, C.R. Webster, E.M. Weinstock and D.L. Wu, 2007, Aura Microwave Limb Sounder upper tropospheric and lower stratospheric H2O and relative humidity with respect to ice validation, *J. Geophys. Res.-Atmospheres* **112**, 29, doi: 10.1029/2007jd008752.

Read, K.A. A.S. Mahajan, L.J. Carpenter, M.J. Evans, B.V.E. Faria, D.E. Heard, J.R. Hopkins, J.D. Lee, S.J. Moller, A.C. Lewis, L. Mendes, J.B. McQuaid, H. Oetjen, A. Saiz-Lopez, M.J. Pilling and J.M.C. Plane, 2008, Extensive halogen-mediated ozone destruction over the tropical Atlantic Ocean, *Nature* **453**, 1232–1235.

Richards, N.A.D. G.B. Osterman, E.V. Browell, J.W. Hair, M. Avery and Q.B. Li, 2008, Validation of Tropospheric Emission Spectrometer ozone profiles with aircraft observations during the intercontinental chemical transport experiment-B, *J. Geophys. Res.-Atmospheres* **113**, D16s29.

Richardson, J.L. J. Fishman and G.L. Gregory, 1991, Ozone Budget over the Amazon - Regional Effects from Biomass-Burning Emissions, *J. Geophys. Res.-Atmospheres* **96**, 13073–13087.

Richter, A. and J.P. Burrows, 2002, Tropospheric NO_2 from GOME measurements. *Adv. Space Res.* **29**, 1673–1683.

Richter, A. F. Wittrock, M. Eisinger and J.P. Burrows, 1998, GOME observations of tropospheric BrO in northern hemispheric spring and summer 1997, *Geophys. Res. Lett.* **25**, 2683–2686.

Richter, A. V. Eyring, J.P. Burrows, H. Bovensmann, A. Lauer, B. Sierk and P.J. Crutzen, 2004, Satellite measurements of NO_2 from international shipping emissions, *Geophys. Res. Lett.* **31**, L23110.

Richter, A. J.P. Burrows, H. Nuss, C. Granier and U. Niemeier, 2005, Increase in tropospheric nitrogen dioxide over China observed from space, *Nature* **437**, 129–132.

Ridley, B.A. T. Zeng, Y. Wang, E.L. Atlas, E.V. Browell, P.G. Hess, J.J. Orlando, K. Chance and A. Richter, 2007, An ozone depletion event in the sub-arctic surface layer over Hudson Bay, Canada, *J. Atmos. Chem.* **57**, 255–280.

Rinsland, C.P. G. Dufour, C.D. Boone, P.F. Bernath and L. Chiou, 2005, Atmospheric Chemistry Experiment (ACE) measurements of elevated Southern Hemisphere upper tropospheric CO, C_2H_6, HCN, and C_2H_2 mixing ratios from biomass burning emissions and long-range transport, *Geophys. Res. Lett.* **32**, 4, doi: 10.1029/2005gl024214.

Rinsland, C.P. C.D. Boone, P.F. Bernath, E. Mahieu, R. Zander, G. Dufour, C. Clerbaux, S. Turquety, L. Chiou, J.C. McConnell, L. Neary and J.W. Kaminski, 2006a, First space-based observations of formic acid (HCOOH): Atmospheric Chemistry Experiment austral spring 2004 and 2005 Southern Hemisphere tropical-mid-latitude upper tropospheric measurements, *Geophys. Res. Lett.* **33**, 6, doi: 10.1029/2006gl027128.

Rinsland, C.P. M. Luo, J.A. Logan, R. Beer, H. Worden, S.S. Kulawik, D. Rider, G. Osterman, M. Gunson, A. Eldering, A. Goldman, M. Shephard, S.A. Clough, C. Rodgers, M. Lampel and L. Chiou, 2006b, Nadir measurements of carbon monoxide distributions by the Tropospheric Emission Spectrometer instrument onboard the Aura Spacecraft: Overview of analysis approach and examples of initial results, *Geophys. Res. Lett.* **33**, L22806.

Rinsland, C.P. P.F. Coheur, H. Herbin, C. Clerbaux, C. Boone, P. Bernath and L.S. Chiou, 2007a, Detection of elevated tropospheric hydrogen peroxide (H_2O_2) mixing ratios in atmospheric chemistry experiment (ACE) subtropical infrared solar occultation spectra, *J Quant Spectrosc Radiat Transf* **107**, 340–348, doi: 10.1016/j.jqsrt.2007.02.009.

Rinsland, C.P. G. Dufour, C.D. Boone, P.F. Bernath, L. Chiou, P.F. Coheur, S. Turquety and C. Clerbaux, 2007b, Satellite boreal measurements over Alaska and Canada during June–July 2004: Simultaneous measurements of upper tropospheric CO, C_2H_6, HCN, CH_3Cl, CH_4, C_2H_2, CH_3OH, HCOOH, OCS, and SF_6 mixing ratios, *Global Biogeochemical Cycles* **21**, 13, doi: 10.1029/2006gb002795.

Rinsland, C.P. M. Luo, M.W. Shephard, C. Clerbaux, C.D. Boone, P.F. Bernath, L. Chiou and P.F. Coheur, 2008, Tropospheric emission spectrometer (TES) and atmospheric chemistry experiment (ACE) measurements of tropospheric chemistry in tropical southeast Asia during a moderate El Nino in 2006, *J Quant Spectrosc Radiat Transf.* **109**, 1931–1942, doi: 10.1016/j.jqsrt.2007.12.020.

Rose, W.I. G.J.S. Bluth, D.J. Schneider, G.G.J. Ernst, C.M. Riley, L.J. Henderson and R.G. McGimsey, 2001, Observations of volcanic clouds in their first few days of atmospheric residence: The 1992 eruptions of Crater Peak, Mount Spurr volcano, Alaska, *J. Geol.* **109**, 677–694.

Rozanov, A. K.U. Eichmann, C. von Savigny, H. Bovensmann, J.P. Burrows, A. von Bargen, A. Doicu, S. Hilgers, S. Godin-Beekmann, T. Leblanc and I.S. McDermid, 2007, Comparison of the inversion algorithms applied to the ozone vertical profile retrieval from SCIAMACHY limb measurements, *Atmos. Chem. and Phys.* **7**, 4763–4779.

Saiz-Lopez, A. J.M.C. Plane and J.A. Shillito, 2004, Bromine oxide in the mid-latitude marine boundary layer, *Geophys. Res. Lett.* **31**, L03111.

Saiz-Lopez, A. J.A. Shillito, H. Coe and J.M.C. Plane, 2006, Measurements and modelling of I_2, IO, OIO, BrO and NO_3 in the mid-latitude marine boundary layer, *Atmos. Chem. and Phys.* **6**, 1513–1528.

Saiz-Lopez, A. K. Chance, X. Liu, T.P. Kurosu and S.P. Sander, 2007a, First observations of iodine oxide from space, *Geophys. Res. Lett.* **34**, L12812.

Saiz-Lopez, A. A.S. Mahajan, R.A. Salmon, S.J.B. Bauguitte, A.E. Jones, H.K. Roscoe and J.M.C. Plane, 2007b, Boundary layer halogens in coastal Antarctica, *Science* **317**, 348–351.

Santee, M.L. A. Lambert, W.G. Read, N.J. Livesey, R.E. Cofield, D.T. Cuddy, W.H. Daffer, B.J. Drouin, L. Froidevaux, R.A. Fuller, R.F. Jarnot, B.W. Knosp, G.L. Manney, V.S. Perun, W.V. Snyder, P.C. Stek, R.P. Thurstans, P.A. Wagner, J.W. Waters, G. Muscari, R.L. de Zafra, J.E. Dibb, D.W. Fahey, P.J. Popp, T.P. Marcy, K.W. Jucks, G.C. Toon, R.A. Stachnik, P.F. Bernath, C.D. Boone, K.A. Walker, J. Urban and D. Murtagh, 2007, Validation of the Aura Microwave Limb Sounder HNO3 measurements, *J. Geophys. Res.-Atmospheres* **112**, 2210, doi: 1029/2007jd008721.

Sauvage, B. R.V. Martin, A. van Donkelaar, X. Liu, K. Chance, L. Jaegle, P.I. Palmer, S. Wu and T.M. Fu, 2007, Remote sensed and *in situ* constraints on processes affecting tropical tropospheric ozone, *Atmos. Chem. and Phys.* **7**, 815–838.

Savage, N.H. K.S. Law, J.A. Pyle, A. Richter, H. Nuss and J.P. Burrows, 2004, Using GOME NO_2 satellite data to examine regional differences in TOMCAT model performance, *Atmos. Chem. and Phys.* **4**, 1895–1912.

Schaub, D. A.K. Weiss, J.W. Kaiser, A. Petritoli, A. Richter, B. Buchmann and J.P. Burrows, 2005, A transboundary transport episode of nitrogen dioxide as observed from GOME and its impact in the Alpine region, *Atmos. Chem. and Phys.* **5**, 23–37.

Schaub, D. D. Brunner, K.F. Boersma, J. Keller, D. Folini, B. Buchmann, H. Berresheim and J. Staehelin, 2007, SCIAMACHY tropospheric NO_2 over Switzerland: estimates of NOx lifetimes and impact of the complex Alpine topography on the retrieval, *Atmos. Chem. Phys.* **7**, 5971–5987.

Schlesinger, W.H. and A.E. Hartley, 1992, A Global Budget for Atmospheric NH_3, *Biogeochemistry* **15**, 191–211.

Schneising, O. M. Buchwitz, J.P. Burrows, H. Bovensmann, M. Reuter, J. Notholt, R. Macatangay and T. Warneke, 2008, Three years of greenhouse gas column-averaged dry air mole fractions retrieved from satellite - Part 1: Carbon dioxide, *Atmos. Chem. and Phys.* **8**, 3827–3853.

Schneising, O. M. Buchwitz, J.P. Burrows, H. Bovensmann, P. Bergamaschi and W. Peters, 2009, Three years of greenhouse gas column-averaged dry air mole fractions retrieved from satellite - Part 2: Methane, *Atmos. Chem. and Phys.* **9**, 443–465.

Schnetzler, C.C. S.D. Doiron, L.S. Walter and A.J. Krueger, 1994, Satellite Measurement of Sulfur-Dioxide from the Redoubt Eruptions of 1989-1990, *J. Volcanology and Geothermal Res.* **62**, 353–357.

Schoeberl, M.R. J.R. Ziemke, B. Bojkov, N. Livesey, B. Duncan, S. Strahan, L. Froidevaux, S. Kulawik, P.K. Bhartia, S. Chandra, P.F. Levelt, J.C. Witte, A.M. Thompson, E. Cuevas, A. Redondas, D.W. Tarasick, J. Davies, G. Bodeker, G. Hansen, B.J. Johnson, S.J. Oltmans, H. Vomel, M. Allaart, H. Kelder, M. Newchurch, S. Godin-Beekmann, G. Ancellet, H. Claude, S.B. Andersen, E. Kyro, M. Parrondos, M. Yela, G. Zablocki, D. Moore, H. Dier, P. von der Gathen, P. Viatte, R. Stubi, B. Calpini, P. Skrivankova, V. Dorokhov, H. de Backer, F.J. Schmidlin, G. Coetzee, M. Fujiwara, V. Thouret, F. Posny, G. Morris, J. Merrill, C.P. Leong, G. Koenig-Langlo and E. Joseph, 2007, A trajectory-based estimate of the tropospheric ozone column using the residual method, *J. Geophys. Res.-Atmospheres* **112**, D24s49.

Schofield, R. P.V. Johnston, A. Thomas, K. Kreher, B.J. Connor, S. Wood, D. Shooter, M.P. Chipperfield, A. Richter, R. von Glasow and C.D. Rodgers, 2006, Tropospheric and stratospheric BrO columns over Arrival Heights, Antarctica, 2002, *J. Geophys. Res.-Atmospheres* **111**, D22310.

Schönhardt, A. A. Richter, F. Wittrock and J.P. Burrows, 2007, First observations of atmopsheric iodine oxide columns from satellite, *Geophys. Res. Abstracts* **9**, Poster 00592.

Schönhardt, A. A. Richter, F. Wittrock, H. Kirk, H. Oetjen, H.K. Roscoe and J.P. Burrows, 2008, Observations of iodine monoxide columns from satellite, *Atmos. Chem. and Phys.* **8**, 637–653.

Schumann, U. and H. Huntrieser, 2007, The global lightning-induced nitrogen oxides source, *Atmos. Chem. Phys.* **7**, 3823–3907.

Seftor, C.J. N.C. Hsu, J.R. Herman, P.K. Bhartia, O. Torres, W.I. Rose, D.J. Schneider and N. Krotkov, 1997, Detection of volcanic ash clouds from Nimbus 7/total ozone mapping spectrometer, *J. Geophys. Res.-Atmospheres* **102**, 16749–16759.

Senten, C. M. De MaziÃre, B. Dils, C. Hermans, M. Kruglanski, E. Neefs, F. Scolas, A.C. Vandaele, G. Vanhaelewyn, C. Vigouroux, M. Carleer, P.F. Coheur, S. Fally, B. Barret, J.L. Baray, R. Delmas, J. Leveau, J.M. Metzger, E. Mahieu, C. Boone, K.A. Walker, P.F. Bernath and K. Strong, 2008, Technical note: New ground-based FTIR measurements at Ile de La Reunion: observations, error analysis, and comparisons with independent data, *Atmos. Chem. Phys.* **8**, 3483–3508.

Sharma, K. S. Blake, S. Self and A.J. Krueger, 2004, SO_2 emissions from basaltic eruptions, and the excess sulfur issue, *Geophys. Res. Lett.* **31**, L13612.

Shephard, M.W. R.L. Herman, B.M. Fisher, K.E. Cady-Pereira, S.A. Clough, V.H. Payne, D.N. Whiteman, J.P. Comer, H. Vomel, L.M. Miloshevich, R. Forno, M. Adam, G.B. Osterman, A. Eldering, J.R. Worden, L.R. Brown, H.M. Worden, S.S. Kulawik, D.M. Rider, A. Goldman, R. Beer, K.W. Bowman, C.D. Rodgers, M. Luo, C.P. Rinsland, M. Lampel and M.R. Gunson, 2008, Comparison of Tropospheric Emission Spectrometer nadir water vapor retrievals with *in situ* measurements, *J. Geophys. Res.-Atmospheres* **113**, D15s24.

Shim, C. Y.H. Wang, Y. Choi, P.I. Palmer, D.S. Abbot and K. Chance, 2005, Constraining global isoprene emissions with Global Ozone Monitoring Experiment (GOME) formaldehyde column measurements, *J. Geophys. Res.-Atmospheres* **110**, D24301.

Shindell, D.T. G. Faluvegi and L.K. Emmons, 2005, Inferring carbon monoxide pollution changes from space-based observations, *J. Geophys. Res.-Atmospheres* **110**, D23303.

Shindell, D.T. G. Faluvegi, D.S. Stevenson, M.C. Krol, L.K. Emmons, J.F. Lamarque, G. Petron, F.J. Dentener, K. Ellingsen, M.G. Schultz, O. Wild, M. Amann, C.S. Atherton, D.J. Bergmann, I. Bey, T. Butler, J. Cofala, W.J. Collins, R.G. Derwent, R.M. Doherty, J. Drevet, H.J. Eskes, A.M. Fiore, M. Gauss, D.A. Hauglustaine, L.W. Horowitz, I.S.A. Isaksen, M.G. Lawrence, V. Montanaro, J.F. Muller, G. Pitari, M.J. Prather, J.A. Pyle, S. Rast, J.M. Rodriguez, M.G. Sanderson, N.H. Savage, S.E. Strahan, K. Sudo, S. Szopa, N. Unger, T.P.C. van Noije and G. Zeng, 2006, Multimodel simulations of carbon monoxide: Comparison with observations and projected near-future changes, *J. Geophys. Res.-Atmospheres* **111**, D19306.

Sierk, B. A. Richter, A. Rozanov, C. Von Savigny, A.M. Schmoltner, M. Buchwitz, H. Bovensmann and J.P. Burrows, 2006, Retrieval and monitoring of atmospheric trace gas concentrations in nadir and limb geometry using the space-borne SCIAMACHY instrument, *Environ. Monit. Assess.* **120**, 65–77.

Simpson, W.R. R. von Glasow, K. Riedel, P. Anderson, P. Ariya, J. Bottenheim, J. Burrows, L.J. Carpenter, U. Frieß, M.E. Goodsite, D. Heard, M. Hutterli, H.-W. Jacobi, L. Kaleschke, B. Neff, J. Plane, U. Platt, A. Richter, H. Roscoe, R. Sander, P. Shepson, J. Sodeau, A. Steffen, A. Wagner and E. Wolff, 2007, Halogens and their role in polar boundary-layer ozone depletion, *Atmos. Chem. Phys.* **7**, 4375–4418.

Singh, H.B. D. Ohara, D. Herlth, J.D. Bradshaw, S.T. Sandholm, G.L. Gregory, G.W. Sachse, D.R. Blake, P.J. Crutzen and M.A. Kanakidou, 1992, Atmospheric Measurements of Peroxyacetyl Nitrate and Other Organic Nitrates at High-Latitudes - Possible Sources and Sinks, *J. Geophys. Res.-Atmospheres* **97**, 16511–16522.

Singh, H.B. M. Kanakidou, P.J. Crutzen and D.J. Jacob, 1995, High-Concentrations and Photochemical Fate of Oxygenated Hydrocarbons in the Global Troposphere, *Nature* **378**, 50–54.

Singh, H.B. W.H. Brune, J.H. Crawford, D.J. Jacob and P.B. Russell, 2006, Overview of the summer 2004 intercontinental chemical transport experiment - North America (INTEX-A), *J. Geophys. Res.-Atmospheres* **111**, D24s01.

Sioris, C.E. C.A. McLinden, R.V. Martin, B. Sauvage, C.S. Haley, N.D. Lloyd, E.J. Llewellyn, P.F. Bernath, C.D. Boone, S. Brohede and C.T. McElroy, 2007, Vertical profiles of lightning-produced NO_2 enhancements in the upper troposphere observed by OSIRIS, *Atmos. Chem. and Phys.* **7**, 4281–4294.

Spichtinger, N. M. Wenig, P. James, T. Wagner, U. Platt and A. Stohl, 2001, Satellite detection of a continental-scale plume of nitrogen oxides from boreal forest fires, *Geophys. Res. Lett.* **28**, 4579–4582.

Spichtinger, N. R. Damoah, S. Eckhardt, C. Forster, P. James, S. Beirle, T. Marbach, T. Wagner, P.C. Novelli and A. Stohl, 2004, Boreal forest fires in 1997 and 1998: a seasonal comparison using transport model simulations and measurement data, *Atmos. Chem. and Phys.* **4**, 1857–1868.

Stajner, I. K. Wargan, S. Pawson, H. Hayashi, L.P. Chang, R.C. Hudman, L. Froidevaux, N. Livesey, P.F. Levelt, A.M. Thompson, D.W. Tarasick, R. Stubi, S.B. Andersen, M. Yela, G. Konig-Langlo, F.J. Schmidlin and J.C. Witte, 2008, Assimilated ozone from EOS-Aura: Evaluation of the tropopause region and tropospheric columns, *J. Geophys. Res.-Atmospheres* **113**, D16s32.

Stavrakou, T. and J.F. Muller, 2006, Grid-based versus big region approach for inverting CO emissions using Measurement of Pollution in the Troposphere (MOPITT) data, *J. Geophys. Res.-Atmospheres* **111**, D15304.

Stavrakou, T. J.F. Muller, K.F. Boersma, I. De Smedt and R.J. van der A, 2008, Assessing the distribution and growth rates of NOx emission sources by inverting a 10-year record of NO_2 satellite columns, *Geophys. Res. Lett.* **35**, L10801.

Steck, T. N. Glatthor, T. von Clarmann, H. Fischer, J.M. Flaud, B. Funke, U. Grabowski, M. Hopfner, S. Kellmann, A. Linden, A. Perrin and G.P. Stiller, 2008, Retrieval of global upper tropospheric and stratospheric formaldehyde (H_2CO) distributions from high-resolution MIPAS-Envisat spectra, *Atmos. Chem. and Phys.* **8**, 463–470.

Steinwagner, J. M. Milz, T. von Clarmann, N. Glatthor, U. Grabowski, M. Hopfner, G.P. Stiller and T. Rockmann, 2007, HDO measurements with MIPAS, *Atmos. Chem. and Phys.* **7**, 2601–2615.

Stiller, G.P. T. von Clarmann, C. Bruehl, H. Fischer, B. Funke, N. Glatthor, U. Grabowski, M. Hopfner, P. Jockel, S. Kellmann, M. Kiefer, A. Linden, M. Lopez-Puertas, G.M. Tsidu, M. Milz, T. Steck and B. Steil, 2007, Global distributions of HO_2NO_2 as observed by the Michelson Interferometer for Passive Atmospheric Sounding (MIPAS), *J. Geophys. Res.-Atmospheres* **112**, D09314.

Stohl, A. H. Huntrieser, A. Richter, S. Beirle, O.R. Cooper, S. Eckhardt, C. Forster, P. James, N. Spichtinger, M. Wenig, T. Wagner, J.P. Burrows and U. Platt, 2003, Rapid intercontinental air pollution transport associated with a meteorological bomb, *Atmos. Chem. and Phys.* **3**, 969–985.

Stohl, A. T. Berg, J.F. Burkhart, A.M. Fjaeraa, C. Forster, A. Herber, O. Hov, C. Lunder, W.W. McMillan, S. Oltmans, M. Shiobara, D. Simpson, S. Solberg, K. Stebel, J. Strom, K. Torseth, R. Treffeisen, K. Virkkunen and K.E. Yttri, 2007, Arctic smoke - record high air pollution levels in the European Arctic due to agricultural fires in Eastern Europe in spring 2006, *Atmos. Chem. and Phys.* **7**, 511–534.

Straume, A.G. H. Schrijver, A.M.S. Gloudemans, S. Houweling, I. Aben, A.N. Maurellis, A.T.J. de Laat, Q. Kleipool, G. Lichtenberg, R. van Hees, J.F. Meirink and M. Krol, 2005, The global variation of CH_4 and CO as seen by SCIAMACHY. *Adv Space Res.* **36**, 821–827.

Tanimoto, H. Y. Sawa, S. Yonemura, K. Yumimoto, H. Matsueda, I. Uno, T. Hayasaka, H. Mukai, Y. Tohjima, K. Tsuboi and L. Zhang, 2008, Diagnosing recent CO emissions and ozone evolution in East Asia using coordinated surface observations, adjoint inverse modeling, and MOPITT satellite data, *Atmos. Chem. Phys.* **8**, 3867–3880.

Tellmann, S. V.V. Rozanov, M. Weber and J.P. Burrows, 2004, Improvements in the tropical ozone profile retrieval from GOME-UV/Vis nadir spectra. *Trace Constituents in the Troposphere and Lower Stratosphere*, Pergamon-Elsevier Science Ltd, Kidlington, 739–743.

Theys, N. M. Van Roozendael, F. Hendrick, C. Fayt, C. Hermans, J.L. Baray, F. Goutail, J.P. Pommereau and M. De Maziere, 2007, Retrieval of stratospheric and tropospheric BrO columns from multi-axis DOAS measurements at Reunion Island (21 degrees S, 56 degrees E), *Atmos. Chem. and Phys.* **7**, 4733–4749.

Theys, N. M. Van Roozendael, B. Dils, F. Hendrick, N. Hao and M. De Maziere, 2009, First satellite detection of volcanic bromine monoxide emission after the Kasatochi eruption, *Geophys. Res. Lett.* **36**, 5, doi: 10.1029/2008gl036552.

Thomas, W. E. Hegels, S. Slijkhuis, R. Spurr and K.I. Chance, 1998, Detection of biomass burning combustion products in Southeast Asia from backscatter data taken by the GOME spectrometer, *Geophys. Res. Lett.* **25**, 1317–1320.

Thomas, W. F. Baier, T. Erbertseder and M. Kastner, 2003, Analysis of the Algerian severe weather event in November 2001 and its impact on ozone and nitrogen dioxide distributions, *Tellus Series B-Chemical and Physical Meteorology* **55**, 993–1006.

Thomas, W. T. Erbertseder, T. Ruppert, M. van Roozendael, J. Verdebout, D. Balis, C. Meleti and C. Zerefos, 2005, On the retrieval of volcanic sulfur dioxide emissions from GOME backscatter measurements, *J. Atmos. Chem.* **50**, 295–320.

Thompson, A.M. D.P. McNamara, K.E. Pickering and R.D. McPeters, 1993, Effect of Marine Stratocumulus on Toms Ozone, *J. Geophys. Res.-Atmospheres* **98**, 23051–23057.

Thompson, A.M. R.D. Diab, G.E. Bodeker, M. Zunckel, G.J.R. Coetzee, C.B. Archer, D.P. McNamara, K.E. Pickering, J. Combrink, J. Fishman and D. Nganga, 1996a, Ozone over southern Africa during SAFARI-92 TRACE A, *J. Geophys. Res.-Atmospheres* **101**, 23793–23807.

Thompson, A.M. K.E. Pickering, D.P. McNamara, M.R. Schoeberl, R.D. Hudson, J.H. Kim, E.V. Browell, V. Kirchhoff and D. Nganga, 1996b, Where did tropospheric ozone over southern Africa and the tropical Atlantic come from in October 1992? Insights from TOMS, GTE TRACE A, and SAFARI 1992, *J. Geophys. Res.-Atmospheres* **101**, 24251–24278.

Thompson, A.M. and R.D. Hudson, 1999, Tropical tropospheric ozone (TTO) maps from Nimbus 7 and Earth Probe TOMS by the modified-residual method: Evaluation with sondes, ENSO signals, and trends from Atlantic regional time series, *J. Geophys. Res.-Atmospheres* **104**, 26961–26975.

Thompson, A.M. B.G. Doddridge, J.C. Witte, R.D. Hudson, W.T. Luke, J.E. Johnston, B.J. Johnston, S.J. Oltmans and R. Weller, 2000, A tropical Atlantic paradox: Shipboard and satellite views of a tropospheric ozone maximum and wave-one in January-February 1999, *Geophys. Res. Lett.* **27**, 3317–3320.

Thompson, A.M. J.C. Witte, R.D. Hudson, H. Guo, J.R. Herman and M. Fujiwara, 2001, Tropical tropospheric ozone and biomass burning, *Science* **291**, 2128–2132.

Thompson, A.M. J.C. Witte, R.D. McPeters, S.J. Oltmans, F.J. Schmidlin, J.A. Logan, M. Fujiwara, V. Kirchhoff, F. Posny, G.J.R. Coetzee, B. Hoegger, S. Kawakami, T. Ogawa, B.J. Johnson, H. Vomel and G. Labow, 2003, Southern Hemisphere Additional Ozonesondes (SHADOZ) 1998-2000 tropical ozone climatology - 1. Comparison with Total Ozone Mapping Spectrometer (TOMS) and ground-based measurements, *J. Geophys. Res.-Atmospheres* **108**, 8238.

Tie, X.X. G.P. Brasseur, C.S. Zhao, C. Granier, S. Massie, Y. Qin, P.C. Wang, G.L. Wang, P.C. Yang and A. Richter, 2006, Chemical characterization of air pollution in Eastern China and the Eastern United States, *Atmospheric Environment* **40**, 2607–2625.

Tie, X.X. S. Chandra, J.R. Ziemke, C. Granier and G.P. Brasseur, 2007, Satellite measurements of tropospheric column O_3 and NO_2 in eastern and southeastern Asia: Comparison with a global model (MOZART-2), *J. Atmos. Chem.* . **56**, 105–125.

Timofeev, Y.M. D.V. Ionov, A.V. Polyakov, N.F. Elanskii, A.S. Elokhov, A.N. Gruzdev, O.V. Postylyakov and E.V. Rozanov, 2000, Comparison between satellite and ground-based NO_2 total content measurements, *Izvestiya Atmos. Ocean Phys.* **36**, 737–742.

Tiwari, Y.K. M. Gloor, R.J. Engelen, F. Chevallier, C. Rodenbeck, S. Korner, P. Peylin, B.H. Braswell and M. Heimann, 2006, Comparing CO_2 retrieved from Atmospheric Infrared Sounder with model predictions: Implications for constraining surface fluxes and lower-to-upper troposphere transport, *J. Geophys. Res.-Atmospheres* **111**, 15, doi: 10.1029/2005jd006681.

Toenges-Schuller, N. O. Stein, F. Rohrer, A. Wahner, A. Richter, J.P. Burrows, S. Beirle, T. Wagner, U. Platt and C.D. Elvidge, 2006, Global distribution pattern of anthropogenic nitrogen oxide emissions: Correlation analysis of satellite measurements and model calculations, *J. Geophys. Res.-Atmospheres* **111**, D05312.

Tupper, A. S. Carn, J. Davey, Y. Kamada, R. Potts, F. Prata and M. Tokuno, 2004, An evaluation of volcanic cloud detection techniques during recent significant eruptions in the western 'Ring of Fire', *Remote Sensing of Environment* **91**, 27–46.

Turquety, S. J. Hadji-Lazaro, C. Clerbaux, D.A. Hauglustaine, S.A. Clough, V. Casse, P. Schlussel and G. Megie, 2004, Operational trace gas retrieval algorithm for the Infrared Atmospheric Sounding Interferometer, *J. Geophys. Res.-Atmospheres* **109**, 19, doi: 10.1029/2004jd004821.

Turquety, S. J.A. Logan, D.J. Jacob, R.C. Hudman, F.Y. Leung, C.L. Heald, R.M. Yantosca, S.L. Wu, L.K. Emmons, D.P. Edwards and G.W. Sachse, 2007, Inventory of boreal fire emissions for North America in 2004: Importance of peat burning and pyroconvective injection, *J. Geophys. Res.-Atmospheres* **112**, D12s03.

Turquety, S. C. Clerbaux, K. Law, P.F. Coheur, A. Cozic, S. Szopa, D.A. Hauglustaine, J. Hadji-Lazaro, A.M.S. Gloudemans, H. Schrijver, C.D. Boone, P.F. Bernath and D.P. Edwards, 2008, CO emission and export from Asia: an analysis combining complementary satellite measurements (MOPITT, SCIAMACHY and ACE-FTS) with global modeling, *Atmos. Chem. Phys.* **8**, 5187–5204.

Uno, I. Y. He, T. Ohara, K. Yamaji, J.I. Kurokawa, M. Katayama, Z. Wang, K. Noguchi, S. Hayashida, A. Richter and J.P. Burrows, 2007, Systematic analysis of interannual and seasonal variations of model-simulated tropospheric NO_2 in Asia and comparison with GOME-satellite data, *Atmos. Chem. and Phys.* **7**, 1671–1681.

Urai, M. 2004, Sulfur dioxide flux estimation from volcanoes using Advanced Spaceborne Thermal Emission and Reflection Radiometer - a case study of Miyakejima volcano, Japan, *J. Volcanology and Geothermal Res.* **134**, 1–13, doi: 10.1016/j.jvolgeores.2003.11.008.

Valks, P.J.M. R.B.A. Koelemeijer, M. van Weele, P. van Velthoven, J.P.F. Fortuin and H. Kelder, 2003a, Variability in tropical tropospheric ozone: Analysis with Global Ozone Monitoring Experiment observations and a global model, *J. Geophys. Res.-Atmospheres* **108**, 4328.

Valks, P.J.M. R.B.A. Koelemeijer, M. van Weele, P. van Velthoven, J.P.F. Fortuin and H. Kelder, 2003b, Variability in tropical tropospheric ozone: Analysis with Global Ozone Monitoring Experiment observations and a global model, *J. Geophys. Res.-Atmospheres* **108**, doi: 10.1029/2002jd002894.

van der A, R.J. R.F. van Oss, A.J.M. Piters, J.P.F. Fortuin, Y.J. Meijer and H.M. Kelder, 2002, Ozone profile retrieval from recalibrated Global Ozone Monitoring Experiment data, *J. Geophys. Res.-Atmospheres* **107**, 4239.

van der A, R.J. D. Peters, H. Eskes, K.F. Boersma, M. Van Roozendael, I. De Smedt and H.M. Kelder, 2006, Detection of the trend and seasonal variation in tropospheric NO_2 over China, *J. Geophys. Res.-Atmospheres* **111**, D12317.

van der A, R.J. H.J. Eskes, K.F. Boersma, T.P.C. van Noije, M. Van Roozendael, I. De Smedt, D. Peters and E.W. Meijer, 2008, Trends, seasonal variability and dominant NOx source derived from a ten year record of NO_2 measured from space, *J. Geophys. Res.-Atmospheres* **113**, D04302.

van Noije, T.P.C. H.J. Eskes, F.J. Dentener, D.S. Stevenson, K. Ellingsen, M.G. Schultz, O. Wild, M. Amann, C.S. Atherton, D.J. Bergmann, I. Bey, K.F. Boersma, T. Butler, J. Cofala, J. Drevet, A.M. Fiore, M. Gauss, D.A. Hauglustaine, L.W. Horowitz, I.S.A. Isaksen, M.C. Krol, J.F. Lamarque, M.G. Lawrence, R.V. Martin, V. Montanaro, J.F. Muller, G. Pitari, M.J. Prather, J.A. Pyle, A. Richter, J.M. Rodriguez, N.H. Savage, S.E. Strahan, K. Sudo, S. Szopa and M. van Roozendael, 2006, Multi-model ensemble simulations of tropospheric NO_2 compared with GOME retrievals for the year 2000, *Atmos. Chem. and Phys.* **6**, 2943–2979.

Van Roozendael, M. T. Wagner, A. Richter, I. Pundt, D.W. Arlander, J.P. Burrows, M. Chipperfield, C. Fayt, P.V. Johnston, J.C. Lambert, K. Kreher, K. Pfeilsticker, U. Platt, J.P. Pommereau, B.M. Sinnhuber, K.K. Tornkvist and F. Wittrock, 2002, Intercomparison of BrO measurements from ERS-2 GOME, ground-based and balloon platforms. *Adv. Space Res*, **29**, 1661–1666.

Velazco, V. J. Notholt, T. Warneke, M. Lawrence, H. Bremer, J. Drummond, A. Schulz, J.R. Krieg and O. Schrems, 2005, Latitude and altitude variability of carbon monoxide in the Atlantic detected from ship-borne Fourier transform spectrometry, model, and satellite data, *J. Geophys. Res.-Atmospheres* **110**, D09306.

Velders, G.J.M. C. Granier, R.W. Portmann, K. Pfeilsticker, M. Wenig, T. Wagner, U. Platt, A. Richter and J.P. Burrows, 2001, Global tropospheric NO_2 column distributions: Comparing three-dimensional model calculations with GOME measurements, *J. Geophys. Res.-Atmospheres* **106**, 12643–12660.

Vogt, R. P.J. Crutzen and R. Sander, 1996, A mechanism for halogen release from sea-salt aerosol in the remote marine boundary layer, *Nature* **383**, 327–330.

Volkamer, R. U. Platt and K. Wirtz, 2001, Primary and secondary glyoxal formation from aromatics: Experimental evidence for the bicycloalkyl-radical pathway from benzene, toluene, and p-xylene, *J. Phys. Chem. A* **105**, 7865–7874.

Volkamer, R. L.T. Molina, M.J. Molina, T. Shirley and W.H. Brune, 2005, DOAS measurement of glyoxal as an indicator for fast VOC chemistry in urban air, *Geophys. Res. Lett.* **32**, doi: 10.1029/2005gl022616.

Volkamer, R. F.S. Martini, L.T. Molina, D. Salcedo, J.L. Jimenez and M.J. Molina, 2007, A missing sink for gas-phase glyoxal in Mexico City: Formation of secondary organic aerosol, *Geophys. Res. Lett.* **34**, L19807.

von Glasow, R. R. von Kuhlmann, M.G. Lawrence, U. Platt and P.J. Crutzen, 2004, Impact of reactive bromine chemistry in the troposphere, *Atmos. Chem. and Phys.* **4**, 2481–2497.

Vrekoussis, M. F. Wittrock, A. Richter and J.P. Burrows, 2009, Temporal and spatial variability of glyoxal as observed from space, *Atmos. Chem. and Phys.* **9**, 4485–4504.

Vukovich, F.M. V. Brackett, J. Fishman and J.E. Sickles, 1996, On the feasibility of using the tropospheric ozone residual for nonclimatological studies on a quasi-global scale, *J. Geophys. Res.-Atmospheres* **101**, 9093–9105.

Vukovich, F.M. V. Brackett, J. Fishman and J.E. Sickles, 1997, A 5-year evaluation of the representativeness of the tropospheric ozone residual at nonclimatological periods, *J. Geophys. Res.-Atmospheres* **102**, 15927–15932.

Wagner, T. and U. Platt, 1998, Satellite mapping of enhanced BrO concentrations in the troposphere, *Nature* **395**, 486–490.

Wagner, T. C. Leue, M. Wenig, K. Pfeilsticker and U. Platt, 2001, Spatial and temporal distribution of enhanced boundary layer BrO concentrations measured by the GOME instrument aboard ERS-2, *J. Geophys. Res.-Atmospheres* **106**, 24225–24235.

Wagner, T. S. Beirle, M. Grzegorski, S. Sanghavi and U. Platt, 2005, El Nino induced anomalies in global data sets of total column precipitable water and cloud cover derived from GOME on ERS-2, *J. Geophys. Res.-Atmospheres* **110**, D15104.

Wagner, T. S. Beirle, M. Grzegorski and U. Platt, 2006, Global trends (1996-2003) of total column precipitable water observed by Global Ozone Monitoring Experiment (GOME) on ERS-2 and their relation to near-surface temperature, *J. Geophys. Res.-Atmospheres* **111**, D12102.

Wagner, T. O. Ibrahim, R. Sinreich, U. Friess, R. von Glasow and U. Platt, 2007, Enhanced tropospheric BrO over Antarctic sea ice in mid winter observed by MAX-DOAS on board the research vessel Polarstern, *Atmos. Chem. and Phys.* **7**, 3129–3142.

Walter, B.P. M. Heimann and E. Matthews, 2001, Modeling modern methane emissions from natural wetlands 2. Interannual variations 1982–1993, *J. Geophys. Res.-Atmospheres* **106**, 34207–34219.

Wang, T. H.L.A. Wong, J. Tang, A. Ding, W.S. Wu and X.C. Zhang, 2006, On the origin of surface ozone and reactive nitrogen observed at a remote mountain site in the northeastern Qinghai-Tibetan Plateau, western China, *J. Geophys. Res.-Atmospheres* **111**, D08303.

Wang, Y.X. M.B. McElroy, R.V. Martin, D.G. Streets, Q. Zhang and T.M. Fu, 2007a, Seasonal variability of NOx emissions over east China constrained by satellite observations: Implications for combustion and microbial sources, *J. Geophys. Res.-Atmospheres* **112**, D06301.

Wang, Y. M.B. McElroy, K.F. Boersma, H.J. Eskes, and J.P. Veefkind, 2007b, Traffic restrictions associated with the Sino-African summit: Reductions of NO_x detected from space, *Geophys. Res. Lett.* **34**, L08814, doi: 10.1029/2007GL029326.

Warneke, T. R. de Beek, M. Buchwitz, J. Notholt, A. Schulz, V. Velazco and O. Schrems, 2005, Shipborne solar absorption measurements of CO_2, CH_4, N_2O and CO and comparison with SCIAMACHY WFM-DOAS retrievals, *Atmos. Chem. and Phys.* **5**, 2029–2034.

Warner, J. M.M. Comer, C.D. Barnet, W.W. McMillan, W. Wolf and E. Maddy, 2007, A comparison of satellite tropospheric carbon monoxide measurements from AIRS and MOPITT during INTEX-A, *J. Geophys. Res.-Atmospheres* **112**, D12s17.

Watson, C.E. J. Fishman and H.G. Reichle, 1990, The Significance of Biomass Burning as a Source of Carbon-Monoxide and Ozone in the Southern-Hemisphere Tropics - a Satellite Analysis, *J. Geophys. Res.-Atmospheres* **95**, 16443–16450.

Wenig, M. N. Spichtinger, A. Stohl, G. Held, S. Beirle, T. Wagner, B. Jahne and U. Platt, 2003, Intercontinental transport of nitrogen oxide pollution plumes, *Atmos. Chem. and Phys.* **3**, 387–393.

Wenig, M. O. A. M. Cede, E. J. Bucsela, E. A. Celarier, K. F. Boersma, J. P. Veefkind, E. J. Brinksma, J. F. Gleason, and J. R. Herman, 2008, Validation of OMI tropospheric NO_2 column densities using direct-Sun mode Brewer measurements at NASA Goddard Space Flight Center, *J. Geophys. Res.* **113**, 45, doi:10.1029/2007JD008988.

Wespes, C. D. Hurtmans, H. Herbin, B. Barret, S. Turquety, J. Hadji-Lazaro, C. Clerbaux and P.F. Coheur, 2007, First global distributions of nitric acid in the troposphere and the stratosphere derived from infrared satellite measurements, *J. Geophys. Res.-Atmospheres* **112**, 10, doi: 10.1029/2006jd008202.

Wittrock, F. A. Richter, H. Oetjen, J.P. Burrows, M. Kanakidou, S. Myriokefalitakis, R. Volkamer, S. Beirle, U. Platt and T. Wagner, 2006, Simultaneous global observations of glyoxal and formaldehyde from space, *Geophys. Res. Lett.* **33**, L16804.

Wolff, M.A. T. Kerzenmacher, K. Strong, K.A. Walker, M. Toohey, E. Dupuy, P.F. Bernath, C.D. Boone, S. Brohede, V. Catoire, T. von Clarmann, M. Coffey, W.H. Daffer, M. De Mazière, P. Duchatelet, N. Glatthor, D.W.T. Griffith, J. Hannigan, F. Hase, M. Höpfner, N. Huret, N. Jones, K. Jucks, A. Kagawa, Y. Kasai, I. Kramer, H. Küllmann, J. Kuttippurath, E. Mahieu, G. Manney, C.T. McElroy, C. McLinden, Y. Mébarki, S. Mikuteit, D. Murtagh, C. Piccolo, P. Raspollini, M. Ridolfi, R. Ruhnke, M. Santee, C. Senten, D. Smale, C. Tétard, J. Urban and S. Wood, 2008, Validation of HNO_3, $ClONO_2$, and N_2O_5 from the Atmospheric Chemistry Experiment Fourier Transform Spectrometer (ACE-FTS), *Atmos. Chem. Phys.* **8**, 3529–3562.

Worden, H.M. J.A. Logan, J.R. Worden, R. Beer, K. Bowman, S.A. Clough, A. Eldering, B.M. Fisher,M.R. Gunson, R.L. Herman, S.S. Kulawik, M.C. Lampel, M. Luo, I.A. Megretskaia, G.B.Osterman and M.W. Shephard, 2007a, Comparisons of Tropospheric Emission Spectrometer (TES) ozone profiles to ozonesondes: Methods and initial results, *J. Geophys. Res.-Atmospheres* **112**, D03309.

Worden, J. X. Liu, K. Bowman, K. Chance, R. Beer, A. Eldering, M. Gunson and H. Worden, 2007b, Improved tropospheric ozone profile retrievals using OMI and TES radiances, *Geophys. Res. Lett.* **34**, L01809.

Worden, H.M. K.W. Bowman, J.R. Worden, A. Eldering and R. Beer, 2008, Satellite measurements of the clear-sky greenhouse effect from tropospheric ozone, *Nat. Geosci.* **1**, 305–308, doi: 10.1038/ngeo182.

Wright, R. S.A. Carn and L.P. Flynn, 2005, A satellite chronology of the May–June 2003 eruption of Anatahan volcano, J Volcanol Geoth Res. **146**, 102–116.

Xiong, X.Z. C. Barnet, E. Maddy, C. Sweeney, X.P. Liu, L.H. Zhou and M. Goldberg, 2008, Characterization and validation of methane products from the Atmospheric Infrared Sounder (AIRS), *J. Geophys. Res.-Biogeosci.* **113**, doi: 1410.1029/2007jg000500.

Yang, K. N.A. Krotkov, A.J. Krueger, S.A. Carn, P.K. Bhartia and P.F. Levelt, 2007, Retrieval of large volcanic SO_2 columns from the Aura Ozone Monitoring Instrument: Comparison and limitations, *J. Geophys. Res.-Atmospheres* **112**, D24s43.

Yudin, V.A. G. Petron, J.F. Lamarque, B.V. Khattatov, P.G. Hess, L.V. Lyjak, J.C. Gille, D.P. Edwards, M.N. Deeter and L.K. Emmons, 2004, Assimilation of the 2000–2001 CO MOPITT retrievals with optimized surface emissions, *Geophys. Res. Lett.* **31**, L20105.

Yumimoto, K. and I. Uno, 2006, Adjoint inverse modeling of CO emissions over Eastern Asia using four-dimensional variational data assimilation, *Atmospheric Environment* **40**, 6836–6845.

Yurganov, L.N. P. Duchatelet, A.V. Dzhola, D.P. Edwards, F. Hase, I. Kramer, E. Mahieu, J. Mellqvist, J. Notholt, P.C. Novelli, A. Rockmann, H.E. Scheel, M. Schneider, A. Schulz, A. Strandberg, R. Sussmann, H. Tanimoto, V. Velazco, J.R. Drummond and J.C. Gille, 2005, Increased Northern Hemispheric carbon monoxide burden in the troposphere in 2002 and 2003 detected from the ground and from space, *Atmos. Chem. and Phys.* **5**, 563–573.

Yurganov, L.N. W.W. McMillan, A.V. Dzhola, E.I. Grechko, N.B. Jones and G.R. van der Werf, 2008, Global AIRS and MOPITT CO measurements: Validation, comparison, and links to biomass burning variations and carbon cycle, *J. Geophys. Res.-Atmospheres* **113**, D09301 Artn d09301.

Zakharov, V.I. R. Imasu, K.G. Gribanov, G. Hoffmann and J. Jouzel, 2004, Latitudinal distribution of the deuterium to hydrogen ratio in the atmospheric water vapor retrieved from IMG/ADEOS data, *Geophys. Res. Lett.* **31**, doi: 410.1029/2004gl019433.

Zeng, T. Y.H. Wang, K. Chance, N. Blake, D. Blake and B. Ridley, 2006, Halogen-driven low-altitude O_3 and hydrocarbon losses in spring at northern high latitudes, *J. Geophys. Res.-Atmospheres* **111**, D17313.

Zerefos, C. K. Ganev, K. Kourtidis, M. Tzortziou, A. Vasaras and E. Syrakov, 2000, On the origin of SO_2 above Northern Greece, *Geophys. Res. Lett.* **27**, 365–368.

Zerefos, C. P. Nastos, D. Balis, A. Papayannis, A. Kelepertsis, E. Kannelopoulou, D. Nikolakis, C. Eleftheratos, W. Thomas and C. Varotsos, 2006, A complex study of Etna's volcanic plume from ground-based, *in situ* and space-borne observations, *Int. J. Remote Sens.* **27**, 1855–1864.

Zhang, Q. D.G. Streets, K. He, Y. Wang, A. Richter, J.P. Burrows, I. Uno, C.J. Jang, D. Chen, Z. Yao and Y. Lei, 2007, NOx emission trends for China, 1995–2004: The view from the ground and the view from space, *J. Geophys. Res.-Atmospheres* **112**, D22306.

Zhang, L. D.J. Jacob, K.F. Boersma, D.A. Jaffe, J.R. Olson, K.W. Bowman, J.R. Worden, A.M. Thompson, M.A. Avery, R.C. Cohen, J.E. Dibb, F.M. Flock, H.E. Fuelberg, L.G. Huey, W.W. McMillan, H.B. Singh and A.J. Weinheimer, 2008, Transpacific transport of ozone pollution and the effect of recent Asian emission increases on air quality in North America: an integrated analysis using satellite, aircraft, ozonesonde, and surface observations, *Atmos. Chem. and Phys.* **8**, 6117–6136.

Zhao, C.S. X.X. Tie, G.L. Wang, Y. Qin and P.C. Yang, 2006, Analysis of air quality in eastern China and its interaction with other regions of the world, *J. Atmos. Chem.* **55**, 189–204.

Zhao, C.S. L. Peng, X.X. Tie, Y.P. Lin, C.C. Li, X.D. Zheng and Y.Y. Fang, 2007, A high CO episode of long-range transport detected by MOPITT, *Water Air and Soil Pollution* **178**, 207–216.

Zheng, Y.G. P.J. Zhu, C.Y. Chan, L.Y. Chan, H. Cui, X.D. Zheng, Q. Zhao and Y. Qin, 2004, Influence of biomass burning in Southeast Asia on the lower tropospheric ozone distribution over South China, *Chinese Journal of Geophysics-Chinese Edition* **47**, 767–775.

Ziemke, J.R. S. Chandra and P.K. Bhartia, 1998, Two new methods for deriving tropospheric column ozone from TOMS measurements: Assimilated UARS MLS/HALOE and convective-cloud differential techniques, *J. Geophys. Res.-Atmospheres* **103**, 22115–22127.

Ziemke, J.R. and S. Chandra, 1999, Seasonal and interannual variabilities in tropical tropospheric ozone, *J. Geophys. Res.-Atmospheres* **104**, 21425–21442.

Ziemke, J.R. and S. Chandra, 2003a, La Nina and El Nino-induced variabilities of ozone in the tropical lower atmosphere during 1970–2001, *Geophys. Res. Lett.* **30**, 1142.

Ziemke, J.R. and S. Chandra, 2003b, A Madden-Julian Oscillation in tropospheric ozone, *Geophys. Res. Lett.* **30**, 2182.

Ziemke, J.R. S. Chandra and P.K. Bhartia, 2000, A new NASA data product: Tropospheric and stratospheric column ozone in the tropics derived from TOMS measurements, *Bull. Amer. Met. Soc.* **81**, 580–583.

Ziemke, J.R. S. Chandra and P.K. Bhartia, 2001, "Cloud slicing": A new technique to derive upper tropospheric ozone from satellite measurements, *J. Geophys. Res.-Atmospheres* **106**, 9853–9867.

Ziemke, J.R. S. Chandra and P.K. Bhartia, 2003, Upper tropospheric ozone derived from the cloud slicing technique: Implications for large-scale convection, *J. Geophys. Res.-Atmospheres* **108**, 4390.

Ziemke, J.R. S. Chandra and P.K. Bhartia, 2005, A 25-year data record of atmospheric ozone in the Pacific from Total Ozone Mapping Spectrometer (TOMS) cloud slicing: Implications for ozone trends in the stratosphere and troposphere, *J. Geophys. Res.-Atmospheres* **110**, D15105.

Ziemke, J.R. S. Chandra, B.N. Duncan, L. Froidevaux, P.K. Bhartia, P.F. Levelt and J.W. Waters, 2006, Tropospheric ozone determined from aura OMI and MLS: Evaluation of measurements and comparison with the Global Modeling Initiative's Chemical Transport Model, *J. Geophys. Res.-Atmospheres* **111**, D19303.

Zingler, J. and U. Platt, 2005, Iodine oxide in the Dead Sea Valley: Evidence for inorganic sources of boundary layer IO, *J. Geophys. Res.-Atmospheres* **110**, D07307.

Chapter 9
Synergistic Use of Retrieved Trace Constituent Distributions and Numerical Modelling

Maria Kanakidou, Martin Dameris, Hendrik Elbern, Matthias Beekmann, Igor B. Konovalov, Lars Nieradzik, Achim Strunk and Maarten C. Krol

9.1 Introduction

Remote sensing of tropospheric constituents from satellite observations of solar irradiance has made significant advances the recent years, opening new horizons in environmental studies and extending observational coverage from individual scarce observations to the global view of short- and long-lived tropospheric constituents. Retrievals of tropospheric trace constituent distributions from satellite observations now provide a concise and global view of the state and of the evolution of the atmosphere. They are valuable for understanding atmospheric responses to natural and human driven emissions, meteorology and climate changes.

For two decades now, the number and type of observations, coupled with improvements in the retrieval algorithms for tropospheric trace constituents and the validation of the retrieved products (trace gases atmospheric columns and profiles, aerosol parameters, fire counts, etc.) has resulted in an increasing confidence in the observations of the Earth's surface and troposphere from space. Thus monitoring air pollution

M. Kanakidou
Environmental Chemical Processes Laboratory, Department of Chemistry, University of Crete, Heraklion, Greece

M. Dameris
Deutsches Zentrum für Luft- und Raumfahrt, Institut für Physik der Atmosphäre, Oberpfaffenhofen, Germany

H. Elbern, L. Nieradzik and A. Strunk
Rhenish Institute for Environmental Research at the University of Cologne, Köln, Germany

M. Beekmann
Laboratoire Interuniversitaire des Systèmes Atmosphériques (LISA) CNRS, Université Paris, Est et Paris 7, Créteil, France

I.B. Konovalov
Institute of Applied Physics, Russian Academy of Sciences, Nizhniy Novgorod, Russia

M.C. Krol
Meteorology and Air Quality, Environmental Sciences Group, Wageningen University, Wageningen, The Netherlands

J.P. Burrows et al. (eds.), *The Remote Sensing of Tropospheric Composition from Space*, Physics of Earth and Space Environments, DOI 10.1007/978-3-642-14791-3_9,

from space is now close to reality and this information can be used for the definition of environmental strategy and control.

Retrieval products of both trace gas and aerosol distributions seen from satellites, are starting to be widely used by the atmospheric modelling community for evaluating models, process studies, improving emissions estimates and estimating environmental and climate impacts occurring in the Earth system due to emissions and chemistry of greenhouse gases and aerosols.

Uncertainties and approximations are associated with both the retrieved data products and the environmental model simulations. These have to be taken carefully into account when using satellite observations jointly with data derived from numerical modelling studies for detecting and quantifying atmospheric changes. In particular, although satellites provide a global view of the atmosphere, this is mostly a composite of several overpasses at specific times of the day and with a specific daily frequency. For instance, SCIAMACHY passes over an area at around 10:30 local time and follows almost the same orbit every 6 days at the equator, whereas GOME-2 monitors an area around 9:30 every 1.5 days and OMI performs more frequent observations with an overpass once a day at 13:30 local time. Similarly the atmospheric models use a variety of horizontal and vertical resolutions ranging for global models from about $5^\circ \times 5^\circ$ to a few tenths of a degree in latitude by longitude and from nine to several tens of vertical levels that also vary in thickness as a function of altitude and location (Stevenson et al. 2006). The model spatial resolutions mostly do not coincide with the spatial resolutions of the satellite sensor observations (Fig. 9.1). This has to be taken into account when comparing satellite data with model results, particularly for short lived species that have a high spatial and diurnal variability such as NO_2, HCHO, and CHOCHO (Vrekoussis et al. 2004; Velasco et al. 2007).

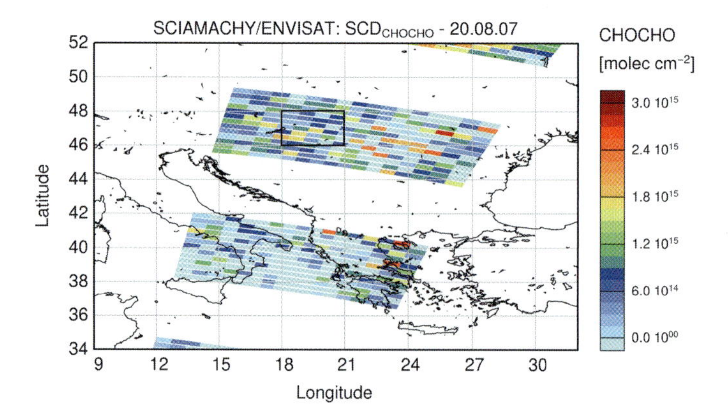

Fig. 9.1 Tropospheric slant column densities of glyoxal, CHOCHO, retrieved from SCIAMACHY sensor observations over south-eastern Europe on 20th August 2007. To demonstrate the non collocation of satellite pixels and model grid boxes, the satellite pixels, which are of variable size, are shown in colour and a model grid box of $2^\circ \times 3^\circ$ resolution is marked in *black* (personal communication from M. Vrekoussis, IUP, University of Bremen).

Using observations of the back scattered solar radiation at the top of the atmosphere and measurements of the extra terrestrial solar irradiance enables the slant column (SC) of a trace gas in a particular wavelength region to be determined. This SC depends on the length of the path of the photons through the atmosphere, the air mass factor (AMF) and the absorptions of the trace gas in a given altitude. So the AMF also depends on the amount of multiple scattering, the fraction of photons reflected at the surface in the direction of the satellite, and therefore the vertical profile of the trace gas (see Chapter 1). The AMF thus provides *a priori* information on the combined effect of all factors that affect the transfer of radiation in the atmosphere and allows conversion of the slant column of atmospheric constituents seen by the satellite sensor to vertical columns or profiles. This information is produced from atmospheric observations and model simulations that take into account the presence of clouds and aerosols, *surface* albedo, the shape of the constituent's vertical profile and temperature in the atmosphere (see in www.iup.uni-bremen.de/E-Learning/section on retrieval procedures and column measurements, and Chapter 1).

One possibility would be to determine in the model the SC and compare model SC with retrieved SC. This would be consistent. However vertical column amounts are easier to comprehend and thus are more likely to be used. It is evident that the quality of the retrievals of the vertical total or tropospheric columns of trace gases and the outcomes of the synergistic use of models with satellite observations strongly depend on the assumptions and *a priori* knowledge being used to determine the AMF.

A similar issue arises when using measurements of the emerging thermal infrared radiation and optimal estimation retrieval techniques. For the retrievals of atmospheric profiles, such as those of CO from the MOPITT instrument, information on the sensitivity of the retrieval to the real profile of the studied atmospheric constituent is required, as well as the *a priori* constituent's profile. This is described by averaging kernel matrixes (Deeter et al. 2003), unique for each retrieval (www.eos.ucar.edu/mopitt/data) that have to be taken into consideration when satellite retrievals are synergistically applied with models.

As an example, Fig. 9.2 shows MOPITT observations of CO columns and compares them with the GEOS-Chem model results sampled along the satellite sensor orbit track and using the MOPITT averaging kernels (Hudman et al. 2004). Asian anthropogenic CO is seen in both distributions. Note that UV/vis/NIR retrievals have no or very limited vertical resolution and so only columns can be retrieved and compared to models.

Provided the sensitivity, for example expressed through the averaging kernels of the retrieved data products, when using DOAS or optimal estimation or other retrieval approaches are appropriately taken into account, satellite observations are and have been synergistically used with modelling for several key objectives:

- To improve our understanding of atmospheric chemistry and its evolution with time under anthropogenic and natural stresses and evaluate models, both on global and smaller scales;

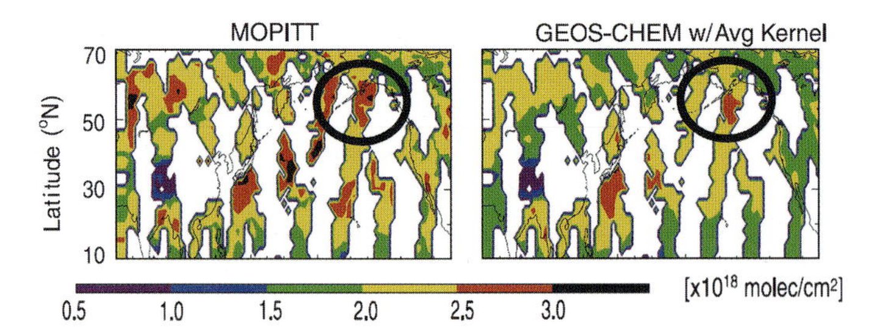

Fig. 9.2 Transpacific Asian pollution event shown by MOPITT observations of CO columns on 4th May 2002 at 00:00 UTC (*left*). The *right panel* shows the corresponding GEOS-Chem model results sampled along the MOPITT orbit tracks and with MOPITT averaging kernels applied. The *black circles* show the plume location (figure adopted from Hudman et al. (2004)).

- To identify the origin and evaluate the environmental consequences of the atmospheric composition changes as seen from space;
- To initiate chemical transport models by using satellite derived distributions of atmospheric constituents as an input for chemistry transport modelling (for instance fire counts observations as a proxy for biomass burning emission distribution; vegetation and chlorophyll-a distributions used to parameterise biogenic emission from the terrestrial and the marine ecosystems, etc.) and thereby to account for processes that are not explicitly resolved in the models;
- To assimilate retrieved data products from satellites to improve model prognostic results;
- To evaluate and improve emission estimates and atmospheric trends (forward and inverse modelling).

This research field, coupling models and satellite data products, is evolving rapidly. In the following sections a flavour of the results obtained from the above scientific applications and their principles rather than an exhaustive list of studies performed is provided. Selected investigations of tropospheric model evaluation, species origin and sources identification are outlined in Section 9.2. Principles, examples and needs for inverse modelling are presented in Section 9.3 and objectives, methods and examples for retrieved data assimilation are discussed in Section 9.4. Overall conclusions on the state of the art in the field and challenges for future research on synergistic use of satellite retrievals and atmospheric models are given in Section 9.5.

9.2 Use of Satellite Data for Process Understanding and Model Evaluation

Consistent global information of the chemical composition and the dynamics of the Earth's atmosphere are provided by space-borne instrument measurements. Satellite data are therefore a major corner stone for better understanding of individual atmospheric processes and feedback mechanisms. In addition, the synergistic use of observations and respective data derived from studies with numerical atmospheric models helps to improve the knowledge of processes driving atmospheric variability and changes on different time scales. Discrepancies between observations and model results can help to identify gaps in our understanding of dynamical, physical and chemical processes in the atmosphere. A detailed evaluation of atmospheric models is necessary to determine their ability to reproduce adequately atmospheric variability and changes. The exact knowledge of strengths and weaknesses of such models is required to enable solid assessments of the future evolutions of atmospheric chemical composition and climate.

9.2.1 Understanding Atmospheric Chemistry

In the recent decade work has been intensified to develop numerical model systems to describe the whole Earth system taking into account interactions, variations and feedbacks of the various compartments of the Earth system, including the atmosphere. Previously, atmospheric models required a number of input parameters, relevant to processes that are not explicitly resolved by these models and which are used as boundary conditions. As a result of the increasing availability of relevant satellite observations, processes in atmospheric models can now be driven by taking the initial data from satellite observed parameters. For example, in the absence of coupled dynamic vegetation/emissions/fire models, satellite retrievals of terrestrial vegetation and fire counts are used to parameterise biogenic VOC emissions or deposition (Guenther et al. 2006), and biomass burning emissions (Mota et al. 2006; Giglio et al. 2006). Synergistic use of products from several satellites allows the construction of long data series. For instance, van der Werf et al. (2006) provided computations of global biomass burning emissions that are widely used by the scientific community. They have used measurements from MODIS in conjunction with ATSR and VIRS satellite data covering a 7-year period. Chlorophyll-a distributions seen from space are also introduced as input into the models to parameterise emissions to the atmosphere from the marine ecosystems (O'Dowd et al. 2008; Arnold et al. 2009; Myriokefalitakis et al. 2010). Satellite retrievals of trace constituents such as total O_3 column, HNO_3 or CO levels in the upper troposphere/lower stratosphere and aerosol optical thickness (Lelieveld and Dentener 2000; Barret et al. 2008; Ito and Penner, 2005) have also been used in models when the driving processes, such as stratospheric chemistry or aerosol dynamics, are not explicitly resolved.

In addition to this one way flow of information from satellite retrievals to models, scientific advances can be achieved through comparisons of model results from targeted simulations with satellite retrievals of trace constituents. Such investigations enable the understanding of atmospheric processes, source identification and quantification, detection and evaluation of the long range transport of pollutants and its impacts as well as model evaluation. Examples of such investigations are discussed in the following sections.

a Formaldehyde, HCHO: A Proxy for VOC Emissions

Formaldehyde, HCHO, is a high-yield oxidation product of numerous VOCs in the atmosphere, including isoprene that is emitted in large amounts from terrestrial vegetation. Due to the short lifetime of HCHO (globally about 4 h) the measured HCHO tropospheric columns are expected to be correlated with the local VOC emissions weighted by the HCHO yield. The signal is smeared out and displaced in the atmosphere due to horizontal transport (Palmer et al., 2003) and diffusion. However, when focusing on areas of the size of a satellite pixel or even larger as those of a global Chemistry-Transport Model (CTM) i.e. with a horizontal resolution of several hundreds of km^2, this displacement should be negligible. Palmer et al. (2003) have developed a methodology based on the synergistic use of the global CTM GEOS-Chem and satellite retrievals of HCHO columns to constrain isoprene emissions from the terrestrial biosphere, taking into account the lifetimes of HCHO and VOC. They assumed that HCHO column variability was mainly linked to isoprene emissions. Model results have been sampled along the ensemble of GOME orbit tracks and used to derive linear relationships between HCHO columns and isoprene emissions over North America in the model. To minimise biases and maximise consistency between retrievals and model results, HCHO columns have been retrieved from GOME based on AMF derived from the model. Clouds, the primary error source in AMF calculations (Millet et al., 2006), have been filtered out using the cloud fraction data from the same sensor. Using GEOS-Chem, Palmer et al. (2003) calculated a mean HCHO molar yield of 1.2 to 1.96 that is consistent with laboratory experiments and with aircraft HCHO and isoprene profile observations (1.6 ± 0.5; Millet et al. (2006)). This ratio allows one to estimate the isoprene emissions from the HCHO columns retrieved from GOME. The retrieval errors, combined with uncertainties in the HCHO yield from isoprene oxidation, have been estimated to result in a 40% (1σ) error in inferring isoprene emissions from HCHO satellite observations (Millet et al., 2006).

Guenther et al. (2006) pointed out the potential importance of biomass burning and anthropogenic emission contributions to HCHO signal that introduces errors to the emission estimates. However, Palmer et al. (2003; 2006) have also shown that VOCs other than isoprene generally either have excessive smearing out or insufficient emission relative to the HCHO detection limit, so that HCHO column observations from space are highly specific to isoprene. Millet et al. (2008) used a similar approach to derive isoprene emissions using the GEOS-Chem model and space observations of HCHO columns by the OMI sensor, with 13×24 km^2 nadir footprint and daily global coverage. They also concluded that the spatial distribution of HCHO columns from OMI follows that of isoprene emission except in few urban locations, like Houston, where anthropogenic hydrocarbon emissions are detectable from space. Fu et al. (2007) performed a similar analysis of GOME retrievals focusing on east and south Asia and found an underestimate of biogenic VOC emissions over China in the current estimates.

Marbach et al. (2009) reported the first detection of signal from ship emissions in the GOME derived tropospheric HCHO columns over the Indian Ocean (Fig. 9.3),

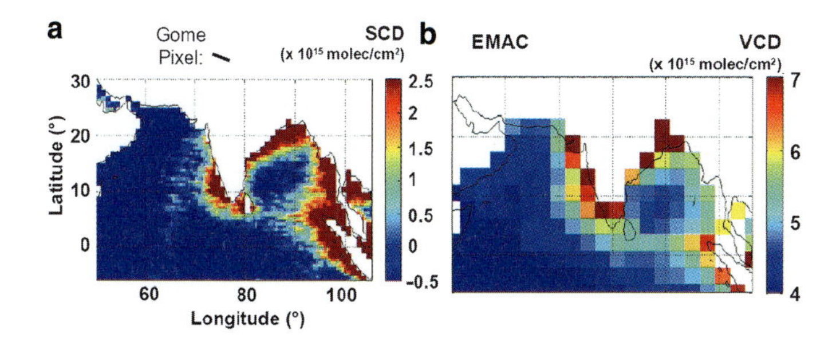

Fig. 9.3 HCHO distribution over the Indian Ocean (land masses are masked out). Figures have been adapted from Marbach et al. (2009). (**a**) GOME SCDs during winter (January to March) 1996–2002 with cloud fractions below 20% are averaged. The ship track is visible from Sri Lanka up to about half the distance to Sumatra. For illustration, size and orientation of a single GOME pixel is displayed above panel a. (**b**) EMAC model results of the mean HCHO VCDs for winter (January to March) 1997–2002, integrated up to 50 hPa (using EMAC local time between 10:00–11:00 a.m.).

where conditions often favour plume detection since ships follow a single narrow track in the same east-west direction as the GOME pixel scanning. From the 7-year composite of cloud free observations, they evaluated a mean HCHO column enhancement over the shipping route of about 2×10^{15} molecules/cm^2. Although the pattern of this enhancement is reproduced by their Climate-Chemistry model EMAC, the HCHO columns are underestimated by a factor of two, when satellite data and model results are similarly sampled and spatially averaged. The discrepancy is tentatively attributed to an underestimate in the emission inventories and their atmospheric dilution as well as to the rather coarse resolution of the model. This is limiting the proper simulation of the fast high NOx ship plume chemistry that enhances oxidation capacity in the marine environment.

b Glyoxal, CHOCHO: Source Apportionment

Glyoxal, CHOCHO, has recently been observed from space (Wittrock et al. 2006). CHOCHO is known to be mostly a product of biogenic VOC oxidation and has been suggested as indicator of secondary aerosol formation in the troposphere (Volkamer et al. 2005). However, a number of anthropogenic hydrocarbons, such as acetylene and aromatics, have been positively identified as CHOCHO precursors. Myriokefalitakis et al. (2008) investigated the contribution of pollution to the CHOCHO levels using the global 3-dimensional (3-D) CTM TM4-ECPL in conjunction with the respective "Vertical Column Amounts or Densities" of CHOCHO retrieved from the SCIAMACHY sensor observations in 2005.

A series of simulations has been performed accounting for various secondary sources as well as a potential primary source of CHOCHO from biomass burning. The simulations have been evaluated by comparison with the retrieved columns of CHOCHO both on an annual (Fig. 9.4) and on seasonal basis. The observations

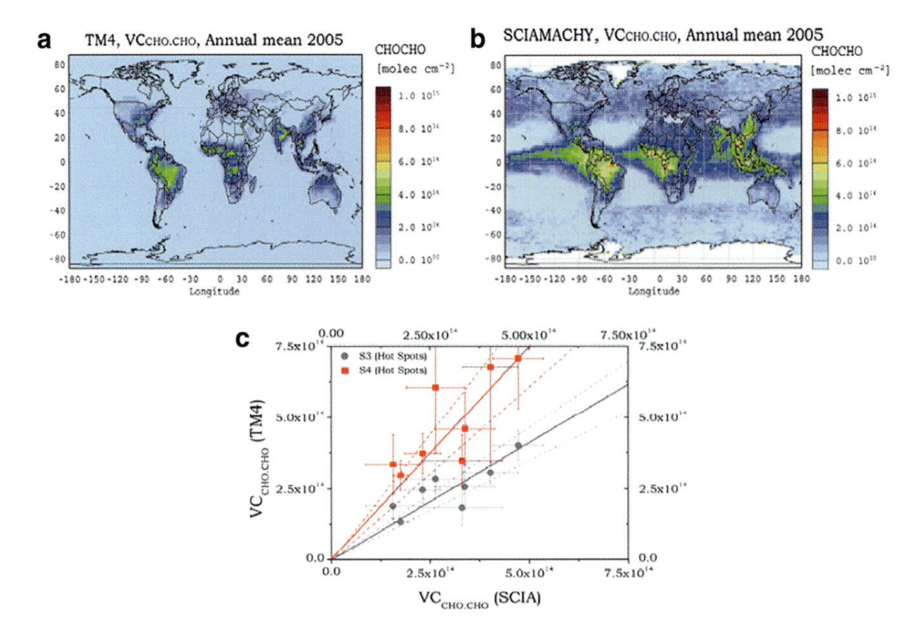

Fig. 9.4 Global annual mean column distribution of glyoxal (CHOCHO) (2° × 3° grid) for the year 2005 (in molecules/cm^2). (**a**) Simulated by TM4-ECPL, taking into account all known photochemical CHOCHO sources; (**b**) Retrieved from the measurements made by the satellite based sensor SCIAMACHY; (**c**) Comparison of annual mean CHOCHO columns from the TM4-ECPL simulations (*black circles*: S3 accounts only for the secondary sources of CHOCHO; *red squares*: S4 accounts for all secondary sources and for a potential primary source of CHOCHO from biomass burning) with the SCIAMACHY data products (in units of molecules/cm^2). Binned data over continental hot spot areas (figure adapted from Myriokefalitakis et al. (2008) where more details can be found).

have been gridded to 2° × 3° in order to fit the model's grids and both the retrieved and the calculated columns have been used. When accounting only for the secondary sources of CHOCHO in the model, the model underestimates CHOCHO columns observed by satellites, and the model fails to simulate the high CHOCHO columns retrieved over tropical oceans. This is tentatively attributed to outflow from the continents and local primary oceanic biogenic or secondary sources of HCHO and CHOCHO that are not taken into account in the model or, alternatively, to an overestimate of CHOCHO columns in the retrievals. Elucidation of this discrepancy between satellite observations and model results requires further targeted experiments as well as forward and inverse modelling investigations.

Using primary emissions of about 7 Tg yr^{-1} of CHOCHO from biomass burning and anthropogenic combustion sources in the model, leads to an overestimate of CHOCHO columns by the model over areas of intensive emissions (Fig. 9.4). For a global mean lifetime for CHOCHO of about 3 h, their model evaluates the global annual mean CHOCHO burden in the model domain at 0.02 Tg equal to the global burden seen by SCIAMACHY over land for the year 2005 (Myriokefalitakis et al., 2008). These results point to the need to understand the presence of CHOCHO over the tropical oceans. Similar conclusions are drawn by Fu et al. (2008) using the GEOS-Chem global CTM. In contrast to HCHO, the secondary anthropogenic contribution from fossil fuel and industrial VOCs emission oxidation to the CHOCHO columns is found to reach 20–70% in the industrialised areas of the northern hemisphere, suggesting that concurrent observations of HCHO and CHOCHO observations over specific locations could provide proxies for VOC emissions (Myriokefalitakis et al., 2008).

Currently there is much interest in the CHOCHO and IO signals over the oceans. There are indications that in the Pacific the regions of elevated CHOCHO, IO, and also possibly HCHO are over areas which contain phytoplankton and in particular diatoms (personal communication, J. Burrows).

c Determining Dominant Chemical Pathways: Air Pollution Impact

Correlations between atmospheric trace constituents provide important information about the dominant atmospheric processes, and this novel capability in satellite remote sensing has important implications not only for investigations of air pollution, but also for air pollution control strategy. For instance, satellite derived HCHO and NO_2 columns can be used to investigate tropospheric O_3 photochemistry. Martin et al. (2004) applied the GEOS-Chem model to evaluate the potential of the ratio of HCHO columns to tropospheric NO_2 columns as an indicator of surface ozone – NO_x ($NO_x = NO + NO_2$) – VOC sensitivity over polluted areas. Relying on these model results, satellite data analysis of the HCHO/NO_2 column ratios have shown the consistency of GOME observations over polluted areas with current understanding of surface O_3 chemistry based on *in situ* observations. The satellite-derived ratios indicate that surface O_3 production is NO_x-limited throughout most continental regions of the northern hemisphere during summer. Exceptions include major urban and industrial centres such as Los Angeles and industrial areas of Germany that tend to be NO_x-saturated (and thus VOC limited). The NO_2 derived from GOME also yields a geographical transition to NO_x-sensitive regime downwind of these centres and a seasonal transition in the autumn when surface O_3 becomes less sensitive to NO_x and more sensitive to VOCs.

The impact of pollution on the photochemical enhancement of O_3 can be also derived from HCHO and NO_2 columns observed from space used in conjunction with chemical box and Langrangian (trajectory) models. For instance, Ladstätter-Weißenmayer et al. (2007) using chemical box calculations associated with the GOME-observed NO_2 and HCHO tropospheric columns, found a potential of daily photochemical enhancement in the tropospheric O_3 columns of about 0.8–1 Dobson Units (DU, equivalent to 2.7×10^{16} molecules/cm^2) and a daily potential of regional photochemical build-up within upwind polluted air masses of about 2–8 DU over Crete in the eastern Mediterranean during spring. At most 10–20 DU of tropospheric O_3 have been attributed to stratosphere-troposphere exchange (STE) whereas the total observed variability in tropospheric O_3 derived from both space and ground based observations was about 25 DU.

CO and aerosol satellite observations also provide information on tropospheric air quality (Chapter 6). Both, CO columns and the aerosol optical depth (AOD) show oxidant-driven seasonal variation since oxidants act both as a source and a sink for CO and a production pathway for secondary aerosols. On global scales, fine mode AOD is driven by sulfate production although carbonaceous particles can be also of importance over several locations (Zhang et al. 2007), particularly during biomass burning events. Edwards et al. (2004) analysed global four year records of concurrent CO and fine mode AOD retrievals from the MOPITT and MODIS observations, both on board the Terra satellite. They concluded that the observed CO and AOD seasonal cycles were several months out of phase, with perturbations occurring during sporadic biomass burning emissions when carbonaceous particles dominate AOD. During such events the retrieved CO columns and AOD are well correlated. Anomalous high pollution observed from space in the northern hemisphere in winter-spring of 2002–2003, has been analysed based on the fire counts from MODIS and on global model simulations with MOZART-2. Artificially releasing pulses of CO over the fire locations in the model and during four other months of the year enabled the evaluation of the persistence of CO in the atmosphere. Edwards et al. (2004) calculated that the build-up of CO in the model for a pulse in October was twice as large as for a pulse in July. This reflects the e-folding time of CO that was calculated to vary over the studied area from about 1.5 months in July to about 3.6 months in October. Thus, the timing of the burning (in late summer-early fall) was favourable for a build-up of CO to anomalously high values in the northern hemisphere in winter compared to other years.

Lelieveld et al. (2009) synergistically used *in situ* aircraft observations of O_3 together with SCIAMACHY and TES satellite sensor observations of NO_2 and O_3, to perform model simulations with the EMAC model to study the origin of observed high O_3 levels over the Persian Gulf. They concluded that the Persian Gulf region is a hot spot of photochemical smog where air quality standards are violated throughout the year. EMAC simulations allow the identification of long distance transport of air pollution from Europe and the Middle East, natural emissions and stratospheric O_3 to the relatively high background O_3 mixing ratios.

d Understanding Differences Between Retrievals and Model Results

Comparison of model results with retrievals of atmospheric constituents from satellite observations is not restricted to the analysis of observations but can also point out deficiencies in the retrieval algorithms and thus initiate their improvements, specifically in the assumptions made for the determination of AMF or the averaging kernels.

For example Martin et al. (2002) have compared the distribution of tropical tropospheric O_3 columns retrieved from TOMS with the GEOS-Chem model results together with additional information from *in situ* observations. They found major discrepancies between model results and TOMS retrievals over northern Africa and southern Asia where the TOMS retrieved O_3 columns did not capture the seasonal enhancements from biomass burning found in the model and in aircraft observations. Martin et al. (2004) attributed this discrepancy to the poor sensitivity of TOMS to Rayleigh scattering that is important for retrieving low troposphere O_3 enhancements by biomass burning. Thus they developed an efficiency correction to the TOMS retrieval algorithm that accounts for the variability of O_3 in the lower troposphere. This correction increased the retrieved O_3 columns over biomass burning regions by 3–5 DU and decreased them by 2–5 DU over oceanic regions, improving the agreement with *in situ* observations. The correction explained about 5 DU of the "tropical Atlantic paradox", i.e. the enhanced tropical tropospheric column of O_3 over the southern tropical Atlantic retrieved during the northern African biomass burning season in December to February. The remainder of the paradox was reproduced by the model; it was attributed to the combination of upper tropospheric O_3 production from lightning NO_x, persistent subsidence over the southern tropical Atlantic and cross-equatorial transport of upper tropospheric O_3 from northern mid-latitudes in the African "westerly duct".

Another recent example is the study by Bergamaschi and Bousquet (2008), who identified a bias in the dry column of CH_4, retrieved by Frankenberg et al. (2008a) from SCIAMACHY data, based on simultaneous assimilation of surface observations and satellite data. A large latitudinal varying bias-correction of the satellite data was required to make these data compatible with surface observations. The bias is, in fact, absent in the most recent CH_4 retrieval algorithm and the change has been attributed to the previously poor knowledge of spectroscopic absorption lines used in the former retrieval algorithm (Frankenberg et al. 2008a), leading to improved CH_4 emission estimates (Frankenberg et al. 2008b). However, it is worth noting that other retrievals of the dry column do not show the same bias (Schneising et al. 2008). Overall the evolution of the accuracy of the retrieval algorithms through improved instrument calibration and validation exercises using measurements and models enables the maximum information content to be retrieved. This process is essential, both for testing models and improving retrieval approaches and algorithms.

9.2.2 Model Evaluations – Comparison with Observation

Recently several tropospheric CTM comparison exercises relied on satellite retrievals to supplement the traditional ground-based and aircraft observational data (Velders et al. 2001; van Noije et al. 2006; Shindell et al. 2006; Dentener et al. 2006a; Textor et al. 2006; Kinne et al. 2006). The year 2000 has been used as base year for several global modelling studies. This year benefits from documented emission inventories (Dentener et al. 2006a; 2006b) for both trace gases and aerosols in the framework of the AEROCOM exercise (Aerosol Comparisons between Observations and Models) that is focussed on aerosols (Textor et al. 2006; Kinne et al. 2006; Schulz et al. 2006). These inventories have been also used for the ACCENT intercomparison exercise that focused on tropospheric O_3,

NO_2 and CO budgets (van Noije et al. 2006; Dentener et al. 2006b; Stevenson et al. 2006; Shindell et al. 2006). The proper comparison of model results with satellite retrievals allows the evaluation both of the emissions used in the model as well as of the parameterisations of their fate in the atmosphere that reflects our understanding of atmospheric processing. In the following we will illustrate model evaluation procedures by comparing with satellite measurements of a series of relevant species.

a NO_2

NO_2 controls tropospheric O_3 production. Its levels in the troposphere show trends that are driven by human activities and these have been observed from space (Richter et al. 2005). The importance of NO_2 for tropospheric O_3 and the consistency of the retrieved tropospheric distributions of NO_2 stimulated their use for model evaluation. Moreover, these satellite observations allow NO_2 global pollution to be evaluated.

Historically the GOME sensor aboard the ERS-2 satellite provided a unique opportunity to compare globally, model calculated NO_2 columns, including, for the troposphere, those from retrievals of remote sensing observations (Lauer et al. 2002). In order to overcome the shortcomings in comparing model results with satellite retrievals outlined in the introduction, van Noije et al. (2006) used 17 different global CTMs that computed daily tropospheric NO_2 column densities for the year 2000. For each model, the computed NO_2 columns were sampled at the satellite overpass time collocated with the measurements to account for sampling biases due to incomplete spatio-temporal coverage by the instrument. The ensemble of the 17 model results has been compared with the mean of the NO_2 columns retrieved from GOME using three different retrieval algorithms (Fig. 9.5).

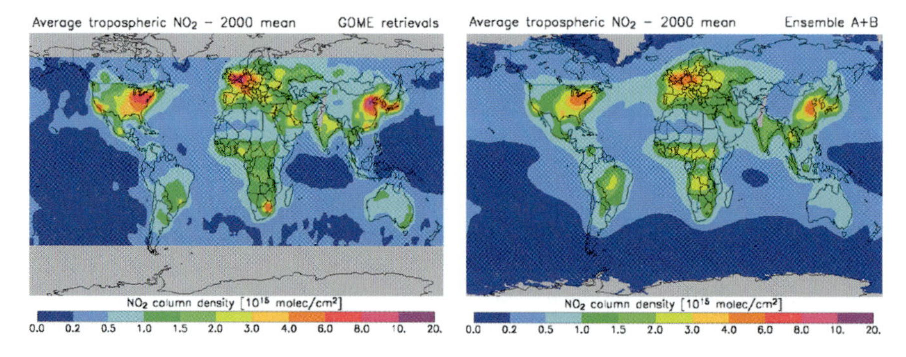

Fig. 9.5 Ensemble average annual mean tropospheric NO_2 column density for three different GOME retrievals (*left panel*) and the full model ensemble (A + B; *right panel*). These quantities have been calculated after smoothing the data to a horizontal resolution of $5^\circ \times 5^\circ$ (figure from van Noije et al. (2006)).

Three major continental regions of high NO_2 tropospheric column densities are indicated in the mean columns derived both from the 17 models and from the three GOME retrievals: North America, western Europe, and China (Fig. 9.5). These regions are subject to high pollution emissions. The averaged model maxima of 6–8 $\times$ 10^{15} molecules/cm^2 are smaller than the GOME observed values, which exceed 10 $\times$ 10^{15} molecules/cm^2. These discrepancies between models and retrievals could neither be explained by *a priori* profile assumptions made in the retrievals, nor by diurnal variations in anthropogenic emissions (van Noije et al. 2006). They have been attributed (Dentener et al. 2006b) to the assumed NO_x emissions that may be unrealistically low in these regions, in particular over the rapidly developing parts of eastern China and South Africa. Similar conclusions for the emissions over Asia were drawn by an earlier and less extensive model intercomparison with GOME observations (Velders et al. 2001). In regions dominated by biomass burning, such as in Africa and South America, the models overestimate the retrievals during the dry season. The comparison improves when using biomass burning emissions specific to the year 2000 instead of a 5-year average inventory used for the base simulations (van Noije et al. 2006), pointing to the importance of the inter-annual variability in the emissions. Another significant finding is that the differences in the GOME retrievals are in many instances as large as the spread in model results (10–50% in the annual mean over polluted regions). This means that in only a few cases, such as China, can robust statements on the underestimation of NO_x emissions be made (Dentener et al. 2006b; van Noije et al. 2006). The findings imply that top-down estimations of NO_x emissions from satellite retrievals of tropospheric NO_2 are strongly dependent on the choice of model and retrieval.

Recently, Boersma et al. (2009) detected diurnal variations of NO_2 over Israel and Egypt synergistically using two different satellite sensor retrievals and analysing them with CTM results: those from SCIAMACHY that observed the atmosphere at 10:00 and those from OMI that overpasses at 13:45. They demonstrated that NO_2 temporal variability over source regions can be followed from space. They derived NO_2 columns about twice as high in winter as in summer and a strong weekly cycle with NO_2 almost twice as low on Saturdays than on weekdays. The diurnal difference between SCIAMACHY (10:00) and OMI (13:45) NO_2 is seen to maximise in summer when SCIAMACHY is up to 40% higher than OMI, and minimise in winter when OMI slightly exceeds SCIAMACHY. The model simulations indicated that a much stronger photochemical loss of NO_2 in summer than in winter is needed to explain these observations.

Another source of NO_x that can be seen from space under certain conditions is that from ships over the remote oceans. Eyring et al. (2007) studied the impact of ship emissions on atmospheric chemistry and climate, using multi-model simulations of Chemistry-Climate Models (CCMs) that have been evaluated for their response to ship emissions. Part of this evaluation used satellite data, particularly the recently observed enhanced tropospheric NO_2 columns over the Red Sea and along the main shipping lane to the southern tip of India, to Indonesia and northwards towards China and Japan (Beirle et al. 2004; Richter et al. 2004).

The tropospheric NO_2 columns derived from SCIAMACHY nadir measurements from August 2002 to April 2004 (Richter et al. 2004) have been compared to ensemble means of the models representing the year 2000. The ensemble mean was derived from eight models that provided tropospheric NO_2 columns at 10:30 a.m. local time, which is close to the overpass time of the ERS-2 satellite. For this comparison, individual model results and SCIAMACHY data were interpolated to a common grid ($0.5° \times 0.5°$). Fig. 9.6 shows that the ensemble of the models reproduces the magnitude and the general pattern of the tropospheric NO_2 columns over the remote ocean observed by SCIAMACHY. However, the shipping signal that is clearly visible in the satellite data with a high horizontal resolution ($30 \times 60\ km^2$), does not appear in the model results with a much lower typical resolution of $5° \times 5°$. In addition, shipping routes in that area are rather close to land and thus the models grid boxes close to the coast are dominated by NO_x emissions from land sources which are much higher (see also Franke et al. (2009)).

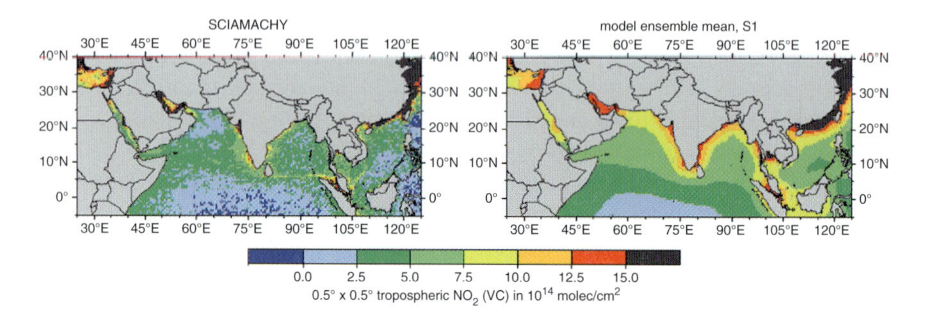

Fig. 9.6 NO_2 tropospheric columns retrieved from SCIAMACHY observations (*left*) and ensemble of model simulations (*right*) interpolated to $0.5° \times 0.5°$grid (figure from Eyring et al. (2007)).

There is clearly a need for model data comparisons with satellite observations over remote oceans. However, such data are inhibited by the distributed nature of the ship emissions over the remote oceans leading to dilution of the emissions, which makes it difficult to distinguish the shipping signal from the effect of long-range transport of polluted air, such as that from the United States towards Europe. The emissions between India and Indonesia present a unique emission pattern. Increasing spatial resolution both of models and of satellite observations might allow us to resolve ship emissions over other oceanic locations.

b CO

In the frame of the ACCENT AT2 global model intercomparison exercise, Shindell et al. (2006) compared near-global satellite observations from the MOPITT instrument and local surface measurements with present-day CO simulations by 26 state-of-the-art atmospheric CTMs and CCMs. For this purpose, they used monthly mean daytime values derived from version 3 retrievals from MOPITT gridded at a resolution of 1° × 1°; the models' coarser grids were sub-sampled at this same resolution for comparison. In addition the models were sampled like MOPITT observations, using space and time varying averaging kernels from the MOPITT retrievals. This procedure minimises potential biases due to the *a priori* information used in the retrievals. Biases may also be induced by the differences between the models' full temporal and spatial averaging and the satellite's limited sampling time and exclusion of cloudy areas. Such biases are expected to be significant for short-lived atmospheric constituents like aerosols but quite small for CO given its relatively long lifetime.

In general, Shindell et al. (2006) pointed out that the models do not adequately capture CO accumulation during the OH-poor winter. The models underestimate both magnitude and seasonality of the CO retrievals throughout the entire extra tropical troposphere in the northern hemisphere, indicating that the biases do not merely reflect an erroneous vertical structure of modelled CO (Fig. 9.7). However, they typically perform reasonably well elsewhere. These results suggest that yearly emissions, probably from fossil fuel burning in eastern Asia and seasonal biomass burning emissions in south-central Africa are greatly underestimated in current inventories such as IIASA and EDGAR3.2. Arellano et al. (2006) performed inverse modelling of CO emissions from various geographical regions and sources from fossil fuel/bio fuel use in Asia based on MOPITT CO data and found that these emissions are almost twice as high as recent bottom-up estimates. The underestimate of the Asian CO emissions in the national estimate-based inventory is also consistent with the under-reporting of NO_x emissions (van Noije et al. 2006). In a more recent study, Arellano and Hess (2006) conducted a sensitivity analysis on the differences in the model treatment of transport on top-down estimates of CO sources. They showed that differences between CO model values are due to atmospheric transport and are of the order of 10–30%, with the highest discrepancies for Indonesia, South America, Europe and Russia.

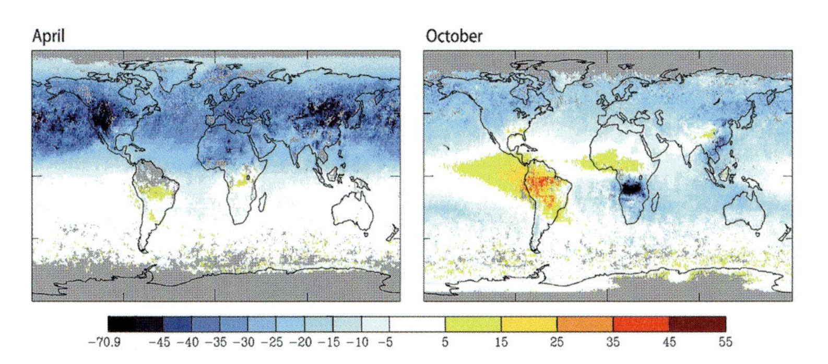

Fig. 9.7 Differences between multi-model average and MOPITT 2000–2004 average CO (ppbv). Values are shown for April (*left*) and October (*right*) for the 500 hPa pressure level (figure from Shindell et al. (2006)).

c Aerosol

Evaluation of aerosol simulations on a global scale is now customarily made through comparison of simulated annual global aerosol optical thickness (AOT) values with those obtained from remote sensing. Fig. 9.8 demonstrates how model simulations for the annual and globally averaged mid-visible AOT (at 550 nm) have changed from the work by Kinne et al. (2003) to the work by Kinne et al. (2006) in the frame of the AEROCOM exercise, and how they compare with data from remote sensing. In the lower panel the number of remote sensing references is reduced to two selections, though of higher quality; a satellite composite, which combines individual satellite retrievals (S*) and an estimate based on statistics at AERONET ground sites (Ae) (Fig. 9.8).

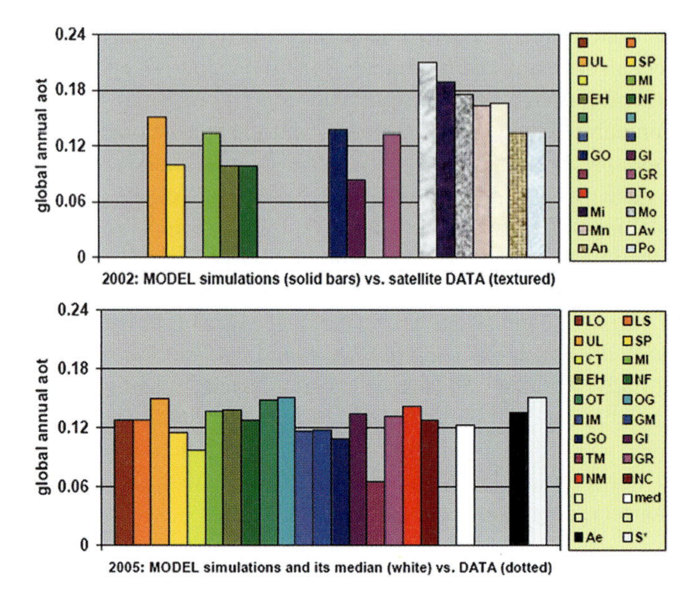

Fig. 9.8 The *upper* panel shows diversity in 2002 among models and satellite data (Kinne et al. 2003). The *lower* panel compares global annual aerosol AOT median value from the 18 models (med) with the satellite data composite (S* – see text) and the Aeronet sunphotometer network (Ae) observations. Other symbols correspond to individual models – for more explanations see text and Kinne et al. (2006).

In the earlier work, fewer models were available and the simulated AOTs exhibited a larger variability between models than in the AEROCOM exercise. The upper panel of Fig. 9.8 presents adjusted global annual averages from TOMS, MISR, MODIS, AVHRR and POLDER retrievals. The composite value (S*) is based on 3° × 3° longitude/latitude monthly averages, where preference is given to year 2000 data. At 0.11–0.14, simulated AOT values are at the lower end of global averages suggested by remote sensing from ground (AERONET about 0.135) and space (satellite composite about 0.15). More detailed comparisons, however, reveal that larger differences in regional distribution and significant differences in compositional mixture remain.

Critical for this exercise has been the production of a composite remote sensing dataset of AOT for comparison. Since all ground based remote sensing data are spatially incomplete, adjustments were needed to make global average AOT values comparable. The adjustments involved the spatially and temporally complete median field from modelling. Details are given in Kinne et al. (2006) who summarise in a table the contributing time-periods, retrieval references and known biases.

9.3 Inverse Modelling

The examples given in the previous section involved a mostly quanlitative analysis of the mismatch between models and satellite observations. To exploit satellite data in a quantitative sense, formal techniques of data assimilation and inverse modelling are required. Solving the *inverse problem* is a common task in many branches of science, where the values of model variables need to be obtained from observations (see 9.6, Appendix). Inverse modelling techniques are in widespread use today in atmospheric science for three major applications: (1) retrieval of atmospheric concentrations from observed radiances (Chapter 4), (2) optimal estimation of atmospheric model parameters and in particular of emissions, and (3) chemical data assimilation. The formal framework for these three problems is similar (Rodgers 2000). Applications presented here will focus on inverse modelling of emissions using satellite observations and also improvements to atmospheric model performances.

9.3.1 Inversions for Short-Lived Species

The use of satellite data to improve emission estimates is illustrated using NO_2 observations from space, which comprise a large part of inverse modelling applications. NO_x finds favour for several reasons. First, NO_2 satellite data products are abundant and, since the NO_2 lifetime is short, the gradients are large and easily observed from space. Second, the relationship between NO_2 columns derived from satellite measurements and NO_x emissions is direct and easy to interpret because NO_x emissions are the major driver of variability on NO_x columns. Finally, the interest for inverse modelling of NO_x emissions is fostered by the fact that these emissions are one of the key factors responsible for air pollution problems.

Many early applications of satellite measurement for estimating NO_x emissions did not explicitly involve any CTM, but used a simple mass balance method which assumed a constant lifetime of the emitted NO_x. For example, Leue et al. (2001) provided estimates of continental and global NO_x emissions, Beirle et al. (2003) investigated weekly variations of anthropogenic NO_x emissions, Beirle et al. (2004)

estimated NO_x emissions from shipping in a specific region of the Indian Ocean, Bertram et al. (2005) investigated daily variations in soil NO_x emissions. Very recently, Hayn et al. (2009) investigated in detail the spatio-temporal patterns of the global NO_2 distribution retrieved from GOME satellite observations using a generalised additive model.

Martin et al. (2003) were first to apply the Bayesian inverse modelling approach to providing inventories for NO_x emissions using satellite observations. Specifically, they performed a probabilistic combination of "top-down" and "bottom-up" emission estimates and provided a global NO_x emission inventory constrained by satellite measurements, which was claimed to be more accurate than the "bottom-up" inventory. They used the global GEOS-Chem CTM in order to define a local linear relationship between NO_x emissions. The transport of NO_x between different grid cells was disregarded. Typically they found correction factors (for *a posteriori* with respect to *a priori* emissions) of 10–20% for most of the regions, with maximum values up to a factor of two. While the use of a global CTM in the inverse modelling scheme may help to improve global emission inventories, models with much higher spatial resolution are needed in order to elaborate the constrained emission inventories for use in air quality studies. Thus, Konovalov et al. (2006a) used a regional CTM with the resolution of $0.5° \times 0.5°$ in combination with GOME and SCIAMACHY measurements to improve NO_x emission estimates on a regular model grid for western Europe (Fig. 9.9). Other novel features of their study were:

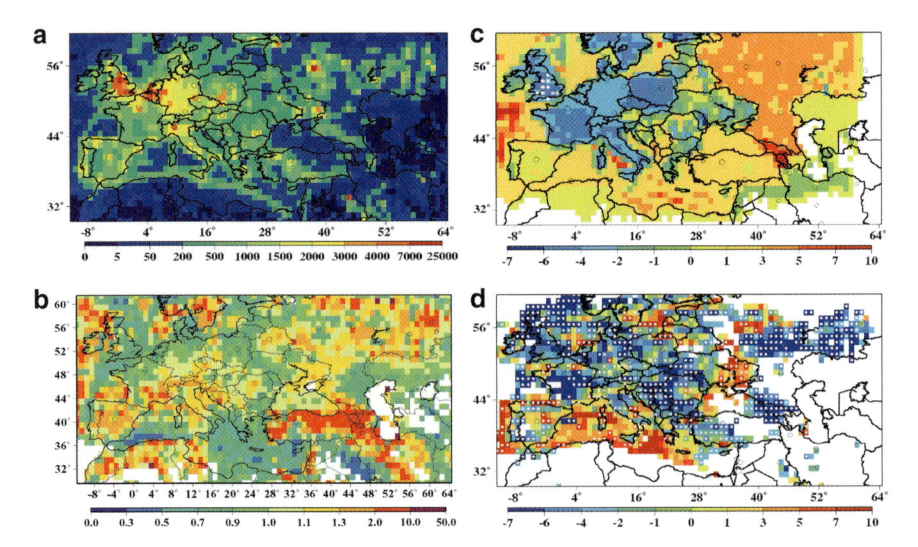

Fig. 9.9 (**a**) *A priori* estimates of summertime anthropogenic NO_x emission rates used in CHIMERE (in 10^8 molecules/cm^2/s); (**b**) *a posteriori* correction factors to them. Decadal trends (percent per year) in summertime anthropogenic NO_x emissions; (**c**) estimated with the EMEP data and (**d**) derived from satellite measurements. *Blank dots* in "**d**" mark grid cells for which the difference between the emission trends is significant in terms of 1σ (from Konovalov et al. (2006b; 2008)).

(1) an original inversion method enabling a partial accounting for the horizontal transport of NO_x, (2) the measurement based estimation of the error variances involved in the Bayesian cost function, and (3) the demonstration that the NO_x emissions constrained by satellite measurements improve the simulated near-surface concentrations of NO_2. In a later study Konovalov et al. (2006b) (Fig. 9.9) found that, on average, the uncertainties in total NO_x emissions are estimated to be about 1.7 in terms of the geometric standard deviation in Europe and about 2.1 outside Europe. The corrected emission estimates provided better agreement of the modelled results with observations for both NO_2 columns and near surface concentrations of O_3.

Although in general satellite data products do not themselves distinguish between anthropogenic and biogenic emissions, it is sometimes possible to use additional sources of information, by taking into account temporal evolution of the measured NO_2 columns and by selecting regions with dominating types of NO_x sources. For example, Jaeglé et al. (2004) estimated NO emissions from soils in Africa by combining inversions of GOME NO_2 columns with space-based observations of fires and bottom-up estimates of fossil fuel and bio fuel emissions. A similar approach was used by Wang et al. (2007) to quantify NO emissions from soils in eastern China. Martin et al. (2007) estimated NO_x emissions from lightning in tropical regions using tropospheric NO_2 columns from SCIAMACHY together with tropospheric O_3 columns from OMI and MLS, and upper tropospheric HNO_3 from ACE-FTS. In their study, the CTM GEOS-Chem was used to identify locations and time periods in which lightning would be expected to dominate the trace gas observations. Multi-annual satellite measurements already available can be used, not only to constrain different sources of emissions and improve their spatial allocation, but also to study their long-term changes. Such studies provide valuable opportunities to verify air pollution control strategies and to monitor changes in air pollution sources in the regions where the ground based monitoring networks are either sparse or absent.

Richter et al. (2005) found a highly significant increase of NO_2 columns over the industrial areas of China in a decadal period from 1996 to 2005 using a combined time series of GOME and SCIAMACHY measurements. They used the MOZART CTM to justify that these changes are caused by similar increases in NO_x emissions. Note that this study is not an inverse modelling study in a classical sense, since it does not involve any inversion of a mathematical relationship between observations and emissions.

Similarly, Kim et al. (2006) and van der A et al. (2006) provided some useful insights into NO_x emission changes in the United States and China, respectively, without employing any inverse modelling technique. Kim et al. (2009) extended their work to the western USA where they found evidence that changing legislation and changing populations have a strong impact on the NO_x emissions in cities in this region. Another true inverse modelling study of multi-annual changes in NO_x emissions was performed by Konovalov et al. (2008). They used the time series of NO_2 columns derived from the GOME and SCIAMACHY measurements in combination with the CHIMERE CTM in a Bayesian inverse modelling scheme, to estimate decadal NO_x emission trends in Europe and the Mediterranean (Fig. 9.9). Instead of looking for deviations from "expert" estimates of emissions (as is common in atmospheric inversion studies), Konovalov et al. (2008) constrained *a priori* only the minimum and maximum values of emission trends in each grid cell. Accordingly, the top-down estimates obtained can be regarded as a measurement-based alternative to trends derived from bottom-up emission cadastres. An even more direct approach involving a simple combination of satellite measurements with CTM simulations was proposed later by Konovalov et al. (2010) to estimate multi-annual NO_x emission trends in megacity regions. NO_x emission reductions in the last decade over western and central Europe are confirmed by the top-down approach, as well as increases in Spain and emissions related to shipping. Differences between the bottom-up and top-down approaches are notable, especially over south-eastern and eastern European countries. Importantly, the estimates of emission trends obtained were found to be consistent with the available surface measurements of NO_x and O_3 (available mainly over western Europe).

Satellite measurements have also been used to estimate emissions of other important short-lived species such as isoprene. Such estimates can be performed by inversion of the modelled relationship between isoprene emissions and HCHO column measurements. For example, Millet et al. (2008) performed the inversion of OMI measurements and found that the derived isoprene emissions in North America are spatially consistent with a normal bottom-up isoprene emission inventory (MEGAN) ($R^2 = 0.48–0.68$) but, on average, lower by 4–25%. Corresponding work was performed for Europe (Dufour et al. 2009) using SCIAMACHY derived HCHO columns, facing the difficulty of much smaller isoprene emissions in Europe compared with the US.

Recently Stavrakou et al. (2009), motivated by the large underestimate from the global CTMs of the observed CHOCHO columns from SCIAMACHY, investigated the possible existence of an additional CHOCHO source of biogenic origin over the land. They performed two inverse modelling scenarios for CHOCHO sources. The first included an additional primary source of CHOCHO over land and the second assumed secondary formation through the oxidation of an unspecified CHOCHO precursor with a lifetime of 5 days. As well as the extra source, the inversion scheme optimised the primary CHOCHO and HCHO emissions as well as their secondary production from other identified NMVOC precursors of anthropogenic, pyrogenic and biogenic origin. The best performance is achieved in the second scenario with the inferred total global continental CHOCHO source estimated to be 108 Tg yr^{-1}, almost twice as high as the global *a priori* source. The extra secondary source is the largest contribution to the global CHOCHO budget (50%), followed by the production from isoprene (26%) and from anthropogenic NMVOC precursors (14%). The updated emissions allowed for a satisfactory agreement of the model with both satellite and *in situ* CHOCHO observations. The large CHOCHO column amounts observed over the tropical oceans are still not explained.

9.3.2 Inversions for CO *and* CH_4

The dense, high-quality CO observations performed by the MOPITT satellite instrument gave rise to several remarkable studies applying the inverse modelling technique to estimate CO emissions. For example, Pétron et al. (2004) used the MOZART CTM and the MOPITT CO retrievals to perform a sequential Bayesian estimation of CO sources in 15 large regions of the world. The largest correction factor (about 50 %) for anthropogenic emissions was needed for eastern Asia. *A posteriori* emissions significantly improved the agreement between the simulated CO distributions and independent ground based measurements. A similar study was presented by Arellano et al. (2006). Their *a posteriori* estimate is also much higher (about a factor of 2) than the corresponding earlier estimate for eastern Asia, and significant differences were found between the earlier and post source emissions in many other regions. Stavrakou and Müller (2006) performed an inversion of MOPITT data using the adjoint of the IMAGE model (Fig. 9.10). The main goal of their study was a comparison of the large region and grid-based Bayesian inversion methods. Both methods gave similar average estimates, but the grid-based approach brings the model columns much closer to the observations because of its better spatial representativity. An interesting novel feature of the method used by Stavrakou and Müller (2006) is optimisation of the sources of the main biogenic VOC compounds simultaneously with the CO sources.

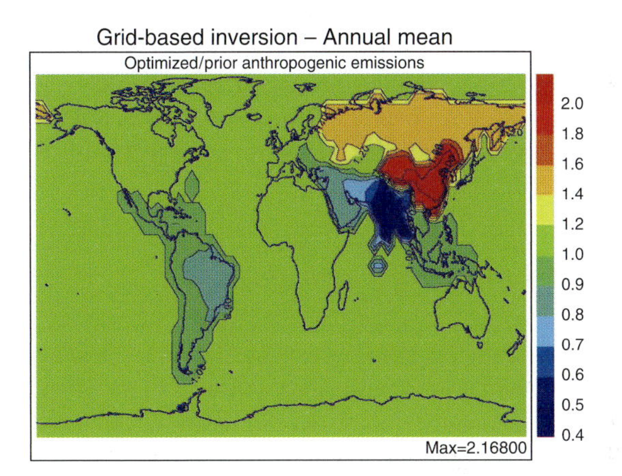

Fig. 9.10 Ratio of optimised to earlier anthropogenic emissions estimated by the inversion of MOPITT observations (adopted from Stavrakou and Müller (2006)).

The CH_4 and CO_2 retrievals are clearly improving but they still differ between different groups (Schneising et al. 2009). As CH_4 has a reasonably uniform distribution over the troposphere, due to its large lifetime (8–9 years), accurate satellite observations, with perhaps less than a few percent of error, are needed to obtain information on CH_4 fluxes, which are primarily emissions. Meirink et al. (2006) applied four-dimensional variational (4D-var) data assimilation method to synthetic measurements of atmospheric CH_4 to investigate the utility of SCIAMACHY observations for CH_4 source estimation. They concluded that SCIAMACHY observations with a precision of 1–2% can contribute considerably to uncertainty reduction in CH_4 source strengths, but that systematic observation errors well below 1% would have a dramatic impact on the quality of the derived emission; thus identification and removal of these biases is crucial. Bergamaschi et al. (2007) presented initial results of a synthesis inversion of coupled surface and satellite CH_4 observations. The use of surface measurements allowed the inverse system to compensate for potential systematic biases in satellite retrievals. They found, in particular, that a coupled inversion yields significantly larger tropical emissions compared to the *a priori* estimates or the inversion estimates based on the surface measurements only. These discrepancies have been reduced with corrected spectroscopic data that lead to more accurate CH_4 retrievals as discussed in Section 9.2.1. Meirink et al. (2008a) demonstrated the advantage of 4D-var in reducing aggregation errors by optimising emissions at the grid scale of the transport model using the 1 year surface observations of CH_4, whereas Meirink et al. (2008b) applied the 4D-var system to analyse SCIAMACHY observations with a focus over South America.

Recently, Bloom et al. (2010) have developed a simple model to combine satellite observations of CH_4 from SCIAMACHY and of gravity anomalies from the Gravity recovery and Climate Experiment (GRACE) satellite, used as proxy for water table depth, together with surface temperature field. Using *a priori* information about rice paddy distribution to isolate wetland regions from their emission estimates they found that tropical wetlands contribute 52–58% of global emissions, with the remainder coming from the outside the tropics. They also estimated a 7% rise in wetland CH_4 emissions over the period 2003–2007, due to warming of mid-latitude and Arctic wetland regions, a figure consistent with recent changes in atmospheric CH_4.

9.3.3 Need for Future Developments

Atmospheric inverse modelling based on satellite data is a new field of research, and it is obvious that the potential for satellite measurements in the given context is yet far from fully exploited. Probable future developments could include an applicable extension of a list of the atmospheric constituents, both gaseous and aerosol, for which sources can be estimated from observations (for example, VOC emissions from CHOCHO and/or HCHO observations, CO_2, SO_2 emissions, etc.). Inverse methods (in particular, Bayesian probabilistic methods) are already widely used in retrievals of satellite data. Future developments should aim at combining satellite retrieval and emission inversion into a coherent framework, because both rely on atmospheric models. Inversions of satellite measurements by means of atmospheric models can also help in estimating parameters of atmospheric processes other than emissions, particularly loss processes in regions without noticeable emissions or when sources have known (say, weekly) patterns of variation. An interesting demonstration of the potential of satellite observations in this sense was provided by Beirle et al. (2003) who estimated the atmospheric lifetime of nitrogen oxides by using GOME NO_2 measurements and a simple box model approach. In addition to chemical sinks, wet and dry deposition rates could also be inverted in future studies. Certainly, observations from a geostationary platform would allow a much finer analysis: in particular, the diurnal variability of emissions or other parameters could be inverted.

9.4 Data Assimilation

9.4.1 *Objectives and State of the Art Approaches*

Coupling models with data in a mathematically sound fashion inevitably requires data assimilation (DA) (Chapter 7). There is a dual requirement in confronting models with data: either, improved forecasting, or control of consistency between observations and model results. Improved forecasting is expected from the assimilation of meteorological and atmospheric chemistry observations, thereby creating a dynamically consistent and complete "movie" of optimal quality in some objective sense of estimation theory. Alternatively, from a scientific viewpoint, evidence is provided about whether model results and measurements are mutually consistent within predefined margins, corroborating or rejecting our system knowledge as sufficient (Bennett 2002).

Advanced DA algorithms incorporate the following subtasks: (1) Filtering the signal from noisy observations, (2) interpolation in space and time, and (3) completion of state variables that are not sampled by the observation network (Cohn, 1997). By doing this, DA serves the classical objective to estimate complete parameter fields from sparse data by chemical and physical laws, estimates forcing-fields acting on the system under investigation, tests scientific hypotheses, helps design optimal observation system configurations, and solves mathematically ill-posed modelling problems (Bennett 1992). DA can further be applied for data validation, field experiments, and climate signal detection.

DA algorithms must be designed to satisfy objective quality criteria. Spatio-temporal DA or inversion techniques are candidates for advanced methods, which are able to combine model information with data in a consistent way, while, at the same time, are able to provide a Best Linear Unbiased Estimate (BLUE), i.e. a linear unbiased estimator of the data to be assimilated by the model having the smallest dispersion matrix. Past attempts to analyse tracer fields were based on monovariate kriging techniques in the troposphere (a basic version of BLUE, Fedorov (1989)), and other purely spatial methods in the stratosphere (Stajner et al. 2001; Struthers et al. 2002).

In many cases, these approaches are equivalent to Optimal Interpolation (OI) (Daley 1991) and satisfy the BLUE property, on the spatial scale but not the temporal scale. The latter fact implies that repeated applications of OI do not force models to maintain a BLUE compatible evolution, which means that chemical imbalances, much larger than observed, cannot be avoided.

Generally, chemical DA methods produce chemical state estimates, frequently referred to as *analyses*, typically on the model grid, after assimilation of observations in model-simulated fields as background information. In contrast to meteorological conditions, chemical DA deals with a vast manifold of chemical species (in the order of 100 species in CTMs) and the tiny number of different observed compounds (less than 5 in most cases: O_3, NO_2, particulate matter, CO, SO_2). Therefore a chemical model to be used as constraint is a key for chemical DA. For this spatio-temporal DA, the BLUE property is provided by two main families of techniques, the 4D-var DA algorithm and the Kalman filter.

For 4D-var, a first successful demonstration was provided by Fisher and Lary (1995) applying a stratospheric chemical box model with a small number of constituents. The authors assessed the applicability of a variational DA method for atmospheric chemistry applications. Eskes et al. (1999) applied the variational method to a two-dimensional model for the assimilation of total satellite columns. For the troposphere, the usefulness of the variational method has been shown by Elbern et al. (1997), applying the box model version of the chemistry mechanism RADM (Regional Acid Deposition Model) (Stockwell et al. 1990). Further, the successful extension to a full chemical 4D-var DA system was demonstrated in the context of identical twin experiments and for an O_3 case study (Elbern and Schmidt 2001), using the University of Cologne EURAD regional CTM. Additional chemistry applications of the 4D-var technique were provided for both the troposphere (Chai et al. 2006) and the stratosphere (Errera and Fonteyn 2001).

Comprehensive DA setups for the troposphere have to account for the fact that, in contrast to stratospheric constituent DA and general meteorological DA, after some time of integration, the evolution of the tropospheric model state is not primarily controlled by the initial state. Rather, emission rates act as strong controlling factors, and exert a direct influence even over short timescales (ranging from seconds close to sources to days in remote areas). Furthermore, emission rates are still not sufficiently well known. Thus, emission rates must be considered as one parameter to be optimised in the DA process. More generally, the parameter to be optimised by DA must be tailored to the simulation objectives. To this end, those parameters must be chosen for optimisation by DA, which exert a strong influence on the simulation or forecast skill and, at the same time, are not sufficiently well known. Given a model parameter, the degree of priority for optimisation is indicated by the product of impact on forecast skill, as quantified by suitable sensitivity tests, and the paucity of knowledge, as quantified by error margins.

For tropospheric chemistry DA, a generalisation with respect to emissions needs to be implemented. This can be done in the incremental formulation of 4D-var by augmenting the state vector by inclusion of deviations from the underlying emission inventory, as well as deviations from a background chemical state, as shown by the work of Elbern and Schmidt (2001) and Elbern et al. (2007) with EURAD-IM. Details on this approach are given in the Appendix, Section 9.6.

An example of a non-chemical implementation of DA, targeted at optimising sources and sinks of CO_2, is provided by the CarbonTracker system (Peters et al., 2007).

9.4.2 Example Results for Tropospheric O_3 assimilation

Since most of the O_3 is in the stratosphere, obtaining accurate tropospheric O_3 measurements from space is challenging. The combination of observations with CTM results is a suitable procedure to obtain global estimates of tropospheric O_3. Two examples are presented here. de Laat et al. (2009) estimated the tropospheric O_3 columns by the subtraction of assimilated O_3 profile observations from total column observations, the so-called Tropospheric O_3 Re-Analysis or TORA method. They evaluated the tropospheric O_3 columns so derived, with space-borne O_3 observations. Six years (1996–2001) of ERS-2 GOME/TOMS total O_3 and GOME O_3 profile observations have been used in the TM5 model with a linearised chemistry parameterisation for the stratosphere.

GOME O_3 profile observations improve the comparisons between model results and ozone-sondes in the tropical UTLS region but slightly degrade the comparisons in the extra-tropical UTLS for both day-to-day variability and monthly means. The large ground pixel size of the GOME O_3 measurements (960×100 km) in combination with retrieval and calibration errors have been suggested to be the main causes of this degradation. Results are expected to improve with higher resolution observations from space.

Stajner et al. (2008) included retrievals from the MLS and the OMI on EOS-Aura in the GEOS-4 O_3 data assimilation system to derive tropospheric O_3. Independent ozone-sondes and MOZAIC data were used for evaluation of the tropospheric O_3 columns. In the troposphere, OMI and MLS provide constraints on the O_3 column, but the O_3 profile shape results from the parameterised O_3 chemistry and the resolved and parameterised transport. Assimilation of OMI and MLS data improves tropospheric column estimates in the Atlantic region but leads to an overestimation in the tropical Pacific, as well as an underestimation in the northern high and middle latitudes in winter and spring. Comparisons of assimilated tropospheric O_3 columns with ozone-sonde data reveal differences of 2.9–7.2 Dobson Units (DU), which are smaller than the model-sonde differences of 3.2–8.7 DU.

Geer et al. (2006) analysed eleven sets of O_3 from seven different DA systems. In most analyses, MIPAS O_3 data are assimilated; two studies assimilate SCIAMACHY observations instead. Analyses are compared to independent O_3 observations (e.g. ozone-sondes) covering the troposphere, stratosphere and lower mesosphere during the period July to November 2003. Where the model results diverge, the main explanation was the way O_3 was modelled. Two analyses used numerical weather prediction (NWP) systems based on general circulation models (GCMs); the other five used CTMs. The systems examined contain either linearised or detailed O_3 chemistry, or no chemistry at all. The result points to the need of further physically based model improvements.

O_3 assimilation also offers the possibility of providing more accurate initial guesses for O_3 retrieval algorithms than are currently available. Other major applications of O_3 assimilation with regard to the troposphere are:

- The provision of vertically resolved global maps of O_3;
- An improvement in the radiative transfer calculations needed to retrieve information from many satellite instruments that require accurate representation of O_3;
- Improvement in the predictions of UV radiation fluctuations at the surface of the Earth, since UV is absorbed by O_3 in the atmosphere; and
- Provision of constraints on other observed constituents that are affected by O_3 chemistry (Rood 2007).

9.4.3 Example Results for NO_2 Tropospheric Column Assimilation

Retrieval results from tropospheric NO_2 columns are ingested into the model by means of averaging kernels, where the observation operator, commonly denoted as **H** (see Appendix, Section 9.6), is constructed by the scalar product of the averaging kernel with the NO_2 molecular density of the model profile (Eskes et al. 2005). The average fraction of the averaging kernel at the surface is roughly 10% of the maximum amplitude in most cases (Fig. 9.11). Recalling the significance of the averaging kernel shape as a sensitivity profile, corrections by the DA procedure are implemented in the same proportion as the sensitivity. The practical meaning is that the modification of

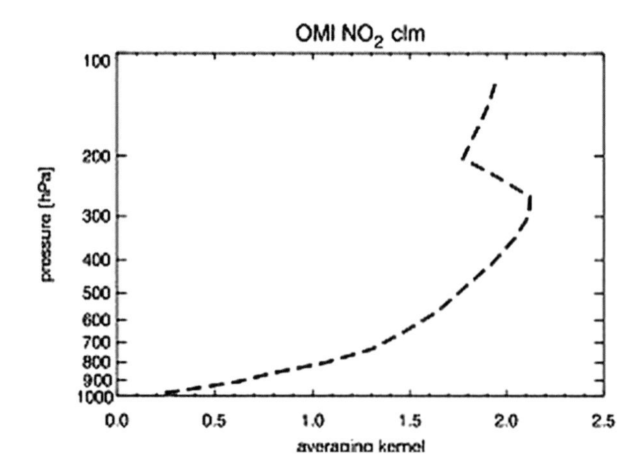

Fig. 9.11 Mean averaging kernel over the European continental scale model domain for a 2 week case study.

the surface layer is hardly affected by this correction in comparison to the free troposphere. This must be remembered when considering the following results.

An example of combined assimilation of SCIAMACHY and OMI NO_2 tropospheric columns with the EURAD-IM CTM for July 2006 is given for illustration. Both SCIAMACHY and OMI satellite retrievals from KNMI were assimilated with averaging kernels, using error information from the appropriate data provider. In an attempt to provide a horizontal model resolution comparable to the minimum OMI 24×13 km^2 footprints, the horizontal model resolution was refined to 15×15 km^2. In this case study, the DA configuration has a time window of 3 h (09:00–12:00 UTC) to include SCIAMACHY satellite sensor data with a late morning overpass over Europe and OMI overpass in the early afternoon over eastern Europe. A longer assimilation window turned out to be unaffordable at 15×15 km^2 horizontal resolution. After assimilation, a 24-h forecast is made, starting at 09:00 UTC. The analysis produced by the assimilation is the initial field, and an emission rate correction factor is applied. Numerical experiments suggest that about 10 iterations are sufficient to ensure convergence to the observations. After assimilation, an *a posteriori* analysis was performed (Talagrand 2003).

The affordable short assimilation interval from 9–12 UTC enforced a fairly disjointed footprint pattern for SCIAMACHY and OMI. While the former covers the western model domain due to its late morning orbit, the latter covers eastern parts due to its early afternoon overpass. Fig. 9.12 illustrates these conditions, along with retrievals (**y**), forecasted retrievals of NO_2 columns ($\mathbf{Hx}_b$), and analysed tropospheric NO_2 columns ($\mathbf{Hx}_a$). The analysis result can be clearly identified as a weighted combination of all information sources, retrievals and forecast acting as background information.

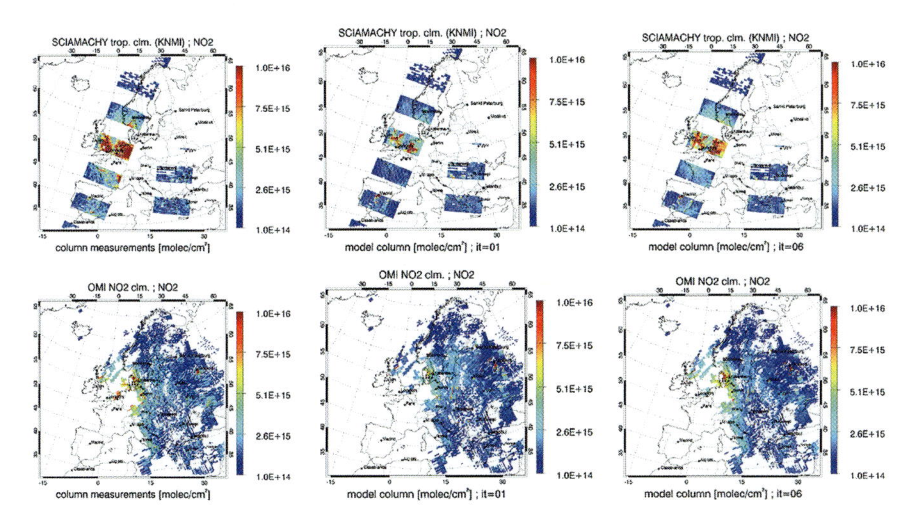

Fig. 9.12 Comparison of NO_2 tropospheric columns in molecules/cm^2 for 6th July 2006. *Left panel* column: KNMI retrieved and assimilated values (**y**); *middle panel* column: EURAD forecasted values ($\mathbf{Hx}_b$); *right panel* column analyses ($\mathbf{Hx}_a$).

The effect of NO_2 column DA for 6th July 2006 is presented in Fig. 9.13. The field obtained is supplemented by difference fields for the tropospheric columns and the concentrations. Clearly major increments can be observed in western England and in the area of north-western Russia. Both these signals are visible for surface concentrations.

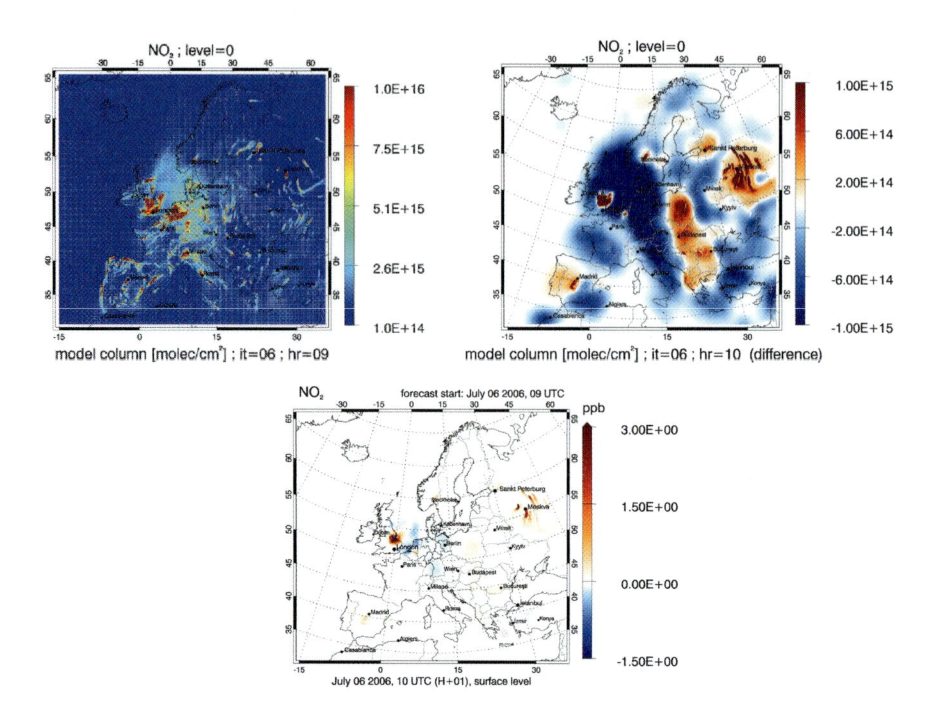

Fig. 9.13 Data assimilation result in terms of tropospheric columns for 6th July 2006. NO_2 model columns based on OMI and SCIAMACHY assimilation within the assimilation interval, 9–12 UTC. Units in molecules/cm^2 (*left panel*). Difference field giving implied changes for tropospheric columns by assimilation (*right panel*), and induced surface concentration changes by NO_2 in ppb (*bottom panel*).

9.4.4 Aerosol Satellite Data Assimilation

A 4D-var approach for aerosol modelling is hampered by the construction of the adjoint model, as the algorithm contains numerous cases which can not be differentiated. Hence, no adjoint of a full-fledged aerosol model with a 3D model is available at present. Sandu et al. (2004) gives a discussion on numerical aspects; however, a 3D-var example is presented here. In addition to difficulties of modelling the variety of processes affecting aerosols, aerosol DA is further hampered by the fact, that the only measurements of lumped particulate matter abundances or aerosol optical depth are possible. Neither quantity is modelled as a space state variable as such.

Enforced by the sparseness of data sources of a single type, the combination of information from remote sensing and *in situ* measurement, that is the use of heterogeneous data sources, is therefore crucial for success. Unlike conditions with satellite retrieved NO_2 tropospheric columns, in satellite retrievals for aerosol data the strongest signals originate from lower atmospheric layers. However, as optical information must be processed, the central problem for DA is bridging information from the optical to the chemical domain. An example of a recent discussion on MODIS data inverse modelling is given in Dubovik et al. (2008). Strictly, related observation operators have to be devised for ingestion of the information.

Aerosol data used in this study are obtained from both satellite retrievals and routine *in situ* observations.

(a) Aerosol data products retrieved from satellite:
SYNergetic Aerosol Retrieval (SYNAER) satellite retrievals: Satellite data are from retrievals based on the SYNAER method (Holzer-Popp et al. 2002a; 2003b). The retrieval procedure has been developed to make a synergistic use of simultaneous GOME and ATSR-2 measurements. It has then been modified to use SCIAMACHY and AATSR data. The retrieval principle is to utilise two complementary properties of (A)ATSR (high spatial resolution) and GOME/SCIAMACHY (high spectral resolution). It is able to deliver various PM_x as integrated values.

(b) *In situ* observations:
The European Environmental Agency collected aerosol data that originate from about 445 routine measurement sites, operated by national and regional environmental protection agencies. From these data, available from the EEA database AirBase, hourly PM_{10} mass concentrations have been used.

The algorithm is highly adaptable to non-isotropic and inhomogeneous aerosol radii of influence, while at the same time, it allows for efficient processing of ensemble model runs for background error covariance matrix estimation. For the study presented here, aerosol dynamics are perturbed to generate ensembles with nine members, modifying gaseous precursors and direct aerosol emission independently.

The demonstration case features 14th July 2003 (Fig. 9.14). The SCIAMACHY footprint track covers parts of the Iberian Peninsula, where the retrievals show elevated PM10 levels, which proved to be biomass burning products from large wildfire events at that time. This case has been selected to demonstrate the capabilities and limits for assimilation under conditions of unpredictable emission events. The special challenge is that extraordinary events like wildfires engender higher level aerosol maxima, which usually differ substantially from modelled profiles, with profile maxima often higher than usual.

Table 9.1 quantifies the beneficial impact of satellite DA in addition to *in situ* DA, with independent *in situ* data, (i.e. direct surface measurements) withheld from the assimilation procedure. A substantial improvement can be claimed for typical conditions (without measured notable wildfires) signified by reduction of root mean

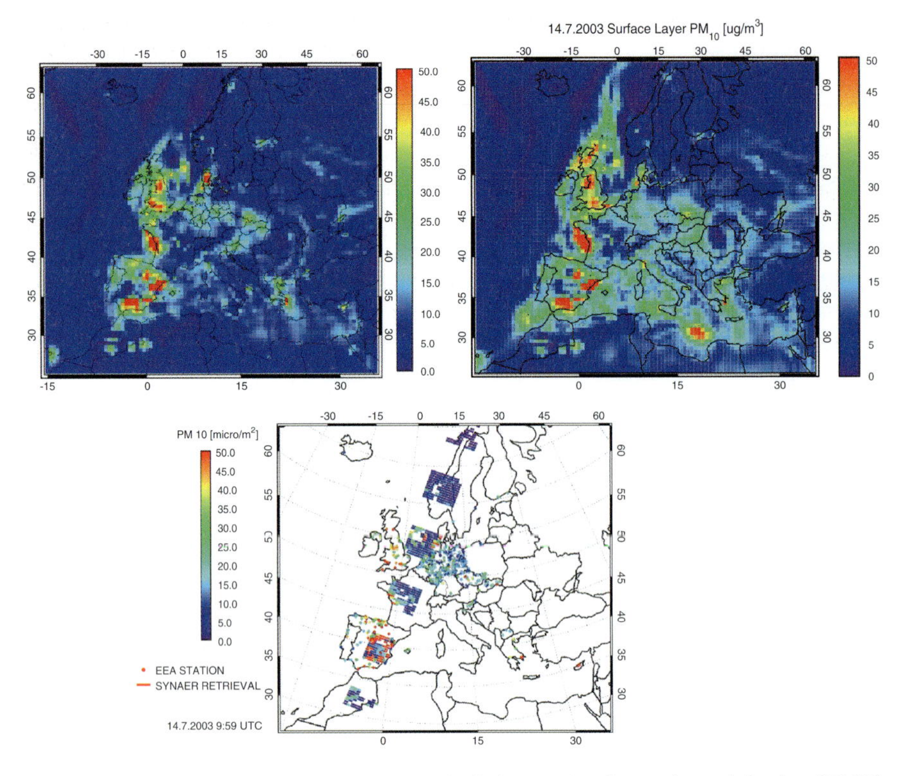

Fig. 9.14 Signal duration of SYNAER data assimilation results. Comparison of simulated PM10 contents: assimilation based analysis after assimilation 14th July 2003, at 10:00 UTC without any previous assimilation (*top left* panel), and with data assimilation processing continuously from 1st to 14th July 2003 (*top right* panel). *In situ* and satellite/SYNAER observation base available for 14th July 2003 (*bottom panel*). The accumulation of information during the 2 weeks is clearly visible from the discrepancies.

Table 9.1 Root mean square errors for assimilation fields for two consecutive dates validated by unassimilated *in situ* observations within satellite footprints. Improvements are given with respect to no assimilation

Date	Number of ground stations withheld	Special characteristics	RMS error [μg m^{-3}] (% improvement)		
			No assimilation	*In situ* data	Satellite + *in situ*
13th July 2003	20	None	6.3	2.5 (61%)	1.9 (70%)
14th July 2003	49	Wildfires	7.3	5.5 (24%)	5.3 (28%)

square errors, especially with SYNAER satellite data. In the case of wildfires, the improvement is markedly reduced, but still clearly visible. In both cases, satellite data from SYNAER demonstrate efficient support for DA improvements.

9.5 Summary: Perspectives

Earth system models, describing dynamical, physical, chemical and biological processes which determine the conditions at the Earth's surface, have been evolving over the past decades. The increasing computer power is facilitating more accurate simulation of current and past conditions and improving our confidence in the accuracy of prediction. The latter is required for policymakers in their quest to achieve legislation facilitating sustainable environmental conditions. Global data sets are required to test the ability of our Earth system models.

In this context the synergistic use of satellite observations with chemistry-transport modelling has opened new horizons in air pollution control and climate change evaluation. However, models and observations from satellite platforms having *higher spatial and temporal resolution* are required in order to attribute better the air pollution to its different sources. The signals from these sources will not be diluted or mixed with other sources in large ground or satellite observations or larger grid boxes of the CTMs. *Improvements* are also needed in the *temporal sampling* of the atmosphere by the satellite based sensors, which has to be taken into account in the models when satellite retrievals are used.

A *continuous dialogue* between satellite retrieved observations of atmospheric trace constituents and model results enables the improvement of: (1) process understanding, (2) the models, and (3) retrieval algorithms, in order to construct a precise picture of our changing atmosphere.

Inverse modelling can contribute to the development of a general observation-based methodology for estimating parameters of the atmosphere that cannot be observed directly. Inverse modelling combining both traditional, i.e. ground-based and airborne (aircraft, balloon) observations and satellite observations can contribute to improving emission inventories of trace constituents, relevant to air pollution and climate change. It also allows verification, by space-based observations, of control strategies for atmospheric emissions.

Data assimilation has a unique role to play with respect to forecasting conditions. It has also been shown that it is important for *filtering* the signal from noisy observations, *interpolation* in space and time, and *completion* of state variables that are not sampled by the observation network. Data assimilation can further be applied for data validation, field experiments, and climate signal detection. The assimilation of satellite observations is a critical step forward in impact modelling for scientific and political decisions with regard to environmental changes. It contributes critically to the improvement of prognostic models, such as those used for weather forecasts and more recently for chemical weather forecasts. Such results are of relevance for society (public services) since they *increase the accuracy of the predictions* and our confidence to them.

9.6 Appendix

Inverse Modelling: Principles

The following equation provides a relationship between $\boldsymbol{y}$, a vector grouping a set of observations, and $\boldsymbol{x}$, the vector of variables to be determined (called *state variables*, e.g. emissions), through applying a *forward* model $\mathbf{F}$ with fixed parameters $\boldsymbol{b}$:

$$\boldsymbol{y} = \mathbf{F}(\boldsymbol{x}, \boldsymbol{b}). \tag{9.1}$$

By inverting this model, analytically or numerically, we can obtain $\boldsymbol{x}$ for a given $\boldsymbol{y}$. As errors are associated with both $\boldsymbol{x}$ and $\boldsymbol{y}$, a probabilistic treatment of the inversion problem is necessary. Its starting point is the Bayesian theorem.

$$P(\boldsymbol{x}|\boldsymbol{y}) = \frac{P(\boldsymbol{y}|\boldsymbol{x})P(\boldsymbol{x})}{P(\boldsymbol{y})}. \tag{9.2}$$

In the context of an inverse modelling problem, $P(\boldsymbol{x})$ and $P(\boldsymbol{y})$ represent the probability distribution functions (pdf) for the state vector $\boldsymbol{x}$ and the observation vector $\boldsymbol{y}$. In Bayesian framework, $P(\boldsymbol{y}|\boldsymbol{x})$ represents the conditional probability to observe a particular vector $\boldsymbol{y}$ for a given value of the set of state variables $\boldsymbol{x}$. While $P(\boldsymbol{x})$ represents the *a priori* information on the state vector (for example an emission inventory with a given uncertainty), then $P(\boldsymbol{x}|\boldsymbol{y})$ represents the *a posteriori* probability that the state vector was $\boldsymbol{x}$ when $\boldsymbol{y}$ was measured reflecting the additional information from measurements. This latter pdf can be used to find a maximum likelihood for an *a posteriori* estimate (MLAP) of $\boldsymbol{x}$ and its uncertainty.

For simplicity, we illustrate the main steps to derive a MLAP estimate for a scalar linear case first. Let x_a be the *a priori* estimate of the state variable, σ_a its uncertainty (assumed to be normal), y an observation with uncertainty σ_e including both the error in observations and in the forward model (due to the model formulation and uncertainty in model parameters b), and k a the forward model linking x and y.

If uncertainties in $\boldsymbol{x}_a$ and $\boldsymbol{y}$ are distributed in accordance to a normal distribution, $P(\boldsymbol{x}|\boldsymbol{y})$ are proportional to an exponential expression combining two terms (Eq. 9.3): the departure of an updated state variable from the *a priori*, and the difference between observations y and simulated values kx for a given x. Combining these two terms, weighed by the respective inverse error variances, is exactly the principle of optimal estimation.

$$P(\boldsymbol{x}|\boldsymbol{y}) \sim \exp\left[-\frac{(x - x_a)^2}{2\sigma_a^2} - \frac{(y - kx)^2}{2\sigma_\varepsilon^2}\right] \tag{9.3}$$

In order to determine the maximum value for $P(x/y)$, we define a *cost function* $J(x)$ given by the terms in brackets in Eq. 9.3. The solution to $\partial J/\partial x = 0$ gives the optimal estimate $\hat{x}$:

$$\hat{x} = x_a + g(y - kx_a). \tag{9.4}$$

where g is the so called gain factor, which weights the contribution of observations y to $\hat{x}$. It is given by:

$$g = \frac{k\sigma_a^2}{k^2\sigma_a^2 + \sigma_\varepsilon^2} \tag{9.5}$$

For $\sigma_\varepsilon/k >> \sigma_a$, g tends to 0, for $\sigma_\varepsilon/k << \sigma_a$, g tends to 1. This latter case is obviously advantageous for inverse modelling because then observations bring much new information significantly reducing the *a posterior* uncertainty, $\hat{\sigma}^2$, given by Eq. 9.6:

$$\frac{1}{\hat{\sigma}^2} = \frac{1}{\sigma_a^2} + \frac{1}{(\sigma_\varepsilon/k)^2}. \tag{9.6}$$

Note however, that $\hat{\sigma}^2 < \sigma_a{}^2$ always holds, so even *uncertain* observations add *some* information.

In the more general case of a set of state variables to be optimised (for example emissions at different locations or times) from a set of observations, we need to reformulate equations 9.3–9.6 in matrix from. The cost function $J(\boldsymbol{x})$ becomes:

$$J(\boldsymbol{x}) = (\boldsymbol{x} - \boldsymbol{x_a})^T \mathbf{S}_\mathbf{a}^{-1}(\boldsymbol{x} - \boldsymbol{x_a}) + (\boldsymbol{y} - \mathbf{K}\boldsymbol{x})^T \mathbf{S}_\varepsilon{}^{-1}(\boldsymbol{y} - \mathbf{K}\boldsymbol{x}) \tag{9.7}$$

Again, an optimal solution for $\boldsymbol{x}$ can be found by differentiating J with respect to $\boldsymbol{x}$. However, several practical problems occur. The *a priori* and observational error terms S_{ij} constituting the matrices $\boldsymbol{S_a}$ and $\boldsymbol{S_\varepsilon}$ are in general not well known. Very often, non-diagonal terms representing error correlations for different measurements or state variables are simply neglected. Subjective estimates for the diagonal error terms are made, which can affect the weighting procedure in the optimal estimates.

In Eq. 9.7, $\mathbf{K}$ represents the so called Jacobian matrix with elements $k_{ij} = \delta y_i/\delta x_j$ indicating the sensitivity of a measurement at i to a state variable at j. It can be solved by running an atmospheric model and varying x_j which needs $n + 1$ runs, n being the dimension of $\boldsymbol{x}$. For the general case of a non-linear model, this procedure has to be repeated several times starting with $\boldsymbol{x_a}$, and iterating for each new update of $\boldsymbol{x}$. Two types of solutions are applied to avoid a large computational burden for large n: (1) the problem is decomposed for sub-domains where each small n, or state variables are aggregated for large regions reducing their number; and (2) use of the adjoint model (Giering and Kaminski 1998) allowing a calculation in one backward run all derivatives $\delta J/\delta x_1 \ldots \delta J/\delta x_n$, and thus to avoiding explicit evaluation of $\boldsymbol{K}$. This procedure has to be applied iteratively (using for example steepest descent algorithms) until an absolute minimum of the cost function J is found. The adjoint method is also known as 4D-var assimilation, because observations can be distributed over a time window, and is in widespread use in meteorology when a task is to optimise initial conditions. For the implementation of the 4D-var approach, a distance function or objective function, which penalises both discrepancies with observations and *a priori* knowledge of emission rates and initial values, also called cost function, may be defined as follows:

$$J(\boldsymbol{x}, \boldsymbol{e}(t)) = \frac{1}{2}(\boldsymbol{x}_b - \boldsymbol{x}(0))^T \mathrm{B}^{-1}(\boldsymbol{x}_b - \boldsymbol{x}(0)) + \frac{1}{2}\int_0^N (\boldsymbol{e}_b(t) - \boldsymbol{e}(t))^T \mathrm{K}^{-1}(\boldsymbol{e}_b(t) - \boldsymbol{e}(t))dt + \frac{1}{2}\int_0^N (\boldsymbol{y}(t) - \mathrm{H}\boldsymbol{x}(t))^T \mathbf{R}^{-1}(\boldsymbol{y}(t) - \mathrm{H}\boldsymbol{x}(t))dt \tag{9.8}$$

where J is a scalar functional defined on the time interval $0 \leq \boldsymbol{t} \leq N$ dependent on the vector valued state variable $\boldsymbol{x(t)}$, and a parameter $\boldsymbol{e(t)}$ to be optimised. Here, observations $\boldsymbol{y}$ are compared with their model equivalent $\mathbf{H}\boldsymbol{x}$ at time $\boldsymbol{t}$, with the operator $\mathbf{H}$ being the forward observation operator. The error covariance matrices of the forecast $\boldsymbol{x}_\mathrm{b}$, the first guess or background emission rates $\boldsymbol{e_b(t)}$ and observations $\boldsymbol{y(t)}$ are denoted $\mathbf{B}$, $\mathbf{K}$ and $\mathbf{R}$, respectively. The CTM with inclusion of emissions is given by $d\boldsymbol{x}/dt = \mathbf{M}(\boldsymbol{x}) + \boldsymbol{e}$, where $\mathbf{M}$ acts as a generally non-linear model operator and $\boldsymbol{e}$ is in our case the vector of emission rates. Both terms uniquely define the state variable $\boldsymbol{x}(t)$ at time t, after an ever fixed initial state $\boldsymbol{x(0)}$ is provided.

The variational chemistry DA algorithm is composed by four components: (1) the forward model, (2) the adjoint of its tangent linear version, (3) the background error covariance matrix, making use of the diffusion paradigm (Weaver and Courtier 2001) for anisotropic and inhomogeneous radii of influence, and (4) the minimisation routine, where the quasi-Newton (L-BFGS) method is selected. Further numerical and implementation details are given in Elbern et al. (2007).

An alternative to the 4D-var approach is Kalman filtering (KF) with the theoretical ability to update the forecast error covariance matrix $\mathbf{B}$ and analysis error covariance matrix $\mathbf{A}$. Given the model and analysis space state dimension of order 10^{6-7} in today's CTMs, direct application of KF is far beyond feasibility (and will always be so), and sophisticated numerical complexity reduction measures must be devised. In sloppy parlance often sequential applications of optimal interpolation are termed "reduced KF", despite the fact that neither $\mathbf{B}$ is updated nor $\mathbf{A}$ computed. In contrast, Kalman filter implementations with sophisticated complexity reduction techniques are still very rare. Advanced implementations are presented by van Loon et al. (2000), where a reduced rank square-root approach (RRSQR-KF) was selected to factorise covariance matrices by a few principal components (Verlaan and Heemink, 1995). Further elaboration on this technique by combination with an ensemble Kalman filter method (En-KF) resulted in additional skill (Hanea et al., 2004). Optimisation parameters include emission rates, photolysis rates, and deposition rates, the correction quantities of which are formally introduced as "noise" parameters in the Kalman filter formulation.

References

Arnold, S. R., D. V. Spracklen, J. Williams, N. Yassaa, J. Sciare, B. Bonsang, V. Gros, I. Peeken, A. C. Lewis, S. Alvain, and C. Moulin, 2009. Evaluation of the global oceanic isoprene source and its impacts on marine organic carbon aerosol, *Atmos. Chem. Phys.*, **9**, 1253–1262.

Arellano, A. F., Jr., and P. G. Hess, 2006. Sensitivity of top-down estimates of CO sources to GCTM transport, *Geophys. Res. Lett.*, **33**, L21807, doi: 10.1029/2006GL027371.

Arellano, A. F., P. S. Kasibhatla, L. Giglio, G. R. van der Werf, J. T. Randerson, and G. J. Collatz, 2006. Time-dependent inversion estimates of global biomass-burning CO emissions using Measurement of Pollution in the Troposphere (MOPITT) measurements, *J. Geophys. Res.*, **111**, D09303, doi: 10.1029/2005JD006613.

Barret B., P. Ricaud, C. Mari, J.-L. Attié, N. Bousserez, B. Josse, E. Le Flochmoën, N. J. Livesey, S. Massart, V.-H. Peuch, A. Piacentini, B. Sauvage, V. Thouret, and J.-P. Cammas, 2008. Transport pathways of CO in the African upper troposphere during the monsoon season: a study based upon the assimilation of spaceborne observations, *Atmos. Chem. Phys.*, **8**, 3231–3246.

Bennett, A. F., 1992. *Inverse methods in physical oceanography*, Cambridge University Press, 347 pp.

Bennett, A. F., 2002. *Inverse modelling of the ocean and the atmosphere*, Cambridge University Press, 234 pp.

Beirle, S., U. Platt, M. Wenig, and T. Wagner, 2003. Weekly cycle of NO_2 by GOME measurements: A signature of anthropogenic sources, *Atmos. Chem. Phys.*, **3**, 2225–2232.

Beirle, S., U. Platt, R. von Glasow, M. Wenig, and T. Wagner, 2004. Estimate of nitrogen oxide emissions from shipping by satellite remote sensing, *Geophys. Res. Lett.*, **31**, L18102, doi: 10.1029/2004GL020312.

Bergamaschi, P., and P. Bousquet, 2008. Estimating Sources and Sinks of Methane: An Atmospheric View, in The Continental-Scale Greenhouse Gas Balance of Europe, edited by A. J. Dolman, et al., pp. 113–133, Springer, New York.

Bergamaschi, P., C. Frankenberg, J. F. Meirink, M. Krol, F. Dentener, T. Wagner, U. Platt, J. O. Kaplan, S. Körner, M. Heimann, E. J. Dlugokencky, A. Goede, 2007. Satellite chartography of atmospheric methane from SCIAMACHY on board ENVISAT 2. Evaluation based on inverse model simulations, *J. Geophys. Res.*, **112**, D02304, doi: 10.1029/ 2006JD007268.

Bertram, T. H., A. Heckel, A. Richter, J. P. Burrows, R. C. Cohen, 2005. Satellite measurements of daily variations in soil NOx emissions, *Geophys. Res. Lett.*, **32**(**24**), L24812.

Bloom, A.A, P. I. Palmer, A. Fraser, D. S. Reay, Ch. Frankenberg, 2010. Large-Scale Controls of Methanogenesis Inferred from Methane and Gravity Spaceborne Data, *Science* **327**, 322–325, DOI: 10.1126/science.1175176

Boersma, K. F., Jacob, D. J., Trainic, M., Rudich, Y., DeSmedt, I., Dirksen, R., and Eskes, H. J., 2009. Validation of urban NO_2 concentrations and their diurnal and seasonal variations observed from the SCIAMACHY and OMI sensors using *in situ* surface measurements in Israeli cities, *Atmos. Chem. Phys.*, **9**, 3867–3879.

Chai, T. F., G. R. Carmichael, A. Sandu, Y. H. Tang, and D. N. Daescu, 2006. Chemical data assimilation of Transport and Chemical Evolution over the Pacific (TRACE-P) aircraft measurements, *J. Geophys. Res.*, **111**, D02301, doi:10.1029/2005JD005883.

Cohn, S., 1997. An introduction to estimation theory, *J. Met. Soc. Japan*, **75**(**1B**), 257–288.

Daley, R., 1991. *Atmospheric Data Analysis*, Cambridge University Press, Cambridge, ISBN: 9780521382151.

Deeter, M. N., L. K. Emmons, G. L. Francis, D. P. Edwards, J. C. Gille, J. X. Warner, B. Khattatov, D. Ziskin, J.-F. Lamarque, S.-P. Ho, V. Yudin, J.-L. Attié, D. Packman, J. Chen, D. Mao, J. R. Drummond, 2003, Operational carbon monoxide retrieval algorithm and selected results for the MOPITT instrument, *J. Geophys. Res.*, **108** (D14), 4399, doi: 10.1029/2002JD003186.

de Laat, A. T. J., R. J. van der A, and M. van Weele, 2009. Evaluation of tropospheric ozone columns derived from assimilated GOME ozone profile observations, *Atmos. Chem. Phys.*

Dentener, F. J., D. Stevenson, K. Ellingsen, T. van Noije, M.Schultz, M. Amann, C. Atherton, N. Bell, D.Bergmann, I. Bey, L. Bouwman, T. Butler, J. Cofala, B. Collins, J. Drevet, R. Doherty, B. Eickhout, H. Eskes, A. Fiore, M. Gauss, D. Hauglustaine, L. Horowitz, I.S.A. Isaksen, B. Josse, M. Lawrence, M. Krol, J.F. Lamarque, V. Montanaro, J.F. Müller, V. H. Peuch, G. Pitari, J. Pyle, S. Rast, J. Rodriguez, M. Sanderson, N.H. Savage, D. Shindell, S. Strahan, S. Szopa, K. Sudo, R.Van Dingenen, O. Wild, G. Zeng, 2006a. The global atmospheric environment for the next generation, *Environ. Sci. Technol.*, **40**, 3586–3594.

Dentener, F., S. Kinne, T. Bond, O. Boucher, J. Cofala, S. Generoso, P. Ginoux, S. Gong, J. J. Hoelzemann, A. Ito, L. Marelli, J. Penner, J.-P. Putaud, C. Textor, M. Schulz, G. R. van der Werf, and J. Wilson, 2006b. Emissions of primary aerosol and precursor gases for the years 2000 and 1750 prescribed data-sets for AeroCom, *Atmos. Chem. Phys.*, **6**, 4321–4344.

Dubovik, O., T. Lapyonok, Y. J. Kaufman, M. Chin, P. Ginoux, R. A. Kahn, and A. Sinyuk, 2008. Retrieving global aerosol sources from satellites using inverse modeling, *Atmos. Chem. Phys.*, **8**, 209–250.

Dufour, G., Szopa, S., Barkley, M. P., Boone, C. D., Perrin, A., Palmer, P. I., and Bernath, P. F.: Global upper-tropospheric formaldehyde: seasonal cycles observed by the ACE-FTS satellite instrument, *Atmos. Chem. Phys.*, **9**, 3893–3910, 2009.

Edwards, D. P., L. K. Emmons, D. A. Hauglustaine, D. A. Chu, J. C. Gille, Y. J. Kaufman, G. Pétron, L. N. Yurganov, L. Giglio, M. N. Deeter, V. Yudin, D. C. Ziskin, J. Warner, J.-F. Lamarque, G. L. Francis, S. P. Ho, D. Mao, J. Chen, E. I. Grechko, and J. R. Drummond, 2004. Observations of carbon monoxide and aerosols from the Terra satellite: Northern Hemisphere variability, *J. Geophys. Res.*, **109**, 24202, doi: 10.1029/2004JD004727.

Elbern, H. and H. Schmidt, 2001. Ozone episode analysis by four-dimensional variational chemistry data assimilation, *J. Geophys. Res.*, **106**, 3569–3590.

Elbern, H., H. Schmidt, and A. Ebel, 1997. Variational data assimilation for tropospheric chemistry modeling, *J. Geophys. Res.*, **102**, 967–985.

Elbern, H., A. Strunk, H. Schmidt, and O. Talagrand, 2007. Emission rate and chemical state estimation by 4-dimensional variational inversion, *Atmos. Chem. Phys.*, **7**, 3749–3769.

Errera, Q. and D. Fonteyn, 2001. Four-dimensional variational chemical assimilation of CRISTA stratospheric measurements, *J. Geophys. Res.*, **106**, 12 253–12 265.

Eskes, H. J., A. J. M. Piters, P. F. Levelt, M. A. F. Allart, and H. M. Kelder, 1999. Variational assimilation of total-column ozone satellite data in a 2D lat-lon tracer-transport model, *J. Atmos. Sci.*, **56**, 3560–3572.

Eskes, H. J., R. J. van der A, E. J. Brinksma, J. P. Veefkind, J. F. de Haan, and P. J. M. Valks, 2005. Retrieval and validation of ozone columns derived from measurements of SCIAMACHY on Envisat, *Atmos. Chem. Phys. Discuss.*, **5**, 4429–4475.

Eyring, V., Stevenson, D. S., Lauer, A., Dentener, F. J., Butler, T., Collins, W. J., Ellingsen, K., Gauss, M., Hauglustaine, D. A., Isaksen, I. S. A., Lawrence, M. G., Richter, A., Rodriguez, J. M., Sanderson, M., Strahan, S. E., Sudo, K., Szopa, S., van Noije, T. P. C., and Wild, O., 2007. Multi-model simulations of the impact of international shipping on Atmospheric Chemistry and Climate in 2000 and 2030, *Atmos. Chem. Phys.*, **7**, 757–780.

Fedorov, V., 1989. Kriging and other estimators of spatial field characteristics (with special reference to environmental studies), *Atmos. Environ.*, **23**, 174–184.

Fisher, M. and D. Lary, 1995. Lagrangian four-dimensional variational data assimilation of chemical species, *Q. J. Roy. Meteor. Soc.*, **121**, 1681–1704.

Franke, K., A. Richter, H. Bovensmann, V. Eyring, P. Jockel, P. Hoor, and J.P. Burrows, 2009. Ship emitted NO_2 in the Indian Ocean: comparison of model results with satellite data, *Atmos. Chem. Phys.*, **9**, 7289–7301.

Frankenberg, C., T. Warneke, A. Butz, I. Aben, F. Hase, P. Spietz, and L. R. Brown, 2008a. Pressure broadening in the 2v3 band of methane and its implication on atmospheric retrievals, *Atmos. Chem. Phys.*, **8**, 5061–5075.

Frankenberg, C., P. Bergamaschi, A. Butz, S. Houweling, J. F. Meirink, J. Notholt, A. K. Petersen, H. Schrijver, T. Warneke, and I. Aben, 2008b. Tropical methane emissions: A revised view from SCIAMACHY onboard ENVISAT, *Geophys. Res. Lett.*, **35**, L15811, doi:10.1029/2008GL034300.

Fu, T.-M., D. J. Jacob, P. I. Palmer, K. Chance, Y. X. Wang, B. Barletta, D. R. Blake, J. C. Stanton, and M. J. Pilling, 2007. Space-based formaldehyde measurements as constraints on volatile organic compound emissions in east and south Asia and implications for ozone, *J. Geophys. Res.*, **112**, D06312, doi: 10.1029/2006JD007853.

Fu, T.-M., D. J. Jacob, F. Wittrock, J. P. Burrows, M. Vrekoussis, and D. K. Henze, 2008. Global budgets of atmospheric glyoxal and methylglyoxal, and implications for formation of secondary organic aerosols, *J. Geophys. Res.*, **113**, D15303, doi: 10.1029/2007JD009505.

Geer, A. J., W. A. Lahoz, S. Bekki, N. Bormann, Q. Errera, H. J. Eskes, D. Fonteyn, D. R. Jackson, M. N. Juckes, S. Massart, V.-H. Peuch, S. Rharmili, and A. Segers, 2006. The ASSET intercomparison of ozone analyses: method and first results, *Atmos. Chem. Phys.*, **6**, 5445–5474.

Giering, R. and T. Kaminski, 1998. Recipes for adjoint code construction, *ACM Trans. on Math. Software*, **24**, 437–474.

Giglio, L., G. R. van der Werf, J. T. Randerson, G. J. Collatz, and P. Kasibhatla, Global estimation of burned area using MODIS active fire observations, *Atmos. Chem. Phys.*, **6**, 957–974, 2006.

Guenther, A., T. Karl, P. Harley, C. Wiedinmyer, P. I. Palmer, and C. Geron, 2006. Estimates of global terrestrial isoprene emissions using MEGAN (Model of Emissions of Gases and Aerosols from Nature), *Atmos. Chem. Phys.*, **6**, 3181–3210.

Hanea, R. G., G. J. M. Velders, and A. Heemink, 2004. Data assimilation of ground-level ozone in Europe with a Kalman filter and chemistry transport, *J. Geophys. Res.*, **109**, D10302, doi: 10.1029/2003JD004283.

Hayn, M., Beirle, S., Hamprecht, F. A., Platt, U., Menze, B. H., and Wagner, T.: Analysing spatio-temporal patterns of the global NO_2-distribution retrieved from GOME satellite observations using a generalized additive model, *Atmos. Chem. Phys.*, **9**, 6459–6477, 2009

Holzer-Popp, T., M. Schroedter, and G. Gesell, 2002a. Retrieving aerosol optical depth and type in the boundary layer over land and ocean from simultaneous GOME spectrometer and ATSR-2 radiometer measurements, 2. Case study application and validation, *J. Geophys. Res.*, **107(D24)**, 4770, doi: 10.1029/2001JD002777.

Holzer-Popp, T., M. Schroedter, and G. Gesell, 2002b. Retrieving aerosol optical depth and type in the boundary layer over land and ocean from simultaneous GOME spectrometer and ATSR-2 radiometer measurements, 1. Method description, *J. Geophys. Res.*, **107(D21)**, 4578, doi: 10.1029/ 2001JD002013,

Hudman, R. C., D. J. Jacob, O. R. Cooper, M. J. Evans, C. L. Heald, R. J. Park, F. Fehsenfeld, F. Flocke, J. Holloway, G. Hubler, K. Kita, M. Koike, Y. Kondo, A. Neuman, J. Nowak, S. Oltmans, D. Parrish, J. M. Roberts, and T. Ryerson, 2004. Ozone production in transpacific Asian pollution plumes and implications for ozone air quality in California, *J. Geophys. Res.*, **109(D23)**, 14, doi: 10.1029/2004JD004974.

Ito, A. and J. E. Penner, 2005. Historical emissions of carbonaceous aerosols from biomass and fossil fuel burning for the period 1870–2000, *Global Biogeochem. Cycles*, **19**, GB2028, doi: 10.1029/2004GB002374.

Jaeglé, L., R. V. Martin, K. Chance, L. Steinberger, T. P. Kurosu, D. J. Jacob, A. I. Modi, V. Yoboué, L. Sigha-Nkamdjou, C. Galy-Lacaux, 2004. Satellite mapping of rain-induced nitric oxide emissions from soils, *J. Geophys. Res.*, **109**, D21310, doi: 10.1029/2004JD004787.

Kim, S.-W., A. Heckel, S. A. McKeen, G. J. Frost, E.-Y. Hsie, M. K. Trainer, A. Richter, J. P. Burrows, S. E. Peckham, G. A. Grell, 2006. Satellite-observed U.S. power plant NOx emission reductions and their impact on air quality, *Geophys. Res. Lett.*, **33**, L22812, doi: 10.1029/2006GL027749.

Kim, S.-W., A. Heckel, G. J. Frost, A. Richter, J. Gleason, J. P. Burrows, S. McKeen, E.-Y. Hsie, C. Granier, and M. Trainer, 2009, NO_2 columns in the western United States observed from space and simulated by a regional chemistry model and their implications for NOx emissions, *J. Geophys. Res.*, **114**, 11301, doi:10.1029/2008JD011343.

Kinne, S., U. Lohmann, J. Feichter, C. Timmreck, M. Schulz, S. Ghan, R. Easter, M. Chin, P. Ginoux, T. Takemura, I. Tegen, D. Koch, M. Herzog, J. Penner, G. Pitari, B. Holben, T. Eck, A. Smirnov, O. Dubovik, I. Slutsker, D. Tanré, O. Torres, M. Mishchenko, I. Geogdzhayev, D.A. Chu, and Y. Kaufman, 2003. Monthly Averages of Aerosol Properties: A Global comparison among models, satellite data and AERONET ground data, *J. Geophys. Res*, **108**, 4634.

Kinne, S., M. Schulz, C. Textor, S. Guibert, Y. Balkanski, S.E. Bauer, T. Berntsen, T.F. Berglen, O. Boucher, M. Chin, W. Collins, F. Dentener, T. Diehl, R. Easter, J. Feichter, D. illmore, S. Ghan, P. Ginoux, S. Gong, A. Grini, J. Hendricks, M. Herzog, L. Horowitz, I. Isaksen, T. Iversen, A. Kirkevåg, S. Kloster, D. Koch, J. E. Kristjansson, M. Krol, A. Lauer, J. F. Lamarque, G. Lesins, X. Liu, U. Lohmann, V. Montanaro, G. Myhre, J. Penner, G. Pitari, S. Reddy, O. Seland, P. Stier, T. Takemura, and X. Tie, 2006. An AeroCom initial assessment – optical properties in aerosol component modules of global models, *Atmos. Chem. Phys.*, **6**, 1815–1834.

Konovalov, I. B., M. Beekmann, A. Richter, J. P. Burrows, 2006a, Inverse modelling of the spatial distribution of NOx emissions on a continental scale using satellite data, *Atmos. Chem. Phys.*, **6**, 1747–1770.

Konovalov, I. B., M. Beekmann, A. Richter, J. P. Burrows, 2006b, The use of satellite and ground based measurements for estimating and reducing uncertainties in the spatial distribution of emissions of nitrogen oxides, arXiv: physics/0612144 (www.arxiv.org).

Konovalov, I. B., M. Beekmann, J. P. Burrows, and A. Richter, 2008. Satellite measurement based estimates of decadal changes in European nitrogen oxides emissions, *Atmos. Chem. Phys.*, **8**, 2623–2641.

Konovalov, I. B., M. Beekmann, A. Richter, A. Hilboll, and J. P. Burrows, 2010. Multi-annual changes of NOx emissions in megacity regions: nonlinear trend analysis of satellite measurement based estimates, *Atmos. Chem. Phys.*, **10**, 8481–8498, doi:10.5194/acp-10-8481-2010, 2010.

Ladstätter-Weißenmayer A., M. Kanakidou, J. Meyer-Arnek, E. V. Dermitzaki, A. Richter, M. Vrekoussis, F. Wittrock, and J. P. Burrows, 2007. Pollution events over the East Mediterranean: Synergistic use of GOME, ground based and sonde observations and models, *Atmos. Environ.*, **41**, 7262–7273, doi:10.1016/j.atmosenv.2007.05.031.

Lauer, A., M. Dameris, A. Richter, J. P. Burrows, 2002. Tropospheric NO_2 columns: A comparison between model and retrieved data from GOME measurements, *Atmos. Chem. Phys.*, **2**, 67–78.

Lelieveld, J. and F. J. Dentener, 2000. What controls tropospheric ozone?, *J. Geophys. Res.*, **105**, 3531–3551.

Lelieveld, J., P. Hoor, P. Jöckel, A. Pozzer, P. Hadjinicolaou, J.-P. Cammas, and S. Beirle, 2009, Severe ozone air pollution in the Persian Gulf region, *Atmos. Chem. Phys.*, **9**, 1393–1406.

Leue, C., M. Wenig, T. Wagner, O. Klimm, U. Platt, and B. Jahne, 2001, Quantitative analysis of NOx emissions from GOME satellite image sequences, *J. Geophys. Res.*, **106**, 5493–5505.

Marbach, T., S. Beirle, U. Platt, P. Hoor, F. Wittrock, A. Richter, M. Vrekoussis, M. Grzegorski, J. P. Burrows, and T. Wagner, 2009. Satellite measurements of formaldehyde from shipping emissions, *Atmos. Chem. Phys.*, **9**, 8223–8234.

Martin, R. V., D. J. Jacob, J. A. Logan, I. Bey, R. M. Yantosca,A. C. Staudt, Q. Li, A. M. Fiore, B. N. Duncan, H. Liu,P. Ginoux, V. Thouret, 2002. Interpretation of TOMS observations of tropical tropospheric ozone with a global model and *in situ* observations, *J. Geophys. Res.*, **107(D18)**, 4351, doi: 10.1029/2001JD001480.

Martin, R. V., D. J. Jacob, K. Chance, T. Kurosu, P. I. Palmer, M. J. Evans, 2003. Global inventory of nitrogen oxide emissions constrained by space-based observations of NO_2 columns, J. Geophys. Res., 108, 4537, doi: 10.1029/2003JD003453.

Martin, R. V., A. M. Fiore, and A. Van Donkelaar, 2004. Space-based diagnosis of surface ozone sensitivity to anthropogenic emissions, *Geophys. Res. Lett.*, **31**, L06120, doi: 10.1029/2004GL019416.

Martin, R. V., B. Sauvage, I. Folkins, C. E. Sioris, C. Boone, P. Bernath, J. Ziemke, 2007. Space-based constraints on the production of nitric oxide by lightning, *J. Geophys. Res.*, **112**, D09309, doi: 10.1029/2006JD007831.

Meirink, J. F., H. J. Eskes, and A. P. H. Goede, 2006. Sensitivity analysis of methane emissions derived from SCIAMACHY observations through inverse modeling, *Atmos. Chem. Phys.*, **6**, 1275–1292.

Meirink J. F., P. Bergamaschi, and M. C. Krol, 2008a. Four-dimensional variational data assimilation for inverse modelling of atmospheric methane emissions: method and comparison with synthesis inversion, *Atmos. Chem. Phys.*, **8(21)**, 6341–635.

Meirink, J. F., P. Bergamaschi, C. Frankenberg, M. T. S. d'Amelio, E. J. Dlugokencky, L. V. Gatti, S. Houweling, J. B. Miller, T. Rockmann, M. G. Villani, and M. Krol, 2008b. Four-dimensional variational data assimilation for inverse modeling of atmospheric methane emissions: Analysis of SCIAMACHY observations, *J. Geophys. Res.*, **113**, D17301, doi:10.1029/2007JD009740.

Millet, D. B., D.J. Jacob, S. Turquety, R.C. Hudman, S. Wu, A. Fried, J. Walega, B. G. Heikes, D. R. Blake, H.B. Singh, B. E. Anderson, and A.D. Clarke, 2006. Formaldehyde distribution over North America: Implications for satellite retrievals of formaldehyde columns and isoprene emission, *J. Geophys. Res.*, **111(24)**, 17, doi: 10.1029/2005JD006853.

Millet, D. B., D. J. Jacob, K. F. Boersma, T. M. Fu, T. P. Kurosu, K. Chance, C. L. Heald, A. Guenther, 2008. Spatial distribution of isoprene emissions from North America derived from formaldehyde column measurements by the OMI satellite sensor, *J. Geophys. Res.*, **113**, D02307, doi: 10.1029/2007JD008950.

Mota, B. W., J. M. C. Pereira, D. Oom, M. J. P. Vasconcelos, and M. Schultz, 2006. Screening the ESA ATSR-2 World Fire Atlas (1997–2002), *Atmos. Chem. Phys.*, **6**, 1409–1424.

Myriokefalitakis, S., M. Vrekoussis, K. Tsigaridis, F. Wittrock, A. Richter, C. Brühl, R. Volkamer, J. P. Burrows, M. Kanakidou, 2008. The influence of natural and anthropogenic secondary sources on the glyoxal global distribution, *Atmos. Chem. Phys.*, **8**, 4965–4981.

Myriokefalitakis, S., E., Vignati, K., Tsigaridis, C., Papadimas, J., Sciare, N., Mihalopoulos, M. C., Facchini, M., Rinaldi, F.J., Dentener, D., Ceburnis, N., Hatzianastasiou, C.D., O'Dowd, M., van Weele, M., Kanakidou, 2010, Global modelling of the oceanic source of organic aerosols, *Adv. Meteo.*, http://www.hindawi.com/journals/amet/aip.939171.html.

O'Dowd, C. D., B. Langmann, S. Varghese, C. Scannell, D. Ceburnis, and M. C. Facchini, 2008. A combined organic-inorganic sea-spray source function, *Geophys. Res. Lett.*, **35**, L01801, doi: 10.1029/2007GL030331.

Palmer, P. I., D. J. Jacob, A. Fiore, R. V. Martin, K. Chance, and T. P. Kurosu, 2003. Mapping isoprene emissions over North America using formaldehyde column observations from space, *J. Geophys. Res.*, **108(D6)**, 4180, doi: 10.1029/2002JD002153.

Palmer P. I., D. S. Abbot, T.-M. Fu, D. J. Jacob, K. Chance, T. P. Kurosu, A. Guenther, Ch. Wiedinmyer, J. C. Stanton, M. J. Pilling, S. N. Pressley, B. Lamb, A.-L. Sumner, 2006. Quantifying the seasonal and interannual variability of North American isoprene emissions using satellite observations of the formaldehyde column, *J. Geophys. Res.*, **111**, D12315, doi:10.1029/2005JD006689.

Peters W., A. R. Jacobson, C. Sweeney, A.E. Andrews, Th. J. Conway, K. Masarie, J.B. Miller, L. M. P. Bruhwiler, G. Pétron, A.I. Hirsch, D. E. J. Worthy, G. R. van der Werf, J. T.

Randerson, P. O. Wennberg, M. C. Krol, and P. P. Tans, 2007. An atmospheric perspective on North American CO_2 exchange: Carbon Tracker, *Proc Natl Acad Sci* U S A. 2007 **104**(**48**): 18925–18930. doi: 10.1073/pnas.0708986104. http://www.esrl.noaa.gov/gmd/ccgg/carbontracker/

Pétron, G., C. Granier, B. Khattatov, V. Yudin, J.-F. Lamarque, L. Emmons, J. Gille, D. P. Edwards, 2004. Monthly CO surface sources inventory based on the 2000–2001 MOPITT satellite data, *Geophys. Res. Lett.*, **31**, L21107, doi: 10.1029/2004GL020560.

Richter, A., H. Nüß, J. P. Burrows, C. Granier, and U. Niemeier, 2004. Long term measurements of NO_2 and other tropospheric species from space, *Proceedings of the XX Quadrennial Ozone Symposium*, 1–8 June 2004, Kos, Greece, 213–214.

Richter, A., J. P. Burrows, H. Nüß, C. Granier, U. Niemeier, 2005. Increase in tropospheric nitrogen dioxide over China observed from Space, *Nature*, **437**, 129–132, doi: 10.1038/nature04092.

Rodgers, C. D., 2000. *Inverse Methods for Atmospheric Sounding*, World Scientific.

Rood R. B., 2007. *Fundamentals of Modeling, Data Assimilation, and High-Performance Computing, in Observing Systems for Atmospheric Composition*, Springer, New York, 10.1007/978-0-387-35848-2.

Sandu, A., W. Liao, G. R. Carmichael, D. Henze, J. H. Seinfeld, T. Chai, D. Daescu, 2004. Computational aspects of data assimilation for aerosol dynamics, *Lecture notes in computer science*, V **3038**, 709–716.

Schneising, O., Buchwitz, M., Burrows, J. P., Bovensmann, H., Bergamaschi, P., and Peters, W., 2009. Three years of greenhouse gas column-averaged dry air mole fractions retrieved from satellite – Part 2: Methane, *Atmos. Chem. Phys.*, **9**, 443-465.

Schulz, M., C. Textor, S. Kinne, Y. Balkanski, S. Bauer, T. Berntsen, T. Berglen, O. Boucher, F. Dentener, S. Guibert, I. S. A. Isaksen, T. Iversen, D. Koch, A. Kirkevåg, X. Liu, V. Montanaro, G. Myhre, J. E. Penner, G. Pitari, S. Reddy, Ø. Seland, P. Stier, and T. Takemura, 2006. Radiative forcing by aerosols as derived from the AeroCom present-day and pre-industrial simulations, *Atmos. Chem. Phys.*, **6**, 5225–5246.

Shindell, D. T., G. Faluvegi, D. S. Stevenson, M. C. Krol, L. K. Emmons, J.-F. Lamarque, G. Petron, F. J. Dentener, K. Ellingsen, M. G. Schultz, O. Wild, M. Amann, C. S. Atherton, D. J. Bergmann, I. Bey, T. Butler, J. Cofala, W. J. Collins, R. G. Derwent, R. M. Doherty, J. Drevet, H. J. Eskes, A. M. Fiore, M. Gauss, D. A. Hauglustaine, L. W. Horowitz, I. S. A. Isaksen, M. G. Lawrence,V. Montanaro, J.-F. Muller, G. Pitari, M. J. Prather, J. A. Pyle, S. Rast, J. M. Rodriguez, M. G. Sanderson, N. H. Savage, S. E. Strahan, K. Sudo, S. Szopa, N. Unger, T. P. C. van Noije, and G. Zeng, 2006. Multimodel simulations of carbon monoxide: Comparison with observations and projected near-future changes, *J. Geophys. Res.*, **111**, D19306, doi: 10.1029/2006JD007100.

Stajner, I., L. P. Riishøjgaard, and R. B. Rood, 2001. The GEOS ozone data assimilation system: Specification of error statistics, *Q. J. Roy. Meteor. Soc.*, **127**, 1069–1094.

Stajner, I., K. Wargan, S. Pawson, H. Hayashi, L.-P. Change, R. C. Hudman, L. Froidevaux, N. Livesey, P. F. Levelt, A. M. Thompson, D. W. Tarasick, R. Stubi, S. B. Andersen, M. Yela, G. Konig-Langlo, F. J. Schmidlin, and J. C. Witte, 2008. Assimilated ozone from EOS Aura: Evaluation of the tropopause region and tropospheric columns, *J. Geophys. Res.*, **113**, D16S32, doi:10.1039/2007JD008863.

Stavrakou, T., J.-F. Müller, 2006. Grid-based versus big region approach for inverting CO emissions using Measurement of Pollution in the Troposphere (MOPITT) data, *J. Geophys. Res.*, **111**, D15304, doi: 10.1029/2005JD006896.

Stavrakou, T., J.-F. Müller, I. De Smedt, M. Van Roozendael, M. Kanakidou, M. Vrekoussis, F. Wittrock, A. Richter, and J. P. Burrows, 2009. The continental source of glyoxal estimated by the synergistic use of spaceborne measurements and inverse modelling, *Atmos. Chem. Phys.*, **9**, 8431–8446.

Stevenson, D. S., F. J. Dentener, M. G. Schultz, K. Ellingsen, T. P. C. van Noije, O. Wild, G. Zeng, H. J. Eskes, A. M. Fiore, M. Gauss, D. A. Hauglustaine, L. W. Horowitz, I. S. A. Isaksen, M. C. Krol, J.-F. Lamarque, M. G. Lawrence, V. Montanaro, J.-F. Müller, G. Pitari, M. J. Prather,

J. A. Pyle, S. Rast, J. M. Rodriguez, M. G. Sanderson, N. H. Savage, D. T. Shindell, S. E. Strahan, K. Sudo, S. Szopa, 2006. Multi-model ensemble simulations of present-day and near-future tropospheric ozone, *J. Geophys. Res.*, **111**, D08301, doi: 10.1029/2005JD006338.

Stockwell, W. R., P. Middleton, and J. S. Chang, 1990. The second generation regional acid deposition model chemical mechanism for regional air quality modeling, *J. Geophys. Res.*, **95**, 16343–16367.

Struthers, H., R. Brugge, W. A. Lahoz, A. O'Neill, and R. Swinbank, 2002. Assimilation of ozone profiles and total column measurements into a General Circulation Model, *J. Geophys. Res.*, **107**, 4438, doi: 10.1029/2001JD000957.

Talagrand, O., 2003. *A posteriori* validation of assimilation algorithms, In: Swinbank, R., V. Shutyaev, and W. A. Lahoz, (eds) Data Assimilation for the Earth System, NATO ASI Series, Kluwer, Dordrecht, The Netherlands.

Textor, C., Schulz, M., Kinne, S., Guibert, S., Balkanski, Y., Bauer, S. E., Berntsen, T., Berglen, T., Boucher, O., Chin, M., Dentener, F., Diehl, T., Easter, R., Feichter, H., Fillmore, D., Ghan, S., Ginoux, P., Gong, S., Grini, A., Hendricks, J., Horowitz, L., Huang, P., Isaksen, I., Iversen, T., Kirkevag, A., Kloster, S., Koch, D., Kristjansson, E., Krol, M., Lauer, A., Lamarque, J. F., Liu, X., Montanaro, V., Myhre, G., Penner, J., Pitari, G., Reddy, S., Seland, Ø., Stier, P., Takemura, T., and Tie, X.: 2006. Analysis and quantification of the diversities of aerosol life cycles within AeroCom, *Atmos. Chem. Phys.*, **6**, 1777–1813.

van der A, R. J. M, D. H. M. U. Peters, H. Eskes, K. F. Boersma, M. van Roozendael, I. De Smedt, H. M. Kelder, 2006. Detection of the trend and seasonal variation in tropospheric NO_2 over China, *J. Geophys. Res.*, **111**, 12317, doi: 10.1029/2005JD006594.

van der Werf, G. R., Randerson, J. T., Giglio, L., Collatz, G. J., Kasibhatla, P. S., and Arellano Jr., A. F., 2006. Interannual variability in global biomass burning emissions from 1997 to 2004, *Atmos. Chem. Phys.*, **6**, 3423–3441.

van Noije, T. P. C., Eskes, H. J., Dentener, F. J., Stevenson, D. S., Ellingsen, K., Schultz, M. G., Wild, O., Amann, M., Atherton, C. S., Bergmann, D. J., Bey, I., Boersma, K. F., Butler, T., Cofala, J., Drevet, J., Fiore, A. M., Gauss, M., Hauglustaine, D. A., Horowitz, L. W., Isaksen, I. S. A., Krol, M. C., Lamarque, J.-F., Lawrence, M. G., Martin, R. V., Montanaro, V., Müller, J.-F., Pitari, G., Prather, M. J., Pyle, J. A., Richter, A., Rodriguez, J. M., Savage, N. H., Strahan, S. E., Sudo, K., Szopa, S., and van Roozendael, M., 2006. Multi-model ensemble simulations of tropospheric NO_2 compared with GOME retrievals for the year 2000, *Atmos. Chem. Phys.*, **6**, 2943–2979.

van Loon, M., P. J. H. Builtjes, and A. J. Segers, 2000. Data assimilation of ozone in the atmospheric transport chemistry model LOTOS, *Environ. Model. Software*, **15**, 603–609.

Velasco, E., Lamb, B., Westberg, H., Allwine, E., Sosa, G., Arriaga-Colina, J. L., Jobson, B. T., Alexander, M. L., Prazeller, P., Knighton, W. B., Rogers, T. M., Grutter, M., Herndon, S. C., Kolb, C. E., Zavala, M., de Foy, B., Volkamer, R., Molina, L. T., and Molina, M. J., 2007. Distribution, magnitudes, reactivities, ratios and diurnal patterns of volatile organic compounds in the Valley of Mexico during the MCMA 2002 & 2003 field campaigns, *Atmos. Chem. Phys.*, **7**, 329–353.

Velders, G. J. M., C. Granier, R. W. Portmann, K. Pfeilsticker, M. Wenig, T. Wagner, U. Platt, A. Richter, and J. P. Burrows, 2001. Global tropospheric NO_2 column distributions: Comparing three-dimensional model calculations with GOME measurements, *J. Geophys. Res.*, **106** (**12**), 12,643–12,660.

Verlaan, M. and A. W. Heemink, 1995. Reduced rank square root filters for large scale data assimilation problems, in: Second International Symposium on Assimilation of Observations in Meteorology and Oceanography.

Volkamer, R., L. T. Molina, M. J. Molina, T. Shirley, and W. H. Brune, 2005. DOAS measurement of glyoxal as an indicator for fast VOC chemistry in urban air, *Geophys. Res. Lett.*, **32**, L08806, doi:10.1029/2005GL022616.

Vrekoussis, M., M. Kanakidou, N. Mihalopoulos, P. J. Crutzen, J. Lelieveld, D. Perner, H. Berresheim, E. Baboukas, 2004. Role of the NO_3 radicals in oxidation processes in the eastern Mediterranean troposphere during the MINOS campaign, *Atmos. Chem. Phys.*, **4**, 169–182.
Wang, Y., M. B. McElroy, R. V. Martin, D. G. Streets, Q. Zhang, and T.-M. Fu, 2007. Seasonal variability of NOx emissions over east China constrained by satellite observations: Implications for combustion and microbial sources, *J. Geophys. Res.*, **112**, D06301, doi: 10.1029/2006JD007538.
Weaver, A. and P. Courtier, 2001. Correlation modelling on the sphere using a generalized diffusion equation, *Q. J. Roy. Meteor. Soc.*, **127**, 1815–1846.
Wittrock, F., A. Richter, H. Oetjen, J. P. Burrows, M. Kanakidou, S. Myriokefalitakis, R. Volkamer, S. Beirle, U. Platt, and T. Wagner, 2006. Simultaneous global observations of glyoxal and formaldehyde from space, *Geophys. Res. Lett.*, **33**, L16804, doi: 10.1029/2006GL026310.
Zhang, Q., J. L. Jimenez, M. R. Canagaratna, J. D. Allan, H. Coe, I. Ulbrich, M. R. Alfarra, A. Takami, A. M. Middlebrook, Y. L. Sun, K. Dzepina, E. Dunlea, K. Docherty, P. F. DeCarlo, D. Salcedo, T. Onasch, J. T. Jayne, T. Miyoshi, A. Shimono, S. Hatakeyama, N. Takegawa, Y. Kondo, J. Schneider, F. Drewnick, S. Borrmann, S. Weimer, K. Demerjian, P. Williams, K. Bower, R. Bahreini, L. Cottrell, R. J. Griffin, J. Rautiainen, J. Y. Sun, Y. M. Zhang, and D. R. Worsnop, 2007. Ubiquity and dominance of oxygenated species in organic aerosols in anthropogenically-influenced Northern Hemisphere midlatitudes, *Geophys. Res. Lett.*, **34**, L13801, doi: 10.1029/2007GL029979.

Chapter 10
Conclusions and Perspectives

John P. Burrows, Ulrich Platt and Peter Borrell

10.1 Introduction: The Need for Satellite Observations

As will have been appreciated from the earlier chapters, key aspects of the Earth's atmosphere and environment are best monitored with instruments mounted on satellites orbiting the Earth in space. The progress towards this aim during the last twenty years has been remarkable. The objectives of the present chapter are to indicate likely future changes and improvements in instruments and techniques, to describe what the scientific community is suggesting and how the national and international space agencies are proposing to meet the challenges in the foreseeable future.

The need for monitoring the atmosphere and environment is well appreciated: the near exponential growth in the Earth's population and the general improvement in standard of living that has brought many societal benefits has been almost literally fuelled by the inexpensive energy made available from the exploitation of fossil fuels. The growth of emissions into the atmosphere has changed pollution from being a matter of local concern to one having a global impact, and thus the Holocene period has been transformed to the Anthropocene, a period characterised by a growing influence of human activities on the global environment. Today, the development of an acceptable strategy to provide a sustainable development and future for the Earth system with an increasing human population, having its traditional aspirations for growth and standard of living, is currently one of the greatest challenges for society.

The changes in atmospheric composition, and other critical parameters determining the nature and behaviour of the Earth system, have been recognised by the

J.P. Burrows (✉)
Institute of Environmental Physics (IUP), University of Bremen, Germany
and
NERC Centre for Ecology and Hydrology, Wallingford, United Kingdom

U. Platt
Institute of Environmental Physics (IUP), University of Heidelberg, Heidelberg, Germany

P. Borrell
P&PMB Consultants, Newcastle-under-Lyme, United Kingdom

J.P. Burrows et al. (eds.), *The Remote Sensing of Tropospheric Composition from Space*, Physics of Earth and Space Environments, DOI 10.1007/978-3-642-14791-3_10,

majority of the active scientific community, and the impact of anthropogenic activity assessed: the state of knowledge and their conclusions have been summarized in a number of international assessments (UNECE LTRAP HTAP 2007; WMO 2007; IPCC 2007; WMO 2010). One successful outcome was the Vienna Convention on Ozone Depleting Substances and its subsequent amendments (UNEP 2009), which instituted a policy to restrict damage to the stratospheric ozone layer. Similarly, in the developed nations, national and international legislation has been successful in improving air quality in cities. However the continuing growth of anthropogenic greenhouse gas emissions, changes in land use, and the release and long range transport and transformation of pollution are now on a scale which challenges the ability of the scientific community to assess their impact and predict the likely consequences.

Although the need for action to maintain the environment, its ecosystem services and biodiversity is generally appreciated, the disappointment at the outcome of the Copenhagen Climate Conference in 2009 and the animated public discussion of the acknowledged minor errors in the IPCC fourth Assessment Report (IPCC 2007), indicate the challenges facing the policymakers in attempting to reach an adequate compromise to control carbon emissions. The impasse re-emphasizes the continuing and growing need for accurate data to assess the Earth's changing environment, if a sustainable future is to be achieved.

A prerequisite for successful global management is a proper knowledge and understanding of the physical, chemical and biological processes and their feedbacks, which determine the conditions of the local, regional and global environment. A positive outcome of the growth of industrial and economic activity has been, over the last 50 years, the birth and growth of space exploration, together with the development of all the techniques and infrastructure required to operate satellite and ground-based systems for monitoring the global and regional atmospheric environment.

International collaboration is essential to achieve global monitoring. The Committee on Earth Observation Satellites, CEOS, coordinates civilian space borne observations of the Earth. It initiated the Integrated Global Observing Strategy, IGOS, which seeks to provide a comprehensive framework to harmonize the common interests of the major space-based and *in situ* systems for global observation of the Earth. The Integrated Global Atmospheric Chemistry Observations theme (IGACO 2004), a component of IGOS, recognised that global measurements of atmospheric trace composition from space-based instrumentation, together with ground-based networks and aircraft, are required to meet the emerging need for accurate data and provide an early warning of deleterious change.

The 2002 World Summit on Sustainable Development and the Group of Eight leading industrialized countries, established the Group on Earth Observations (GEO), which coordinates efforts to build a Global Earth Observation System of Systems (GEOS 2005); it now incorporates the objectives of IGOS. The European Global Monitoring of Environment and Security (GMES) is the European contribution to Group on Earth Observations. While it is clear that some progress is being achieved, the question remains whether the progress will be sufficient to meet the goals: that is delivery of the evidence needed to underpin an international environmental policy to provide sustainable development, before the changes in the Earth system become irreversible.

10.2 Some Scientific Highlights

The principal chapters of this book have documented the revolution in knowledge and the most important aspects of the evolution of the technology of measurement and retrieval theory for the remote sensing of tropospheric constituents from space. The resulting measurements of the tropospheric composition are challenging and revolutionise our understanding of the chemical processing and dynamics within the atmosphere, of exchange at the surface and at the tropopause and of the biogeochemical cycles.

Perhaps the most important highlight is the unique capability to make measurements needed for numerical weather prediction and to measure tropospheric pollution, its transport and transformation.

The observations have resulted in several discoveries of both natural phenomena and the results of anthropogenic activity: large clouds of bromine monoxide, BrO, seen at high latitudes, the global distribution of iodine monoxide, IO, the global observations of ammonia, NH_3, and the increase in nitrogen dioxide, NO_2, ozone, O_3, and sulfur dioxide, SO_2, over Asia, together with the impact of biomass burning. These have led to a new understanding of our atmospheric environment and have revolutionised research in atmospheric chemistry.

The following section picks out some highlights from Chapters 2, 3, 4, 6 and 8.

10.2.1 Observed Compounds

Perhaps the most notable development in the field is the sheer number of species that can now be observed from space (Table 8.3). The prominent pollutants, ozone, O_3, nitrogen dioxide, NO_2, carbon monoxide, CO, and sulfur dioxide, SO_2, are observed and it is now also possible to observe selected volatile organic compounds, VOCs, methane, CH_4, acetylene, C_2H_2, ethane, C_2H_6, and ethene, C_2H_4. Also observable are some of the products from photo-oxidation such as methanol, CH_3OH, formaldehyde, HCHO, formic acid, HCOOH, glyoxal, CHOCHO, acetone, CH_3COCH_3, PAN, peroxy nitric acid (HO_2NO_2), hydrogen cyanide, HCN, nitric acid, HNO_3, hydrogen peroxide, H_2O_2 and carbonyl sulfide, OCS. Another problem compound observed, resulting from intensive agriculture, is ammonia, NH_3.

The greenhouse gases carbon dioxide, CO_2, methane, CH_4, nitrous oxide, N_2O and water vapour, H_2O are observable. Partially deuterated water, HDO, has also been seen. In addition all three phases of H_2O, ice, liquid and vapour can be observed with microwave sensors.

The interest in stratospheric ozone depletion is reflected in observations in the upper troposphere of the chlorofluorocarbons, CFC-11 ($CFCl_3$), CFC-12 ($CFCl_2$), CFC-113 (Cl_2FC-$CClF_2$), HCFC-142b (ClF_2C-CH_3) and HCFC-22 ($CHClF_2$). Some other halogen compounds seen in this region are sulfur hexafluoride (SF_6), methyl chloride, CH_3Cl, and hydrogen chloride, HCl.

The observations of reactive halogen free radicals BrO and IO have demonstrated the global importance of tropospheric halogen chemistry. The sudden explosive spring growth in the concentrations of BrO in the Arctic and Antarctic, together with the measurements of the depletion of O_3 at the surface, have triggered much interest and speculation about unexpected chemical reactions on the surface of the ice sheets.

The observation of cloud and aerosol optical parameters from space based instrumentation has also seen a remarkable development in the last two decades. As each compound or parameter has its own peculiar spectral characteristics and unique sources and sinks, not all the compounds or parameters can be observed under all conditions.

10.2.2 The Multiple Roles of NO_2

NO_2 is a prominent product from industrial and domestic energy production, biomass burning and transport. NO_2 also plays a key role in a number of pollution problems: acidification of terrestrial ecosystems, eutrophication of lakes and the marine environment, formation of tropospheric O_3 and effects on human health and agricultural productivity (Grennfelt et al. 1994). NO_2 is a key photochemical species in the troposphere, which participates as a chain carrier in catalytic cycles, producing and destroying tropospheric O_3. It is perhaps fortunate then that NO_2 can readily be observed from space in the UV/vis region.

The retrieval of tropospheric NO_2 is providing global knowledge of NO_2 columns. During the period of observations a large increase of NO_2 over urban areas in Asia has been observed, as has the simultaneous decrease over Europe, attributed to legislative demands. Weekly cycles of NO_2 in urban areas were discovered as well as changes in NO_2 due to the fertilisation of soils.

Emissions from shipping show the NO_2 ship tracks over the ocean and indicate the growing impact of pollution from shipping on ocean ecosystems. Observations of the formation of NO_2 from lightning have improved the estimation of nitrogen budgets.

10.2.3 Industrial Emissions and Biomass Burning

The observation of the tropospheric columns of CO and the aldehydes, HCHO and CHOCHO, are providing valuable information about the oxidation of VOC of both natural and anthropogenic origin. These help to enlarge our knowledge of fossil fuel combustion, and anthropogenic, biogenic and biomass burning emissions of VOC and CO.

The extensive CO observations have been particularly well exploited as an indicator of industrial pollution and biomass burning; the long range transport in the southern hemisphere being one focus of activity. The recent measurements of HNO_3 and, in particular NH_3, should add to the possibilities and further constrain the complex models. Measurements in the upper atmosphere from limb observations have also shown the global importance of biomass burning and enabled exchange between the troposphere and stratosphere to be studied.

The increase of air pollutants such as NO_2, SO_2 and aerosol parallels the rapid economic development in eastern Asia. The observation of the transport of pollution from Asia to North America, from North America to Europe and from Europe into the Arctic has huge implications for transboundary pollution control.

10.2.4 *Ozone,* O_3

Appreciable improvements have been made in the retrieval of tropospheric O_3, particularly in the tropics, where stratospheric O_3 is spatially uniform. However as about 90% of the O_3 lies above the tropopause in the stratosphere, TIR radiation must pass once through this layer and solar backscatter twice before detection. Thus accurate knowledge of the stratospheric O_3 is required in order to retrieve the tropospheric O_3, and particularly boundary layer O_3, with any accuracy.

In the sub-tropics the rapid changes in lower stratospheric O_3 and tropopause height, resulting from streamers and the movement of frontal systems, presents a challenge and requires more research to improve our knowledge of global tropospheric O_3. In spite of the progress made in the retrieval of tropospheric O_3, both in the UV/vis and TIR regions, determination of accurate boundary layer O_3 remains a challenging target for tropospheric remote sensing. However the combined simultaneous use of different spectral features of O_3 and different viewing geometries offers potentially a synergetic solution for this issue.

10.2.5 *Greenhouse Gases*

H_2O is the most important greenhouse gas and a key component of the hydrological cycle, and the measurements of water profiles and columns have made much progress in recent years. The measurement of the dry columns of the greenhouse gases, CO_2, CH_4 from space has now begun in earnest. Nadir sounding TIR instruments are providing mid- and upper-tropospheric columns and solar back scattered instrumentation in the near and short wave infrared spectral regions are yielding the dry columns of these gases. Seasonal maps of CO_2 concentrations confirm the seasonal features in the Mauna Loa records (Keeling et al. 1976; 2003), and there are some excellent animations using the changes in the dry column of CO_2 to show the "breathing" Earth. However measurements of high accuracy are needed here to assess both anthropogenic emission and the response of the biogeochemical carbon cycle to change. Recent studies have demonstrated the feasibility of measuring CO_2 and CH_4 remotely – and these now pose a considerable challenge for the next generation of instrumentation and systems.

10.2.6 Water Vapour, and Other Hydrological and Cloud Parameters

Parameters, which determine the transport within the hydrological cycle, have been a very successful target for remote sensing. Gas phase H_2O is retrieved successfully by instruments from the microwave to the visible regions; the rotational and ro-vibrational spectra of H_2O are being exploited in both emission and absorption. In addition liquid and ice phase parameters are now also retrieved from both IR and microwave spectral regions.

An important focus of the use of wide-swath sounding, passive and active sensors operating at microwave and mm wavelengths, see Chapter 4, has been the retrieval of meteorological parameters. In this context the accurate retrieval of H_2O and temperature profiles have been critical for the development and improvement of numerical weather prediction.

In limb sounding, retrievals using data from some microwave instruments, although primarily aimed at the retrieval of stratospheric parameters, yield O_3, CO and HNO_3, as well as the established water vapour and cloud-ice observations in the upper troposphere.

The retrieval of liquid water and ice from clouds represent significant Earth observation milestones, and the use of radar will be increasingly exploited for precipitation applications in the future.

10.2.7 Aerosol and Cloud Parameters

The growth in the measurement of aerosol parameters has been a feature in the past few years. These have been facilitated by the launch of new dedicated LEO satellite instruments as well as GEO instruments. There is now a long data record of nadir viewing at selected wavelengths, which yield long term changes in aerosol parameters.

Currently instruments using multiple views and observing different polarisations are providing much improved data products, and the pioneering active systems will give reliable profile data (Chapter 6). A major recent enhancement has been the successful deployment in space of lidar for the observation of cloud and aerosol parameters. Building on LITE and GLAS, the combination of CALIOP and the improved passive remote sounding instruments is providing valuable new insights to enhance our understanding of aerosol amount and transport.

A topical example is provided by the recent eruption of the volcano under the Eyjafjallajoekull glacier in Iceland. Eruptions began unexpectedly in March 2010, the first time since 1821. A strong eruption on the 14th April pushed gas and ash into the upper troposphere. The volcanic plume can be seen in Fig. 10.1 (top) as the

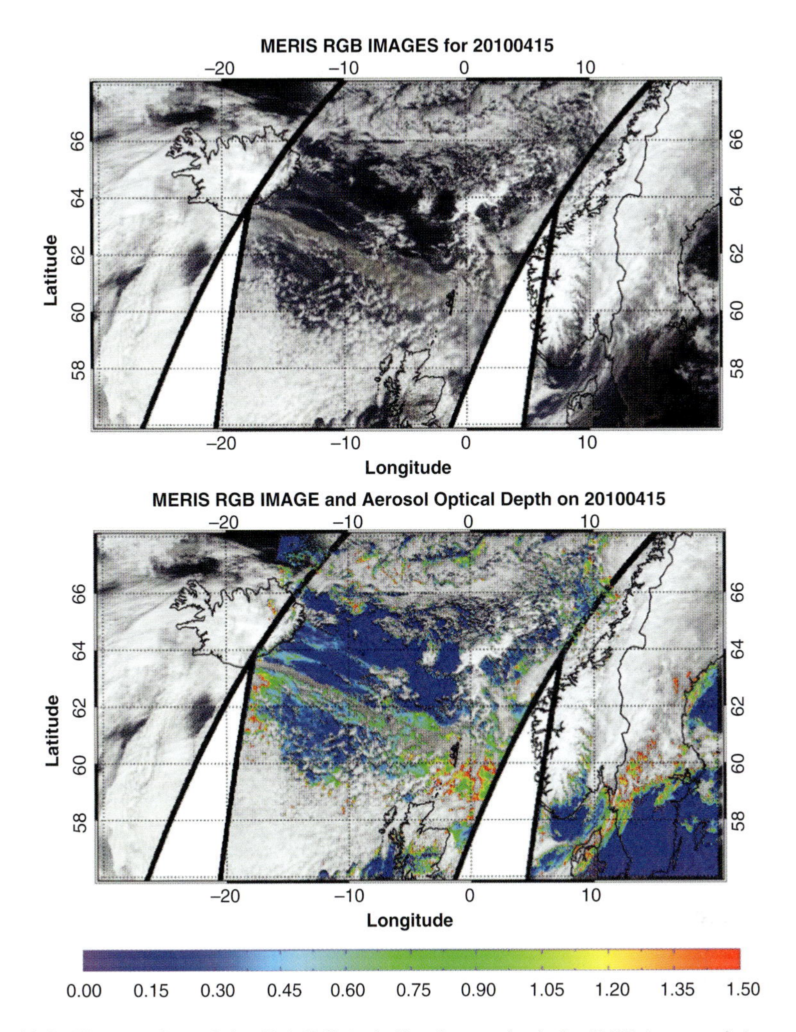

Fig. 10.1 The eruption of the Eyjafjallajoekull volcano: (*top*) the RGB image of three orbit segments (left orbit: 42475, middle orbit: 42476, right orbit: 42477) of MERIS taken close to 10 a.m. local time on the 15th April 2010. In the middle orbit the ash cloud is visible after one day of transport; (*bottom*) as above but with an overlay of the aerosol optical depth; the highest aerosol optical depths are found close to the Shetland Islands with values of about 1.5. AOD derived from the Bremen AErosol Retrieval (BAER). C. Schlundt, W. von Hoyningen Huene, M. Vountas, and J.P. Burrows, Instiute of Environemntal Physics, University of Bremen.

brownish cloud originating from the south of Iceland and transported in an ESE direction while dispersing; it spread the particulate matter over northern Europe leading to closure of air space and the total disruption of air traffic.

An overlay of the optical density of the fine particles has been added in Fig. 10.1 (bottom) and shows the spread of fine particulates over much of the region viewed. The pictures illustrate the assistance that can be gained in practical air traffic management from regular remote sensing measurements with their high spatial and temporal sampling.

Finally for the adequate scientific interpretation of global remote sensing atmospheric data products, results for a number of other parameters are required. Among these are night-time light pollution, vegetation distribution, fire counts and lightning counts; these are illustrated as global maps in Appendix B. They provide practical examples of the synergetic use of multiple data products to improve retrievals and understanding (Section 10.3).

10.2.8 Volcanic Emissions

As shown in Fig 1.11, SO_2 is emitted not only by pollution sources but also, in many cases continuously, by volcanoes. Their releases result in the formation of aerosol and cloud condensation nuclei, which in turn impact on cloud amount and precipitation patterns.

In addition, as Fig. 1.10 shows, volcanic eruptions are a major source of aerosol and particulate matter. This was highlighted in spring 2010 when the inability of aircraft to fly had a large negative impact on sectors of the economy.

10.3 Scientific Needs

The authors of each chapter have provided a future perspective by indicating likely or needed advances in their respective areas. While the chapters should be consulted for the detailed recommendations, a number of common themes emerge.

- Present passive remote sensing instruments in the UV/vis/NIR and the TIR are demonstrating performance close to the achievable limits for the data retrieval, so larger detector areas are needed to improve the signal to noise ratio. Instruments having improved spatial and temporal sampling are required to cope with the accuracy needed to observe the detailed changes in concentrations of long-lived and short lived climatically active gases, and also the concentrations of O_3 and its precursors in the all-important planetary boundary. Better resolution ground coverage is needed in most studies (Sections 2.8.1, 3.5, and 8.4).

- To ensure that measurements are made often enough under cloud free conditions at mid and low latitudes, GEO observations are needed. However the polar regions are not probed by GEO, and combinations of instruments on GEO and LEO satellites with orbits optimised to observe the polar regions are desirable (Sections 2.8.1, 3.5, 6.15, and 10.8).
- More extensive multi viewing, hyperspectral, coupled with polarisation measurements, are needed to improve the determination of aerosol properties and a knowledge of tropospheric radiative transfer (Sections 2.8.1, 6.15, and 10.6.1).
- The determination of tropospheric height-profile information for all atmospheric constituents needs improvement. For selected constituents this might best be provided by the deployment of active systems (Sections 2.8.1, 4.7, and 6.15). The use of simultaneous multiple viewing, passive, remote sensing instruments requires further study for trace gas observations in the lower troposphere.
- While retrievals are under continuous improvement, improved reference input data is needed. For example, absorption cross sections and line parameters are not known adequately well, even for common molecules (Section 2.8.2). Similarly the knowledge of surface spectral reflectivity and emissivity is inadequate.
- Synergistic studies involving instruments mounted on different satellites in different orbits offer the potential to deduce the physical and chemical state of the troposphere including, for example, studying the development of pollution episodes in a more systematic way. However they will require more extensive cross-calibration (Sections 2.8.3, 3.5, 4.8, 8.4, 9.5, and 10.4.2).
- Data assimilation, an essential tool in meteorology for numerical weather prediction, is now being developed for the numerical prediction of chemical composition. However much more development and experience is required in applications such as inverse modelling (Sections 4.8, 9.5, and 10.4.3).
- Although much needed, the production of long term consolidated and consistent data sets of atmospheric parameters is challenging, principally because of the cost. Unfortunately the issue is coupled with the fact that fewer missions designed to study trace species in the troposphere are planned (see the timelines in Appendix D). As a result of the time taken to build and commission space-based instrumentation for Earth observation, there are likely to be problems in the future when more detailed measurements and understanding will be required (Section 3.5, Appendix D). A good example of this recently was the eruption of the volcano under the Eyjafjallajoekull glacier (Section 10.2.7), which resulted in all air traffic over Europe being stopped. The airline industries argue that insufficient data was available and are now asking governments to compensate them for their losses. One solution would be to levy the energy and transport industries to provide the funding needed for an adequate measurement system.
- Improvements in instruments and retrievals are dealt with in more detail in Sections 10.5 and 10.6.

10.4 Further Interpretation of Data from Current Instrumentation

Much of the current generation of space instrumentation was not developed specifically for the determination of tropospheric composition. It was often a by-product of instrumentation designed to observe the stratosphere or to deliver parameters for numerical weather prediction. However much progress has been made in the retrieval of trace constituents and parameters using the current generation of instrumentation, and many more data products have been delivered than expected.

Overall there has been an exceptionally high return on the investment in the hardware and ground segment for research. Operational test services using the new capability have been initiated: for example in Europe, ESA, EUMETSAT and ECMWF have initiated upstream and downstream service elements. To evolve, these will require improved systems, based on the successes of the first generation of instrumentation. However it is also essential to improve instrument performance such as temporal and spatial sampling. Nevertheless the improved evaluation of data from existing sensors in space still holds much promise for better tropospheric data products.

For some chemical species, such as BrO, the maximum information content has been achieved because the optimal spectral window has been used. However for molecules such as NO_2 and O_3 many more spectral features are available. The simultaneous use of such features, often from different spectral regions, has been recognised but its use is still sparse. Similarly the use of multiple measurements for the determination of cloud and aerosol parameters is in its infancy. One important point to note is that improved knowledge of cloud and aerosol will also improve the general accuracy of trace gas retrievals, using data recorded simultaneously.

Limiting factors for the simultaneous use of multiple spectral regions, or combinations of different instruments, are a knowledge of the calibration of the measurements, the accuracy of data bases of the reference spectra and the completeness of the radiative transfer programmes used as forward models in the retrieval process. Further progress using the current and archived set of measurements and data products is likely to come from the following areas.

10.4.1 Retrieval Algorithm Developments

For cloud free scenes in the spectral ranges from the UV to the mid-IR, improved radiation transport modelling has the potential to improve the retrieval of tropospheric trace gases. This could be achieved by:

- having improved simultaneous knowledge of aerosol properties, and
- the explicit inclusion of surface effects, e.g. consideration of (weak) narrow band structures in surface reflectivity and emissivity.

For cloudy and partially cloudy ground scenes, the retrieval of cloud parameters, trace gas and aerosol properties, and also the adequacy of the forward models used in the retrieval algorithms is of central importance.

Presently all operational cloud retrieval algorithms rely on a homogeneous, single-layered cloud model, while real clouds are inhomogeneous objects. Scattered cloud fields, situations with extensive vertical convection or, for thin clouds, both cloud inhomogeneity and the underlying surface bi-directional reflectance function, must all be accounted for. Another issue is the retrievals of ice clouds, which at the moment relies on *a priori* models of crystal shapes mostly based on empirical evidence. The retrievals of mixed phase cloud properties, such as the ice/water fraction, are not yet developed satisfactorily. In general improving the description of the scattering within the atmosphere and at the surface will reduce uncertainty in the retrieved data products.

10.4.2 The Use of Multiple Observations

The retrieval of data products depends on accurate knowledge of the atmospheric conditions. Although the potential is recognised, the synergetic use of simultaneous or near simultaneous observation of trace gas features in the visible, infrared, and microwave regions has not been systematically explored for the retrieval of tropospheric data products. Investigations are necessary to meet this challenge and to show how to benefit from the complementary sensitivities of the different wavelengths to each atmospheric parameter, and thus help to constrain the inversion problems.

Combining data products from different sensors (on the same satellite or on different satellites) potentially yields:

- an increased number of retrieved data products, noting that several trace gases may only be retrieved in selected wavelength ranges, such as IO and BrO in the UV, and NH_3 or CH_3OH in the mid-IR (see Chapters 2 and 3). Combining these data leads to a better characterization and thus to improved constraints of the chemical modelling of a probed air mass;
- improved vertical resolution can be achieved by combining the different sensitivities of trace gas absorption in different spectral regions as a function of altitude, e.g. CO measurements in the short-wave IR and the TIR, or the combination of limb and nadir observation of the same species;
- the combination of simultaneous or near simultaneous knowledge of chemical composition and dynamical parameters.

10.4.3 Data Assimilation

As is well recognised by the operational meteorological community, data assimilation has a special role to play in the improvement of the prediction. The area of environmental and climate prediction is a growing area which meets many of the needs of the policymaking community. Data assimilation combines the knowledge from model and measurement. The method is also valuable in filling in regions where data is sparse, and thus provides an approach for the comparison of retrieved data products which are not measured simultaneously. The resultant hybrid data sets have many scientific uses.

However for the detection of changes and trends, and to ensure unforeseen processes and feedbacks are recognised, it is advisable that there should always be (alternative) data products, which are as free from any modelling information as possible.

10.5 Idealised Requirements for the Evolution of Instrumentation

The type and quality of the data obtained from remote sensing necessarily depend on the scientific problem being tackled. As in all physical experiments the sampling and the signal to noise ratio of the data products must be sufficient to reveal true behaviour of the process being investigated.

However there are different priorities for the data products determining atmospheric composition: numerical weather prediction, numerical environmental prediction, global climate and tropospheric pollution and atmospheric chemistry; all need different temporal and spatial sampling scales. Nevertheless there are common themes which include the following.

(a) The improvement of the vertical resolution for trace gas and aerosol measurements. Three-dimensional fields of trace gas concentrations are desirable at vertical resolutions much better than the atmospheric scale height (7 ± 1 km).
(b) Sufficient horizontal resolution of the measurements, in particular for the retrieval of trace gas column densities and concentrations. In order to resolve structures in biomass burning events, volcanic plumes, and mega cities spatial resolutions better than 4 km are required.
(c) Temporal resolution of the measurements, which is enough to resolve diurnal cycles of trace gas concentrations, aerosol and cloud parameters.
(d) Extension of the number of gaseous species that can be observed from space. Desirable species include speciated VOC, especially alkanes and olefins. Presently only CH_4 and a few oxygenated species are measurable in the boundary layer from space.
(e) Improved measurements of the key species of atmospheric chemistry such as tropospheric O_3, and NO_2. RO_x radicals (OH, HO_2, RO_2 where R is an organic radical) are required to understand the processing on short time scales in air masses.
(f) Improved sensitivity for a series of species to study background levels of NO_x, hydrocarbons, and tropospheric O_3.

For many of these requirements, there is not yet any mature technological solution. This is particularly the case for improved vertical resolution. Improved sensitivity, spatial, and temporal resolution can be achieved using either instruments with a larger aperture (i.e. able to collect more photons in a given period of time), or constellations of instruments (see below). These will provide a challenge for instrument engineering and sampling logistics.

10.6 Perspectives for the Improvement of Instrument Technology

We are in the middle of an ongoing revolution in technology for remote sensing; the progress in the miniaturisation of computer and electronic control systems and in the increase of the rate of data transmission and storage will continue to have a major impact on the development of remote sensing.

For passive remote sensing, the performance is usually determined by the detector system, the limit ideally being determined by the shot noise limit. For active remote sensing, the emitter and receivers would both benefit from technical improvements. However, as a result of the strict quality control used in space hardware and the long development time, the equipment used in space is, in fact, surprisingly old fashioned.

The first generations of satellite instruments capable of probing the troposphere from space necessarily had or have deficiencies and limitations, resulting from technical issues, from the lack of sufficient funding and sometimes from limited vision.

Technical issues, such as the polarization sensitivity of the instruments, narrow band spectral structures of the instrument, spectral under-sampling, spectral structures in the reflectivity of the diffuser plate, degradation and icing of instruments in the space environment, are all examples of known problems. Thus an obvious way to improve future instruments would be to address these issues and to avoid, or minimise, the known deficiencies in future designs.

For instruments relying on solar radiation, improved knowledge of the solar spectrum, the Fraunhofer structure and the small but significant variations, is likely to improve the accuracy of data products in these spectral regions.

In the TIR, sub-mm and microwave spectral regions, improved detector technology will undoubtedly result in improved performance. Similarly improved antennae design and constellations of instruments are likely to meet the evolving needs better.

When considering the likely evolution of technology our vision encompasses the following suggestions.

10.6.1 Polarisation Measurements

Instruments having better defined and calibrated polarization sensitivities are needed. These would provide improved information on the atmospheric radiative transfer and potentially add tropospheric profile information for O_3 (Chapter 2). In addition, the observation of the same ground scene under a variety of viewing angles yields improved discrimination of aerosol scattering and surface reflection.

10.6.2 Measurements for Tomographic Reconstruction

For specific atmospheric conditions having strong gradients, tomographic inversion techniques could be applied to resolve vertical structures in trace gas and aerosol distributions, provided the signal to noise ratio is adequate. Tomography has already been successfully applied to multiple measurements made in the upper atmosphere and in some microwave applications. Multiple observation of the same scene in the troposphere, where chemical lifetimes are often relatively short compared to the upper atmosphere, opens the opportunity for tomographic retrievals.

10.6.3 Multi-Wavelength Hyper-Spectral Measurements

For the optimal vertical resolution for troposphere, many trace gases will be observed by combining passive solar and TIR measurements. This will require both improved data sets of atmospheric absorption cross sections and line parameters, and an adequate, consolidated and consistent radiative transfer model for the different spectral regions.

10.6.4 Multi-Instrument Measurements

Modern micro-optical techniques permit the development of instruments consisting of multiple identical units (e.g. spectrometers or interferometers), each one observing a particular direction or wavelength interval. These would appreciably improve the signal to noise ratio.

10.6.5 Microwave and Sub-mm Spectral Region

One important potential growth area is the new cooled superconductor-insulator-superconductor receivers (SIS) which have the potential to yield measurements with dramatically improved precision and spatial resolution (Chapter 4).

10.6.6 Active Systems

Active remote instruments for trace gas measurements offer the advantage of good vertical resolution (see Section 1.11.2). For aerosol, improved vertical resolution is achieved by evolving the lidar space based systems. DIAL (Differential Absorption lidar) systems are capable of measuring and mapping concentrations and mass emissions of various molecules in the lower atmosphere from the ground upwards. Instruments such as CALIOP need to be extended to multi-wavelength observations. DIAL may also yield tropospheric trace gas observations with unprecedented vertical resolution, albeit with reduced spatial coverage. These techniques for trace constituents are promising for a selected number of gases but probably not generally applicable. Radar systems are being, and will be further, improved to provide better cloud parameters and precipitation measurements. However an adequate constellation of instruments is required to provide global observations for predictive purposes.

10.7 Current and Future Planned Missions

The diagrams in Appendix D give time lines for the various present and accepted future missions to study tropospheric gases and aerosols. The time lines present the information in terms of the tropospheric species being studied, divided into three groups: reactive gases, greenhouse gases and aerosols. Each of these is divided into nadir, occultation and geostationary observations. The diagrams can be used to gauge the likely extent of coverage in the future and highlight the unfortunate gaps that seem likely to develop.

10.7.1 LEO Satellite Instruments

The current generation of NASA low earth orbit research satellites having tropospheric capability comprises the large platforms, such as Terra, Aqua and Aura, coupled with the other smaller missions participating in the A Train. These and the European earth observation satellite, ENVISAT were conceived in the mid-1980s and challenged the engineering experience of the day. Since this time, focussed parts of these missions are being evolved into operational missions by EUMETSAT and NPOESS. The latter system was planned to monitor a wide range of parameters in Earth observation and provide data for long-range weather and climate forecasts. Unfortunately, as a result of cost overruns, a significant fraction the instrumentation aimed at tropospheric remote sounding, has been reduced in scope.

The EUMETSAT ESA MetOp satellite series is the first of the European operational meteorological satellites in low earth orbit. It was initiated with the launch of MetOp A in late 2006 and is planned to measure up to about 2020. MetOp includes the instruments GOME-2, IASI, and AVHRR, which yield concentrations of tropospheric trace constituents. EUMETSAT and ESA are now planning the post-MetOp platforms and payloads to make measurements in the 2020–2030 period. For tropospheric remote sensing of trace constituents, two instruments, UV/vis/NIR sounding mission (UVNS), and the infrared instrument (IRS) are relevant. UVNS builds on the heritage of SCIAMACHY and GOME and has been funded, as the GMES Sentinel 5, through an EU contribution to ESA and EUMETSAT. The IRS builds on the IASI heritage. Instruments for cloud and aerosol parameters such as the visual infrared imager (VII) and multi-channel, multi-viewing, multi-polarisation imager (3MI) are also proposed. VII and 3MI build on the heritage of MERIS and POLDER. ESA has initiated a Sentinel 5 precursor mission, which will have an OMI and an IR instrument focused on CO measurements.

10.7.2 GEO Satellite Instruments

In order to measure air quality, high spatial and temporal measurements are needed, which yield the changing composition of the planetary boundary layer. Geostationary measurements provide high sampling and maximise the number of cloud free scenes for a given spatial resolution.

Sir Arthur C. Clarke's 1945 prediction in the *Wireless World* magazine about the possibility and value of geostationary satellites became reality in 1965 with the launch of INTELSAT I *Early Bird,* the first commercial geostationary communication satellite. The use of GEO orbits for communications and satellite television has blossomed since then. The usefulness of GEO for operational meteorology was recognised in 1974 with the US Synchronous Meteorological Satellite (SMS-1) which became the first operational meteorological geostationary satellite. In 1977 a global system was established with the Japanese Geostationary Meteorological Satellite (GMS-1) and European Meteosat-1. Later EUMETSAT took operational control of the European Meteosat second generation satellites: MSG-1 came into operational service in 2004.

GEO offers a number of possibilities for tropospheric sounding. The fate of most short-lived atmospheric trace constituents and pollutants is regulated by the diurnal cycle of day and night and these crucial variations are lost in single LEO measurements. The smaller signal intensity at the distance of the GEO orbit can be compensated for with larger optics and some signal integration. The principal advantages of GEO instrument platforms are the high temporal coverage and the relatively high probability of observations under cloud free conditions for a given spatial resolution.

A number of proposals to make tropospheric constituent measurements from geostationary orbits have been made by the scientific community.

- Krueger and colleagues (NASA) proposed a geostationary satellite to observe the volcanic ring of fire using a TOMS like instrument aimed at SO_2 and O_3.
- Fishman and colleagues (NASA) proposed the measurement of O_3 and CO in the GeoTropSat initiative in the middle of the 1990s (Little et al. 1997).
- The Geostationary Imaging Fourier Transform Spectrometer (Smith et al. 2006) has been proposed for trace gas measurements primarily aimed at parameters for NWP.

- In Europe the GeoSCIA and GeoTROPE initiatives and proposals were evolved between 1997 and 2004 (Bovensmann et al. 2002; 2004; Burrows et al. 2004). GeoTROPE was planned to combine a SCIAMACHY-like instrument, GeoSCIA, with an IASI-like FTIR instrument, GeoFIS, to make simultaneous measurements from a geostationary orbit.

Although these various proposals clearly indicated the advantages of measuring air quality from a geostationary orbit and provided an impetus for such an initiative, they were not accepted by the agencies at that time.

However ESA and EUMETSAT are now currently preparing the third generation of Meteosat, which comprises two three-axis stabilized platforms, to be launched in between 2015 and 2018. This has been incorporated in the GMES programme (http://www.esa.int/esaLP/SEM3ZT4KXMF_LPgmes_0.html). As a result, the GMES Sentinel 4, which has its heritage in the GeoSCIA concept and the later EUMESAT initiatives to build on this concept, has been selected to fly as part of MTG. Sadly, cost considerations, rather than scientific goals and operational services, seem to be restricting its measurements to Europe rather than Europe and Africa, as proposed for GeoSCIA. This will fly together with an infrared spectrometer, which has some capability for trace gases but is primarily aimed at the provision of parameters for NWP. This combination is approaching the objective of GeoTROPE.

NASA, as part of the NRC Decadal Study, has proposed a GeoCAPE mission (Fishman et al. 2008; NRC 2007). This builds on the GeoTROPSAT and GeoTRACE initiatives and will make measurements similar to GMES Sentinel 4. Recently the JAXA has announced plans for a GEO mission, similar in intention to and building on the GeoSCIA concept. South Korea is also considering a similar mission. The various missions are not aimed at the full Earth disc measurements, principally to minimise cost; this strategy seems likely to be a mistake.

Overall one hopes that the future, envisaged in the IGOS/IGACO strategy (IGACO 2004) and the GeoSCIA initiatives, will be realized with a network of geostationary satellites coupled with similar LEO instruments (Section 10.8). Such a system would provide optimal spatial and temporal sampling for many short lived tropospheric gases (Appendix D).

10.7.3 Greenhouse Gases

A recent achievement is the measurement of the dry columns of the greenhouse gases CO_2 and CH_4 which were shown to be feasible by SCIAMACHY and IASI measurements (Sections 8.2.8 and 8.2.9). The JAXA GOSAT, a dedicated CO_2 and CH_4 platform, was launched in 2009 and data products are now becoming available. A new flight opportunity is available for a second OCO after the failure of launch of the first OCO. However, under current planning, all these experiments will finish around 2015 and, in spite of the recognised need for highly accurate measurements of the two most important long lived greenhouse gases, no follow-on missions have been approved, although CarbonSat mission and CarbonSat constellation have been proposed to meet the need.

An active mission ASCENDS is being considered by NASA, but the proposed active European mission ASCOPE has not been selected. It appears therefore that long lifetime space based laser missions are considered too challenging at the present time. A Franco-German mission MERLIN for the active measurement of CH_4 by DIAL from space is in pre-phase A.

In conclusion, as pointed out by the GMES Working Group 4, there is a clear gap in the measurement of CO_2 and CH_4 after 2014 and there is a need for new and improved measurements of dry columns of greenhouse gases (Appendix D).

10.7.4 Observations from the Lagrange Point

Also being considered by NASA is the resurrection of The Deep Space Climate Observatory (DSCOVR), previously called Triana. DSCOVR was intended to exploit the unique observation opportunities at the Earth's L1 Lagrangian point, at a distance of 1.5 million kilometers, to observe the Earth. At this location a continuous view of the sun-lit side of the Earth yields full diurnal coverage. Although the satellite was constructed nearly 10 years ago, it has been in storage since then. It is now hoped that it will be allocated a launch vehicle in the near future.

10.8 Future Monitoring of the Troposphere from Space

An essential application of trace gas measurements from space will be the continuous monitoring of tropospheric pollution levels by satellite instruments, which, like today's meteorological satellites, would provide maps of the *chemical weather*. To achieve this goal, the frequency of measurements at any point on the Earth has to be increased from once every one to three days, to perhaps a measurement every 1–2 h. Global observation, which would have an adequate temporal resolution of the order of 15–30 min, could be achieved in several ways.

1. The deployment of a number of smaller satellites in polar orbit that pass over any point on the Earth following each other in a determined temporal pattern (see Fig. 10.2).
2. The deployment of an instrument in a GEO orbit (Fig. 10.3) yields the diurnal variation of the atmospheric composition over about one third of the Earth's surface from a single satellite. Three such satellites, positioned appropriately, would be sufficient to observe the entire globe.
3. The large distance of the GEO orbit to the earth of about 36,000 km, compared to ~800 km for LEO, can be compensated for by somewhat larger optics. Also the time for observation is much longer in GEO than in LEO and offers the

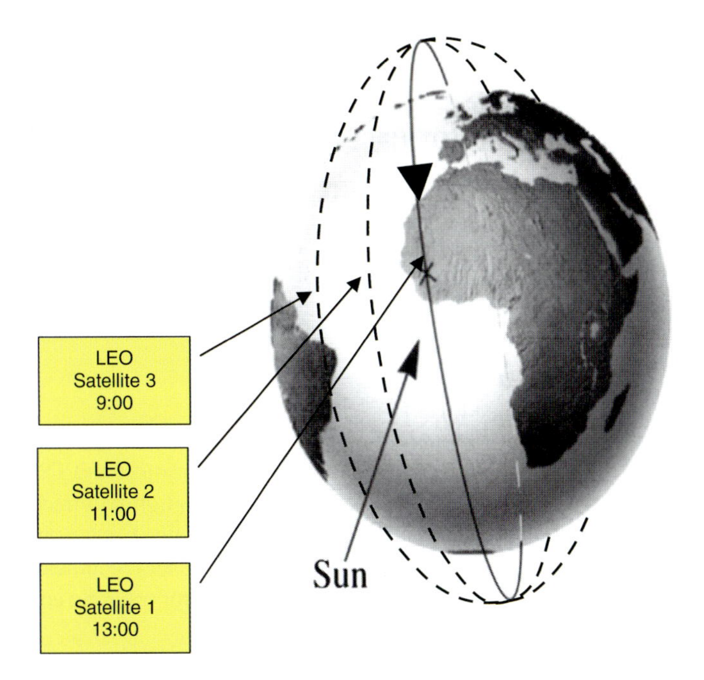

Fig. 10.2 A series of LEO satellites on sun-synchronous (low earth) orbits with suitably staged equator crossing times can provide diurnal profiles and global coverage at the same time. Compared to satellites in geostationary orbits the global coverage of polar regions would be excellent.

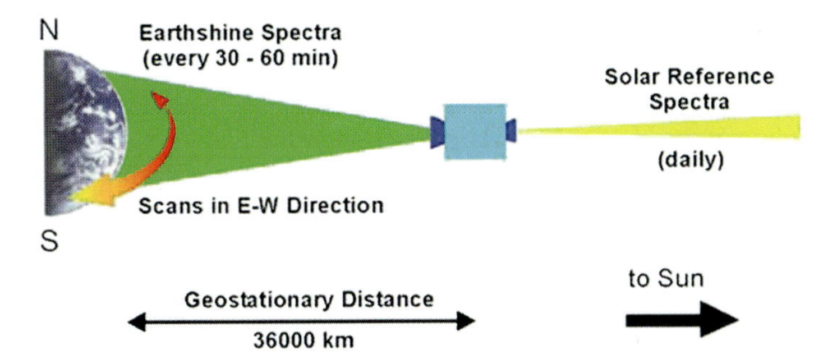

Fig. 10.3 Orbiting the Earth once a day, a GEO satellite maintains a constant position above a point on the Earth's surface. A complete scan of the viewed portion of the Earth's surface would be made every 30 min or so. The GEO altitude (~36,000 km) contrasts with that of the normal LEO satellites at ~800 km above the Earth. Figure: J.P. Burrows and S. Noel, IUP, Bremen.

possibility of time averaging to improve the signal/noise ratio. Several missions are now being considered as described above.

4. Placing a satellite in a halo orbit around the first Lagrange point, L1, about 1.5 million km above the Earth surface, is an interesting possibility. The optical requirements (see point 2, above) are still more demanding than for a geostationary

satellite instrument. DSCOVR, mentioned above, may be the first L1 instrument; however more capability is required that that envisaged in the old design.

In summary, a network of *chemical weather* satellites is needed to measure and monitor our changing environment. The increasing world population will have an increasing thirst for energy, so that an adequate system is required within the next decade. Such a system will be required for the foreseeable future to test our knowledge of the Earth system and its response to anthropogenic activity, the impact of international legislation aimed at achieving sustainable development.

10.9 Conclusions

This book, which has been made possible by the ACCENT project and TROPOSAT scientific community, provides the reader with the information required to understand the remote sensing of tropospheric constituents in the Earth's atmosphere from space. It describes some of the remarkable scientific and technical developments in the emerging scientific area of remote sensing of tropospheric composition that have been achieved in the last three decades.

Less than a generation ago, the global observation of tropospheric trace constituents by remote sensing was considered impossible. Now a set of research instruments and the first operational instruments are demonstrating the benefits for Earth science, and are providing the evidence for the development, monitoring, verification and transparency of international environmental policy.

The evolution of global observing capability for the troposphere has required a rapid development of instrument concepts, atmospheric radiative transfer and retrieval theory, by the scientific community. The new data has expanded our understanding of the biogeochemical processing of the Earth system, and the new capabilities provide the evidence needed to identify and assess the impact of natural processes and anthropogenic activity.

One challenge for the future of tropospheric remote sensing is the logistics of coordinating multiple simultaneous remote sensing and *in situ* measurements to provide useful data products. Such information has strategic importance as the Anthropocene develops.

The operational meteorological agencies, which established the need for NWP, provide a valuable model. A similar approach and commitment is needed to address the needs of environmental remote sensing, which focuses on the identification and quantification of environmental and climate change. This investment has been lacking up to now.

The variety of international initiatives including IGOS and GEO, have been developed to stimulate the satellite missions needed and there is general agreement that more missions are needed. However the rate of generation of new and focussed missions does not meet the need, as the timelines in Appendix D show. NPOESS has aspirations with respect to providing the space segment of instrumentation for both

NWP and numerical environmental prediction. Similarly EUMETSAT includes both NWP and climate. However, in practice, budgets for these organisations are limited and NWP currently dominates the priorities for missions. More resources are needed for the scientific and technological aspects in order to achieve a focussed global measurement strategy that is fit for purpose, such as the GMES space segment for MACC (Monitoring Atmospheric Composition and Climate).

The space agencies and industry are successful but they are expensive: their mandates, given by national governments, often focus on support for improving the national technological infrastructure rather than on clear scientific or policy objectives. In addition the creation of the necessary administrative structure has an appreciable financial overhead.

Scientific understanding and exploitation is the key to the overall development and use of Earth observation but, unfortunately, it is still the poor relative in the overall space programme, despite the undoubted nature of the deleterious effects of the changing atmosphere. Overall governments are to be congratulated on their investments in Earth observation, but only time will tell whether the provision of data will be adequate to meet the future challenges in the detection, attribution and adaptation to global climate change.

The provision of Earth observation, of which the remote sensing of tropospheric composition is a part, is a facet of the greening of the industrial base of key strategic importance and of the knowledge and information revolution. However there is a need for a faster evolution of missions and systems to provide the proper monitoring of the atmospheric environment so that environmental change can be recognised soon enough to take any necessary action. New missions would yield data products which monitor both the health of the environment and its ecosystems as well as the pollution. They would further enhance our understanding of the Earth system and improve the predictive capability of our models. The data products from such missions would provide the evidence needed for the monitoring, reporting, verification and transparency of international environmental agreements.

The scientific community has shown the leadership by initiating the evolution of Earth observation instrumentation in the field of tropospheric composition. This would not have been possible without massive public investment in the infrastructure. We are convinced that the scientific community will continue to propose missions, which not only challenge our understanding of the atmospheric composition but also provide the evidence that the Earth is being managed successfully. Hopefully the space agencies supported by government will continue to facilitate these visions and improve our knowledge of environmental change, thereby supporting sustainable economic development.

References

Bovensmann H., S. Noel, P. Monks, A.P.H. Goede, J.P. Burrows, 2002, The Geostationary Scanning Imaging Absorption Spectrometer (GEOSCIA) Mission: Requirements and Capabilities, *Adv. Space Res.*, **29**, 1849–1859.

Bovensmann H., K.U. Eichmann, S. Noel, J.M. Flaud, J. Orphal, P.S. Monks, G. K. Corlett, A.P. Goede, T. von Clarmann, T. Steck, V. Rozanov and J.P. Burrows, 2004, The Geostationary scanning imaging absorption spectrometer (GeoSCIA) as part of the Geostationary pollution explorer (GeoTROPE) mission: requirements concepts and capabilities. *Adv. Space Res.* **34**, 694–699.

Burrows J.P., H. Bovensmann, G. Bergametti, J.M. Flaud, J. Orphal, S. Noel, P.S. Monks, G.K. Corlett, A.P. Goede, T. von Clarmann, T. Steck, H. Fischer and F. Friedl-Vallon, 2004, The geostationary tropospheric pollution explorer (GeoTROPE) missions: objects, requirements and mission concept, *Adv. Space Res.* **34**, 682–687, doi:10.1016/j.asr.2003.08.067.

Fishman J., K.W. Bowman, J.P. Burrows, A. Richter, K.V. Chance, D.P. Edwards, R.V. Martin, G.A. Morris, R. Bradley Pierce, J.R. Ziemke, J.A. Al-Saadi, J.K. Creilson, T.K. Schaack and A.M. Thompson, 2008, Remote Sensing of Tropospheric Pollution from Space, *Bull Amer. Met. Soc.* **89**(6), 805–821.

GEOS, 2005, System Capabilities and the Role for U.S. EPA, Recommendations of a Community Panel. EPA/600/R-05-009, U.S. EPA.

Grennfelt, P., Ø Hov R. G. Derwent, 1994, Second Generation Abatement Strategies for NOx, NH_3, SO_2 and VOC, *Ambio*, **23**, 425–433.

IGACO, 2004, The Changing Atmosphere: an Integrated Global Atmospheric Chemistry Observation Theme for the IGOS Partnership: Report, September 2004. ESA SP-1282, No. 159 (WMO TD No. 1235) ftp://ftp.wmo.int/Documents/PublicWeb/arep/gaw/gaw159.pdf.

IPCC, 2007, Climate Change 2007: Synthesis Report. Contribution of Working Groups I, II and III to the Fourth Assessment Report of the Intergovernmental Panel on Climate Change, Geneva, Switzerland, pp 104.

Keeling C.D., R.B. Barcastow, A.E. Bainbridge, C.A. Ekdahl, P.R. Guenther and L.S. Waterman, 1976, Atmospheric carbon dioxide variations at Mauna Loa observatory, Hawaii, *Tellus* **28**, 538–551.

Keeling R.F., S.C. Piper, A.F. Bollenbacher and J.S. Walker, 2003, CDIAC, doi: 10.3334/CDIAC/atg.035

Little, A. D., D.O. Neil, G.W. Sachse, J. Fishman and A. Krueger, 1997, Remote sensing from geostationary orbit: GEO TROPSAT, a new concept for atmospheric remote sensing, Sensors, Systems, and Next-Generation Satellites, *SPIE Proc.* **3221**, 480–488. Aerospace Remote Sensing, London.

NRC, 2007, Earth Science and Applications from Space: National Imperatives for the Next Decade and Beyond Committee on Earth Science and Applications from Space: A Community Assessment and Strategy for the Future, National Research Council ISBN: 0-309-66714-3, 456. http://www.nap.edu/catalog/11820.html

Smith W. L., H.E. Revercomb, D.K. Zhou, G.E. Bingham, W.F. Feltz, H.L. Huang, R.O. Knuteson, A.M. Larar, X. Liu, R. Reisse and D.C. Tobin, 2006, Geostationary Imaging Fourier Transform Spectrometer (GIFTS): science applications, *SPIE Soc.* **6405**, F4050–F4050.

UNECE LTRAP HTAP, 2007, Hemispheric Transport of Air Pollution 2007 Air Pollution Studies No. 16: Interim report prepared by the Task Force on Hemispheric Transport of Air Pollution acting within the framework of the Convention on Long-Range Transboundary Air Pollution. United Nations New York and Geneva, 2007, United Nations Publications, Sales No. E.08.II.E.5 ISSN 1014-4625 ISBN 978-92-1-116984-3 Copyright © United Nations. The next HTAP report will be completed in 2010. http://www.htap.org/.

UNEP Ozone Secretariat, 2009, Handbook for the Vienna Convention for the Protection of the Ozone Layer, 8th Edition, UNEP, Nairobi pp 79.

WMO, 2007, Scientific Assessment of Ozone Depletion: 2006, Global Ozone Research and Monitoring Project—Report No. 50, pp 572. Geneva, Switzerland.

WMO, 2010, Scientific Assessment of Ozone Depletion: 2010, Global Ozone Research and Monitoring Project WMO Geneva, Switzerland. The Ozone Assessment will be published in 2011.

Appendices

Appendix A: Satellite Instruments for the Remote Sensing in the UV, Visible and IR

Tables of microwave instruments are given in Tables 4.1 and 4.2. However MLS on Aura is included here as well.

Abbreviations Used in the Table

Aerosol Properties: *AOD* aerosol optical depth, *AOT* aerosol optical thickness, *AE* Ångström exponent, *FMF* fine mode fraction, *CMF* coarse mode fraction, *NS* non-spherical, *PSD* particle size distribution, type, *AAI* absorbing aerosol index, *ssa* single scattering albedo

Cloud Properties: *CA* cloud albedo, *CER* cloud effective radius, *CF* cloud fraction, *COT* cloud optical thickness, *CP* cloud phase, *CPP* cloud particle phase, *CPS* cloud particle size, *CTH* cloud top height, *CTP* cloud top pressure, *CTT* cloud top temperature, *LWP* liquid water path, *DZ* droplet size, *CZ* crystal size

Viewing: *N* nadir, *L* limb, *O* occultation

Sounding: *Tot* total column, *Sp* stratospheric profile, *Tp* tropospheric profile, *Tc* tropospheric column, *Sc* stratospheric column, *Mp* mesospheric profile

Spectral region: UV/vis/NIR, ultraviolet visible and near infrared; *IR* infrared; continuous spectrum taken or selected spectral channels. For IR Fourier Transform Spectrosocopy (FTS), the resolution is given as the optical path difference, OPD

J.P. Burrows et al. (eds.), *The Remote Sensing of Tropospheric Composition from Space*, Physics of Earth and Space Environments, DOI 10.1007/978-3-642-14791-3,

AATSR	Advanced Along-Track Scanning Radiometer *Satellite*; *lifetime*: ESA ENVISAT; 2002 – present; *re-visit period*: 5 days; *equator crossing time:* 10.00 ascending; *species:* sea surface temperature; *cloud properties*: CF, COT, CP, CPS, CTH, CTP, CTT, LWP, DZ, CZ; *aerosol properties*: AOD, AE, aerosol mixing ratio; *viewing*: N and 55° forward; *sounding* Tot; *footprint*: 1 × 1 km, swath 512 km; *spectral region*: vis/IR, 2 views; *channels*: 555, 659, 865, 1,600, 3,700, 11,000, 12,000 nm; *resolution*: 20 nm (1–3), 300 nm (4–5), 1,000 nm (6–7)
ACE-FTS	Atmospheric Chemistry Experiment – FTS *Satellite*; *lifetime*: CSA SCISAT-1; 2003 – present; *species:* H_2O, CO_2, CH_4, N_2O, O_3, CO, CFC-11, CFC-12, $ClNO_3$, HCl, HF, HNO_3, NO_2, NO, N_2O_5 and more; *viewing*: O; *sounding* Tp (upper); *spectral region*: IR; FTS, contin. spectrum; *resolution*: OPD 25 cm
AIRS	Atmospheric Infrared Sounder *Satellite*; *lifetime*: NASA Aqua, 2002 – present; *re-visit period*: twice a day; *equator crossing time:* 13.30; *species:* H_2O, CO_2, CH_4, O_3, CO; *viewing*: N + scan; *sounding* Tot, Tc; *footprint*: 13.5 × 13.5 km; *spectral region*: IR; 650–1,136, 1,216–1,613, 2,170–2,674 cm^{-1}; *resolution*: $\lambda/\Delta\lambda = 1{,}200$
ATMOS	Atmospheric Trace Molecule Spectroscopy *Satellite*; *lifetime*: NASA Spacelab, 1985, ATLAS: 1992, 1993, 1994; *re-visit period and equator crossing time:* not applicable; *species:* O_3, NO_x, N_2O_5 $ClONO_2$, HCl, HF, CH_4, CFCs; *viewing*: O; *sounding* Sc, Tc (upper); *spectral region*: IR; continuous; *resolution*: 0.01–1 cm^{-1}
ATSR-2	Along Track Scanning Radiometer *Satellite*; *lifetime*: ESA ERS-1, 2; 1991–2002; *re-visit period*: 5 days; *equator crossing time:* 10.00; *species:* sea surface temperature; *cloud properties*: CF, COT, CP, CPS, CTH, CTP, CTT, LWP, DZ, CZ; *aerosol properties*: AOD, AE, aerosol mixing ratio; *viewing*: N; *sounding* Tot; *footprint*:1 × 1 km; *spectral region*: vis/IR, 2 views; channels: 555, 659, 865, 1,600, 3,700, 11,000, 12,000 nm; resolution: 20 nm (1–3), 300 nm (4–5), 1,000 nm (6–7)
AVHRR	Advanced Very High Resolution Radiometer *Satellite*; *lifetime*: NASA TIROS-N, NOAA-6, NOAA 15 Metop A; 1978 – present; *re-visit period*: 2days; *equator crossing time:* 06:00 to 10:00 and 09:30; *species:* fire, vegetation, aerosol properties; *cloud properties*: CTH, COT, CTT, LWP, DZ, CZ; *viewing*: N; *sounding* Tot; *footprint*: 1.25 km × 1.25 km, 5 km × 5 km, and 25 km × 25 km; *spectral region*: vis/IR; 0.58–0.68 μm, 0.725–1.0 μm; IR 1.58–1.64 μm, 3.55–3.93 μm, 10.3–11.3 μm, 11.5–12.5 μm
BUV	Backscatter Ultraviolet Ozone Experiment *Satellite*; *lifetime*: NASA Nimbus 4; 1970–1974; *re-visit period*: 6 days; *species:* O_3; *viewing*: N; *sounding* Sp, Tc; *footprint*:230 km × 230 km; *spectral region*: UV; *resolution*: 1–5 nm
CALIOP	Cloud-Aerosol Lidar with Orthogonal Polarization *Satellite*; *lifetime*: NASA CALIPSO (A TRAIN); 2006; *equator crossing time:* 13.30, ascending; *species: aerosol properties*: see Table 6.4; *viewing*: N; *sounding* Tot; *footprint*: 330 × 100 m; vert. 30–60 m; *spectral region*: lidar; 532 (polarised), 1,064 nm

(*continued*)

CLAES	Cryogenic Limb Array Etalon Spectrometer *Satellite*; *lifetime*: NASA UARS; 1991–1993; e*quator crossing time:* asynchronous; *species:* temperature, pressure, O_3, H_2O, CH_4, N_2O, NO, NO_2, N_2O_5, HNO_3 , $ClONO_2$, HCl, CFC-11, CFC-12, and aerosol absorption coefficients; *viewing*: L; s*ounding* Sp; *spectral channels*: IR; 780, 792, 843, 879, 925, 1,257, 1,605, 1,897, 2,843 cm^{-1}; *resolution*: 0.19, 0.25, 0.26, 0.22, 0.22, 0.26, 0.39, 0.47, 0.65 cm^{-1}
GLAS	Geoscience Laser Altimeter System *Satellite* ICESat; *viewing*: N; *sounding*: altimeter
GOME	Global Ozone Monitoring Experiment *Satellite*; *lifetime*: ESA ERS-1, 1995 – present; *re-visit period*: 3 days; e*quator crossing time:* 10.30; *species:* O_3, NO_2, H_2O, BrO, OClO, SO_2, HCHO, CHOCHO, IO, H_2O, O_2, O_4; *cloud properties*: CTH, CF, COT; CA; and aerosols; *viewing*: N; s*ounding* Tot, Tc; *footprint*: 320 × 40 km; *spectral region*: UV/vis; continuous; *resolution*: 0.2 nm
GOME-2	Global Ozone Monitoring Experiment-2 *Satellite*; *lifetime*: ESA EUMETSAT MetOp-A, 2006 – present; *re-visit period*: 1.5 days; e*quator crossing time:* 09.30; *species:* O_3, NO_2, H_2O, BrO, OClO, SO_2, HCHO; *cloud properties*: CTH, CF, COT, CA; and aerosols; *sounding* Tot; *viewing*: N; *footprint*: 80 × 40 km; *spectral region*: UV/vis; continuous; *resolution*: ~0.2 nm
GOMOS	Global Ozone Monitoring by Occultation of Stars *Satellite*; *lifetime*: ESA ENVISAT; 2002 – present; e*quator crossing time:* 10:00; *species:* O_3, NO_2, NO_3, H_2O, O_2; *viewing*: O; *sounding:* Tp; Sp, Tot, Me; *spectral regions*: UV/vis; 248–693 nm, NIR; 750–776 nm, 915–956 nm; *resolution*: 1.2 nm (UV/vis), 0.2 nm (NIR)
HALOE	Halogen Occultation Experiment *Satellite*; *lifetime*: NASA UARS; 1991–2005; *species:* CO_2, H_2O, O_3, NO_2, HF, HCl, CH_4, NO; *viewing*: O; *sounding* Sp, (Tp); *spectral region*: several channels from 2.45 to 10.04 μm; *resolution*: gas correlation radiometer
HIRDLS	The High Resolution Dynamics Limb Sounder *Satellite; lifetime*: NASA Aura, 2004 – present; *re-visit period*: twice a day; e*quator crossing time:* 01.43; *species:* H_2O, CH_4, N_2O, O_3, CFC-11, CFC-12, $ClONO_2$, NO_2, N_2O_5, HNO_3, T; and cloud properties; *viewing*: L + scan; *sounding:*Sp ;*spectral region*: 21 channels from 6.12 to 17.76 μm; *resolution*: channel dependent
IASI	Infrared Atmospheric Sounding Interferometer *Satellite; lifetime*: EUMETSAT MetOp; 2006 – present; *re-visit period*: twice a day; e*quator crossing time:* 09.30; *species:* H_2O, HDO, CO_2, CH_4, N_2O, O_3, CO, CFC-11, CFC-12, HCFC22, HNO_3, SO_2, NH_3, C_2H_4, HCOOH, CH_3OH *et al. viewing*: N + scan; s*ounding* Tot, Tc, Tp; *footprint*: 12 km diameter; *spectral region*: IR, Michelson interferometer; continuous; *resolution*: OPD 2 cm
ILAS I, II	Improved Limb Atmospheric Spectrometer *Satellite; lifetime*: I: NASDA ADEOS; 1996–1997; II ADEOS II; 1999; *species:* O_3, NO_2, N_2O, H_2O, CF_3Cl, CH_4, $ClONO_2$, T, P; *viewing*: O; s*ounding* Sp (Tp); *footprint*: 2 × 2 km (NIR) and 2 × 13 (TIR); *spectral region*: 753–784 nm (NIR), 6.21–11.77 μm (TIR); *resolution*: 0.15 nm (NIR), 0.12 μm (TIR)

(*continued*)

IMG	Atmospheric Infrared Sounder *Satellite*; *lifetime*: NASDA ADEOS, 1996–1997; *re-visit period*: 10 days; *equator crossing time:* 10.30 descending; *species:* H_2O, CO_2, CH_4, O_3, CO; *viewing*: N; *sounding* Tot, Tc, Tp; *footprint*: 8 × 8 km; *spectral region*: IR; continuous; *resolution*: OPD = 10 cm
ISAMS	Improved Stratospheric and Mesospheric Sounder *Satellite; lifetime*: NASA UARS; 1991–1992; *species:* CO_2, H_2O, CO, N_2O, CH_4, NO, NO_2, N_2O_5, HNO_3, O_3; *viewing*: L + scan ; *sounding* Sp, Mp; *footprint*: 2.6 × 13 km; *spectral region*: 605–2,257 cm^{-1} (14 Bands); *resolution*: gas correlation
LIMS	Limb Infrared Monitor of the Stratosphere *Satellite; lifetime*: NASA Nimbus 7; 1978–1979; *species:* CO_2, HNO_3, O_3, H_2O, NO_2; *viewing*: L + scan; *sounding* Sp; *spectral regions*: 637–673, 579–755, 844–917, 926–114, 1,237–1,560, 1,560–1,630 cm^{-1}
LITE	Lidar In-space Technology Experiment *Satellite* Space Shuttle Discovery*; lifetime*: 9 days; *species*: aerosols, clouds; *viewing*: N; *sounding* Tot, Np; *footprint*: 300 m; *spectral region*: 355 nm, 532 nm, 1,064 nm
LRIR	Limb Radiance Inversion Radiometer *Satellite; lifetime*: NASA Nimbus 7; *equator crossing time:* local noon; *species:* CO_2, O_3; *viewing*: L; *sounding* Sp; *spectral regions*: IR; 14.6–15.9 μm, 14.2–17.3 μm, 8.8–10.1 μm, 20–25 μm
MAPS	Measurement of Air Pollution from Satellites *Satellite; lifetime*: NASA Space Shuttle; 1981, 1984, 1994; *species:* CO; *viewing*: N; *sounding* Tc; *spectral method*: Gas Correlation (uses CO and N_2O as reference)
MAS	Millimeter Wave Atmospheric Sounder *Satellite*; *lifetime*: Shuttle ATLAS 1, 2 and 3:1992, 1993, 1994; *species:* ClO, O_3, H_2O; *viewing*: L + scan; *sounding* Sp; *spectral bands*: 60 GHz, 183 GHz, 184 GHz, 204 GHz
MERIS	Medium Resolution Imaging Spectrometer for Passive Atmospheric Sounding *Satellite*; *lifetime*: ESA-ENVISAT; 2002 – present; *re-visit period*:1–2 days; *equator crossing time:* 10.00 ascending; *species:* H_2O; *aerosol properties:* AOD, AE; *cloud properties*: CA, COT, CTH, CTP; *viewing*: N; *sounding* Tc; *footprint*: 0.3 × 0.3 km *swath*: 1,150 km; *spectral channels*: vis/NIR: 412.5, 442.5, 490, 510, 560, 620, 665, 681.25, 705, 753.75, 760, 775, 865, 890, 900 nm; *resolution*: 1.8 nm
MIPAS	Michelson Interferometer for Passive Atmospheric Sounding *Satellite*; *lifetime*: ESA ENVISAT, 2002 – present; *re-visit period*: 6 days; *equator crossing time:* 10.00 ascending; *species:* H_2O, CO_2, CH_4, N_2O, O_3, CO, CFC-11, CFC-12, ClO, $ClONO_2$, OClO, HNO_3, C_2H_6, SF_6, NO_2, NO, NH_3, OCS, SO_2; *viewing*: L; *sounding* Tc (upper); *spectral region*: IR; FTS continuous; *resolution*: OPD: 20–8 cm
MLS (UARS)	Microwave Limb Sounder *Satellite*; *lifetime*: NASA UARS; 1994–2001; *species:* ClO, CH_3CN, H_2O, HNO_3, O_3, SO_2, temp., geopotential height, ice water content, ice water path, relative humidity with respect to ice; *viewing*: L; *sounding:* Sp; *spectral bands*, 63 GHz, 183 GHz, 205 GHz

(*continued*)

MLS (Aura)	Microwave Limb Sounder *Satellite*; *lifetime*: NASA (A-TRAIN) Aura; 2004 – present ; *equator crossing time:* 13.38; *species:* BrO, CH_3CN, ClO, CO, H_2O, HCl, HCN, HNO_3, HO_2, HOCl, N_2O, O_3, OH, SO_2, temp., geopotential height, ice water content, ice water path, relative humidity with respect to ice; *viewing*: L; *sounding:* Sp; *spectral bands*, 118, 190, 240, 640, 2,250 GHz
MODIS	Moderate Resolution Imaging Spectroradiometer *Satellite*; *lifetime*: Terra 1999 – present; Aqua (A TRAIN); 2002 – present; *re-visit period*:1–2 days; *equator crossing:* 10.30 descending (T); 13.30 ascending (A); *aerosol properties*: AOD, FM AOD, CM AOD, type, psd (over ocean); *cloud properties*: CER, CIWP, COT, CPP, CTP, CTH, CTT, LWP, DZ, CZ; *viewing*: N; *sounding* Tot; *footprint*: bands 1–2: 0.25×0.25 km; bands 3–7: 0.5×0.5 km; bands 8–36: 1×1 km; swath 2,330 km; *spectral channels*: UV/vis/NIR; 412.5, 443, 469, 488, 531, 551, 555, 645, 667, 678, 748, 858, 869.5, 905, 936, 940, 1,240, 1,375, 1,640, 2,130, 3,750, 3,859, 4,050, 4,465, 4,516, 6,715, 7,325, 8,550, 9,730, 11,030, 12,020, 13,335, 13,635, 13,935, 14,235 nm; *resolution*: variable
MOPITT	Measurement of Pollution in the Troposphere *Satellite*; *lifetime*: NASA Terra 1999 – present; *re-visit period*: 3 days; *equator crossing time*: 10.30 descending; *species*: CO; *viewing*: N + scan; *sounding*: Tc, Tp; *footprint*: 22×22 km; *spectral region*: IR; correlation radiometer, 3 bands, 8 channels; *resolution*: 0.04 cm^{-1} (effective); length and pressure modulated correlation spectrometer
OMI	Ozone Monitoring Instrument *Satellite*; *lifetime*: NASA Aura (A TRAIN), 2004 – present; *re-visit period*: 1 day; *equator crossing time:* 13.00 ascending; *species:* O_3, NO_2, SO_2, BrO, OClO, HCHO, CHOCHO, O_4; *aerosol properties*: AOD, AAI, ssa; *cloud properties*: CF, CP, CTH; *viewing*: N; *sounding* Tot, Tc; *footprint*: 24×13 km swath: 2,600 km; *spectral region*: UV/vis; continuous; *resolution*: ~0.5 nm
OSIRIS/IRI	Optical Spectrograph and Infrared Imaging System *Satellite*; *lifetime*: Swedish ODIN, 2001 – present; *equator crossing time:* 06.00; *species:* O_3, BrO; *viewing*: L ; *sounding* Sp; *spectral regions*: 280–800 nm (OSIRIS); 2 bands, 1.27 μm and 1.53 μm (IRI); *resolution*: 1 nm (OSIRIS)
POLDER (PARASOL)	Polarization and Anisotropy of Reflectances for Atmospheric Science coupled with Observations from a LIDAR *Satellite*; *lifetime*: NASA PARASOL (A TRAIN); 2004 – present; *re-visit period*: 1 day; *equator crossing time:* 13.30, ascending; *aerosol properties*: AOD, FM AOD, CF AOD, NS AOD; *cloud properties*: CER, CF, COT, CP, CTH, CTP, LWP, SW albedo; *viewing*: N, multi-directional; *sounding:* Tot; *footprint*: 6×6 km; *swath*: 2,400 km *spectral region*: vis/NIR; *polarised channels*: 443, 490, 565, 670, 763, 765, 865, 910, 1,020 nm
POAM-II	Polar Ozone and Aerosol Measurement II *Satellite*; *lifetime*: SPOT-3, 1993–1996; *equator crossing time:* 10.30 descending; *species:* O_3, H_2O, NO_2, aerosol properties, temperature; *viewing*: O; *sounding* Sp; *orbit:* sun-synchronous polar; *spectral region*: UV/vis/NIR; *channels*, 353.0, 442.0, 448.3, 600.0, 760.8, 780.0, 920.0, 935.5, 1,059.0 nm; *resolution*: 2–16 nm

(*continued*)

POAM-III	Polar Ozone and Aerosol Measurement III *Satellite*; *lifetime*: SPOT-4, 1998–2005; e*quator crossing time:* 10.30 decending; *species:* O_3, H_2O, NO_2, aerosol properties, temperature; *viewing*: O; s*ounding* Sp; *orbit:* sun-synchronous polar; *spectral channels*: UV/vis/NIR; 354.0, 439.6, 442.2, 603.0, 761.3, 779.0, 922.4, 935.9, 1,018.0 nm; *resolution*: 2–16 nm
SAGE-1	Stratospheric Aerosol and Gas Experiment I *Satellite*; *lifetime*: NASA Atm. Explorer, 1979–1981; *species:* O_3, NO_2, aerosol properties; *viewing*: O; s*ounding* Sp, Tp (upper); *spectral region*: vis/NIR; *channels*: 380, 450, 600, 1,000 nm; *resolution*: 2–20 nm
SAGE-2	Stratospheric Aerosol and Gas Experiment II *Satellite*; *lifetime:* NASA Earth Radiation Budget,1984 – present; *species:* O_3, NO_2, H_2O, aerosol properties; *viewing*: O; s*ounding* Sp, Tp (upper); *spectral region*: vis/NIR; *channels*: 385 nm, 448 nm, 453 nm, 525 nm, 600 nm, 940 nm, 1,020 nm; *resolution*: 2–20 nm
SAGE-3	Stratospheric Aerosol and Gas Experiment III *Satellite*; *lifetime*: NASA Meteo 3M, 1999; ISS, 2002; *species:* O_3, NO_2, OClO, BrO, NO_3, aerosol properties; *viewing*: O; *sounding* Sp, Tp (upper); *spectral channels*: vis/NIR; 290, 385, 430–450, 525, 600, 740–780, 920–960, 1,020, 1,500 nm; *resolution*: 2–20 nm
SAM II	Stratospheric Aerosol Measurement II *Satellite*; *lifetime*: Nimbus 7; 1979–1990; e*quator crossing time:* local noon; *species:* aerosol properties; *viewing*: O; s*ounding* Sp/Tp; *footprint*: 30-arc-second circle; *spectral region*: IR; 1 channel at 1 μm with 0.038 μm bandpass
SAMS	Stratospheric and Mesospheric Sounder *Satellite*; *lifetime*: NASA Nimbus 7; 1979–1990; e*quator crossing time:* local noon; *species:* CO_2, H_2O, CO, N_2O, CH_4, NO, temperature; *viewing*: L + scan; s*ounding* Sp, Mp (20–100 km); *footprint*: 10 H × 100 L km; *spectral channels*: IR; 4.3, 5.3, 7.7, 15, 100 μm, 25–100 μm; *resolution*: gas pressure modulation technique
SBUV	Solar Backscatter Ultraviolet Ozone Experiment *Satellite*; *lifetime*: NASA Nimbus 7; 1979–1990; e*quator crossing time:* 12:00; *species:* O_3; *viewing*: N; s*ounding:* SP; *footprint*: 200 × 200 km; *spectral channels*: UV; 252, 273, 283, 288, 292, 298, 302, 306, 312, 318, 331, 340 nm; *resolution*: ~1 nm
SBUV-2	Solar Backscatter Ultraviolet Ozone Experiment *Satellite*; *lifetime*: NOAA-9, NOAA- 11, NOAA-14 (1985 – present); e*quator crossing time:* variable due to the satellite drift; *species:* O_3; *viewing*: N; s*ounding:* SP; *footprint*: 200 × 200 km; *spectral region*: UV; *channels*: 252, 273, 283, 288, 292, 298, 302, 306, 312, 318, 331, 340 nm; *resolution*: ~1 nm
SCIAMACHY	Scanning Imaging Absorption Spectrometer for Atmospheric Cartography *Satellite*; *lifetime*: ESA-ENVISAT, 2002 – present; *re-visit period*: 6 days; e*quator crossing time:* 10.00 ascending; *species:*O_3, NO, N_2O, NO_2, CO, CO, CO_2, CH_4, BrO, OClO, HCHO, SO_2, CHOCHO, IO, H_2O, O_2, O_4, and aerosols ; *cloud properties*: CTH, COT, LWP, DZ, CZ; *viewing*: N, L, O; s*ounding* Tot, Sp, Tp, Me; *footprint*: 60 × 30 km; *spectral region*: UV/vis/NIR; continuous: 240–1,750 nm, 1,940–2,040 nm, 2,265–2,380 nm; *resolution*: 0.2–1.5 nm

(*continued*)

SEVIRI	Spinning Enhanced Visible and InfraRed Imager *Satellite*; *lifetime:* MSG (Meteosat 2nd Gen.); 2005 – present; *geostationary, scan repeat*: 15 min; *species: aerosol properties*: AOD, *cloud properties*: CTT, CTP; *viewing:* GEO*; sounding:* Tot*; footprint:* 1 × 1 km (high resolution vis channel); 3 × 3 km (IR and other vis channels); *spectral region:* vis/IR; *channels* 635, 810, 1,640, 3,920, 6,250, 7,350, 8,700, 9,660, 10,800, 12,000, 13,400 nm
TANSO	Thermal and short wave infra-red Sensor for observing greenhouse gases *Satellite*; *lifetime*: JAXA GOSAT, 2009; *re-visit period*: 3 days; e*quator crossing time:* 13.00; *species:* O_2, CO_2, CH_4, H_2O; cloud and aerosol properties; *viewing*: N; s*ounding* Tot; *footprint*: 10 × 10 km; *spectral bands*: vis/NIR/IR; 750–780, 1,560–1,730, 1,920–2,090, 5,500–14,300 nm, *resolution*: 0.015–4 nm
SCR	Selective Chopper Radiometer *Satellite*; *lifetime*: NASA Nimbus 4,5; 1970–1975; *species:* CO_2 temperature profile, water vapour, ice; *viewing*: nadir; s*ounding* Tot ; *footprint*: 25 × 25 km; *spectral region*: IR(1) 4 CO_2 channels between 13.8 and 14.8 μm, (2) four channels at 15.0 μm, (3) an IR window channel at 11.1 μm, H_2O at 18.6 μm, two channels at 49.5 and 133.3 μm, and (4) four channels at 2.08, 2.59, 2.65,and 3.5 μm
SeaWiFS	Sea-viewing Wide Field-of-View Sensor *Satellite*; *lifetime*: SeaStar, August 1997 – present; *re-visit period*: 1 day; *equator crossing time:* 12.20 descending; *swath:* 2,801 km; *species:* AOT (at 865 nm); *viewing*: N; *footprint*: 1.1 × 4.5 km; *spectral channels*: UV/vis; 412, 443, 490, 510, 555, 670, 765, 865 nm; *bandwidths (FWHM)*: 20, 20, 20, 20, 20, 20, 40, 40 nm; *other features*: hyperspectral image, normalized water leaving radiance, attenuation coefficient, Angstrom coefficient, photosynthetically active radiation, land reflectance
SME	Solar Mesospheric Experiment(SME was a mission consisting of 5 single instruments) *Satellite; lifetime*: NASA SME: 1983; e*quator crossing time:* 03.00–15.00 Sun-synchronous orbit; *species:* O_3, $O_2(^1\Delta_g)$, NO_2 s*ounding* Sp, Me; *spectral region*: UV/vis
TES	Tropospheric Emission Spectrometer *Satellite*; *lifetime*: NASA Aura, 2003 – present; *re-visit period*: several days; e*quator crossing time:* 01.45; *species:* H_2O, CH_4, N_2O, O_3, CO, NO, NO_2, HNO_3; *viewing*: N, L; s*ounding* Tot, Tc, Tp; *footprint*: 5 × 8 km; *spectral region*: IR; FTS continuous spectrum; *resolution:* OPD: 8.45 cm
TOMS	Total Ozone Monitoring Spectrometer *Satellite*; *lifetime*: NASA Nimbus 7, 1979–1992; ADEOS, 1996–1997; Meteor, 1992–1094; Earth Probe, 1996 to present; *re-visit period*: 1.5 days; e*quator crossing time:* 12.00; *species:* O_3; SO_2; *viewing*: N; s*ounding* Tot, Tc; *footprint*: 50 × 50 km; *spectral region*: UV; *channels*: 379.95, 359.88, 339.66, 331.06, 317.35, 312.34 nm; *resolution*: ~1 nm

Appendix B: Atlas of Ancillary Global Data

Steffen Beirle
Max-Planck-Institut für Chemie, Mainz, Germany

(a) Cloud-free composite of the Earth's view from space (*MODIS/NASA*).
(b) Night-time light pollution derived from DMSP measurements.
Data processing by NOAA's National Geophysical Data Center. DMSP data collected by US Air Force Weather Agency.
(c) Normalized Differenced Vegetation Index for August 2007 from the NASA instrument MODIS.
Terra/MODIS measurements; http://neo.sci.gsfc.nasa.gov/.
(d) Fires (absolute fire counts on a 1° grid) 2003–2005, produced from ESA remote sensing data.
ATSR World Fire Atlas, received from the ESA Data User Element.
(e) Lightning flash climatology (flashes per km^2 per year) derived from LIS/OTD.
The v1.0 gridded satellite lightning data were produced by the NASA LIS/OTD Science Team; http://ghrc.msfc.nasa.gov.

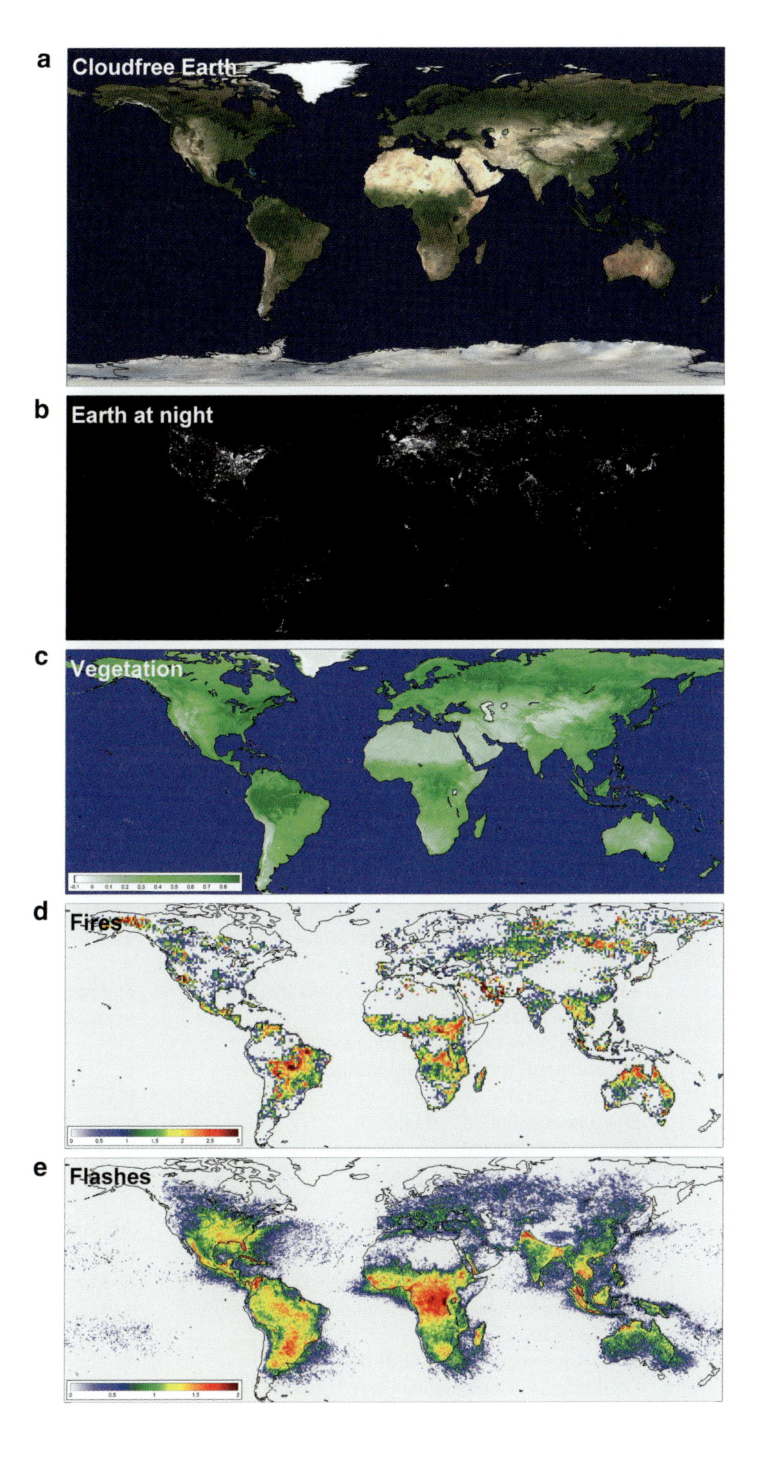
a
Cloudfree Earth
b
Earth at night
c
Vegetation
-0.1 0 0.1 0.2 0.3 0.4 0.5 0.6 0.7 0.8
d
Fires
0 0.5 1 1.5 2 2.5 3
e
Flashes
0 0.5 1 1.5 2

Appendix C: Abbreviations and Acronyms

A list of chemical names and molecular formulae is given just before the first chapter

		Chapters
AAI	Aerosol absorbing index	Appendix D
AATSR	Advanced along-track scanning radiometer	5, 6, Appendix A
ACCENT	Atmospheric Composition Change/The European Network of Excellence	5, 6, 10
ACE	Aerosol characterisation experiment	4
ACE-FTS	Atmospheric chemistry experiment – FTS	3, Appendix A
ADEOS	Advanced earth observation satellite	3
ADV	AATSR dual view algorithm	6
AEROCAN	Canadian aerosol network	6
AERONET	Aerosol robotic network – a ground based network	6, 7
AIRS	Atmospheric infrared sounder	3, 4, Appendix A
AMAERO	OMI multi-wavelength aerosol algorithm	6
AMF	Air mass factor	2, 9
AMS	American Meteorological Society	4
AMSR	Advanced microwave scanning radiometer	4
AMSUA, B	Advanced microwave sounding unit – A, B	4
AMV	Atmospheric motion vectors	4
AOD	Aerosol optical depth	6, 7, 9, Appendix D
AOS	Acousto-optical-spectrometers	4
AOT	Aerosol optical thickness	6, 7, 9, Appendix D
APS	Aerosol polarimetry sensor	6
ARM	Atmospheric radiation measurement site	5
AROME	Application of research to operations in meso-scale	4
ARTS	Atmospheric radiative transfer simulator	4
ASCAT	MetOp's advanced scatterometer	4
AT2	ACCENT-TROPOSAT-2	1
ATLAS	Atmospheric laboratory for application and science	4
ATM	Atmospheric transmission at microwaves	4
ATMOS	Atmospheric trace molecule spectroscopy	3, Appendix A
ATSR-2	Along track scanning radiometer	6, Appendix A
AVHRR	Advanced very high resolution radiometer	4, 5, 6, Appendix A
BAER	Bremen aerosol retrieval algorithm	6
BB	Biomass burning	8
BLUE	Best linear unbiased estimate	9
BRDF	Bi-directional distribution function	2
BRDF	Bi-directional reflection function	6
BT	Brightness temperature	5
BUV	Backscatter ultraviolet ozone experiment	Appendix A
CA	Cloud albedo	Appendix A
CALIOP	Cloud-aerosol lidar with orthogonal polarization	6, Appendix A
CALIPSO	Cloud-aerosol lidar and infrared pathfinder satellite observations	4, 6
CAMA	Toolkit for validation of OMI data	7
CARIBIC	Civil aircraft for the regular investigation of the atmosphere based on an instrument container	7
CCM	Chemistry climate model	9

(*continued*)

		Chapters
CCN	Cloud concentration nuclei	1, 6
CEOS	Committee on Earth observation satellites	7, 10
CERES	Cloud and Earth's radiant energy system	4
CESAR	Cabauw experimental site for atmospheric research	6
CF	Cloud fraction	Appendix A
CFC	Chloroflurocarbon	3
CHAMP	Challenging mini-satellite payload	4
CIMEL	Commercial sun photometer	7
CIW	Cloud ice water	4
CIWSIR	Cloud ice water sub-millimeter imaging radiometer	4
CL	Chemiluminescence	7
CLAES	Cryogenic limb array etalon spectrometer	Appendix A
CLW	Cloud liquid water	4
CNES	Centre National d'Etudes Spatiales	3
COSMIC	Constellation observing system for meteorology ionosphere and climate	4
CoSSIR	Conical scanning submillimeter wave imaging radiometer	4
COT	Cloud optical thickness	5, Appendix A
CPR	Cloud profiling radar	4
CRD	Cloud radiation database	4
CRL	Communications Research Laboratory, Japan	4
CRM	Cloud resolving model	4
CRTM	Community radiative transfer model	4
CSA	Canadian Space Agency	3, 4
CSU	Colorado State University	4
CT	Cloud temperature	Appendix A
CTH	Cloud top height	5, Appendix A
CTM	Chemical transport model	8, 9
CZ	Crystal size	Appendix A
CZCS	Coastal zone colour scanner	5
DA	Data assimilation	9
DDA	Discrete dipole approximation	4
DIAL	Differential absorption lidar	1, 7, 10
DMSP	Defence meteorological satellite program	4
DOAS	Differential optical absorption spectroscopy	1
DOFS	Degrees of freedom for signal	3
DOIT	Discrete ordinate iterative method	4
DPR	Dual-frequency (Ku/Ka-band) precipitation radar	4
DSD	Droplet size distribution	4
DU	Dobson unit	2, 8, 9
DUE	Data users element	6
DUP	Data users program	6
DZ	Droplet size	Appendix A
EARLINET	European research lidar network	6, 7
ECC	Electrochemical concentration cell	7
ECMWF	European Centre for Medium Range Weather Forecasting	3, 4, 10
EGPM	European GPM	4
E-GVAP	EUMETNET GPS water vapour programme	4
ELDO	European Launcher Development Organisation	1

(continued)

		Chapters
EMAC	ECHAM/MESSy atmospheric chemistry model	9
ENVISAT	Environmental satellite	3, 5, 6
EOS	Earth observing system	6
EPS	EUMETSAT's polar system	4
ERTS-1	Earth resources technology satellite	6
ESA	European Space Agency	1, 3, 4, 5, 6, 10
ESRO	European Space Research Organisation	1
EUCAARI	European integrated project on aerosol cloud climate air quality interactions	6
EUMETNET	The Network of European meteorological services	4
EUMETSAT	European Organisation for the Exploitation of Meteorological Satellites	1, 6, 10
EURAD	European Air Pollution Dispersion model system	9
EUSAAR	European supersites for atmospheric aerosol research	6
FFTS	Fast Fourier transform spectrometer	4
FIRSC	Far infrared sensor for cirrus	4
FMI	Finnish Meteorological Institute	6
FT	Fourier transform	1
FTIR	Fourier transform infra red	7
FTS	Fourier transform spectrometer	3, 4, 5
FWHM	Full width half maximum	1, 3
GADS	Global aerosol data set	6
GAW-PFR	Global Atmospheric Watch – precision filter radiometer	6
GCE	Goddard cumulus ensemble	4
GCM	Global climate model	8
GCM	General circulation model	9
GCOS	Global climate observing system	
GEISA	A spectroscopic database	3
GEMS	Global and regional Earth-system monitoring using satellite and *in situ* data	6
GEO	Geostationary Orbit	1, 4, 10
GEO	Group on Earth Observations	
GEOS	Goddard Earth observing system model	4
GEOS-Chem	Goddard Earth observing system-chemistry model	9
GEOSS	Global observing system of systems	1, 10
GLAS	Geoscience laser altimeter system	6
GMAO	Global modelling and assimilation office	6
GMES	Global Monitoring of Environment and Security	1, 6, 10
GMI	GPM microwave imager	4
GOMAS	Geostationary observatory for microwave atmospheric sounding	4
GOME-2	Global ozone monitoring experiment-2	Appendix A
GOMOS	Global ozone monitoring by occultation of stars	Appendix A
GOSAT	Greenhouse gases observing satellite	5
GPM	Global precipitation measurement	4
GPS	Global positioning system	4
GPSRO	GPS radio occultation	4
GRAS	Global navigation satellite system receiver for atmospheric sounding	4
GSFC	Goddard Space Flight Center	4

(*continued*)

		Chapters
GVS	Ground validation system	4
HALOE	Halogen occultation experiment	Appendix A
HERA	Hybrid extinction algorithm	6
HIRDLS	High resolution dynamics limb sounder	3, Appendix A
HIRS	High-resolution infrared radiation sounder	4
HITRAN	A spectroscopic database	3
HITRAN	High resolution transmission model	4
HSB	Humidity sensor for Brasil	4
HTAP	Hemispheric transport of air pollution	1
IASI	Infrared atmospheric sounding interferometer	3, Appendix A
ICESat	Ice, cloud, and land elevation satellite	6
IDL	Language used in the CAMA toolkit	7
IGACO	Integrated Global Atmospheric Chemistry Observations	10
IGOS	Integrated Global Observing System	1, 10
IGS	International GPS service	4
IGY	International geophysical year	1
IIR	Imaging infrared radiometer	6
ILAS I, II	Improved limb atmospheric spectrometer	Appendix A
ILS	Instrumental line shape	3
IMG	Interferometric monitor for greenhouse gases	3, AA
IPCC	Inter-governmental panel on climate change	1, 10
IR	Infrared	1, 7, 10
ISAMS	Improved stratospheric and mesospheric sounder	Appendix A
ISCCP	International satellite cloud climatology project	4, 5
ISO	International Organisation for Standardization	7
ISS	International space station	4
ITCZ	Inter tropical convergence zone	5
JAXA	Japanese Aerospace Space Agency	1, 4, 5, 10
JCSDA	Joint centre for satellite data assimilation	4
JEM-SMILES	The Japanese experiment module superconducting submillimeter-wave limb-emission sounder	4
JMA	Japan Meteorological Agency	4
JPL	Jet propulsion laboratory	3, 4
KNMI	Royal Netherlands Meteorological Institute	6
LEO	Low earth orbit	1, 4, 10
Lidar	Light detection and ranging	1, 5, 10
LIF	Laser induced fluorescence	7
LIMS	Limb infrared monitor of the stratosphere	4, Appendix A
LIS	Lightning imaging sensor	4, 8
LITE	Lidar in space technology experiment	6
LOS	Line of sight	2
LOWTRAN	Low resolution transition model database	1
LRIR	Limb radiance inversion radiometer	4, Appendix A
LRT	Long range transport	8
LRTAP	Long range transboundary air pollution	1
LT	Lower troposphere	8
LTE	Local thermal equilibrium	3
LUT	Look-up table	5–7
LWP	Liquid water path	5

(*continued*)

		Chapters
MACC	Monitoring atmospheric composition and climate	6, 10
MAN	Maritime aerosol network	6
MAPS	Measurements of atmospheric pollution from satellites	3, Appendix A
MAS	The millimeter-wave atmospheric sounder	4
MAS	Microwave atmospheric sounder	Appendix A
MAXDOAS	Multi-axis differential optical absorption spectroscopy	7
MEGAPOLI	Megacities: emissions, urban, regional and global atmospheric pollution and climate effects, and integrated tools for assessment and mitigation	6
MERIS	Medium resolution imaging spectrometer	5, 6, Appendix A
METEOSAT	Meteorological satellite	6
METOP	MetOp is a series of three satellites, forming a segment of EPS	4, 6
MHS	Microwave humidity sounder	4
MIPAS	Michelson interferometer for passive atmospheric sounding	3, Appendix A
MIR	Millimeter-wave imaging radiometer	4
MISR	Multiangle imaging spectro-radiometer	4, 5, 6
MJO	Madden-Julian oscillation	4
MLS	Microwave limb sounder	4, Appendix A
MODIS	Moderate-resolution imaging spectro-radiometer	4, 5, 6, Appendix A
MOPITT	Measurements of pollution in the troposphere	3, Appendix A
MOZAIC	Measurements of ozone and water vapour by in-service Airbus aircraft	4, 7
MOZART	Model for ozone and related chemical tracers	9
MSC	Meteorological Service of Canada	4
MSG	Meteosat second generation	6
MSS	Multi spectral scanner	6
MSU	Microwave sounding unit	4
MTPE	Mission to planet Earth	4
MWHS	Microwave humidity sounder	4
MWMOD	Microwave model	4
MWRI	Microwave radiation imager	4
MWTS	Microwave-tomographic imaging	4
NADCC	Network for the detection of atmospheric composition change	7
NAO	North Atlantic oscillation	8
NASA	National Aeronautics and Space Administration (USA)	1, 3–6, 10
NASDA	National Space Development Agency in Japan	4
NCEP	National centers for environmental prediction (NOAA)	4
NDVI	Normalised difference vegetation index	6
NEMS	Nimbus-E microwave sensor	4
NESR	Noise-equivalent spectral radiance	3
NEXRAD	Next generation weather radar	4
NH	Northern hemisphere	8, 9
NiCT	National Institute of Information and Communications Technology, Japan	4
NIR	Near infrared	2, 6
NIS	NEXRAD in Space	4

(*continued*)

		Chapters
NMHC	Non-methane hydrocarbons	8
NMVOC	Non-methane volatile organic compounds	8
NOAA	National Oceanic and Atmospheric Administration (USA)	1, 4, 5
NPOESS	National Polar-orbiting Operational Environmental Satellite System	1, 4, 10
NPP	NPOESS preparatory project	
NSF	National Science Foundation	10
NWP	Numerical weather prediction	1,4, 10
NWP-SAF	NWP- Satellite Application Facility	4
OE	Optimal estimation	3
OI	Optimal interpolation	9
OMI	Ozone monitoring instrument	6, Appendix A
OPAC	Optical properties of aerosols and clouds package	6
OPD	Optical path difference	3
OSE	Observing system experiment	4
OSIRIS	Optical spectrograph and infrared imaging system	Appendix A
OSSE	Observing system simulation experiments	4
PAH	Polyaromatic hydrocarbons	1
PARASOL	Polarization and anisotropy of reflectances for atmospheric sciences coupled with observations from a lidar	5, 6, Appendix A
PGR	Polarisation gain ratio	6
PHOTONS	European aerosol network	6, 7
PIA	Path integrated attenuation	4
PM10, PM2.5	Particulate matter (diameter less than 10 μm) (diameter less than 2.5 μm)	6
POAM-II, III	Polar ozone and aerosol measurement II, III	Appendix A
POLDER	Polarization and directionality of the Earth's reflectances	5, 6
POP	Persistent organic pollutants	1
PPS	Precipitation processing system	4
PR	Precipitation radar	4
PREMIER	Process exploration through measurements of infrared and millimeter-wave emitted radiation	4
PW	Precipitable water	4
RADM	Regional acid deposition model	9
RH	Relative humidity	6
RMS	Root mean square	7, 9
RMSD	Root mean square deviation	6
RPV	Raman-Pinty-Verstraete mode	6
RRS	Rotational Raman scattering	1
RTE	Radiative transfer equation	4
RTM	Radiation transfer modelling	2
RTM	Radiation transfer model	6, 7
RTTOV	Radiative transfer for TOVS	4
RVRS	Rotational-vibrational Raman scattering	1
SACURA	Semi-analytical cloud retrieval algorithm	5
SAGE – 1, 2, 3	Stratospheric aerosol and gas experiment 1, 2, 3	Appendix A
SAM, II	Stratospheric aerosol measurement instrument, II	6, Appendix A

(*continued*)

		Chapters
SAMS	Stratospheric and mesospheric sounder	4, Appendix A
SBUV, - 2	Solar backscatter ultraviolet ozone experiment, -2	Appendix A
SCA	Scene classification algorithm	6
SCAMS	Scanning microwave sounder	4
SCD	Slant column density	2
SCIAMACHY	Scanning imaging absorption spectrometer for atmospheric cartography	5, Appendix A
SCR	Selective chopper radiometer	Appendix A
SD	Slant delay	4
SEVIRI	Spinning enhanced visible and infrared imager	6, Appendix A
SH	Southern hemisphere	8, 9
SHADOZ	Southern hemisphere additional ozone sondes	7
SHDOM	Spherical harmonic discrete ordinate method	4
SIBYL	Selective interactive boundary algorithm	6
SIRICE	Submillimeter infrared radiometer ice cloud experiment	4
SIS	Superconductor-insulator-superconductor	4
SLST	Sea and land surface temperature	6
SME	Solar mesospheric experiment	Appendix A
SMLS	Scanning microwave limb sounder	4
SMR	The Odin sub-millimeter radiometer	4
SOA	Secondary organic aerosol	8
SOD	Slant optical density	2
SRT	Surface reference technique	4
SSA	Single scattering albedo	6, Appendix D
SSM/T-1, -2	Special sensor microwave/temperature-1, -2	4
SSMI/S	Special sensor microwave imager/sounder	4
SST	Sea surface temperature	4
SSU	Stratospheric sounding unit	4
STEAM-R	The stratosphere troposphere exchange and climate monitor radiometer	4
SZA	Solar zenith angle	2
TANSO	Thermal and short wave infra-red sensor for observing greenhouse gases	Appendix A
TEMIS	Tropospheric emission monitoring internet service	6
TEOM-FDMS	Tapered element oscillating microbalance with filter dynamics measurement system	6
TES	Tropospheric emission spectrometer	3, 7, Appendix A
THV	Threshold value	5
TIR	Thermal infrared	5, 6, 10
TIROS	Television Infrared Observation Satellite	4, 5
TM4-ECPL	Chemistry transport model version 4 – Environmental chemical processes laboratory	9
TM5	Chemistry transport model version 5	9
TMI	Tropical rainfall measuring mission microwave imager	4
TNO	Netherlands Organisation for Applied Scientific Research	6
TOA	Top of atmosphere	2, 5, 6, 7
TOMS	Total ozone mapping spectrometer	5, 6, Appendix A
TORA	Tropospheric ozone re-analysis method	9

(*continued*)

		Chapters
TOVS	TIROS operational vertical sounder	4, 5
TP	Tikhonov – Phillips regularisation	3
TRMM	Tropical rainfall measuring mission	4
TROPOSAT	Use and usability of satellite data for tropospheric research	1, 10
TTL	Tropopause transition layer	4, 8
UAE2	United Arab Emirates Unified Aerosol Experiment	6
UARS	Upper atmosphere research satellite	4
UNECE	United Nations Economic Commission for Europe	1, 10
UNEP	United Nations Environment Programme	1
UNFCCC	United Nations Framework Convention on Climate Change	1
UT	Upper troposphere	8
UTH	Upper tropospheric humidity	4
UTLS	Upper troposphere/lower stratosphere	3, 9
UV	Ultra-violet	6
UV/vis	UV/visible	1, 2, 7
UW-NMS	University of Wisconsin non-hydrostatic modelling system	4
VCD	Vertical column density	2
VFM	Vertical feature mask	6
VIM	International vocabulary for metrology	7
VIRS	Visible and infrared scanner	4
VMR	Volume mixing ratio	8
VOC	Volatile organic compound	1, 8, 9
VRS	Vibration rotation spectra	1
WALES	Water vapour lidar experiment in space	1
WF	Weighting function	2
WFC	Wide field camera	6
WMO	World Meteorological Organisation	1, 10
WOUDC	World ozone and UV radiation data centre	7
ZTD	Zenith total delay	4

Appendix D: Timelines for Present and Future Missions

John P. Burrows and Stefan Noel, IUP, Bremen

These figures are based on the best available 2010 information.

D.1 Tropospheric Reactive Gases

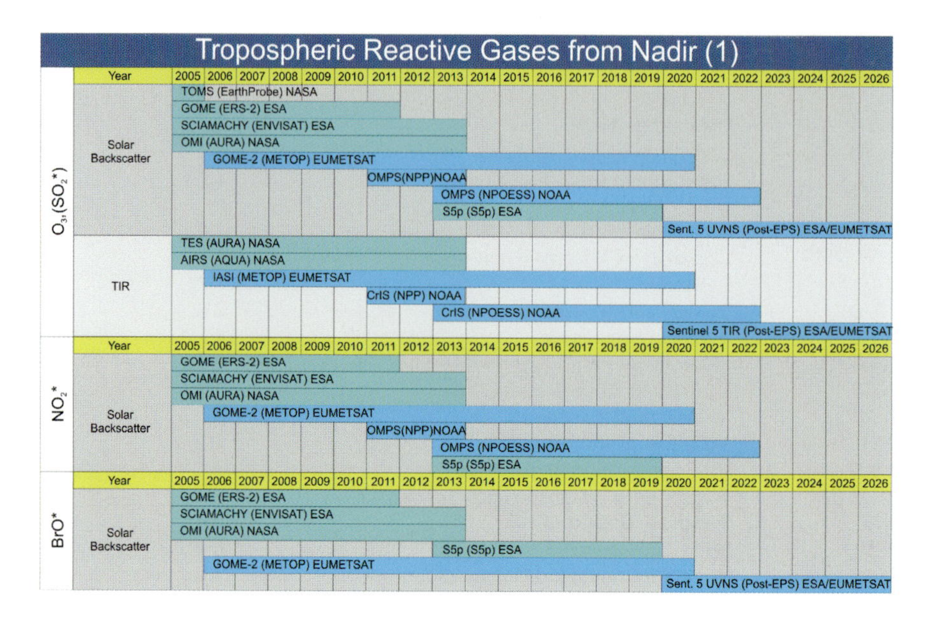

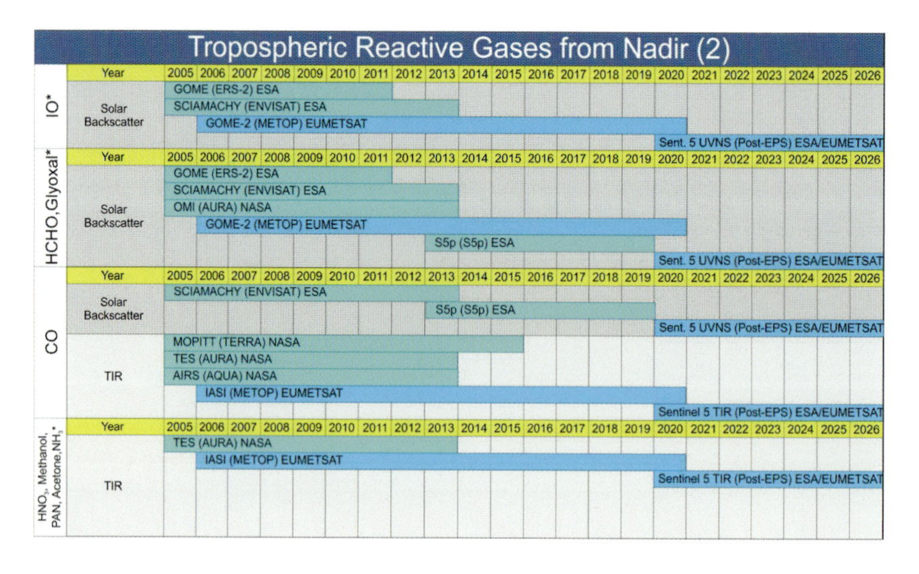

Upper Tropospheric Reactive Gases from Limb/Occultation

Gas	Technique	2005	2006	2007	2008	2009	2010	2011	2012	2013	2014	2015	2016	2017	2018	2019	2020	2021	2022	2023	2024	2025	2026
O_3	Solar Backscatter	SCIAMACHY (ENVISAT) ESA																					
								OMPS(NPP)NOAA															
	TIR	ACE (SCISAT-1) CSA																					
		MIPAS, GOMOS (ENVISAT) ESA																					
	MW	SMR (ODIN) SSC																					
		MLS (AURA) NASA																					
NO_2*	Solar Backscatter	SCIAMACHY (ENVISAT) ESA																					
								OMPS(NPP)NOAA															
	TIR	ACE (SCISAT-1) CSA																					
		MIPAS, GOMOS (ENVISAT) ESA																					
SO_2*	Solar Backscatter							OMPS(NPP)NOAA															
BrO	Solar Backscatter	SCIAMACHY (ENVISAT) ESA																					
CO	TIR	ACE (SCISAT-1) CSA																					
		MIPAS (ENVISAT) ESA																					
HNO_3, Methanol, PAN, Acetone, NH_3*	TIR	ACE (SCISAT-1) CSA																					
		MIPAS (ENVISAT) ESA																					

Tropospheric Reactive Gases from Geostationary Orbit

Gas	Technique	2005	2006	2007	2008	2009	2010	2011	2012	2013	2014	2015	2016	2017	2018	2019	2020	2021	2022	2023	2024	2025	2026
O_3, (SO_2*)	Solar Backscatter													UVN (GEO-CAPE) NASA									
															UVN (MTG) ESA/EUMETSAT								
															GEMS (MP-GEOSAT) KARI								
																UVNS Instr. (GEO-ASIA) JAXA							
	TIR														IRS (MTG) EUMETSAT								
NO_2*	Solar Backscatter													UVN (GEO-CAPE) NASA									
															UVN (MTG) ESA/EUMETSAT								
															GEMS (MP-GEOSAT) KARI								
																UVNS Instr. (GEO-ASIA) JAXA							
HCHO, Glyoxal*	Solar Backscatter													UVN (GEO-CAPE) NASA									
															UVN (MTG) ESA/EUMETSAT								
															GEMS (MP-GEOSAT) KARI								
																UVNS Instr. (GEO-ASIA) JAXA							
CO	TIR													IR (GEO-CAPE) NASA									
															IRS (MTG) EUMETSAT								
															GEMS (MP-GEOSAT) KARI								
																UVNS Instr. (GEO-ASIA) JAXA							

*Implies short lived gases with highly variable amounts. These are measured when column amounts are above the instrumental detection limit e.g. SO_2 from volcanoes for TIR detection and from tropospheric pollution at the ground retrieved from solar backscattered measurements

Notes

1. Concerning the retrieval of data products for tropospheric trace gases:
   (a) In the TIR, the averaging kernel depends on the temperature difference between the atmosphere and the earth's surface. If this difference is low

there is then little sensitivity to the lower troposphere and the information content in the observation is primarily in the middle and upper troposphere. However measurements can be made both by day and night.

(b) The retrieval of trace gases using solar backscatter is sensitive to the lowermost troposphere as UV, visible and near IR radiation reaches the surface. However the sensitivity is reduced in the ultraviolet as a result of multiple scattering, and no measurements can be made at night.

2. Colouring: darker blue implies existing or funded operational meteorological satellites and instrumentation; lighter blue implies funded space agency research missions; the hatched pale blue implies missions under study but not yet funded.

D.2 Greenhouse Gases: CH_4, CO_2

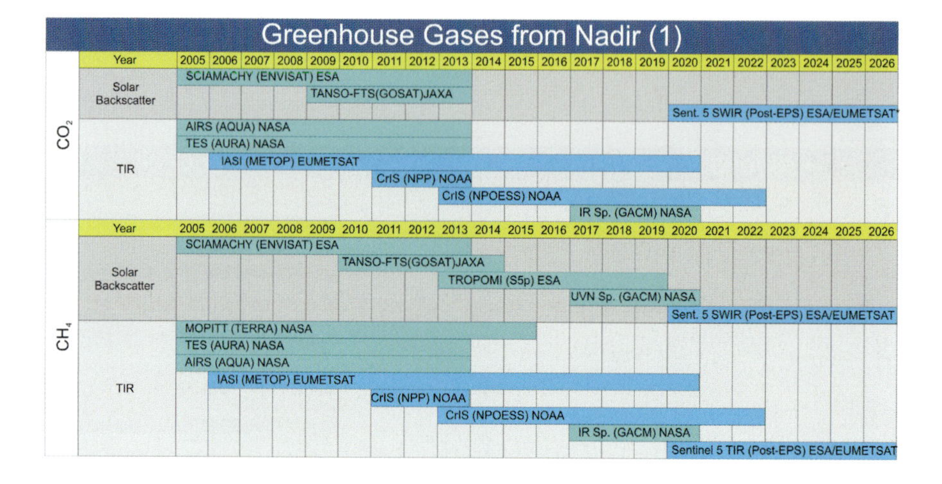

*Used for CH_4 normalisation; not optimised for CO_2 retrieval

Upper Trop. Greenhouse Gases from Limb/Occultation (1)

Year 2005 2006 2007 2008 2009 2010 2011 2012 2013 2014 2015 2016 2017 2018 2019 2020 2021 2022 2023 2024 2025 2026

CO_2 — Solar Backscatter: SCIAMACHY (ENVISAT) ESA

CO_2 — TIR: MIPAS (ENVISAT) ESA

CH_4 — Solar Backscatter: SCIAMACHY (ENVISAT) ESA

CH_4 — TIR: HIRDLS (AURA) NASA; MIPAS (ENVISAT) ESA

Greenhouse Gases from Geostationary Orbit (1)

Year 2005 2006 2007 2008 2009 2010 2011 2012 2013 2014 2015 2016 2017 2018 2019 2020 2021 2022 2023 2024 2025 2026

CO_2 — TIR: IRS (MTG) EUMETSAT

Notes

1. Concerning the retrieval of data products for tropospheric trace gases:
   (a) In the TIR, the averaging kernel depends on the temperature difference between the atmosphere and the Earth's surface. If this difference is low there is then little sensitivity to the lower troposphere and the information content in the observation is primarily in the middle and upper troposphere. However measurements can be made both day and night.
   (b) The retrieval of trace gases using solar backscatter is sensitive to the lowermost troposphere as UV, visible and near IR radiation reaches the surface. However the sensitivity is reduced in the ultraviolet as a result of multiple scattering, and no measurements can be made at night.
2. Colouring: darker blue implies existing or funded operational meteorological satellites and instrumentation; lighter blue implies funded space agency research missions; the hatched pale blue implies missions under study but not yet funded.

D.3 Greenhouse Gases: Water Vapour

Greenhouse Gases from Nadir (2): Water Vapour

	Year	2005–2026
H_2O	Solar Backscatter	AVHRR (NOAA15-19) NOAA
		GOME (ERS-2) ESA
		SCIAMACHY, MERIS, AATSR (ENVISAT) ESA
		MODIS (AQUA) NASA
		MODIS (TERRA) NASA
		AVHRR, GOME-2, HIRS (METOP) EUMETSAT
		UVNS,VII,3MI/3MI'(Post-EPS)ESA/EUMETSAT
	TIR	HIRS (NOAA15-19) NOAA
		AIRS (AQUA) NASA
		IASI (METOP) EUMETSAT
		CrIS, VIIRS (NPP) NOAA
		CrIS, VIIRS (NPOESS) NOAA
	MW	AMSU (NOAA15-19) NOAA
		AMSU (AQUA) NASA
		SSM/I(S), SSM/T2 (DMSP) NOAA
		AMSU (METOP) EUMETSAT
		ATMS (NPP) NOAA
		ATMS (NPOESS) NOAA

Greenhouse Gases from Limb/Occultation (2): Water Vapour

	Year	2005–2026
H_2O	Solar Backscatter	SAGE-3(METEOR-3M)NASA
		SCIAMACHY (ENVISAT) ESA
	TIR	HALOE(UARS)NASA
		ACE (SCISAT-1) CSA
		SMR (ODIN) SSC
		MIPAS, GOMOS (ENVISAT) ESA
		MLS (AURA) NASA
	GPS	various (see Notes)

Greenhouse Gases from Geostationary Orbit (2): Water Vapour

	Year	2005–2026
H_2O	Solar Backscatter	SEVIRI (MSG) EUMETSAT
		FCI (MTG) EUMETSAT
	TIR	Imager (GOES) NOAA
		SEVIRI (MSG) EUMETSAT
		IRS (MTG) EUMETSAT

Notes

1. GPS refers to all Global Positioning System satellites, which are used for H_2O retrieval. These includes those already delivering water vapour products, e.g. GRAS on METOP and COSMIC, a constellation flown by the South Korean Space Agency in collaboration with NCAR. Further missions which are expected to deliver water vapour products are GALILEO, GLONASS and RO on PostEPS.
2. Concerning the retrieval of data products for tropospheric trace gases:
   (a) In the TIR, the averaging kernel depends on the temperature difference between the atmosphere and the earth's surface. If this difference is low there is then little sensitivity to the lower troposphere and the information content in the observation is primarily in the middle and upper troposphere. However measurements can be made both day and night.
   (b) The retrieval of trace gases using solar backscatter is sensitive to the lowermost troposphere as UV, visible and near IR radiation reaches the surface. However the sensitivity is reduced in the ultraviolet as a result of multiple scattering, and no measurements can be made at night.
3. Colouring: darker blue implies existing or funded operational meteorological satellites and instrumentation; lighter blue implies funded space agency research missions; the hatched pale blue implies missions under study but not yet funded.

D.4 Tropospheric Aerosol

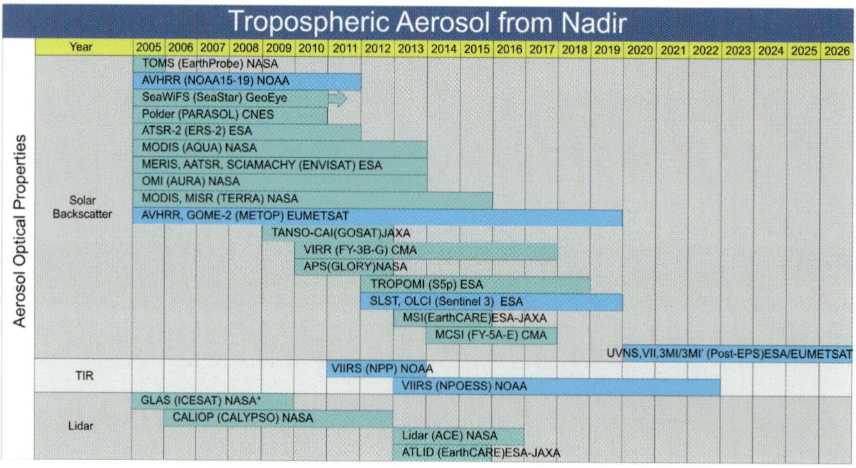

Tropospheric Aerosol from Nadir

Aerosol Optical Properties

Year 2005 2006 2007 2008 2009 2010 2011 2012 2013 2014 2015 2016 2017 2018 2019 2020 2021 2022 2023 2024 2025 2026

Solar Backscatter
TOMS (EarthProbe) NASA
AVHRR (NOAA15-19) NOAA
SeaWiFS (SeaStar) GeoEye
Polder (PARASOL) CNES
ATSR-2 (ERS-2) ESA
MODIS (AQUA) NASA
MERIS, AATSR, SCIAMACHY (ENVISAT) ESA
OMI (AURA) NASA
MODIS, MISR (TERRA) NASA
AVHRR, GOME-2 (METOP) EUMETSAT
TANSO-CAI(GOSAT)JAXA
VIRR (FY-3B-G) CMA
APS(GLORY)NASA
TROPOMI (S5p) ESA
SLST, OLCI (Sentinel 3) ESA
MSI(EarthCARE)ESA-JAXA
MCSI (FY-5A-E) CMA
UVNS,VII,3MI/3MI' (Post-EPS)ESA/EUMETSAT

TIR
VIIRS (NPP) NOAA
VIIRS (NPOESS) NOAA

Lidar
GLAS (ICESAT) NASA*
CALIOP (CALYPSO) NASA
Lidar (ACE) NASA
ATLID (EarthCARE)ESA-JAXA

*Intermittent operation

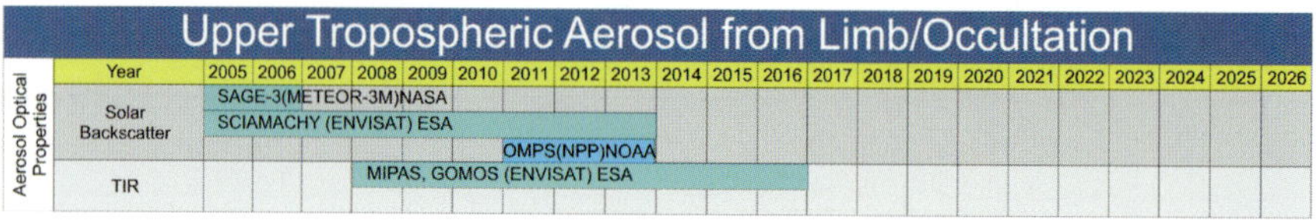

Upper Tropospheric Aerosol from Limb/Occultation

Aerosol Optical Properties

Year 2005 2006 2007 2008 2009 2010 2011 2012 2013 2014 2015 2016 2017 2018 2019 2020 2021 2022 2023 2024 2025 2026

Solar Backscatter
SAGE-3(METEOR-3M)NASA
SCIAMACHY (ENVISAT) ESA
OMPS(NPP)NOAA

TIR
MIPAS, GOMOS (ENVISAT) ESA

Tropospheric Aerosol from Geostationary Orbit

Year: 2005 2006 2007 2008 2009 2010 2011 2012 2013 2014 2015 2016 2017 2018 2019 2020 2021 2022 2023 2024 2025 2026

Aerosol Optical Properties	Missions
Solar Backscatter	SEVIRI (MSG) EUMETSAT
	OCI (COMS) KARI
	FCI (MTG) EUMETSAT
	UVN&IR (GEO-CAPE) NASA
	UVN (MTG) ESA/EUMETSAT
	GEMS (MP-GEOSAT) KARI
	UVNS (GEO-ASIA) JAXA
TIR	FCI (MTG) EUMETSAT

Notes

1. The aerosol optical proprieties in the figure description refer to the broad range of aerosol data products generated. These include: aerosol optical thickness, AOT, or aerosol optical depth, AOD, single scattering albedo, SSA, aerosol absorbing index, AAI, size distribution where available, size discrimination, coarse and fine.
2. Colouring: darker blue implies existing or funded operational meteorological satellites and instrumentation; lighter blue implies funded space agency research missions; the hatched pale blue implies missions under study but not yet funded.

*Intermittent operation

D.5 Clouds

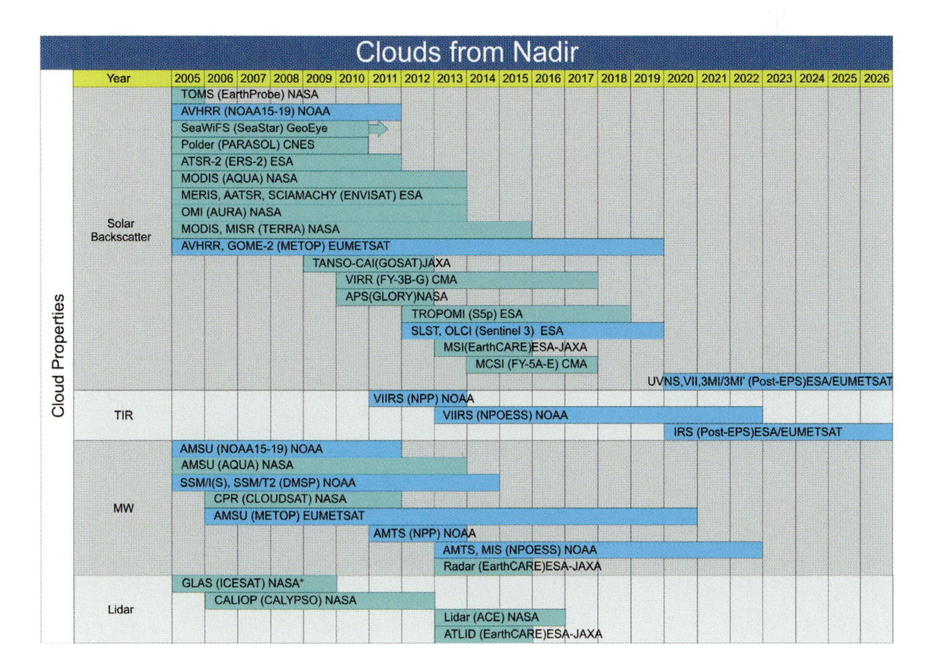

Clouds from Limb/Occultation

	Year	2005	2006	2007	2008	2009	2010	2011	2012	2013	2014	2015	2016	2017	2018	2019	2020	2021	2022	2023	2024	2025	2026
Cloud Properties	Solar Backscatter	SAGE-3(METEOR-3M)NASA																					
		SCIAMACHY (ENVISAT) ESA																					
	TIR	ACE (SCISAT-1) CSA																					
		SMR (ODIN) SSC																					
		MIPAS, GOMOS (ENVISAT) ESA																					
		MLS (AURA) NASA																					

Clouds from Geostationary Orbit

	Year	2005	2006	2007	2008	2009	2010	2011	2012	2013	2014	2015	2016	2017	2018	2019	2020	2021	2022	2023	2024	2025	2026
Cloud Properties	Solar Backscatter	SEVIRI (MSG) EUMETSAT																					
							MeteoImager (COMS) KARI																
												FCI (MTG) EUMETSAT											
														UVN&IR (GEO-CAPE) NASA									
															UVN (MTG) ESA/EUMETSAT								
															GEMS (MP-GEOSAT) KARI								
																UVNS (GEO-ASIA) JAXA							
	TIR	Imager (GOES) NOAA																					
															IRS (MTG) EUMETSAT								

Notes

1. Instruments/missions often deliver different cloud products; see the individual missions for details.
2. Colouring: darker blue implies existing or funded operational meteorological satellites and instrumentation; lighter blue implies funded space agency research missions; the hatched pale blue implies missions under study but not yet funded.

Index

A list of satellite instruments is given in Appendix A.
A full list of abbreviations and acronyms is given in Appendix C.
A list of chemical names and molecular formulae is given on page xxxi.

Printing and Binding: Stürtz GmbH, Würzburg